Spectrum Sharing in Wireless Networks

Fairness, Efficiency, and Security

OTHER COMMUNICATIONS BOOKS FROM CRC PRESS

TO ORDER
Call: 1-800-272-7737 • Fax: 1-800-374-3401 • E-mail: orders@crcpress.com

Spectrum Sharing in Wireless Networks

Fairness, Efficiency, and Security

Edited by
John D. Matyjas • Sunil Kumar • Fei Hu

CRC Press
Taylor & Francis Group
Boca Raton London New York

CRC Press is an imprint of the
Taylor & Francis Group, an **informa** business

CRC Press
Taylor & Francis Group
6000 Broken Sound Parkway NW, Suite 300
Boca Raton, FL 33487-2742

First issued in paperback 2020

© 2017 by Taylor & Francis Group, LLC
CRC Press is an imprint of Taylor & Francis Group, an Informa business

No claim to original U.S. Government works

Version Date: 20160706

ISBN 13: 978-0-367-57410-9 (pbk)
ISBN 13: 978-1-4987-2635-1 (hbk)

Contents

SECTION III: MODELING ISSUES   369

List of Figures

List of Tables

Preface

Spectrum sharing is one of the hottest R&D topics in wireless communications today. Electromagnetic spectrum, as part of the global commons, is a critical medium and resource that underpins international commerce, personal and private-sector communications, public safety, national security, and defense. The explosion of mobile devices and social media content has increased the competitiveness for and value of spectrum while straining availability. For example, in 2015, 25MHz of spectrum sold for $45 billion dollars in the United States. While this type of exclusive licensed access is favorable revenue-wise to a government licensing authority and companies that "own" it, this is not a sustainable long-term approach for a finite resource and inhibits efficacy in its access and use. This has motivated the quest for new approaches in real-time, negotiated spectrum sharing that cut across the public & private sector use-cases and their varying geo-spatial and temporal dimensions. Forthwith, new research challenges and innovation opportunities are surfacing that cut across sensing, antennas, signal processing & interference mitigation, co-existence strategies, policies/rulesets, and security/enforcement therein.

Some of the most important issues in spectrum sharing include: (1) *Fairness*: how do we ensure that different (prioritized) wireless users can share the spectrum in an equitable manner? Users with similar spectrum access privileges should have nearly equal opportunities to access the available spectrum. If some users have higher priority over others, they should be able to occupy the spectrum more often and/or be able to pre-empt lower priority users. (2) *Efficiency*: Efficient use of spectrum resources among different users at any given time must come with minimal control/coordination overhead, where it is critical not to waste any spectrum in space and time. (3) *Security*: Spectrum sharing methodologies need a robust trust model that can detect the overuse of the spectrum or falsified reports on users' occupied spectrum along with policies to prevent selfish users from overusing the spectrum. Distributed resource verification and data falsification safeguards are required herein.

The spectrum sharing paradigm can be briefly described as follows: Assume that there are two types of users in a wireless network, higher-priority (e.g., primary users or PUs), and lower-priority (e.g., secondary users or SUs). Though PUs and SUs are more related to cognitive radio networks, we use these terms to differentiate between two user types. There may exist 3 spectrum sharing models between them: (1) *Non-cooperative*: The SU can only transmit data in the absence of PUs' signals, where there is no cooperation between PUs and SUs; (2) *Cooperative - Concurrent Transmission*: PUs and SUs can concurrently transmit data, as long as the interference caused by SUs to PUs does not exceed a certain threshold (called interference temperature); (3) *Cooperative – SU Assisting PU*: The SU assists PU by relaying its traffic. In exchange, the PU can release small portions of its bandwidth for SU's data transmission.

An example of spectrum sharing is the apportionment of government-held spectrum with non-government users in the form of TV Whitespaces (TVWS). The cost of clearing and re-allocating spectrum is very high and can be avoided if effective sharing can be achieved. It is becoming widely recognized that spectrum sharing is an essential ingredient in freeing up enough spectrum for use by mobile broadband. To be suitable for spectrum sharing, radio units must have some or all of the following attributes or abilities: Operate in a range of channel widths in the same equipment; Adjust transmit power; Operate with a range of waveforms, most commonly OFDM or one of its derivatives, with a choice of modulation depths and error coding schemes; Perform sensing or radio environment monitoring, which can be in-band (intra-frequency) or out-of band (inter-frequency); Process user data using scheduling and queueing for different qualities of service (QoS).

Furthermore, there are many challenging R&D issues posed by integrated hardware/software radio platform designs suitable for efficient and fair spectrum sharing. A few examples include: Using advanced algorithms (such as game theory) to coordinate PUs and SUs to maximize mutual benefit; Determining the spectrum licensing, pricing and auction schemes; Using information theory to determine how long an SU can help to relay the PU's traffic; Integration of cooperative communication with network coding to improve throughput; Balancing the tradeoff of PU's capacity improvement and leased bandwidth when asking SU to also act as a relay.

Until now, there have been no technical books that cover the aforementioned issues with sufficient depth. This book provides a comprehensive "big picture" on the topic of spectrum-sharing that strategically deep-dives into key technical details, whetting the appetite of many academic researchers, and covers efficient spectrum sharing strategies under different network environments (such as cellular networks, ad hoc networks, etc.) of keen interest to industry engineers.

We have invited subject matter experts from all over the world to provide chapter contributions. They have spent over one year on the writing and editing of their chapters. The book editors have carefully arranged the chapters into the following seven parts:

- Part I. Big Picture: This part consists of two chapters on the basic concept of spectrum sharing, hardware/software function requirements for efficient sharing, and future trends of sharing strategies.

- Part II. Approaches to Spectrum Sharing: This is the main part of the book consisting of 11 chapters that discuss different efficient spectrum sharing approaches. Especially, we first introduce a new coexistence and sharing scheme for multi-hop network; then we describe the space-time sharing concept. LTE-U scheme is introduced and sharing in broadcast and unicast hybrid cellular network is also described, followed by chapters on different PU/SU cooperation strategies to achieve mutual benefits for PUs and SUs. Some protocols (such as Routing) are also discussed in spectrum sharing context, and different game theory models are provided between PUs and SUs.

- Part III. Modeling Issues: This part consists of four chapters that model the interactions of PUs and SUs, describe efficient calculation methods to find out the available spectrum, and the scheduling schemes to achieve efficient SU traffic delivery.

- Part IV. MIMO-oriented Design: Directional antennas and MIMO antennas greatly enhance the wireless network performance. This part consists of three chapters that describe the capacity/rate calculation for MIMO as well as the beamforming issues.

- Part V. Power Control: Chapters in this part discuss the interference-aware power allocation schemes among cognitive radio users, and power control issues for spectrum sharing.

- Part VI. Security: Security is very important in spectrum sharing. The chapters in this part comprehensively discuss different types of spectrum sharing attacks and threats, as well as the corresponding countermeasure schemes.

- Part VII. Military Considerations: This component describes some issues on current sharing schemes and policies in military applications along with a construct for future spectrum-dependent capabilities. For example, how do non-civilian tasks protect their data flows when sharing the spectrum with civilian applications?

Target audiences: This book targets both academia and industry, as it identifies some interesting research problems in this critically important and emerging field with an opportunity to learn about promising solutions. Graduate students will be informed by emerging research that can be leveraged for their thesis or dissertation topics. Researchers will be inspired by the models and protocols considered and discussed in various chapters. Engineers/practitioners from industry will be exposed to engineering design trades and will benefit from an awareness of corresponding practical solutions covered in select chapters.

About the Editors

Dr. John D. Matyjas earned his PhD in electrical engineering from State University of New York at Buffalo in 2004. Currently, he is serving as the Connectivity & Dissemination Core Technical Competency Lead at the Air Force Research Laboratory (AFRL) in Rome, NY. His research interests include dynamic multiple-access communications and networking, software defined RF spectrum mutability, statistical signal processing and optimization, and neural networks. He serves on the IEEE Transactions on Wireless Communications Editorial Advisory Board.

Dr. Matyjas is the recipient of the 2014 Air Force Scientific Management Award, the 2012 IEEE R1 Technology Innovation Award, the 2013 AFRL Harry Davis Award for "Excellence in Basic Research," and the 2010 IEEE Int'l Communications Conf. Best Paper Award. He is an IEEE Senior Member, chair of the IEEE Mohawk Valley Signal Processing Society, and member of Tau Beta Pi and Eta Kappa Nu.

Dr. Fei Hu is currently an associate professor in the Department of Electrical and Computer Engineering at the University of Alabama, Tuscaloosa, Alabama, USA. He earned his PhD degrees at Tongji University (Shanghai, China) in the field of Signal Processing (in 1999), and at Clarkson University (New York, USA) in Electrical and Computer Engineering (in 2002). He has published over 200 journal/conference papers and books. Dr. Hu's research has been supported by U.S. National Science Foundation, Cisco, Sprint, and other sources. His research expertise can be summarized as *3S: Security, Signals, Sensors:* (1) Security: This is about how to overcome different cyber attacks in a complex wireless or wired network. Recently he focuses on cyber-physical system security and medical security issues. (2) Signals: This mainly refers to *intelligent signal processing*, that is, using machine learning algorithms to process sensing signals in a smart way in order to extract patterns (i.e., pattern recognition). (3) Sensors: This includes micro-sensor design and wireless sensor networking issues.

Dr. Sunil Kumar is currently a professor and Thomas G. Pine Faculty Fellow in the Electrical and Computer Engineering Department at San Diego State University (SDSU), San Diego, California, USA. He earned his PhD in Electrical and Electronics Engineering from the Birla Institute of Technology and Science (BITS), Pilani (India) in 1997. From 1997 to 2002, Dr. Kumar was a Postdoctoral Researcher and Adjunct Faculty at the University of Southern California, Los Angeles. He also worked as a consultant in industry on JPEG2000 and MPEG-4 related projects, and was a member of US delegation in JPEG2000 standardization activities. Prior to joining SDSU, Dr. Kumar was an assistant professor at Clarkson University, Potsdam, NY (2002–2006). He was an ASEE Summer Faculty Fellow at the Air Force Research Lab in Rome, NY during the summer of 2007 and 2008, where he conducted research in Airborne Wireless Networks. Dr. Kumar is a senior member of IEEE, and has published more than 125 research articles in international journals and conferences, including three books/book chapters. His research has been supported by grants/awards from the National Science Foundation, U.S. Air Force Research Lab, Department of Energy, California Energy Commission, and industry. His research areas include wireless networks, cross-layer and QoS-aware wireless protocols, and error-resilient video compression.

Contributors

Andreas Achtzehn
Institute for Networked Systems
RWTH Aachen University
Aachen, Germany

Ahmed M. Ahmed
School of Electrical Engineering and Computer
    Science (EECS)
University of Ottawa
Ottawa, Canada

Petri Ahokangas
Oulu Business School
Oulu, Finland

Ala Abu Alkheir
School of Electrical Engineering and Computer
    Science (EECS)
University of Ottawa
Ottawa, Canada

Mohamed-Slim Alouini
Computer, Electrical, and Mathematical
    Science of Engineering Division
King Abdullah University of Science and
    Technology (KAUST)
Thuwai, Saudi Arabia

Ahmad Alsharoa
Department of Electrical and Computer
    Engineering
Iowa State University (ISU)
Ames, Iowa, USA

Alagan Anpalagan
Department of Electrical and Computer
    Engineering
Ryerson University
Toronto, Canada

Guoan Bi
School of Aeronautics and Astronautics
University of Electronic Science and
    Technology of China
Chengdu, China

Jesse Bourque
Senior OSD Consultant (Policy & CIO)
AECOM
USA

Timothy X Brown
Department of Electrical and Computer
    Engineering
Carnegie Mellon University in Rwanda
Kigali, Rwanda, USA

Qian Chen
Institute for Infocomm Research (I^2R)
Singapore

Chunxiao Chigan
Department of Electrical and Computer
    Engineering
University of Massachusetts Lowell One
    University Avenue
Lowell, Massachusetts, USA

Francois Chin
WiMedia Alliance
Institute for Infocomm Research (I^2R)
Agency for Science, Technology and Research
    (A*STAR)
Singapore

Daniel B. da Costa
Department of Computer Engineering
Federal University of Ceara (UFC)
Sobral, Ceará, Brazil

Haiyang Ding
State Key Lab of Integrated Services Networks
Xidian University
Xian, China

Zhiyong Feng
School of Information and Communication
    Engineering
Beijing University of Posts and
    Telecommunications
Beijing, China

Michael Fitch
British Telecom Research
Martlesham Heath, UK

Ethan Gaebel
Complex Networks and Security Research
    (CNSR) Lab
Virginia Polytechnic Institute and State
    University
Blacksburg, Virginia, USA

Jianhua Ge
State Key Lab of Integrated Services Networks
Xidian University
Xian, China

Hakim Ghazzai
Computer, Electrical, and Mathematical
    Science of Engineering (CEMSE) Division
King Abdullah University of Science and
    Technology (KAUST)
Thuwai, Saudi Arabia

Daesik Hong
School of Electrical and Electronic Engineering
Yonsei University
Seoul, Korea

Y. Thomas Hou
Complex Networks and Security Research
    (CNSR) Lab
Bradley Department of Electrical and Computer
    Engineering
Virginia Tech
Blacksburg, Virginia, USA

Yongwei Huang
Department of Electronic and Computer
    Engineering
Hong Kong University of Science and
    Technology
Clear Water bay, Kowloon, Hong Kong

Brian Jalaeian
Complex Networks and Security Research
    (CNSR) Lab
Bradley Department of Electrical and Computer
    Engineering
Virginia Tech
Blacksburg, Virginia, USA

Sastry Kompella
Naval Research Laboratory
Washington, D.C., USA

Hongxiang Li
Xian Communication Institute
Xian, China

Lei Li
Department of Electrical and Computer
    Engineering
University of Massachusetts Lowell One
    University Avenue
Lowell, Massachusetts, USA

Wulin Liu
Xian Communication Institute
Xian, China

Wenjing Lou
Complex Networks and Security Research
    (CNSR) Lab
Department of Computer Science
Virginia Tech
Blacksburg, Virginia, USA

Shan Luo
School of Aeronautics and Astronautics
University of Electronic Science and
    Technology of China
Chengdu, China

Petri Mähönen
Institute for Networked Systems
RWTH Aachen University
Aachen, Germany

Farokh Marvasti
Department of Electrical Engineering
Sharif University of Technology
Tehran, Iran

Marja Matinmikko
VTT Technical Research Centre of Finland
Oulu, Finland

Scott F. Midkiff
Virginia Tech
Blacksburg, Virginia, USA

Mohammad Robat Mili
Department of Electrical Engineering
Sharif University of Technology
Tehran, Iran

Hussein T. Mouftah
School of Electrical Engineering and Computer
    Science (EECS)
University of Ottawa
Ottawa, Canada

Miia Mustonen
VTT Technical Research Centre of Finland
Oulu, Finland

Gosan Noh
Electronics and Telecommunications Research
    Institute (ETRI)
Daejeon, Korea

Xiaoming Peng
Institute for Infocomm Research (I^2R)
Agency for Science, Technology and Research
    (A*STAR)
Singapore

Barbaros Preveze
Department of Electrical-Electronics
    Engineering
Cankaya University
Ankara, Turkey

Haythem Bany Salameh
Telecommunications Engineering Department
Yarmouk University
Irbid, Jordan

Shamik Sengupta
Department of Computer Science and
    Engineering
University of Nevada, Reno (UNR)
Reno, Nevada, United States

Douglas C. Sicker
Department of Engineering and Public Policy
Carnegie Mellon University
Pittsburgh, Pennsylvania, USA

Ali Tajer
Department of Electrical, Computer, and
    Systems Engineering
Rensselaer Polytechnic Institute
Troy, New York, USA

Chee Wei Tan
Department of Electrical and Computer
    Engineering
National University of Singapore
Singapore

Shensheng Tang
Department of Engineering Technology
Missouri Western State University
Saint Joseph, Missouri, USA

Deepak K. Tosh
Department of Computer Science and
    Engineering
University of Nevada, Reno (UNR)
Reno, Nevada, United States

Xiaodong Wang
Electrical Engineering Department
Columbia University
New York City, New York, USA

Zhiqing Wei
School of Information and Communication
Engineering
Beijing University of Posts and
Telecommunications
Beijing, China

David Tung Chong Wong
Institute for Infocomm Research (I^2R)
Agency for Science, Technology and Research
(A*STAR)
Singapore

Yong Xiao
Department of Electrical and Computer
Engineering
University of Houston
Houston, Texas, USA

Tangwen Xu
State Key Lab of Integrated Services Networks
Xidian University
Xian, China

Yuhua Xu
College of Communications Engineering
PLA University of Science and Technology
Nanjing, China

Seppo Yrjölä
Nokia Networks
Oulu
Finland

Pu Yuan
Department of Electrical and Computer
Engineering
University of Houston
Houston, Texas, USA

Xu Yuan
Complex Networks and Security Research
(CNSR) Lab

Bradley Department of Electrical and Computer
Engineering
Virginia Tech
Blacksburg, Virginia, USA

Xiangping Zhai
Department of Computer Science and
Technology
Nanjing University of Aeronautics and
Astronautics
Nanjing, China

Ning Zhang
Department of Electrical and Computer
Engineering
University of Waterloo
Waterloo, Ontario, Canada

Qixun Zhang
School of Information and Communication
Engineering
Beijing University of Posts and
Telecommunications
Beijing, China

Yani Zhang
Northwestern Polytechnical University
Xian, China

Yinfa Zhang
Xian Communication Institute
Xian, China

Liang Zheng
Department of Electrical Engineering
Princeton University
Princeton, New Jersey, USA

Junni Zou
School of Communications and Information
Engineering
Shanghai University
Shanghai, China

BIG PICTURE

I

Chapter 1

Physical Aspects of Spectrum Sharing

Dr Michael Fitch

CONTENTS

1.1 Introduction

Spectrum sharing means different things to different people. To the national administrations it means to make more spectrum available for services whose growth is in the national interest, without upsetting too much the existing users of the spectrum. To network operators with a diverse spectrum portfolio it means sharing that portfolio among its customers to maximize business value. Other actors in the value chain, such as service providers and end users, don't care about spectrum; they are interested in receiving sufficient service at acceptable cost. We restrict ourselves here to dynamic spectrum access (DSA), whereby sharing is organized among users, and the allocation can change in time depending on the demands of the systems that are sharing. This is distinct from co-existence, whereby a fixed provision is made for users of the same or different spectrum so that they don't cause each other harmful interference. In addition to the division into national and corporate sharing, we can also divide spectrum sharing according to the licence conditions, and here we make two classes, which are licence-exempt sharing (LES) and licensed assisted access (LAA). Both LES and LAA can be on a national or corporate basis and we have to consider all the combinations when thinking about licensed shared access (LSA).

Because radio transmitters and receivers are not perfect, all users will transmit some energy into spectrum that they do not intend to, and all users will be vulnerable to interference from spectrum that they are not supposed to be receiving from. The level of unwanted emissions and the level of susceptibility to interference are specified for each type of radio technology in the form of spectral masks which are generally harmonized across world regions and enforced by the national regulators in the licence conditions. In the case where different radio technologies are sharing the same band using DSA (co-channel), it is necessary to set limits on transmit power and / or on sharing distance so that the technologies do not cause harmful interference to each other. In the case where different radio technologies are not sharing the same band, it is necessary also to set limits so that the unwanted emissions from one do not cause harmful interference to the other. These in-band and out-of-band limits will be different according to the mix of radio technologies involved, and it is necessary for the spectrum manager to take account of them. There are also 'grey" areas whereby a degree of interference can be tolerated which degrades the quality of service (QoS) and the manager can administer a portfolio of spectrum which likely includes QoS as a parameter.

An example of spectrum being made available in the national interest, and being LES, is the sharing of government-held spectrum with non-government users in the form of TV whitespace (TVWS). In this case, the unlicensed users share the spectrum with licensed users, where the licensed users have priority. They are the "'primary" users, and the unlicensed sharing users are the "secondary" users; they share on the condition that they obey certain rules laid down by the national regulator to make sure they do not cause harmful interference to the primary users. In this way, the secondary users have "opportunistic" use of the spectrum and the regulator does not care to protect these users from one another or from the primary users. So the spectrum manager has to, by law, incorporate the rules specified by the regulator, but in order to be effective it also has to incorporate mechanisms to optimize the secondary user experience. This latter area is one of many fertile places for innovation.

Trials of TVWS have been carried out in the US, Europe and Asia, and commercialization of TVWS spectrum managers is planned to take place in late 2015 at least in the US and the UK. Both the FCC and Ofcom in the UK have stated a policy whereby other spectrum will follow suit, and it is becoming widely recognized that spectrum sharing is an essential ingredient in freeing up enough spectrum for use by mobile broadband.

The US NTIA has the model city initiative where federal users will share with non-federal, and this initiative has its roots in the PCAST report about making more efficient use of government held spectrum[1]. In memorandum dated June 2013 to the NTIA and FCC, the president of the US called for speeding up the progress towards spectrum sharing to support the administration's vision of 500 MHz being freed up for mobile communications by 2020. The activity is targeting spectrum below 3 GHz , and the US administration recognizes that spectrum sharing will be necessary to achieve this vision. The cost of clearing and re-allocating spectrum is very high, and can be avoided if effective sharing can be achieved. Other administrations are coming to the same conclusions; the European Commission is promoting spectrum sharing in its Digital Agenda for Europe[2], and they say that meeting growing spectrum needs for wireless connectivity is constrained by lack of vacant spectrum and by the high price associated with re-allocating spectrum to new uses, in terms of cost, delays and the occasional need to switch off incumbent users.

The physical aspects of spectrum sharing take into account the necessary technical aspects such as waveform and tuning flexibility, and also the rules that enable sharing without causing harmful interference. The rest of the chapter attempts to describe these aspects.

1.2 Radio Technology Aspects of Sharing

In order to be suitable for spectrum sharing, radio units must have some or all of the following attributes or abilities:

- Tune over a range of frequencies without re-starting. In practice this is limited by the power amplifier bandwidth and the extent to which filters can be electronically tuned. A typical maximum range over which a radio unit can be tuned without component changes is an octave (doubling) in frequency, and often is much less,

- Operate in a range of channel widths, from say 1 to 20 MHz in the same equipment. Carrier aggregation can be used to utilize more than one channel at one time, and such aggregation can use adjacent channels or non-adjacent channels. Narrow band radio units can use narrower channels, down to <10kHz.

- Ability to adjust transmit power, for example from less than 1mW to more than 20W for some equipment,

- Utilize frequency division duplex (FDD) or time-division duplex (TDD), ideally being able to switch between them,

- Operate with a range of waveforms, most commonly orthogonal frequency division multiplexing (OFDM) or one of its derivatives, with a choice of modulation depths and error coding schemes,

- Perform sensing or radio environment monitoring, which can be in-band (intra-frequency) or out-of band (inter-frequency),

- Process user data using scheduling and queuing for different QoS.

Figure 1.1 shows a notional block diagram of a radio unit suitable for dynamic spectrum sharing.

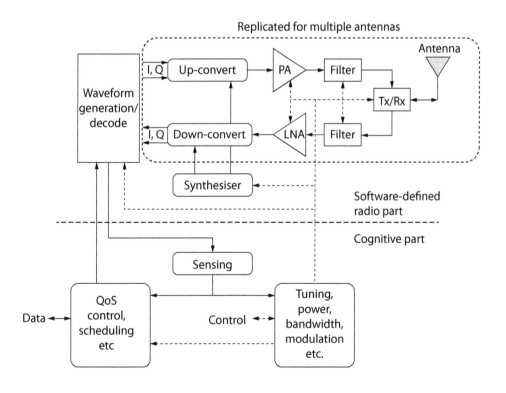

Figure 1.1: Notional block diagram of a radio unit suitable for DSA.

Figure 1.1 is divided into two major parts, which are a software-defined radio (SDR) part and a cognitive part. The SDR part is passive in the sense that it has no intelligence or algorithms running within it, but it takes instructions from the cognitive part. The dashed lines in Figure 1.1 represent the control plane; the function of the control plane is to tune the radio by setting the synthesizer (local oscillator source), to adjust the transmit power, to tune the filters, and to control the Tx / Rx switching, the waveform generation and the scheduling and queueing parameters.

The sensing block could be in either part, but is shown in the cognitive part because ideally it would adapt to different sensing methods, for example feature detection or energy detection.

With regard to tuning, different parts of the radio unit have different abilities. The blocks capable of tuning over the widest range are the synthesizer, converters and sensing. Typically, semiconductor devices are available for these functions that cover over a decade in frequency, from say 450 MHz to 6 GHz . Antennas can be designed to cover a large range, typically an octave say from 2 – 4 GHz , but the compromise is large size and low gain. A small printed antenna will be band-specific with ¡10% bandwidth. Some designs have a separate wideband antenna for sensing. The low noise amplifier (LNA), transmit power amplifier and the filters are probably the worst performing in terms of tuning and bandwidth. These typically have less than 50% tuning range, say from 2 – 3 GHz .

It is the analog components of the radio unit that are currently restricting its tuning range, and if the unit is required to have a range that exceeds any of them we have to fit multiple parallel analog stages, each operating at a different centre-frequency. To avoid this high-cost method of achieving greater range, the design and fabrication of wide-band amplifiers, filters and antennas are currently fertile areas for research.

High bandwidth is as important as wide tuning range for some services, which puts more demand on the analog parts of the radio unit. For example, a common LTE bandwidth is 20 MHz, and if the radio unit is required to cover the tuning range of 100 to 1000 MHz, then the range from the lowest possible signal frequency to the highest is 90 to 1010 MHz. LTE-A uses carrier aggregation as a way of increasing the available bandwidth, using up to 5 * 20 MHz carriers. It is obviously best if these carriers are all within the bandwidth and tuning range of a single set of analog stages in the radio unit, which is easiest if the carries are all adjacent. But the likelihood of them being adjacent is very low, given the current allocation of spectrum between operators, and it is highly likely that carriers to be aggregated will be in different bands, for example one at 1800 MHz and one at 2600 MHz. Using carrier aggregation where the carriers are far apart like this means that the radio units have to employ parallel analog transmit and receive chains.

Interference can occur in three ways. First the interfering signal can block the wanted signal when the two signals are not occupying the same spectrum, second the interfering signal can be produced unintentionally by a transmitter (i.e., unwanted emissions) that falls within the wanted signal spectrum, and third the interfering signal can intentionally be on the same channel. Let us look at these three mechanisms.

With blocking, a strong signal that is not on the same frequency as the wanted signal causes the receiver to overload or become non-linear. Figure 1.2 shows normal operation, where the radio receiver is designed to select the required frequency (Frequency Selected – fs) and the receiver LNA and down-conversion functions can amplify and process the signal correctly.

Figure 1.2 shows that fs just gets bigger. In Figure 1.3 a strong unwanted interfering signal (fi) is introduced which is not perfectly filtered at the receiver input, and therefore gets passed to the LNA and down-conversion functions, which are having problems in handling its large amplitude.

The interferer (fi) will suppress the wanted signal (fs) and generate more unwanted signals (3rd order products). The resulting lower wanted signal, and the presence of additional spectral products, will impact the ability of the receiver to decode the wanted signal.

With interference caused by unwanted emissions, the interferer is also on a different frequency to the wanted signal, but unwanted emissions from it extend across the wanted signal spectrum. This is illustrated in Figure 1.4.

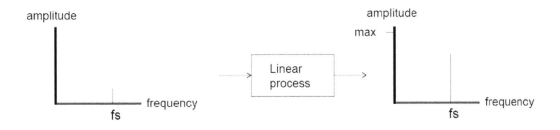

Figure 1.2: RF front end requires the signal being received to work in the linear range.

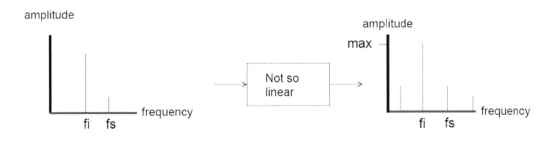

Figure 1.3: Presence of a strong unwanted signal can drive a receiver front end into non-linearity.

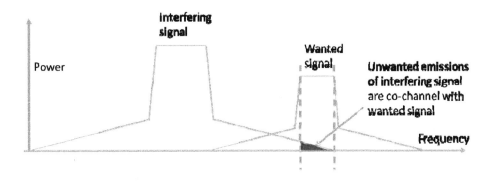

Figure 1.4: Interference caused by unwanted emissions from the interfering signal appearing in the same spectrum (co-channel) as the wanted signal.

In this case, the situation cannot be improved by receiver filtering or by increasing the linearity of the receiver, since the interference is on the same frequency as the wanted signal. It can be improved by better filtering on the interfering transmitter, or by setting a minimum distance between the two. Table 1.1 below summarizes the interference mechanisms and the ways in which they can be mitigated. In addition to these ways, all cases of interference can be reduced by maintaining a minimum distance between the interfering transmitter and the wanted signal receiver.

Table 1.1   Interference Mechanisms and Their Mitigation.

Interference mechanism	*Mitigation*
Blocking	Improving receiver front-end filtering and / or linearity
Unwanted emissions from interferer	Improving filtering and / or linearity on interfering signal transmitter

1.3   Types of Spectrum Sharing

Digital Europe [3] defines two types of spectrum sharing, which we mentioned in the introduction to this chapter. The two are (LSA) otherwise known as authorized shared access (ASA), and (LES).

A radio network operator has the following options to satisfy use demand as far as technically and commercially feasible:

1. Allocating spectrum from his licensed portfolio (LSA),

2. Allocating spectrum from the licensed portfolio of another operator (LSA), within a commercial and political framework agreement between operators similar to a roaming agreement,

3. Using entirely spectrum that is designated as unlicensed (LES).

4. Allocating spectrum from both his licensed portfolio (e.g., for control traffic or for some user traffic) and also using some unlicensed spectrum (e.g., for use traffic only),

5. Using spectrum that is owned by another operator on an opportunistic basis (LES).

Options 3 and 4 involve the use of unlicensed spectrum, or strictly, it is that unlicensed equipment can use the spectrum. Spectrum allocations for unlicensed equipment that are below 1 GHz are shown in Table 1.2, generally known as Industrial Scientific and Medical (ISM) bands, which are largely harmonized across Europe:

As these are narrow-band channels they support only low bit-rates, typically less than 100kbit/s. Some are suited to infrequent bursts of data and most of them have restrictions on transmit power and / or duty cycle. The operating power and bandwidth of the 870 MHz – 876 MHz band is uncertain at time of writing.

Above 1 GHz there the wide-band ISM bands at 2.4 GHz , 5.4 GHz and 5.8 GHz used commonly for Wi-Fi, and these can support higher bit-rates of several hundred Mbit/s.

The ISM bands offer no guarantee of service and have varying degrees of ability to cope with contention for the channel. Some do not have any mechanism for dealing with contention and take their chance ("fire and forget"), while others such as Wi-Fi___33 have sophisticated random back-off mechanisms and automatic selection of free channels using listen before talk (LBT). Any sharing

Table 1.2   Spectrum for unlicensed equipment that is below 1 GHz .

Frequency band	EIRP	Duty	Channel	
(MHz)	(dBm)	cycle	bandwidth	Comments
458.5–459.5	+20	No limit	25 kHz	Intented for short-range devices
863–865	+10	No limit	<300 kHz	Wireless microphones
868–868.6	+14	< 1%	No limit	
868.7–869.2	+14	< 0.1%	No limit	
869.3–869.4	+10	No limit	< 25 kHz	Access protocol required
869.4–869.65	+27	< 10%	< 25 kHz	Channels may be combined
869.7–870	+7	No limit	No limit	
870–876		< 10% possible		SRD, RFID, Smart metering. Need to co-ordinate with GSM R.

that we put in place that uses these ISM bands must also comply with the regulatory conditions such as those in Table 1.2, and any contention mechanisms that apply.

Options 1 and 2 using LSA will need to comply with the conditions attached to the licences, and also comply with any relevant standards applicable to the country or region, such as the ETSI standards. These conditions and standards are put in place to control the amount of interference that one system can impose on others, and also to remain below safe transmission radiation limits. This is best illustrated with a couple of examples. The first example is the 2.6 GHz frequency band that was auctioned in the UK two years ago and is designated for use with LTE. This spectrum band is allocated to three operators as shown in Figure 1.5.

Going from left to right in Figure 1.5, this band consists of an frequency division duplex (FDD) uplink portion a time-division duplex (TDD) portion and an FDD downlink portion. The FDD uplink and downlink portions need to have a separation in the centre because of the need to provide adequate isolation between them in the radio units, and most of this isolation is provided by the analog filtering as discussed in the previous section. In this center gap is placed TDD carriers, from 2570 MHz to 2615 MHz. At the upper frequency end of the band is a 10 MHz guard band between the top of the allocated band at 2690 MHz and a radar band that starts at 2700 MHz. This band is used for the approach radar systems at London's Heathrow airport and at many other airports throughout Europe. In order that EE can use this top band without interfering with the radar systems, Ofcom imposed the out of band limits shown in Table 1.3.

LTE uses an OFDM-based physical layer, which cannot on its own achieve −45 dBm/ MHz unwanted emissions if the top LTE channel is used. The 10 MHz guard band is not sufficient for the OFDM spectrum to fall that far, if reasonable transmitter powers are to be used. The limit of -45 dBm/ MHz is possible to achieve by utilizing additional filtering after the transmitter power

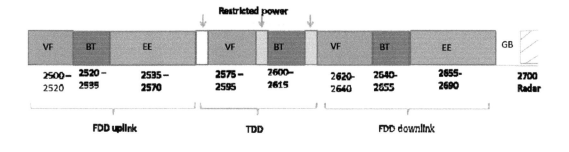

Figure 1.5: The 2.6 GHz LTE band UK allocations: VF = Vodafone, BT = British Telecom, EE = Everything Everywhere, GB = Guard band.

Table 1.3   Unwanted Emissions Limits at Upper End of the 2.6 GHz LTE Band.

Frequency	Unwanted emission limit	Notes
>2700 MHz	−45 dBm/ MHz	This is stricter than the 3GPP limit which is −30 dBm/ MHz
>2720 MHz	−65 dBm/ MHz	This is needed if exclusion zones are to be avoided

amplifier, which adds some cost and mass to the base-stations. But the limit of -65 dBm beyond 2720 MHz is more difficult, it requires more expensive and bulky filters, and the additional cost of this filtering has to be weighed against the alternative approach, which is to establish exclusion zones around the radar systems.

In line with the mitigation methods in Table 1.3, the radar operators had to fit filters to their receivers to prevent LTE from blocking them. The cost of fitting the filters was borne by the UK government, as part of the costs of making the 2.6 GHz LTE band available.

The other aspect of interest in this band is the placing of TDD in between the FDD uplink and downlink. Guard bands are used here also, to help isolate the systems from one another, and a further guard band is places between the two TDD operators. This is needed again so that adequate filtering can be provided, but this time to prevent a TDD transmitter from interfering too much with FDD receivers, or with TDD receivers on the other system that may not use the same uplink: downlink ratio. These guard bands are 5 MHz each and need not be absolutely clear of transmissions; they are known as restricted power bands, where the operator sitting above the guard band can use it at reduced power, typically +25 dBm/5 MHz. The transmission power in the main parts of the band is typically +60 dBm/5 MHz.

Any sharing of this band needs to take into account these regulatory conditions and any exclusion zones.

The second example we use is the 800 MHz allocation to LTE, also in the UK. In this case, a re-allocation was made that transferred some of the TV broadcast spectrum to mobile communications use. The part of the TV spectrum that was above 790 MHz – 820 MHz was transferred, along with 832 – 862 MHz to form an FDD pair of bands. At the lower end of this allocation, there is a thin 1 MHz guard band between the upper end of the TV broadcast band and the start of the new mobile band as shown in Figure 1.6.

The technical challenge with this band is interference from and into TV services. The 700 MHz band works the reverse way around to the other mobile bands in that the base-station transmit is in the lower part of the FDD spectrum and the UE transmit is in the upper part. This was done so that the TV transmitters are closer in frequency to the UE receivers than they are to the base-station receivers, and as the UEs are normally at less height, the problem is reduced. However the other side of this issue is that the LTE base-stations are transmitting at frequencies close to the TV receivers at the top channel (channel 60), especially the base-stations owned by 3. The UK government made available low-pass filters, for free, that could be fitted to TV sets to solve the blocking problems that might arise.

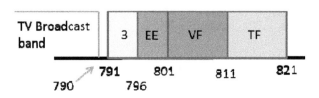

Figure 1.6: 700 MHz allocation in the UK.

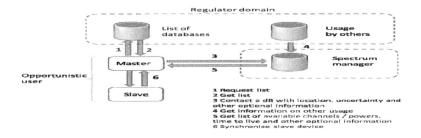

Figure 1.7: Database structure example for opportunistic (secondary) user sharing spectrum with primary users.

1.4 Spectrum Management

All of the options for sharing dynamically can be managed through the use of a spectrum management database, or using sensing, or both. Figure 1.7 shows a possible arrangement for sharing where an opportunistic user is sharing spectrum with a primary (licensed) user. This is the database structure used with TVWS.

Figure 1.7 shows a case where the spectrum manager must be aware of the primary users of the spectrum and avoid causing them harmful interference. The primary users in this case are TV broadcast, and also wireless microphones. The national regulator took exhaustive steps in this case to protect the primary users, and so they installed a system whereby spectrum manager providers would need to certify their database operations before being allowed to go live. The regulator keeps a list of certified managers, and when a secondary system wants spectrum, it must first get one from the list. Then it contacts the spectrum manager, which returns a list of available channels and powers that the secondary devices can use. The spectrum manager must also accept input from the regulator relating to operational areas of the primary user.

Sensing is not mandatory in the structure shown in Figure 1.7, but it can be used to enrich the information that the secondary user receives. The information that would be useful to the secondary system is

- a choice of spectrum to use;

- allowable transmit power and bandwidth;

- restrictions on duty cycle;

- amount of pollution on the channel.

The first three items on this list could come from one or more databases that are held either by a regulator (as is the case with TVWS) or by the organization that holds the licence. The last item would come from sensing, whereby sharing users would report channel conditions back to the manager.

Two-stage spectrum managers have been studied [4] where one manager is responsible for maintaining a spectrum portfolio that could belong to an organization or to a government, and a second manager is used to allocate the spectrum to individual radio links, a resource manager. The portfolio manager would be centralized and the resource manager would be distributed, as shown in Figure 1.8.

Figure 1.8 shows a two-stage spectrum manager approach and gives typical information that needs to flow between the master radio unit (the base-station) and the resource manager, and further up to the portfolio manager. There will generally be one portfolio manager and multiple co-operating resource managers, and this approach has been simulated to be scalable to large numbers of master

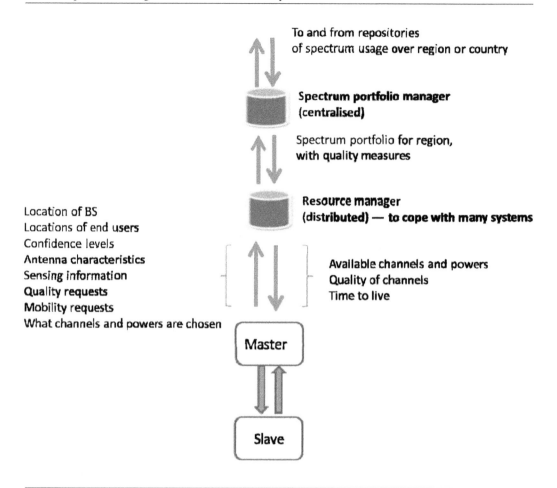

To and from repositories
of spectrum usage over region or country

**Spectrum portfolio manager
(centralised)**

Spectrum portfolio for region,
with quality measures

**Resource manager
(distributed) — to cope with many systems**

Location of BS
Locations of end users
Confidence levels
Antenna characteristics
Sensing information
Quality requests
Mobility requests
What channels and powers are chosen

Available channels and powers
Quality of channels
Time to live

Master

Slave

Figure 1.8: Two-stage spectrum manager.

units. There must exist a protocol between master and slave devices to prevent slaves from transmitting until the master has found a channel. If the slave wants to initiate a connection, it can do so by emitting a short low-power burst before the master has been allocated a channel.

Spectrum aggregation may be used in any combination of usage we have discussed, to give higher bit-rates to the user. To show the type situation where spectrum aggregation can give advantage, have a look at Figure 1.9, which is the 900 MHz mobile spectrum allocation to Vodafone and Telefonica in the UK at the time of writing.

Figure 1.9 shows only the uplink portion of this FDD band. The downlink band is of identical pattern between 925.1 MHz and 959.9 MHz. Due to some unfortunate history of how this band was divided up, Vodafone and Telefonica have allocations of between 4.6 and 7.6 MHz alternatively across the band. Also, there is a 0.4 MHz gap in both the uplink and downlink bands that neither operator is using.

Currently, this band is used for 2G, but if we try and operate LTE in the band, the efficiency would not be good, because the LTE bandwidth options of 1.4 MHz, 3 MHz, 5 MHz, 10 MHz, 15 MHz and 20 MHz are all a bad fit. It would only be possible to run 2 * 5 MHz and 1 * 3 MHz LTE channels across the three allocations belonging to each operator. However, if the spectrum could be aggregated, then users would obtain a much greater bit-rate than if they were only able to utilize one LTE channel at a time. Aggregation like this, which is using fragments of spectrum in the same

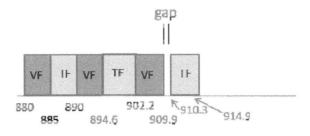

Figure 1.9: UK 900 MHz mobile spectrum. Uplink only shown. VF = Vodafone, TF = Telefonica.

band, is easier to achieve than aggregation across different bands, because it uses just one receiver and one transmitter chain.

If Vodafone and Telefonica could agree with the regulator to re-allocate the band so each had contiguous spectrum in two halves, they would have 17.5 MHz each and could run the 15 MHz LTE option, giving them a 15% increase in bit-rate over the aggregated case. This example shows an extreme allocation pattern, but generally the allocation of spectrum is fragmented to some degree over usually more than two operators in every band, because the bands become available at different times and are divided up among operators usually through an auction process. The allocation is fragmented to a certain degree in every band in every country. Rationalization at a later date is disruptive, takes a long time, and is expensive, and is almost never done. So sharing, including aggregation, will be of benefit for the foreseeable future.

1.5 Traffic Patterns

Dynamic spectrum sharing is one aspect of radio resource management, and the holding time of the connection is a function of the traffic that is being carried.

What we are really interested in doing is giving users what they want, at the price they are willing to pay, wherever they are located. Ideally this could involve choices, for example the user can elect to pay more for a movie to download in 1 minute rather than wait 10 minutes. This kind of 'premium" service comes with higher QoS and could involve usage of spectrum reserved for the purpose.

The traffic demand is bursty, and the burstiness comes about from user behaviour. A broadband user at home may usually request less than 50Mbit/s but every few hours download a game or a movie that is 7.5 Gbyte, with the expectation that it arrives in less than a minute, which is a download rate of 1 Gbit/s. Even the lower rate traffic may be bursty if he is viewing catch-up TV or web-browsing. If he downloads a game that interacts with other users, he may request a very low latency while playing that may not involve large volumes of traffic.

The above figures are what the user requests, but in practice the experience he gets depends on the equipment he has, how much he is willing to pay, and where he is located. Bit-rates of over 500Mbit/s can currently be achieved over wireless only if the users are very close to an LTE-A base-station that utilizes Multiple Input Multiple Output (MIMO) and carrier aggregation. The 5G systems are being dimensioned for a minimum of 1Gbit/s and it will be a challenge to provide this over wide geographical areas. Whatever the technology, it will be several years before these high rates are available in rural areas.

We can classify the traffic types as follows:

- Long-term traffic (say >30 minutes) due to applications like web-browsing, shopping, video conferencing or catch-up TV, with a mean and a peak value for bit-rate and a moderate latency (say 10 – 100ms)

- Long-term traffic which although of moderate volume is sensitive to latency, such as gaming (say 1ms)

- Short-term traffic (say <10 minutes) due to large file transfers like movies and games, which require a burst peak rate of say 50x the mean bit-rate

The bursty nature of the traffic, coupled with the knowledge that not all users will need to burst together, will further increase the need for a flexible and dynamic spectrum sharing solution. Perhaps one way of measuring the effectiveness of the system to satisfy the burst requirements is its responsiveness. If a user is connected via a radio link that gives 8 Mbit/s, and a request is made for a burst of 100 Mbit/s for 10 seconds, how quickly does the system respond by changing or modifying the link to accommodate the request? We propose that the performance parameters should be extended beyond the well-known ones of bit-rate and latency. We should add the difference between delivered and requested burst performance, the rate of change of bits/s as well as limits of bits/s, and the minimum wastage of radio resource when switched around bursty users. These are some of the key performance indicators that we hope become part of 5G systems.

References

[1] Promoting spectrum sharing in the wireless broadband era. National Telecommunications and Information Administration. January 9th 2015. Available here: http://www.ntia.doc.gov/blog/2015/promoting-spectrum-sharing-wireless-broadband-era. Retrieved May 1, 2015

[2] Promoting the shared use of radio spectrum resources in the internal market. Communication from the Commission to the European Parliament, the Council, the European Economic and Social Committee and the Committee of the Regions. Brussels 3.9.2012 COM(2012) 478 Final. Available here: https://ec.europa.eu/digital-agenda/sites/digital-agenda/files/com-ssa.pdf. Retrieved May 1, 2015.

[3] Digital Europe Positioning Paper on Licence Shared Access, February 14, 2013. Available here: http://www.digitaleurope.org/DesktopModules/Bring2mind/DMX/Download.aspx? Command=Core_Download&EntryId=519&PortalId=0&TabId=353. Retrieved May 3, 2015.

[4] Integrated final functional specification of spectrum management framework and procedures. QoSMOS deliverable 6.7. Available here: http://cordis.europa.eu/docs/projects/cnect/4/248454/080/deliverables/001-D67final.pdf. Retrieved May 3, 2015.

Chapter 2

Perspective on the Design of Opportunistic Spectrum Sharing

Haythem Bany Salameh

CONTENTS

One major challenge in the design of opportunistic cognitive radio networks (CRNs) is how to simultaneously provide efficient spectrum utilization (network throughput) while protecting the performance of licensed legacy primary radio networks (PRNs). Efficient medium access control and careful spectrum assignment have a great potential to achieve this goal. In this chapter, we investigate the issue of spectrum assignment in CRNs and examine various opportunistic spectrum access approaches proposed in the literature. We provide insight into the efficiency of such approaches and their ability to attain their design objectives. We discuss the factors that impact the selection of the appropriate operating channel(s), including the important interaction between the cognitive link-quality conditions and the time-varying nature of PRNs. Protocols that consider such interaction are described. We argue that using best quality channels does not achieve the maximum possible throughput in CRNs (does not provide the best spectrum utilization). The impact of guard bands on the design of opportunistic spectrum access protocols is also investigated. Various complementary techniques and optimization methods are underlined and discussed, including the utilization of variable-width spectrum assignment, resource virtualization, full-duplex capability, cross-layer design, beamforming and multiple input multiple output (MIMO) technology, cooperative communication, network coding, discontinuous-OFDM (orthogonal frequency division multiplexing) technology, and software defined radios. Finally, we highlight several directions for future research in this field.

2.1   Introduction

The design of spectrum sharing and access mechanisms for cognitive radio networks (CRNs) has attracted significant attention in the last few years. The interest in CRNs is mainly attributed to their ability to enable efficient spectrum utilization and to provide wireless networking solutions in scenarios where the un-licensed spectrum bands are heavily utilized. Furthermore, due to their cognitive nature, CRNs are more spectrum efficient and robust than their non-cognitive counterparts against spectrum unavailability, and have the capability to utilize different frequency bands and adapt their operating parameters based on the surrounding radio frequency (RF) environment. Specifically, CR is considered as the key technology to effectively address the inefficient spectrum utilization in legacy licensed wireless communication systems[1] by providing opportunistic on-demand access [1–5] (e.g., Figure 2.1 shows actual spectrum measurements taken in downtown Berkeley [6] that indicate a vast under-utilization in the licensed spectrum). CR technology enables unlicensed users to opportunistically utilize the idle PR channels (so-called spectrum holes). The spectrum holes represent the PR channels that are currently under-utilized. In order to utilize these spectrum opportunities without interfering with the PRNs, CR users should perform accurate spectrum sensing, through which idle channel lists are identified. In addition, the CR users should be flexible enough to quickly vacate the operating channel when a PR user reclaims it. In this case, CR users should quickly and seamlessly switch their operating channel(s).

---

[1]Spectrum measurement reports by the FCC and other organizations have indicated huge geographical and temporal variations in the utilization of the licensed portions of the spectrum, ranging from 15% to 85%.

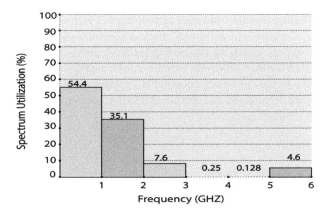

Figure 2.1: Spectrum utilization measurements [6].

While wide-scale deployment of CRNs is still to come, extensive research attempts are currently underway to improve the effectiveness of spectrum sharing protocols and improve the spectrum management and operation of such networks [5, 7–21]. Two of the most crucial challenges in deploying CRNs are the needs to maximize spectrum efficiency and minimize the caused interference to PRNs. On other words, providing efficient communication and spectrum access protocols that provide high throughput while protecting the performance of licensed PRNs are the crucial design challenge in CRNs. We now mention the main capabilities and limitations of a CR device that need to be considered when designing efficient CR communication protocols:

- Spectrum sensing capability: The main purpose of spectrum sensing is to allow CR users identifying spectrum opportunities.

- Spectrum sharing capability: The main purpose of spectrum sharing is to enable CR users to effectively share the available spectrum with other CR nodes without interfering with PRNs.

- Distributed coordination capability: The main purpose of distributed coordination is to allow CR users to establish common control channel and to organize their transmissions.

- Re-configurability capability: The main purpose of this capability is to allow CR users to adapt their transmission parameters (e.g., transmit power, carrier frequency, channel bandwidth, waveform shape, transmission technology) according to the sensed RF information [5].

- Spectrum mobility: This capability allows CR users to switch between channels in response to PR activities.

- Transmit power constraints: A CR user should use regulated transmit powers in order to prevent degrading the reception of the PR users.

The main objective of this chapter is to overview and analyze the key schemes and protocols for spectrum access/sharing/managemnent that have been developed for CRNs in the literature. Furthermore, we briefly highlight a number of opportunistic spectrum sharing and management schemes and explain their operation details. As confirmed later, it follows logically that cross-layer design, link quality/channel availability tradeoff, and interference management are the key design principles for providing efficient spectrum utilization in CRNs. We start by describing the main CRN architectures and operating environment. Then, the spectrum sharing problem is stated. The various objectives used to formulate the spectrum sharing problem in CRNs are summarized. We then point out the several design challenges in designing efficient spectrum sharing and access mechanisms. The tradeoffs in selecting the operating channel(s) in CRNS are discussed. A number of spectrum sharing design categories are then surveyed. Various complementary approaches, new technologies, and optimization methods that have great potential in facilitating the design of efficient CRN communication protocols are highlighted and discussed. Finally, concluding remarks are provided with several open research challenges.

2.2 Network Architecture

2.2.1 Cognitive Radio Network Model

Typical CRN environment consists of N different types of PRNs and one or several CRNs. The PR and CR networks geographically co-exist within the same area. In terms of network topology, two basic types of CRNs are proposed: centralized multi-cell CRNs and infrastructure-less ad hoc CRNs. Figure 2.2 depicts a composition view of a CRN operating environment consisting of an ad hoc CRN and a multi-cell centralized CRN that coexist with two different types of PRNs. The different PRNs have license to transmit over orthogonal non-overlapping spectrum bands, each with a different licensed bandwidth. PR users of a given PRN operate over the same set of licensed channels. CR users can opportunistically utilize the entire PR licensed and unlicensed spectrum. For ad hoc multi-hop CRNs without centralized entity, it is necessary to provide distributed spectrum access protocols that allow each CR user to separately access and utilize the available spectrum. On the other hand, for centralized multi-cell CRNs, it is desirable to provide (1) centralized spectrum allocation protocols that allocate the available channels to the different CR cells, and (2) centralized channel assignment mechanisms that enable efficient spectrum reuse inside each cell.

We note here that the IEEE 802.22 WRAN (wireless regional area network) is the first multi-cell centralized CR system that enables commercial applications based on CR technology. According to the IEEE 802.22, each CR cell consists of a base-station (BS) and a group of CR users. To provide accurate channel availability information, the CR users sense the spectrum availability in their locality and periodically share their sensing measurements with the associated BS. The IEEE 802.22 standard utilizes the idle channels in the VHF/UHF TV broadcast systems, ranging from 54 MHz to 862 MHz. The IEEE 802.22 standard provides a broadband (high data-rate) access to hard-to-reach areas with low population density, which makes it suitable for rural areas.

2.2.2 PR ON/OFF Channel Model

In general, the channel availability model of each PR channel in a given locality is described by a two-state ON/OFF Markov process. This model describes the evolution between idle (OFF) and busy (ON) states (i.e., the ON state of a PR channel indicates that the PR channel is busy, while the OFF state reveals that the PR channel is idle). The model is further described by the stochastic distribution of the busy and idle periods, which are generally distributed. The distributions of the idle and busy states depend on the PR activities. We note here the ON and OFF periods of a given channel are independent random variables. For a given channel i, the average idle and busy periods are $\overline{T}_I$ and $\overline{T}_B$, respectively. Based on this model, the idle and busy probabilities of a PR channel i are, respectively, given by $P_I^{(i)} = \frac{\overline{T}_I}{\overline{T}_I + \overline{T}_B}$ and $P_B^{(i)} = \frac{\overline{T}_B}{\overline{T}_I + \overline{T}_B}$. Figure 2.3 shows a transition diagram of a 2-state busy/idle Markov model of a given PR channel. We note here that neighboring CR users typically have similar views to spectrum availabilities, while non-neighboring CR users have different channel availability conditions.

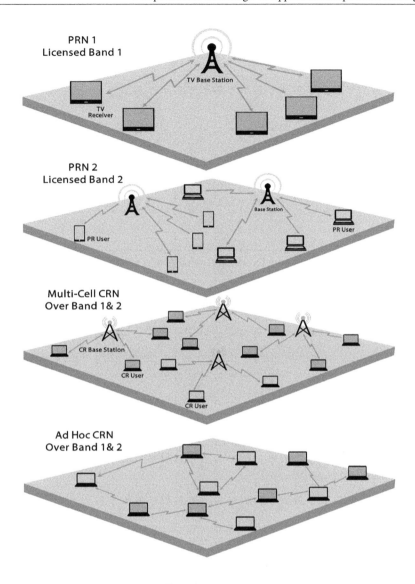

PRN 1
Licensed Band 1

PRN 2
Licensed Band 2

Multi-Cell CRN
Over Band 1& 2

Ad Hoc CRN
Over Band 1& 2

Figure 2.2: Generic architecture of a CRN environment.

2.3   Spectrum Sharing Problem Statement and Objectives

Spectrum sharing problem (including spectrum management and decision) can be stated as follows: " Given the output of spectrum sensing, the main goal is to determine which channel(s) to use, at what powers, and at what rates, such that a given performance metric (objective function) is optimized. This is often a joint optimization problem that is very difficult to solve (often it constitutes an NP-hard problem). Recently, several spectrum assignment strategies have been proposed for CRNs [20–54]. These strategies are designed to optimize a number of performance metrics including:

- Maximizing the CR throughput (individual users or network-level) based on Shannon capacity or a realistic staircase rate-SINR (signal to interference and noise ratio) function (e.g., [22–25]).

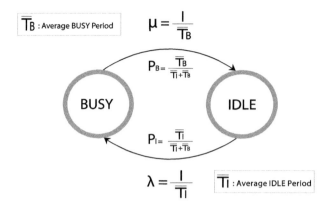

Figure 2.3: Two-state Markov channel availability model of a given PR channel.

- Minimizing the number of assigned channels for each CR transmission (e.g., [20, 26]).

- Maximizing the CR load balance over the different PR channels (e.g., [27]).

- Minimizing the probability of PR disruption, i.e., minimizing the PR outage probability (e.g., [21, 28]).

- Minimizing the average holding time of selected channels (minimizing the PR disruption) (e.g., [29, 30]).

- Minimizing the frequency of channel switching due to PR appearance by selecting the channel with maximum residual idle time, i.e., minimizing the CR disruption in terms of forced-termination rate (e.g., [31–33]).

- Maximizing the CR probability of success (e.g., [34, 36]).

- Minimizing the spectrum switching delay for CR users (e.g., [37–39]).

- Minimizing the expected CR waiting time to access a PR channel (e.g., [40, 41]).

- Minimizing CRN overhead and providing CR quality of service (QoS) (e.g., [42]).

- Minimizing the overall energy consumption (e.g., [43–46]).

- Achieving fair spectrum allocation and throughput distribution in the CRN (e.g., [47–51]).

- Maintaining CRN connectivity with predefined QoS requirements (e.g., [52–54]).

We note here that the spectrum sharing problem for any of the aforementioned objectives is, in general, NP-hard. Therefore, several heuristics algorithms and approximations have been proposed to provide suboptimal solutions for the problem in polynomial-time. These heuristics and approximations can be classified based on their adopted optimization method as: graph theory-based algorithms (e.g., [56, 56]), game theory-based algorithms (e.g., [38]), genetic-based algorithm (e.g., [57]), linear programming relaxation-based algorithms (e.g., [58]), fuzzy logic-based algorithms (e.g., [59]), dynamic programming-based algorithms (e.g., [60]), and sequential-fixing-based algorithms (e.g., [20]).

2.4 Issues in Designing Spectrum Sharing Mechanisms

2.4.1 Interference Management and Co-Existence Issue

The coexistence problem is one of the most limiting factors in achieving efficient CR communications. In a CRN environment, there are three kinds of harmful interference that should be considered: PR-to-CR/CR-to-PR interference (the so-called PR coexistence) and CR-to-CR interference (the so-called self-coexistence). While several mechanisms have been proposed to effectively deal with the PR-to-CR interference problem based on cooperative (e.g., [61–64]) or noncooperative (e.g., [65–67]) spectrum sensing, the CR-to-CR and CR-to-PR interference problems are still challenging issues.

2.4.1.1 Self-Coexistence Management

To address the CR-to-CR interference problem in ad hoc CRNs, several channel allocation and self-coexistence management mechanisms have been proposed based on either (1) exclusive channel assignment or (2) joint channel assignment and power control. On the contrary, the CR-to-CR interference problem has been addressed in multi-cell centralized CRNs based on either fixed channel allocation [68–72] or adaptive traffic-aware/spectrum-aware channel allocation [73–75].

2.4.1.2 Providing Performance Guarantees to PR Users

It has been shown that the CR-to-PR interference is the most crucial interference in CRN environment, because it has a direct effect on the performance of PRNs. Hence, the transmission power of CR users over the PR channels should be adaptively computed such that the performance of the PRNs is protected. Based on the outcomes of spectrum sensing, two different power control strategies can be identified: binary and multilevel transmission power strategies. According to the binary-level strategy (the most widely used power control strategy in CRNs), CR users can only transmit over idle channels with no PR activities. Specifically, for a given PR channel, a CR transmits 0 power if the channel is busy, and uses the maximum possible power if the PR channel is idle. While this strategy ensures collision-free spectrum sharing between the CR and PR users, it requires perfect spectrum sensing. Worse yet, the binary-level strategy can lead to non-optimal spectrum utilization. On the other hand, using a multi-level adaptive frequency-dependent transmission power strategy allows the CR and PR users to simultaneously share the available spectrum in the same locality, which can significantly improve spectrum utilization. By allowing CR users to utilize both idle and partially-occupied PR channels, much better spectrum utilization can be achieved. The multi-level power strategy can also be made time-dependent to capture the dynamics of PR activities. Under this strategy, controlling the CR-to-PR interference is nontrivial. In addition, computing the appropriate multi-level power strategy is still a challenging issue, which has been studied under some simplified assumptions. Specifically, the authors in [26] proposed an adaptive multi-level frequency- and locality-dependent CR transmission power strategy that provides a soft guarantee on PRNs' performance. This adaptive strategy is dynamically determined according to the PR traffic activities and interference margins.

2.4.2 Distributed Coordination Issue

In this section, we review several well-known distributed coordination mechanisms designed for CRNs. We note that control channel designs for CRNs can be loosely classified into seven different categories [5, 76]:

- Dedicated pre-specified out-of-band common control channel (CCC) design [21, 77–80].

- Non-dedicated pre-specified in-band CCC design [8, 9, 36, 81, 82].

- Hybrid out-of-band and in-band CCC design [83].

- Hopping-based control channel design [84–90].

- Hybrid in-band CCC and hopping-based control channel design [91].

- Spread-spectrum-based control channel design [92, 93].

- Cluster-based local CCC design [94–100].

Despite the fact that using a dedicated out-of-band CCC is straightforward, it contradicts the opportunistic behavior of CRNs, may result in a single-point-of-failure (SPOF) and performance bottleneck due to CCC saturation under high CR traffic loads. Similarly, using a pre-specified non-dedicated in-band CCC is not a practical solution due to spectrum heterogeneity and, if exists, such solution can result in a SPOF, become a performance bottleneck, and introduce security challenges. Another approach that can effectively deal with the CCC saturation issue (bottleneck problem) is to use a hybrid out-of-band and in-band CCC (simultaneous control communications over in-band PR channels and dedicated out-of-band CCCs). This approach exploits the strengths of out-of-band and in-band signaling and, hence, can significantly enhance the performance of multi-hop CRNs. Using a hopping-based control channel can address the SPOF, bottleneck, and security issues. However, in such type of solutions, the response to PR appearance is challenging, as CR users cannot use a PR channel once reclaimed by PR users. In addition, this type of solution is generally spectrum unaware. Another key design issue in such solutions is the communication delay that heavily depends on the time to rendezvous. A hybrid control channel design of in-band CCC and frequency-hopping schemes can reduce the required time to rendezvous and maintain connectivity between CR users by adopting multiple control channels. Using cluster-based coordination solutions, where neighboring CR users are dynamically grouped into clusters and establish local CCCs, can provide reliable distributed coordination in CRNs [5, 76]. However, adopting this type of solutions in a multi-hop CRN is limited by several challenges, such as providing reliable inter-cluster communication (i.e., different cluster may consider different CCCs), maintaining connectivity, broadcasting control information, identifying the best/optimal cluster size, and maintaining time-synchronization [5]. Finally, using spread-spectrum-based distributed coordination is a promising solution to most of the aforementioned design challenges, but the practicality and design issues of such a solution need to be further investigated. According to this solution, the control information is spread over a huge PR bandwidth with a very low transmission power level (below the noise level). Consequently, with a proper design, an efficient CCC design can be implemented using spread spectrum with minor effect on PRNs' performance. In conclusion, various distributed coordination mechanisms have been developed to provide reliable communications for CRNs, none of which are totally satisfactory. Hence, designing efficient distributed coordination schemes in CRNs should be based on novel coordination mechanisms along with effective transmission technologies that enable effective, robust, and efficient control message exchanges.

2.5   Tradeoffs in Selecting the Operating Channel

The spectrum (channel) assignment problem in CRNs has been extensively studied in the literature. Existing channel assignment/selection solutions can loosely be classified into three categories: best link-quality schemes, larger availability-period schemes, and joint link-quality and channel-availability-aware schemes. It has been shown (e.g., [21, 33, 101]) that using the best link-quality schemes in CRNs, where the idle channel(s) with the highest transmission rate(s) are selected, can only provide good performance under relatively static PR activities with average PR channel idle durations that are much larger than the needed transmission times for CR users [9,21,33,34,36]. Under highly dynamic PR activities, this class of schemes can result in increasing the CR termination rate, leading to a reduction in CRN performance, as a CR user may transmit over a good-quality PR channel with relatively short availability time (short channel-idle period). On the other hand, employing the larger availability-period schemes in CRNs (e.g., [102]) can result in increasing the CR forced-termination rate as an idle PR channel of very poor link-quality (low transmission rate) may be chosen, resulting in a significant reduction in CRN performance. We note here that the interaction between the CRN and PRNs is fundamental for conducting channel assignment in CRNs.

The above discussion presents sufficient motivation to jointly consider the link-quality and average idle durations of PR channels when assigning operating channels to CR users. However, several open questions in this domain still need to be addressed; perhaps the most challenging one is how to jointly consider the link-quality and average idle durations into one metric to perform channel assignment. Other important questions are: How can a CR user estimate the distribution of the idle periods of the different PR channels? What are the implications of the interaction between the CRN and the PRNs? How can a CR user determine the link-quality conditions over the various (large number) PR channels? Some of these questions have been addressed in [33, 34, 36] by introducing the CR packet success probability metric. This metric is derived based on stochastic models of the time-varying PR behaviors. The probability of success over a given channel is a function of both the link-quality condition and the average-idle period of that PR channel. It has been proven that it is necessary to jointly consider the link-quality conditions and availability times of available PR channels to improve the overall network performance [33].

2.6   State-of-the-Art Spectrum Sharing Protocols in CRNs

There are several attempts that have been made to design spectrum sharing protocols with the objective of improving the overall spectrum utilization while protecting the performance of licensed PRNs. Existing spectrum sharing/access protocols and schemes for CRNs can loosely be categorized into four main classes based on: the number of radio transceivers per CR user (single-transceiver, dual-transceiver, or multiple transceiver), their reaction to PR behavior (reactive, proactive, or interference-based threshold), their spectrum allocation behavior (exclusive or non-exclusive spectrum occupancy model), and the guardband considerations (guardband-aware or guardband-unaware).

2.6.1   Number of Radio Transceivers and Assigned Channels

Spectrum sharing protocols and schemes for CRNs can also be categorized based on the number of radio transceivers per a CR user (i.e., single transceiver [10, 13–16, 108–112], dual transceivers [17, 113], and multiple transceivers [9, 21, 34, 114–116]). Using multiple (or dual) transceivers greatly simplifies the task of spectrum access design and significantly improve system performance. This is because a CR user can simultaneously utilize multiple channels (the potential benefits of utilizing multi-channel parallel transmission in CRNs were demonstrated in [20, 117]). In addition, the spectrum access issues such as hidden/exposed terminals, transmitter deafness, and connectivity can be easily overcome as one of the transceivers can be switched to the assigned control channel (i.e., CR users can always receive control packet over the CCC even when they are operating over the data channels). However, the achieved performance gain of using multiple transceivers (multi-channel parallel transmission) comes at the expense of extra hardware. Worse yet, the optimal joint channel assignment and power control problem in multi-transceiver CRNs is, in general, NP-Hard. On the other hand, it has been shown that the design of efficient channel assignment schemes for single-transceiver single-channel low-cost CRNs is simpler than that of the multi-transceiver counterpart [9]. While single-transceiver designs can greatly simplify the task of finding the optimal channel assignment, the aforementioned channel access issues are not trivial, and the performance is limited to the capacity of the selected channel.

2.6.2 Reaction to PR Appearance

Spectrum sharing schemes in the CRNs can also be classified based on their reaction to the appearance of PR users into three main groups: (1) proactive (e.g., [118–120]), (2) reactive (e.g., [121–123]), and (3) interference threshold-based (e.g., [20, 26, 117, 124]). In reactive schemes, the active CR users switch channels after the PR appearance. On the other hand, in proactive schemes, the CR users predict the PR appearance and switch channels accordingly. The threshold-based schemes allow the CR users to share the spectrum (both idle and partially-occupied PR channels) with PR users as long as the interference introduced at the PR users is within acceptable values. Existing threshold-based schemes attempt at reducing the impacts of the un-controllable frequency-dependent PR-to-CR interference on CRN performance through proper power control based on either (1) the instantaneous sensed interference [20], (2) the average measured PR interference [117], or (3) using stochastic PR interference models [26].

2.6.3 Spectrum Sharing Model

The spectrum sharing model represents the type of interference model used to solve the channel and power assignment problem. There are two different spectrum sharing models: protocol (interference avoidance) and physical (interference) models [5]. The former employs an exclusive channel occupancy strategy, which eliminates the CR-to-CR interference and simplifies the management of the CR-to-PR interference [20, 21]. However, it does not support concurrent CR transmissions over the same channel, which may reduce the spectrum efficiency. On the other hand, the overlay physical model allows for multiple concurrent interference-limited CR transmissions to simultaneously proceed over the same channel in the same locality, which improves spectrum efficiency [125]. However, the power control issue (CR-to-CR and CR-to-PR interference management) under this model is not trivial. Worse yet, using this model requires a distributed iterative power adjustment for individual CR users, which was shown that it results in slow protocol convergence [125].

2.6.4 Guard-Band Considerations

Most of existing spectrum sharing protocols for CRNs were designed assuming orthogonal channels, where the adjacent channel interference (ACI) is ignored (e.g., [8–10, 21, 36, 80, 103]). However, this requires using ideal sharp transmit and receive filters, which is practically not feasible. In practice, frequency separation (guard bands) between adjacent channels is needed to mitigate the effects of ACI and protect the performance of ongoing PR and CR users operating over adjacent channels. It has been shown that introducing guard bands can significantly impact the spectrum efficiency, and, hence, it is very important to account for the guard-band constraints when designing spectrum sharing protocols for CRNs.

Few number of CRN spectrum access and sharing protocols have been designed while accounting for the guard band issue [20, 104–107]. Guard band-aware strategies enable effective and safe spectrum sharing, have a great potential to enhance the spectral efficiency, and protect the receptions of the ongoing CR and PR transmissions over adjacent channels. The need for guard band-aware spectrum sharing mechanisms and protocols was discussed in [20]. Specifically, the authors, in [20], have investigated the ACI problem and proposed guard-band-aware spectrum access/sharing protocols for CRNs. The main objective of their proposed mechanism is to minimize the total number of reserved guard-band channels, such that the overall spectrum utilization is maximized. In [104], the authors showed that selecting the operating channels on per block (group of adjacent channels) basis instead of per channel basis (unlike the work in [20]) provides better spectrum efficiency. The work in [104] attempts at selecting channels, such that at most one guard band is introduced for each new CR transmission. In [105], the authors proposed two guard-band spectrum sharing mechanisms for CRNs. The first mechanism is a static single-stage channel assignment that is suitable for distributed multi-hop CRNs. The second one is an adaptive two-stage channel assignment that is suitable for centralized CRNs. The main objective of the proposed mechanisms is to maximize spectrum efficiency while providing soft guarantees on CR performance in terms of a pre-specified rate demand.

2.7 Complementary Approaches and Optimizations

In this section, we discuss several approaches that interact with spectrum sharing protocols to further enhance spectrum utilization in CRNs.

2.7.1 Resource Virtualization in CRNs

The resource virtualization concept has been extensively discussed in the literature, which refers to the process of creating a number of logical resources based on the set of all available physical resources. This concept allows the users to utilize the logical resources in the same way they are using the physical resources. This leads to a better utilization of the physical resources as virtualization allows more users to share the available physical resources. In addition, virtualization introduces an additional layer of security as a user's application cannot directly control the physical resources. The concept of virtualization was originally used in computer systems to better utilize the available physical resources (e.g., processors, memory, storage units, and network interfaces). These resources are virtualized into separate sets of logical resources, and each set of these virtual resources can be assigned to different users. Using system virtualization can achieve: (1) users' isolation, (2) customized services, and (3) improved resource efficiency. Virtualization was also been introduced in wired networks by introducing the framework of virtual private networks (VPNs).

Recently, several attempts have been made to implement the virtualization concept in wireless CRNs. We note here that employing virtualization in CRNs is daunted by several challenges including: spectrum sharing, limited infrastructure, different geographical regions, self co-existence, PR co-existence, dynamic spectrum availability, spectrum heterogeneity, and users' mobility [126]. In [127], a single cell CRN virtualization framework was introduced. According to this framework, a network with one BS and M physical radio nodes (PNs) with varying sets of resources are considered. The resources include the number of radio interfaces at each PN, the set of orthogonal idle channels at each PN, and the employed coding schemes. Each PN hosts a set of virtual nodes (VNs). The VNs located in the different PNs can communicate with each other. To facilitate such communications, VNs request resources from their hosting PNs. Simulation results have demonstrated the effectiveness of using network virtualization in improving network performance. In [128], the authors have proposed a virtualization framework for multi-channel multi-cell CRNs. In this work, a virtualization based semi-decentralized resource allocation mechanism for CRNs using the concept of multilayer hypervisors was proposed. The main objective of this work is to reduce the overall CR control overhead by minimizing the CR users' reliance on the base-station in assigning spectrum resources. Simulation results have indicated significant improvement in CRN performance (in terms of control overhead, spectrum utilization, and blocking rate) is achieved by the virtualized framework compared to non-virtualized resource allocation schemes.

2.7.2 Full Duplex Communications

The problem of computing the optimal spectrum access strategy for CR users has been well investigated in [129–131], but for CR users that are equipped with half-duplex (HD) transceivers. It has been shown that using HD transceivers can significantly reduce the achieved network performance [132]. Motivated by the recent advances in full-duplex (FD) communications and self-interference suppression (SIS) techniques (including Antenna Cancellation (AC), Radio Frequency Interference Cancellation (RIC) And Digital Interference Cancellation (DIC) [133–136]), several attempts have been made to exploit the FD capabilities and SIS techniques in designing communication protocols for CRNs [132, 137, 138]. The main objective of these protocols is to improve the overall spectrum efficiency by allowing simultaneous transmission, sensing and reception (over the same channel or over different channels) at each CR user. These protocols, however, require additional hardware support (i.e., duplexers). The practical aspects of using FD radios in CRNs need to be further investigated. The design of effective channel/power/rate assignment schemes for FD-based CRNs is still an open problem.

2.7.3 Beamforming Techniques

Beamforming techniques are another optimization that can enable efficient spectrum sharing [139–144]. According to beamforming, the transmit and receive beamforming coefficients are adaptively computed by each CR user such that the achieved CR throughput is maximized while minimizing the introduced interference at the CR and PR users. Furthermore, the performance gain achieved by using beamforming in CRNs can be significantly improved by allowing for adaptive adjustment of the allocated powers to the transmit beamforming weights [143]. The operation details of such an approach need to be further explored.

2.7.4 Software Defined Radios and Variable Spectrum-Width

The use of variable channel widths through channel aggregation and bonding is another promising approach in improving spectral efficiency. However, this approach has not given enough attention. Based on its demonstrated excellent performance (compared to fixed bandwidth channels), variable channel widths has been chosen as an effective spectrum allocation mechanism in cellular mobile communication systems, including the recently deployed 4G wireless systems. Thus, it is very important to use variable-bandwidth channels in CRNs. More specifically, in CRNs, assigning variable bandwidth to different CR users can be achieved through channel bonding and aggregation. This has a great potential in improving spectrum efficiency. The use of variable bandwidth transmission in CRNs is not straightforward due to the dynamic time-variant behavior of PR activities and the hardware nature of most of existing CR devices [5], which make it very hard to control the channel bandwidth [20].

So far, most of CR systems have been designed with the assumption that each CR user is equipped with one or several radio transceivers. Using hardware radio transceivers can limit the number of possibly assigned channels to CR users and cannot fully support variable-width channel assignment. One possible approach to enable variable-width spectrum assignment and increase network throughput is to employ software defined radios (SDRs). The use of the SDRs enables the CR users to bond and/or aggregate any number of channels, thus enabling variable spectrum-width CR transmissions. Thus, SDRs support more efficient spectrum utilization, which significantly improves the overall CRN performance and provides QoS guarantees to CR users.

2.7.5 Cross-Layer Design Principle

Cross-layer design is essential for efficient operation of CRNs. Spectrum sharing protocols for CRNs should select the next-hop and the operating PR frequency channel(s) using a cross-layer design that incorporates the network, medium access control (MAC), and physical layers. A cross-layer routing metric called the maximum probability of success (MPoS) was proposed in [36]. The MPoS incorporates the link quality conditions and the average availability periods of PR users to improve the CRN performance in terms of the network throughput. The metric assigns operating channels to the candidate routes so that a route with the maximum probability of success and minimum CR forced termination-rate is selected. The main drawback of the MPoS approach is its requirement of known PR channel availability distributions (the probability density function of idle periods of the PR channels).

2.7.6 *Discontinuous-OFDM Technology*

Based on the spectrum availability conditions and to enable efficient CRN operation, a CR user may need to utilize multiple adjacent (contiguous) idle PR channels (the so-called spectrum bonding) or non-adjacent (non-contiguous) idle PR channels (the so-called spectrum aggregation). Spectrum bonding and aggregation can be realized using either the traditional frequency division multiplexing (FDM) or the discontinuous-orthogonal frequency division multiplexing (D-OFDM) technology [18, 20, 145]. The former technology requires several HD transceivers and tunable filters at each CR user, where each assigned channel will use one of the available transceivers. While this approach is simple, it requires a large number of transceivers and does not provide the enough flexibility needed to implement channel aggregation and bonding at a large-scale. The D-OFDM is a novel wireless radio technology that allows a CR transmission to simultaneously take place over several (adjacent or non-adjacent) channels using one HD OFDM transceiver. According to D-OFDM, each channel includes a distinct equal-size group of adjacent OFDM sub-carriers. According to D-OFDM, spectrum bonding and aggregation with any number of channels can be realized through power control, in which the sub-carriers of a non-assigned channel will be assigned 0 power and all the sub-carries of a selected channel will be assigned controlled levels of powers. We note here that the problem of assigning different powers to different OFDM symbols within the same channel is still an open issue.

2.7.7 *Spectrum Sharing for MIMO-Based Ad Hoc CRNs*

Multiple Input Multiple Output (MIMO) is considered as a key technology to increase the achieved wireless performance. Specifically, MIMO can be used to improve spectrum efficiency, throughput performance, wireless capacity, network connectivity, and energy efficiency. The majority of previously proposed works on MIMO-based CRNs (e.g., [146–149]) have focused on the physical layer and addressed a few of the challenging issues at the upper layers, but certainly more effort is still required to investigate the achieved capacity of MIMO-based CRNs, the design of optimal channel/power/rate assignment for such CRNs, the interoperability with the non-MIMO CRNs, and many other challenging issues.

2.7.8 *Cooperative CR Communication (Virtual MIMO)*

One of the main challenges in the design of CRNs communication protocols is the time-varying nature of the wireless channels due to the PR activities and the multi-path fading. Cooperative communication is a promising approach that can deal with the time-varying nature of the wireless channels, and, hence, improve the CRN performance. Cooperative communication can create a virtual MIMO system by allowing CR users to assist each other in data delivery (by relaying data packets to the receiver). Hence, the received data packets at the CR destination traverse several independent paths achieving diversity gains. Cooperative communication can also extend the coverage area. The benefits of employing cooperative communication, however, are achieved at the cost of an increase in power consumption, an increase in computation resources and an increase in system complexity. It has been shown that cooperation may potentially lead to significant long-term resource savings for the whole CRN. An important challenge in this domain is how to design effective cooperative MAC protocols that combine the cooperative communication with CR multiple-channel capability such that the overall network performance is improved. The CR relay selection is another challenging problem that needs to be further investigated. Therefore, new cooperative CRN MAC protocols and relay selection strategies are needed to effectively utilize the available resources and maximize network performance.

2.7.9   Network Coding

Network coding in CRNs is another interesting approach that has not yet explored in CRNs. Based on its verified excellent performance in wireless networks [150], it is natural to consider it in the design of cooperative-based CRNs. The packet relaying strategies in cooperative communication are generally implemented on a per packet basis, where a store-and-forward (SF) technique is used (the received packets at the CR relays are received, stored and retransmitted towards the receiver). While this type of relaying mechanisms is simple, it has been shown that it provides a sub-optimal performance in terms of the overall achieved CRN throughput (especially, in multi-cast scenarios). Instead of using SF, network coding can be used to maximize the CRN performance. With network coding, the intermediate relay CR users can combine the incoming packets using mathematical operations (additions and subtractions over finite fields) to generate the output packets.

One drawback in using network coding is that the computational complexity increases as the finite field size increases. The higher the field size, the better is the network performance. However, the tradeoff should be further investigated and more efforts are required to identify and study the benefits and drawbacks of increasing the field-size in CRNs. In addition, the performance achieved through network coding can be further enhanced in CRNs by dynamically adapting the total number of coded packets that need to be sent by the source CR user. Such adaptation adjustment is yet to be explored, which should be based on the PR activities, link loss rates, link correlations, and nodes' reachability.

2.8   Summary and Open Research Problems

CR technology has a great potential to enhance the overall spectrum efficiency. In this chapter, we first highlighted the main existing CRN architectures. Then, we described the unique characteristics of their operating RF environment that need to be accounted for in designing efficient communication protocols and spectrum assignment mechanisms for these networks. We then surveyed several spectrum sharing approaches for CRNs. We showed that these approaches differ in their design objectives. Ideally, one would like to design a spectrum sharing solution that maximizes spectrum efficiency while causing no harmful interference to PR users. We showed that interference management (including self-coexistence and the PR coexistence) and distributed coordination are the main crucial issues in designing efficient spectrum sharing mechanisms. The key idea in the design of effective spectrum sharing and assignment protocols for CRNs is to jointly consider the PR activities and CR link-quality conditions.

The reaction to PR appearance is another important issue in designing spectrum sharing schemes for CRNs. Currently, most of spectrum sharing schemes are either reactive or proactive schemes. Interference threshold-based schemes are very promising, where more research should be conducted to explore their advantages and investigate their complexities. Another crucial and challenging problem is the incorporation of the guard-band constraints in the design of spectrum sharing schemes for CRNs. A huge amount of interference is leaked into the adjacent channels when guard bands are not used. This can significantly reduce spectrum efficiency and cause harmful interference to PR users. The effect of introducing guard-bands on the spectrum sharing design has not been well explored.

Many interesting open design issues still to be addressed. Variable-width spectrum sharing approach is quite promising, but their design assumptions and feasibility should be carefully investigated. Resource virtualization is another important concept that can significantly improve the overall spectrum utilization. Beamforming and MIMO technology have recently been proposed as a means of maximizing spectrum efficiency. The use of beamforming in CRNs with MIMO capability can achieve significant improvement in spectrum efficiency. However, the spectrum sharing problem becomes more challenging due to the resurfacing of several design issues, such as the determination of the beamforming weights, the joint channel assignment and power control, etc., which need to be further addressed. Research should focus also on the cooperative CR communication and cross-layer concepts. Using FD radios versus using HD radios is another interesting issue. Moreover, utilizing network coding is very promising in improving the CRN's performance. Finally, we showed that channel bonding and aggregation can be realized through the use of D-OFDM technology. This technology allows a CR user to simultaneous transmit or receive over multiple channels using a single radio transceiver.

References

[1] B. Wang and K. J. R. Liu, Advances in cognitive radio networks: A survey, *IEEE Journal of Selected Topics in Signal Processing*, vol. 5, no. 1, pp. 5–23, 2011.

[2] Federal Communications Commission, ET Docket No. 03-322 Notice of Proposed Rule Making and Order, Dec 2003.

[3] Federal Communications Commission, "Spectrum Policy Task Force," ET Docket No. 02-135, Nov., 2002.

[4] I. F. Akyildiz et al., NeXt Generation/Dynamic Spectrum Access/Cognitive Radio Wireless Networks: A Survey, *Comp. Networks J.*, vol. 50, Sept. 2006, pp. 212–759.

[5] H. Bany Salameh and M. Krunz, "Channel access protocols for multihop opportunistic networks: challenges and recent developments," *IEEE Network*, vol. 23, no. 4, 2009.

[6] D. Cabric, S. Mishra, D. Willkomm, R. Brodersen, and A. Wolisz, "A cognitive radio approach for usage of virtual unlicensed spectrum," in Proceedings of the 14th IST Mobile and Wireless Communications Summit, June 2005.

[7] M. T. Hassan, E. Ahmed, J. Qadir, and A. Baig, "Quantifying the multiple cognitive radio interfaces advantage," in Proceedings of the 27th IEEE International Conference on Advanced Information Networking and Applications Workshops (WAINA), 2013, pp. 511–516.

[8] H. Bany Salameh, "Throughput-oriented channel assignment for opportunistic spectrum access networks," *Mathematical and Computer Modelling*, vol. 53, Iss. 11–121, pp. 2108–2118, June 2011.

[9] H. Bany Salameh, "Rate-maximization channel assignment scheme for cognitive radio networks," in Proceedings of the IEEE GLOBECOM Conference, Florida, 2010.

[10] L. Tan and L. Le, "Channel assignment with access contention resolution for cognitive radio networks," *IEEE Transactions on Vehicular Technology*, vol. 61, no. 6, pp. 2808–2823, 2012.

[11] J. Wang and Y. Huang, "A cross-layer design of channel assignment and routing in cognitive radio networks," in Proceedings of the 3rd IEEE International Conference on Computer Science and Information Technology (ICCSIT10), China, 2010, pp. 542–547.

[12] R. Irwin, A. MacKenzie, and L. DaSilva, "Resource minimized channel assignment for multi-transceiver cognitive radio networks," *IEEE Journal on Selected Areas in Communications*, vol. 31, no. 3, pp. 442–450, 2013.

[13] P.-K. Tseng, H. Chen, and W.-H. Chung, "Joint design on energy efficiency and throughput for non-infrastructure based cognitive radio networks," in Proceedings of the 8th International Wireless Communications and Mobile Computing Conference (IWCMC12), Cyprus, 2012, pp. 642–647.

[14] P.-K. Tseng, W.-H. Chung, H. Chen, and C.-S. Wu, "Distributed energy-efficient cross-layer design for cognitive radio networks," in Proceedings of the 23rd International Symposium on Personal Indoor and Mobile Radio Communications (PIMRC12), Sydney, Australia, 2012, pp. 161–166.

[15] K. Bian and J.-M. Park, "Segment-based channel assignment in cognitive radio ad hoc networks," in Proceedings of the 8th International Conference on Cognitive Radio Oriented Wireless Networks and Communications (CrownCom07), Orlando, Florida USA, 2007, pp. 327–335.

[16] J. D. Ser, M. Matinmikko, S. Gil-Lopez, and M. Mustonen, "Centralized and distributed spectrum channel assignment in cognitive wireless networks: A harmony search approach," *Applied Soft Computing*, vol. 12, no. 2, pp. 921–930, 2012.

[17] E. Anifantis, V. Karyotis, and S. Papavassiliou, "A markov random field framework for channel assignment in cognitive radio networks," in Proceedings of the IEEE International Conference on Pervasive Computing and Communications Workshops (PERCOM Workshops12), Lugano, Switzerland, 2012, pp. 770–775.

[18] H. Bany Salameh, M. Krunz, and D. Manzi, "An efficient guard-band-aware multi-channel spectrum sharing mechanism for dynamic access networks," in Proceedings of the IEEE GLOBECOM Conference, Dec. 2011.

[19] M. Ahmadi, Y. Zhuang, and J. Pan, "Distributed robust channel assignment for multi-radio cognitive radio networks," in Proceedings of the 76th IEEE Vehicular Technology Conference (VTC12-Fall), Quebec City, Canada, 2012, pp. 1–5.

[20] H. Bany Salameh, M. Krunz, and D. Manzi, "Spectrum bonding and aggregation with guard-band awareness in cognitive radio networks," IEEE Transactions on Mobile Computing, 2014.

[21] H. Bany Salameh, M. Krunz, and O. Younis, "Dynamic spectrum access protocol without power mask constraints," in Proceedings of the IEEE INFOCOM Conference, Brazil, April 2009.

[22] Q. Zhao, L. Tong, A. Swami, and Y. Chen, "Decentralized cognitive MAC for opportunistic spectrum access in ad hoc networks: A POMDP framewrok," *IEEE Journal on Selected Areas in Communications*, vol. 25, no. 3, pp. 589–600, 2007.

[23] Q. Xiao, Y. Li, M. Zhao, S. Zhou, and J. Wang, "Opportunistic channel selection approach under collision probability constraint in cognitive radio systems," *Computer Communications*, vol. 32, no. 18, pp. 1914–1922, 2009.

[24] Q. Zhao, and S. Geirhofer, L. Tong, and B. M. Sadler, "Opportunistic spectrum access via periodic channel sensing," *IEEE Transactions on Signal Processing*, vol. 56, no. 2, pp. 785–796, 2008.

[25] D. Sahu, and A. Trivedi, "A Bayesian approach using M-QAM modulated primary signals for maximizing spectrum utilization in cognitive radio," in Proceedings of the International Conference on Advances in Computing, Communications and Informatics (ICACCI), 2014, pp. 1183–1187.

[26] H. Bany Salameh, M. Krunz, and O. Younis, "MAC protocol for opportunistic cognitive radio networks with soft guarantees," *IEEE Transactions on Mobile Computing*, vol. 8, no. 10, 2009.

[27] L.-C. Wang, C.-W. Wang, and F. Adachi, "Load-balancing spectrum decision for cognitive radio networks," *IEEE Journal on Selected Areas in Communications*, vol. 29, no. 4, pp. 757–769, April 2011.

[28] P. A. K. Acharya, S. Singh, and H. Zheng, "Reliable open spectrum communications through proactive spectrum access," in Proceedings of the TAPAS Conference, 2006.

[29] M. Bkassiny and S. K. Jayaweera, "Optimal channel and power allocation for secondary users in cooperative cognitive radio networks," in Proceedings of the 2nd International Conference on Mobile Lightweight Wireless Systems (MOBILIGHT), 2010.

[30] H. Wang, J. Ren, and T. Li, "Resource allocation with load balancing for cogntive radio networks," in Proceedings of the IEEE GLOBECOM Conference, 2010.

[31] S.-U. Yoon and E. Ekici, "Voluntary spectrum handoff: A novel approach to spectrum management in CRNs," in Proceedings of the IEEE ICC Conference, 2010.

[32] Y. Song and J. Xie, "Common hopping based proactive spectrum handoff in cogntive radio ad hoc networks," in Proceedings of the GLOBECOM Conference, 2010.

[33] O. Badarneh and H. Bany Salameh, "Quality-aware Routing in Cognitive Radio Networks under Dynamically Varying Spectrum Opportunities," *Computers and Electrical Engineering Journal*, vol. 38, Iss. 6, pp. 1731–1744, November 2012.

[34] H. Bany Salameh, "Resource management with probabilistic performance guarantees in opportunistic networks," *International Journal of Electronics and Communications AEU*, vol. 67, Iss. 7, pp. 632–636, 2013.

[35] H. Bany Salameh, "Probabilistic Spectrum Assignment for QoS-constrained Cognitive Radios with Parallel Transmission Capability," in Proceedings of the IFIP Wireless Days Conference, Dublin, Ireland, Nov. 2012

[36] H. Bany Salameh and O. Badarneh, "Opportunistic medium access control for Maximizing packet delivery rate in dynamic access networks," *Journal of Network and Computer Applications*, vol. 36, Iss. 1, pp. 523–532, 2013.

[37] W.-y. Lee, S. Member, and I. F. Akyildiz, "A Spectrum decision framework for cognitive radio networks," *IEEE Transaction on Mobile Computing*, vol. 10, no. 2, pp. 161–174, 2011.

[38] I. Malanchini, M. Cesana, and N. Gatti, "On spectrum selection games in cognitive radio networks," in Proceedings of the IEEE GLOBECOM Conference, 2009, pp. 1–7.

[39] Q. D. Xue Feng, S. Guangxi, and L. Yanchun, "Smart channel swiching in cognitive radio networks," in Proceedings of the CISP Conference, 2009.

[40] A. C.-C. Hsu, D. S.-L. Wei, and C.-C. J. Kua, "A cognitive MAC protocol using statistical channel allocation for wireless ad-hoc networks," in Proceedings of the IEEE WCNC Conference, 2007.

[41] R.-T. Ma, Y.-P. Hsu, and K.-T. Feng, "A POMDA-based spectrum handoff protocol for partially observable cognitive radio networks," in Proceedings of the IEEE WCNC Conference, 2009.

[42] P. Zhu, J. Li, and X. Wang, "A new channel parameter for cognitive radio," in Proceedings of the CrownCom Conference, 2007.

[43] L. Yu, C. Liu, and W. Hu, "Spectrum allocation algorithm in cognitive ad-hoc networks with high energy efficiency," in Proceedings of the International Conference on Green Circuits and Systems (ICGCS), 2010.

[44] S. Byun, I. Balasingham, and X. Liang, "Dynamic spectrum allocation in wireless cognitive sensor networks: Improving fairness and energy efficiency," in Proceedings of the 68th IEEE Vehicular Technology Conference (VTC 2008-Fall), 2008, pp. 1–5.

[45] S. Gao, L. Qian, and D. Vaman, "Distributed energy efficient spectrum access in wireless cognitive radio sensor networks," in Proceedings of the IEEE WCNC Conference, 2008, pp. 1442–1447.

[46] X. Li, D. Wang, and J. McNair, "Residual energy aware channel assignment in cognitive radio sensor networks," in Proceedings of the IEEE WCNC Conference, 2011, pp. 398–403.

[47] T. Zhang, B. Wang, and Z. Wu, "Spectrum assignment in infrastructure based cognitive radio networks," in Proceedings of the IEEE National Aerospace and Electronics Conference (NAECON), 2009, pp. 69–74.

[48] C. Peng, H. Zheng, and B. Y. Zhao, "Utilization and fairness in spectrum assignment for opportunistic spectrum access," *Mobile Networks and Applications*, vol. 11, no. 4, pp. 555–576, May 2006.

[49] L. Le and E. Hossain, "Resource allocation for spectrum underlay in cognitive radio networks," *IEEE Transactions on Wireless Commuications*, vol. 7, no. 12, pp. 5306–5315, 2008.

[50] G. Yildirim, B. Canberk, and S. Oktug, "Enhancing the performance of multiple IEEE 802.11 network environment by employing a cognitive dynamic fair channel assignment," in Proceedings of the 9th IFIP Annual Mediterranean Ad Hoc Networking Workshop (Med-Hoc-Net), 2010, pp. 1–6.

[51] Y. Ge, J. Sun, S. Shao, L. Yang, and H. Zhu, "An improved spectrum allocation algorithm based on proportional fairness in Cognitive Radio networks," in Proceedings of the 12th IEEE International Conference on Communication Technology (ICCT), 2010, pp. 742–745.

[52] Y. Li, Z. Wang, B. Cao, and W. Huang, "Impact of spectrum allocation on connectivity of cognitive radio ad-hoc networks," in Proceedings of the IEEE GLOBECOM Conference , pp. 1–5, Dec. 2011.

[53] M. Ahmadi and J. Pan, "Cognitive wireless mesh networks: A connectivity preserving and interference minimizing channel assignment scheme," in Proceedings of the IEEE Pacific Rim Conference on Communications, Computers and Signal Processing, pp. 458–463, Aug. 2011.

[54] H. M. Almasaeid and A. E. Kamal, "Receiver-based channel allocation for wireless cognitive radio mesh networks," in Proceedings of the IEEE Symposium on New Frontiers in Dynamic Spectrum (DySPAN), pp. 1–10, Apr. 2010.

[55] C. Zhao, B. Shen, T. Cui, and K. Kwak, "Graph-theoretic cooperative spectrum allocation in distributed cognitive networks using bipartite matching," in Proceedings of the IEEE 3rd International Conference on Communication Software and Networks (ICCSN), 2011, pp. 223–227.

[56] A. Hoang, Y.-C. Liang, "Maximizing spectrum utilization of cognitive radio networks using channel allocation and power control," in Proceedings of the 64th IEEE Vehicular Technology Conference (VTC-2006 Fall), 2006, pp. 1–5.

[57] F. Ye, R. Yang, and Y. Li, "Genetic algorithm based spectrum assignment model in cognitive radio networks," in Proceedings of the 2nd IEEE Information Engineering and Computer Science International Conference (ICIECS10), China, 2010, pp. 1–4.

[58] F. Hou and J. Huang, "Dynamic channel selection in cognitive radio network with channel heterogeneity," in Proceedings of the IEEE GLOBECOM Conference, Florida, 2010.

[59] P. Kaur, M. Uddin, and A. Khosla, "Adaptive bandwidth allocation scheme for cognitive radios," *International J. Advancements in Computing Technology*, vol. 2, no. 2, pp. 35–41, 2010.

[60] L. Gao, S. Cui, "Power and rate control for cognitive radios: A dynamic programming approach," in Proceedings of the 3rd International Conference on Cognitive Radio Oriented Wireless Networks and Communications (CrownCom 2008), pp. 1–7, 2008.

[61] E. Peh, Y-C Liang, Y. Guan, Y. Zeng, "Optimization of cooperative sensing in cognitive radio networks: a sensing-throughput trade off view", *IEEE Transactions on Vehicular Technology*, vol. 58, pp. 5294–5299, 2009.

[62] W. Wang, J. Cai, B. Kasiri, and A.S. Alfa, "Channel assignment of cooperative spectrum sensing in multi-channel cognitive radio networks," in Proceedings of the IEEE International Conference on Communications (ICC), 2011, pp. 1–5.

[63] Y. Zeng, Y-C Liang, E. Peh, A. Hoang, "Cooperative covariance and eigenvalue based detections for robust sensing", in Proceedings of the IEEE GLOBECOM Conference, December 2009, Hawaii, USA.

[64] J. Unnikrishnan, V. Veeravalli, "Cooperative sensing for primary detection in cognitive radio", *IEEE Journal on Selected Topics in Signal Processing*, Vol 2, No. 1, pp. 18–27, 2008.

[65] K. Seshukumar, R. Saravanan, M. Suraj, "Spectrum sensing review in cognitive radio," in Proceedings of the IEEE International Conference on Emerging Trends in VLSI, Embedded System, Nano Electronics and Telecommunication System Conference (ICEVENT), pp. 1–4, Jan. 2013.

[66] S. Haykin, D.J. Thomson, J.H. Reed, "Spectrum sensing for cognitive radio," *Proceedings of the IEEE*, vol. 97, no. 5, pp. 849–877, 2009.

[67] Z. Quan, S. Cui, A.H. Sayed, H.V. Poor, "Optimal multiband joint detection for spectrum sensing in cognitive radio networks," *IEEE Transactions on Signal Processing*, vol. 57, no. 3, pp. 1128–1140, March 2009.

[68] K. Bian and J.-M. Park, "A coexistence-aware spectrum sharing protocol for 802.22 WRANs," in Proceedings of the 18th Internatonal Conference on Computer Communications and Networks (ICCCN), Aug. 2009, pp. 1–6.

[69] V. Gardellin, S.K. Das, and L. Lenzini, "A fully distributed game theoretic approach to guarantee self-coexistence among WRANs," in Proceedings of the IEEE INFOCOM Conference, March 2010, pp. 1–6.

[70] G.-Z. Ko, A.A. Franklin, S.-J. You, J.-S. Pak, M.-S. Song, and C.-J. Kim, "Channel management in IEEE 802.22 WRAN systems," *IEEE Communications Magazine*, vol. 48, no. 9, pp. 88–94, 2011.

[71] P. Camarda, C. Cormio, and C. Passiatore, "An exclusive self-coexistence (ESC) resource sharing algorithm for cognitive 802.22 networks," in Proceedings of the 5th IEEE International Symposium on Wireless Pervasive Computing (ISWPC), May 2010, pp. 128–133.

[72] W. Hu, D. Willkomm, M. Abusubaih, J. Gross, G. Vlantis, M. Gerla, and A. Wolisz, "Cognitive radios for dynamic spectrum access - dynamic frequency hopping communities for efficient IEEE 802.22 operation," *IEEE Communications Magazine*, vol. 45, no. 5, pp. 80–87, May 2007.

[73] H. Bany Salameh, Y. Jararweh, T. Aldalgamouni, and A. Khreishah, "Traffic-driven exclusive resource sharing algorithm for mitigating selfcoexistence problem in WRAN systems," in Proceeding of the IEEE WCNC Conference, April 2014, pp. 1933–1937.

[74] S. Debroy and M. Chatterjee, "Intra-cell channel allocation scheme in IEEE 802.22 networks," in Proceedings of the 7th IEEE conference on Consumer communications and networking conference (CCNC), 2010, pp. 284–289.

[75] M. Bani Hani, H. Bany Salameh, Y. Jararweh, and A. Bousselham, "Traffic-aware self-coexistence management in IEEE 802.22 WRAN systems," in Proceeding of the 7th IEEE GCC Conference and Exhibition, Nov 2013, pp. 507–510.

[76] B. F. Lo, "A survey of common control channel design in cognitive radio networks," *Physical Communication*, vol. 4, no. 1, pp. 26–39, 2011.

[77] K. Chowdhury and I. Akyildiz, "OFDM based common control channel design for cognitive radio ad hoc networks," *IEEE Transactions on Mobile Computing*, vol. 10, pp. 228–238, 2011.

[78] H. Su and X. Zhang, "Cross-layer based opportunistic MAC protocols for QoS provisionings over cognitive radio wireless networks," *IEEE Journal on Selected Areas in Communications*, pp. 118–129, 2008.

[79] Y. Yuan, P. Bahl, R. Chandra, T. Moscibroda, and Y. Wu., "Allocating dynamic time-spectrum blocks in cognitive radio networks," in Proceedings of the ACM International Symposium on Mobile and Ad-Hoc Networking and Computing (MobiHoc), Sept. 2007.

[80] J. Jia, Q. Zhang, and X. Shen, "HC-MAC: A Hardware-Constrained Cognitive MAC for Efficient Spectrum Management," *IEEE Journal on Selected Areas in Communications*, vol. 26, no. 1, pp. 106–117, Jan. 2008.

[81] A. Masri, C.-F. Chiasserini, and A. Perotti, "Control information exchange through UWB in cognitive radio networks," in Proceedings of the 5th IEEE International Symposium on Wireless Pervasive Computing Conference (ISWPC), 2010, pp. 110–115.

[82] M.E. Sahin and H. Arslan, "System design for cognitive radio communications," in Proceedings of the 1st International Conference on Cognitive Radio Oriented Wireless Networks and Communications Conference (CrownCom), 2006, pp. 1–5.

[83] J. Marinho, E. Monteiro, "CORHYS: Hybrid Signaling for Opportunistic Distributed Cognitive Radio", Computer Networks, February 2015.

[84] J.-M. J. Park, R. Chen, and K. Bian, "Control channel establishment in cognitive radio networks using channel hopping," *IEEE Journal on Selected Areas in Communications*, vol. 29, no. 4, pp. 689–703, 2011.

[85] U. Tefek, and T. Lim, "Channel-hopping on multiple channels for full rendezvous diversity in cognitive radio networks," in Proceedings of IEEE GLOBECOM Conference, pp. 4714–4719, Dec. 2014.

[86] J.-M. J. Park, R. Chen, and K. Bian, "A quorum-based framework for establishing control channels in dynamic spectrum access networks," in Proceedings of the 15th annual international conference on Mobile computing and networking Conference (MobiCom), 2009, pp. 25–36.

[87] N. Baldo, M. Zorzi, and A. Asterjadhi, "A distributed network coded control channel for multihop cognitive radio networks," *IEEE Network*, vol. 23, no. 4, 2009, pp. 26–32.

[88] N. Baldo, A. Asterjadhi, and M. Zorzi, "Dynamic spectrum access using a network coded cognitive control channel," *IEEE Transactions on Wireless Communications*, vol. 9, no. 8, pp. 2575–2587, 2010.

[89] C. Cormio and K. R. Chowdhury, "Common control channel design for cognitive radio wireless ad hoc networks using adaptive frequency," *Ad Hoc Networks*, vol. 8, no. 4, pp. 430–438, 2010.

[90] L. A. DaSilva and I. Guerreiro, "Sequence-based rendezvous for dynamic spectrum access," in Proceedings of the IEEE DySPAN Conference, 2008, pp. 1–7.

[91] Z. Htike, J. Lee, C-S. Hong, "A MAC protocol for cognitive radio networks with reliable control channels assignment," in Proceedings of the International Conference on Information Networking (ICOIN), 2012, pp. 81–85.

[92] H. Bany Salameh, "Spread spectrum-based coordination design for multi-hop spectrum-agile wireless networks," in Proceedings of the IEEE 81th Vehicular Technology Conference (VTC'15-Spring), Scotland, 2015, pp. 1–5.

[93] S. Perez-Salgado, E. Rodriguez-Colina, M. Pascoe-Chalke, and A. Prieto-Guerrero, "Underlay control channel using adaptive hybrid spread spectrum techniques for dynamic spectrum access," in Proceedings of the International Symposium on Performance Evaluation of Computer and Telecommunication Systems, July 2013, pp. 99–106.

[94] J. Zhao, H. Zheng, and G.-H. Yang, "Distributed coordination in dynamic spectrum allocation networks," in Proceedings of the IEEE DySPAN Conference, 2005, pp. 259–268.

[95] T. Chen, H. Zhang, G. M. Maggio, and I. Chlamtac, "CogMesh: A cluster-based cognitive radio network," in Proceedings of the IEEE DySPAN Conference, 2007, pp. 168–178.

[96] T. Chen, H. Zhang, M. D. Katz, and Z. Zhou, "Swarm Intelligence Based Dynamic Control Channel Assignment in CogMesh," in Proceedings of the IEEE ICC Workshops, 2008, pp. 123–128.

[97] L. Lazos, S. Liu, and M. Krunz, "Spectrum opportunity-based control channel assignment in cognitive radio networks," in Proceedings of the IEEE SECON Conference, 2009, pp. 1–9.

[98] S. Liu, L. Lazos, and M. Krunz, "Cluster-based control channel allocation in opportunistic cognitive radio networks," *IEEE Transactions on Mobile Computing*, vol. 11, no. 10, pp. 1436–1438, 2012.

[99] V. Gardellin, S. Das, and L. Lenzini, "Coordination problem in cognitive wireless mesh networks," *Pervasive and Mobile Computing*, vol. 9, no. 1, pp. 18–34, 2013.

[100] H. Yu, and K.-F. Ssu, "Spatially varied control channel assignment in cognitive radio ad hoc networks," in Proceedings of the Sixth International ICST Conference on Cognitive Radio Oriented Wireless Networks and Communications (CROWNCOM), 2011, pp. 311–315.

[101] T. Clancy, "Achievable capacity under the interference temperature model", in Proceedings of the IEEE INFOCOM Conference, 2007, pp. 794–802.

[102] S. Jones, N. Merheb, and I.-J. Wang, "An experiment for sensing-based opportunistic spectrum access in csma/ca networks," in Proceedings of the IEEE DySPAN Conference, 2005, pp. 593–596.

[103] J. Jia, J. Zhang, and Q. Zhang, "Cooperative relay for cognitive radio networks," in Proceedings of the IEEE INFOCOM Conference, 2009.

[104] G. Uyanik, M. Abdel-Rahman, and M. Krunz, "Optimal guard-band-aware channel assignment with bonding and aggregation in multi-channel systems," in Proceedings of the IEEE GLOBECOM Conference, Atlanta, USA, 2013, pp. 4898–4903.

[105] M. Abdel-Rahman, F. Lan, and M. Krunz, "Spectrum-efficient stochastic channel assignment for opportunistic networks," in Proceedings of the IEEE GLOBECOM Conference, Atlanta, USA, 2013, pp. 1272–1277.

[106] L. Yuan, Z. Feng, Z. Feng, Q. Zhang, and B. Liu, "A guard-band-aware channel allocation algorithm for multi-channel cognitive radio networks," in Proceedings of the 78th IEEE Vehicular Technology Conference (VTC Fall), 2013, pp. 1–5.

[107] M.J. Abdel-Rahman, M. Krunz, "Stochastic Guard-band-aware Channel Assignment with Bonding and Aggregation for DSA Networks," IEEE Transactions on Wireless Communications, 2015.

[108] H. Bany Salameh and M. Krunz, "Adaptive power-controlled MAC protocols for improved throughput in hardware-constrained cognitive radio networks," *Ad Hoc Networks Journal*, vol. 9, no. 7, pp. 1127–1139, Sep. 2011.

[109] J. R. Gallego, M. Canales, and J. Ortin, "Flow allocation with joint channel and power assignment in multihop cognitive radio networks using game theory," in Proceedings of the 9th IEEE International Symposium on Wireless Communication Systems (ISWCS12), Paris, France, 2012, pp. 91–95.

[110] J. Wu, Y. Dai, and Y. Zhao, "Effective channel assignments in cognitive radio networks," *Computer Communications*, vol. 36, no. 4, pp. 411–420, 2013.

[111] C. Zheng, R. P. Liu, X. Yang, I. B. Collings, Z. Zhou, and E. Dutkiewicz, "Maximum flow-segment based channel assignment and routing in cognitive radio networks," in Proceedings of the 73rd Vehicular Technology Conference (VTC11-Spring), Budapest, Hungary, 2011, pp. 1–6

[112] Y. Dai and J. Wu, "Efficient channel assignment under dynamic source routing in cognitive radio networks," in Proceedings of IEEE 8th International Conference on Mobile Adhoc and Sensor Systems (MASS11), Valencia, Spain, 2011, pp. 550–559.

[113] P. Junior, M. Fonseca, A. Munaretto, A. Viana, and A. Ziviani, "ZAP: a distributed channel assignment algorithm for cognitive radio networks," *EURASIP Journal on Wireless Communications and Networking*, vol. 2011, no. 1, pp. 1–11, 2011.

[114] Y. Xing, R. Chandramouli, S. Mangold, S. Shankar, "Dynamic spectrum access in open spectrum wireless networks", *IEEE Journal on Selected Areas in Communications*, vol. 24, no. 3, pp. 626–637, 2006.

[115] Y. Xing, C. Mathur, M. Haleem, R. Chandramouli, and K. Subbalakshmi, "Dynamic spectrum access with QoS and interference temperature constraints", *IEEE Transactions on Mobile Computing*, vol. 6, no. 4, pp. 423–433, 2007.

[116] S. Sankaranarayanan, P. Papadimitratos, A. Mishra, and S. Hershey, "A bandwidth sharing approach to improve licensed spectrum utilization", in Proceedings of the IEEE DySPAN Conference, 2005, pp. 279–288.

[117] T. Shu, S. Cui, and M. Krunz, "Medium access control for multi-channel parallel transmission in cognitive radio networks," in Proceedings of GLOBECOM Conference, pp. 1–5, 2006.

[118] J. Vartiainen, M. Hoyhtya, J. Lehtomaki, and T. Braysy, "Priority channel selection based on detection history database," in Proceedings of the CROWNCOM Conference, 2010.

[119] L. Yang, L. Cao, and H. Zheng, "Proactive channel access in dynamic spectrum networks," *Elsevier Physical Communications Journal*, vol. 1, pp. 103–111, 2008.

[120] T. C. Clancy and B. D. Walker, "Predictive dynamic spectrum access," in Proceedings of the SDR Forum Technical Conference, Florida, USA, 2006.

[121] A. Mishra and R. Brodersen, "Cooperative sensing among cognitive radios," in Proceeding of the IEEE ICC Conference, 2006.

[122] X. Liu and S. N., "Sensing-based opportunistic channel access," *Mobile Networks and Applications*, vol. 11, pp. 577–591, 2006.

[123] S. Fourati, S. Hamouda, and S. Tabbane, "RMC-MAC: A reactive multi-channel MAC protocol for opportunistic spectrum access," in Proceedings of the 4th IFIP International Conference on New Technologies, Mobility and Security (NTMS), 2011, pp. 1–5.

[124] G. D. Nguyen and S. Kompella, "Channel sharing in cognitive radio networks," in Proceedings of the MILCOM Conference, 2010.

[125] F. Wang, M. Krunz, and S. Cui, "Price-based spectrum management in cognitive radio networks," *IEEE Journal of Selected Topics on Signal Processing*, vol. 2, no. 1, Feb. 2008, pp. 74–87.

[126] X. Wang, P. Krishnamurthy, and D. Tipper, "Wireless network virtualization", in Proceedings of the IEEE International Conference on Computing, Networking and Communications (ICNC), pp. 818–822, 2013.

[127] Y. Jararweh, M. Al-Ayyoub, A. Doulat, A. Abed Al Aziz, H. Bany Salameh, and A. Khreishah, "Software defined cognitive radio network framework: Design and evaluation", International Journal of Grid and High Performance Computing (IJGHPC), 2014.

[128] A. Doulat, A. Abed Al Aziz, M. Al-Ayyoub, Y. Jararwah, H. Bany Salameh, and A. Khreishah: "Software defined framework for multi-cell cognitive radio networks," in Proceedings of the IEEE 10th International Conference on Wireless and Mobile Computing, Networking and Communications (WiMob), Larnaca, Cyprus, 2014.

[129] Q. Zhao and J. Ye, "Quickest detection in multiple on-off processes," *IEEE Transactions on Signal Processing*, vol. 58, no. 12, pp. 5994–6006, Dec. 2010.

[130] W. Afifi, A. Sultan, and M. Nafie, "Adaptive sensing and transmission durations for cognitive radios," in Proceedings of the IEEE DySPAN11 Conference, 2011, pp. 380–388.

[131] S. Huang, X. Liu, and Z. Ding, "Optimal sensing-transmission structure for dynamic spectrum access," in Proceedings of the IEEE INFOCOM09 Conference, April 2009, pp. 2295–2303.

[132] W. Afifi and M. Krunz, "Adaptive transmission-reception-sensing strategy for cognitive radios with full-duplex capabilities," on Proceedings of the IEEE DySPAN 2014 Conference, McLean, VA, April 2014.

[133] Quellan Inc., "Qhx220 narrowband noise canceller ic.," http://www.quellan.com/products/qhx220 ic.php.

[134] R. Bozidar, R. Dinan, K. Peter, P. Alexandre, S. Nikhil, B. Vlad, and D. Gerald, "Rethinking indoor wireless mesh design: Low power, low frequency, full-duplex," in Proceedings of the Fifth IEEE Workshop on Wireless Mesh Networks (WIMESH 2010), 2010, pp. 16.

[135] S. Gollakota and D. Katabi, "Zigzag decoding: combating hidden terminals in wireless networks," in Processinds of the ACM SIGCOMM 2008 Conference on Data Communications, New York, NY, USA, 2008, pp. 159–170.

[136] J. Choi, J. Mayank, S. Kannan, L. Philip, and K Sachin, "Achieving single channel, full duplex wireless communication," in Processinds of the 16th ACM MOBICOM Conference, Chicago, USA, Sep. 2010.

[137] W. Afifi and M. Krunz, "Exploiting self-interference suppression for improved spectrum awareness/efficiency in cognitive radio systems." in Proceedings of the IEEE INFOCOM13 Conference, Turin, Italy, 2013, pp. 1258–1266.

[138] W. Cheng, X. Zhang, and H. Zhang, "Full duplex spectrum sensing in non-time-slotted cognitive radio networks," in Proceedings of the MILCOM11 Conference, Nov. 2011, pp. 1029–1034.

[139] M.A. Beigi, S.M. Razavizadeh, "Cooperative beamforming in Cognitive Radio networks," in Proceedings of the 2nd IFIP Wireless Days (WD), 2009.

[140] M. Hassan, Md.J. Hossainr, "Cooperative beamforming for cognitive radio systems with asynchronous interference to primary user," *IEEE Transactions on Wireless Communications*, vol. 12, no. 11, pp. 5468–5479, 2013.

[141] C. Zhang, L. Guo, R. Hu, J. Lin, "Opportunistic distributed beamforming in cognitive radio networks with limited feedback," in Proceedings of the IEEE WCNC Conference, 2004, pp. 893–897.

[142] J. Zhang, L. Guo, T. Kang, P. Zhang, "Cooperative beamforming in cognitive radio network with two-way relay," in Proceedings of the IEEE 79th Vehicular Technology Conference (VTC Spring), pp. 1–5, May 2014.

[143] J.-H. Noh, S.-J. Oh, "Beamforming in cognitive radio with partial channel state information," in Proceedings of the IEEE GLOBECOM Conference, pp. 1–6, 2011.

[144] H. Bany Salameh, T. Hailat, "An Iterative Beamforming Algorithm for Improved Throughput in Multi-cell Multi-antenna Wireless Systems", IET Communications, 2015.

[145] J. Poston and W. Horne, "Discontiguous OFDM considerations for dynamic spectrum access in idle TV channels," in Proceedings of the IEEE DySPAN Conference, 2005, pp. 607–610.

[146] G. Scutari, D. Palomar, and S. Barbarossa, "Cognitive MIMO radio," *IEEE Signal Procesing Magazine*, vol. 25, no. 6, pp. 46–59, Nov. 2008.

[147] L. Bixio, G. Oliveri, M. Ottonello, M. Raffetto, and C. S. Regazzoni, "Cognitive radios with multiple antennas exploiting spatial opportunities," *IEEE Transaction on Signal Processing*, vol. 58, no. 8, pp. 4453–4459, Aug. 2010.

[148] S. Hua, H. Liu, M. Wu, and S. Panwar, "Exploiting MIMO antennas in cooperative cognitive radio networks," in Proceedings of the IEEE INFOCOM Conference, pp. 2714–2722, Shanghai, 2011.

[149] D. Nguyen and M. Krunz, "Heterogeneous spectrum sharing with rate demands in cognitive MIMO networks," in Proceedings of the IEEE GLOBECOM 2013 Conference, Atlanta, Dec. 2013.

[150] A. Khreishah, I. Khalil, and J. Wu, "Distributed network coding-based opportunistic routing for multicast," in Proceedings of the ACM Mobihoc Conference, June 2012.

APPROACHES TO SPECTRUM SHARING

Chapter 3

New Coexistence and Sharing Paradigms for Multihop Secondary Networks

Xu Yuan, Brian Jalaian, Y. Thomas Hou, Wenjing Lou, Scott F. Midkiff, and Sastry Kompella

CONTENTS

3.1 Introduction

The last decade has witnessed a rapid advance in the research and development of spectrum-sharing technologies. A recent report by the President's Council of Advisors on Science and Technology (PCAST) [13] called for the sharing of 1 GHz of the federal government radio spectrum with non-government entities in order to spur economic growth. This report further accelerated the pace of commercialization of innovative spectrum-sharing technologies. This report also motivates us to pursue a much more aggressive and bold vision for enhancing spectrum utilization.

    The current prevailing spectrum-sharing paradigm is that secondary nodes (typically equipped with cognitive radios (CRs)) are allowed to use a spectrum channel allocated to the primary nodes only when such a use will not cause interference to the primary nodes. This can be done by having the secondary nodes exploit transmission opportunities in time, space, and spectrum domains. This is also called "interweave" paradigm in [4]. The rational behind this paradigm is that secondary nodes should not produce interference that may be harmful to primary nodes. Under this paradigm, the wireless networking community has invested significant research efforts in algorithm design and protocol implementation to optimize secondary CR users' performance while ensuring that their activities will not interfere with the primary users [3, 6, 15, 23].

Although the interweave paradigm is simple, there are two major limitations:

- A secondary network can only exploit spectrum holes in the time, space, and frequency domains. It cannot be active simultaneously with the primary network in time, space, and frequency domains.

- The primary and secondary networks are completely independent, in the sense that there is no cooperation and sharing between the two networks for data forwarding.

To address these two major limitations, in this chapter, we propose two novel paradigms and discuss them in the context of multihop networks.

- **Transparent Coexistence Paradigm** To address the first limitation of the interweave paradigm, we propose a new paradigm called transparent coexistence (TC). Under this paradigm, the secondary nodes are allowed to use the same spectrum simultaneously with the primary network as long as they can cancel their interference to the primary nodes in such a way that the primary nodes do not feel the presence of the secondary nodes. In other words, activities by the secondary network are made transparent (or "invisible") to the primary network. Such transparency is accomplished through a systematic interference cancellation (IC) by the secondary nodes without any impact on the primary network. Although the idea of TC paradigm has been explored in the information theory (IT) community (known as "underlay" in [4]), results from IT community only focused on either cellular networks or have limited its scope to very simple network settings, e.g., several nodes or link pairs, all for *single-hop* communications [1, 5, 10, 26, 28]. TC paradigm is still not well understood in the wireless networking community, particularly for multihop networks.

- **Cooperative Sharing Paradigm** To address the second limitation of interweave paradigm, we envision a cooperative sharing (CS) paradigm that allows cooperation between the primary and secondary networks to relay each other's traffic. There are some previous efforts on having secondary network help relay traffic for the primary network [8, 9, 11, 12, 19, 21, 27]. But there is no consideration of the converse (i.e, primary helping the secondary). This was called "overlay" in [4]. Different from overlay paradigm, our proposed CS paradigm allows cooperation in both directions. It allows to pool together the resources from both the primary and secondary networks so that users in each network can access a much richer network resources from the combined network. Such cooperation could vary from unilateral cooperation (i.e., only secondary nodes help relay primary user traffic but not vice versa), bilateral cooperation, constrained cooperation, or other customized policy based on particular application needs or requirements. There are many potential benefits of CS, such as much improved network topology, opportunity of better power control, more flexibility in link layer scheduling and network layer routing, and a much richer set of service offerings for users in the primary and secondary networks.

The remainder of this chapter is organized as follows. In Sections 3.2 and 3.3, we explore the TC and CS paradigm, respectively, for multihop networks. For each paradigm, we discuss their benefits and challenges, and develop the mathematical models for analysis. Through separate case studies, we demonstrate that the TC paradigm is a potential solution to address the first limitation

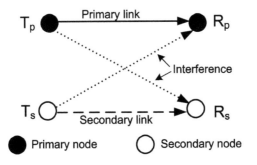

Figure 3.1: A simple example illustrating the TC paradigm.

of the interweave paradigm, while the CS paradigm is a potential solution to address the second limitation of the interweave paradigm. Section 3.4 summarizes this chapter.

3.2 Transparent Coexistence Paradigm

3.2.1 Overview

The TC paradigm aims to achieve simultaneously activation of both primary and secondary networks so as to enhance access to the radio spectrum. Under the TC paradigm, primary nodes use their spectrum just as they do under the current interweave paradigm. They do not feel the presence of the secondary nodes, even though the secondary nodes may be using the same spectrum at the same time. The burden of IC rests solely upon the secondary nodes. Such an aggressive spectrum sharing paradigm may be realized by a number of advances at the physical layer.

We use a simple example to illustrate this concept. In Figure 3.1, suppose T_p and R_p are a pair of transmit and receive nodes in the primary network, while T_s and R_s are a pair of transmit and receive nodes in the secondary network. Assume that all nodes share the same channel. Suppose T_p is transmitting 1 data stream to R_p. Under the interweave paradigm, secondary transmit node T_s is prohibited from transmission on the same channel, as it will interfere with primary receive node R_p. However, when multiple input multiple output (MIMO) is employed on the secondary nodes, simultaneous transmissions can be achieved under the TC paradigm. Assume secondary nodes T_s and R_s are each equipped with 4 antennas (4 DoFs). T_s can use 1 of its DoFs to cancel its interference to R_p so that R_p can receive its 1 data stream correctly from T_p. At node R_s, R_s can use 1 of its DoFs to cancel interference from T_p. After IC, both T_s and R_s still have 3 DoFs remaining, which can be used for SM of 3 data stream from T_s to R_s.

To achieve TC, all IC responsibility should rest upon the secondary nodes. Specifically, a secondary transmit node needs to cancel its interference to all neighboring primary receive nodes that are interfered by this secondary transmitter. A secondary receive node also needs to cancel interference from all neighboring primary transmit nodes that interfere with this secondary receiver. It is important for the secondary nodes to have accurate channel state information (CSI) for IC. The problem is: How can a secondary node obtain the CSI between itself and its neighboring primary nodes while remaining transparent to the primary nodes? We propose the following solution to resolve this problem.

For each primary node, it typically sends out a pilot sequence (training sequence) to its neighboring primary nodes so that those primary nodes can estimate the CSI for communication. This is the practice for current cellular networks, and we assume such a mechanism is available for a pri-

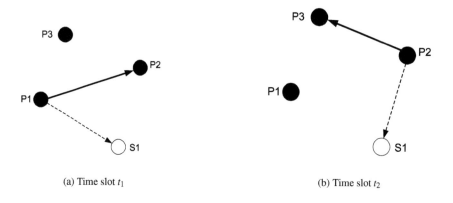

(a) Time slot t_1 (b) Time slot t_2

Figure 3.2: CSI estimation at secondary node S_1.

mary network. Since we consider a multi-hop network, where each node will act as a transmitter in one time slot but as a receiver in another time slot. Then, each secondary node can *overhear* the pilot sequence signal from the primary node while staying transparent. For example, in Figure 3.2(a), in time slot t_1, when P_1 is transmitting the pilot sequence, a secondary node S_1 can overhear this sequence from P_1. Likewise, in Figure 3.2(b), in time slot t_2, when P_2 is transmitting its pilot sequence, the secondary node S_1 can overhear this pilot sequence from P_2. Suppose the pilot sequence from the primary nodes is publicly available (as in cellular networks) and is known to the secondary nodes. Then the secondary node S_1 can use this information and the actual received pilot sequence signal from the primary nodes for channel estimation. Based on the reciprocity property of a wireless channel [20], a secondary node S_1 will be able to estimate the CSI in both directions to/from P_1 and P_2. Likewise, the secondary node may use the same approach to derive CSI among the secondary nodes.

To realize TC, we need to successfully address the following challenges:

- **Channel/time slot scheduling** In a secondary network, an intermediate relay node is both a transmitter and a receiver. Assuming half-duplex at each node, a node cannot transmit and receive on the same channel within the same time slot. Therefore, scheduling (either in time slot or channel) is needed. Scheduling can be performed both in time slot and channel allocation (time and frequency domains). Note that scheduling transmission/reception at a secondary node will lead to a particular interference relationship among the primary and secondary nodes in the underlying time slot and channel. This joint time/channel scheduling plays an integral role for IC in the network.

- **Inter-network IC** A secondary transmitter needs to cancel its interference to its neighboring primary receivers while a secondary receiver needs to cancel the interference from its neighboring primary transmitters.

- **Intra-network IC** In addition to inter-network IC, interference from a secondary node may also interfere with another secondary node within their own network (i.e., "intra-network" interference). Such an interference must also be canceled properly (either by a secondary transmitter or receiver) to ensure successful data communications inside the secondary network.

It is important to realize that the above three key challenges are not independent, but deeply intertwined with each other (see Figure 3.3). In particular, channel/time slot scheduling at a secondary node is directly tied to the interference relationship between primary and secondary nodes as well as interference among the secondary nodes. Therefore, a mathematical model of TC paradigm must capture all these components jointly.

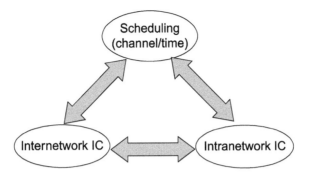

Figure 3.3: Coupling relationship among scheduling, inter-network IC, and intra-network IC.

3.2.2 Mathematical model

In this subsection, we develop a mathematical model for the TC paradigm under which a multi-hop secondary network can access the same spectrum as a primary network (see Figure 3.4). Referring to Figure 3.4, we consider a secondary multi-hop network consisting of a set of nodes S that is co-located with a primary multi-hop network consisting of a set of nodes P. Assume there is a set of channels B available for the primary network. Suppose that there are T time slots in each time frame. For the primary network, there is no special requirement on the primary nodes and we assume that each primary node is a single-antenna node. A primary node may transmit and receive on any channel and time slot as needed. Suppose there is a set of sessions $\tilde{F}$ within the primary network P. For a given routing for each session, denote $\tilde{L}$ as the set of active links in the primary network (shown in solid arrow lines in Figure 3.4). Denote $\tilde{z}^b_{(\tilde{l})}(t)$ as the number of data streams over primary link $\tilde{l} \in \tilde{L}$ on channel b in time slot t. Then due to single antenna on each primary node, $\tilde{z}^b_{(\tilde{l})}(t) = 1$ if link $\tilde{l}$ is active (on channel b and time slot t) and 0 otherwise.

For the secondary network, we assume MIMO capability at each node. Denote A_i as the number of antennas on a secondary node $i \in S$. Suppose there is a set of multi-hop sessions F in S. For a given routing for each session, denote L as the set of secondary links (shown in dashed arrow line in Figure 3.4).

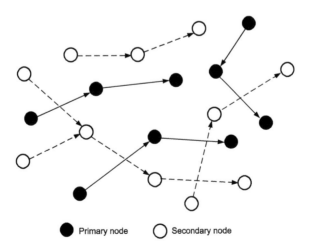

Primary node   Secondary node

Figure 3.4: A multi-hop secondary network co-located with a multi-hop primary network.

To model scheduling at a secondary node for transmission or reception, we denote $x_i^b(t)$ and $y_i^b(t)$ ($i \in \mathcal{S}, b \in \mathcal{B}$ and $1 \leq t \leq T$) as whether node i is a transmitter or receiver on channel b in time slot t, respectively. We have

$$x_i^b(t) = \begin{cases} 1 & \text{if node } i \text{ is a transmitter on } b \in \mathcal{B} \text{ and } 1 \leq t \leq T; \\ 0 & \text{otherwise.} \end{cases}$$

$$y_i^b(t) = \begin{cases} 1 & \text{if node } i \text{ is a receiver on } b \in \mathcal{B} \text{ and } 1 \leq t \leq T; \\ 0 & \text{otherwise.} \end{cases}$$

To consider half-duplex (a node cannot transmit and receive on the same channel in the same time slot), we have the following constraint on $x_i^b(t)$ and $y_i^b(t)$:

$$x_i^b(t) + y_i^b(t) \leq 1 \qquad (i \in \mathcal{S}, b \in \mathcal{B}, 1 \leq t \leq T). \tag{3.1}$$

Node ordering for IC in secondary network. Recall that the secondary network is solely responsible for "inter-network" IC (as well as "intra-network" IC). To avoid unnecessary duplication in DoF allocation for IC, it has been shown in [17] that node-ordering based IC is very effective. Under this scheme, all secondary nodes are put into an ordered list. DoF allocation at each secondary node for IC is based on the position of the node in the list. It was shown in [17] that such a disciplined approach can ensure that there is no duplication in IC (and, thus, no waste of DoF resources) while the final DoF allocation is feasible at the physical layer. We will describe the specific rules for DoF allocation at a secondary node for IC (depending on whether it is a transmitter or receiver) shortly. But first, we give a mathematical model for the node ordering concept.

Denote $\pi^b(t)$ as an ordered list of the secondary nodes in the network on $b \in \mathcal{B}$ and $1 \leq t \leq T$, and denote $\pi_i^b(t)$ as the position of node $i \in \mathcal{S}$ in $\pi^b(t)$. Therefore, $1 \leq \pi_i^b(t) \leq S$, where $S = |\mathcal{S}|$. For example, if $\pi_i^b(t) = 3$, then it means that node i is the third node in the list $\pi^b(t)$.

To model the relative ordering between any two secondary nodes i and j in $\pi^b(t)$, we use a binary variable $\theta_{j,i}^b(t)$ and define it as follows:

$$\theta_{j,i}^b(t) = \begin{cases} 1 & \text{if node } j \text{ is before node } i \text{ in } \pi^b(t) \text{ for } b \in \mathcal{B} \text{ and } 1 \leq t \leq T; \\ 0 & \text{otherwise.} \end{cases}$$

It was shown in [17] that the following relationships hold among $\pi_i^b(t), \pi_j^b(t)$ and $\theta_{j,i}^b(t)$.

$$\pi_i^b(t) - S \cdot \theta_{j,i}^b(t) + 1 \leq \pi_j^b(t) \leq \pi_i^b(t) - S \cdot \theta_{j,i}^b(t) + S - 1, \tag{3.2}$$

where $i, j \in \mathcal{S}, b \in \mathcal{B}$, and $1 \leq t \leq T$.

Constraints at a secondary transmitter. At secondary transmitter i, it needs to expend DoFs for SM, IC to neighboring primary receivers, and IC to a subset of its neighboring secondary receivers based on their orders in the node list.

- **DoF for SM.** For SM, denote $z_{(l)}^b(t)$ and $\mathcal{L}_i^{\text{Out}}$ as the number of data streams over link $l \in \mathcal{L}$ and the set of outgoing links from secondary node i. Then the number of DoFs at secondary node $i \in \mathcal{S}$ for SM is $\sum_{l \in \mathcal{L}_i^{\text{Out}}} z_{(l)}^b(t)$ for $b \in \mathcal{B}$ and $1 \leq t \leq T$.

- **DoF for IC to neighboring primary receivers.** To ensure transparent coexistence, a secondary transmitter needs to cancel its interference to neighboring primary receivers. Recall that if a primary receiver $p \in \mathcal{P}$ is within the interference range of node i, the number of DoFs at node i that is used for canceling the interference to node p is equal to the number of data stream that are received at node p. Denote $\tilde{\mathcal{L}}_p^{\text{In}}$ as the set of incoming primary links to node p. Denote $\tilde{\mathcal{I}}_i$ as the set of primary nodes that are located within the interference range of secondary transmitter i. For node $p \in \tilde{\mathcal{I}}_i$, the number of DoFs used at node i for canceling interference to node p is $\sum_{\tilde{l} \in \tilde{\mathcal{L}}_p^{\text{In}}} \tilde{z}_{(\tilde{l})}^b(t)$ for $b \in \mathcal{B}$ and $1 \leq t \leq T$. Now for all primary receive nodes in $\tilde{\mathcal{I}}_i$, the number of DoFs used at node i to cancel interference to these nodes is $\left(\sum_{p \in \tilde{\mathcal{I}}_i} \sum_{\tilde{l} \in \tilde{\mathcal{L}}_p^{\text{In}}} z_{(\tilde{l})}^b(t) \right)$ for $b \in \mathcal{B}$ and $1 \le t \le T$.

- **DoF for IC to secondary receivers.** For IC within the secondary network, this secondary transmitter i only needs to cancel its interference to a subset (instead of all) of its neighboring secondary receivers based on the node ordering list [17]. Specifically, this secondary transmitter i only needs to expend DoFs to null its interference to neighboring secondary receivers that are *before* itself in the ordered secondary node list $\pi^b(t)$. Node i does not need to expend any DoF to null its interference to those secondary receivers that are *after* itself in the ordered node list $\pi^b(t)$. This is because the interference from node i to those secondary receivers (that are after this node in $\pi^b(t)$) will be nulled by those secondary receivers later (when we perform DoF allocation at those nodes). This is the key to avoid duplication in IC.

Recall that if a secondary receiver $j \in \mathcal{S}$ is within the interference range of secondary transmit node i, the number of DoFs required at transmit node i to cancel its interference to node j is equal to the number of data-stream that are being received at node j. Denote $\mathcal{L}_j^{\text{In}}$ as the set of the incoming links to node j. Denote $\mathcal{I}_i$ as the set of secondary nodes that are located within the interference range of node i. For secondary receive node $j \in \mathcal{I}_i$, the number of DoFs used at secondary transmit node i for canceling its interference to node j is $(\theta_{j,i}^b(t) \cdot \sum_{k \in \mathcal{L}_j^{\text{In}}}^{\text{Tx}(k) \neq i} z_{(k)}^b(t))$. Note that we are using the indicator variable $\theta_{j,i}^b(t)$ to consider only those secondary receive nodes that are before node i in the ordered node list $\pi^b(t)$. Now for all secondary receive nodes in $\mathcal{I}_i$, the number of DoFs used at node i to cancel interference to these nodes is $\sum_{j \in \mathcal{I}_i} (\theta_{j,i}^b(t) \cdot \sum_{k \in \mathcal{L}_j^{\text{In}}}^{\text{Tx}(k) \neq i} z_{(k)}^b(t))$ for $b \in \mathcal{B}$ and $1 \le t \le T$.

Putting all these DoF consumptions together at a secondary transmitter i, we have the following constraints:

- If this secondary transmit node i is active, i.e., $x_i^b(t) = 1$, we have

$$x_i^b(t) \le \sum_{l \in \mathcal{L}_i^{\text{Out}}} z_{(l)}^b(t) + \left(\sum_{p \in \tilde{\mathcal{I}}_i} \sum_{\tilde{l} \in \tilde{\mathcal{L}}_p^{\text{In}}} \tilde{z}_{(\tilde{l})}^b(t) \right) + \sum_{j \in \mathcal{I}_i} \left(\theta_{j,i}^b(t) \cdot \sum_{k \in \mathcal{L}_j^{\text{In}}}^{\text{Tx}(k) \neq i} z_{(k)}^b(t) \right) \le A_i, \quad (3.3)$$

which means that the DoF consumption at node i cannot be more that the total number of its antennas.

- If node i is not active, i.e., $x_i^b(t) = 0$, we have

$$\sum_{l \in \mathcal{L}_i^{\text{Out}}} z_{(l)}^b(t) = 0. \quad (3.4)$$

To incorporate $x_i^b(t)$ into mathematical constraints, we rewrite (3.3) and (3.4) into the following two constraints:

$$x_i^b(t) \le \sum_{l \in \mathcal{L}_i^{\text{Out}}} z_{(l)}^b(t) + \sum_{p \in \tilde{\mathcal{I}}_i \tilde{l} \in \tilde{\mathcal{L}}_p^{\text{In}}} \tilde{z}_{(\tilde{l})}^b(t) + \sum_{j \in \mathcal{I}_i} \theta_{j,i}^b(t) \sum_{k \in \mathcal{L}_j^{\text{In}}}^{\text{Tx}(k) \neq i} z_{(k)}^b(t) \le A_i x_i^b(t) + (1 - x_i^b(t))M, \quad (3.5)$$

$$\sum_{l \in \mathcal{L}_i^{\text{Out}}} z_{(l)}^b(t) \le x_i^b(t) \cdot A_i, \quad (3.6)$$

where M is a large constant, which is an upper bound of $\sum_{p \in \tilde{\mathcal{I}}_i} \sum_{\tilde{l} \in \tilde{\mathcal{L}}_p^{\text{In}}} \tilde{z}_{(\tilde{l})}^b(t) + \sum_{j \in \mathcal{I}_i} \theta_{j,i}^b(t) \cdot \sum_{k \in \mathcal{L}_j^{\text{In}}}^{\text{Tx}(k) \neq i} z_{(k)}^b(t)$ when $x_i^b(t) = 0$. For example, we can set $M = \sum_{j \in \mathcal{I}_i} A_j + \sum_{p \in \tilde{\mathcal{I}}_i} \sum_{\tilde{l} \in \tilde{\mathcal{L}}_p^{\text{In}}} \tilde{z}_{(\tilde{l})}^b(t)$.

To see that (3.5) and (3.6) can replace (3.3) and (3.4), note that (i) when $x_i^b(t) = 1$, (3.5) becomes (3.3) and (3.6) holds trivially; (ii) when $x_i^b(t) = 0$, (3.4) and (3.6) are equivalent, and (3.5) holds trivially.

Since (3.5) has a nonlinear term $(\theta_{j,i}^b(t) \cdot \sum_{k \in \mathcal{L}_j^{\text{In}}}^{\text{Tx}(k) \neq i} z_{(k)}^b(t))$, we can use *Reformulation-Linearization Technique* (RLT) [7, Chapter 6] to reformulate this nonlinear term by introducing new variables and adding new linear constraints. We define a new variable $\lambda_{j,i}^b(t)$ as follows:

$$\lambda_{j,i}^b(t) = \theta_{j,i}^b(t) \cdot \sum_{k \in \mathcal{L}_j^{\text{In}}}^{\text{Tx}(k) \neq i} z_{(k)}^b(t) \,,$$

where $i \in \mathcal{S}, j \in \mathcal{I}_i, b \in \mathcal{B}$, and $1 \leq t \leq T$. For binary variable $\theta_{j,i}^b(t)$, we have the following associated constraints:

$$\theta_{j,i}^b(t) \geq 0 \,,$$
$$(1 - \theta_{j,i}^b(t)) \geq 0 \,.$$

For $\sum_{k \in \mathcal{L}_j^{\text{In}}}^{\text{Tx}(k) \neq i} z_{(k)}^b(t)$, we have

$$\sum_{k \in \mathcal{L}_j^{\text{In}}}^{\text{Tx}(k) \neq i} z_{(k)}^b(t) \geq 0 \,,$$

$$A_j - \sum_{k \in \mathcal{L}_j^{\text{In}}}^{\text{Tx}(k) \neq i} z_{(k)}^b(t) \geq 0 \,.$$

We can multiply each of the two constraints involving $\theta_{j,i}^b(t)$ by each of the two constraints involving $\sum_{k \in \mathcal{L}_j^{\text{In}}}^{\text{Tx}(k) \neq i} z_{(k)}^b(t)$, and replacing the product term $(\theta_{j,i}^b(t) \cdot \sum_{k \in \mathcal{L}_j^{\text{In}}}^{\text{Tx}(k) \neq i} z_{(k)}^b(t))$ by $\lambda_{j,i}^b(t)$. Then (3.5) can be replaced by the following linear constraints:

$$x_i^b(t) \leq \sum_{l \in \mathcal{L}_i^{\text{Out}}} z_{(l)}^b(t) + \left(\sum_{p \in \tilde{\mathcal{I}}_i} \sum_{\tilde{l} \in \tilde{\mathcal{L}}_p^{\text{In}}} \tilde{z}_{(\tilde{l})}^b(t) \right) + \sum_{j \in \mathcal{I}_i} \lambda_{j,i}^b(t) \leq A_i x_i^b(t) + (1 - x_i^b(t))M, \quad (3.7)$$

$$\lambda_{j,i}^b(t) \geq 0, \quad (3.8)$$

$$\lambda_{j,i}^b(t) \leq \sum_{k \in \mathcal{L}_j^{\text{In}}}^{\text{Tx}(k) \neq i} z_{(k)}^b(t), \quad (3.9)$$

$$\lambda_{j,i}^b(t) \leq A_j \cdot \theta_{j,i}^b(t), \quad (3.10)$$

$$\lambda_{j,i}^b(t) \geq A_j \cdot \theta_{j,i}^b(t) - A_j + \sum_{k \in \mathcal{L}_j^{\text{In}}}^{\text{Tx}(k) \neq i} z_{(k)}^b(t), \quad (3.11)$$

where $i \in \mathcal{S}, j \in \mathcal{I}_i, b \in \mathcal{B}$, and $1 \leq t \leq T$.

Constraints at a secondary receiver. At a secondary receiver i, it needs to expend DoFs for SM, canceling interference from neighboring primary transmitters, and canceling interference from a subset of its neighboring secondary transmitters based on their orders in the node list.

- **DoF for SM.** For SM, the number of DoFs consumed at a secondary receiver $i \in \mathcal{S}$ is $\sum_{k \in \mathcal{L}_i^{\mathrm{In}}} z_{(k)}^b(t)$ for $b \in \mathcal{B}$ and $1 \leq t \leq T$.

- **DoF for IC from neighboring primary transmitters.** A secondary receiver needs to cancel the interference from neighboring primary transmitters. If a primary transmitter $p \in \mathcal{P}$ is within the interference range of secondary receive node $i \in \mathcal{S}$, the number of DoFs at node i required for canceling this interference from node p is equal to the number of data-stream that are being transmitted by node p. Denote $\tilde{\mathcal{L}}_p^{\mathrm{Out}}$ as the set of outgoing links from primary node p. For $p \in \tilde{\mathcal{I}}_i$, the number of DoFs used at node i for canceling interference from node p is $\sum_{\tilde{l} \in \tilde{\mathcal{L}}_p^{\mathrm{Out}}} \tilde{z}_{(\tilde{l})}^b(t)$. Now for all primary transmit nodes in $\tilde{\mathcal{I}}_i$, the number of DoFs used at node i to cancel interference from these nodes is $(\sum_{p \in \tilde{\mathcal{I}}_i} \sum_{\tilde{l} \in \tilde{\mathcal{L}}_p^{\mathrm{Out}}} \tilde{z}_{(\tilde{l})}^b(t))$ for $b \in \mathcal{B}$ and $1 \leq t \leq T$.

- **DoF for IC from secondary transmitters.** For IC within the secondary network, this secondary receiver i only needs to null the interference from a subset (instead of all) of its neighboring secondary transmitters based on node ordering list. Specifically, this secondary receiver i only need to expend DoFs to null the interference from neighboring secondary transmitters that are *before* itself in the ordered secondary node list $\pi^b(t)$. Node i does not need to expend any DoF to null the interference from these secondary transmitters that are *after* itself in the ordered node list $\pi^b(t)$. This is because the interference to node i from those secondary transmitters (that are after this node $\pi^b(t)$) will be nulled by those secondary transmitters later (when we perform DoF allocation at those nodes).

  Recall that if node i is within the interference range of a secondary transmit node $j \in \mathcal{S}$, the number of DoFs at node i that is used for canceling the interference from node j is equal to the number of data-stream that are being transmitted at node j. For a secondary transmit node $j \in \mathcal{I}_i$, the number of DoFs used at secondary receive node i for canceling interference from node j is $(\theta_{j,i}^b(t) \cdot \sum_{l \in \mathcal{L}_j^{\mathrm{Out}}}^{\mathrm{Rx}(l) \neq i} z_{(l)}^b(t))$. Now for all other secondary transmit nodes in $\mathcal{I}_i$, the number of DoFs used at node i to cancel interference from those nodes is $\sum_{j \in \mathcal{I}_i} (\theta_{j,i}^b(t) \cdot \sum_{l \in \mathcal{L}_j^{\mathrm{Out}}}^{\mathrm{Rx}(l) \neq i} z_{(l)}^b(t))$ for $b \in \mathcal{B}$ and $1 \leq t \leq T$.

We can put all DoF consumption at a secondary receiver as follows:

$$y_i^b(t) \leq \sum_{k \in \mathcal{L}_i^{\mathrm{In}}} z_{(k)}^b(t) + \sum_{p \in \tilde{\mathcal{I}}_i} \sum_{\tilde{l} \in \tilde{\mathcal{L}}_p^{\mathrm{Out}}} \tilde{z}_{(\tilde{l})}^b(t) + \sum_{j \in \mathcal{I}_i} \theta_{j,i}^b(t) \sum_{l \in \mathcal{L}_j^{\mathrm{Out}}}^{\mathrm{Rx}(l) \neq i} z_{(l)}^b(t) \leq A_i y_i^b(t) + (1 - y_i^b(t)) N, \quad (3.12)$$

$$\sum_{k \in \mathcal{L}_i^{\mathrm{In}}} z_{(k)}^b(t) \leq y_i^b(t) \cdot A_i, \quad (3.13)$$

where N is a large constant, which is an upper bounder of $\sum_{p \in \tilde{\mathcal{I}}_i} \sum_{\tilde{l} \in \tilde{\mathcal{L}}_p^{\mathrm{Out}}} \tilde{z}_{(\tilde{l})}^b(t) + \sum_{j \in \mathcal{I}_i} (\theta_{j,i}^b(t) \cdot \sum_{l \in \mathcal{L}_j^{\mathrm{Out}}}^{\mathrm{Rx}(l) \neq i} z_{(l)}^b(t))$ when $y_i^b(t) = 0$. For example, we can set $N = \sum_{j \in \mathcal{I}_i} A_j + \sum_{p \in \tilde{\mathcal{I}}_i} \sum_{\tilde{l} \in \tilde{\mathcal{L}}_p^{\mathrm{Out}}} \tilde{z}_{(\tilde{l})}^b(t)$.

Following the same token as in the last section, we can use RLT to linearize the nonlinear term

$(\theta_{j,i}^b(t) \cdot \sum_{l \in \mathcal{L}_j^{\text{Out}}}^{\text{Rx}(l) \neq i} z_{(l)}^b(t))$ in (3.12). Denote $\mu_{j,i}^b(t)$ as $(\theta_{j,i}^b(t) \cdot \sum_{l \in \mathcal{L}_j^{\text{Out}}}^{\text{Rx}(l) \neq i} z_{(l)}^b(t))$. Then (3.12) can be replaced by the following linear constraints:

$$y_i^b(t) \leq \sum_{k \in \mathcal{L}_i^{\text{In}}} z_{(k)}^b(t) + \left(\sum_{p \in \tilde{\mathcal{I}}_i} \sum_{\tilde{l} \in \mathcal{L}_p^{\text{Out}}} \tilde{z}_{(\tilde{l})}^b(t) \right) + \sum_{j \in \mathcal{I}_i} \mu_{j,i}^b(t) \leq A_i y_i^b(t) + (1 - y_i^b(t))N, \qquad (3.14)$$

$$\mu_{j,i}^b(t) \geq 0, \qquad (3.15)$$

$$\mu_{j,i}^b(t) \leq \sum_{l \in \mathcal{L}_j^{\text{Out}}}^{\text{Rx}(l) \neq i} z_{(l)}^b(t), \qquad (3.16)$$

$$\mu_{j,i}^b(t) \leq A_j \cdot \theta_{j,i}^b(t), \qquad (3.17)$$

$$\mu_{j,i}^b(t) \geq A_j \cdot \theta_{j,i}^b(t) - A_j + \sum_{l \in \mathcal{L}_j^{\text{Out}}}^{\text{Rx}(l) \neq i} z_{(l)}^b(t), \qquad (3.18)$$

where $i \in \mathcal{S}, j \in \mathcal{I}_i, b \in \mathcal{B}$, and $1 \leq t \leq T$.

Obviously, the node ordering is very important for IC, both for a transmit node and a receive node. A natural question arises as follows: What kind of ordering should we use? We find that there is no better solution other than putting the ordering problem as part of the overall optimization problem. According to the specific optimization objective, the optimal solution will give an optimal ordering.

3.2.3 Case Study

Using the above mathematical model for the TC paradigm, various problems could be investigated. In this section, we study a throughput optimization problem in the secondary network. Denote $r(f)$ as the rate of session $f \in \mathcal{F}$. Then at any link $l \in \mathcal{L}$ in the network, the aggregate throughput rate among the flows that traverse this link cannot exceed the link's scheduling capacity (over a time frame). That is,

$$\sum_{f \in \mathcal{F}}^{f \text{ traversing } l} r(f) \leq c \cdot \frac{1}{T} \sum_{b \in \mathcal{B}} \sum_{t=1}^{T} z_{(l)}^b(t) \qquad (l \in \mathcal{L}), \qquad (3.19)$$

where c is the data volume carried by a datastream.

For the throughput maximization problem, suppose we are interested in maximizing the minimum throughput rate among all secondary sessions. Then the problem can be formulated as follows:

> OPT-TC
> max $\quad r_{\min}$
> s.t $\quad r_{\min} \leq r(f) \quad (f \in \mathcal{F})$;
> Half-duplex constraints: (3.1);
> Node ordering constraints: (3.2);
> Transmitter DoF constraints: (3.6)–(3.11);
> Receiver DoF constraints: (3.13)–(3.18);
> Link capacity constraints:(3.19).

Table 3.1   Location of Each Node for the 20-Node Primary Network and 30-Node Secondary Network

		Primary Network			
Node	*Location*	*Node*	*Location*	*Node*	*Location*
P_1	(10, 10)	P_8	(15, 50)	P_{15}	(20, 80)
P_2	(30, 30)	P_9	(40, 70)	P_{16}	(31, 48)
P_3	(50, 30)	P_{10}	(60, 90)	P_{17}	(35, 85)
P_4	(75, 50)	P_{11}	(85, 90)	P_{18}	(90, 80)
P_5	(90, 20)	P_{12}	(40, 10)	P_{19}	(3, 35)
P_6	(90, 45)	P_{13}	(70, 10)	P_{20}	(6, 97)
P_7	(75, 65)	P_{14}	(55, 55)		
		Secondary Network			
Node	*Location*	*Node*	*Location*	*Node*	*Location*
S_1	(23, 66)	S_{11}	(55, 60)	S_{21}	(88, 62)
S_2	(3, 89)	S_{12}	(8, 56)	S_{22}	(70, 20)
S_3	(42, 41)	S_{13}	(3, 78)	S_{23}	(76, 74)
S_4	(19, 37)	S_{14}	(62, 2)	S_{24}	(84, 30)
S_5	(10, 70)	S_{15}	(92, 92)	S_{25}	(22, 92)
S_6	(29, 6)	S_{16}	(36, 94)	S_{26}	(60, 40)
S_7	(8, 25)	S_{17}	(82, 4)	S_{27}	(28, 16)
S_8	(51, 10)	S_{18}	(35, 60)	S_{28}	(99, 3)
S_9	(63, 75)	S_{19}	(76, 40)	S_{29}	(98, 38)
S_{10}	(65, 98)	S_{20}	(48, 21)	S_{30}	(47, 85)

In this formulation, $r_{\min}, r(f), x_i^b(t), y_i^b(t), z_{(l)}^b(t), \pi_i^b(t), \lambda_{j,i}^b(t), \mu_{j,i}^b(t)$ and $\theta_{j,i}^b(t)$ are optimization variables, and $A_i, M, N, \bar{z}_{(l)}^b(t)$ and c are given constants. This optimization problem is in the form of a mixed-integer linear program (MILP). Although the theoretical worst-case complexity to a general MILP problem is exponential [14], there exist highly efficient optimal/approximation algorithms (e.g., branch-and-bound with cutting planes [16]) and heuristics (e.g., sequential fixing algorithm [6, 25]) to solve it.

In the following, we present some numerical results for OPT-TC to illustrate how it actually works in a multihop secondary network, and show the tremendous benefits (in terms of spectrum access and throughput gain) of the TC over the interweave paradigm. We consider a 20-node primary network and a 30-node secondary network randomly deployed in the same 100×100 area. For the ease of scalability and generality, we normalize all units for distance, bandwidth, and throughput with appropriate dimensions. The location for each node (both primary and secondary) is generated at random and is listed in Table 3.1. We assume that there are four antennas on each secondary node, and all nodes' transmission range and interference range are 30 and 50, respectively.[1] There are two channels owned by the primary network ($B = 2$). A time frame is divided into four time slots ($T = 4$). For simplicity, we assume the data rate of one data stream in a time slot is 1 unit ($c = 1$).

We assume there are three active sessions in the primary network and four active sessions in the secondary network (see Table 3.2). For simplicity, we assume that minimum-hop routing is used for the primary and secondary sessions, although other routing methods will also work here. Further, the channel and time slot allocation on each hop for each primary session is known *a priori* and is shown in Figure 3.5, where (b,t) means this link is transmitting on channel b in time slot t. The solid arrows represent the links in the primary network, while the dashed arrows represent the links in the secondary network.

---

[1] For an in-depth study on how to set interference range, we refer readers to our previous work in [18].

Table 3.2 Source and Destination Nodes of Each Session in the Primary and Secondary Networks

	Primary Network	
Session	*Source Node*	*Destination Node*
1	P_1	P_{14}
2	P_5	P_7
3	P_{11}	P_{15}
	Secondary Network	
Session	*Source Node*	*Destination Node*
1	S_7	S_{25}
2	S_{21}	S_{17}
3	S_{14}	S_3
4	S_{30}	S_{23}

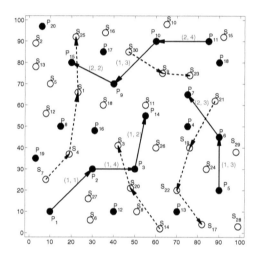

Figure 3.5: Active sessions in the primary and secondary networks.

For this network setting, we solve OPT-TC and the obtained objective value is 1.0. The channel and time slot scheduling on each link for each secondary session is shown in the shaded box as in Figure 3.6, where (b, t) on each secondary link represents that this link transmits on channel b in time slot t. The details of DoFs used for SM on each channel in each time slot on each link in the secondary network are shown in Table 3.3. The link rate (i.e., total number of DoFs used for SM averaged over a 4-time-slot frame) on a link is also shown in this table.

To see how the secondary node can be active simultaneously with the primary nodes while remaining transparent, consider $(b, t) = (1, 2)$ (channel 1, time slot 2) in Figure 3.6. Here, link $P_3 \rightarrow P_{14}$ in the primary network is active; links $S_{14} \rightarrow S_{20}$, $S_{22} \rightarrow S_{17}$, $S_{21} \rightarrow S_{19}$, $S_{30} \rightarrow S_9$, and $S_4 \rightarrow S_1$ in the secondary network are also active. Based on a node's interference range, the interference relationships among the nodes associated with these active links are shown in Figure 3.7, where the dotted arrow lines show the interference from a (primary or secondary) transmitter to an unintended (primary or secondary) receiver. Table 3.4 shows the DoF allocation at each secondary node for SM, IC to/from primary nodes, and IC within the secondary network for $(b, t) = (1, 2)$.

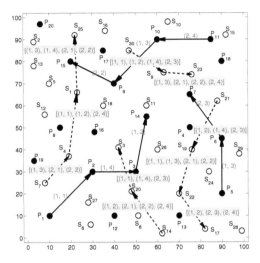

Figure 3.6: Channel and time slot scheduling on each link for the secondary sessions by our solution algorithm. Channel and time slot scheduling on each link for the primary sessions are given in Figure 3.5.

- First, we check whether there is any interference to primary receiver P_{14}. Note that there are four potential interference from secondary transmitters, i.e., S_4, S_{21}, S_{22}, and S_{30}. Since each of these secondary transmitter uses one DoF to cancel its interference to primary receiver P_{14} (fifth column in Table 3.4), all interference on the primary receiver P_{14} is effectively nulled. Therefore, the primary receiver P_{14} is not interfered by the simultaneous activation of its neighboring secondary transmitters.

- Next, we check whether the interference from the primary transmitter is nulled properly at its neighboring secondary receivers ("internetwork" interference). Note that primary transmit node P_3 is interfering its neighboring secondary receive nodes S_1, S_{20}, S_{17}, S_{19}, and S_9. Since each of these secondary receive nodes uses one DoF to cancel this interference (fifth column in Table 3.4), this interference from primary transmit node P_3 is effectively nulled at these secondary receive nodes.

- Finally, we check whether the interference within the secondary network ("intranetwork" interference) is nulled properly by the secondary nodes themselves. The IC within the secondary network follows the node ordering, which is shown in the third column of Table 3.4. The number of DoFs used for IC to/from other secondary nodes is shown in the last column of Table 3.4. As an example, consider node S_{22}, which is a transmit node. Referring to Table 3.4, S_{22} only needs to cancel its interference to those receive nodes that are before itself in the ordered node list and within S_{22}'s interference range, i.e., node S_{19}. Table 3.4 (last column) shows that S_{22} indeed uses one DoF to cancel its interference to S_{19}. For its interference to the secondary receive node S_{20}, which is also in S_{22}'s interference range, S_{22} does not need to do anything, as S_{20} is after node S_{22} in the ordered list. This interference to S_{20} will be canceled by S_{20} (as shown in Table 3.4, last column).

It can be easily verified that for all interference among the active secondary nodes are properly canceled. Further, at each active secondary node, the DoFs used for SM, IC to/from the primary nodes, IC within the secondary network is not more than its total DoFs (i.e., 4).

Table 3.3   Channel and Time Slot Scheduling on Each Link, DoF Allocation for SM, and Throughput on Each Link for the Secondary Sessions

Session	Link	(Channel, Time Slot) Scheduling	DoF for SM	Link rate
1	$S_7 \longrightarrow S_4$	(1, 3)	2	1.0
		(2, 1)	1	
		(2, 2)	1	
	$S_4 \longrightarrow S_1$	(1, 1)	1	1.0
		(1, 2)	1	
		(2, 4)	2	
	$S_1 \longrightarrow S_{25}$	(1, 3)	1	1.0
		(1, 4)	1	
		(2, 1)	1	
		(2, 2)	1	
2	$S_{21} \longrightarrow S_{19}$	(1, 2)	1	1.0
		(1, 4)	2	
		(2, 3)	1	
	$S_{19} \longrightarrow S_{22}$	(1, 1)	1	1.0
		(1, 3)	1	
		(2, 1)	1	
		(2, 2)	1	
	$S_{22} \longrightarrow S_{17}$	(1, 2)	1	1.0
		(2, 3)	1	
		(2, 4)	2	
3	$S_{14} \longrightarrow S_{20}$	(1, 2)	1	1.0
		(2, 1)	1	
		(2, 2)	1	
		(2, 4)	1	
	$S_{20} \longrightarrow S_3$	(1, 1)	2	1.0
		(1, 4)	2	
		(2, 3)	1	
4	$S_{30} \longrightarrow S_9$	(1, 1)	1	1.0
		(1, 2)	1	
		(1, 4)	1	
		(2, 3)	1	
	$S_9 \longrightarrow S_{23}$	(1, 3)	1	1.0
		(2, 1)	1	
		(2, 2)	1	
		(2, 4)	1	

The above illustration is for $(b,t) = (1,2)$ (i.e., channel 1, time slot 2), the results for the other channel and time slots (i.e., $(1,1)$, $(1,4)$, $(1,3)$, $(2,2)$, $(2,3)$, and $(2,4)$) are similar and are omitted to conserve space.

To see the benefits of the TC paradigm, we compare it to the interweave paradigm. Under the interweave paradigm, a secondary node is not allowed to transmit (receive) on the same channel at the same time when a nearby primary receiver (transmitter) is using this channel. Therefore, the set of available channel and time slots that can be used by secondary nodes is smaller. The problem formulation for this paradigm is similar to (but simpler than) OPT-TC. In particular, we can remove the second term ($\sum_{p \in \tilde{\mathcal{I}}_i} \sum_{\tilde{l} \in \tilde{\mathcal{L}}_p^{\text{In}}} \tilde{z}_{(\tilde{l})}^b(t)$ and $\sum_{p \in \tilde{\mathcal{I}}_i} \sum_{\tilde{l} \in \tilde{\mathcal{L}}_p^{\text{Out}}} \tilde{z}_{(\tilde{l})}^b$) in constraints (3.5) and (3.12) in OPT-TC that are used for secondary nodes to cancel interference to/from the primary nodes.

Table 3.4   DoF Allocation for SM and IC on $(b,t) = (1,2)$ At Each Node in the Secondary Network

Node i	TX/RX	$\pi_i^1(2)$	DoF for SM	DoF for IC to/from primary nodes	DoF for IC within secondary network
S_{19}	RX	1	1	1 from P_3	0
S_{14}	TX	2	1	0	1 to S_{19}
S_{22}	TX	4	1	1 to P_{14}	1 to S_{19}
S_{21}	TX	5	1	1 to P_{14}	0
S_{17}	RX	6	1	1 from P_3	1 from S_{14}
S_{20}	RX	8	1	1 from P_3	1 from S_{22}
S_{30}	TX	9	1	1 to P_{14}	0
S_9	RX	11	1	1 from P_3	1 from S_{21}
S_4	TX	12	1	1 to P_{14}	1 to S_{20}
S_1	RX	13	1	1 from P_3	1 from S_{30}

Following the same setting as above, we solve the optimization problem under the interweave paradigm. Note that the available channels and time slot resources at each node are only a subset of 2 channels and 4 time slots, vs. full 2 channels and 4 time slots for each secondary node in the TC paradigm. The obtained objective value is 0.5 (compared to 1.0 under TC paradigm). The channel and time slot scheduling on each link of each secondary session is shown in Figure 3.8. Comparing Figures 3.6 and 3.8, we find that the set of channels and time slots used by each secondary link under interweave paradigm is smaller.

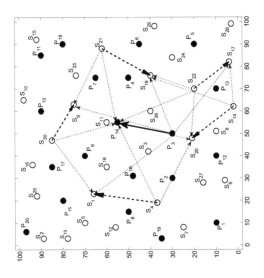

Figure 3.7: Illustration of interference relationships among the primary and secondary links on channel 1 in time slot 2 in the case study.

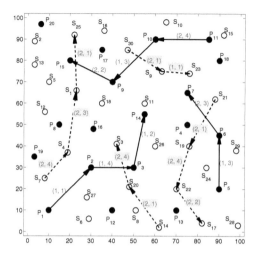

Figure 3.8: Channel and time slot scheduling on each link for the secondary sessions under the interweave paradigm.

3.3 Cooperative Sharing Paradigm

3.3.1 Overview

To address the second limitation associated with the interweave paradigm, we propose a CS paradigm that allows cooperation between the primary and secondary networks on the data plane in terms of relaying *each other*'s traffic. Such cooperation policies could vary from unilateral cooperation (i.e., only having secondary nodes help relay primary user traffic but not vice versa), bilateral cooperation, constrained cooperation, or other customized policy based on particular application needs or requirements.

As a concrete example, we consider the UPS policy [24], which is the abbreviation of United cooperation of Primary and Secondary networks. UPS allows primary nodes to help relay traffic for the secondary network and vice versa. UPS policy pools all the resources from the primary and secondary networks together to enable users in each network to access much richer network resources from the combined network. Figure 3.9 illustrates the concept of UPS, where the nodes in the two networks interact to form one combined network. It is not hard to see that there are

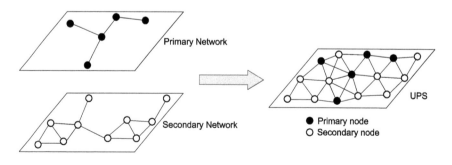

Figure 3.9: An illustration of the UPS paradigms.

many potential benefits associated with the UPS policy. We briefly describe some of the benefits as follows:

- **Topology.** Comparing to the primary or secondary network in isolation, the combined network allows both primary and secondary networks to have a much better node connectivity, with nodes from both networks.

- **Power Control.** By adjusting its transmission power, a node has the flexibility in choosing its next hop node. This flexibility can be exploited for different upper layer performance requirements (connectivity, scheduling) or objectives (throughput).

- **Link Layer.** The improved physical topology allows more opportunities (as well as challenges) at the link layer for medium access. Both the primary and secondary networks can better coordinate with each other in transmission and interference avoidance. Further, the potential impact associated with node or link failure can now be easily investigated as compared to the case when the primary and secondary networks operate independently.

- **Network Diversity.** The combined network offers more routing opportunities to users in both networks. This directly translates into improved throughput and delay performance for user sessions.

- **Services and Applications.** The UPS architecture (combining both the primary and secondary networks) allows to offer much richer services and applications than those services explored in [8,9,11,12,19,21,27]. Although the two networks are combined into one physical network, constrained services and applications offered to users in the primary and secondary networks can still be supported at the transport level, by implementing certain traffic engineering policies. In other words, the combined network does not infer that service guarantee to the primary network will be lost. On the contrary, by specifying the desired resource management policy appropriately, one can easily achieve various service differentiation objectives and application goals.

3.3.2   *Mathematical model*

In this section, we develop a mathematical model for the UPS policy. Suppose that there is a set of sessions in the primary network, with each session having certain rate requirement. In the secondary network, suppose there is also a set of sessions, with each session having an elastic traffic requirement. By "elastic," we mean that each secondary session does not have a stringent rate requirement as the primary session. Instead, each secondary session will be supported on a best-effort basis and will transmit as much as the remaining network resource allows. The goal is to have the combined network to support the rate requirements of the primary sessions while maximizing the throughput of the secondary sessions. There are a number of technical challenges that one must address:

- **Guaranteed service for the primary traffic.** In the above problem, each primary session has hard rate requirement, and the combined network should support it at all possibility. This problem alone may not be challenging. What is challenging (and interesting) is that should there are multiple ways to support primary sessions' rate requirements, we should find such a way that the rates for the secondary sessions are maximized in the combined network.

- **Relay selection.** To meet the service requirement (guaranteed service for primary traffic) and to optimize the objective (maximize the rates of the secondary sessions), relay node selection along a route (for either a primary or secondary session) is a key problem.

- **Scheduling.** To maximize the rates of the secondary sessions while guaranteeing the rates of the primary sessions, scheduling in each time slot is not a trivial problem. In particular, in addition to addressing traditional self-interference (half-duplex) and mutual interference problems, primary network must be cooperative so as to help the secondary sessions to achieve their optimization objective in the combined network.

Denote $\mathcal{N}$ as the combined set of nodes consisting of both the set of primary nodes $\hat{\mathcal{N}}_P$ and the set of secondary nodes $\mathcal{N}_S$, i.e., $\mathcal{N} = \hat{\mathcal{N}}_P \bigcup \mathcal{N}_S$. In the combined network, denote $\mathcal{T}_i$ as the set of nodes (including both primary and secondary nodes) located within a nodes i's transmission range, where i can be either a primary or secondary node (i.e., $i \in \mathcal{N}$). Denote $\mathcal{I}_j$ as the set of nodes (including both primary and secondary nodes) located within a node j's interference range, where j can be either a primary or secondary node. For a primary session $l \in \hat{\mathcal{L}}$, we assume it has a hard requirement on its data rate, which we denote as $\hat{R}(l)$. For a secondary session $m \in \mathcal{L}$, we assume that it does not have a rate requirement. Instead, the data rate $r(m)$ on $m \in \mathcal{L}$ is supported on a best-effort basis and will be an optimization variable in the problem formulation.

Guaranteed service for the primary sessions. Under the UPS policy, the primary sessions consider the combined network $\mathcal{N}$ as their communication resources. For flexibility and load balancing, we allow flow splitting in the network. That is, the flow rate of a session may split and merge inside the network in whatever loop-free manner as long as it can help support the given rate requirement $\hat{R}(l)$ of session $l \in \hat{\mathcal{L}}$. Denote $\hat{f}_{ij}(l)$ as the data rate on link (i, j) that is attributed to primary session $l \in \hat{\mathcal{L}}$, where $i \in \mathcal{N}$ and $j \in \mathcal{T}_i$. Denote $\hat{s}(l)$ and $\hat{d}(l)$ as the source and destination nodes of primary session $l \in \hat{\mathcal{L}}$, respectively. We have the following flow balance constraints:

- If node i is the source node of primary session $l \in \hat{\mathcal{L}}$ (i.e., $i = \hat{s}(l)$), then

$$\sum_{j \in \mathcal{T}_i} \hat{f}_{ij}(l) = \hat{R}(l) \qquad (l \in \hat{\mathcal{L}}). \tag{3.20}$$

- If node i is an intermediate relay node for primary session l (i.e., $i \neq \hat{s}(l)$ and $i \neq \hat{d}(l)$), then

$$\sum_{\substack{j \in \mathcal{T}_i}}^{j \neq \hat{s}(l)} \hat{f}_{ij}(l) = \sum_{\substack{k \in \mathcal{T}_i}}^{k \neq \hat{d}(l)} \hat{f}_{ki}(l) \qquad (l \in \hat{\mathcal{L}}, i \in \hat{\mathcal{N}}_P). \tag{3.21}$$

- If node i is the destination node of primary session l (i.e., $i = \hat{d}(l)$), then

$$\sum_{k \in \mathcal{T}_i} \hat{f}_{ki}(l) = \hat{R}(l) \qquad (l \in \hat{\mathcal{L}}). \tag{3.22}$$

It can be easily verified that once (3.20) and (3.21) are satisfied, then (3.22) must be satisfied. As a result, it is sufficient to list only (3.20) and (3.21) in the formulation.

Best effort service for secondary sessions. Under the UPS policy, the primary sessions have priority in accessing the combined network resources (in the form of guaranteed services). While the primary sessions are supported, the secondary sessions may use as much as the remaining resources of the combined network. How the primary and secondary sessions interact in the combined network is part of our optimization problem. Denote $f_{ij}(m)$ as the data rate on link (i, j) that is attributed to secondary session $m \in \mathcal{L}$. Denote $s(m)$ and $d(m)$ as the source and destination nodes of secondary session $m \in \mathcal{L}$, respectively. Similar to that for the primary sessions, we allow flow splitting for the secondary sessions. We have the following flow balance constraints:

- If node i is the source node of secondary session $m \in \mathcal{L}$ (i.e., $i = s(m)$), then we have

$$\sum_{j \in \mathcal{T}_i} f_{ij}(m) = r(m) \qquad (m \in \mathcal{L}). \tag{3.23}$$

- If node i is an intermediate relay node for secondary session m (i.e., $i \neq s(m)$ and $i \neq d(m)$), then

$$\sum_{\substack{j \in \mathcal{T}_i}}^{j \neq s(m)} f_{ij}(m) = \sum_{\substack{k \in \mathcal{T}_i}}^{k \neq d(m)} f_{ki}(m) \qquad (m \in \mathcal{L}, i \in \mathcal{N}_S). \tag{3.24}$$

- If node i is the destination node of secondary session m (i.e., $i = d(m)$), then

$$\sum_{k \in \mathcal{T}_i} f_{ki}(m) = r(m) \qquad (m \in \mathcal{L}). \tag{3.25}$$

Again, to avoid redundancy, it is sufficient to list only (3.23) and (3.24) in the formulation.

Note that although (3.23)–(3.25) are similar to (3.20)–(3.22), there is an important difference between them: Unlike $\hat{R}(l)$ for primary session $l \in \hat{\mathcal{L}}$, which is a given *constant*, secondary session rate $r(m)$, $m \in \mathcal{L}$, is an optimization *variable*. Therefore, we will only need to optimize the flow path in (3.20)–(3.22), while we need to both optimize the routes and maximizing the objective $r(m)$ in (3.23)–(3.25).

Self-interference constraints. We assume scheduling is done in time slot on a frame-by-frame basis, with each frame consisting of T time slots. We use a binary variable $x_{ij}[t], i, j \in \mathcal{N}$ and $1 \leq t \leq T$, to indicate whether node i transports data to node j. That is, if node i transports data to node j, $x_{ij}[t] = 1$; otherwise, $x_{ij}[t] = 0$.

Since each primary or secondary session is unicast, a node i only needs to transmit to or receive from one node in a time slot. We have

$$\sum_{j \in \mathcal{T}_i} x_{ij}[t] \leq 1 \qquad (i \in \mathcal{N}, 1 \leq t \leq T), \tag{3.26}$$

$$\sum_{k \in \mathcal{T}_i} x_{ki}[t] \leq 1 \qquad (i \in \mathcal{N}, 1 \leq t \leq T). \tag{3.27}$$

To account for half-duplex at each node i, we have

$$x_{ij}[t] + x_{ki}[t] \leq 1 \qquad (i \in \mathcal{N}, j, k \in \mathcal{T}_i, 1 \leq t \leq T). \tag{3.28}$$

These three constraints in (3.26), (3.27), and (3.28) can be replaced by the following single and equivalent constraint.

$$\sum_{j \in \mathcal{T}_i} x_{ij}[t] + \sum_{k \in \mathcal{T}_i} x_{ki}[t] \leq 1 \qquad (i \in \mathcal{N}, 1 \leq t \leq T). \tag{3.29}$$

To see this, note that in (3.29), if node i is receiving data from some node in $\mathcal{T}_i$ in time slot t, we must have $\sum_{j \in \mathcal{T}_i} x_{ij}[t] = 0$, i.e., node i cannot transmit in the same time slot. This is exactly the half-duplex constraint. In this case, (3.29) also becomes (3.27). On the other hand, if node i is transmitting to some node in $\mathcal{T}_i$ in time slot t, then $\sum_{k \in \mathcal{T}_i} x_{ki}[t] = 0$, i.e., node i cannot receive in the same time slot. Again, this is the half-duplex constraint, obviously. In this case, (3.29) becomes (3.26).

Mutual interference constraints. For any primary or secondary node $j \in \mathcal{N}$ that is receiving data in time slot t, it shall not be interfered by another (unintended) transmitting node $p \in \mathcal{I}_j$ in the same time slot. We have the following mutual interference constraint.

$$x_{ij}[t] + x_{pk}[t] \leq 1, \tag{3.30}$$

where $i \in \mathcal{T}_j, p \in \mathcal{I}_j, k \in \mathcal{T}_p, j \in \mathcal{N}, j \neq k$, and $1 \leq t \leq T$.

Following the same token in (3.29), the three constraints in (3.26), (3.27), and (3.30) can be replaced by the following single and equivalent constraint.

$$\sum_{i \in \mathcal{T}_j} x_{ij}[t] + \sum_{k \in \mathcal{T}_p} x_{pk}[t] \leq 1 , \tag{3.31}$$

where $p \in \mathcal{I}_j, j \in \mathcal{N}, j \neq k$, and $1 \leq t \leq T$.

Link capacity constraints. For each link (i, j), denote the link capacity as C_{ij}. Since the aggregate flow rate from primary and secondary sessions on each link (i, j) cannot exceed the average link rate (over T time slots), we have

$$\sum_{l \in \hat{\mathcal{L}}}^{j \neq \hat{s}(l), i \neq \hat{d}(l)} \hat{f}_{ij}(l) + \sum_{m \in \mathcal{L}}^{j \neq s(m), i \neq d(m)} f_{ij}(m) \leq \frac{1}{T} \sum_{t=1}^{T} C_{ij} \cdot x_{ij}[t]. \tag{3.32}$$

3.3.3   Case Study

In the combined network, our goal is to offer guaranteed support for the primary sessions (each with a given rate requirement) while maximize the throughput for the secondary sessions, whose traffic are assumed to be elastic. To ensure fairness among the sessions, we set our objective function to maximize the minimum session rate among all secondary sessions. We define r_{min} as the minimum data rate among all sessions. The optimization problem can be written as follows:

> OPT-UPS
>
> max $\quad r_{min}$
>
> s.t. $\quad r_{min} \leq r(l) \quad (l \in \mathcal{L})$;
> Guaranteed service for primary sessions: (3.20), (3.21);
> Best effort service for secondary sessions: (3.23), (3.24);
> Self-interference constraints: (3.29);
> Mutual interference constraints: (3.31);
> Link capacity constraints: (3.32).

In this formulation, $\hat{R}(l)$ and C_{ij} are constants, $x_{ij}[t]$ are binary variables, and $\hat{f}_{ij}(l), f_{ij}(m)$, and $r(m)$ are continuous variables. This problem is in the form of MILP. Although the theoretical worst-case complexity to a general MILP problem is exponential, we found that OPT-UPS can be solved by CPLEX efficiently, due to fact that all integer variables $x_{ij}[t]$ are binary.

In the following, we present some numerical results to demonstrate the capabilities and advantages of the UPS policy, and have a close look at how the primary and secondary nodes help each other in the UPS policy. We consider a UPS network where both the primary and the secondary nodes are randomly deployed in a 100×100 area. For generality, we normalize the units for distance, bandwidth, power, and data rate with appropriate dimensions. We assume the bandwidth of the channel allocated to the primary network is $B = 10$. The transmission power spectral density Q_i for each node $i \in \mathcal{N}$ is 1. We assume the transmission range and interference range at all nodes are 30 and 50, respectively. The number of time slots in a frame is $T = 10$. A link's capacity is calculated by $C_{ij} = B \log_2(1 + \frac{Q_i d_{ij}^{-4}}{N_0})$, where d_{ij} is the distance between nodes i and j, and N_0 is the ambient Gaussian noise density. We assume $N_0 = 10^{-6}$.

We consider a 30-node network, with 15 primary nodes and 15 secondary nodes randomly deployed in a 100×100 area (see Figure 3.10). We assume that there are two primary sessions in the primary network and two secondary sessions in the secondary network. The source and destination

Table 3.5 The Source and Destination Nodes for Each Sessions in the 30-Node Network

Session	Source	Destination
Primary session 1	P_{10}	P_7
Primary session 2	P_{15}	P_1
Secondary session 1	S_6	S_{15}
Secondary session 2	S_{12}	S_3

nodes for each session are randomly chosen in each network and are shown in Table 3.5. Denote the rate requirements of the two primary sessions as $\hat{R}(1)$ and $\hat{R}(2)$, respectively. We gradually increase the rate requirements of $\hat{R}(1)$ and $\hat{R}(2)$ and examine (i) whether such rates can be supported under the UPS policy and the interweave paradigm, respectively, and (ii) the objective value of secondary session rate in our optimization problem under the two paradigms. The optimization problem for maximizing the minimum secondary session rate under interweave paradigm can be formulated following a similar token to OPT-UPS and is omitted here to conserve space.

Table 3.6 summarizes the results of this study. The second column represents increasing rate requirements for the primary sessions (i.e., $\hat{R}(1)$ and $\hat{R}(2)$). For ease of explanation, we break this table into five regions, with each region representing an operating behavior for comparison under the two paradigms. The third and fourth columns show the performance under the UPS policy. Specifically, the third column shows whether the rate requirements of the two primary sessions are feasible in the primary network (abbreviated as "PN" in the table); the fourth column shows the maximized minimum data rate between the two secondary sessions (abbreviated as "SS" in the table), with 0 indicating zero rates for the secondary sessions and "N/A" indicating not applicable, as the corresponding network cannot even support the rate requirements of the primary sessions. The fifth and sixth columns show the performance under the interweave paradigm, which are to be compared to the third and fourth columns under the UPS paradigm, respectively.

Region 1 This region represents the scenario where the rate requirements of the primary sessions can be supported under both paradigms *and* the rates of the secondary sessions are positive. Comparing columns four and six, we can find that the secondary sessions always achieve higher performance under the UPS policy than that under the interweave paradigm. This confirms our expectation that the UPS policy can offer higher throughput for the secondary sessions.

As an example, consider the case when the two primary sessions have rate requirements (2.0, 3.0). The objective value achieved for the secondary sessions under the UPS and the interweave paradigms are 4.0019 and 1.2461, respectively. Under UPS policy, the flow routing and scheduling for the primary and secondary sessions are shown in Figure 3.10. The number in the box on each link represents the active time slots for this link. Note that primary nodes P_4, P_5, P_6, P_{11}, and P_{13} are helping relay secondary sessions' data while secondary nodes S_2, S_8, and S_{13} are helping relay the primary sessions' data. Under the interweave paradigm, the flow routing and scheduling for primary network is shown in Figure 3.11, which can satisfy rate requirements for primary sessions. According to the time slots used by the primary network, the secondary network calculate the available time slot at each node and use them to maximize their minimum data rate among all sessions. The flow routing and scheduling for the secondary sessions under interweave paradigm are also shown in Figure 3.11. As expected, there is no cooperation at the node level between the two networks in terms of relaying each other's data.

Region 2 This region represents the scenario where the rate requirements of the primary sessions can be supported under both paradigms, while the secondary sessions can only be supported under the UPS policy but not under the interweave paradigm (with $r_{\min} = 0$). This region shows that the

Table 3.6   Performance Comparison between the UPS and the Interweave Paradigms for Different Primary Session Rate Requirements

	Rate Requirements $(\hat{R}(1), \hat{R}(2))$	UPS Feasible in PN	SS rate	Interweave Paradigm Feasible in PN	SS rate
Region 1	(0, 0)	Yes	6.0073	Yes	4.8277
	(0.2, 0.3)	Yes	5.5229	Yes	3.4457
	(0.4, 0.6)	Yes	5.5229	Yes	3.4457
	(0.6, 0.9)	Yes	5.5.5229	Yes	3.4457
	(0.8, 1.2)	Yes	5.5229	Yes	3.4457
	(1.0, 1.5)	Yes	4.6263	Yes	3.378
	(1.2, 1.8)	Yes	4.6263	Yes	3.378
	(1.4, 2.1)	Yes	4.6263	Yes	2.4923
	(1.6, 2.4)	Yes	4.6263	Yes	1.7958
	(1.8, 2.7)	Yes	4.0019	Yes	1.2461
	(2.0, 3.0)	Yes	4.0019	Yes	1.2461
Region 2	(2.2, 3.3)	Yes	2.9815	Yes	0
	(2.4, 3.6)	Yes	2.7795	Yes	0
	(2.6, 3.9)	Yes	2.7795	Yes	0
	(2.8, 4.2)	Yes	2.7795	Yes	0
	(3.0, 4.5)	Yes	2.7795	Yes	0
	(3.2, 4.8)	Yes	2.7795	Yes	0
Region 3	(3.4, 5.1)	Yes	2.7795	No	N/A
	(3.6, 5.4)	Yes	1.4907	No	N/A
Region 4	(3.8, 5.7)	Yes	0	No	N/A
	(4.0, 6.0)	Yes	0	No	N/A
	(4.2, 6.3)	Yes	0	No	N/A
	(4.4, 6.6)	Yes	0	No	N/A
	(4.6, 6.9)	Yes	0	No	N/A
	(4.8, 7.2)	Yes	0	No	N/A
	(5.0, 7.5)	Yes	0	No	N/A
	(5.2, 7.8)	Yes	0	No	N/A
	(5.4, 8.1)	Yes	0	No	N/A
	(5.6, 8.4)	Yes	0	No	N/A
	(5.8, 8.7)	Yes	0	No	N/A
	(6.0, 9.0)	Yes	0	No	N/A
	(6.2, 9.3)	Yes	0	No	N/A
	(6.4, 9.6)	Yes	0	No	N/A
	(6.6, 9.9)	Yes	0	No	N/A
Region 5	(6.8, 10.2)	No	N/A	No	N/A

combined network can offer more to the secondary sessions than the isolated networks under the interweave paradigm.

Region 3   This region represents the scenario where the rate requirements of the primary sessions can be supported under the UPS policy but not so under the interweave paradigm. For secondary sessions, there is still remaining resource to support them under the UPS policy. For fairness in comparison, we do not consider the achieved rate of the secondary sessions under the interweave paradigm (marked as "N/A"). The region shows the definitive advantage of using a combined network to support the primary sessions over an independent primary network.

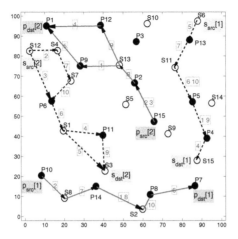

Figure 3.10: A Region 1 example that shows the flow routing topologies for the primary and secondary sessions in the UPS policy, where solid line segments are for the primary sessions, while dashed line segments are for the secondary sessions.

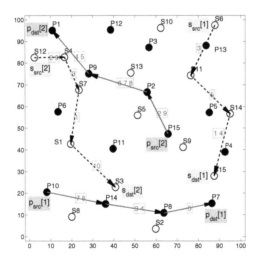

Figure 3.11: Region 1 example that shows the flow routing and scheduling for each primary session and secondary session under the interweave paradigm.

Region 4   This region represents the scenario where the rate requirement of the primary sessions can be satisfied under the UPS policy but not so under the interweave paradigm. The secondary sessions can no longer be supported under the UPS policy (with $r_{min} = 0$). For fairness in comparison, we do not consider the achieved data rate of the secondary sessions under the interweave paradigm (marked as "N/A"), as even the rate requirements for the primary sessions cannot be supported.

Similar to Region 3, Region 4 shows the advantage of using a combined network to support the primary sessions over an independent primary network.

Region 5 As the rate requirements of the primary sessions continue to increase, even the UPS policy will no longer be able to support them after a certain point. This is shown in Region 5.

3.4 Summary

In this chapter, we revealed two fundamental limitations associated with the traditional interweave paradigm. The first limitation is that a secondary network cannot be active simultaneously with the primary network in time, space, and frequency domains. The second limitation is that the primary and secondary networks are completely independent, in the sense that there is no cooperation and sharing between the two networks for data forwarding. To address the first limitation, we proposed the TC paradigm, under which the secondary nodes are allowed to use the same spectrum simultaneously with the primary network as long as they can cancel their interference to the primary nodes. To address the second limitation, we proposed the CS paradigm that allows cooperation between the primary and secondary networks to relay each other's traffic. For both paradigms, we discussed their properties and benefits. We also developed mathematic models in the context of multi-hop primary and secondary networks. Through case studies and numerical results, we showed that the TC and CS paradigms can offer significant improvement in spectrum access and throughput performance over the interweave paradigm.

3.5 Acknowledgments

This work was supported in part by the national science foundation (NSF) and office of naval research (ONR). The work of Dr. S. Kompella was supported in part by the ONR. Part of Prof. W. Lou's work was completed while she was serving as a program director at the NSF. Any opinion, findings, and conclusions or recommendations expressed in this chapter are those of the authors and do not reflect the views of the NSF.

References

[1] O. Bakr, M. Johnson, R. Mudumbai, and K. Ramchandran, "Multi-antenna interference cancellation techniques for cognitive radio applications," in *Proc. IEEE WCNC,* Budapest, Hungary, 6 pages, April 2009.

[2] K. Chowdhury and I.F. Akyildiz, "CRP: A routing protocol for cognitive radio ad hoc networks," *IEEE Journal on Selected Areas in Communications,* vol. 29, no. 4, pp. 794–804, April 2011.

[3] S. Geirhofer, L. Tong, and B.M. Sadler, "Dynamic spectrum access in the time domain: Modeling and exploiting white space," *IEEE Communications Magazine,* vol. 45, no. 5, pp. 66–72, May 2007.

[4] A. Goldsmith, S.A. Jafar, I. Maric, and S. Srinivasa, "Breaking spectrum gridlock with cognitive radios: An information theoretic perspective," *Proceedings of the IEEE,* vol. 97, no. 5, pp. 894–914, May 2009.

[5] F. Gao, R. Zhang, Y.-C. Liang, and X. Wang, "Design of learning-based MIMO cognitive radio systems," *IEEE Transactions on Vehicular Technology,* vol. 59, no. 4, pp. 1707–1720, May 2010.

[6] Y.T. Hou, Y. Shi, and H.D. Sherali, "Spectrum sharing for multi-hop networking with cognitive radios," *IEEE Journal on Selected Areas in Commun.,* vol. 26, no. 1, pp. 146–155, Jan. 2008.

[7] Y.T. Hou, Y. Shi, and H.D. Sherali, *Applied Optimization Methods for Wireless Networks,* Cambridge University Press, 2014, ISBN-13: 978-1107018808.

[8] S. Hua, H. Liu, M. Wu, and S.S. Panwar, "Exploiting MIMO antennas in cooperative cognitive radio networks," in *Proc. IEEE INFOCOM,* pp. 2714–2722, Shanghai, China, April 10-15, 2011.

[9] S.K. Jayaweera, M. Bkassiny, and K.A. Avery, "Asymmetric cooperative communication based spectrum leasing via auctions in cognitive radio networks," *IEEE Trans. on Wireless Commun.,* vol. 10, no. 8, pp. 2716–2724, August 2011.

[10] S.-J. Kim and G.B. Giannakis, "Optimal resource allocation for MIMO ad hoc cognitive radio networks," *IEEE Transactions on Information Theory,* vol. 57, no. 5, pp. 3117–3131, May 2011.

[11] R. Manna, R.H.Y. Louie, Y. Li, and B. Vucetic, "Cooperative spectrum sharing in cognitive radio networks with multiple antennas," *IEEE Trans. on Signal Processing,* vol. 59, no. 11, pp. 5509–5522, Nov. 2011.

[12] T. Nadkar, V. Thumar, G. Shenoy, A. Mehta, U.B. Desai, and S.N. Mechant, "A cross-layer framework for symbiotic relaying in cognitive radio networks," in *Proc. IEEE DySPAN,* pp. 498–509, Aachen, Germany, May. 3–6, 2011.

[13] President's Council of Advisors on Science and Technology (PCAST), "Report to the President — Realizing the Full Potential of Government-held Spectrum to Spur Economic Growth," July 2012, available online: http://www.whitehouse.gov/sites/default /files/microsites/ostp/pcast_spectrum_report _final_july_20_2012.pdf.

[14] A. Schrijver, *Theory of Linear and Integer Programming,* Wiley Interscience, New York, NY, 1986.

[15] S. Sengupta and K.P. Subbalakshmi, "Open research issues in multi-hop cognitive radio networks," *IEEE Communication Magazine,* vol. 52, no. 4, pp. 168–176, April 2013.

[16] S. Sharma, Y. Shi, Y.T. Hou, H.D. Sherali, S. Kompella, and S.F. Midkiff, "Joint flow routing and relay node assignment in cooperative multi-hop networks," *IEEE Journal on Selected Areas in Communications,* vol. 30, no. 2, pp. 254–262, Feb. 2012.

[17] Y. Shi, J. Liu, C. Jiang, C. Gao, and Y.T. Hou, "A DoF-based link layer model for multi-hop MIMO networks," *IEEE Trans. on Mobile Computing,* vol. 12, issue 7, pp. 1395–1408, 2014.

[18] Y. Shi, Y.T. Hou, J. Liu, and S. Kompella, "Bridging the gap between protocol and physical models for wireless networks," *IEEE Trans. on Mobile Computing,* vol. 12, issue 7, pp. 1404–1416, July 2013.

[19] O. Simone, I. Stanojev, S. Savazzi, Y. Bar-Ness, U. Spagnolini, and R. Pickholtz, "Spectrum leasing to cooperating secondary ad hoc networks," *IEEE Journal on Selected Areas in Commun.,* vol. 26, no. 1, pp. 203–213, Jan. 2008.

[20] G.S. Smith, "A direct derivation of a single-antenna reciprocity relation for the time domain," *IEEE Trans. on Antennas and Propagation,* vol. 52, no. 6, pp. 1568–1577, June 2004.

[21] W. Su, J.D. Natyjas, and S. Batalama, "Active cooperation between primary users and cognitive radio users in cognitive ad-hoc network" in *Proc. IEEE ICASSP,* pp. 3174–3177, March. 14–19, 2010.

[22] X. Wang and H.V. Poor, "Iterative (Turbo) soft interference cancellation and decoding for coded CDMA," *IEEE Transactions on Communications,* vol. 47, no. 7, pp. 1046–1061, July 1999.

[23] A.M. Wyglinski, M. Nekovee, and Y.T. Hou, *Cognitive Radio Communications and Networks: Principles and Practice.* Academic Press/Elsevier, 2010. ISBN: 978-0-12-374715-0.

[24] X. Yuan, Y. Shi, Y.T. Hou, W. Lou, and S. Kompella, "UPS: A united cooperative paradigm for primary and secondary networks," in *Proc. IEEE MASS,* pp. 78–85, Hangzhou, China, Oct. 14–16, 2013.

[25] X. Yuan, C. Jiang, Y. Shi, Y.T. Hou, W. Lou, S. Kompella, and S.F. Midkiff, "Toward transparent coexistence for multi-hop secondary cognitive radio networks," *IEEE Journal on Selected Areas in Communications,* vol. 33, no. 5, pp. 958–971, May 2015.

[26] R. Zhang and Y.-C. Liang, "Exploiting multi-antennas for opportunistic spectrum sharing in cognitive radio networks," *IEEE Journal of Selected Topics in Signal Processing,* vol. 2, no. 1, pp. 88–102, February 2008.

[27] J. Zhang and Q. Zhang, "Stackelberg game for utility-based cooperative radio network," in *Proc. ACM MobiHoc,* pp. 23–32, New Orleans, LA, USA, May 18–21, 2009.

[28] Y.J. Zhang and A.M.-C. So, "Optimal spectrum sharing in MIMO cognitive radio networks via semidefinite programming," *IEEE Journal on Selected Areas in Commun.,* vol. 29, no. 2, pp. 362–373, February 2011.

Chapter 4

Space-Time Spectrum Sharing in Cognitive Radio Networks

Zhiqing Wei, Zhiyong Feng, and Qixun Zhang

CONTENTS

4.1 Introduction

Cognitive radio (CR) [1] is one of the most promising technologies for efficient spectrum utilization. It enables flexible and comprehensive usage of available spectrum [2], and allows the optimization of radio resource utilization by exploiting spectrum holes, which are the available spectrum bands for secondary users (SUs). Notice that spectrum holes exist in space-time dimensions and SUs at different locations or time may experience different spectrum occupancies. For example, SUs that are far away from primary users (PUs) can transmit simultaneously with PUs, i.e., these SUs have spatial spectrum opportunities. However, SUs that are close to PUs can only utilize the spectrum not currently being used by PUs, namely, these SUs have temporal spectrum opportunities.

The exploration and exploitation of available spectrum holes can increase spectrum opportunities for SUs and improve spectrum utilization. To explore spectrum holes, space-time spectrum sensing is proposed to detect spectrum opportunities in space-time dimensions. Tandra *et al.* in [3]–[4] designed a space-time spectrum sensing scheme and proposed a new metric to evaluate the performance of space-time spectrum sensing. Furthermore, various space-time spectrum sensing schemes had been addressed in [5]–[8]. Do *et al.* in [5] proposed a two-stage spectrum sensing scheme, where SUs detect temporal spectrum opportunities in the first stage, if there are no temporal spectrum opportunities, SUs will then switch to the second stage to detect spatial spectrum opportunities. Ding *et al.* in [6]–[7] studied spatial-temporal opportunity detection in cognitive radio networks and proposed a two-dimensional sensing (TDS) framework to improve the opportunity detection performance [7]. They also studied cooperative spatial-temporal spectrum opportunity detection in [8]. However, theoretical calculation on the locations of space-time spectrum holes in [5]–[8] are still insufficient.

In terms of the locations of space-time spectrum holes, Vu *et al.* in [9] analyzed the region where spatial spectrum holes exist and proposed the notion of primary exclusive region (PER). However, temporal spectrum holes were not considered, thus SUs in [9] do not exploit temporal spectrum opportunities. Actually, there are temporal spectrum holes in part of PER, where SUs can opportunistically access vacant spectrum. With the exploitation of temporal spectrum opportunities, spectrum utilization can be improved, because more SUs are deployed in the entire region. The follow-up works in [10]–[11] extended PER concept by considering the shadowing effect and network-level performance analysis.

How to discover and calculate the locations of space-time spectrum holes is still a challenge for efficient spectrum sharing in cognitive radio networks (CRNs). To solve these problems, we propose to define three regions for space-time spectrum sharing in CRNs that consists of black region, gray region, and white region. Geographically, black region is surrounded by gray region, and gray region is surrounded by white region. As illustrated in Figure 4.1, PUs inside black region have exclusive right to use the licensed spectrum, and SUs are not allowed to operate in black region. Temporal spectrum holes exist in gray region, and SUs can opportunistically access the licensed spectrum with interweave[1] spectrum sharing. In white region, SUs taking advantage of their long distances from PUs can exploit spatial spectrum opportunities and transmit with maximum power at any time without causing severe interference to PUs.

In order to exploit spectrum holes in space-time dimensions efficiently, this chapter is focusing on the calculation of the bounds of three regions. First, we analyze the constraints of temporal spectrum sensing to determine the bound of gray region. Then the bound of white region is determined by estimating the aggregate interference from SUs to PUs and considering the outage constraint of PUs. The optimal radius of black region is determined by solving an optimization problem in the view of dynamic spectrum leasing.

---

[1]In the interweave spectrum sharing manner, SU utilizes the spectrum not currently being used by PU [19] [20] [21].

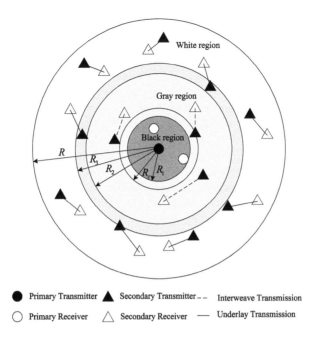

Figure 4.1: Black region is the disc with radius R_1, where only PUs are allowed to transmit. The gray region is the annulus with inner radius R_1 and outer radius R_2, where SUs can access the licensed spectrum with interweave spectrum sharing. White region is the annulus with inner radius R_3 and outer radius R, where SUs can transmit at any time with maximum transmit power. The annulus between gray region and white region is the transition zone, which can be eliminated under some conditions. R_c is the coverage radius of primary transmitter, out of which PUs will be in outage. The annulus with inner radius R_1 and outer radius R_c is the sacrificed region of PUs.

The rest of this chapter is organized as follows. The system model is introduced in Section II. Temporal spectrum opportunities in gray region are analyzed in Section III. Spatial spectrum opportunities in white region are analyzed in Section IV. In Section V, the bounds of the proposed three regions are derived in closed form. The numerical results are presented in Section VI. Finally, Section VII summarizes this chapter.

4.2   System Model

As illustrated in Figure 4.1, we consider a network model where PUs and SUs are located in a disc with radius R. A primary transmitter, for example, a TV station[2], is located in the center of the disc and transmits with power P_0. PUs, such as TV receivers, are distributed uniformly around primary transmitter within the coverage region. The entire disc is divided into three regions, i.e., black region with outer radius R_1, gray region with outer radius R_2, and white region with inner radius R_3 and outer radius R, where R can be infinite. The values of R_1, R_2, and R_3 are calculated in this chapter. SUs are distributed uniformly in gray and white regions, with density of λ users per unit area and transmit with power P_s. A channel model with only path loss is considered. Given a distance L between a transmitter and its receiver, the channel power gain is $g = \frac{V}{L^\alpha}$, where V is a frequency

---

[2]Notice that the primary network in this chapter is not limited to TV broadcasting network, as in the standard of IEEE 802.22.

dependent constant and α is the path loss factor. We normalize V to 1 for simplicity and consider $\alpha \geq 2$, which is typical in practice [9].

When $R_2 < R_3$, a transition zone between gray region and white region exists, which is illustrated as the annulus with inner radius R_2 and outer radius R_3 in Figure 4.1. In transition zone, SUs need to control their transmit power to mitigate the interference to PUs. Since SUs exploiting space-time spectrum opportunities must have a higher sensitivity than regular primary receivers [7] [14], temporal spectrum opportunities are proved to exist between black region and white region, which has not been taken into account in [9]. Moreover, spectrum utilization has been improved when compared with [9] by deploying SUs in gray region, because there are more SUs deployed in the entire region than those in [9]. The black region is within the coverage region of primary transmitter with radius R_c in Figure 4.1. The primary receivers outside the coverage region cannot decode the signal from primary transmitter even with no interference from SUs. PUs within black region are protected from the interference generated by SUs, thus the annulus with inner radius R_1 and outer radius R_c is a "sacrificed" region of primary networks [3], where quality of service (QoS) of PUs degrades as they are located outside black region.

In order to determine the (outer) radius of gray region (R_2)[3], key factors of both missed detection and false alarm are considered. On one hand, constraints of missed detection probability are provided in current spectrum sharing standards, such as IEEE 802.22. On the other hand, if the probability of false alarm is too high, SUs will not use the licensed spectrum band because SUs have scarce transmission opportunities in this situation and the benefit cannot offset the cost due to the time-energy consumption in spectrum sensing. Considering both missed detection and false alarm constraints simultaneously, when the distance between primary transmitter and secondary detector exceeds a threshold, at least one of the constraints cannot be satisfied, and this threshold is the outer radius of gray region (R_2). The inner radius of white region (R_3) is calculated by satisfying outage constraints of PUs on the edge of black region. To estimate the interference suffered by PUs, both accurate and supplement methods are designed to calculate the aggregate interference.

The radius of black region (R_1) can be determined by solving an optimization problem in the view of dynamic spectrum leasing. We have discovered that the expansion of black region will protect more PUs while spectrum opportunities of SUs are reduced. On the other aspect, the shrink of black region will increase spectrum opportunities of SUs, however, the interference suffered by PUs is more severe. Thus the fluctuation of the bound of black region reveals an interesting trade-off between the benefits of SUs and PUs. However, there is always a minimum protected region to separate SUs from PUs for PUs that are related with public security. For example, when PUs are ground-based or airborne radar systems, FCC recommends exclusion zones to protect the radar systems [13]. In this case, the minimum protected region of PUs must be considered in the interaction between the radii of black region and white region. Key parameters involved in this chapter are listed in Table 4.I.

4.3 Temporal Spectrum Opportunity In Gray Region

In gray region, since SUs operate in the interweave spectrum sharing manner, they are required to detect the spectrum bands which are not occupied by PUs. In this chapter, energy detection scheme is used to detect primary users. Let $x(t)$, h and $w(t)$ denote the band-pass primary signal with carrier frequency f_c and bandwidth W, the channel gain, and the additive white Gaussian noise (AWGN), respectively. When a primary signal $x(t)$ is transmitted through wireless channel with channel gain h, it is sampled at sampling frequency f_s. Thus, during the sensing time τ, the number of samples is $N = \lfloor \tau f_s \rfloor$. For a band-pass signal $x(t)$, the nth sample of the received signal $y(t)$ in the detector is

---

[3]In this chapter, we often refer to "outer radius of gray region" as "radius of gray region," and "inner radius of white region" as "radius of white region" for the ease of description.

Table 4.1 Key Parameters

Symbol	Description
R	The radius of the entire region
R_1	The radius of black region
R_2	The (outer) radius of gray region
R_3	The (inner) radius of white region
R_c	The radius of coverage area of primary transmitter
P_0	Transmit power of PU
P_s	Transmit power of SU
λ	Density of SUs in gray and white regions
λ_p	Density of PUs in black region
α	Path loss factor
σ_w^2	Power density function of AWGN
P_m, ξ_m	Missed detection probability and its upper bound
P_f, ξ_f	False alarm probability and its upper bound
r	SNR at the energy detector
N	Number of samples of energy detector
ε	Detection threshold of energy detector
I_0	Aggregate interference from SUs to PU

$y[n]$ that follows a binary hypothesis: $\mathcal{H}_0$ (primary signal absent) and $\mathcal{H}_1$ (primary signal present), as follows [22].

$$y[n] = \begin{cases} w[n] & : \mathcal{H}_0 \\ hx[n] + w[n] & : \mathcal{H}_1 \end{cases}, \tag{4.1}$$

where $x[n]$ and $w[n]$ are the samples of primary signal and noise process, respectively. $w[n]$ is assumed to be a Gaussian random variable with zero mean and variance σ_w^2, namely, $w[n] \sim \mathcal{N}(0, \sigma_w^2)$. Similarly, we assume $x[n] \sim \mathcal{N}(0, \sigma_s^2)$ [16]. Because we only consider path loss in the channel model, h is only related to the geographic location of SU. Thus we assume h remains unchanged in a sensing duration. Besides, $x[n]$, h, $w[n]$ and $y[n]$ are assumed to be real variables. In the sequel, we take into account of the constraints of missed detection and false alarm, and first evaluate the performance limits of spectrum sensing with precise result to derive the radius of gray region (R_2). Because there is no closed form of R_2 in the precise result, we evaluate the performance limits of spectrum sensing using Central Limit Theorem (CLT) and similarly derive the radius of gray region with closed form.

4.3.1 Precise Result

In this section, the test statistic is $Y = \sum_{i=1}^{N} (y[i])^2$. According to [17], the received signal of energy detector follows a noncentral chi-square distribution under hypothesis $\mathcal{H}_1$ and follows a chi-square distribution under hypothesis $\mathcal{H}_0$. The probability density functions of Y under $\mathcal{H}_1$ and $\mathcal{H}_0$ are

$$f_{Y|\mathcal{H}_1}(y) = \frac{1}{2\sigma_w^2} \left(\frac{y}{2r} \right)^{(v-1)/2} e^{-\frac{2r+y}{2\sigma_w^2}} I_{v-1} \left(\frac{\sqrt{2ry}}{\sigma_w^2} \right), \tag{4.2}$$

$$f_{Y|\mathcal{H}_0}(y) = \frac{1}{2^v \sigma_w^{2v} \Gamma(v)} y^{v-1} e^{-\frac{y}{2\sigma_w^2}}, \tag{4.3}$$

where $r = \frac{\sigma_s^2}{\sigma_w^2}$ is the signal to noise ratio (SNR) at the detector, N denotes the number of samples, and $v = N/2$. $I_v(*)$ is the vth-order modified Bessel function of the first kind. $\Gamma(*)$ is the Gamma function. The probability of detection $P_d = \Pr\{\mathcal{H}_0|\mathcal{H}_1\} = \Pr\{Y > \varepsilon|\mathcal{H}_1\}$ and the probability of false alarm $P_f = \Pr\{\mathcal{H}_1|\mathcal{H}_0\} = \Pr\{Y > \varepsilon|\mathcal{H}_0\}$ are as follows:

$$P_d = \int_\varepsilon^\infty f_{Y|\mathcal{H}_1}(y)dy = Q_v\left(\sqrt{\frac{2r}{\sigma_w^2}}, \sqrt{\frac{\varepsilon}{\sigma_w^2}}\right), \tag{4.4}$$

$$P_f = \int_\varepsilon^\infty f_{Y|\mathcal{H}_0}(y)dy = \frac{\Gamma\left(v, \frac{\varepsilon}{2\sigma_w^2}\right)}{\Gamma(v)}, \tag{4.5}$$

where ε is the detection threshold, $Q_v(*,*)$ represents the generalized Marcum Q-function, and $\Gamma(\alpha,x)$ denotes the upper incomplete Gamma function with definition $\Gamma(\alpha,x) = \int_x^\infty e^{-t}t^{\alpha-1}dt$. With definition $P_m = 1 - P_d$, from the perspective of PUs, the lower P_m is, the less interference PUs will receive from SUs. From the perspective of SUs, the lower P_f is, the more spectrum opportunities SUs can exploit. Both P_m and P_f should be as small as possible. But the receiver operating characteristic (ROC) declares that P_m and P_f cannot be reduced simultaneously. Thus, we bound P_m and P_f as follows:

$$P_f = \frac{\Gamma\left(v, \frac{\varepsilon}{2\sigma_w^2}\right)}{\Gamma(v)} \le \xi_f \tag{4.6}$$

$$P_m = 1 - P_d = 1 - Q_v\left(\sqrt{\frac{2r}{\sigma_w^2}}, \sqrt{\frac{\varepsilon}{\sigma_w^2}}\right) \le \xi_m. \tag{4.7}$$

We define ε_f and ε_m as follows:

$$\varepsilon_f = 2\sigma_w^2 \Gamma^{-1}\left(v, \xi_f \Gamma(v)\right) \tag{4.8}$$

$$\varepsilon_m = \sigma_w^2\left(Q_v^{-1}\left(\sqrt{\frac{2r}{\sigma_w^2}}, 1 - \xi_m\right)\right)^2. \tag{4.9}$$

As to constraint (4.6), we have Theorem 4.1.

Theorem 4.1
When $\varepsilon \ge \varepsilon_f$, the constraint for the probability of false alarm is satisfied, where $\Gamma^{-1}(\alpha,x)$ is the inverse upper incomplete Gamma function of x with α as a constant.

Proof 4.1   The derivative of P_f is

$$\frac{\partial P_f}{\partial \varepsilon} = \frac{1}{\Gamma(v)}\frac{\partial \Gamma\left(v, \frac{\varepsilon}{2\sigma_w^2}\right)}{\partial \varepsilon} = -\frac{e^{-\frac{\varepsilon}{2\sigma_w^2}}\left(\frac{\varepsilon}{2\sigma_w^2}\right)^{v-1}}{2\sigma_w^2 \Gamma(v)} < 0. \tag{4.10}$$

Thus, P_f is a decreasing function of ε. Solving inequality (4.6), we have $\varepsilon \ge 2\sigma_w^2 \Gamma^{-1}\left(v, \xi_f \Gamma(v)\right) \overset{\Delta}{=} \varepsilon_f$. Since $\Gamma(v, \frac{\varepsilon}{2\sigma_w^2})$ is a monotonic function of ε, its inverse function exists. ■

As to constraint (4.7), we have Theorem 4.2.

Theorem 4.2
When $\varepsilon \le \varepsilon_m$, the constraint for the probability of missed detection is satisfied, where $Q_v^{-1}(\alpha,x)$ is the inverse generalized Marcum Q-function of x with α as a constant.

Proof 4.2 The derivative of P_m is

$$\frac{\partial P_m}{\partial \varepsilon} = \frac{\partial \int_0^\varepsilon f_{Y|\mathcal{H}_1}(y)dy}{\partial \varepsilon} = f_{Y|\mathcal{H}_1}(\varepsilon) \geq 0. \tag{4.11}$$

Thus, P_m is an increasing function of ε. Solving inequality (4.7), we have $\varepsilon \leq \sigma_w^2 \left(Q_v^{-1} \left(\sqrt{\frac{2r}{\sigma_w^2}}, 1 - \xi_m \right) \right)^2 \triangleq \varepsilon_m$. Because the generalized Marcum Q-function $Q_v(\alpha, x)$ is a monotonic function of x, its inverse function exists. ■

Since the inverse functions of upper incomplete Gamma function and generalized Marcum Q-function do not have closed form, ε_f and ε_m do not have closed form either. However, their values can be obtained numerically. We explore the relation between ε_m and the distance between primary transmitter and secondary detector, and have Theorem 4.3.

Theorem 4.3
ε_m *is a decreasing function of the distance between primary transmitter and secondary detector.*

Proof 4.3 Since ε_m is an implicit function of SNR r, the relation between ε_m and r is

$$F(\varepsilon_m, r) = 1 - Q_v\left(\sqrt{\frac{2r}{\sigma_w^2}}, \sqrt{\frac{\varepsilon_m}{\sigma_w^2}}\right) - \xi_m = 0. \tag{4.12}$$

Thus, we have

$$\frac{d\varepsilon_m}{dr} = -\frac{\frac{\partial F(\varepsilon_m, r)}{\partial r}}{\frac{\partial F(\varepsilon_m, r)}{\partial \varepsilon_m}}, \tag{4.13}$$

where $\frac{\partial F(\varepsilon_m, r)}{\partial \varepsilon_m} > 0$, which is verified in the proof of Theorem 4.2. Thus, we need to prove $\frac{\partial F(\varepsilon_m, r)}{\partial r} \leq 0$.

Another form of the probability of missed detection is

$$\begin{aligned}
P_m &= \int_0^{\varepsilon_m} f_{Y|\mathcal{H}_1}(y)dy \\
&\overset{(a)}{=} \frac{1}{2\sigma_w^2} \int_0^{\varepsilon_m} \left(\frac{y}{2r}\right)^{\frac{v-1}{2}} e^{-\frac{2r+y}{2\sigma_w^2}} \left(\frac{ry}{2\sigma_w^4}\right)^{\frac{v-1}{2}} \sum_{k=0}^{\infty} \frac{\left(\frac{ry}{2\sigma_w^4}\right)^k}{k!\Gamma(v+k)} dy \\
&\overset{(b)}{=} e^{-\frac{r}{\sigma_w^2}} \sum_{k=0}^{\infty} \int_0^{\frac{\varepsilon_m}{2\sigma_w^2}} \left(\frac{r}{\sigma_w^2}\right)^k e^{-x} x^{k+v-1} \frac{1}{k!\Gamma(v+k)} dx \\
&\overset{(c)}{=} e^{-\frac{r}{\sigma_w^2}} \sum_{k=0}^{\infty} \left(\frac{r}{\sigma_w^2}\right)^k \frac{\gamma(v+k, \frac{\varepsilon_m}{2\sigma_w^2})}{k!\Gamma(v+k)},
\end{aligned} \tag{4.14}$$

where (a) takes the expansion of the modified Bessel function of the first kind, which is as follows [30, Section 8.445],

$$I_{v-1}\left(\frac{\sqrt{2ry}}{\sigma_w^2}\right) = \left(\frac{\sqrt{2ry}}{2\sigma_w^2}\right)^{v-1} \sum_{k=0}^{\infty} \frac{\left(\frac{ry}{2\sigma_w^4}\right)^k}{k!\Gamma(v+k)}, \tag{4.15}$$

(b) changes the order of integration and summation, and (c) takes the form of lower incomplete

Gamma function with definition $\gamma(\alpha,x) = \int_0^x e^{-t}t^{\alpha-1}dt$. The form of P_m in (4.14) is not applicable in computation, but it is useful in analysis. Substituting (4.14) into (4.12), we have

$$F(\varepsilon_m, r) = e^{-\frac{r}{\sigma_w^2}} \sum_{j=0}^{\infty} \frac{(\frac{r}{\sigma_w^2})^j}{j!} \frac{\gamma(j+v, \frac{\varepsilon_m}{2\sigma_w^2})}{\Gamma(j+v)} - \xi_m = 0. \tag{4.16}$$

Then we have

$$\sigma_w^2 e^{\frac{r}{\sigma_w^2}} \frac{\partial F(\varepsilon_m, r)}{\partial r}$$

$$= \sum_{j=1}^{\infty} \frac{j(\frac{r}{\sigma_w^2})^{j-1}}{j!} \frac{\gamma(j+v, \frac{\varepsilon_m}{2\sigma_w^2})}{\Gamma(j+v)} - \sum_{j=0}^{\infty} \frac{(\frac{r}{\sigma_w^2})^j}{j!} \frac{\gamma(j+v, \frac{\varepsilon_m}{2\sigma_w^2})}{\Gamma(j+v)}$$

$$\stackrel{(d)}{=} \sum_{j=0}^{\infty} \frac{(\frac{r}{\sigma_w^2})^j}{j!} \frac{\gamma(j+v+1, \frac{\varepsilon_m}{2\sigma_w^2})}{\Gamma(j+v+1)} - \sum_{j=0}^{\infty} \frac{(\frac{r}{\sigma_w^2})^j}{j!} \frac{\gamma(j+v, \frac{\varepsilon_m}{2\sigma_w^2})}{\Gamma(j+v)} \tag{4.17}$$

$$= \sum_{j=0}^{\infty} \frac{(\frac{r}{\sigma_w^2})^j}{j!} \left(\frac{\gamma(j+v+1, \frac{\varepsilon_m}{2\sigma_w^2})}{\Gamma(j+v+1)} - \frac{\gamma(j+v, \frac{\varepsilon_m}{2\sigma_w^2})}{\Gamma(j+v)} \right),$$

where (d) is obtained by changing the starting index of the summation from 1 to 0. The expansion of the lower incomplete Gamma function is as follows:

$$\frac{\gamma(n,x)}{\Gamma(n)} = 1 - \exp(-x) \sum_{j=0}^{n-1} \frac{x^j}{j!}. \tag{4.18}$$

Thus, $\frac{\gamma(n,x)}{\Gamma(n)}$ is a decreasing function of n. Therefore, we have $\frac{\gamma(j+v, \frac{\varepsilon_m}{2})}{\Gamma(j+v)} \geq \frac{\gamma(j+v+1, \frac{\varepsilon_m}{2})}{\Gamma(j+v+1)}$ and $\frac{\partial F(\varepsilon_m, r)}{\partial r} \leq 0$ in (4.17). According to (4.13), we have $\frac{d\varepsilon_m}{dr} \geq 0$. Note that SNR r is a decreasing function of the distance between primary transmitter and secondary detector, which is denoted as L. Thus, ε_m is also a decreasing function of L, because $\frac{d\varepsilon_m}{dL} = \frac{d\varepsilon_m}{dr} \frac{dr}{dL} \leq 0$. ∎

SUs take the detection threshold ε from the interval $[\varepsilon_f, \varepsilon_m]$, where ε_f is a constant and ε_m is a decreasing function of the distance L. As L increases, there must be an L, where $\varepsilon_m < \varepsilon_f$ and the interval $[\varepsilon_f, \varepsilon_m]$ is empty. Then SUs cannot find a detection threshold for spectrum sensing. Therefore $\varepsilon_m = \varepsilon_f$ reveals the limit of temporal spectrum sensing. Solving this equation yields the bound of gray region as follows:

$$2\sigma_w^2 \Gamma^{-1}\left(v, \xi_f \Gamma(v)\right) = \sigma_w^2 (Q_v^{-1}(\sqrt{\frac{2r}{\sigma_w^2}}, 1 - \xi_m))^2$$

$$\Rightarrow 1 - \xi_m = Q_v(\sqrt{\frac{2r}{\sigma_w^2}}, \sqrt{2\Gamma^{-1}(v, \xi_f \Gamma(v))}), \tag{4.19}$$

which can be solved numerically to get the solution of r. We can get the candidate[4] radius of gray region as $L = (\frac{P_0}{r\sigma_w^2})^{1/\alpha} \triangleq R_2^p$ by solving the equation $\frac{P_0/L^\alpha}{\sigma_w^2} = r$. Since the solution of r in (4.19) does not have a closed form, we do not have a closed form of R_2^p using the precise form of the probabilities of false alarm and missed detection. Therefore, we derive the closed form of the candidate radius of gray region using CLT, which is denoted as R_2^c.

---

[4]The relation between the *candidate* radius of gray region R_2^c and the radius of gray region R_2 is $R_2 = \min\{R_2^p, R_3\}$, because SUs prefer spatial spectrum holes in white region to temporal spectrum holes in gray region.

4.3.2 Result with CLT

With the assumption that both $x[n]$ and $w[n]$ are real-valued Gaussian random variables, according to CLT, the test statistic $Y = \frac{1}{N} \sum_{i=1}^{N} (y[i])^2$ can be approximated by a Gaussian random variable with the following distributions [16]:

$$Y \sim \begin{cases} \mathcal{N}\left(\sigma_w^2, \frac{2}{N}\sigma_w^4\right) & : \mathcal{H}_0 \\ \mathcal{N}\left((r+1)\sigma_w^2, \frac{2}{N}(r+1)^2\sigma_w^4\right) & : \mathcal{H}_1. \end{cases} \tag{4.20}$$

The probabilities of missed detection $P_m = \Pr\{\mathcal{H}_0|\mathcal{H}_1\}$ and false alarm $P_f = \Pr\{\mathcal{H}_1|\mathcal{H}_0\}$ are as follows:

$$P_m = \Pr\{Y < \varepsilon|H_1\} = Q\left(\left(1 - \frac{\varepsilon}{\sigma_w^2(r+1)}\right)\sqrt{\frac{N}{2}}\right)$$

$$P_f = \Pr\{Y > \varepsilon|H_0\} = Q\left(\left(\frac{\varepsilon}{\sigma_w^2} - 1\right)\sqrt{\frac{N}{2}}\right), \tag{4.21}$$

where $Q(x) = \frac{1}{\sqrt{2\pi}} \int_x^\infty \exp\left(-\frac{t^2}{2}\right) dt$ is the tail probability of standard normal distribution. Similarly, we bound P_m and P_f as follows:

$$P_m(\varepsilon, N) \leq \xi_m$$
$$P_f(\varepsilon, N) \leq \xi_f, \tag{4.22}$$

where ξ_m and ξ_f are the constraints of P_m and P_f. Substituting (4.21) into (4.22), we rewrite the constraints as follows:

$$Q\left(\left(1 - \frac{\varepsilon}{\sigma_w^2(r+1)}\right)\sqrt{\frac{N}{2}}\right) \leq \xi_m$$

$$Q\left(\left(\frac{\varepsilon}{\sigma_w^2} - 1\right)\sqrt{\frac{N}{2}}\right) \leq \xi_f \tag{4.23}$$

Solving the inequalities above, we have two constraints of the detection threshold ε as follows:

$$\varepsilon \leq \sigma_w^2(r+1)\left(1 - Q^{-1}(\xi_m)\sqrt{\frac{2}{N}}\right) \triangleq \varepsilon_m \tag{4.24}$$

$$\varepsilon \geq \sigma_w^2\left(Q^{-1}(\xi_f)\sqrt{\frac{2}{N}} + 1\right) \triangleq \varepsilon_f, \tag{4.25}$$

where $Q^{-1}(*)$ is the inverse Q function. Two key parameters ε_m and ε_f determine the range of ε. The detection threshold ε must be lower than ε_m to satisfy the constraint of P_m, and ε must be higher than ε_f to satisfy the constraint of P_f. Therefore, the detection threshold $\varepsilon \in [\varepsilon_f, \varepsilon_m]$. Notice that ε_f is a function of N, while ε_m is a function of N and r. If we fix N as a large number, for example, $N = 100$, then ε_f is a constant and ε_m is a function of r. As to ε_m, we have the following theorem.

Theorem 4.4
ε_m is a decreasing function of the distance between primary transmitter and secondary detector for sufficiently large N.

Proof 4.4   The derivative of ε_m is

$$\frac{d\varepsilon_m}{dr} = \sigma_w^2 \left(1 - Q^{-1}(\xi_m)\sqrt{\frac{2}{N}}\right). \qquad (4.26)$$

For sufficiently large N, namely, when $N > 2\left(Q^{-1}(\xi_m)\right)^2$, we have $Q^{-1}(\xi_m)\sqrt{\frac{2}{N}} < 1$ and $\frac{d\varepsilon_m}{dr} > 0$. For example, let $\xi_m = 10^{-3}$, then when $N \geq 20$, we have $\frac{d\varepsilon_m}{dr} > 0$. Actually, (4.21) is derived by the CLT [16], and to satisfy CLT, N must be sufficiently large.

Since SNR $r = \frac{P_0/L^\alpha}{\sigma_w^2}$ is a decreasing function of L, we have $\frac{d\varepsilon_m}{dL} = \frac{d\varepsilon_m}{dr}\frac{dr}{dL} < 0$. And this theorem is proved.   ■

The calculation of gray region is similar to the previous section. Since ε_m is a decreasing function of distance L and ε_f is a constant as we fix the value of N, when L increases, there must be an $L = R_2^c$, where $\varepsilon_m = \varepsilon_f$. Further, if $L > R_2^c$, we have $\varepsilon_m < \varepsilon_f$, and there is no detection threshold that can satisfy the constraints of false alarm and missed detection in (4.22) simultaneously. Then R_2^c is the candidate radius of gray region. To get R_2^c, we solve the equation as follows:

$$\varepsilon_m = \varepsilon_f$$

$$\Rightarrow r = \frac{Q^{-1}(\xi_f)\sqrt{\frac{2}{N}} + 1}{1 - Q^{-1}(\xi_m)\sqrt{\frac{2}{N}}} - 1. \qquad (4.27)$$

Substituting $r = \frac{P_0/L^\alpha}{\sigma_w^2}$ into (4.27), we have the candidate radius of gray region as follows:

$$L = \left(\frac{P_0}{\sigma_w^2}\frac{\sqrt{\frac{N}{2}} - Q^{-1}(\xi_m)}{Q^{-1}(\xi_f) + Q^{-1}(\xi_m)}\right)^{1/\alpha} \triangleq R_2^c. \qquad (4.28)$$

Note that R_2^c is an increasing function of P_0, because when the transmit power of PU increases, the SNR at SU's detector increases and the performance of spectrum sensing is improved, hence, gray region expands. R_2^c is also an increasing function of the number of samples N. As N increases, the performance of spectrum sensing is improved and gray region correspondingly expands. The R_2^p in previous section is satisfied for any value of N. However, it does not have a closed form. Although R_2^c in (4.28) is only satisfied for sufficiently large N^5, we still use R_2^c as the candidate radius of gray region in the following sections because it has a closed form. In Figure 4.2, we compare R_2^p and R_2^c with different number of samples N, which has verified that when N is big enough, the result with CLT is very close to the precise result.

4.3.3   Impact of Noise Uncertainty on R_2^c

In practice, there is uncertainty in the estimation of noise variance. Thus, we investigate the impact of noise uncertainty on the radius of gray region. The distributional uncertainty of noise can be summarized in a single interval $\sigma_w^2 \in \left[\frac{\sigma_n^2}{\rho}, \rho\sigma_n^2\right]$, where σ_n^2 is the nominal noise power and $\rho > 1$ is a parameter that quantifies the size of uncertainty [22]. To achieve the target P_f and P_m, the P_f and P_m are modified as follows:

---

[5] In practice, this condition can be easily satisfied.

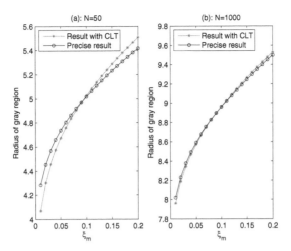

Figure 4.2: The comparison between the precise result (R_2^p) and the result with CLT (R_2^c). For $\xi_f = 0.05$, $\alpha = 3$, and $\sigma_w^2 = 1$, $P_0 = 100$.

$$P_f = \max_{\sigma_w^2 \in \left[\frac{\sigma_n^2}{\rho}, \rho\sigma_n^2\right]} Q\left(\left(\frac{\varepsilon}{\sigma_w^2} - 1\right)\sqrt{\frac{N}{2}}\right)$$

$$= Q\left(\left(\frac{\varepsilon}{\rho\sigma_n^2} - 1\right)\sqrt{\frac{N}{2}}\right) \tag{4.29}$$

$$P_m = \max_{\sigma_w^2 \in \left[\frac{\sigma_n^2}{\rho}, \rho\sigma_n^2\right]} Q\left(\left(1 - \frac{\varepsilon}{\sigma_w^2(r+1)}\right)\sqrt{\frac{N}{2}}\right)$$

$$= Q\left(\left(1 - \frac{\varepsilon\rho}{\sigma_n^2(r+1)}\right)\sqrt{\frac{N}{2}}\right). \tag{4.30}$$

Considering the constraints for P_f and P_m, we have the following relations:

$$P_f = Q\left(\left(\frac{\varepsilon}{\rho\sigma_n^2} - 1\right)\sqrt{\frac{N}{2}}\right) \leq \xi_f \tag{4.31}$$

$$P_m = Q\left(\left(1 - \frac{\varepsilon\rho}{\sigma_n^2(r+1)}\right)\sqrt{\frac{N}{2}}\right) \leq \xi_m. \tag{4.32}$$

With some manipulations, the constraints for the detection threshold are achieved as follows:

$$\varepsilon \geq \rho\sigma_n^2\left(Q^{-1}(\xi_f)\sqrt{\frac{2}{N}} + 1\right) \triangleq \varepsilon_f^n \tag{4.33}$$

$$\varepsilon \leq \frac{\sigma_n^2(r+1)}{\rho}\left(1 - Q^{-1}(\xi_m)\sqrt{\frac{2}{N}}\right) \triangleq \varepsilon_m^n. \tag{4.34}$$

Following the same rule as in Section III-B, we let $\varepsilon_f^n = \varepsilon_m^n$ and have the following value of r:

$$r = \frac{(\rho^2 - 1)\sqrt{\frac{N}{2}} + \rho^2 Q^{-1}(\xi_f) + Q^{-1}(\xi_m)}{\sqrt{\frac{N}{2}} - Q^{-1}(\xi_m)}. \tag{4.35}$$

To achieve the radius of gray region robustly, we substitute $r = \frac{P_0/L^\alpha}{\rho \sigma_n^2}$ into (4.35), and have the candidate radius of gray region as follows:

$$R_2^c = \left(\frac{P_0}{\rho \sigma_n^2} \frac{\left(\sqrt{\frac{N}{2}} - Q^{-1}(\xi_m)\right)}{(\rho^2 - 1)\sqrt{\frac{N}{2}} + \rho^2 Q^{-1}(\xi_f) + Q^{-1}(\xi_m)} \right)^{1/\alpha}. \tag{4.36}$$

When $\rho = 1$ and $\sigma_w^2 = \sigma_n^2$, uncertainty of σ_w^2 does not exist. Comparing (4.36) with (4.28), we notice that gray region shrinks when noise uncertainty exists. Besides, R_2^c is a decreasing function of ρ.

4.4 Spatial Spectrum Opportunity In White Region

In this section, the bound of white region is analyzed by considering the interference from SUs to PUs. To estimate the interference suffered by PUs, both accurate and supplement methods are designed to calculate the aggregate interference.

4.4.1 Accurate method

As illustrated in Figure 4.3, PUs on the edge of black region suffer from the worst interference generated by SUs, which is proved in the following Theorem 4.5. The interference experienced by the PU on the edge of black region from a SU in white region is

$$I(l, \theta) = \frac{P_s}{[d(l, \theta)]^\alpha} = \frac{P_s}{\left(l^2 + R_1^2 - 2R_1 l \cos \theta\right)^{\alpha/2}}, \tag{4.37}$$

where $d(l, \theta)$ is the distance from this SU to the PU on the edge of black region. Since SUs are distributed uniformly, both l and θ are random variables with probability density functions as follows [9]:

$$f_l(l) = \frac{2l}{R^2 - R_x^2}, R_x < l \leq R \tag{4.38}$$

$$f_\theta(\theta) = \frac{1}{2\pi}, 0 < \theta \leq 2\pi, \tag{4.39}$$

where R_x is the radius of white region[6]. As the density of SUs is λ users per unit area, the average number of SUs in white region is $n = \lambda \pi (R^2 - R_x^2)$. Assuming that $R_x > 1$, then the singular point is removed from the integral. Denote I_0 as the aggregate interference inflicted on PU, whose

---

[6]In this chapter, R_3 is the lower bound of R_x. Technically, any radius of $R_x \geq R_3$ can be the radius of white region.

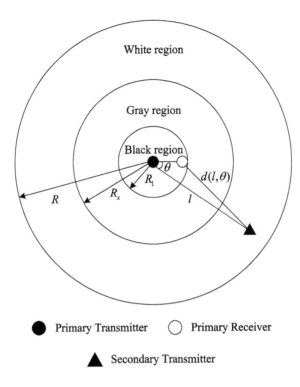

Figure 4.3: A scenario with the worst interference from SUs to PU.

expectation is given by the following equations:

$$E[I_0] = n \int_{R_x}^{R} \int_{0}^{2\pi} I(l,\theta)\, f_l(l) f_\theta(\theta) d\theta dl$$

$$= \int_{R_x}^{R} \int_{0}^{2\pi} \frac{\lambda l P_s}{\left(l^2 + R_1^2 - 2R_1 l \cos\theta\right)^{\alpha/2}} d\theta dl. \tag{4.40}$$

When $\alpha = 2k$ with k as a positive integer, a closed form expression on the expectation of the aggregate interference can be derived as (4.41).

$$E[I_0] = \frac{\lambda \pi P_s}{R_1^{2k-2}} \sum_{j=0}^{k-1} \left(\frac{(k+j-1)!}{(j!)^2 (k-j-1)!} \frac{R_1^{2(k+j-1)}}{k+j-1} \right.$$

$$\left. \times \left(\frac{1}{\left(R_x^2 - R_1^2\right)^{k+j-1}} - \frac{1}{\left(R^2 - R_1^2\right)^{k+j-1}} \right) \right) \tag{4.41}$$

Specifically, when $\alpha = 2$, we have

$$E[I_0] = \pi \lambda P_s \left[\ln\left(R^2 - R_1^2\right) - \ln\left(R_x^2 - R_1^2\right) \right]. \tag{4.42}$$

Since $\lim_{R \to \infty} E[I_0] = \infty$, $\alpha = 2$ should not be considered for an infinite area. When $\alpha = 4$, we have

$$E[I_0] = \pi \lambda P_s \left[-\frac{R^2}{\left(R^2 - R_1^2\right)^2} + \frac{R_x^2}{\left(R_x^2 - R_1^2\right)^2} \right]. \tag{4.43}$$

And $\lim\limits_{R \to \infty} E[I_0] = \frac{\pi \lambda P_s R_x^2}{\left(R_x^2 - R_1^2\right)^2}$. The interference for general path loss factor is not available in the accurate method because (4.40) is nonintegrable for real number α. The unavailability of closed form expressions for $E[I_0]$ with general path loss factor motivates the search for its upper bounds, which we will discuss in the following section. Before proceeding further, we present Theorem 4.5 as follows.

Theorem 4.5
Primary user on the edge of black region receives the worst interference when $R \to \infty$.

Proof 4.5 In (4.41), we replace R_1 with x and fix the value of R_x, then we find the monotonicity of $E[I_0]$ as a function of $x \in [0, R_1]$. When $R \to \infty$, with some manipulations, the monotonicity of $E[I_0]$ is the same as the following function.

$$f(x) = \frac{x^{2j}}{\left(R_x^2 - x^2\right)^{k+j-1}}, \tag{4.44}$$

which is a monotonic increasing function of x. Thus, when $x = R_1$, the interference of PU is largest, namely, the PU on the edge of black region receives the worst interference. ■

Theorem 4.5 reveals that the PU on the edge of black region receives the most severe interference. Thus, in order to protect PUs within black region, we only need to protect PUs on the edge of black region.

4.4.2 Supplement Method

Similar to [9], we recenter the network at the PU on the edge of black region, and fill the annulus whose inner radius is $R_x - R_1$ and outer radius is $R_1 + R$ with SUs. The density of these SUs is still λ users per unit area. Assume that $R_x - R_1 > 1$ and the singular point can be removed from the integral. We get the upper bound of the interference suffered by PUs and its limit as follows:

$$E[I_0] \le 2\pi P_s \lambda \int_{R_x - R_1}^{R+R_1} l^{1-\alpha} dl$$

$$= \frac{2\pi P_s \lambda}{\alpha - 2} \left(\frac{1}{(R_x - R_1)^{\alpha-2}} - \frac{1}{(R+R_1)^{\alpha-2}} \right) \tag{4.45}$$

$$\lim\limits_{R \to \infty} E[I_0] \le \frac{2\pi P_s \lambda}{\alpha - 2} \frac{1}{(R_x - R_1)^{\alpha-2}}. \tag{4.46}$$

When R_1 is big, the upper bound of interference calculated by supplement method is loose. (4.46) can also be used in finding the (inner) radius of white region.

4.4.3 Radius of White Region

The outage constraints of PUs are adopted to bound the radius of white region R_3. Assuming the data rate of PU is T_0, the outage event occurs when T_0 falls below a threshold data rate C_0. To guarantee the QoS of PU, the outage probability cannot exceed β, which is formulated as follows:

$$p_{out} = \Pr[T_0 \le C_0] \le \beta. \tag{4.47}$$

We derive the outage constraints of PUs, which suffer from the worst interference, i.e., PUs on the edge of black region [9], as follows.

$$
\begin{aligned}
p_{out} &= \Pr\left[\log\left(1 + \frac{P_0/R_1^\alpha}{I_0 + \sigma_w^2}\right) \le C_0\right] \le \beta \\
&= \Pr[I_0 \ge \frac{P_0/R_1^\alpha}{2^{C_0} - 1} - \sigma_w^2] \le \beta,
\end{aligned}
\tag{4.48}
$$

where σ_w^2 is the noise power spectral density in primary receiver. Define $I_{th} = \frac{P_0/R_1^\alpha}{2^{C_0}-1} - \sigma_w^2$, which is an interference threshold. When $I_0 \ge I_{th}$, an outage of PU occurs. Note that $I_{th} \ge 0$ and the radius of black region (R_1) has to satisfy the following condition:

$$
R_1 \le \left(\frac{P_0}{\sigma_w^2(2^{C_0} - 1)}\right)^{1/\alpha} \triangleq R_c,
\tag{4.49}
$$

where R_c is the radius of coverage area of primary network. When $R_1 > R_c$, the primary receiver is in outage due to noise, because the primary receiver is too far away from primary transmitter and merely noise can result in outage. R_c is the upper bound of R_1, therefore the radius of black region is

$$
R_1 < R_c.
\tag{4.50}
$$

Generally, the detector of SU is more sensitive than the detector of PU [7] [14], thus, $R_2 > R_c$ and $R_2 > R_1$. The radius of white region (R_3) can be derived by the outage constraints of PUs. Due to Markov inequality $\Pr[x \ge \varepsilon] \le \frac{E[x]}{\varepsilon}$, we have

$$
p_{out} \le \frac{E[I_0]}{I_{th}} = \frac{E[I_0]}{\frac{P_0/R_1^\alpha}{2^{C_0}-1} - \sigma_w^2}.
\tag{4.51}
$$

When $\alpha = 4$ and $R \to \infty$, we have

$$
p_{out} \le \frac{\pi\lambda P_s R_x^2}{\left(R_x^2 - R_1^2\right)^2 \left(\frac{P_0/R_1^\alpha}{2^{C_0}-1} - \sigma_w^2\right)}.
\tag{4.52}
$$

Using β to bound the right-hand side of the inequality above, the radius of white region can be derived as follows:

$$
\frac{\pi\lambda P_s R_x^2}{\left(R_x^2 - R_1^2\right)^2 \left(\frac{P_0/R_1^\alpha}{2^{C_0}-1} - \sigma_w^2\right)} \le \beta
$$

$$
\Rightarrow R_x \ge \underbrace{\frac{\sqrt{\pi\lambda P_s} + \sqrt{\pi\lambda P_s + 4\beta R_1^2\left(\frac{P_0/R_1^\alpha}{2^{C_0}-1} - \sigma_w^2\right)}}{2\sqrt{\beta\left(\frac{P_0/R_1^\alpha}{2^{C_0}-1} - \sigma_w^2\right)}}}_{R_3},
\tag{4.53}
$$

where the right-hand side of the inequality above is defined as R_3. The R_3 in (4.53) is derived by using the $E[I_0]$ in the accurate method. However, it is only applicable for the case of $\alpha = 4$. For the case of general path loss factor, we use the aggregate interference result in (4.46) with supplement method to derive the radius of white region. Substituting the upper bound of $E[I_0]$ in (4.46) into (4.51), we get the value of R_3 with supplement method as follows:

$$R_3 < \left(\frac{2\pi P_s \lambda}{\beta(\alpha-2)\left(\frac{P_0/R_1^\alpha}{2^{C_0}-1} - \sigma_w^2\right)} \right)^{1/(\alpha-2)} + R_1, \tag{4.54}$$

which is the upper bound of R_3 for general path loss factor. Notice that the R_3's in (4.53) and (4.54) are increasing functions of P_s and R_1, because when P_s or R_1 increases, SUs should be farther away from PUs to reduce the interference to PUs and R_3 thus increases.

4.4.4 Impact of Noise Uncertainty on R_3

In practice, there is uncertainty in the estimation of noise variance. Thus we investigate the impact of noise uncertainty on the radius of white region. As in Section III-C, the variance of noise is $\sigma_w^2 \in \left[\frac{\sigma_n^2}{\rho}, \rho\sigma_n^2\right]$, where σ_n^2 is the nominal noise power and $\rho > 1$ is a parameter that quantifies the size of uncertainty [22]. To achieve the radius of white region robustly, we substitute $\sigma_w^2 = \rho\sigma_n^2$ into (4.53) and (4.54), and get R_3's as follows:

- R_3 with accurate method is

$$R_3 = \frac{\sqrt{\pi\lambda P_s} + \sqrt{\pi\lambda P_s + 4\beta R_1^2\left(\frac{P_0/R_1^\alpha}{2^{C_0}-1} - \rho\sigma_n^2\right)}}{2\sqrt{\beta\left(\frac{P_0/R_1^\alpha}{2^{C_0}-1} - \rho\sigma_n^2\right)}} \tag{4.55}$$

- R_3 with supplement method is

$$R_3 = \left(\frac{2\pi P_s \lambda}{\beta(\alpha-2)\left(\frac{P_0/R_1^\alpha}{2^{C_0}-1} - \rho\sigma_n^2\right)} \right)^{1/(\alpha-2)} + R_1 \tag{4.56}$$

Notice that R_3's increase when uncertainty of noise variance exists. Besides, R_3's are increasing functions of ρ.

4.5 Bounds of Three Regions

In this section, we first provide the radii of three regions for single primary network. Then we address the radii of three regions for multiple primary networks. Since the increase in R_1 will enlarge the protected area of PUs and reduce spectrum opportunities of SUs in white region, the R_1 in (4.50) can be derived by solving an optimization problem, which is addressed in Section V-C. Finally, we investigate into the existence condition of transition zone.

4.5.1 Radii of Three Regions for Single Primary Network

SUs prefer spectrum opportunities in white region to those in gray region, because SUs in white region can transmit continuously and would not be interrupted by PUs. Thus, the radius of gray region takes the minimum of R_2^c and R_3. The radii of three regions are summarized as follows:

- The radius of black region is $R_1 < R_c$.
- The radius of white region is R_3, given by (4.53) and (4.54).
- The radius of gray region is $R_2 = \min\{R_2^c, R_3\}$, where R_2^c is provided in (4.28).

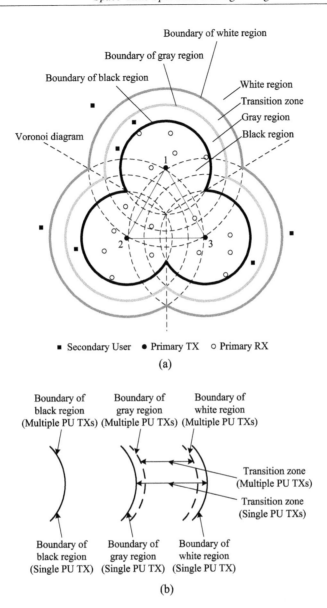

Figure 4.4: (a): The black, gray, and white regions for multiple primary networks. (b): Boundaries of three regions for multiple primary networks (multiple primary transmitters) and single primary network (single primary transmitter).

4.5.2 Radii of Three Regions for Multiple Primary Networks

As to the bounds of three regions of multiple primary networks, we consider the scenario that primary networks are operating on the same spectrum band[7]. In this situation, the black region of multiple primary networks is the union of each black region of primary networks, whose outer boundary is depicted in Figure 4.4(a).

---

[7]For the scenario that the primary networks are operating on different spectrum bands, the primary networks can be regarded as independent of each other, and three regions for multiple primary networks are the same as those of multiple isolated primary networks.

For the scenario of multiple primary networks, SU receives aggregate signal of multiple primary transmitters. Thus, the performance of spectrum sensing is better than that in single primary network and the gray region of multiple primary networks will expand, which is illustrated in Figure 4.4(b). The lower bound of the boundary of gray region for multiple primary networks is the envelope of each gray region of primary networks, which is depicted in Figure 4.4(a).

In the scenario of multiple primary networks, the number of SUs that generate interference to PUs is smaller than those in single primary network. Thus the white region of multiple primary networks will invade inward, which is illustrated in Figure 4.4(b). Therefore the upper bound of the boundary of white region is the envelope of each white region of primary networks, which is depicted in Figure 4.4(a).

Thus, we summarize the bounds of three regions for multiple primary networks as follows.

1) The black region of multiple primary networks is the union of each black region of primary networks.

2) The *lower bound* of the boundary of gray region for multiple primary networks is the envelope of each gray region of primary networks.

3) The *upper bound* of the boundary of white region for multiple primary networks is the envelope of each white region of primary networks.

4) The transition zone of multiple primary networks is between the envelopes of each gray region and white region of primary networks.

The envelopes of three regions of primary networks can be regarded as the boundaries of three regions of multiple primary networks. Thus, the results of three regions for single primary network can be directly applied to the scenario of multiple primary networks without violating the constraints of spectrum sensing and interference to PUs. But this will expand the transition zone. As illustrated in Figure 4.4(b), the transition zone of single primary network is wider than the real transition zone of multiple primary networks.

4.5.3   Optimal Radius of Black Region

First, we reveal the relation among R_3, R_1, and R_c by substituting (4.49) into (4.53), and we have

$$R_3 = \frac{\sqrt{\pi\lambda P_s} + \sqrt{\pi\lambda P_s + 4\beta\sigma_w^2 R_1^2\left(\frac{R_c^\alpha}{R_1^\alpha} - 1\right)}}{2\sqrt{\beta\sigma_w^2\left(\frac{R_c^\alpha}{R_1^\alpha} - 1\right)}}. \tag{4.57}$$

Notice that R_3 is an increasing function of R_1. Besides, when $R_1 \to R_c$, we have $R_3 \to \infty$, which means that there are no spatial spectrum opportunities for SUs in this situation. Note that the increase in R_1 will expand the protected area of PUs and protect more PUs. On the contrary, spectrum opportunities of SUs in white region are reduced because R_3 increases with R_1. Thus, we can optimize R_1 in the view of dynamic spectrum leasing, namely, PUs have an incentive (e.g., monetary rewards, such as leasing payments) to allow SUs to operate in their licensed spectrum band [23]. Hence, R_1 does not have to be R_c, because in this situation, PUs cannot get rewards from SUs in white region via dynamic spectrum leasing. We define the utility function of primary network in (4.58). For simplicity, we assume transition zone is eliminated. Assuming the density of PUs is λ_p users per unit area, the utility function is defined as the total revenue of primary network from PUs and SUs.

$$U_p(R_1) = \underbrace{c_p \lambda_p \pi R_1^2}_{D} + \underbrace{\int_{R_1}^{R_3} c_s w(l) p_0 \lambda 2\pi l \, dl}_{E}$$

$$+ \underbrace{\int_{R_3}^{\infty} c_s w(l) \lambda 2\pi l \, dl}_{F}, \tag{4.58}$$

where D is the revenue of primary network from its own users. Notice that we do not distinguish PUs by their locations. Each PU contributes revenue of c_p to the primary network. E and F are the revenue of primary network by leasing spectrum to SUs in gray region and white region, respectively. In gray region, the SU with distance l away from primary transmitter pays $c_s w(l) p_0$ to primary network, where $p_0 \in (0, 1]$ is the discount rate of spectrum price because the quality of spectrum in gray region is worse than that in white region. In this chapter, we assume p_0 is the probability that primary signal is absent. $w(l) = A \exp(-\kappa l)$ [3, eq. (5)] is the weighting function. In white region, the SU with distance l away from primary transmitter pays $c_s w(l)$ to primary network. Note that the payment of SUs to primary network is decreasing with the increase in l. After some manipulations, the utility of primary network is as follows:

$$U_p(R_1) = c_p \lambda_p \pi R_1^2 + 2A c_s \lambda \pi \frac{e^{-\kappa R_3}(1 + \kappa R_3)}{\kappa^2}$$

$$+ 2A c_s \lambda p_0 \pi \frac{e^{-\kappa R_1}(1 + \kappa R_1) - e^{-\kappa R_3}(1 + \kappa R_3)}{\kappa^2}. \tag{4.59}$$

Then R_1 can be determined by maximizing $U_p(R_1)$. Because of the complexity of $U_p(R_1)$, the closed form of the optimal R_1 is not available. In the numerical results, we will present the optimal R_1 and the impact of A, κ, and p_0 on it.

4.5.4 Existence Condition of Transition Zone

The transition zone can be eliminated by tuning system parameters. And we have a theorem as follows.

Theorem 4.6
The existence condition of transition zone is

$$f(\xi_m, \xi_f) = \left(R_3^\alpha + \frac{P_0}{\sigma_w^2} \right) Q^{-1}(\xi_m) + R_3^\alpha Q^{-1}(\xi_f) > C, \tag{4.60}$$

where $C = \frac{P_0}{\sigma_w^2} \sqrt{\frac{N}{2}}$.

Proof 4.6   When $R_2 < R_3$, i.e., $R_2^c < R_3$, a transition zone between gray region and white region exists. Otherwise, the outer radius of gray region and the inner radius of white region are the same and transition zone is eliminated. Substituting the expressions of R_2^c in (4.28) and R_3 in (4.53) into the relation $R_2^c < R_3$, we can get this theorem. ■

Therefore, transition zone disappears when

$$f(\xi_f, \xi_m) \leq C. \tag{4.61}$$

Notice that $f(\xi_f, \xi_m)$ is a decreasing function of ξ_f and ξ_m, which means that when ξ_f or ξ_m is

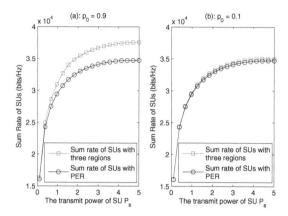

Figure 4.5: Sum rate improvement of the architecture of three regions. For $R = 400$m, $R_1 = 100$m, $C_0 = 0.1$bps/Hz, $\sigma_w^2 = 10^{-6}$W/Hz, $\beta = 0.3$, $\alpha = 4$, $N = 100$, $\xi_m = \xi_f = 0.1$. **Figure 4.5(a) is for the case of $p_0 = 0.9$ and Figure 4.5(b) is for the case of $p_0 = 0.1$.**

big enough, (4.61) will be satisfied and transition zone will be eliminated, namely, the relaxation of the constraints of missed detection or false alarm can help to eliminate transition zone. Besides, C is an increasing function of P_0 and N, therefore, when the value of P_0 or N is increased, (4.61) can be satisfied and transition zone will disappear.

In transition zone, temporal spectrum sensing fails because the probability of missed detection is so large that the performance of spectrum sensing violates the constraints in some protocols, such as IEEE 802.22. Note that we derive R_3 using Markov inequality, therefore, R_3 is an upper bound and there is still margin for SUs in transition zone to transmit with power control. Thus, SUs in transition zone need to control their power to mitigate the interference to PUs.

The transition zone can be eliminated by relaxing the constraints of missed detection and false alarm. However, this will bring interference to PUs and reduce spectrum utilization. Another way to eliminate transition zone is turning transition zone into gray region, namely, SUs in gray region can share spectrum sensing results with SUs in transition zone (e.g., via a public database), and SUs in transition zone can exploit temporal spectrum holes. In this way, gray region can be expanded to the inner boundary of white region and transition zone is eliminated. This scheme, however, requires the cooperation between SUs in gray region and transition zone, which may complicate the design of cognitive radio networks. Moreover, if the status of spectrum band is not updated in time, the transmission of SUs in transition zone may bring interference to PUs because they are not operating in white region.

4.6 Numerical Results

4.6.1 Performance of Three Regions

The performance improvement of the architecture of three regions is provided in Figure 4.5, where the relation between the sum rate and transmit power of SUs is provided. The sum rate is calculated by aggregating data rate of SUs in the disc with radius $R = 400$m in Figure 4.1. As illustrated in Figure 4.5, the sum rate is increasing with P_s. However, when P_s is large enough, the improvement is not significant because R_3 correspondingly increases. As illustrated in Figure 4.5(a), when the probability that PUs are idle, namely, p_0 is large, the performance improvement of three regions is significant when compared with the PER scheme in [9]. However, when p_0 is small, the performance

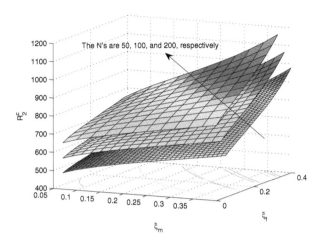

Figure 4.6: Relation between R_2^c and ξ_m, ξ_f, N. For $\alpha = 3$, $P_0 = 100$W and $\sigma_w^2 = 10^{-6}$W/Hz. The N's are 50, 100, and 200, respectively, for three surfaces.

improvement is negligible, because temporal spectrum opportunities are rare in this case, which is illustrated in Figure 4.5(b).

4.6.2 Radius of Gray Region

The relations between the (candidate) radius of gray region R_2^c and ξ_m, ξ_f, N are illustrated in Figure 4.6, where the upper, middle, and bottom surfaces are the results with $N = 200$, $N = 100$, and $N = 50$, respectively. Note that the radius of gray region increases with N, because the increase in N can improve the accuracy of spectrum sensing. Besides, R_2^c increases with ξ_m and ξ_f, which means that the relaxation of the constraints for the probabilities of missed detection and false alarm can expand gray region.

We investigate the impact of noise uncertainty on the radius of gray region in Figure 4.7, where the relation between R_2^c and N is illustrated. Note that R_2^c is decreasing with the increase in ρ, which means that when there is more uncertainty in noise variance, the gray region will shrink to detect PUs robustly.

4.6.3 The radius of white region

Figure 4.8 compares the radii of white regions in (4.53) and (4.54). The R_3 in (4.53) is derived by substituting the accurate value of the aggregate interference in (4.41) into (4.51). However, the R_3 in (4.53) is only applicable for even α. In contrast, the R_3 in (4.54) is applicable for general α. (4.54) is derived by substituting the upper bound of the aggregate interference of (4.46) into (4.51). Thus the R_3 in (4.54) is bigger than the R_3 in (4.53), which is verified by Figure 4.8.

Figure 4.9 shows the relation between R_3 and R_1 with P_s as the reference variable. R_3 is increasing with R_1, because SUs in white region need to make a concession to mitigate the interference to PUs when the black region expands. Note that the increase in R_1 will expand the protected area of PUs and reduce spectrum opportunities of SUs because R_3 correspondingly increases. Thus, the relation between R_3 and R_1 reveals the tradeoff between the benefits of PUs and SUs. Notice that although we assume R_1 can be as small as possible, there is always a minimum protected region for PUs that are related with public security [13]. Figure 4.10 shows the relation between R_3 and

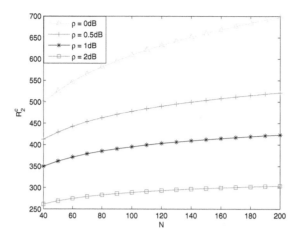

Figure 4.7: Relation between R_2^c and N with noise uncertainty. For $\alpha = 3$, $\xi_f = \xi_m = 0.1$, $P_0 = 100\text{W}$, and $\sigma_w^2 = 10^{-6}\text{W/Hz}$.

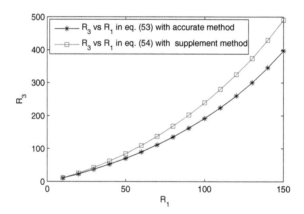

Figure 4.8: Comparison between the radii of white regions in (4.53) and (4.54). For $\lambda = 0.01\text{Users/m}^2$, $C_0 = 0.1\text{bps/Hz}$, $\sigma_w^2 = 10^{-6}\text{W/Hz}$, $\beta = 0.3$, $\alpha = 4$, $P_s = 5\text{W}$, and $P_0 = 200\text{W}$.

R_1, with P_0 as the reference variable. R_3 decreases with the increase in P_0, due to the fact that when the signal strength of PUs increases, the interference tolerance of PUs also improves and the white region can expand inward.

We investigate the impact of noise uncertainty on the radius of white region in Figure 4.11, where the relation between R_3 and R_1 is illustrated. Notice that R_3 increases with ρ in Figure 4.11, which means that when there is more uncertainty in noise variance, R_3 will be bigger to mitigate the interference from SUs to PUs robustly.

4.6.4 Transition Zone

The existence condition of transition zone is illustrated in Figure 4.12, where a family of curves with different values of C's are presented. When the pair (ξ_f, ξ_m) is below the curve, transition

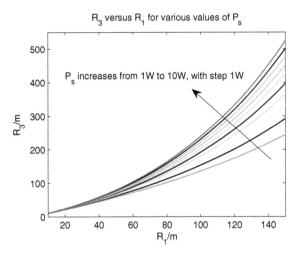

Figure 4.9: Relation between R_3 and R_1 for various values of P_s. For $\lambda = 0.01 \mathrm{Users/m^2}$, $P_0 = 200\mathrm{W}$, $C_0 = 0.1\mathrm{bps/Hz}$, $\sigma_w^2 = 10^{-6}\mathrm{W/Hz}$, **and** $\beta = 0.3$, $\alpha = 4$.

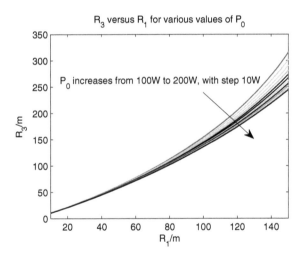

Figure 4.10: Relation between R_3 and R_1 for various values of P_0. For $\lambda = 0.01 \mathrm{Users/m^2}$, $P_s = 1\mathrm{W}$, $C_0 = 0.1\mathrm{bps/Hz}$, $\sigma_w^2 = 10^{-6}\mathrm{W/Hz}$, **and** $\beta = 0.3$, $\alpha = 4$.

zone exists. Take $(\xi_f, \xi_m) = (0.2, 0.1)$ as an example, when $C = 4.5$, transition zone is eliminated, however, when $C = 4$, transition zone exists. Therefore, transition zone is easier to be eliminated with a bigger C. Because $C = \frac{P_0}{\sigma_w^2} \sqrt{\frac{N}{2}}$, the increase in P_0 or N makes transition zone easier to be eliminated. Generally, the elimination of transition zone can simplify the system architecture, since SUs in transition zone need to control their power to mitigate the interference to PUs. Transition zone can be eliminated by tuning the parameters of spectrum sensing, such as N, ξ_f and ξ_m. Figure 4.12 shows that the increase in ξ_f or ξ_m can eliminate transition zone. However, this will bring more interference to PUs and reduce spectrum opportunities of SUs.

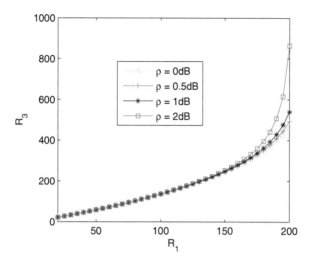

Figure 4.11: Relation between R_3 and R_1 with noise uncertainty. For $\lambda = 0.01 \text{Users}/\text{m}^2$, $P_s = 1\text{W}$, $P_0 = 200\text{W}$, $C_0 = 0.1 \text{bps}/\text{Hz}$, $\sigma_w^2 = 10^{-6}\text{W}/\text{Hz}$, **and** $\beta = 0.3$, $\alpha = 4$.

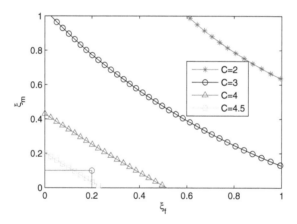

Figure 4.12: Existence condition of transition zone. The expression of C is given in (4.60).

The relation between the width of transition zone and P_s is illustrated in Figure 4.13. Note that the width of transition zone is increasing with P_s, because when P_s increases, the interference from SUs is worse than before, and transition zone needs to be expanded to protect PUs.

The relation between the width of transition zone and P_0 is illustrated in Figure 4.14. The width of transition zone is decreasing with the increasing in P_0, because the enhancement of P_0 will expand gray region outward and expand the white region inward simultaneously. Notice that transition zone can be eliminated by tuning parameter P_s and P_0, i.e., when P_s is sufficiently small or P_0 is big enough, transition zone will disappear, which is illustrated in Figure 4.13 and Figure 4.14.

We investigate the impact of noise uncertainty on transition zone. In Figure 4.13 and Figure 4.14, transition zone expands with ρ, because when ρ increases, R_2^c will decrease and R_3 will increase, which makes transition zone wider than before.

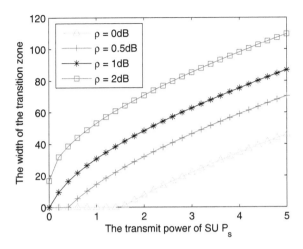

Figure 4.13: Relation between the width of transition zone and P_s with noise uncertainty. For $R_1 = 100$m, $C_0 = 0.1$bps/Hz, $\sigma_w^2 = 10^{-6}$W/Hz, $\beta = 0.3$, $\alpha = 4$, $N = 100$, and $\xi_m = \xi_f = 0.1$.

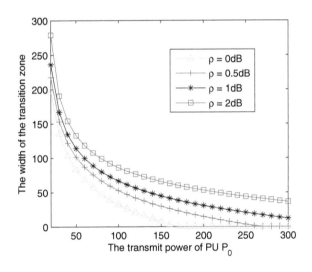

Figure 4.14: Relation between the width of transition zone and P_0 with noise uncertainty. For $R_1 = 100$m, $C_0 = 0.1$bps/Hz, $\sigma_w^2 = 10^{-6}$W/Hz, $\beta = 0.3$, $\alpha = 4$, $N = 100$, and $\xi_m = \xi_f = 0.1$.

4.6.5 Optimal Radius of Black Region

In Figure 4.15, the relations between U_p and R_1 are illustrated with various values of A. With the increase in A, U_p is improved because the price of spectrum rises and the revenue of primary network from dynamic spectrum leasing correspondingly increases. In Figure 4.15, U_p's have maximum values, and the filled dots denote the maximum values of U_p's and the corresponding optimal R_1's. Note that when A is bigger, namely, the price of spectrum is higher, the optimal R_1 will be smaller, which means that PUs are willing to lease more spectrum opportunities to SUs due to the higher revenue from dynamic spectrum leasing.

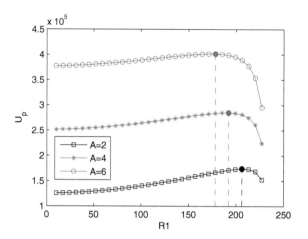

Figure 4.15: Utility of primary network versus R_1 for various values of A. For $\lambda_p = 0.05\text{Users}/\text{m}^2$, $\lambda = 0.01\text{Users}/\text{m}^2$, $c_p = 10$, $c_s = 1$, $\kappa = 0.001$, $p_0 = 0.5$, $C_0 = 0.1\text{bps}/\text{Hz}$, $\sigma_w^2 = 10^{-6}\text{W}/\text{Hz}$, $\beta = 0.3$, **and** $\alpha = 4$.

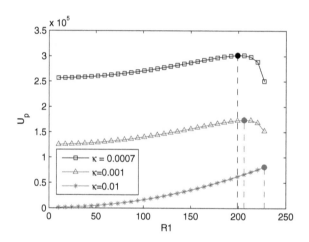

Figure 4.16: Utility of primary network versus R_1 for various values of κ. For $\lambda_p = 0.05\text{Users}/\text{m}^2$, $\lambda = 0.01\text{Users}/\text{m}^2$, $c_p = 10$, $c_s = 1$, $A = 4$, $p_0 = 0.5$, $C_0 = 0.1\text{bps}/\text{Hz}$, $\sigma_w^2 = 10^{-6}\text{W}/\text{Hz}$, $\beta = 0.3$, **and** $\alpha = 4$.

In Figure 4.16, the relations between U_p and R_1 are illustrated with various values of κ. Note that U_p is increasing with the decrease in κ, because smaller value of κ means higher price of spectrum, and primary network can get more revenue from SUs via dynamic spectrum leasing. When κ increases, the price of spectrum will decrease. Hence, when κ is sufficiently large, for example, $\kappa = 0.01$ in Figure 4.16, primary network is not willing to lease spectrum to SUs in white region[8], because the revenue from dynamic spectrum leasing cannot offset the loss of PUs outside of black region. Similar to Figure 4.15, when κ decreases, the price of spectrum rises and PUs are willing to lease more spectrum to SUs and the optimal R_1 will decrease.

---

[8]In this situation, primary network is still willing to lease spectrum to SUs in gray region.

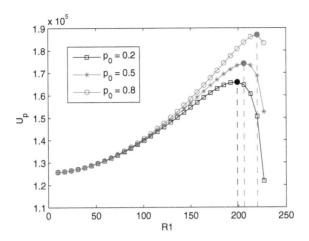

Figure 4.17: Utility of primary network versus R_1 for various values of p_0. For $\lambda_p = 0.05\text{Users}/\text{m}^2$, $\lambda = 0.01\text{Users}/\text{m}^2$, $c_p = 10$, $c_s = 1$, $\kappa = 0.001$, $A = 2$, $C_0 = 0.1\text{bps}/\text{Hz}$, $\sigma_w^2 = 10^{-6}\text{W}/\text{Hz}$, $\beta = 0.3$, and $\alpha = 4$.

In Figure 4.17, the relations between U_p and R_1 are illustrated with various values of p_0. Note that U_p increases with p_0 because the price of spectrum in gray region rises, and the revenue of primary network is improved. However, when R_1 is small, U_p's are nearly the same because the area of gray region is small in this situation. Notice that p_0 is the discount rate of spectrum price in gray region. Thus, when p_0 increases, the primary network can get considerable revenue from SUs in gray region. Besides, because $c_p > c_s$[9], the increase in R_1 will lead to more revenue from PUs and SUs in gray region, with small loss of the revenue from SUs in white region. Thus, we have obtained an interesting conclusion that if SUs in gray region pay less to PUs, namely, the discount rate p_0 is smaller, primary network will decrease R_1 to provide SUs with more spatial spectrum opportunities to gain more revenue from SUs in white region via dynamic spectrum leasing, which is illustrated in Figure 4.17, where the optimal R_1 decreases with decreasing p_0.

4.6.6   *Typical Application Scenario of IEEE 802.22*

We apply the architecture of three regions in an IEEE 802.22 scenario. The primary transmitter is a TV tower that operates on channels DS-33 and transmits with power of 200KW. SUs are cognitive femtocells, which are the cognitive radio wireless regional area networks that aim to exploit the temporal and spatial spectrum opportunities in TV networks. The transmit power of cognitive femtocell varies from 10mW to 100mW [28]. In this scenario, the relations between R_3 and R_1 are illustrated in Figure 4.18. Notice that R_3 increases with R_1, which is the same as previous simulations. Besides, white region is very close to black region because the transmit power of SUs is very small. In this situation, there are much more spatial spectrum opportunities than temporal spectrum opportunities.

---

[9]For primary network, we assume its own users (PUs) are the main sources of revenue. The revenue from SUs is only supplementary.

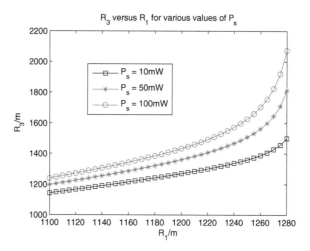

Figure 4.18: The relation between R_3 and R_1 for various values of P_s in an IEEE 802.22 scenario in Beijing. For $\lambda = 0.01\text{Users/m}^2$, $P_0 = 200\text{KW}$, $C_0 = 0.1\text{bps/Hz}$, $\sigma_w^2 = 10^{-6}\text{W/Hz}$, **and** $\beta = 0.05$ **to protect PUs,** $\alpha = 4$.

4.7 Conclusion

In this chapter, we have designed three regions for space-time spectrum sharing, including black region, gray region, and white region. The bound of gray region is determined by the constraints of missed detection and false alarm. The bound of white region is derived by analyzing the impact of aggregate interference from SUs to PUs. And the optimal bound of black region can be determined in the view of dynamic spectrum leasing. We have also investigated the bounds of three regions for multiple primary networks. Besides, a transition zone between gray region and white region may exist, and its existence condition is obtained. Therefore, three regions can guide space-time spectrum sharing in CRNs. Finally, numerical results are provided to show key relations in three regions.

References

[1] J. Mitola, "Cognitive radio: An integrated agent architecture for software defined radio," PhD dissertation, KTH Royal Inst. of Technol., Stockholm, Sweden, 2000.

[2] B. Wang and K. J. R. Liu, "Advances in cognitive radio networks: A survey," *IEEE Journal of Selected Topics in Signal Processing*, vol. 5, no. 1, pp. 5–23, Feb. 2011.

[3] R. Tandra, S. M. Mishra, and A. Sahai, "What is a spectrum hole and what does it take to recognize one?" *Proceedings of the IEEE*, vol. 97, no.5, pp. 824–848, May 2009.

[4] R. Tandra, A. Sahai and V. Veeravalli, "Unified space-time metrics to evaluate spectrum sensing," *IEEE Communications Magazine*, vol. 49, no. 3, pp. 54–61, Mar. 2011.

[5] T. Do and B. L. Mark, "Joint spatial-temporal spectrum sensing for cognitive radio networks," *IEEE Transactions on Vehicular Technology*, vol. 59, no. 7, pp. 3480–3490, Sep. 2010.

[6] G. Ding, Q. Wu, Y.-D. Yao, et al., "Kernel-based learning for statistical signal processing in cognitive radio networks: Theoretical foundations, example applications, and future directions," *IEEE Signal Processing Magazine*, vol. 30, no.4, pp. 126–136, July 2013.

[7] Q. Wu, G. Ding, J. Wang, et al., "Spatial-temporal opportunity detection for spectrum-heterogeneous cognitive radio networks: Two-dimensional sensing," *IEEE Transactions on Wireless Communications*, vol. 12, no. 2, pp. 516–526, Feb. 2013.

[8] G. Ding, J. Wang, Q. Wu, et al, "Spectrum sensing in opportunity-heterogeneous cognitive sensor networks: How to cooperate?" *IEEE Sensors Journal*, vol. 13, no. 11, pp. 4247–4255, Nov. 2013.

[9] M. Vu, N. Devroye and V. Tarokh, "On the primary exclusive region of cognitive networks," *IEEE Transactions on Wireless Communications*, vol. 8, no. 7, pp. 3380–3385, July 2009.

[10] A. Bagayoko, P. Tortelier, and I. Fijalkow, "Impact of shadowing on the primary exclusive region in cognitive networks," *European Wireless Conference*, pp. 105–110, Apr. 2010.

[11] J. Dricot, G. Ferrari, F. Horlin, et al., "Primary exclusive region and throughput of cognitive dual-polarized networks," *IEEE International Conference on Communications (ICC) Workshops*, pp. 1–5, May 2010.

[12] X. Hong, C., X. Wang, H., H. Chen, and Y. Zhang, "Secondary spectrum access networks," *IEEE Vehicular Technology Magazine*, vol.4, pp. 36–43, Jun. 2009.

[13] FCC, Enabling Innovative Small Cell Use In 3.5 GHZ Band NPRM & Order, Doc. 12-148, Dec. 12, 2012. Avaliable at: http://www.fcc.gov/document/enabling-innovative-small-cell-use-35-ghz-band-nprm-order.

[14] J. Ma, G. Li, and B. H. Juang, "Signal processing in cognitive radio," *Proceedings of the IEEE*, vol 97, no. 5, pp. 805–823, May 2009.

[15] G. Ganesan, Y. Li, B. Bing, and S. Li, "Spatiotemporal sensing in cognitive radio networks," *IEEE Journal on Selected Areas in Communications*, vol. 26, no. 1, pp. 5–12, Jan. 2008.

[16] Y. C. Liang, Y. Zeng, E. C. Y. Peh, et al., "Sensing-throughput tradeoff for cognitive radio networks," *IEEE Transactions on Wireless Communications*, vol. 7, no. 4, pp. 1326–1337, Apr. 2008.

[17] F. F. Digham, M. S. Alouini and M. K. Simon, "On the energy detection of unknown signals over fading channels," *IEEE Transactions on Communications*, vol.55, pp. 21–24, 2007.

[18] M. Vu and V. Tarokh, "Scaling laws of single-hop cognitive networks," *IEEE Transactions on Wireless Communications*, vol. 8, no. 8, pp. 4089–4097, Aug. 2009.

[19] J. Lee, H. Wang, J. G. Andrews, et al., "Outage probability of cognitive relay networks with interference constraints," *IEEE Transactions on Wireless Communications*, vol. 10, no. 2, pp. 1326–1337, Feb. 2011.

[20] M. Song, C. Xin, Y. Zhao, and X. Cheng, "Dynamic spectrum access: From cognitive radio to network radio," *IEEE Wireless Communications Magazine*, vol. 19, no. 1, pp. 23–29, Feb. 2012.

[21] E. Axell, G. Leus, E. G. Larsson, and H. V. Poor, "Spectrum sensing for cognitive radio: State-of-the-art and recent advances," *IEEE Signal Processing Magazine*, vol. 29, no. 3, pp. 101–116 , May 2012.

[22] R. Rahul Tandra and A. Anant Sahai, "SNR walls for signal detection," *IEEE Journal of Selected Topics in Signal Processing*, vol. 2, no. 1, pp. 4–17, Feb. 2008.

[23] S. K. Jayaweera and T. Li, "Dynamic spectrum leasing in cognitive radio networks via primary-secondary user power control games," *IEEE Transactions on Wireless Communications*, vol. 8, no. 6, pp. 3300–3310, Jun. 2009.

[24] S. K. Jayaweera, G. Vazquez-Vilar, C. Mosquera, "Dynamic spectrum leasing: A new paradigm for spectrum sharing in cognitive radio networks," *IEEE Transactions on Vehicular Technology*, vol. 59, no. 5, pp. 2328–2339, Jun. 2010.

[25] M. Barrie, S. Delaere, G. Sukareviciene, et al., "Geolocation database beyond TV white space? Matching applications with database requirements," *IEEE International Symposium on Dynamic Spectrum Access Networks (DySPAN)*, pp. 467–478, 2012.

[26] V. Petrini and H.R. Karimi, "TV white space databases: Algorithms for the calculation of maximum permitted radiated power levels," *IEEE International Symposium on Dynamic Spectrum Access Networks (DySPAN)*, pp. 552–560, 2012.

[27] J. Heo, G. Noh, S. Park, et al., "Mobile TV white space with multi-region based mobility procedure," *IEEE Wireless Communications Letters*, vol. 1, no. Dec. 6, 2012, pp. 569–572.

[28] 3GPP TR 36.814, "Further advancements for E-UTRA physical layer aspects (Release 9)," 2010.

[29] Z. Wei, Z. Feng, Q. Zhang, and W. Li, "Three regions for space-time spectrum sensing and access in cognitive radio networks," *IEEE Globecom 2012*, pp. 1283–1288, Dec. 2012

[30] I. S. Gradshteyn and I. M. Ryzhik, Table of Integrals, Series, and Products, 7th edition. Academic Press, 2007.

Chapter 5

LTE in the Unlicensed Band (LTE-U)

David Tung Chong Wong, Qian Chen, Francois Chin, and Xiaoming Peng

CONTENTS

5.1   Introduction

Cellular networks have evolved from the first generation (1G) systems to the current fourth generation (4G) systems with higher data rates in the downlink and the uplink from one generation to the next generation, leveraging on new technologies. Similarly, wireless local area networks or Wi-Fi have also evolved from one generation of standard to the next generation of standard. In terms of spectrum, cellular networks traditionally operate in the licensed band, while wireless local area networks traditionally operate in the unlicensed band. However, there is a trend to have (3.9)th generation (3.9G)/4G cellular networks operating in the unlicensed band to increase its throughput, while coexisting with wireless local area networks in recent years. Although Wi-Fi operates in the 2.4 GHz, 5 GHz and 60 GHz unlicensed band, the focus of having 3.9G/4G cellular networks in the unlicensed band is in the 5 GHz band initially. long-term evolution (LTE) is the 3.9 G cellular network, while LTE-Advanced is the 4G cellular network. The first name coined for such a scenario of cellular networks coexisting with Wi-Fi networks is LTE in the unlicensed band or LTE-U. Later, another name for this coexistence scenario in the unlicensed band is used—LTE licensed-assisted access (LAA).

This chapter has three main parts. The first part covers the traditional LTE/LTE-Advanced cellular networks in the licensed band, while the second part covers the traditional Wi-Fi wireless local area networks (WLANs) in the unlicensed band. The third part covers the LTE in the unlicensed band (LTE-U) coexisting with Wi-Fi WLANs.

Traditional LTE/LTE-A cellular networks operate in the licensed band, while traditional Wi-Fi WLANs operate in the unlicensed band. Since 2013, Qualcomm and Ericsson have proposed to use LTE in the unlicensed band called LTE-U. Since then, many companies have jumped into the bandwagon. Through Carrier Aggregation technique, the throughput of LTE can be increased via LTE in the licensed band and LTE-U in the unlicensed band. To allow for spectrum sharing in the unlicensed band for LTE-U with Wi-Fi, coexistence mechanisms are needed. These mechanisms can either use a listen-before-talk protocol or without it. The focus of this chapter is on the coexistence mechanisms for LTE-U with and without the listen-before-talk protocol. This spectrum sharing in the unlicensed band differs from the traditional spectrum sharing in the licensed band, where the cellular network is the incumbent or primary users and Wi-Fi is the secondary users. Priority for the two types of networks, LTE-U and Wi-Fi, in the unlicensed bands can be expressed in terms of the Mean Duty Cycles of the On and Off Periods of LTE-U. The traditional LTE networks include LTE and LTE-A, while the traditional Wi-Fi networks include IEEE 802.11a/b/g/n/ac. Scenarios, coexistence mechanisms, spectrum regulations, benefits, and some results of LTE-U are presented in details. Furthermore, LTE-U is planned to be released in 3GPP Release 13.

The objective of this chapter is to provide a good foundation of LTE-U with an industrial favor. The approach in this chapter is to explain the networks in the traditional licensed and unlicensed bands, leading to the coexistence of the two types of networks, both cellular networks and WLANs, in the unlicensed band. Our focus in this chapter is on the LTE-U in the unlicensed band. Although, the focus in this chapter is on LTE-U using the 5 GHz unlicensed band, it can be extended to the sub-GHz, the 2.4 GHz, and 60 GHz unlicensed bands, as well as any other unlicensed bands. IEEE 802.11ah, IEEE 802.11ad and IEEE 802.11aj WLANs can be included in the future too. Thus,

LTE-U stretches the imagination of spectrum usage of cellular networks that we know today. The techniques developed in LTE-U can also be used for WiMAX cellular systems and Home eNodeBs or femtocells in the unlicensed bands, as well as future cellular network technologies. Synergies of new network architectures for heterogeneous networks (HETNETs) and millimeter wave (mmWave) backhaul for 5G cellular networks with LTE-U are certainly new areas to explore.

5.2   Overview of Traditional LTE in the Licensed Band

A brief overview of traditional LTE in the licensed band is presented in this section. LTE is a 3.9G all-Internet Protocol (all-IP) cellular system, while LTE-A is a 4G all-IP cellular system. This section briefly describes the network architectures, physical layer air interfaces and medium access control scheduling and traditional spectrum for LTE and LTE-A cellular networks. References [1–16] are referred and consulted for the descriptions of LTE and LTE-A cellular systems in this section. For more details on LTE and LTE-A, the reader is referred to [1–16].

5.2.1   *Network Architectures*

This subsection describes the network architectures for LTE and LTE-A. The evolved packet system, evolved UTRAN, IP-multimedia subsystem and external networks are briefly described.

5.2.1.1   *LTE*

Figure 5.1 shows the LTE network architecture. In the evolved universal mobile telecommunications service (UMTS) evolution, the blocks are the Evolved UTRAN (E-UTRAN) and the evolved packet core (EPC) [2]. Evolved UMTS evolution is also known as evolved packet system (EPS), while E-UTRAN is also known as evolved access network. EPC is also known as evolved packet core network. From references [1–16], E-UTRAN consists of a network of evolved nodeBs (eNodeBs) or base stations (BSs) with no centralized controller. Thus, the E-UTRAN architecture is known to be flat, unlike previous generations of cellular networks. The eNodeBs or BSs are connected via an X2 interface. The eNodeBs or BSs are connected to the mobility management entity (MME) via an S1-MME interface and to the serving gateway (GW) via a S1-U interface. The protocols that are executed between the eNodeBs or BSs and the mobile user (MT) (or user equipment (UE)) are known as the access stratum (AS) protocols.

The E-UTRAN is responsible for all radio-related functions [2]. These functions include radio resource management (RRM), header compression (HC), security and connectivity to the EPS. The RMM is responsible for all functions concerning the radio bearers. These functions include radio bearer control, radio admission control, scheduling, and dynamic allocation to MTs or UEs in both directions—the downlink and uplink. The function of the HC is to compress IP packet headers. This is done so that significant overhead for small packets, such as voiceover IP (VoIP), is avoided.

For security, all data are encrypted for transmissions over the radio interface [2]. The connectivity to the EPC consists of the signaling in the direction of the MME and the bearer path in the direction of the serving GW.

The components of the EPC are the MME, serving gateway GW, packet data network (PDN) GW, and the policy and charging rules function (PCRF) [2]. All functions in the control plane are the responsibility of the MME. These functions concern the subscriber and session management, security procedures, terminal-to-network session handling, and idle terminal location management. The MME is connected to the home subscriber server (HSS) via a S6 interface. The HSS is made up of the home location register (HLR) and the authentication center (AuC) in third generation (3G) cellular networks. A database with all subscription information is supported by the HSS. On the other hand, the serving GW is the end point of the packet data interface in the direction of the

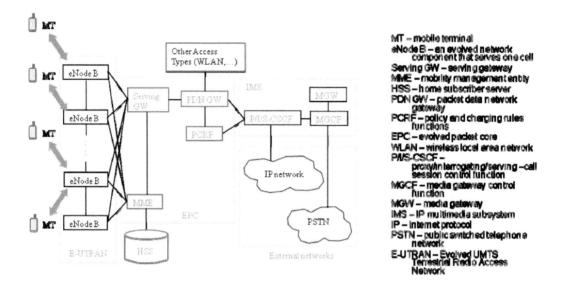

Figure 5.1: Long term evolution (LTE) network architecture.

E-UTRAN. It also functions as a local mobility anchor when the MTs (UEs) move between the eNodeBs or BSs. Packets are exchanged via this point for intra-E-UTRAN mobility and mobility with other 3GPP technologies such as second generation (2G) global system for mobile communications (GSM) and 3G UMTS cellular networks. In addition, the PDN GW is the end point of the packet data interface in the direction of the PDN.

It is also the anchor point for sessions in the direction of the PDN. On top of it, it also supports policy enforcement features. These features include applying operator-defined rules for resource allocation and usage, packet filtering, like deep packet inspection for virus signature detection, and evolved charging like URL charging. A URL is an address of a Web page on the World Wide Web (WWW). Furthermore, policy control decision-making and controlling the flow-based charging functionalities in the PDN GW are the responsibility of the PCRF. It also has provision for quality-of-service (QoS) authorization of data flow through the PDN GW and ensures the user's subscription profile is valid.

The IP multimedia subsystem (IMS) makes provision for IP-based multimedia services [2]. Within the IMS, the call session control function (CSCF) is very crucial and important in the IMS architecture. The CSCF consists of three types. They are the proxy, interrogating, and serving types. The CSCF can start, end and change an IMS session. Another crucial function within the IMS is the multimedia gateway control function (MGCF). Call control protocol conversion, media gateway (MGW) and Interrogation CSCF are supported by MGCF. On the other hand, media conversion, bearer control, and payload processing like codec, echo canceller, etc., are the responsibility of the MGW.

The external networks are connected to the IMS via IP networks and via the public switched telephone network (PSTN).

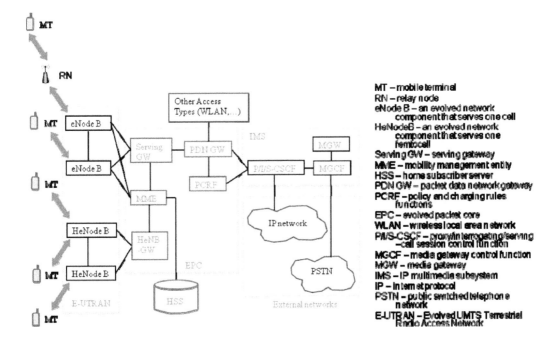

Figure 5.2: Long term evolution (LTE) advanced network architecture.

5.2.1.2   IEEE LTE-Advanced

The E-UTRAN for LTE-A can support home eNodeB (HeNodeB). HeNodeB is also known as a femtocell. A HeNodeB is basically an eNodeB or BS of lower cost for indoor coverage improvement. The connectivity of a HeNodeB to the ECP can be directly or via a HeNodeB GW. This HeNodeB GW allows support for a large number of HeNodeBs or indoor base stations. The E-UTRAN for LTE-A also supports relay nodes and enhanced relaying strategies for increased coverage. Higher data rates, better QoS performance, and fairness for different users are also its advantages.

5.2.2   Overview of Physical Layer (PHY) Air Interfaces

This subsection briefly describes the physical layers for LTE and LTE-A. The basic data rates and the main factors in the uplink and downlink access techniques are briefly mentioned. The LTE system is not backward-compatible with the 3G systems, while the LTE-A system is backward-compatible with the LTE system.

5.2.2.1   LTE

The QoS in LTE is greatly improved by having a high peak data rate and a low latency. The peak data rate in the downlink is 300 Mbps, while the peak data rate in the uplink is 75 Mbps. Furthermore, LTE uses multiple-input multiple-output (MIMO) technique (i.e., it uses multiple antennas). The downlink is based on orthogonal frequency division multiple access (OFDMA) at the physical layer, while the uplink is based on single carrier frequency division multiple access (SC-FDMA). SC-FDMA allocates carriers across a continuous block of spectrum and therefore limits the flexibility in scheduling. In terms of radio frame structure of LTE, each generic radio frame is 10 ms long and can be divided into subframes of 1 ms and slots of 0.5 ms duration. From this frame structure, a physical resource block (PRB) is a time-frequency block with 0.5 ms duration and 180 kHz in width. In terms of modulation for LTE, the modulation schemes supported are quadrature phase shift keying (QPSK), 16-quadrature amplitude modulation (QAM) and 64-QAM. LTE supports scalable bandwidths of 1.4MHz, 3MHz, 5MHz, 10MHz, 15MHz and 20 MHz. However, Carrier aggregation is not supported. Nevertheless, LTE supports both time-division duplex (TDD) and frequency-division duplex (FDD).

5.2.2.2 IEEE LTE-A

The quality of service in LTE-A is even more greatly improved by having a higher peak data rate and a lower latency. The peak data rate in the downlink is 1 Gbps, while the peak data rate in the uplink is 500 Mbps. Furthermore, LTE-A uses enhanced MIMO techniques. The downlink is also based on OFDMA at the physical layer, while the uplink is based on clustered SC-FDMA. Clustered SC-FDMA supports frequency-selective scheduling of component carriers and, therefore, gives better link performance. In terms of radio frame structure for LTE-A, each generic radio frame is also 10 ms long and can be divided into subframes of 1 ms and slots of 0.5 ms duration. In terms of modulation for LTE-A, the modulation schemes supported are also QPSK, 16-QAM, and 64-QAM. One of the differences between LTE-A and LTE is that LTE-A allows carrier aggregation across multiple component carriers in the licensed bands up to 100 MHz; thus, carrier aggregation increases bandwidth and directly increases the bit rate. LTE-A also support both TDD and FDD. In addition, the LTE-A network is backward-compatible with the LTE network.

5.2.3 Medium Access Control Scheduling

This subsection briefly describes the medium access control (MAC) scheduling for LTE and LTE-A. A packet scheduler in the eNodeB or BS is responsible for allocation of radio resources to MTs or UEs. It decides the time/frequency resource blocks in which the MTs or UEs transmit on. Note that the scheduling algorithms (e.g., [8–10]) are *not standardized* in LTE and LTE-A systems. The scheduling algorithms are for the downlink from the eNodeB or BS to the MT or UE and for the uplink from the MT or UE to the eNodeB or BS.

5.2.3.1 LTE

The downlink packet scheduling in a LTE system tightly interacts with the link adaptation and hybrid automatic repeat request (ARQ) [2]. The decision to multiplex within a subframe for user transmissions may depend on the following parameters: (1) minimum and maximum data rate, (2) available power to share among mobiles, (3) basal electrical rhythm (BER) target requirements according to the service, (4) latency requirement, depending on the service, (5) QoS parameters and measurements, (6) payload buffered in the eNodeB ready for scheduling, (7) pending retransmissions, (8) channel quality indicator reports from the UEs, (9) UE capabilities, (10) UE sleep cycles and measurement gaps/periods, and (11) system parameter such as bandwidth and interference level patterns, etc. These parameters are important and critical for the decisions made in downlink packet scheduling.

On the other hand, uplink scheduling in an LTE system depends on the states of buffers inside the mobiles, which are unknown to the eNodeB. Scheduling cannot be based on the type of information as in the downlink. However, some time-frequency resources can be allocated for contention-based access. Within these time-frequency resources, MTs or UEs can transmit without first being scheduled. As a minimum, contention-based access should be used for random access and for request-to-be scheduled signaling. As mentioned before, carrier aggregation is not supported. More details on the downlink and uplink scheduling for LTE can be found in [8] and [9], respectively. Detailed description of different scheduling schemes for LTE is beyond the scope of this chapter and note that scheduling is *not standardized* in LTE.

5.2.3.2 LTE-A

From references [1–16], the components of the scheduler in the LTE-A system are the resource scheduling and the packet scheduling. Carrier aggregation across carrier components is supported up to 100 MHz. The carrier components can be contiguous or discontiguous for the carrier aggregation. The resource scheduling is based on the following parameters [11]: (1) channel quality indication (CQI), (2) dynamic subcarrier assignment (DSA), (3) adaptive modulation and coding (AMC), (4) adaptive power control (APC), and (5) multi-antenna: MIMO/beamforming. These parameters are important and critical for the decisions made in resource scheduling.

On the other hand, the packet scheduling is based on the following parameters [11]: (1) resource partitioning between BS and relay node (RN), (2) QoS priorities and substrategies, and (3) buffer/queue management. More details on the downlink and uplink scheduling for LTE-A can be found in [10] and [9], respectively. Detailed description of different scheduling schemes for LTE-A is beyond the scope of this chapter and note that scheduling is *not standardized* in LTE-A.

5.2.4 Traditional Spectrum

This subsection briefly describes the traditional spectrum allocated for LTE and LTE-A.

5.2.4.1 LTE

3GPP defined the operating bands for TD-LTE and FDD-LTE in Table 5.1. The LTE system supports six different kinds of bandwidth options: 1.4 MHz, 3 MHz, 5 MHz, 10 MHz, 15 MHz, and 20 MHz. However, different operating bands allow only a subset of the six possible bandwidth cases, as listed in Table 5.1. Typically, an LTE system operates in a certain band, and its channel bandwidth is allocated in terms of resource blocks (RBs), which is shown in Table 5.2.

5.2.4.2 IEEE LTE-A

LTE-A operates in the same frequency bands with LTE as given in Table 5.1. However, the LTE-A system has a capability of carrier aggregation (CA), which allows aggregating up to 5 component carriers in the same band or in the different band to achieve a maximum bandwidth of 100 MHz.

5.3 Overview of Traditional Wi-Fi in the Unlicensed Band

A brief overview of traditional Wi-Fi or WLAN in the unlicensed band is presented in this section. This section briefly describes the network architectures of IEEE 802.11, physical layers of IEEE 802.11a/b/g/n/ac, contention-based medium access control of IEEE 802.11, IEEE 802.11e, and IEEE 802.11ac, and traditional spectrum for IEEE 802.11 WLANs. References [17–34] are referred and consulted for the descriptions of Wi-Fi WLANs in this section. For more details of IEEE 802.11a/b/g/n/ac Wi-Fi WLAN, the reader is referred to [12–32], while for more details of IEEE 802.11ad WLAN in the 60 GHz band, the reader is referred to [33] and the references therein.

5.3.1 Network Architectures

IEEE 802.11 has basically three types of network architectures. They are infrastructure mode, ad hoc mode and wireless mesh mode network architectures. The smallest basic unit of an IEEE 802.11 WLAN is a basic service set (BSS) [17]. From reference [17], a BSS consists of a number of stations (STAs) using the same MAC protocol and competing for access to the same shared wireless medium. A BSS may be standalone or it may be linked to a backbone distribution system (DS) via an access

Table 5.1   LTE Operating Bands

LTE Operating Band	Uplink (UL), MHz F_{UL_low}—F_{UL_high}	Downlink (DL), MHz F_{DL_low}—F_{DL_high}	Duplex Spacing, MHz	Duplex Mode	Channel Bandwidth, MHz
1	1920 – 1980	2110 – 2170	190	FDD	5, 10, 15, 20
2	1850 – 1910	1930 – 1990	80	FDD	1.4, 3, 5, 10, 15, 20
3	1710 – 1785	1805 – 1880	95	FDD	1.4, 3, 5, 10, 15, 20
4	1710 – 1755	1805 – 1880	400	FDD	1.4, 3, 5, 10, 15, 20
5	824 – 849	869 – 894	45	FDD	1.4, 3, 5, 10
6	830 – 840	875 – 885	45	FDD	5, 10
7	2500 – 2570	2620 – 2690	120	FDD	5, 10, 15, 20
8	880 – 915	925 – 960	45	FDD	1.4, 3, 5, 10
9	1749.9 – 1784.9	1844.9 – 1879.9	95	FDD	5, 10, 15, 20
10	1710 – 1770	2110 – 2170	400	FDD	5, 10, 15, 20
11	1427.9 – 1447.9	1475.9 – 1495.9	48	FDD	5, 10
12	699 – 716	729 – 746	30	FDD	1.4, 3, 5, 10
13	777 – 787	746 – 756	31	FDD	5, 10
14	788 – 798	758 – 768	30	FDD	5, 10
15	Reserved	Reserved		FDD	
16	Reserved	Reserved		FDD	
17	704 – 716	734 – 746	30	FDD	5, 10
18	815 – 830	860 – 875	45	FDD	5, 10, 15
19	830 – 845	875 – 890	45	FDD	5, 10, 15
20	832 – 862	791 – 821	41	FDD	5, 10, 15, 20
21	1447.9 – 1462.9	1495.9 – 1510.9	48	FDD	5, 10, 15
22	3410 – 3490	3510 – 3590	100	FDD	5, 10, 15, 20
23	2000 – 2020	2180 – 2200	180	FDD	1.4, 3, 5, 10, 15, 20
24	1626.5	1660.5	101.5	FDD	5, 10
25	1850 – 1915	1930 – 1995	80	FDD	1.4, 3, 5, 10, 15, 20
26	814 – 849	859 – 894	45	FDD	1.4, 3, 5, 10, 15
27	807 – 824	852 – 869	45	FDD	1.4, 3, 5, 10
28	703 – 748	758 – 803	55	FDD	3, 5, 10, 15, 20
29	N/A	717 – 728	N/A	FDD	3, 5, 10
30	2305 – 2315	2350 – 2360	45	FDD	5, 10
31	452.5 – 457.5	462.5 – 467.5	10	FDD	1.4, 3, 5
32	N/A	1452 – 1496	N/A	FDD	5, 10, 15, 20
33	1900 – 1920		N/A	TDD	5, 10, 15, 20
34	2010 – 2025		N/A	TDD	5, 10, 15
35	1850 – 1910		N/A	TDD	1.4, 3, 5, 10, 15, 20
36	1930 – 1990		N/A	TDD	1.4, 3, 5, 10, 15, 20
37	1910 – 1930		N/A	TDD	5, 10, 15, 20
38	2570 – 2620		N/A	TDD	5, 10, 15, 20
39	1880 – 1920		N/A	TDD	5, 10, 15, 20
40	2300 – 2400		N/A	TDD	5, 10, 15, 20
41	2496 – 2690		N/A	TDD	5, 10, 15, 20
42	3400 – 3600		N/A	TDD	5, 10, 15, 20
43	3600 – 3800		N/A	TDD	5, 10, 15, 20
44	703 – 803		N/A	TDD	3, 5, 10, 15, 20

Table 5.2   The Number of RBs in Different Kinds of Channels

Channel Bandwidth, MHz	1.4	3	5	10	15	20
Number of Resource Blocks	6	15	25	50	75	100

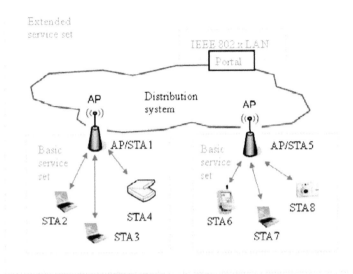

AP – access point
STA – station
LAN – local area network

Figure 5.3: IEEE 802.11 WLAN (infrastructure mode) network architecture.

point (AP). The AP functions as a bridge. A bridge is a network device that links multiple network segments and acts in the first two layers of the International Organization for Standardization/Open System Interconnection (ISO/OSI) model. The MAC protocol may be fully distributed or controlled by a central coordination function in the AP. More details on the MAC protocols are presented in Subsections III.C. The BSS is like a cell with a coverage area. The coverage area is roughly circular for omni-directional MAC protocols. IEEE 802.11ad MAC protocols are directional in nature. The DS can be a switch, a wired network, or a wireless network. Most of the time, the DS is either a switch or a wired network. Figure 5.7 shows the simplest configuration, where each station belongs to a single BSS. That is, each station is within the wireless range only of the other stations within the same BSS. Two BSSs can also overlap geographically, such that a single station can be in more than one BSS. Furthermore, the association between a station and a BSS is dynamic. Association can be done actively by sending out a probe frame, getting a response frame or beacon from the AP to obtain the AP's capabilities, and then sending an association request frame or listening to the beacon sent out by the AP to obtain the AP's capabilities before out an association request frame. An extended service set (ESS) consists of two or more BSSs interlinked by a DS. Typically, the DS is a wired backbone local area network (LAN) or any other wireline or wireless network. The ESS appears as a single logical LAN to the logical link control (LLC) level. Figure 5.3 shows the AP implemented as part of a station. The AP is the logic within a station that provides access to the DS by providing DS services, on top of being a station. A portal is used to link the IEEE 802.11 architecture with a traditional wired LAN (IEEE 802.x). The portal is implemented within a bridge or a router, that is, part of the wired LAN, and is also linked to the DS. A router is a network device that links multiple network segments and acts in the first three layers of the ISO/OSI model.

In the ad hoc network architecture, as shown in Figure 5.4, stations are connected directly to each other in an ad hoc manner without an AP. This is like a mesh network topology, or sometimes known as peer-to-peer network topology.

Furthermore, this mode of operation is also known as an independent BSS (IBSS).

In the wireless mesh network topology, as shown in Figure 5.5, the distribution system can be a wireless mesh network among the APs.

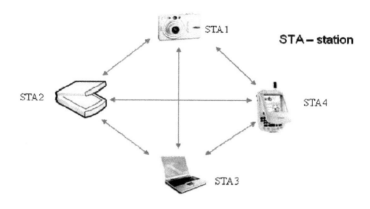

Figure 5.4: IEEE 802.11 WLAN (ad hoc mode) network architecture.

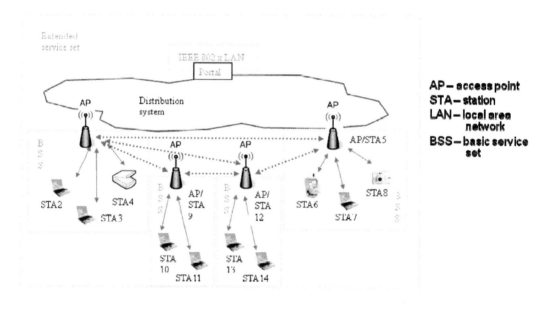

Figure 5.5: IEEE 802.11 WLAN (wireless mesh mode) network architecture.

5.3.2 *Overview of Physical Layers (PHYs) of IEEE 802.11a/b/g/n/ac*

IEEE 802.11b supports a data rate of up to 11 Mbps, while both IEEE 802.11a and IEEE 802.11g support a data rate of up to 54 Mbps. IEEE 802.11b and IEEE 802.11g operate in the 2.4 GHz band, while IEEE 802.11a operates in the 5 GHz band. IEEE 802.11b uses direct-sequence spread spectrum (DSSS), while IEEE 802.11a and IEEE 802.11g use orthogonal frequency division multiplexing (OFDM) for the PHY. IEEE 802.11n supports a data rate of up to 600 Mbps using OFDM PHY and MIMO technology to enhance diversity. IEEE 802.11n operates on both the 2.4 GHz and 5 GHz bands. IEEE 802.11ac is a very high throughput WLAN. IEEE 802.11ac WLAN supports a data rate of up to 6933.3 Mbps. IEEE 802.11ac operates below the 6 GHz band, except for the 2.4 GHz band. The very high throughput PHY of IEEE 802.11ac uses OFDM PHY. The IEEE 802.11ad WLAN operates in the 60 GHz band and has a maximum data rate of up to 6756.75 Gbps. It uses control PHY, single-carrier PHY, and OFDM PHY. Thus, only IEEE 802.11a, IEEE 802.11n, and IEEE 802.11ac can operate in the 5 GHz unlicensed band.

5.3.3   *Medium Access Control (MAC)*

5.3.3.1   *IEEE 802.11 Distributed Coordination Function (DCF)*

The 802.11 distributed coordination function (DCF) MAC uses the CSMA/CA MAC protocol [18]. There are two access methods in Carrier Sense Multiple Access/CA MAC. They are the basic access method and the request-to-send/clear-to-send (RTS/CTS) access method. The basic access method is a two-way handshaking as shown in Figure 5.6, while the RTS/CTS access method is a four-way handshaking as shown in Figure 5.7. In the former access method, the source station sends its frame to the destination station in the data transmission phase. After correctly receiving the frame, the destination station sends an acknowledgement to the source station in the acknowledgement phase. Thus, this process completes the two-way handshaking. In the latter access method, the source station sends an RTS frame to the destination station. If the destination station receives the RTS frame correctly and is available for reception, it replies with a CTS frame. Then the source station sends its data frame to the destination station.

Upon correctly receiving the data frame, the destination station acknowledges receipt of the data frame with an acknowledgement frame. This completes the four-way handshaking. If the payload is below a certain threshold, the basic access method is used; otherwise, the RTS/CTS access method is used. Figure 5.8 shows the channel access in IEEE 802.11 DCF MAC protocol.

The CSMA/CA MAC works as follows.

- If the channel is idle for more than a distributed coordination function inter-frame space time (DIFS) and the backoff counter is zero, a station can transmit immediately.

- If the channel is busy, the station will generate a random backoff period. This random backoff period is uniformly selected from zero to the current contention window size. The backoff counter decrements by one if the channel is idle for each time slot and freezes if the channel is sensed busy. The backoff counter is reactivated to count down when the channel is sensed idle for more than a DIFS time. At the initial backoff stage, the current contention window size is set at the minimum contention window size.

- If the backoff counter reaches zero, the station attempts to transmit its frame. If it is successful, the destination station sends an acknowledgement after a short interframe space (SIFS), and the current contention window size is reset to the minimum contention window size. If it is not successful, it increases the current contention window size by doubling it and adding one until a maximum contention window size is reached in the next backoff stage and a new random backoff period is selected as before.

- This process repeats itself until the frame is successfully transmitted or until the maximum retry limit is reached. If the frame is still not successfully transmitted, then it is dropped.

- If a station does not receive an acknowledgment within an acknowledgment timeout period

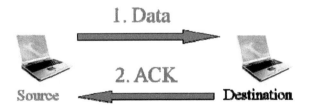

Figure 5.6: Basic access for IEEE 802.11 DCF MAC protocol.

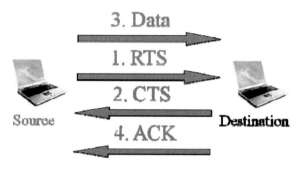

Figure 5.7: RTS/CTS access for IEEE 802.11 DCF MAC protocol.

Figure 5.8: Channel access in IEEE 802.11 DCF MAC protocol.

after a frame is transmitted, it continues to attempt to retransmit the frame according to the backoff algorithm.

- In the RTS/CTS access method, if a station does not receive a CTS frame within a CTS timeout period after sending an RTS frame, it attempts to retransmit the frame according to the RTS/CTS access method and the backoff algorithm.

An example of the exponential increase in the contention window size is shown in Figure 5.9, while a Markovian state transition diagram for IEEE 802.11 DCF MAC is shown in Figure 5.10. An analytical model for saturated throughput and delay can be obtained from this diagram. However, the analytical model is beyond the scope of this chapter.

For more details of a similar approach, the reader is referred to [34].

5.3.3.2 IEEE 802.11e Enhanced Distributed Channel Access (EDCA)

The contention-based IEEE 802.11e [19–21] uses carrier sense multiple access with collision avoidance (CSMA/CA) similar to that in IEEE 802.11. The main differences are that it supports multiple classes and allows the transmission of several data frames at one go with block acknowledgment. There are eight priority classes and they are mapped into four access, categories (ACs). The four ACs are for background, best effort, video, and voice traffic. The channel access for this traffic is differentiated by using different arbitration interframe spaces (AIFSs) and the minimum and maximum contention window sizes. The shorter the AIFS, the higher the priority for these access categories. Figure 5.11 shows the channel access for IEEE 802.11e, while Table 5.3 shows the arbitration inter-frame space number (AIFSN), the minimum contention window (CWmin) and the maximum contention window (CWmax) for background, best effort, video, and voice traffic.

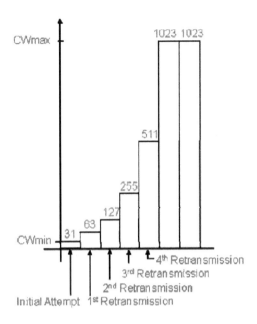

Figure 5.9: Example of exponential increase in the contention window.

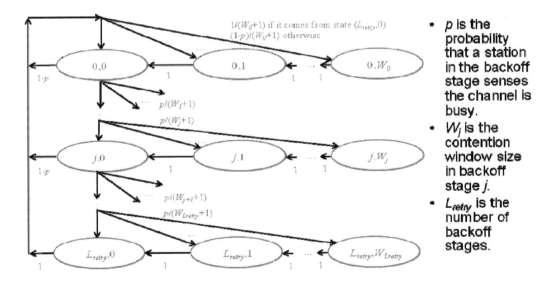

Figure 5.10: State transition diagram for IEEE 802.11 DCF MAC.

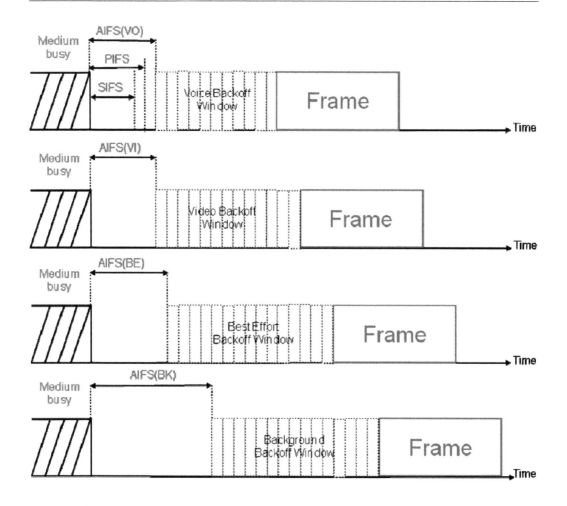

Figure 5.11: Channel access in IEEE 802.11e enhanced distributed channel access (EDCA) MAC.

5.3.3.3 IEEE 802.11n Reverse Direction Protocol (RDP)

There is a reverse direction MAC protocol in IEEE 802.11n standard as shown in Figure 5.12. After successful RTS and CTS frames, the reverse direction MAC protocol allows data transmissions in both directions between two stations at one successful attempt for initial data transmission from Station A to Station B, and the subsequent reverse data transmission from Station B to Station A is piggy-backed on this successful attempt. There are multiple MAC protocol data units (MPDUs) within the data transmissions, and there is a block acknowledgment request (BAR) embedded at the end of each data transmissions. These BARs are acknowledged by block acknowledgments

Table 5.3   AIFSN and Minimum and Maximum Contention Window Sizes for IEEE 802.11e EDCA MAC

Traffic Class	AIFSN	CWmin	CWmax
Background	7	aCWmin	aCWmax
Best Effort	3	aCWmin	aCWmax
Video	2	(aCWmin+1)/2−1	aCWmin
Voice	2	(aCWmin+1)/4−1	(aCWmin+1)/2−1

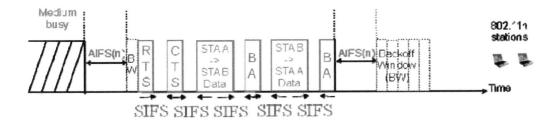

Figure 5.12: Channel access in IEEE 802.11n reverse direction protocol (RDP) MAC.

(BAs) after a SIFS time from the data transmissions. This protocol cuts down access delay for the destination station and improves turnaround time for the destination station to respond to the source station. This is particularly important for voice application, which is a bidirectional traffic. For more details of IEEE 802.11n, the reader is referred to [1].

5.3.3.4   IEEE 802.11ac Dynamic Channel Bandwidth Access (DCBA)

The contention-based IEEE 802.11ac uses CSMA/CA with dynamic channel bandwidth. This dynamic channel bandwidth access (DCBA) is similar to DCF, except that it can have channel bonding of static 80 MHz or dynamic 20/40/80 MHz. These two modes of accesses are shown in Figures 5.13 and 5.14. The figures are for illustration purpose only; they are not drawn to scale. In the static 80 MHz mode of access, all channels—the primary channel and the three secondary channels—must be sensed idle for a period of point coordination function inter-frame space (PIFS) time before the IEEE 802.11ac station's backoff counter reaches zero. This is on top of the AIFS for the IEEE 802.1ac (AIFS(ac)) station and its backoff window (BW). If all the channels are idle, then the data frames can be sent across the 80 MHz channels, and BAs are sent across all channels. Otherwise, the IEEE 802.11ac station cannot send the data frames. Other IEEE 802.11a/n stations in the secondary channels can access their shared channels as per their normal MAC protocols, including the AIFS for the IEEE 802.11a/n (AIFS(a/n)) and its BW. Note that the AIFSs for IEEE 802.11a stations and IEEE 802.11n are different, and the duration of the backoff slot times for them are also different.

For the dynamic 20/40/80 MHz mode of access, the number of channels that are available for the IEEE 802.11ac station is/are used for data transmission(s). One channel, two channels, or four channels is/are used depending on availability of the channels. Similarly, the available channel(s) must be sensed for the PIFS period before the IEEE 802.11ac station's backoff counter reaches zero. If the available channel(s) is/are idle, then the data frame(s) can be sent across the 20/40/80 MHz channel(s), and BAs are sent across the 20/40/80 channels. Dynamic channel bandwidth access e-Business Communication Association (EBCA) can be extended for Static 160 MHz and dynamic 20/40/80/160 MHz. This is optional. More details of IEEE 802.11ac can be found in [22–32].

5.3.3.5   IEEE 802.11ad Directional MAC Access Protocols in the Service Periods (SPs) and Contention-Based Periods (CBAPs)

The IEEE 802.11ad basic beacon interval (BI) structure is shown in Figure 5.15. Each BI consists of four parts [33]: a beacon time interval (BTI), association beamforming training time (A-BFT), announcement time interval (ATI), and a data transfer time interval (DTI). Service periods (SPs) and contention-based periods (CBAPs) are within the DTI. There is also MAC protocols in the SPs and CBAPs. A directional MAC protocol in the SP is used for data transmission between two devices, while a MAC protocol in the CBAP is used for data transmissions among devices via a modified directional EDCA MAC protocol. IEEE 802.11ad operates in the 60 GHz unlicensed band. For more details of IEEE 802.11ad, the reader is referred [33].

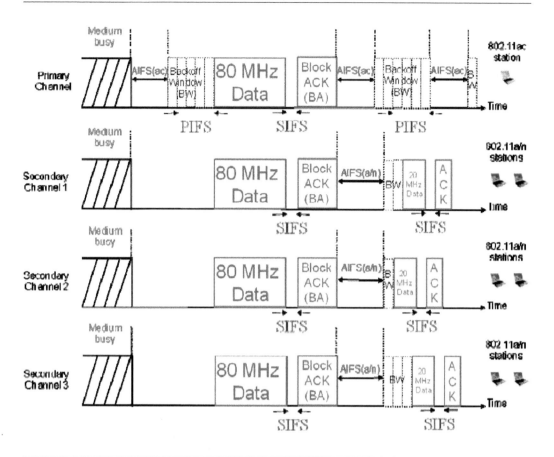

Figure 5.13: Dynamic channel bandwidth access for static 80 MHz.

5.3.4 Traditional Spectrum

This subsection briefly describes the traditional spectrum allocated for IEEE 802.11 WLANs.

The IEEE 802.11 working group defined more than 7 frequency bands for IEEE 802.11 WLANs: 2.4 GHz, 3.6 GHz, 4.9 GHz, 5 GHz, 5.9 GHz, 45 GHz, and 60 GHz. Currently, the 3.6 GHz frequency band is used as a licensed band in the United States for IEEE 802.11y, the 4.9 GHz frequency band is used by public safety entities in the United States, and the 5.9 GHz frequency band is used as licensed intelligent transportation system by IEEE 802.11p. Moreover, the 45 GHz frequency band is used as both licensed and unlicensed bands in China, where the unlicensed band plan is published as shown in Figure 5.16. Except for these bands, many current and forthcoming wireless applications are concentrated in the 2.4GHz, 5GHz, and 60 GHz frequency bands. We will introduce each of them one at a time in the remaining Subsection.

First, IEEE 802.11b/g/n operates in the 2.4 GHz frequency band that contains a total of 14 channels, as listed in Table 5.4. These channels are separated by 5 MHz in most cases, but the bandwidth is set to 22 MHz. Therefore, the channel overlaps with each other, as shown in Figure 5.17.

To mitigate the interference, the adjacent IEEE 802.11b/g/n WLANs are required to operate in the nonoverlapping channels. As a result, we see that there are five combinations of available nonoverlapping channel sets in Figure 5.18. Typically, the set of channels 1, 6, and 11 is widely used since the most Wi-Fi routers set Channel 6 as default.

Moreover, if IEEE 802.11n WLAN operates in the 2.4 GHz band, it is allowed to use the channel

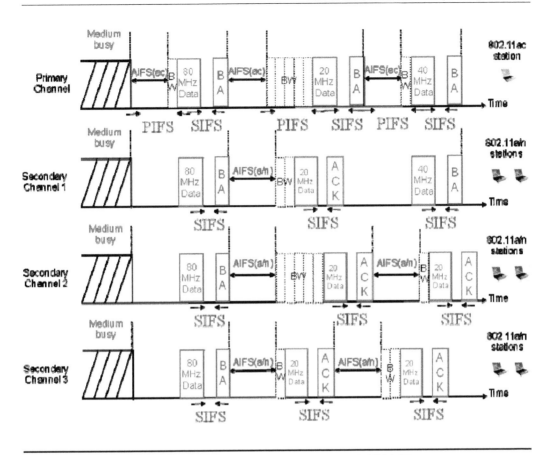

Figure 5.14: Dynamic channel bandwidth access for dynamic 20/40/80 MHz.

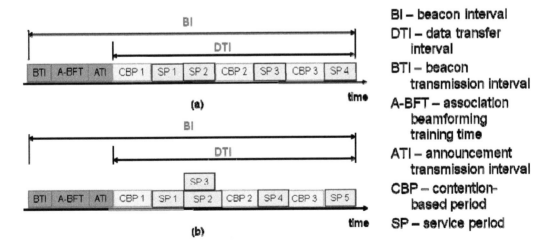

BI – beacon interval

DTI – data transfer interval

BTI – beacon transmission interval

A-BFT – association beamforming training time

ATI – announcement transmission interval

CBP – contention-based period

SP – service period

Figure 5.15: IEEE 802.11ad beacon interval (BI) structure (a) without spatial reuse, and (b) with spatial reuse.

Bandwidth (MHz)	Channel Number n	Center Frequency (GHz)
1080	1, 2, 3, 4	43.065+1.08(n-1)
	5	47.800
540	6, 7, 8, 9, 10, 11, 12, 13	42.795+0.54(n-6)
	14, 15	47.530+0.54(n-14)

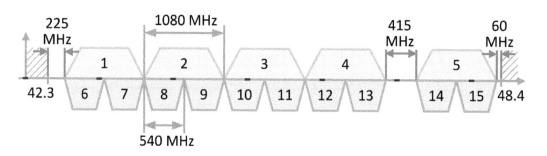

Figure 5.16: Unlicensed 45 GHz band plan in China.

Table 5.4   GHz Band Plan

Channel Number	Lower Frequency MHz	Center Frequency MHz	Upper Frequency MHz
1	2401	2412	2423
2	2404	2417	2428
3	2411	2422	2433
4	2416	2427	2438
5	2421	2432	2443
6	2426	2437	2448
7	2431	2442	2453
8	2436	2447	2458
9	2441	2452	2463
10	2446	2457	2468
11	2451	2462	2473
12	2456	2467	2478
13	2461	2472	2483
14	2473	2484	2495

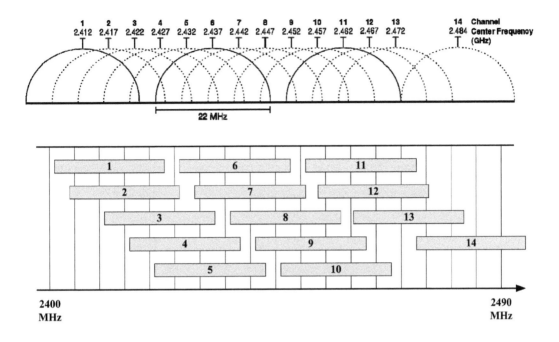

Figure 5.17: WLAN channels in 2.4 GHz band.

bandwidth of 40 MHz for the purpose of high data rate. In this case, the number of nonoverlapping channels will be reduced. For example, if Channel 3 and Channel 11 are used by an IEEE 802.11n WLAN, there will be no available channel for other WLANs.

Second, as the 2.4 GHz band becomes more crowded, many wireless applications choose to use the 5 GHz frequency band, e.g., IEEE 802.11a/h/j/n/ac. Different countries have different spectrum regulation, and we list the available operating channels for Europe, North America, and Japan in Table 5.5. The channel bandwidth is set to 20 MHz, and two adjacent channels can form a 40 MHz channel, e.g., Channel 36 and Channel 40.

Channel Number	Low Freq. (GHz)	Center Freq. (GHz)	High Freq. (GHz)	3 dB BW (MHz)	Roll-Off Factor
1	57.240	58.320	59.400	1728	0.25
2	59.400	60.480	61.560	1728	0.25
3	61.560	62.640	63.720	1728	0.25
4	63.720	64.800	65.880	1728	0.25

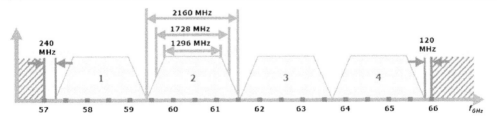

Figure 5.18: IEEE 802.11ad WLAN channels in 60 GHz band.

Table 5.5   GHz Band Plan and Spectrum Regulation in Different Countries

Channel Number	Center Frequency MHZ	Europe (ETSI)	North America (FCC)	Japan
36	5180	Indoors	Yes	Yes
40	5200	Indoors	Yes	Yes
44	5220	Indoors	Yes	Yes
48	5240	Indoors	Yes	Yes
52	5260	Indoors / DFS / TPC	DFS	DFS / TPC
56	5280	Indoors / DFS / TPC	DFS	DFS / TPC
60	5300	Indoors / DFS / TPC	DFS	DFS / TPC
64	5320	Indoors / DFS / TPC	DFS	DFS / TPC
100	5500	DFS / TPC	DFS	DFS / TPC
104	5520	DFS / TPC	DFS	DFS / TPC
108	5540	DFS / TPC	DFS	DFS / TPC
112	5560	DFS / TPC	DFS	DFS / TPC
116	5580	DFS / TPC	DFS	DFS / TPC
120	5600	DFS / TPC	No Access	DFS / TPC
124	5620	DFS / TPC	No Access	DFS / TPC
128	5640	DFS / TPC	No Access	DFS / TPC
132	5660	DFS / TPC	DFS	DFS / TPC
136	5680	DFS / TPC	DFS	DFS / TPC
140	5700	DFS / TPC	DFS	DFS / TPC
149	5745	SRD	Yes	No Access
153	5765	SRD	Yes	No Access
157	5785	SRD	Yes	No Access
161	5805	SRD	Yes	No Access
165	5825	SRD	Yes	No Access

Note: DFS = dynamic frequency selection; TPC = transmit power control; SRD = short range devices 25 mW max power.

Each channel has the bandwidth of 2.16 GHz, and different countries have different spectrum regulations, e.g., only Channel 2 and Channel 3 are available in China.

Third, IEEE 802.11ad WLAN operates in the 60 GHz band and four of operating channels are given in Figure 5.18.

Last, to solve the scarcity problem of the operating channels of 60 GHz band in China, IEEE 802.11aj working group proposed a way that allows splitting each 2.16 GHz channel into two 1.08 GHz channels. Thus, IEEE 802.11aj WLAN can operate in either a 2.16 GHz channel or a 1.08 GHz channel, which is illustrated in Figure 5.19.

5.4   LTE in the Unlicensed Band

This section describes the scenarios, co-existence mechanisms, spectrum regulations at the 5 GHz unlicensed band, benefits and some results for LTE-U. For more details of LTE-U, the reader is referred to [35–39].

5.4.1   Scenarios

In this subsection, the scenarios in LTE-U are presented. There are two main prioritized scenarios for LTE-U deployment [35]. The scenario in the first phase to be considered is the collocated case

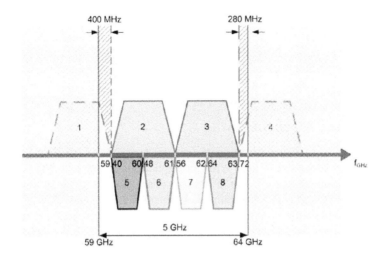

Figure 5.19: IEEE 802.11aj WLAN channels in 60 GHz band.

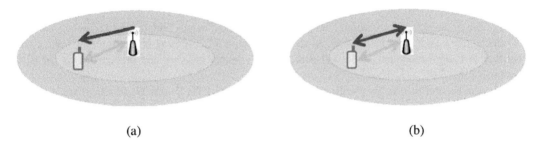

Figure 5.20: Collocated LTE base station and Wi-Fi AP for (a) downlink only in the unlicensed band, and (b) both downlink and uplink in the unlicensed band.

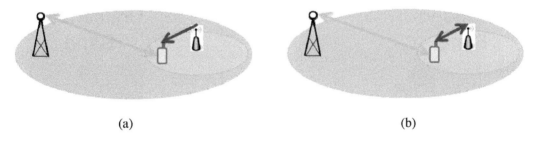

Figure 5.21: Non-Collocated LTE Base Station and Wi-Fi AP for (a) Downlink only in the Unlicensed Band, and (b) Both Downlink and Uplink in the Unlicensed Band.

for the LTE base station and Wi-Fi AP, as shown in Figure 5.20, while the scenario in the second phase to be considered is the noncollocated case for the LTE base station and the Wi-Fi AP, as shown in Figure 5.21.

In each case, the usage of the unlicensed band can be for downlink transmissions only or can be for both downlink and uplink transmissions. The coverage of the LTE in licensed band assumed to be larger than that of the Wi-Fi in the unlicensed band. Thus, the bandwidth of LTE-U can be aggregated via the licensed bands in LTE/LTE-A and the unlicensed bands in Wi-Fi. This is a new form of carrier aggregation or spectrum aggregation, similar to those techniques in the licensed bands only.

5.4.2   Coexistence Mechanisms

In this subsection, the coexistence mechanisms for LTE-U are presented. There are three mechanisms for coexistence between LTE and Wi-Fi in the unlicensed band for LTE-U without listen-before-talk (LBT) requirements [36]. The three mechanisms are channel selection, carrier-sensing adaptive transmission (CSAT), and opportunistic supplemental downlink (SDL). For LTE-U with LBT requirement, modifications to the LTE physical layer and MAC scheduling are needed. These changes will be discussed in upcoming 3GPP meetings. Countries like the United States, China, and South Korea do not have regulatory requirements for LBT waveform in the unlicensed band, while Europe, Japan, and India have regulatory requirements for LBT waveform. Thus, this chapter focuses on LTE-U with and without LBT requirement. Furthermore, only downlink transmissions in the unlicensed band are considered. Extensions to support both downlink and uplink transmissions in the unlicensed bands are needed in the future.

For the first mechanism without LBT requirements, the cleanest channel in the unlicensed band is selected for LTE SDL transmission, as shown in Figure 5.22. If there is interference in the channel that is being scanned, LTE-U will search for a cleaner channel and switched to it for SDL transmission.

For the second mechanism without LBT requirements, LTE-U and Wi-Fi can share the unlicensed band using on and off duty cycles of LTE-U, as shown in Figure 5.23. During the on cycle of LTE-U, LTE-U can transmit its uplink (UL) or downlink (DL) packets, while during the off cycle of LTE-U, Wi-Fi can transmit its UL or DL packets for infrastructure mode of network architecture. The priority of LTE-U or Wi-Fi depends on the mean duty cycles of LTE-U in the unlicensed band. The larger the mean on cycle of LTE-U, the higher the priority for LTE-U in terms of mean duty

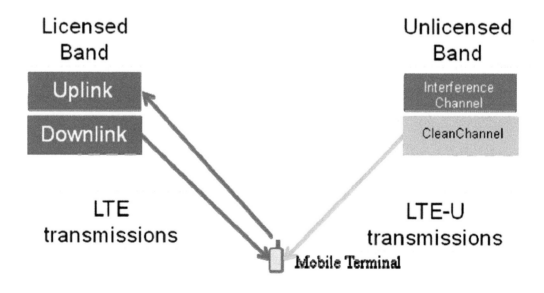

Figure 5.22: Channel Selection for LTE-U transmissions.

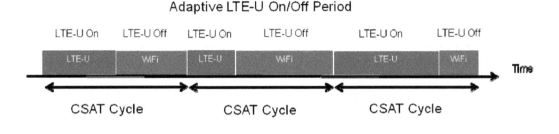

Figure 5.23: CSAT mechanism allows LTE-U and Wi-Fi to share the channel in a time division manner.

cycle; similarly, the larger the mean off cycle of LTE-U, the higher the priority for Wi-Fi. This spectrum sharing in the unlicensed band differs from the traditional spectrum sharing in the licensed band, where the cellular network is the incumbent or primary users, and Wi-Fi is the secondary users. Priority for the two types of networks, LTE-U and Wi-Fi, in the unlicensed bands can be expressed in terms of the mean duty cycles of on and off periods of LTE-U.

For the third mechanism without LBT requirement, when there are active LTE users in the unlicensed band's coverage area and the downlink traffic load exceeds a threshold, the SDL is turned on for data transmissions. On the other hand, when there is no user in the coverage area, the SDL carrier is turned off.

For the coexistence mechanism with LBT requirement, an example of an LBT protocol is presented in [36]. A listening period known as the clear channel assessment (CCA) is used by LTE-U before attempting to transmit a downlink frame. In between LTE-U frame transmissions, Wi-Fi can transmit their data frames using CSMA/CA MAC protocols, as shown in Figure 5.24. This can also be extended to cases for LTE-U with IEEE 802.11ac with DCBA MAC protocols, using multiple channels.

Throughput for LTE can be improved via carrier aggregation of LTE transmissions in the licensed band and LTE-U transmissions in the unlicensed band. For more details for LTE-U coexistence mechanisms, the reader is referred to [35–37].

5.4.3 Spectrum Regulations at the 5 GHz Unlicensed Band

The spectrum regulations at the 5 GHz unlicensed band is presented in Subsection III.D. The reader is referred to Subsection III.D for more details.

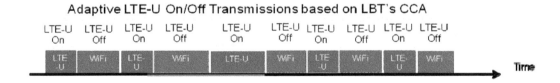

Figure 5.24: CCA mechanism for listen-before talk protocol in LTE-U to allow LTE-U and Wi-Fi to co-exist in a channel.

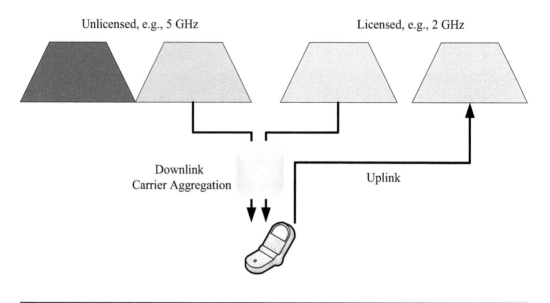

Figure 5.25: Carrier aggregation in LTE-U.

5.4.4 LTE-U Benefits

In this subsection, the quantitative benefits of LTE-U are presented. LTE-U is considered as a performance booster in operator-deployed networks. The traditional licensed band can guarantee the control and management packet transmission. The data packet transmission can occur in the licensed or unlicensed band through carrier aggregation, as illustrated in Figure 5.25, so that the throughput performance is dramatically improved. This is also called LTE (LAA), as mentioned earlier in this chapter.

Carrier Wi-Fi is another way to offload the cellular traffic to the unlicensed spectrum. However, such efforts may not always achieve the expected network performance improvement or cost reduction.

The reasons are various, such as the investment on the backhaul and core network, in addition to the existing cellular infrastructure, the inferior performance of Wi-Fi technology, and the lack of good coordination between the cellular system and the Wi-Fi system, which eventually results in the low-efficient use of spectrum and poor user experience.

As compared to the carrier Wi-Fi, LTE-U can provide higher spectrum efficiency, reliability, and quality and robust fallback. The improvement is derived from the following aspects: (1) LTE-U has a better and more robust structure for mobility; (2) LTE-U's coordinated and synchronized architecture makes the best use of resources by managing and mitigating the interference; and (3) the mandatory anchor in the licensed spectrum ensures that the control signaling is always efficiently delivered and seamless mobility is achieved.

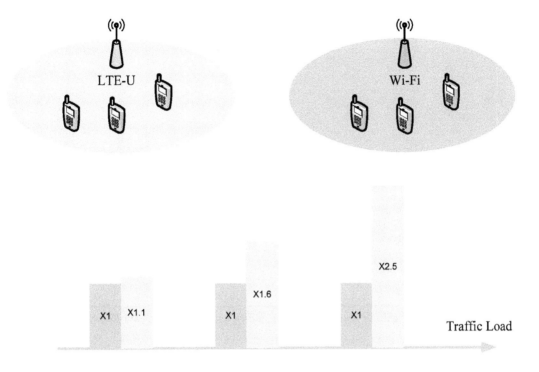

Figure 5.26: Spectrum efficiency comparison between Wi-Fi and LTE-U.

The unified network structure is also an advantage for LTE-U, which is easier to be deployed and coordinated. All the existing backhaul, core network, and even sites deployed for licensed LTE carriers can be reused for the operation of unlicensed spectrum, with updates only in eNodeBs or base stations. Therefore, the whole network has unified operation and management between the licensed band and the unlicensed spectrum.

5.4.5   *Some Results*

In this subsection, some results from Huawei and Qualcomm are presented to illustrate the performance benefits of LTE-U.

In Huawei's report [37], they consider an isolated deployment scenario that only LTE-U or Wi-Fi exists, as illustrated in Figure 5.26 and obtain the conclusion that LTE-U achieves up to 2.5 times spectrum efficiency of that of Wi-Fi by its inherent interference mitigation mechanism.

Moreover, in Qualcomm's report [38], they consider another scenario, such that LTE-U coexists with Wi-Fi within the same channel. Their simulation results show that both LTE-U and Wi-Fi system, improve their capacity as compared, to the pure two Wi-Fi systems, which is shown in Figure 5.27.

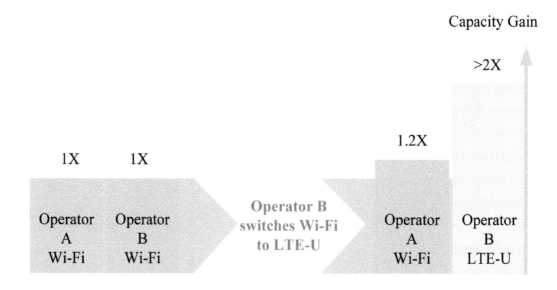

Figure 5.27: Capacity gain in both operators A and B, when operator B Wi-Fi switches to LTE-U.

5.5   Conclusions

LTE-U is described and presented in this chapter. An overview of LTE/LTE-A in the licensed band and Wi-Fi in the unlicensed band is presented. For LTE\LTE-A in the licensed band, the network architectures, physical layers, MAC scheduling, and traditional spectrum are briefly described, Similarly, for Wi-Fi in the unlicensed band, the network architectures, physical layer MAC protocols, and traditional spectrum are also briefly described. For detailed descriptions of LTE/LTE-A and Wi-Fi, the reader is referred to [1–34].

The focus of this chapter is on LTE-U or LTE LAA. Scenarios, coexistence mechanisms, spectrum regulations, benefits, and some results of LTE-U are presented in detail.

Three coexistence mechanisms for LTE-U without listen-before-talk protocol and a coexistence mechanism for LTE-U with listen-before talk protocol are described. Four coexistence mechanisms with and without listen-before-talk protocol are also presented, based on [36]. For LTE-U with listen-before-talk requirements, modifications to the LTE physical layer and MAC scheduling are needed. These changes will be discussed in upcoming 3GPP meetings. For more understanding of the status of many companies participating in LTE-U, the reader is referred to the presentations and summary of the 3GPP Workshop on LTE in Unlicensed Spectrum in [35, 37, 38]. For a recent article on LTE-U, the reader is referred to [39].

IEEE 802.11ah, IEEE 802.11aj, and IEEE 802.11ax WLANs will be standardized in the future. IEEE 802.11ah WLAN operates in sub-GHz unlicensed band to provide extended range for a large number of stations or sensors [40–42]. On the other hand, IEEE 802.11aj WLAN is for the Chinese millimeter wave bands in the 59–64 GHz and the 45 GHz to provide more logical channels, multi-Gbps throughput, and lower power [43]. Whereas, IEEE 802.11ax WLAN [44] operates in the 5 GHz unlicensed band and is four times the throughput of IEEE 802.11ac WLAN [45]. LTE-U can also be extended via new carrier aggregation or spectrum aggregation of the licensed bands of LTE/LTE-A and the unlicensed bands at the sub-GHz bands, the 2.4 GHz bands, the 5 GHz bands, and the 60 GHz bands in the future, as well as other unlicensed bands, not just the 5 GHz unlicensed bands that are being considered currently. LTE-U's coexistence with other technologies, like IEEE 802.15.4 Zigbee and Bluetooth, in the unlicensed bands needs to be considered in the future, too, not just the coexistence of LTE-U with Wi-Fi. Thus, LTE-U or LTE LAA is the next exciting technology to watch out for in the unlicensed band with a new twist in the spectrum sharing arena. The coexistence techniques developed in LTE-U can also be applied to worldwide interoperability for microwave access (WiMAX) cellular system in the unlicensed bands in the future, as well as future cellular network technologies. These techniques can also be applied to HeNodes in the future. HeNodes are also known as femtocells.

Thus, LTE in the unlicensed band is certainly good for LTE by expanding its throughput via carrier aggregation of spectrum in the licensed bands, as well as a portion of the spectrum in the unlicensed band. On the other hand, there are concerns in the Wi-Fi industry on sharing the traditional unlicensed band with LTE. These issues need to be addressed. To the customers or users and the LTE and Wi-Fi industries, the following questions need to be answered as well. Will the customers or users on both sides of the technologies, LTE and Wi-Fi, truly benefit from LTE-U, or will the customers or users who are sometime on both sides of the technologies truly benefit from LTE-U, or how will LTE-U finally roll out? Only time will give answers to these questions.

Another interesting networking scenario is heterogeneous networks (HETNETs). In HETNETs, a mobile user of cellular networks and Wi-Fi networks can switch seamlessly from these networks without user intervention and without interruption in his/her applications. LTE-U may play an important role in HETNETs in the future. Thus, LTE-U could synergize with HETNETs to give an excellent QoE for the end users in the future.

Furthermore, another exciting networking scenario is mmWave backhaul for fifth generation (5G) cellular networks. Coupling mmWave backhaul for 5G cellular networks with LTE-U certainly creates novelty in new network architecture and is a new vision for future 5G cellular networks. There are also tremendous interests in mmWave for 5G cellular networks from both academia and industry. Thus, this is a new exciting area to watch and work on for at least the next few years to come.

Finally, it must be stressed that WiFi and LTE-U need to coexist together harmoniously to ensure the success of LTE-U and the continued usage of WiFi concurrently in the unlicensed band.

References

[1] D.T.C. Wong et al., *Wireless Broadband Networks*, John Wiley and Sons, March 2009.

[2] P. Lescuyer and T. Lucidarme, *Evolved Packet System (EPS)*, John Wiley and Sons, 2008.

[3] S. Sesia, I. Toufik, and M. Baker, *LTE — The UMTS Long Term Evolution: From Theory to Practice*, John Wiley and Sons, 2009.

[4] A. Gosh, J. Zhang, J.G. Andrews, and R. Muhamed, *Fundamentals of LTE*, Prentice Hall, 2011.

[5] M. Sauter, *From GSM to LTE-Advanced:An Introduction to Mobile Networks and Mobile Broadband*, Second Edition, Wiley, 2014.

[6] C. Cox, *An Introduction to LTE: LTE, LTE-Advanced, SAE, VoLTE and 4G Mobile Communications*, Second Edition, Wiley, 2014.

[7] E. Dahlman, S. Parkvall, and J. Skold, *4G: LTE/LTE Advanced for Mobile Broadband*, Academic Press, 2014.

[8] T. Dikamba, "Downlink Scheduling in 3GPP Long Term Evolution (LTE)," M.Sc. Thesis, Delft University of Technology, 2011.

[9] N. Abu-Ali, et al., "Uplink scheduling in LTE and LTE-advanced: tutorial, survey and evaluation framework," *IEEE Communications Surveys and Tutorials*, vol. 16, no. 3, pp. 1239–1265, December 18, 2013.

[10] M. Chmiel, J. Shi, and D.X. Zhou, "Downlink scheduling and rate capping for LTE-Advanced carrier aggregation," *Communications and Networks*, vol. 5, no. 3b, pp. 1–5, September 2013.

[11] Rainer Schoenen, "MAC layer considerations for 4G multihop cellular wireless systems," I^{2}R Seminar, Singapore, November 22, 2010.

[12] K. Higuchi, "Introduction of evolved UTRA and UTRAN," I^{2}R Seminar, Singapore, 25 March 2008.

[13] J. Zyren, "Overview of the 3GPP long term evolution physical layer," white paper, Freescale Semiconductor, July 2007.

[14] T. Godfrey, "Long-term evolution protocol: How the standard impacts media access control," Online dated June 26, 2007.

[15] 3GPP TS 36.300 V8.2.0 (2007–09).

[16] "3GPP long term evolution", Qualcomm, online dated January 2008.

[17] W. Stallings, *Wireless Communications and Networks*, Prentice Hall, 2002.

[18] Yang Xiao and Yi Pan (Editors), *Emerging Wireless LANs, Wireless PANs, and Wireless MANs*, John Wiley and Sons, 2009.

[19] "IEEE Standard 802.11e," September 2005.

[20] G. Bianchi, "Performance Analysis of the IEEE 802.11 distributed coordination function," *IEEE Journal on Selected Areas in Communications*, vol. 18, no. 3, March 2000.

[21] Z.N. Kong, D.H.K. Tsang, and B. Bensaou, "Performance analysis of IEEE 802.11e contention-based channel access," *Journal on Selected Areas in Communications*, vol. 22, no. 10, December 2004.

[22] IEEE P802.11ac Standard "Part 11: Wireless LAN medium access control (MAC) and physical layer (PHY) specifications amendment 4: Enhancements for very high throughput for operations in bands below 6 GHz," December 2013.

[23] E. Perahia and R. Stacey, *Next Generation Wireless LANs: 802.11n and 802.11ac*, Second Edition, Cambridge Press, 2013.

[24] M.S. Gast, *802.11ac: A Survival Guide*, O'Reilly Press, 2013.

[25] E. Perahia and M.X. Gong, "Gigabit wireless LAN: An overview of IEEE 802.11ac and 802.11ad," *ACM SigMobile Mobile Computing and Communications Review*, 2011.

[26] E. Charfi, L. Chaari and L. Kamoun, "PHY/MAC enhancements and QoS mechanisms for very high throughput WLANs: A survey," *IEEE Communication Surveys and Tutorials*, vol. 15, no. 4, 2013.

[27] S.J. Vaughan-Nichols, "Gigabit Wi-Fi is on its way," *Computer*, November 2010.

[28] L. Garber, "Wi-Fi Races to a faster future," *Computer*, March 2012.

[29] L. Verma, M. Fakharzadek, and S. Choi, "Wi-Fi on steroids: 802.11ac and 802.11ad," *IEEE Wireless Communications Magazine*, December 2013.

[30] E. Perahia and M.X. Gong, "Gigabit wireless LANs: An overview of IEEE 802.11ac and 802.11ad," *SIGMOBLE Mobile Computing and Communications Review*, November 2011.

[31] M.X. Gong, B. Hart, L. Xia, and R. Want, "Channel bonding and MAC protection mechanisms for 802.11ac," IEEE Globecom 2011.

[32] M. Park, "IEEE 802.11ac: Dynamic bandwidth channel access," IEEE International Conference on Communications, 2011.

[33] D.T.C. Wong, F. Chin, X. Peng, and Q. Chen, "IEEE 802.11ad wireless local area network and its MAC performance," to be published in *Wireless Network Performance Enhancement via Directional Antennas: Models, Protocols, and Systems*, J.D. Matyjas, F. Hu, and S. Kumar, Eds., CRC Press, 2015.

[34] Q. Ni, et al., "Saturated throughput analysis of error-prone 802.11wireless networks," *Wireless Communications and Mobile Computing*, vol.5, pp. 945–956, 2005.

[35] 3GPP workshop on LTE in unlicensed spectrum presentations, Sophia Antipolis, France, June 13, 2014. http://www.3gpp.org/news-events/3gpp-news/1603-lte_in_unlicensed, dated January 6, 2015.

[36] Qualcomm, "Qualcomm research — LTE in the unlicensed spectrum: Harmonious coexistence with Wi-Fi," online, June 2014.

[37] Huawei, "U-LTE: unlicensed spectrum utilization of LTE," http://www.huawei.com/ilink/en/download/HW_327803, dated March 6, 2015.

[38] Qualcomm, "Extending LTE advanced to unlicensed spectrum," December 2013, online, dated March 6, 2015.

[39] F.M. Abinader et al., "Enabling the coexistence of LTE and Wi-Fi in unlicensed bands," *IEEE Communications Magazine*, pp. 54–61, November 2014.

[40] Wikipedia online on IEEE 802.11ah, http://en.wikipedia.org/wiki/IEEE_802.11ah, dated June 30, 2014.

[41] E. Khorov et al., "A survey on IEEE 802.11ah: An enabling networking technology for smart cities," *Computer Communications*, in press.

[42] M. Park, "IEEE 802.11ad: Energy efficient MAC protocols for long range wireless LAN," IEEE International Conference on Communications, 2014.

[43] IEEE 802.11aj website, http://www.ieee802.org/11/Reports/tgaj_update.htm, dated June 30, 2014.

[44] Wikipedia on IEEE 802.11, http://en.wikipedia.org/wiki/IEEE_802.11, dated July 3, 2014.

[45] IEEE P802.11ac Standard "Part 11: Wireless LAN medium access control (MAC) and physical layer (PHY) specifications amendment 4: Enhancements for very high throughput for operations in bands below 6 GHz," December 2013.

Chapter 6

Licensed Shared Access to spectrum

Seppo Yrjölä, Marja Matinmikko, Petri Ahokangas, and Miia Mustonen

CONTENTS

Need for the mobile broadband is growing at increasing pace, placing emergent demands on the scarce radio spectrum resources. Licensed Shared Access (LSA) is a new complementary licensing method to traditional exclusive licensing and unlicensed operations for spectrum access that allows current incumbent spectrum users to share their spectrum with additional LSA licensees, such as mobile network operators (MNOs), according to this LSA regulatory framework issued by a National Regulatory Authority. The advantage of LSA with conditions that resemble exclusive licensing is that the rights of both the Incumbent and MNO spectrum users are guaranteed and Quality of Service (QoS) fully supported. This chapter presents the LSA spectrum sharing framework, concept and system architecture based on the recent developments in regulation and standardization. We discuss LSA system concept, stakeholders and their relations in the LSA use cases. Next results of the recent LSA demonstrations and trials are introduced with a focus on indentifying key enabling technologies. Finally, spectrum sharing business model scenarios are shared showing how

LSA can become an important solution to gain access to new mobile broadband spectrum on time while enabling new innovative business opportunities for existing and new stakeholders.

6.1   Introduction

Spectrum is one of the most in-demand resources in our digitalizing information economy. We have witnessed the exponential growth of wireless services to access information, enjoy content, and conduct commerce from mobile devices anywhere, anytime. The number of Mobile Broadband (MBB) subscribers, number and variety of devices and the amount of data used per user is set to grow significantly over the coming years [1] leading to increasing spectrum demand. The US President's Council of Advanced Science & Technology (PCAST) report [2] three years ago painted the urgency for new thinking within wireless industry to meet the growing spectrum crisis in spectrum allocation, utilization and management. The significance of spectrum sharing was highlighted to find a balance between the different domains, systems and services with different spectrum needs and dynamics. At the same time in the Europe, the European Commission's (EC) Radio Spectrum Policy Group (RSPG) promoted the shared use of radio spectrum resources in the internal market and in particular showed growing interest on the Licensed Shared Access (LSA) as a complementary spectrum tool to cope with the growing spectrum demand for wireless broadband. The RSPG opinion in 2013 stated that *"To meet the growing demand for spectrum the industry and administrations are under pressure to introduce new technologies and regulatory mechanisms to optimize the use of the limited frequency resources. In this context, the promotion of the shared use of radio spectrum resources is a valuable means to offer additional spectrum access to broadband communications, for license exempt but also licensed usage, which is a new paradigm referred to as Licensed Shared Access."* [3].

For a spectrum sharing concept, where several radio systems operate in the same spectrum, to be a feasible and attractive, close cooperation between business, policy and technology domains is essential. Without active contribution from the key business stakeholders along the ecosystem, these system concepts will not become deployed in the commercial services. Industry created user stories, requirements and sound business model designs for all the key stakeholders are critical success factors for any concept to succeed. In the wireless ecosystems, spectrum policy and regulation has played central role in triggering current multibillion MBB operator businesses via exclusive spectrum usage rights, guaranteeing QoS for end users through the minimization of interference and stability for MNOs through long license durations. The role of spectrum regulation as triggering innovations as well as setting frames will become even stronger with shared spectrum initiatives as typically the government is the biggest single user of spectrum, with licenses issued and administered by the National Regulatory Authority (NRA). Governmental spectrum can be maximized to provide additional spectrum for commercial use, but only if the government is willing to examine how ministries use their allocated spectrum and offers incentives for more efficient use. Furthermore, industry and research should collaborate early on in innovating, testing and trialing of the applicability of the enabling technologies, new concepts and business models.

Therefore, only a subset of the conducted research on spectrum sharing has ever entered into regulatory and business domains, an example being the early research on cognitive radio with spectrum sensing as the only interference mitigation technique. Additionally, there are several spectrum sharing models, widely studied and supported by NRAs and standardization, that have remained niche in the wireless market, the TV White Space (TVWS) being the latest example [4] and [5]. Following a decade of intense unlicensed TVWS studies in the US and the Europe, licensing founded and database interference mitigation based sharing models have recently emerged and are currently under regulatory discussion. The most well known topical spectrum sharing concepts under study in the technology, policy and business domains are the LSA [6] from the Europe and the three tier Citizens Broadband Radio Service (CBRS) from the US [7].

Although the LSA system concept and framework could be deployed across different wireless communication domains, the main focus is to apply it first to MBB. In this scenario a Mobile Network Operator (MNO) would be allowed to access spectrum identified for International Mobile Telecommunication (IMT) systems by sharing with the Incumbent spectrum user of other industry type. In this chapter we apply the LSA system and framework for MNOs to enhance and complement their future spectrum options to cope with the growing mobile broadband demand with novel business opportunities and models. We review the system concept and regulatory framework status of the LSA and their future evolution. In particular we analyze the sharing based novel business opportunities and business models designs on the LSA concept for future MBB networks. To make this happen, the purpose of this research is to identify the characteristics needed for the LSA business model to scale and to be able to leverage sharing economy within the MBB ecosystem. More specifically, the objectives of this chapter are as follows:

1. Describe and illustrate the LSA system concept and regulatory framework;

2. Discuss recent LSA field trials and assess the role of key enabling technologies;

3. Assess business opportunities and business model design characteristics of the LSA for future mobile broadband.

The rest of the chapter is organized as follows. First the underlying LSA spectrum sharing system concept is identified followed by an introduction to the regulatory framework and standardization status. Next, LSA field trials are introduced and key enabling technologies discussed. Then, business model design scalability and sharing economy criteria is introduced and business model opportunities derived and evaluated. Finally, conclusions are drawn.

6.2   LSA system concept and regulatory framework

LSA is a complementary regulatory framework that provides access to spectrum resources otherwise unavailable for the MBB use. LSA enables predictable QoS and protection from harmful emissions for the spectrum right holders, the Incumbent and the LSA Licensee. Furthermore the LSA framework ensures voluntary participation for the stakeholders and continuous access rights to the Incumbent's required spectrum [8]. B*andwidth Expansion for Mobile Network Operator* is the major use case for the LSA concept defined by European Telecommunications Standards Institute (ETSI) [8]. An MNO operating LTE in the licensed band in a region applies for an individual authorization to use radio frequency within the 2.3 GHz spectrum in the same region under the LSA regime. Authorization for the MNO LSA Licensee plans to access the agreed part of the spectrum may be location and/or time based. Access to additional LSA spectrum will require modifications to existing mobile broadband network operations. In this section we are translating the latest regulatory and standardization requirements and studies into the reference LSA system and architecture concept and discuss functionalities of each system element. The LSA regulatory and standardization time line with referred key results and documents are discussed in details in the subsequent section.

6.2.1   LSA functional system model

In managing and integrating additional LSA spectrum resources dynamically into MNO's MBB network, novel LTE Heterogonous Network (HetNet) and Self Organizing Networks (SON) features are essential. MNO's Operations Support Systems (OSS) Network Management System (NMS) enables relevant evolved NodeB (eNB) LTE Base Stations (BS) to start transmission in the LSA band, when permitted, and load balancing and traffic steering algorithms in the LTE network move users to the LSA band and balance load between LSA and possible other spectrum band and Radio Access Networks (RAN). When the LSA band needs to be evacuated due to the appearance of an Incumbent, the OSS reconfigures relevant eNBs, and OSS algorithms move users to other networks in order to protect the Incumbent. The LSA spectrum resource could be considered as a part of MNO's overall HetNet consisting of several spectrum band, Radio Access Technologies (RAT) and layers like macro cells or small cells. In the LSA use case scenario the MNO assesses the network conditions against the technical and business optimization criteria and determines whether additional LSA spectrum resource is needed and how to optimally utilize it.

The key players in the LSA concept are the Incumbent, the Regulator, and the MNO LSA Licensee, as shown in Figure 6.1 [9]. In the LSA concept [3] and [6] spectrum sharing is allowed between an Incumbent spectrum user and a Licensee in a binary way so that both have exclusive individual access to a spectrum at a given time and location. The spectrum Regulator is responsible for identifying LSA spectrum to be licensed and defining the *sharing framework* [6] consisting of rules and conditions for sharing as well as granting the *license* to the LSA Licensee. Based on the national framework the Incumbent and LSA Licensee negotiate the private commercial *sharing agreement* [9] under the permission and governance of the Regulator. In the voluntary LSA framework and agreement the Incumbent spectrum user defines the part of its spectrum that can be used for sharing with the LSA concept, the license duration, geographical area and the required evacuation time. Compared to traditional auctioning & licensing approach, the LSA framework concept enables faster, lower cost and flexible localized access to spectrum through the avoidance of the lengthy and expensive re-farming process enforced by regulators. The long term predictability and availability of the spectrum resource are key antecedents of the framework and the agreement for the use case to succeed and guarantee the MNO to invest in the LSA concept deployment.

The LSA concept introduces two new logical elements to the RAN architecture, the *LSA Controller (LC)* and the *LSA Repository (LR)* [10]. These new elements are required to support the dynamic LSA spectrum resource availability, guarantee interference free operation and the rights of the Incumbent. The LSA architecture reference model and mapping of the high level functions [10] are shown in Figure 6.2 and its integration into MBB architecture is suggested in Figure 6.3.

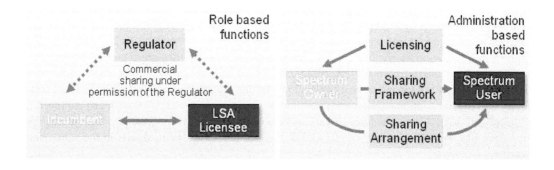

Figure 6.1: The key roles in the LSA concept.

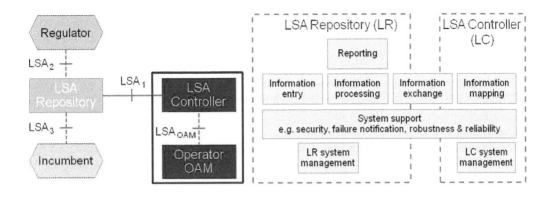

Figure 6.2: The LSA architecture reference model and mapping of the high level functions.

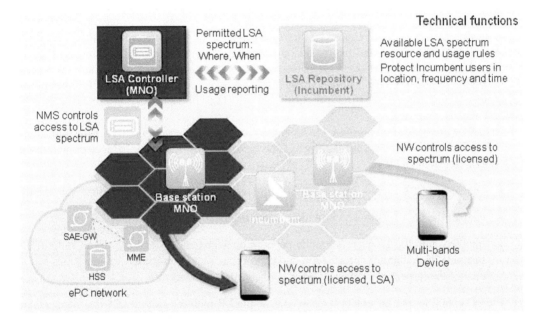

Figure 6.3: The LSA technical function integration into MBB architecture.

The *LSA Repository* (LR) is a database with the following key functions [10]:

- the entry and storage of information describing Incumbent's usage and protection requirements,

- it conveys availability information to authorized LCs,

- it receives and stores acknowledgement information received from the LCs,

- it provides means for NRA to monitor operation of the LSA system, and to provide the system with information on the sharing framework and the LSA Licensees and

- it ensures that the LSA system operates in conformance with the sharing framework, and may in addition implement any non-regulatory details of the sharing arrangement.

Based on this information, protected areas are defined based on the underlying regulatory requirements. These protected areas are *exclusion zones* within which LSA Licensees are not allowed to have active radio transmitters, *protection zones* where Incumbent receivers will not be subject to harmful interference caused by LSA Licensees' transmissions or *restriction zones*, where LSA Licensees are allowed to operate radio transmitters, under certain restrictive conditions, e.g., maximum Equivalent Isotropically Radiated Power (EIRP) limits and/or constraints on antenna parameters [9]. The Incumbent user is requested to make a specific *LSA Spectrum Resource Availability Notification* to enable the LR to send *LSA Spectrum Resource Availability Information* (LSRAI) to the LC. It can be used to send either specific immediate notifications, or periodic updates of the overall LSA spectrum resource availability information related to this LC. In addition Licensee could send *the LSA Spectrum Resource Availability Information Request* to make a request for LSA spectrum resource availability information. This procedure can be used to initiate LSA operation, or to synchronize information between LR and LC during LSA operation. The LR stores information describing Incumbent's usage and conveys availability information to authorized LCs when the information changes. The LR is able to communicate with several LCs. Regulators may monitor spectrum usage via the LR, which monitors the LSA system for possible exception situations such as the unavailability of LC or unconfirmed protection request. Notification will be sent to Regulator immediately if failure occurs. It is critical to guarantee that the data exchange between different stakeholders is efficient, secure and reliable. Transport layer mechanisms utilize transmission control protocol (TCP) and user datagram protocol (UDP), and depending on the security requirements, use of Internet protocol security architecture (IPSec), transport layer security (TLS) or datagram transport layer security (DTLS) may be applicable.

The *LSA Controller (LC)* provides the Licensee with means to access the LSA spectrum and to react on the spectrum resource availability and the Incumbent user activity. LC located within the LSA Licensee's domain [10]:

- enables the LSA Licensee to obtain spectrum resource availability information from the LR

- enables the LSA Licensee to provide acknowledgment information to the LR

- interacts with the Licensee's mobile network management system in order to support the mapping of availability information into appropriate radio transmitter configurations and to receive the respective confirmations from the mobile network.

The LC combines the information received from the LR with the current network management information in order for HetNet OSS to be able to configure and optimize the use of LSA spectrum resource. As the OSS requires full knowledge of the network layout and access to potentially business sensitive information, the LC is assumed to be under purview of the MNO as shown in Figure 6.2 and Figure 6.5. Output data of the LC work flow are configuration parameters for the RAN BSs. Operations robustness could be checked by the connectivity check procedure at application initiated by the LR and/or the LC. E.g., in the case the LC has sent a Connectivity Check Request to LR but does not receive a Connectivity Check Response, a typical behavior would be to repeat the check attempts a number of times before further recovery action agreed in the sharing agreement is invoked. In case the LC failure management detects system operation malfunction or outage, it initiates actions, e.g., inform Incumbent or LSA Licensee, generate alarm message for LSA system management, change LSA Spectrum resource availability information and trigger emergency evacuation to guarantee the Incumbent and the LSA Licensee protections. The Incumbent and the LSA Licensee protections are considered to be specified in the Sharing Framework and/or Sharing Arrangement including the case of LSA system failures.

An essential part for allowing the coexistence between the LSA network and the Incumbent is to define a criterion and algorithms which guarantees an interference-free operation of the LSA Licensee and Incumbent transmissions. In the reference LC implementations two basic algorithms have been used. *Minimum Separation Distance* (MSD) protection algorithm calculates the minimum required distance between the Incumbent and the LSA transmitter taking into account both the Incumbent and Licensee radio transmission parameters and in particular the cell sector antenna configuration, such as direction and down tilt angels, to calculate the MSDs to specific geographical directions. The MSD Incumbent protection methodology for the different Program Making and Special Event (PMSE) use case protections, such as Cordless camera, a Mobile video link and a Portable video corresponding to the worst case scenarios, is presented in [11].

However, since the MBB Network (MN) is an interference-limited system where multiple spatially separated BSs are transmitting simultaneously on the same frequency band, the aggregate field strength created by the MN at the Incumbent receiver can result in intolerable interference. Therefore, more advanced protection criteria is needed like, e.g., the Protection Zone Optimization (PZO) algorithm [12]. Even if the MSDs of all individual BSs are satisfied the interference created by the MN can be higher than allowed, resulting in MSD shorter than MSD of any single LSA transmitter, that is, the aggregate interference from all BSs of the network can exceed the protection zone limit even if none of the BSs exceeds it alone. This limit is defined by the Incumbent receiver sensitivity, noise floor, and additional interference margin. The PZO method computes the cumulative interference created by the MN. Specifically, convex optimization methods and accurate propagation modeling could be used to determine the individual cells which are required to be re-configured so that the resulting aggregate field strength at the Incumbent receiver remains below the Protection zone limit. This allows the MNO to operate its network at full viable capacity while satisfying the criteria for interference-free operation of the co-existing Incumbent. The LC algorithm outputs two lists: 1) BS cells which experience interference and should be re-configured if sectors are active and 2) cells that are not interfering with at least one of the Incumbent users and are possible candidates for re-configuration/activation. However, a cell can be activated only if the same cell is not included to the other Incumbents' lists and the sector is currently off air.

Apart from the two new logical elements complementing the LTE MBB network and their interfaces, no change is needed to the existing LTE network consisting of User Equipments (UE), eNBs, evolved Packet Core (ePC) and OSS NMS as shown in Figure 6.3. On the contrary, several existing LTE and LTE-Advanced (LTE-A) technologies could be leveraged in providing a solid base for the implementation of the additional features required for the LSA system work flow optimization in activation, operation and de-activation phases as shown in Figure 6.4 [13].

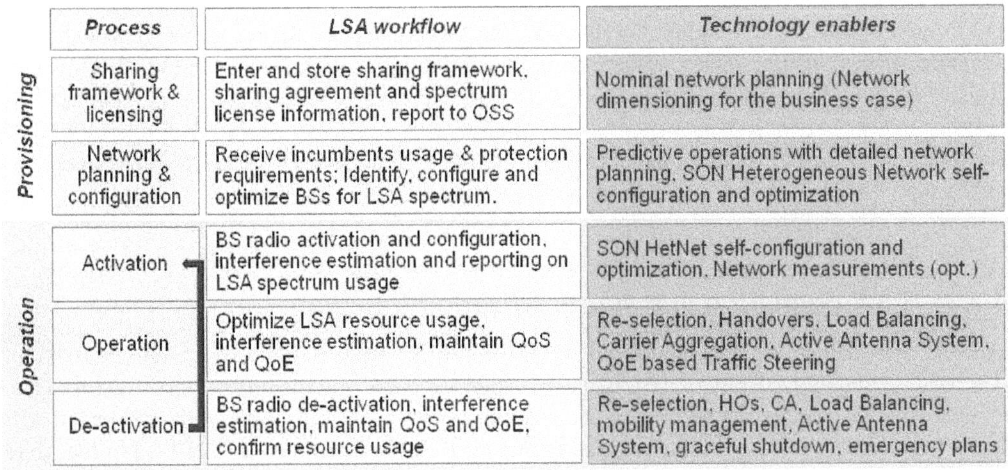

	Process	LSA workflow	Technology enablers
Provisioning	Sharing framework & licensing	Enter and store sharing framework, sharing agreement and spectrum license information, report to OSS	Nominal network planning (Network dimensioning for the business case)
	Network planning & configuration	Receive incumbents usage & protection requirements; Identify, configure and optimize BSs for LSA spectrum.	Predictive operations with detailed network planning, SON Heterogeneous Network self-configuration and optimization
Operation	Activation	BS radio activation and configuration, interference estimation and reporting on LSA spectrum usage	SON HetNet self-configuration and optimization. Network measurements (opt.)
	Operation	Optimize LSA resource usage, interference estimation, maintain QoS and QoE	Re-selection, Handovers, Load Balancing, Carrier Aggregation, Active Antenna System, QoE based Traffic Steering
	De-activation	BS radio de-activation, interference estimation, maintain QoS and QoE, confirm resource usage	Re-selection, HOs, CA, Load Balancing, mobility management, Active Antenna System, graceful shutdown, emergency plans

Figure 6.4: the LSA work flow and key technology enablers.

The RAN comprises complex combinations of radio cells, frequencies, technologies and layers that require smart network management and optimization. SON features automates the configuration, healing and optimization of such networks. By automating the management of HetNets, SON enhances their interworking and mobility. With tools to manage and interoperate multiple layers and technologies, SON ensures small cells interwork with the macro layer, even in a multivendor environment. Other key SON functions related to LSA reference implementation are load balancing, traffic steering and mobility management. The Load balancing [14] is LTE SON self optimization feature, allowing monitored and controlled terminals to switch between, e.g., the FDD-LTE and the LSA TD-LTE networks on demand. Load balancing aims to even out the load generated across the network by moving users from one cell to another in order to improve QoS. LSA enabled BSs can be used as an additional capacity layer, providing more capacity to balance the load and optimized connectivity experience for users. The nature of LSA spectrum availability leads to considerations on which user segments can be best served and are least affected by possible evacuation. Traffic steering directs traffic to a particular RAT or layer to enable operators to optimize their resources, improve the Quality of Experience (QoE) services and additionally minimize power consumption. Traffic steering works hand-in-hand with mobility management to ensure a reasonable number of handovers and eliminate radio link failures (RLF). It also considers other factors such as the capabilities of the terminals and network and the load in different RATs and layers. Today, most network operating processes are well established and it can be hard to identify the right time and place to intervene by starting to implement automation in order to raise efficiency and reduce complexity. Process integrated SON enables SON support for operator process sub-entities like site creation and LSA [15].

Consistent QoS and QoE of end users is one of the key requirement for any practical implementation of the LSA system in case LSA spectrum resource availability changes abruptly. The users connected to interfering evacuated cells will experience a RLF if the cells are locked abruptly via hard shutdown. In order to reduce the number of RLF, the shutdown or the modification of the Tx power and or antenna downtilt could be done during a certain period, allowing users to be handed over to other cells via *Graceful Shutdown*. In the case of HetNet implementation, the MNO should consider which alternative network layer and resource to use for the back off handovers. For example, the alternative network resource could be of lower capacity or it could be congested, leading to lowered QoS for the end users after the LSA evacuation process. The LTE-A Carrier Aggregation (CA) feature [16] could be utilized proactively to combine a LSA carrier to a carrier on another licensed band at the device side to increase the end user rates across the cell coverage area and to smooth potential transitions. In this way, the MNO can use LSA resources to provide additional capacity to its users, without the risk of connection break caused by changing LSA resource availability. Supplemental Downlink (SDL), as a special case of CA, allows leveraging the LSA resource to boost down link capacity in order to cope with increasing downlink uplink asymmetry in MBB networks. The integration scenario of the LC with the MNO's OAM is illustrated in the Figure 6.5.

6.2.2   LSA regulation and standardization timeline

Efficient and scalable implementation of the Incumbent protection in the LSA system introduces new requirements for information exchange between the NRA, the Incumbents and the Licensee's mobile network, which calls regulation and standardization. This concept was initially called ASA (authorized shared access), and introduced within CEPT by Qualcomm early 2011. In order to response to growing industry interest in spectrum sharing the EC RSPG initially introduced the LSA concept November 2011 in their assessment of different sharing concepts [17]. In 2012 the EC issued a standardization mandate to ETSI for Reconfigurable Radio Systems (RRS) [18] and more specifically requested RSPG opinion on the spectrum regulations and economic aspects of the LSA

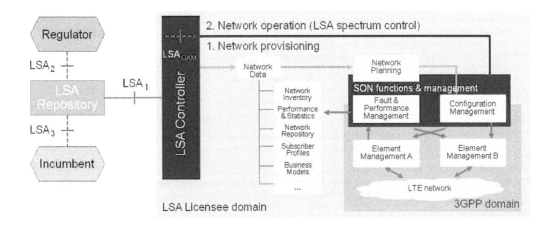

Figure 6.5: The LSA Controller integration with network's OAM.

[19]. The opinion published in November 2013 defined the LSA concept through describing main features and implementation options [20]. CEPT defined LSA as general regulatory framework and assessed it related to the current regulatory practices on the use of the 2.3GHz spectrum band [21]. An ECC Decision on the harmonized technical conditions [22] and an ECC Recommendation on the cross-border coordination in the 2.3 GHz band [23] appeared in 2014. Next mandated by the EC in April 2014 [21] the European Conference of Postal and Telecommunications Administrations (CEPT) started to develop harmonized technical conditions and guidelines for the sharing framework at the 2.3GHz band and published reports in 2015 discussing Incumbent usage cases on the band and related trial implementation examples [24], followed by more focused study on the PMSE use case and sharing framework guidelines for NRAs [25].

Even though spectrum sharing and LSA is a national matter, global spectrum harmonization is essential antecedent for any radio innovation to scale and succeed. International Telecommunication Union Radio communication sector (ITU-R) works on international agreements and recommendations defining allocation of spectrum to different services. In order to define needed technical parameters and coordination ITU-R conducts sharing studies. 2014 published studies has recognized LSA as a possible cognitive radio solution for the vertical sharing [26], future trend for the IMT systems [27] and as best practice and innovative regulatory tool for the shared use of spectrum [28].

In parallel with above regulatory actions ETSI RRS according to the above discussed EC mandate from 2011 has worked on more detailed requirements and architecture to ensure interoperability, harmonization and scale for the concept. In July 2013 report [8] the LSA concept was introduced in high level with the primary use case, operational features, functions and performance requirements. Followed by high level functional and performance system requirements for the mobile broadband systems on the band [9] concluding the standardization stage 1 (*Requirements*) in 2014. At stage 2 (*Architecture + Interfaces + High Level Procedures*) ETSI RRS worked on the architecture reference model, more detailed functional descriptions and information flows between the system elements [10] succeeded by a liaison statement to the Third Generation Partnership Project (3GPP) Service and System Aspects Telecom Management working group (3GPP SA5) in April 2015 [29] with a study item (SI) on LSA. At present the ETSI RRS is working on the information elements and protocols for the interface between LC and LR. The SA5 is cooperating with the ETSI RRS in order to identify how the solution and architecture in [10] based on [9] requirements may provide a global solution also supported by the 3GPP network management architecture defined in [30]. The SI also analyzed the LSA functionalities and the information flow in [10] and studied the impacts on network internal interfaces as shown in Figure 6.2 and Figure 6.5. supporting both static and semi-static spectrum sharing scenarios and reference use case as defined by the ETSI RRS.

6.3   LSA field trials

The LSA is being considered in several research and trial projects, e.g., the European projects CORE+ [31], CoMoRa [32] and METIS [33] as well as NRA initiated proof of concepts. The *Finnish Cognitive Radio Trial Environment* (CORE) program with its unique end-to-end ecosystem consortium is to the authors' knowledge the only existing live field trial environment of LSA. The CORE program has actively contributed to LSA and 5G development especially regarding regulation and standardization [34] and [35], technical field trials [36] and [37] and technology [38], [39], [40] and [41]. The CORE program has been at the cutting edge of the development of LSA by providing future business scenarios[42], identifying opportunities [43] and [44] and exploring business models [45] and [46] for the key stakeholders in the LSA evolution. The LSA concept has been field trialed the first time by CORE consortium in April 2013 [36] followed by iteratively updated features demonstrated in April 2014 [47] and November 2014 [48]. The CORE+ LSA trial environment consists of the following key elements as shown in Figure 6.6:

- Commercial available heterogeneous LTE-A network of TDD and FD LTE macro and small cell evolved Node B (eNB) Base Stations (BSs), Evolved Packet Core (EPC) core network, network management system (NMS) and end user equipments (UEs),

- PMSE incumbent spectrum users with LR and the Incumbent Spectrum Manager,

- LC research platform based on cognitive engine

- LC utilizing commercially available OSS NMS and SON platforms and interfaces with incumbent protection algorithms and SON features.

Incumbent spectrum user in the trial is selected according to the national Finnish LSA use case to be the employees of a media or broadcasting company using PMSE services in program making

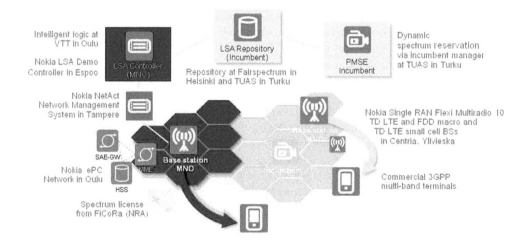

Figure 6.6: The Finnish LSA CORE+ trial environment.

on the 3GPP band 40 (2.3-2.4 GHz), as defined in [49] and [11]. CEPT has used the Finnish LSA trial system as an example of the technical reference implementation of the LSA concept [24]. In May 2015 LSA trial the CORE environment was further enhanced by introducing first time LSA controller implemented as SON solution fully integrated into commercial OSS with advanced MSD and PZO algorithms needed to optimize protection zones to protect the Incumbent's business while maximizing availability for the Licensee. Performance validation was conducted by measuring the duration of the spectrum evacuation workflow steps in releasing the LSA band due to Incumbent's immediate spectrum resource availability notification. The measured average end-to end evacuation time of 51 seconds revealed that the evacuation operation can be done in a way that fulfills typical PMSE service incumbent's requirements in the Finnish sharing use case and wider in a static and a semi-static LSA use cases. Comparing results to previous research platform based LC demonstrations OSS integrated LSA controller reduced overall LC operations delay approximately 85% [12].

In the *CoMoRa* project real time LSA RAN emulation and traffic simulation were done and the Season LSA Software Emulation tool with e2e radio HW lab environment used to demonstrate base technologies enabling spectrum sharing, e.g., Carrier Aggregation [32]. Demonstration was showcased at the GSMA Mobile World Congress (MWC) 2013 and 2014 by Nokia, Qualcomm and Intel [50]. The EU FP7 *METIS* project has focused on future spectrum needs and usage principles towards 5G considering spectrum sharing scenarios, their technical enablers and potential impact to mobile ecosystem. Spectrum sharing toolbox with general functional architecture applicable for the LSA regulation was introduced [33]. In the Feb 2015 at MWC *RED technologies* demonstrate LSA options facilitating sharing between Mobile Services and PMSE video links both using the 2.3 GHz band. The demonstration was based on the study on the coexistence of PMSE video links with Mobile Services done by the Spanish regulator SETSI in collaboration with RED Technologies and reported to the latest CEPT FM PT 52 technical work on LSA [51]. The *Joint Research Centre* (JRC), the in-house science service of the EC, has conducted simulation studies on LSA usage scenarios to assess the QoS as seen by the end-user under different wireless technologies. In their demo the traffic data was simulated in MATLAB and UEs in LabView [52]. Furthermore, a regulatory pilot for the LSA was launched in Rome in 2015 by the Italian Ministry for Economic Development and the Joint Research Centre of the European Commission to validate the technical conditions for spectrum sharing in the 2.3 GHz band. Following the CEPT approach to LSA, the pilot has defined a sharing framework for LSA between variety of the incumbents (PMSE, fixed links and government users) and the mobile broadband [53].

6.4 Spectrum sharing economy - Business model design in Licensed Shared Access

Business model can be defined as a vehicle that is built to explore and exploit a business opportunity [54], [55] and connecting the firm with its external business environment, customers, competitors, and society [56]. To remain competitive, firms must continuously develop and reinvent their business models in order to create and capture value by and from their business activities. Wirtz et al. [57] presented four business models of the web 2.0 era, i.e., the Connection, Content, Context and Commerce business models that are relevant also when providing MBB with LSA:

- At the connection layer the service provider offers connectivity to one or several networks,

- At the content layer the service provider offers any content the customer should want or need,

- At context layer the service provider offers information about alternative connections, content, context services and commerce platforms available, and

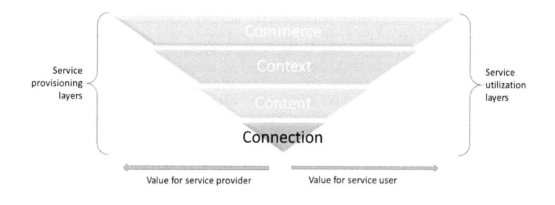

Figure 6.7: The layered 4C business model framework for MBB with LSA.

- At the commerce layer the service provider offers all stakeholders a platform for trading alternative connectivity solutions, content or context information.

These four models can be seen as layered (4C model) and forming an ecosystem where the lower level business models serve as enablers and value levers for the higher layers as depicted in Figure 6.7. Each of these business models can be offered alone or as bundled, and the business potential of the whole ecosystem depends on the ecosystem players' synergies when providing their services.

In the development of new spectrum sharing concepts, it is essential early on to consider the underlying business opportunities that are attractive and feasible for all the key stakeholders, thus bringing our attention to value co-creation and co-capture in sharing economy based business models. Spectrum sharing confront the wireless industry with increasing strategic environmental changes, such as emerging competitive market structures, policy and regulatory changes as well as technology progress and complexity, which all require companies to adapt or reinvent one or more aspects of their capabilities, competitive advantages and business models. In the following we examine how business models and related ecosystem roles could evolve, grow and scale in response to novel spectrum sharing models, LSA specifically.

6.4.1 LSA business model scalability

Business model scalability has been shown to be the primary factor for the venture growth [58] and attractiveness towards investors [59]. Chrisman [60] identified and categorized the antecedents of business model scalability into five mutually exclusive factors and Stampfl further defined in the explorative business model scalability model [61].

The emerging sharing economy concept has leveraged these scalability factors with focus on resource efficiency. Stephany [62] has recently defined *Sharing Economy* as "*the value in taking the underutilized assets and making them accessible online to a community, leading to a reduced need for ownership of those assets.*" Utilization of the concept has lately evolved from collaborative individual peer-to-peer community consumption to corporations and governments participating the ecosystem as buyers, sellers or lenders [63]. What characterizes the sharing economy business opportunities is that the opportunities can be seen as two-sided, i.e., simultaneously related to the provisioning and utilization of resources. Proposed scalability antecedent factors within sharing economy used in assessing business opportunities and business model design characteristics of the LSA for future mobile broadband are:

a) *Technology*: scalability of technical infrastructure, automation of processes; and platform for online accessibility

b) *Cost structure*: superior value proposition with low initial costs; and reduced need for the ownership

c) *Revenue structure*: generate sustainable and continuous revenue early; leveraging underutilized assets

d) *Adaptability to different legal, regulatory and policy regimes*: have potential to create strong barrier in terms of scalability.

e) *Network externalities*: Network effect 'lock-ins' and critical mass; utilizing communities

f) *Value and user orientation*: Uniqueness and mission critical user driven 'need pull'. Simplicity of the offer to solve real problem built around existence of user knowledge

Next, we analyze and compare the LSA spectrum sharing primary use case and the model introduced in the LSA system concept section against the business model scalability and sharing economy criteria discussed above. Key LSA enabling and framing business model features are summarized in Table 6.1.

a) Technology scalability and accessibility

Global LTE ecosystem with its scale and harmonization will be the key technology scalability factor for LSA. Users in the LSA system could start using LTE RAN for the targeted spectrum band and OSS with off-the-shelf technology. In the LSA model MNOs are able to fully utilize existing MBB infrastructure and network management assets considering LC as added SON functionality to network OSS system. In the convergence of technologies and businesses, unified technology platforms and off-the-shelf technologies capable for offering versatile services, enable the exploitation of new collaborative (two-sided) opportunities between the MNO and commercial/ governmental incumbents, e.g., through utilizing LTE for broadcasting/PMSE, public safety or defence. In the LSA system where spectrum resource control via the LC is inside the MNO domain, diffusion towards Cognitive Networks (CN) in large could also be retained within MNOs control. In the LSA evolution, managing a increasing volume of dynamic transactions, online accessible spectrum management services based on big data analytics capabilities could become a competitive advantage. Focusing on high-density capacity areas and small cell layer LSA Licensees could utilize their fixed optical infrastructure assets in urban small cell sites and backhauling. On a longer term, the LSA concept has potential to reduce the need for parallel network infrastructure when spectrum and infra are tradable and shared.

b) Cost structure and reduced need for the ownership

For an MNO, the LSA model offers access to lower cost spectrum without coverage obligations when and where needed, with QoS secured by traditional exclusive licensing based model. At the same time, for a greenfield, or a new challenger operator, the related up-front lump-sum spectrum license payment combined with needed new infrastructure continues to set an entry barrier. Utilizing extra capacity, established MNOs could create differentiated value propositions around QoS and QoE. In the future the spectrum sharing regulatory approach has potential to unbundle investments in spectrum, network infrastructure and services. Faster access to low cost spectrum with lower initial investments enable local 'pro-competitive' deployments and could further expand sharing mechanism for pooling spectrum and infrastructure resources between operators. Incumbents are not essentially in need of new LSA spectrum resources, but the cost pressures faced by both the commercial and governmental incumbents increase their need for efficiency and induce sharing in many forms. Incumbents may seek internal efficiency, seek to share infrastructure, or seek to utilize alternative commercial technologies. By allowing sharing, incumbents could continue their operations in the spectrum and in case of a governmental incumbent fulfill their obligations defined

Table 6.1   LSA Spectrum sharing business model scalability factors

Antecedents	LSA business model features
a) Technology enablers: scalability of technical infrastructure and automation of processes; *Platform for online accessibility*	• Fully utilize existing MBB infrastructure and network management assets and scale of the LTE ecosystem • Unified technology platform and off-the-shelf technologies for new converged services: media, public safety, defence • Simple LSA repository function • LSA controller as added SON functionality to network OSS system • Heterogenous infrastructure and MNO driven diffusion to CR • Utilization of big data analytics capabilities to manage higher complexity with dynamic sharing • Reduces on a longer term the need for parallel network infrastructure when spectrum and infra are tradable and shared
b) Cost structure: superior value proposition with low initial costs; *Reduced need for the ownership*	Faster access to lower cost spectrum without coverage obligations – when and where needed • Based on traditional exlusive licensing model with up front lump sum payment • Build on existing MNO infrastructure with radio upgrades • Unbundles investment in spectrum, network infrastructure and services • Enables local 'pro-competitive' deployments • Expands sharing mechanism for pooling spectrum and infrastructure resources • Transaction cost related to sharing framework and system
c) Revenue structure: generate sustainable and continuous revenue early; *Leveraging underutilized assets*	MNO model as is with differentiation possibilites through extra data capacity and higher speed (enables QoS and QoE pricing) • Capacity wholesale service opportunity • Spectrum and small cell hosted solution as a service • Advertisement & transaction based models • Introduces new roles with LSA system evolution (repositories & controllers) • Incumbent incentives and preserve future use of spectrum asset
d) Adaptability to different legal, regulatory and policy regimes	Legal certainty/stability and security with existing regulatory framework • Uncertainty with long term availability of the spectrum • Need regulation and standardization with Incumbent ecosystem • 'Pro-competition' with lower entry barrier Intial European focus– need adaptability to other regimes, e.g., the US CBRS
e) Network externalities: Network effect 'lock-ins' and critical mass; *Utilizing communities*	Utilizing existing 3GPP ecosystem scale and harmonization • MNOs consumer ownership through connectivity • Small cell ecosystem introduce new players • With evolution towards LSA phase 2 and CBRS extends to internet 'innovation' ecosystems with consumer and customer data ownership on apps and services
f) *Value* and user value orientation: Uniqueness and mission critical user driven 'need pull'. Simplicity of the offer to solve real problem built around existence of user knowledge	Additional quality data capacity to serve customers with improved QoS and QoE • MNOs' existing customer billing relationship and data • Flexible regulatory framework allows faster efficient access to new systems and services • Wider and faster access to media and internet services • Enables serving heterogeneous customer profiles • Local new business models and services with evolution

by the society with minimum additional investment. The discussed more dynamic market mechanisms, needed security and added flexibility might increase overall system complexity and could potentially increase transaction costs.

c) Revenue structure and underutilized assets

The LSA model deployment enables MNOs to timely respond to the increasing MBB traffic with QoS and QoE pricing differentiation through enhanced data capacity and high speed. Furthermore, possible extra spectrum resource opens up a capacity wholesale service opportunity for the converging ICT and media industries. In addition to offloading and nomadic WiFi type of internet access services, in particular to dense urban environment hot spots, new business model designs and revenue structures could emerge around spectrum and small cell solutions. E.g., underutilized spectrum resources and local utilities could be bundled as hosted solution-as-a-service, advertisement & transaction based models and enabling new vertical segments in Internet of Things (IoT). Also, incumbents could become interested to take up a new role by starting to offer, e.g., mobile services or bundling of service offerings by utilizing underutilized assets themselves as allowed by regulation. LSA evolution with higher complexity introduces new independent or integrated roles to the ecosystem, e.g., related to repository and controller service provisioning. In the initial LSA model LRs act as a basic databases with support the entry and storage of information and conveys availability information to authorized LCs creating value for the whole ecosystem but with limited capabilities to capture the value, monetize it. Whereas the LSA evolution with spectrum pooling and towards the US CBRS [64] system with its enhanced dynamics enables new roles with value creation and value capture potential. In addition to basic spectrum availability information evolved repository could offer value added interference mitigation services and facilitate spectrum aggregation and brokerage market place.

d) Adaptability to different legal regimes

The LSA offers predictability and security with existing regulatory framework and relatively high administrative burden which on one hand protects the turf for established players but on the other hand still continues to limit the scalability as an entry barrier. Scale and global harmonization have been key enabling factors for the successful 3GPP ecosystem growth from GSM to LTE. In spite of the technology synergies, the scalability of LSA could become limited due to fragmented national incumbent use cases and related regulatory differences. In the LSA model the detailed procedure of defining a sharing framework has to be defined at national level in Europe and needs adaptability to other authorization regimes. Regulatory and standardization actions needed with Incumbents' ecosystem will potentially further limit the scalability. In particular in public and governmental services, the value of spectrum is a political decision. With LSA evolution towards higher dynamism and with regulatory 'pro competition' stances of NRAs, LSA is targeted at lowering administrative burden and entry barriers particularly for the challenger MNOs and novel types of operators.

e) Network externalities and communities

In network externalities LSA business model design represents a co-opetitive (simultaneous competition and collaboration) situation between MBB and the incumbent as well as with novel operator types, e.g., from the Internet domain. Traditional MNOs deploying additional LSA spectrum resources could utilize the existing 3GPP ecosystem scale, harmonization and customer base to achieve direct network effect, critical mass and use existing consumer ownership on connectivity for customer lock-in. New LSA operators could leverage Internet 'innovation' ecosystems and consumer and customer data ownership on apps and services in creating their critical mass and network effects. The switching cost between the service provider and the customer could be widened in relation to other stakeholders. E.g., in the dense urban LSA small cell deployment there are many new critical capabilities and assets from legal and real estate aspects to radio planning and site cam-ouflaging due to the fact that small cells will attach to structures and building not owned by the traditional MNO. To seize related new opportunities, technology vendors and MNOs could form partnerships with various specialist companies like infrastructure owners and providers, real estate and street furniture owners, utility service companies and backhaul providers. The initial binary LSA regulatory framework do not fully meet the community criteria of the sharing economy as it could be more defined as discussed above an ecosystem with supply and demand. Further evolution in LSA with possible spectrum pooling between MNOs and the US CBRS model with opportunis-tic *license-by-rule* General Authorized Access (GAA) third layer will expand the concept further towards community and furthermore increase the scalability.

f) Value and user orientation

MNO access to excess spectrum resource facilitated by the deployment of the LTE-A and SON technologies could lead to improved QoS, QoE and strengthened customer satisfaction. Offering could fulfill existing need pull with familiar services and simplicity of the offer built around ex-isting customer data and Customer Experience Management (CEM) systems. The more flexible regulatory approach of the LSA opens up potential to innovate new unique local business model designs, application and services. For MNOs it enables serving heterogeneous customer profiles and novel operator types allow faster efficient access to new systems and services. Internet players could build their unique offering around their extensive user knowledge. For incumbents the LSA is an opportunity to show their cooperative attitude and societal responsibility by improving the efficiency of spectrum use and offering new secure critical services for new kind of partners, for example within healthcare or energy sector, provided that regulatory schemes allow it.

6.4.2   LSA business model structure

An alternative, complementary perspective to the 4C business model categorization discussed earlier in the texts, is provided by Messerschmitt & Szyperski [65], who discussed ICT ecosystems and presented a model of the roles within ICT businesses. There are two basic ways how to approach the ICT ecosystems as a concept. The traditional approach in which an ecosystem is based on technical infrastructure, a platform, to which other players of the ecosystem integrate. This technical approach is widely used but captures only one side of the coin. The other approach is from business perspective in which an ecosystem could be defined to consist of synergistic business models that enable simultaneous value creation and capture through the business models employed within that ecosystem. To root this to the Messerschmitt & Szyperski's logic, the value creation can be seen across the technology life cycle that has the following phases: research, technology, product, system and service. At the same time, the value capture dimension is attempted through applications or infrastructure provided by the players as presented in Figure 6.8.

The business models employed with the ICT sector can be grouped into three generic busi-ness model categories. The first traditional generic business model is the *vertical business model* employed, for example, by most infrastructure and technology providers. These companies believe that to be competitive they need to create value for their customers, thereby living in a "value cre-ation economy" and being trapped inside their own selected verticals. The second traditional generic business model is the *horizontal business model* employed by most service-oriented and consumer business companies. These companies believe that to be competitive they need to serve and hook a wide clientele and reach across different segments - and try to capture as much value from their customers as possible. This is why for example MNOs pay so much attention to Average Revenue Per User (ARPU) as a measure of their success. These companies live in a "value capture economy" where their task is to milk the customers and defend their existing position against competition, thus becoming extremely cost-aware and not-so-innovative as they used to be at earlier phases of

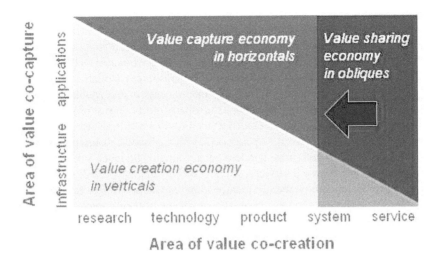

Figure 6.8: Business model design evolution towards oblique business model framework.

their development. The third recently emerged generic business model is what we call the *oblique business model,* employed by fast-growing and service-oriented companies that utilize the resources of third parties in their business and thus seek two-sided business opportunities. Many apps and web service providers has used this *sharing economy* based strategy to enter the market. Apple's iPod was among the first ones to create an oblique business model by basically combining memory stick (product) to content (service) distributed to masses: cheap hardware with very versatile content, bypassing completely the more old-fashioned music distribution logic employed by the music industry. However, with the emergence of the sharing economy concept, where resource efficiency plays a crucial role, the oblique business models really started to thrive. These companies live in the value sharing economy and turn an ecosystem's underutilized assets to a more efficient or better use - thus generating themselves revenue by that means. The number of oblique business models is increasing fast, transforming and converging whole industries, winning market share, and jeopardizing the established or Incumbent companies' horizontal and vertical business models.

In sharing economy framework shared spectrum assets are no longer being sought after only by the established MNOs. The earlier categorization of connection, content, context and commerce business models is becoming blurred or fuzzy at the firm level, as companies seek bundled or hybrid business models that combine or aggregate services from different layers of the 4C model. Also, with the introduction of more dynamic localized sharing approaches nontraditional players like utilities, railways, private enterprises and service companies are now getting into the spectrum fora, considering novel hybrid business models and ecosystem roles to strengthen the core of their business model. Unbundling investment in spectrum resource, network infrastructure and services creates new opportunities related to context and commerce of the spectrum asset. In particular with higher frequencies like IMT bands on 2.3 GHz and 3.5 GHz the key focus will be on dense urban area and in-building coverage. This is a transformational change to MNOs with a vast increase in radios and locations which are sited in venues traditionally not owned or controlled by the operator. New 'as-a-service' business models are emerging where investments can be efficiently shared across multiple providers, avoiding a long term high upfront parallel network infrastructure investments and wasteful duplication. To date hosted *Small Cell as a Service* (SCaaS) model proof of concepts have focused on particular parts of the existing value chain and their combinations: antennas, radios, core network, electricity, backhaul, site acquisitions, site ownership, leveraging existing asset ownership, to deliver cost savings. LSA will introduce a needed spectrum resource to complement these models and enable them to scale and utilize sharing economy business model innovations. Novel use cases like: connected venues, enterprises and routes, each with differing requirements, could include coordination of deployments through a neutral governance model with shared operations combining licensed, unlicensed and shared spectrum resources. Collaboration along ecosystem and community is essential in making SCaaS solutions successful and enables SCaaS operators to emerge from different angles: a venue owner or a third-party utility service provider, e.g., companies with attachment rights, fixed & cable Internet Service Providers (ISPs), tower companies, advertising agencies or Mobile Virtual Network Operators (MVNOs). Small cell equipment vendors coming from a mobile broadband or enterprise wireless background could enter leveraging their expertise in system integration and managed services, while telecom vendors with e2e offering will be in the best position to provide operating, management, and maintenance services for the technology. Additionally major infrastructure vendors could take advantage of their complete e2e HetNet product and service portfolio and customer intimacy build on outsourcing and managed services to facilitate their hosted SCaaS offerings enhanced with shared spectrum.

6.5 Conclusions

Mobile broadband networks are facing a tremendous increase in data traffic volumes through growth of wireless services with a wide range of diverse devices and applications. In order to meet this need, large amounts of spectrum will be a key prerequisite for any wireless network evolution. To satisfy the demand, MNOs will need new spectrum allocations on the one hand and, on the other, ways of utilizing spectrum more efficiently. Spectrum sharing techniques can be used to optimize spectrum utilization in the heterogeneous networks and, more importantly, to provide opportunities for network operators to access additional spectrum, which is typically allocated to other radio services and thus not available via traditional exclusive licensing. This way different spectrum sharing options are complementing existing network capacity based on exclusively licensed and license-exempt spectrum.

The role of shared spectrum is likely to increase in the future, as the new means to respond to the growing traffic demand in a scalable and timely fashion. The novel LSA regulatory framework enables the LSA Licensee to gain faster access to licensed and QoS guaranteed local spectrum with lower costs and without lengthy and costly traditional re-farming process while guaranteeing the Incumbent spectrum users' rights. On the other hand improved spectrum use efficiency via sharing allows the Incumbent to continue the use of spectrum for current service with possible incentives from sharing. Furthermore the LSA framework concept can strengthen the global spectrum regulatory harmonization efforts and the ECC harmonization measure by taking harmonized IMT spectrum assets into mobile broadband use by sharing with existing non-MBB Incumbent users. In Europe, only very few countries would be able to open access to the 3GPP band 40 without the LSA resulting in low interest for major MNOs to invest in networks and further technology vendors to manufacture Europe region handsets for the spectrum.

This chapter has identified the key technical, administrative and role based functions of the LSA concept and considered in detail the architecture and key technology enablers. The LSA system concept can be put into action utilizing existing 3GPP LTE networks with feasible and practical modifications to the OSS infrastructure and regulatory framework. Moreover, the system concept has been demonstrated in over the air field trials and incorporated into the European regulatory and standardization framework. LSA standardization effort is ongoing and would be leveraged in the recently started parallel the US three tier CBRS regulatory model in which, optimally the LSA could be fully aligned with the two highest tiers, Incumbent Access and Priority Access. The LSA framework was shown to offer scalable business opportunities for the key stakeholders utilizing sharing economy antecedents with novel "oblique" business model designs. Future research directions include an analysis of the impact of the dense urban small cell environment on spectrum sharing and alignment and co-evolution of the European LSA and the US CBRS concepts toward the practical implementation.

6.6 Acknowledgement

This work is supported by Tekes – the Finnish Funding Agency for Technology and Innovation in 5thGear programme. The authors would like to acknowledge the CORE++ project consortium.

References

[1] Cisco white paper, Cisco Visual Networking Index: Global Mobile Data Traffic Forecast Update, 2014–2019. [Online]. Available: https://www.cisco.com/c/en/us/solutions/collateral/service-provider/visual-networking-index-vni/white_paper_c11-520862.pdf, Feb. 2015.

[2] The White House, Realizing the Full Potential of Government-Held Spectrum to Spur Economic Growth, President's Council of Advisors on Science and Technology (PCAST) Report, July 2012.

[3] RSPG Opinion on Licensed Shared Access. RSPG13-538, Radio Spectrum Policy Group, November 2013.

[4] FCC, White Spaces, [Online] http://www.fcc.gov/topic/white-space

[5] Ofcom, TV White Spaces Pilot. [Online]. Available: http://www.fcc.gov/topic/white-space

[6] ECC, Licensed Shared Access (LSA), ECC Report 205, Feb. 2014.

[7] FCC, Report and order and second further notice of proposed rulemaking (in the 35GHz band). [Online]. Available: http://transition.fcc.gov/Daily_Releases/Daily_Business/2015/db0421/FCC-15-47A1.pdf, Apr 2015.

[8] ETSI, Mobile Broadband services in the 2300-2400 MHz frequency band under Licensed Shared Access regime. ETSI TR 103.113 v 1.1.1, July 2013.

[9] ETSI, System requirements for operation of Mobile Broadband Systems in the 2300 MHz -2400 MHz band under LSA. ETSI TS 103 154, Sept. 2014.

[10] ETSI, System Architecture and High Level Procedures for operation of Licensed Shared Access (LSA) in the 2300 MHz-2400 MHz band. TS 103 235 v 0.0.9, Apr. 2015.

[11] CEPT, "Broadband Wireless Systems Usage in 2300-2400 MHz," ECC Report 172, March 2012.

[12] S.Yrjölä et al., "Licensed Shared Access (LSA) field trial using LTE network and Self Organized Network LSA Controller," WInnComm-Europe, Oct. 2015, unpublished.

[13] M. Mustonen, Tao Chen, H. Saarnisaari, M. Matinmikko, S. Yrjola and M. Palola, "Cellular architecture enhancement for supporting the European licensed shared access concept," IEEE Wireless Commun., vol. 21, no. 3, pp. 37–43, 2014.

[14] 3GPP, "SON Policy and Optimization Function Definitions," TS 32.522 V11.7.0, Sept. 2013.

[15] Nokia whitepaper, "Intelligent Self Organizing Networks (iSON)," [Online]. Available: http://networks.nokia.com/sites/default/files/document/nokia_ison_white_paper.pdf

[16] 3GPP technical report, "Evolved Universal Terrestrial Radio Access (E-UTRA); Carrier Aggregation; Base Station (BS) radio transmission and reception," TR 36.808, June 2012.

[17] RSPG, "Report on Collective Use of Spectrum (CUS) and other spectrum sharing approaches", RSPG11-392, European Commission, Radio Spectrum Policy Group, Nov. 2011.

[18] EC, "Standardization Mandate to CEN, CENELEC and ETSI for Reconfigurable Radio Systems," M/512, Nov. 2012.

[19] EC, "Request for Opinion on Licensed Shared Access (LSA)," RSPG12-424, Nov. 2012.

[20] RSPG, "RSPG Opinion on Licensed Shared Access," RSPG13-538, Nov. 2013.

[21] EC, "Mandate to CEPT to develop harmonized technical conditions for the 2300-2400MHz ('2.3GHz') frequency band in the EU for the provision of wireless broadband electronic communication services," DG CONNECT/B4, Apr. 2014.

[22] CEPT, Harmonized technical and regulatory conditions for the use of the band 2300-2400 MHz for Mobile/Fixed Communications Networks (MFCN), ECC Decision (14)02, June 2014.

[23] CEPT Cross-border coordination for mobile/fixed communications networks (MFCN) and between MFCN and other systems in the frequency band 2300-2400 MHz," ECC Recommendation (14)04, May 2014.

[24] CEPT, "Technological and regulatory options facilitating sharing between Wireless broadband applications (WBB) and the relevant Incumbent services/applications in the 2.3 GHz band," CEPT Report 56, March 2015.

[25] CEPT, "Technical sharing solutions for the shared use of the 2300-2400 MHz band for WBB and PMSE", Draft CEPT Report 58, May 2015.

[26] ITU-R, "Cognitive radio systems in the land mobile service", Report M.2330, ITU-R, Nov. 2014.

[27] ITU-R, "Future technology trends of terrestrial IMT systems", Report M.2320, Nov 2014.

[28] ITU-R, "Innovative regulatory tools to support enhanced shared use of the spectrum", draft new Report ITU-R SM, June 2014.

[29] 3GPP Work Item, " 670028 (FS_OAM_LSA) Study on OAM support for Licensed Shared Access (LSA) [Rel-13]," April 2015.

[30] 3GPP TS, "Telecommunication management; Principles and high level requirement," 3GPP TS 32.101 V12.0.0 [Rel-12], Sept. 2014.

[31] CORE+; Cognitive Radio Trial Environment +. [Online]. Available: http://core.willab.fi/

[32] CoMoRa; Cognitive Mobile Radio. [Online]. Available: http://www.comora.de

[33] METIS; Mobile and wireless communications Enablers for the Twentytwenty Information Society. [Online]. Available: https://www.metis2020.com/

[34] M. Matinmikko, M. Mustonen, D. Roberson, J. Paavola, M. Höyhtyä, S. Yrjölä, and J. Röning. Overview and comparison of recent spectrum sharing approaches in regulation and research, IEEE DySPAN, McLean, VA, 1–4 April 2014.

[35] M. Mustonen, M. Matinmikko, D. Roberson & S. Yrjölä. Evaluation of recent spectrum sharing models from the regulatory point of view. 5G for Ubiquitous Connectivity (5GU) conference, Levi, Finland, 26.-28. Nov. 2014.

[36] M. Matinmikko et al., "Cognitive Radio Trial Environment: First Live Authorised Shared Access (ASA) based Spectrum Sharing Demonstration," IEEE Veh. Technol. Mag., vol. 8, no. 3, pp. 30–37, Sept. 2013.

[37] M. Palola, M. Matinmikko, J. Prokkola, M. Mustonen, M. Heikkilä, T. Kippola, S. Yrjölä, V. Hartikainen, L. Tudose, A. Kivinen, J. Paavola, and K. Heiska, "Live field trial of Licensed Shared Access (LSA) concept using LTE network in 2.3 GHz band", in the 7th IEEE Symposium on New Frontiers in Dynamic Spectrum Access Networks (DySPAN), McLean, Virginia, USA, Apr. 1st-4th, 2014.

[38] M. Mustonen, M. Matinmikko, M. Palola, S. Yrjölä, K. Horneman. An Evolution Towards Cognitive Cellular Systems: Licenced Shared Access (LSA) for Network Optimization. IEEE Communications Magazine. May 2015.

[39] M. Matinmikko, H. Okkonen, M. Palola, S. Yrjölä, P. Ahokangas & M. Mustonen. Spectrum sharing using Licensed Shared Access (LSA): The concept and its work flow for LTE-Advanced networks. IEEE Wireless Communications Magazine, vol. 21, no. 2, pp. 72–79, April 2014.

[40] M. Matinmikko, H. Okkonen, M. Palola, S. Yrjölä, P. Ahokangas & M. Mustonen. Spectrum sharing using Licensed Shared Access (LSA): The concept and its work flow for LTE-Advanced networks. IEEE Wireless Communications Magazine, vol. 21, no. 2, pp. 72–79, April 2014.

[41] M. Mustonen, M. Matinmikko, M. Palola, S. Yrjölä, J. Paavola, A. Kivinen & J. Engelberg. Considerations on the Licensed Shared Access (LSA) Architecture from the Incumbent Perspective. Conference on Cognitive Radio Oriented Wireless Networks and Communications (CrownCom) 2014, Oulu, Finland, 2-4 June 2014, pp. 150–155.

[42] P. Ahokangas, M. Matinmikko, S. Yrjölä, M. Mustonen, E. Luttinen & A. Kivimäki, "Business Scenarios for Incumbents in Licensed Shared Access (LSA)," CrownCom 2014, Oulu, Finland, 2–4 June 2014.

[43] M. Matinmikko, H. Okkonen, S. Yrjölä, P. Ahokangas, M. Mustonen, M. Palola, V. Gonçalves, A. Kivimäki, E. Luttinen & J. Kemppainen. Business benefits of Licensed Shared Access (LSA) for key stakeholders. In O. Holland, H. Bogucka & A. Medeisis (eds.) Opportunistic Spectrum Sharing and White Space Access: The Practical Reality. John Wiley & Sons, pp. 407–424. May 2015.

[44] P. Ahokangas, M. Matinmikko, S. Yrjölä, H. Okkonen & T. Casey. "Simple rules" for mobile network opera-tors' strategic choices in future cognitive spectrum sharing networks. IEEE Wireless Communications, vol. 20, no. 2, pp. 20–26, April 2013.

[45] P. Ahokangas, M. Matinmikko, I. Atkova, L. F. Minervini, S. Yrjölä & M. Mustonen, "Coopetitive Business Models in Future Mobile Broadband," 6th Workshop on Coopetition Strategy - "Coopetition Strategy and Practice", Umeå, Sweden, 22–23 May 2014.

[46] P. Ahokangas, M. Matinmikko, S. Yrjölä, M. Mustonen, H. Posti, E. Luttinen & A. Kivimäki. Business models for mobile network operators in Licensed Shared Access (LSA). IEEE DyS-PAN, McLean, VA, 1–4 April 2014.

[47] M. Palola, M. Matinmikko, J. Prokkola, M. Mustonen, M. Heikkila, T. Kippola, S. Yrjola, et al., "Live field trial of Licensed Shared Access (LSA) concept using LTE network in 2.3 GHz band," in Proc. IEEE DySPAN, Washington D.C., pp. 38–47, April 2014,

[48] ETSI RRS Plenary, [Online]. Available: http://www.etsi.org/index.php/news-events/ events/807-etsi-rrs-workshop-2014, Dec. 2014.

[49] ERC Report 38, "Handbook on Radio Equipment and Systems Video Links for ENG/OB use," May 1995.

[50] Nokia, "NSN and Qualcomm bring ASA to life, opening up new spectrum" [online]. Available: https://blog.networks.nokia.com/mobile-networks/2014/03/05/nsn-and-qualcomm-bring-asa-to-life-opening-up-new-spectrum/, Mar. 2014.

[51] CEPT Project Team FM PT 52, FM52(15)13 April 2015

[52] P. Chawdhry, "Licensed Shared Access - A new approach to efficient use of spectrum," ETSI workshop on Reconfigurable Radio Systems, [Online]. Available: http://www.etsi.org/ index.php/news-events/events/807-etsi-rrs-workshop-2014, Dec. 2014.

[53] JRC MiSE PR May 2015. Unpublished

[54] C. Zott and R. Amit, "Business model design: An activity system perspective". Long Range Planning, vol. 43, no. 2-3, pp. 216–226, 2010.

[55] R. McGrath,. "Business models: A discovery driven approach." Long Range Planning, vol. 43, no. 2-3, pp. 247–261, 2010.

[56] D. Teece, "Business models, business strategy and innovation," Long Range Planning, vol. 43, no. 2-3, pp. 172–194, 2010.

[57] B. Wirtz,, O. Schilke, and S. Ullrich, "Strategic development of business models. Implications of the Web 2.0 for creating value on the internet," Long Range Planning, vol. 43, no. 2-3, pp. 272–290, 2010.

[58] L. Berry, V. Shankar, J. Parish, S. Cadwallader and T. Dotzel, "Creating new markets through service innovation," MIT Sloan management Review, Vol. 47, No. 2, 2006, pp. 56–63.

[59] N. Franke, M. Gruber, D. Harhoff and J. Henkel, "Venture capitalists' evaluations of startup teams: trade-offs, knock-out criteria, and the impact of VC experience," Entrepreneurship: Theory & Practice, Vol. 32, No. 3, pp. 459–483, 2008.

[60] J. Chrisman, C. Hofer and W. Boulton, "Toward a system for classifying business strategies," The Academy of Management Review, Vol. 13, No. 3, pp. 413–428, 1988.

[61] G. Stampfl, R. Prügl and V. Osterloh, "An explorative model of business model scalability," Int. J. Product Development, Vol. 18, Nos. 3/4, pp. 226–248, 2013.

[62] A. Stephany, The Business of Sharing: Making it in the New Sharing Economy, Palgrave and Macmillan, 2015.

[63] A. Sundararajan, "From Zipcar to the Sharing Economy". January 3, 2013. HBR, June 2013.

[64] FCC: Report and Order and second FNPRM to advance availability of 3550-3700 MHz band for wireless broadband, 2015.

[65] D. Messerschmitt and C. Szyperski, Software Ecosystem: Understanding an Indispensable Technology and Industry, Cambridge, MA: MIT Press, 2003.

Chapter 7

Spectrum Sharing in Broadcast and Unicast Hybrid Cellular Network

Hongxiang Li

CONTENTS

Rarely have wireless innovations changed everyday life as widely and profoundly as broadcast TV network and mobile cellular network. Since its inception, broadcast TV has served the public for a century and penetrated into almost every household; however, viewership in the US has decreased significantly over the last three decades. Meanwhile, the number of mobile connected devices exceeded the global population in 2013, and forecasters [1] predicted that mobile video traffic will increase 13-fold between 2014 and 2019, reaching 17.5 exabytes per month and accounting for nearly three-fourths of the world's mobile data traffic by 2019. This huge demand for mobile broadband service is occurring while radio spectrum, a necessary ingredient for this service, is becoming increasingly more scarce. Even with the increased availability of unlicensed spectrum (for Wi-Fi offload) and the denser deployment of infrastructure, we still face the need for additional spectrum for mobile services.

As two of the most prominent wireless infrastructure networks, broadcast TV networks and mobile cellular networks have historically evolved along distinct trajectories due to their inherent differences. The broadcast TV network is a one-way communication network that delivers common information (TV programs) to all receivers in a large geographic area. In recent years, analog broadcast TV has been converted to digital broadcast TV in many parts of the world for better TV quality and more efficient use of the spectrum. On the other hand, the mobile cellular network is largely a unicast network that delivers private information to individual receivers, and requires a bidirectional channel. In recent years, cellular networks have evolved from voice telephony networks to data access networks, including mobile video services. Compared to broadcast, unicast allows on-demand user interaction and thus has the flexibility of consuming resources only when a user is actively using the network service. However, the unfavorable scaling behavior of unicast [2] can prove problematic; the network resource is quickly depleted when many users are requesting video services at the same time (e.g., the Super Bowl). In this case, broadcast is much more resource (spectrum and energy) efficient, since a single transmission will simultaneously accommodate all users. As the cellular network faces the dual challenges of supporting large data volumes and seeking severely limited and expensive radio spectrum, the evolved multimedia broadcast multicast service (eMBMS) was introduced in 3rd Generation Partnership Project Long-Term Evolution (3GPP LTE) standard [3] to provide capacity offload from unicast transmissions.

On the other hand, to address spectrum scarcity for mobile broadband, the wireless research community has focused on dynamic spectrum access (DSA) under the name of cognitive radio (CR) [4, 5], where the secondary users can opportunistically use the primary network's licensed spectrum through such methods as sensing or geolocation. CR approaches require fast and reliable spectrum sensing or conservative geolocation-based power restrictions to avoid interference with primary users. This seemingly simple straightforward task is actually notoriously challenging in practice due to the large variations in the dynamic range and bandwidth of signals to be detected. In fact, quantitative analysis from recent studies has revealed that TV whitespace [6] is not suitable for secondary systems providing wide-area coverage due to the interference constraint for primary TV receivers; only short-range systems with smaller interference footprints can exploit the local secondary spectrum opportunity [7,8]. More importantly, the conservative federal spectrum policies have stifled the possibility of interaction and collaboration between the broadcast TV network and the mobile cellular network. Another more radical approach is to repurpose the broadcast TV bands for mobile broadband systems and to distribute the traditional broadcast TV contents over cellular infrastructures as one of many services.

In this chapter, we consider a broadcast (BC) and unicast (UC) hybrid cellular network and discuss new nonorthogonal spectrum sharing technology that maximizes the spectrum utilization efficiency. Note that broadcast and unicast systems have historically been distinct because of their inherent differences, and have evolved along distinct trajectories. With increasing user demands for mixed services, cellular hybrid is becoming an intriguing choice for wireless data providers to simultaneously offer broadband Internet and broadcast multimedia services [10]– [14]. The hybrid cellular concept defines the natural evolution path for converging one-way broadcast and two-way unicast networks. Unlike the cognitive radio approach, the hybrid architecture investigated here collaboratively deliver both broadcast and unicast information on a common platform. Such coordination is expected to significantly enhance the aggregate capacity, as compared to uncoordinated individual networks.

We first consider a single cell hybrid system with the aim of finding the suitable modem that approaches the system capacity. Toward this end, we introduce a new performance metric called hybrid capacity region, which incorporates broadcast *outage capacity* and unicast *Shannon capacity*. Then, we derive resource allocation schemes to approach the hybrid capacity region. For the case of multicell hybrid, we evaluate the additional hybrid capacity and broadcast coverage gains due to multicell collaborative transmission.

7.1 System Model

For wireless multimedia applications, most of the broadband systems (Digital Video Broadcasting-Terrestria/Handheld (DVB-T/H), Worldwide Interoperability for Microwave Access (WiMAX), long term evolution (LTE)) are already orthogonal frequency division multiplexing (OFDM) based [15]- [20]. In this chapter, we consider a multicarrier broadcast and unicast hybrid system in which broadcast and unicast signals are overlaid across the entire frequency band.

This work considers the hybrid transmission of broadcast and unicast information from the base station to two separate sets (at least logically) of users. In this case, a fundamental question is how shall broadcast and unicast signals share the available radio spectrum? In 3GPP eMBMS, multicast, and unicast signals are multiplexed in time domain (i.e. Time division multiple access (TDMA)) [9], where the configuration of radio resources used for eMBMS can be determined dynamically, within a wide range (0.3% to 60%) of downlink unicast capacity. On the other hand, it is well known that TDMA is suboptimal in terms of spectrum utilization efficiency [27]. Furthermore, we also proved that even the prevailing OFDMA technology (currently adopted by a number of standards including 4G LTE) is almost always suboptimal in achieving the unicast multiuser channel capacity, and the performance loss can be significant in some practical scenarios [30, 34]. To achieve the true multiuser channel capacity, nonorthogonal approaches must be used. As will become clear in the remainder of this chapter, this technique enables the exploitation of multiuser diversity while maintaining interference-free reception for unicast receivers, therefore offering considerable gains in total system capacity without added complexity at the user side.

Over the past few years, transmitter pre-coding technique, i.e., dirty paper coding (DPC), has attracted a lot of attention, since it was proved to be capacity-achieving in multiple input multiple output (MIMO) downlink channels [9, 10, 21, 24, 26, 37, 38]. The basic principle of DPC is illustrated in Figure 7.1(a). Assume v is the desired signal to be transmitted, s is the interference, and n is the additive white gaussian noise (AWGN) noise. If the interference s is non-causally known at the transmitter, by adding a smart precoder at the transmitter, the receiver can demodulate source v as if the interference were not present. That is, without increasing the transmitting power, the capacity of interference channel is the same as that of the AWGN channel without interference. Figure 7.1(b) shows the nonorthogonal spectrum sharing between two signals.

The broadcast and unicast signals in the hybrid system can be modeled as "known" interference to each other. Casting hybrid transmission into the DPC framework, we arrive at a multicarrier modem structure illustrated in Figure 7.2.

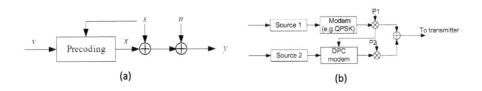

(a)   (b)

Figure 7.1: (a) Dirty paper precoding. (b) Nonorthogonal spectrum sharing via DPC.

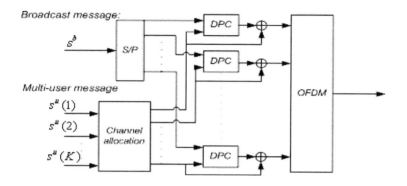

Figure 7.2: Hybrid modem.

The hybrid transmitter possesses the following characteristics:

- OFDM modem: Multicarrier modulation is used to support broadband unicast and broadcast applications. The total system bandwidth B is divided into N frequency flat parallel subchannels. The subchannels are orthogonal, which is advantageous in practical implementation.

- Broadcast signals overlaid on top of unicast signals across all subchannels: The broadcast signals and unicast signals are no longer delivered through separate infrastructures or orthogonal channels — they are superimposed in both time and frequency from the same transmitter. For multimedia applications, the common information may be layered (i.e., carrying source signals of different qualities) as well.

- DPC precancellation at the transmitter side: Without preprocessing, the superimposed broadcast and unicast signals will obviously interfere with each other. Although signal separation can be performed at the receiver by superposition code with successive interference cancellation (SC-SIC) [35, 36], it incurs a significant burden to all receivers, both in computation and memory costs. More importantly, the delay due to SIC may be intolerable to certain applications.

The DPC-based hybrid networks takes advantage of the fact that the unicast and broadcast signals are transmitted from the same transmitter. As a result, each subchannel is degraded in the sense that both signals arrive at any given receiver through the same wireless channel. Consequently, interference free broadcast can potentially be achieved through aggregation of dirty-paper precoded common information. Alternatively, interference free unicast can be delivered with the same strategy. Obviously, the hybrid system reduces to a regular OFDM broadcast system (e.g., DVB-T) when no power is allocated to unicast, and a regular multicarrier unicast system (e.g., WiMAX) when no power is allocated to broadcast. The hybrid modem represents a distinct change from the set of problems previously addressed in this area with the following benefits:

- Nonintrusive overlay: Although broadcast and unicast signals are superimposed in both frequency and time, DPC precancellation at the transmitter enables interference free broadcast or unicast reception with no added cost or delay at receivers.

- Increased unicast data rate and broadcast coverage area: Since the hybrid system converges two networks into one platform, the guard band between two isolated networks is eliminated. In addition, one can expect an increased unicast data rate and a better broadcast coverage area over the traditional time division multiplexing/frequency division multiplexing (TDM/FDM) scheme, as will be shown in the ensuing sections.

By relaxing the orthogonality constraint, the objective therein is to maximize the achievable unicast sum rate and the broadcast coverage region. We invoke the following assumptions and notation conventions throughout the paper:

1. For unicast, perfect knowledge of the instantaneous channel state information (CSI) is available at the transmitter through uplink feedback.

2. For broadcast, only the channel statistics are available at the transmitter. In other words, the broadcast users are passive receivers without the transmitting capability.

3. Different receivers experience independent block Rayleigh fading channels.

N	the total number of subcarriers
P	the total transmitting power of the hybrid system
P^b	the total broadcast power
p_n^b	broadcast power on subcarriers n
h_n^b	BC channel gain on subcarrier n of the worst receiver
K	the total number of unicast users
P^u	the total unicast transmitting power
$p_n^u(k)$	transmission power for unicast user k on subcarriers n
$h_n^u(k)$	channel gain for unicast user k on subchannel n
B	the total system bandwidth
B_n	subchannel bandwidth
$\max\limits_{x}\{f(x)\}$	maximum value of $f(x)$ maximized overall x
$\arg\max\limits_{x}\{f(x)\}$	value of x that maximizes the function $f(x)$

It is worth pointing out that Assumption 1 is not a requirement for the proposed single cell hybrid operation — however, the availability of the unicast CSI can be used to improve the spectrum efficiency. Also we do not specify the channel model in Assumption 2. We begin our discussion with the single cell case. A key element of the hybrid scheme is the DPC-based transmitter that enables interference-free reception. As such, the DPC principle and its implementation is reviewed first.

7.2 Hybrid System Analysis

7.2.1 Performance Metric

In order to quantify the performance of the hybrid system, we must first define the performance metric. For a broadcast channel where no instantaneous CSI is available at the transmitter, there are two channel capacity definitions that are relevant to the system design [27]: the ergodic capacity (also called the Shannon capacity) and the outage capacity. The ergodic capacity defines the maximum data rate that can be sent to the receiver with asymptotically small error probability through all the fading states. Obviously, it is not suitable for applications, such as TV broadcast with delay constraints. The outage capacity defines the maximum data rate that can be transmitted with certain outage probability that the received data cannot be decoded with negligible error probability. If the received signal-to-noise ratio (SNR) is above the threshold corresponding to the outage probability, the transmitted data can be decoded with negligible probability of error; otherwise, the transmission is in outage. By allowing some outage, the broadcast receiver can decode the message during each fading state and thus, meet the tight delay constraint. In this chapter, we use the outage capacity as the figure of merit for broadcasting.

For the unicast application under consideration, no outage capacity is needed because the transmitter has the instantaneous CSI. Based on unicast channel realization, the transmitter dynamically sends unicast information at a rate that can always be decoded correctly.

In light of the two different applications, we define the hybrid performance metric as follows:

Definition 7.1 (i) $r^o = \begin{bmatrix} r_1^o & \cdots & r_N^o \end{bmatrix}$: the broadcast *common information rate vector* over subcarriers; (ii) $R^o = \sum_{n=1}^{N} r_n^o$: the total broadcast *common information rate*; (iii) $q(l)$: the outage probability (package loss rate) associated with the broadcast receiver l.

Definition 7.2 (i) $r^u(k) = \begin{bmatrix} r_1^u(k) & \cdots & r_N^u(k) \end{bmatrix}$: the private information rate vector to unicast user k over all subcarriers; (ii) $R^u(k) = \sum_{n=1}^{N} r_n^u(k)$: the total private information rate to unicast user k.

Remark 6.1 (1) Throughout the chapter, we assume independent channel coding across subchannels, i.e., a separate capacity-achieving code is used over each of the subchannels. (2) Joint coding across subchannels, while more practical, is difficult to analyze due to its dependency on fading channel models, the size of the coding block, and the interleaver mechanisms. This problem will be studied separately in our future work. (3) All broadcast receivers receive the same common information. Depending on the fading statistics, broadcast receivers decode the common information with different outage probabilities. (4) Unlike broadcast, the unicast users receive different rates.

Using the above definitions, our goal in the hybrid network design is to find transmission schemes that maximize the broadcast and unicast network performance jointly. Specifically, the objective for unicast application is to maximize the achievable rate region $\begin{bmatrix} R^u(1) & \cdots & R^u(K) \end{bmatrix}$. On the other hand, there are two interchangeable ways to evaluate the broadcast performance:

1. Given a broadcast rate R^o and an outage probability q^o, find the maximum coverage area A, such that any broadcast receiver within A has outage probability $q \leq q^o$.

2. Given a coverage area A and an outage probability q^o, find the maximum rate R^o, such that any broadcast receiver within A has outage probability $q \leq q^o$.

Remark 6.2 Let L_A be the index of a broadcast receiver that has the worst channel gain statistics in A. Given any channel path loss and fading model, receiver L_A always receives the common information, with the highest outage probability among all broadcast receivers in A. Therefore, the following two constraints are equivalent:

$$\forall \, l \in A, q(l) \leq q^o \text{ iff } q(L_A) \leq q^o, \tag{7.1}$$

where the worst receiver L_A is typically on the edge of area A. Consequently, we only need to consider the farthest receiver L_A in broadcast optimization.

Note that the above two broadcast optimization problems are equivalent in maximizing the common information transmission rate from base station to receiver L_A. For convenience, we fix q^o and

A and try to maximize R^o. Accordingly, we define the capacity region of the hybrid network as:

$$\begin{bmatrix} R^o & \mathbf{R}^u \end{bmatrix} = \begin{bmatrix} R^o & ; & R^u(1) & \cdots & R^u(K) \end{bmatrix} \tag{7.2}$$

$$s.t. \; P^b + P^u \leq P \tag{7.3}$$

$$q(L_A) \leq q^o, \tag{7.4}$$

which is the closure of all achievable broadcast and unicast rate set under the power and outage probability constraints.

7.2.2 Hybrid Capacity

Having defined the hybrid capacity region (7.2), our goal is to find transmission strategies and power loading schemes that approach the boundary point of the capacity region. Since we only need to consider the farthest receiver L_A in broadcasting, the capacity region defined in (7.2) can be cast into the capacity region of $K + 1$ users.

To simplify the problem, we assume the same outage probability q^o over all subcarriers. This allows us to replace the unknown broadcast channel gain with a channel gain threshold and arrive at the following theorem:

Theorem 7.1
Define broadcast channel threshold on subcarrier n as:

$$\left\{ |h_n^T| \Big| \Pr \left[|h_n^b| \leq |h_n^T| \right] = q^o \right\}, \tag{7.5}$$

where $|h_n^b|$ is the unknown random channel gain of broadcast receiver L_A on subcarrier n. The capacity region $\begin{pmatrix} R^o & \mathbf{R}^u \end{pmatrix}$ is the set of rate pairs

$$C_{hybrid} = \bigcup_{\left\{ \sum_{n=1}^N \left(p_n^b + \sum_{k=1}^K p_n^u(k) \right) = P \right\}} \tag{7.6}$$

$$\left\{ R^o ; R^u(1) \cdots R^u(K) \right\},$$

$$where \begin{cases} R^o = \sum_{n=1}^N B_n \log_2 \left(1 + \dfrac{|h_n^T|^2 p_n^b}{N_o B_n + |h_n^T|^2 J_n^b} \right) \\ R^u(k) = \sum_{n=1}^N B_n \log_2 \left(1 + \dfrac{|h_n^u(k)|^2 p_n^u(k)}{N_o B_n + |h_n^u(k)|^2 J_n^u(k)} \right), \end{cases}$$

$$with \begin{cases} J_n^b = \sum_{j=1}^K p_n^u(j) \mathbf{1}[|h_n^u(j)| > |h_n^T|] \\ J_n^u(k) = p_n^b \mathbf{1}[|h_n^T| > |h_n^u(k)|] + \\ \qquad \sum_{j=1}^K p_n^u(j) \mathbf{1}[|h_n^u(j)| > |h_n^u(k)|], \end{cases}$$

where $\mathbf{1}[\cdot]$ denotes the indicator function.

Proof 7.1 Because the transmitter does not have the instant broadcast channel information, it has to transmit the common information at a fixed rate during each unicast channel realization. For any outage rate r_n^o and power allocation $(p_n^b, p_n^u(k))$ on subcarrier n, the broadcast outage probability is given by

$$q^o = \Pr \left[B_n \log_2 \left(1 + \dfrac{|h_n^b|^2 p_n^b}{N_o B_n + |h_n^b|^2 J_n^b} \right) \leq r_n^o \right], \tag{7.7}$$

where J_n^b is the interference from unicast users that are not precancelled by DPC. Note that in (7.7) p_n^b and J_n^b are all determined after power loading, and h_n^b is the only random variable during each unicast channel realization, the outage probability q^o is simply determined by the distribution of $\left|h_n^b\right|$. Using the fact that $B_n \log_2\left(1 + \frac{\left|h_n^b\right|^2 p_n^b}{N_0 B_n + \left|h_n^b\right|^2 J_n^b}\right)$ monotonically increases with $\left|h_n^b\right|$, we can uniquely determine a channel gain threshold $\left|h_n^T\right|$, such that $r_n^o = B_n \log_2\left(1 + \frac{\left|h_n^T\right|^2 p_n^b}{N_0 B_n + \left|h_n^T\right|^2 J_n^b}\right)$, with $\Pr\left[\left|h_n^b\right|^2 \leq \left|h_n^T\right|^2\right] = q^o$. Since the distribution of $\left|h_n^b\right|$ is available at the base station, the channel threshold $\left|h_n^T\right|$ is known. Therefore, the capacity region $\left(\begin{array}{cc} R^o & \mathbf{R}^u \end{array}\right)$ is essentially equivalent to the $K + 1$ user capacity region with informed transmitter. By [28, Theorem 1] and [29, Theorem 2.1], Equation (7.6) is the capacity region. ■

Theorem 6.1 converts the hybrid capacity region into an equivalent $K + 1$ dimensional capacity region, with h_n^T being the effective channel gain of broadcast receiver L_A. Obviously, C_{hybrid} is a function of the outage probability q^o.

While Theorem 6.1 provides the capacity region of the hybrid system, the high dimensional capacity region is nontrivial to compute when K is large. For tractability, we focus the optimization criterion on the two-dimensional rate region $\left(\begin{array}{cc} R^o & R_{sum}^u \end{array}\right)$, where the scalar $R_{sum}^u = \sum_{k=1}^{K} R^u(k)$ is the unicast sum-rate. Specifically, under the total power and bandwidth constraints, our objective is

1. to characterize the achievable rate region of $\left(\begin{array}{cc} R^o & R_{sum}^u \end{array}\right)$; and

2. to determine the suitable power allocation schemes to achieve or approach the maximum capacity region.

In general, base station can transmit signals to all the $K + 1$ users on every subcarrier. In order to reach the capacity region, we have the following theorem regarding the optimal subcarrier allocation:

Theorem 7.2
The boundary point of the capacity region $\left(\begin{array}{cc} R^o & R_{sum}^u \end{array}\right)$ is achieved only if

1. OFDMA [30] is used for unicasting;

2. subcarrier n is shared by broadcasting as well as unicasting to the user with the highest channel gain $\max_{1 \leq k \leq K}\left\{\left|h_n^u(k)\right|^2\right\}$.

The proof of Theorem 2 can be found in Appendix A. Theorem 2 is a pleasant surprise, because it suggests that each subcarrier should be assigned exclusively to one unicast user who has the channel gain $\max_{1 \leq k \leq K}\left\{\left|h_n^u(k)\right|^2\right\}$ — only broadcast signal can share any given subcarrier with the best unicast user. The theorem reveals an important result regarding the hybrid design — the capacity of the hybrid system can be achieved with only two overlaid signals, one broadcasting and one unicasting, on each subcarrier. Combining Theorems 1 and 2, we have the following corollary:

Corollary 7.1
The capacity region $\begin{pmatrix} R^o & R^u_{sum} \end{pmatrix}$ is the set of rate pairs

$$C_{hybrid} = \bigcup_{\{\sum_{n=1}^{N} p_n^b + p_n^u = P\}} \tag{7.8}$$

$$\left\{ R^o = \sum_{n=1}^{N} B_n \log_2 \left(1 + \frac{|h_n^T|^2 p_n^b}{N_o B_n + |h_n^T|^2 p_n^u \mathbf{1}[|h_n^u| > |h_n^T|]} \right), \right.$$

$$\left. R^u_{sum} = \sum_{n=1}^{N} B_n \log_2 \left(1 + \frac{|h_n^u|^2 p_n^u}{N_o B_n + |h_n^u|^2 p_n^b \mathbf{1}[|h_n^T| > |h_n^u|]} \right) \right\},$$

where $|h_n^u| = \max_{1 \leq k \leq K} |h_n^u(k)|$ is the best channel gain of unicast users on subcarrier n.

7.2.3 Power Loading in Hybrid Transmission

In principle, the precancellation order in Corollary 6.1 does not have to be fixed (i.e., either the broadcast or the unicast signal can be regarded as interference). For all practical purposes, however, it is desirable to precancel the broadcast signal for unicast receivers. Note that the unicast channel gain $|h_n^u| = \max_{1 \leq k \leq K} |h_n^u(k)|$ is usually higher than $|h_n^T|$ because broadcast receiver L_A is far from the base station, and the outage probability q^o is set small. In the following analysis, we focus on precancellation of broadcast signal from unicasting for the purpose of maximizing the overall hybrid information rate.

Corollary 7.2
Assuming the broadcast signal is precanceled using DPC, the capacity region $\begin{pmatrix} R^o & R^u_{sum} \end{pmatrix}$ is the set of rate pairs

$$C_{hybrid} = \bigcup_{\{\sum_{n=1}^{N} p_n^b + p_n^u = P\}} \tag{7.9}$$

$$\left\{ R^o = \sum_{n=1}^{N} B_n \log_2 \left(1 + \frac{|h_n^T|^2 p_n^b}{N_o B_n + |h_n^T|^2 p_n^u} \right), \right.$$

$$\left. R^u_{sum} = \sum_{n=1}^{N} B_n \log_2 \left(1 + \frac{|h_n^u|^2 p_n^u}{N_o B_n} \right) \right\}.$$

From Corollary 6.2, the unicast private information can be decoded at the user end as if the overlaid broadcast signal does not exist. On the other hand, the unicast signal constitutes a pure interference to the broadcast signal, which needs to be coped with despreading.

The capacity region (7.9) is the convex hull of the union of all rate pairs over all power allocation (P^b, P^u) satisfying the total power constraint. The optimal power allocation scheme that achieves any boundary point of the capacity region was introduced in [29, Theorem 2.1] and [31] and it is essentially multiuser water filling. The optimal power loading has a greedy interpretation and is somewhat computationally expensive when N is large. Note that the precancellation order is fixed for all subcarriers in (7.9), we present two simple power loading schemes that have similar performance with the optimal power allocation.

7.2.3.1 Power Loading Scheme 1

Since the base station does not have the instant broadcast CSI, it assumes the same broadcast channel threshold $|h_n^T|$ on all subcarriers and equally distributes the broadcast power P^b over frequency. On the other hand, since the broadcast signal is precancelled by DPC and the unicast CSI is available at the base station, unicast power P^u can be allocated over subcarriers by single user water filling, regardless of the broadcast power allocation. The scheme is illustrated in Figure 7.3a, where $z_n^u = \frac{N_0 B_n}{|h_n^u|^2}$ is the unicast effective noise on subcarrier n.

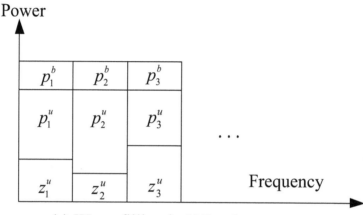

(a) Water filling in UC only

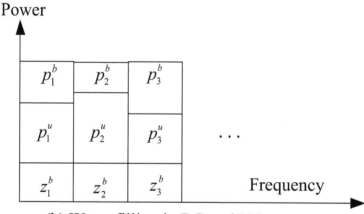

(b) Water filling in BC and UC

Figure 7.3: Hybrid power loading.

7.2.3.2 Power Loading Scheme 2

Despite the absence of CSI, power loading with water filling can be performed for broadcasting as well to deliver better achievable rates. This is possible because the unicast transmitting power, which is an interference to broadcast, is indeed available at the base station. As a result, broadcast power loading based on the unicast power distribution becomes an necessity, as shown in Figure 7.3b.

Given power pair (P^b, P^u), Corollary 6.2 suggests the following multicarrier power loading algorithm:

Step 1: Allocate unicast power P^u over N subchannels by single user water filling with effective noise $\frac{N_0 B_n}{|h_n^u|^2}$;

Step 2: Allocate broadcast power P^b over N subchannels by single user water filling with effective noise $\frac{N_0 B_n}{|h_n^T|^2} + p_n^u$.

Unicast power loading is performed first because its signals are free from interference. Once Step 1 is complete, the interference to broadcast is known and another water-filling for broadcast can follow. This decoupled unicast and broadcast power loading algorithm yields additional capacity gains. While intuitively sounding, the above algorithm does not achieve the boundary of the capacity region promised in (7.9).

Remark 6.3   Given (P^b, P^u), the unicast power loading in step 1 is suboptimal in achieving the boundary of capacity region (7.9). However, provided that the optimal unicast power loading is already achieved, the broadcast power loading in step 2 is optimal in achieving the boundary of (7.9). The reason is that unicast power loading causes interference to broadcast and, thus, affects the outage rate R^o. On the other hand, the broadcast power loading does not affect unicast sum rate because it has been precancelled.

Compared with the optimal multiuser water filling, single user water filling in the above algorithm reduces the computational complexity from $o(2N)$ to $o(N)$. As will be shown in what does this refer to?, the achievable rate region using above power loading schemes is close to the optimal capacity region in most cases.

7.3   Collaborative Hybrid System

In a multicell network, each hybrid cell can operate independently as in most regular cellular systems with frequency planning. On the other hand, hybrid transmission can also be employed at macro-level through multiple base stations that cooperate with each other in delivering both common and private information. The collaborative multicell is essentially a multiple input and single output (MISO) system, which is mathematically identical to a "super" base station equipped with multiple geographically dispersed antennas. For unicast, the multicell cooperation can increase the sum capacity [32]. For broadcast, distributed transmission provides additional diversity and forms an extended ellipse coverage beyond the superposition of individual cells [33], as shown in Figure 7.4.

The hybrid capacity region becomes much more complicated in the multicell setting — Theorems 6.1 and 6.2 no longer hold in general, even for unicast-only systems [34]. In this section, we will limit our discussion to interference-free OFDMA based unicasting without addressing the optimality issue. Specifically, we investigate the achievable rate region by using DPC precancellation and analyze the maximum broadcast coverage in a multicell environment.

7.3.1   DPC Precancellation

Since the unicast CSI is available at the transmitter, on each subcarrier, we can apply DPC to precancel the broadcast signals and achieve interference-free unicast transmission. For unicast transmission, the private information is beamformed from multiple base stations to maximize the transmission rate, i.e., the same unicast signal is weighted by a complex scale and sent from all col-

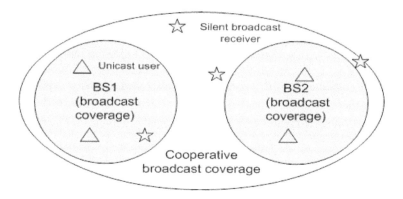

Figure 7.4: Cooperative transmission for broadcast coverage extension.

laborating cells. For broadcast transmission, collaborating cells send space-time coded common information to improve the transmitter diversity, that is, the broadcast signals sent from multiple base stations are independent.

Let vector s_n^b and scalar s_n^u be the broadcast and unicast signals on subcarrier n, respectively, and M_t be the number of cooperation cells. The transmitting signal on subcarrier n is given by:

$$x_n = w_n^b \odot s_n^b + w_n^u s_n^u,$$

where w_n^b and w_n^u are $M_t \times 1$ complex weight vectors for signals s_n^b and s_n^u, respectively, and "$\odot$" is the element by element multiplication operator. Note that the weights must satisfy $||w_n^b||^2 = ||w_n^u||^2 = 1$ to meet the total transmitting power constraint. By judiciously choosing the weight vector (w_n^b, w_n^u), additional performance gains over single cell transmission can be achieved.

At the receiver side, each user receives the superposition of these signals. The received signals on subcarrier n are given by:

$$\text{BC:} \qquad y_n^b = (h_n^b)^H \left(w_n^b \odot s_n^b \right) + (h_n^b)^H w_n^u s_n^u \qquad (7.10)$$

$$\text{UC:} \qquad y_n^u = (h_n^u)^H \left(w_n^b \odot s_n^b \right) + (h_n^u)^H w_n^u s_n^u, \qquad (7.11)$$

where h_n^b and h_n^u are $M_t \times 1$ complex channel gain vectors for broadcast and unicast, respectively.

Because the unicast CSI is known at the transmitter, we choose the UC weight vector as

$$w_n^u = \frac{(h_n^u)^*}{||h_n^u||}, \qquad (7.12)$$

which is the maximum ratio beamforming vector that maximizes the unicast rate.

Since the broadcast CSI is not available at the transmitter, we evenly divide the broadcast transmitting power among the M_t cells, i.e., the broadcast input covariance matrix on subcarrier n is

$$Q_n^b = E \left[s_n^b \left(s_n^b \right)^H \right] = \frac{p_n^b}{M_t} \mathbf{I},$$

where $\mathbf{I}$ is the identity matrix.

Note that regardless of the selection of w_n^b, the interference term in (7.11), $(h_n^u)^H \left(w_n^b \odot s_n^b \right)$, is known to the unicast transmitter. Therefore, we can still apply DPC to precancel it without knowledge of the broadcast channel, yielding the following achievable capacity region [27, pp. 483]:

$$C_{\text{hybrid}} = \bigcup_{\{\sum_{n=1}^{N} p_n^b + p_n^u = P\}}$$

$$\left\{ \begin{array}{c} R^o = \sum_{n=1}^{N} B_n \log_2 \left(1 + h_n^T p_n^b\right), \\ R_{sum}^u = \sum_{n=1}^{N} B_n \log_2 \left(1 + \frac{||h_n^u||^2 p_n^u}{N_o B_n}\right), \end{array} \right\} \tag{7.13}$$

$$\text{with } h_n^T : \Pr \left[\frac{\frac{||h_n^b||^2}{M_t}}{N_o B_n + \left|(h_n^b)^H w_n^u\right|^2 p_n^u} \geq h_n^T \right] = q^o.$$

The optimal power loading scheme for multicell hybrid transmission is nontrivial. However, we can compute the achievable rate region using the power loading algorithm in Section III.C with the broadcast effective noise on subcarrier n as $\frac{1}{|h_n^T|^2}$.

7.3.2 Broadcast Coverage Gain

In practical broadcast applications, the common information rate is usually prefixed and, thus, the objective of hybrid design is to maximize broadcast coverage area and unicast sum rate simultaneously. Compared to the single cell transmission, the DPC precancellation scheme can significantly increase the broadcast outage rate R^o and unicast sum rate R_{sum}^u simultaneously. Alternatively, if we keep the R^o and q^o the same in Equation (7.13), the broadcast coverage area can be expanded.

Note that the maximum broadcast coverage area is achieved when all power is assigned to broadcast network. For convenience, we set $p_n^u = 0$ in Equation (7.13) in order to quantify the broadcast gains:

$$\begin{aligned} R^O &= \sum_{n=1}^{N} R_n^o \left| \Pr \left[B_n \log_2 \left(1 + \frac{\frac{||h_n^b||^2}{M_t} p_n^b}{N_o B_n}\right) \leq R_n^o \right] = q^o \right. \\ &= \sum_{n=1}^{N} B_n \log_2 \left(1 + \frac{y_0 p_n^b}{N_o B_n}\right) \left| \Pr \left[\frac{||h_n^b||^2}{M_t} \leq y_0 \right] = q^o. \right. \end{aligned}$$

We can see that the maximum broadcast rate (coverage area) depends solely on the distribution of $\left|\left|h_n^b\right|\right|^2$. In general, when $p_n^u \neq 0$, the received signal-to- interference-plus-noise (SNIR)

$\frac{\frac{||h_n^b||^2}{M_t} p_n^b}{N_o B_n + \left|(h_n^b)^H w_n^u\right|^2 p_n^u}$ does not monotonically increase with $\left|\left|h_n^b\right|\right|^2$. However, we can approximate the

received SNIR as $\frac{\frac{|h_n^b|^2}{M_t} p_n^b}{N_o B_n + \frac{|h_n^b|^2}{M_t} p_n^u}$, which strictly monotonically increase with $\left|\left|h_n^b\right|\right|^2$. The approxima-

tion allows us to calculate the broadcast coverage area based on the distribution of $\left|\left|h_n^b\right|\right|^2$. Assume the average path loss is simply a function of the distance between base station and broadcast receivers, we know the worst receiver L_A is always on the edge of A. Under the Rayleigh fading channel model, we exemplify the coverage gain by analyzing three collaborative cells as follows.

First, we need to derive the distribution of $\frac{||h_n^b||^2}{M_t}$. Let $X_1 = \frac{|h_n^b(1)|^2}{M_t}$, $X_2 = \frac{|h_n^b(2)|^2}{M_t}$, $X_3 = \frac{|h_n^b(3)|^2}{M_t}$,

and $Y = \frac{||h_n^b||^2}{M_t}$, and we have $Y = X_1 + X_2 + X_3$.

Since the collaborating cells are separated far enough, we can consider X_1, X_2, and X_3 are independent chi-square R.V.s. The distribution of scaled chi-square R.V. is given by:

$$PDF \quad : \quad f(x) = \frac{1}{2\sigma^2} \exp(-\frac{x}{2\sigma^2}) \tag{7.14}$$

$$CDF \quad : \quad F(x) = 1 - \exp(-\frac{x}{2\sigma^2}). \tag{7.15}$$

The cumulative distribution function (CDF) of Y can be obtained by convoluting (7.14) and (7.15) as:

$$F_Y(y) = f_{X_1}(y) * f_{X_2}(y) * F_{X_3}(y) = \tag{7.16}$$

$$
\begin{cases}
\text{if } \sigma_1 = \sigma_2 = \sigma_3 = \sigma: \\
1 - \exp(\frac{-y}{2\sigma^2})\left(1 + \frac{y}{2\sigma^2} + \frac{y^2}{8\sigma^4}\right) \\
\text{if } \sigma_1 = \sigma_2 = \sigma \neq \sigma_3: \\
1 + \frac{\sigma^2(2\sigma_3^2 - \sigma^2)}{(\sigma_3^2 - \sigma^2)^2}\exp\left(\frac{-y}{2\sigma^2}\right) - \frac{\sigma_3^4}{(\sigma_3^2 - \sigma^2)^2}\exp\left(\frac{-y}{2\sigma_3^2}\right) \\
\quad + \frac{y}{2(\sigma_3^2 - \sigma^2)}\exp\left(\frac{-y}{2\sigma^2}\right) \\
\text{if } \sigma_1 \neq \sigma_2 \neq \sigma_3: \\
1 + \frac{\sigma_1^4}{(\sigma_2^2 - \sigma_1^2)(\sigma_1^2 - \sigma_3^2)}\exp\left(\frac{-y}{2\sigma_1^2}\right) + \frac{\sigma_2^4}{(\sigma_1^2 - \sigma_2^2)(\sigma_2^2 - \sigma_3^2)}\exp(\frac{-y}{2\sigma_2^2}) \\
\quad + \frac{\sigma_3^4}{(\sigma_1^2 - \sigma_3^2)(\sigma_3^2 - \sigma_2^2)}\exp(\frac{-y}{2\sigma_3^2}),
\end{cases}
$$

where σ_i is determined by the pass loss from cell i to the receiver.

Then Equation (7.16) can be used to determine the multicell broadcast coverage area. Specifically, in order to maximize the coverage area, three cells should be separated equally in space by symmetry. Figure 7.5 shows the polar coordinates of the cell locations: $(\frac{D}{\sqrt{3}}, \frac{\pi}{2})$, $(\frac{D}{\sqrt{3}}, \frac{7\pi}{6})$, and $(\frac{D}{\sqrt{3}}, \frac{11\pi}{6})$, where D is the distance between any two cells. Note that for path loss, σ_i is simply a function of the distance between cell i and the receiver. The base station fixes the common information rate $r_n^o = B_n \log_2(1 + \frac{y_0 p_n^b}{N_o B_n + y_0 p_n^u})$ on subcarrier n. For each direction ($0 \leq \theta \leq 360$) from the origin, we calculate a radius $\rho(\theta)$, such that $Y = \frac{||h_n^b||^2}{M_t}$ at (ρ, θ) satisfies $\Pr[Y(\rho, \theta) \geq y_0] = q^o$, i.e., we find the coverage edge in all directions. Then the maximum coverage area can be numerically computed. An example will be shown in the next section.

7.4 Numerical Results

In this section, we provide numerical results to evaluate the performance of the hybrid network. We assume that the channel gain is Rayleigh distributed, i.e., the envelope of the complex channel gain has the following distribution:

$$f(|h|) = \frac{|h|}{\sigma^2} \exp\left(-\frac{|h|^2}{\sigma^2}\right), \tag{7.17}$$

where $E[|h|^2] = 2\sigma^2$ is determined by path loss.

For the single cell hybrid transmission, we consider the achievable rate region (7.9) under different power loading schemes in Section III.C and compare them against the traditional TDM scheme. With fixed transmitting power P and the broadcast coverage area A, we compute the hybrid capacity regions with different $\frac{|h_n^u|^2}{|h_n^T|^2}$, N and q^o, where $\frac{|h_n^u|^2}{|h_n^T|^2}$ indicates the channel quality difference between

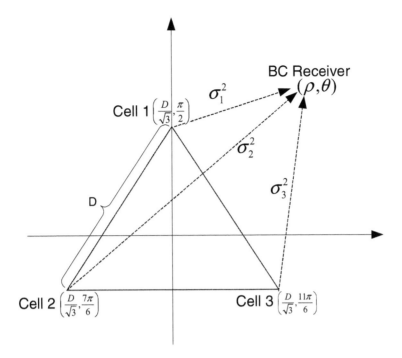

Figure 7.5: Multi-cell broadcast coverage.

unicasting and broadcasting. Figure 7.6 illustrates the capacity region of the hybrid system over the traditional TDM with $N = 4$. The solid curves show the achievable rate region by different power loading schemes described in Section III.C, and the dashed and dotted lines show the capacity region of the time-sharing scheme. While R^o is clearly a function of the outage probability, the rate regions of the hybrid system are considerably higher than those of the TDM benchmark. In all cases, we observe that the achievable rates obtained from our power loading schemes are almost the same as the optimal capacity region. Note that the performance gain of the hybrid system over TDM depends on $\frac{|h_n^u|^2}{|h_n^T|^2}$. The DPC-based hybrid system benefits the most when $\frac{|h_n^u|^2}{|h_n^T|^2}$ is large.

For the multicell collaboration, we apply the DPC precancellation scheme described in Section IV and compare its performance with that of TDM. Figure 7.7 shows the achievable hybrid rate region for $M_t = 3$. We can see the hybrid scheme clearly outperforms the TDM in all cases. Because of the multicell cooperation, the hybrid capacity region is increased substantially. To quantify the broadcast coverage gain, we evaluate a multicell scenario with the following settings.

For large scale path loss, the Hata model is the most common model for signal prediction in large urban macro-cells [27]. This model is applicable over distances of 1 km–100 km and frequency ranges of 150MHz–1500MHz. The standard formula for empirical path loss in urban areas under the Hata model is

$$P_L(i) \text{ dB} = 69.55 + 26.16\log_{10}(f_c) - 13.82\log_{10}(h_t) \tag{7.18}$$
$$-a(h_r) + (44.9 - 6.55\log_{10}(h_t))\log_{10}(d_i),$$

where d_i is the distance between base station i and the broadcast receiver, f_c is the carrier frequency, h_t/h_r is the transmitter/receiver antenna height, and d is the distance between transmitter and receiver. For larger cities at frequencies $f_c > 300$ MHz, the correction factor $a(h_r)$ is given by

$$a(h_r) = 3.2(\log_{10}(11.75h_r))^2 - 4.97 \text{ dB}.$$

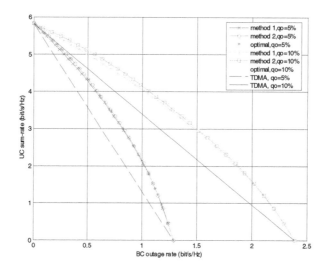

Figure 7.6: Single-cell hybrid capacity region with $N = 4$.

From (7.18), the average received power from cell i on subcarrier n is given by:

$$2\sigma_i^2 = p_n^b 10^{P_L(i)/10}.$$

Without loss of generality, we set $r_n^o = 515$ Kbps, $h_r = 1$ m and $p_n^b = 1$ watt in single cell transmission, such that $d_0 = 1$ km is the benchmark distance, with outage probability $q^o(d_0) = 5\%$.

In the multicell transmission with $M_t = 3$, we use Equation (7.16) to calculate the broadcast coverage area. We assume the same total power constraint on both single cell and multicell cases. Under the fixed broadcast common information rate $r_n^o = 515$ Kbps, Figure 7.8 shows the maximum broadcast coverage area for both single-cell and multicell transmissions, with $q^o = 1\%$. As pointed out in Section IV.B, multiple cells need to be separated evenly in space to achieve the best performance. The circle in the centre indicates the single input, single output (SISO) coverage area. The outer region is the extended coverage area with multicell cooperation, and the three small circles

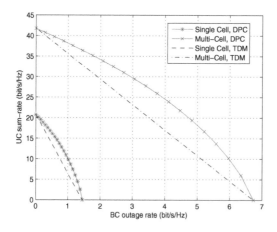

Figure 7.7: Capacity region for $Mt = 3$, $N = 16$, and $p^o = 5\%$.

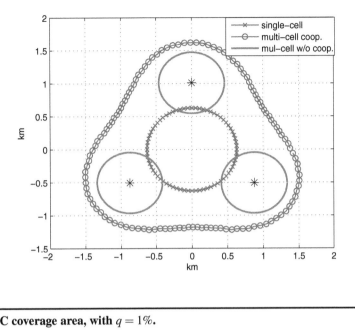

Figure 7.8: BC coverage area, with $q = 1\%$**.**

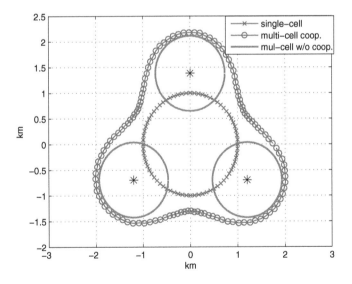

Figure 7.9: BC coverage area, with $q = 5\%$**.**

around base stations are the multicell coverage area when they do not collaborate. Figure 7.9 shows similar results for $q^o = 5\%$. Compared to the single cell case, the multicell cooperation coverage gains are 498% and 315% for outage probability 1% and 5%, respectively.

In order to find the optimal cell separation, we numerically calculate the coverage area as a function of cell separation distance for different q^os. As shown in Figure 7.10, the optimal cell separation is 1.8kms, 2.4kms, and 2.8 kms to achieve the maximum coverage area for outage probability 1%, 5%, and 10%, respectively.

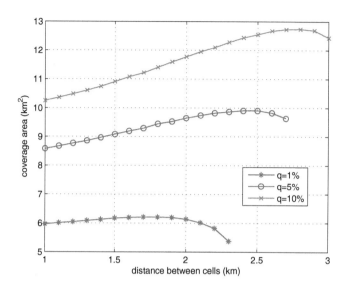

Figure 7.10: BC coverage area vs. cell distance.

7.5 Structured Dirty Paper Coding

The key in DPC based collaborative transmission is to design a simple but efficient precoding scheme. At high SNR, DPC can be approximated by the Tomlinson-Harashima precoding (THP) [40, 41]. Unfortunately, THP has been shown to suffer from a 1.53dB shaping loss in high SNR regime [42]. In low SNR regime, where broadcasting normally operates, the THP performance loss is even more severe, typically in the range of 4 dB, 5 dB [43]. The cause of this degradation is quantization (i.e., modulo operation loss) at the transmitter and the receiver. Recently, several trellis precoding based algorithms have been developed to recover the losses of THP [43–46]. However, the added complexity makes them less desirable in practical applications.

In multiuser wireless communications, the network performance is mostly affected by those users with poor channel conditions (i.e., receivers in low SNR regime). Therefore, we seek to develop low-complexity DPC scheme that can reduce or eliminate the THP modulo loss at low SNR regime. Toward this end, we observe that, in cellular and WiFi applications, the modulation structures of different datastreams are preknown and, thus, this information is available to all receivers. By exploiting this known structure, we arrive at a new precoding algorithm named structured DPC (SDPC) [24]. To see how the SDPC works, let's compare the THP and SDPC using an example shown in Figure 7.11. Since only low-order constellations are feasible in low SNR regime, both the desired signal v and the interfering signal s are assumed to be quadrature phase shift keying (QPSK). The different types of dots in Figure 7.11 represent different source symbols at the receiver end. Figure 7.11(a) can be viewed as the constellation of y at the input of the THP receiver. The received signal is then folded (modulo) into the dashed rectangular box before performing detection at the THP receiver.

We notice that the constellation of y at the THP receiver is generally an expanded version of the constellation of v due to the modulo operation at the transmitter. As a result, we should be able to directly demodulate y based on its constellation without performing the modulo operation, as in THP. Along the same line, if we take advantage of the structure information of interference and accordingly design the precoder, a more receiver friendly y constellation can be achieved. Specifically, in SDPC, source information is directly demodulated from y using region-based (minimum

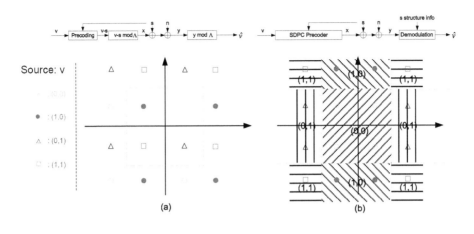

Figure 7.11: (a) Constellation of *y* **at THP receiver. (b) Constellation of** *y* **at SDPC receiver.**

distance) detection. Interestingly, we can rearrange the constellation of the received signal *y* for better performance through SDPC precoder. It can be verified that the mapping shown in Figure 7.11(b) yields the best performance [24]. As a matter of fact, its Bit Error Rate (BER) performance is comparable to that of QPSK with the same minimum distance. Then the SDPC precoding rule is designed accordingly.

i) *Precoding:* The precoder modulates 2-bit/symbol based on the following rule:

$$\begin{cases} x = v - s; & |v| > |s| \\ x = sign(s)(|2i \cdot v - |v||) - s; & |v| \le |s| \end{cases} \quad i = \left\lfloor \frac{1}{2} \left(\frac{|s|}{|v|} + 1 \right) \right\rfloor, \tag{7.19}$$

Here, $|\cdot|$ denotes the amplitude and $\lfloor \cdot \rfloor$ is the floor operator. The precoding rule is applied to both dimensions of the QPSK signal. The power of *x* is the average power of random signal $\pm[2(i-1) \cdot |v| - |s|]$ and $\pm[2(i+1)|v| - |s|]$, which are bounded.

ii) *Decoding:* The decoder detects the 2-bit symbol based on the location of the received signal (relative to four decision regions). For $|v| > |s|$, the decision regions are the same as that of the QPSK. For $|v| \le |s|$, the four decision regions are asymmetric, as illustrated in Figure 7.11(b). Nevertheless, a direct mapping from the source *v* to *y* can be established. By removing the modulo operation, the noise folded into the modulo interval around the origin is eliminated.

Note that, in contrast to prior works that assume arbitrary interference, the SDPC takes advantage of the constellation structure of the interference to approach the promised DPC channel capacity, with demodulation complexity similar to that of a regular quadrature amplitude modulation (QAM) demodulator. To get real world performance, we have implemented the SDPC and SC-SIC on hardware testbed [24], which confirms that SC-SIC requires higher complexity and a larger receiver buffer compared to SDPC. In our setup, both the source and interference signals are QPSK MPEG video streams with data rate of 5 Mbps, which are generated by an MPEG encoder residing in a PC. The MPEG datastreams are processed in the field-programmable gate array (FPGA) before OFDM transmission. Specifically, they are first channel encoded with a rate 1/2 (7,5) turbo encoder, followed by our SDPC precoder. At the receiver end, the demodulated intermediate frequency (IF) signal is sampled and then sent into the FPGA, where SDPC decoding and channel decoding are conducted. In the turbo decoder, eight iterations are performed using the log-map decoding algorithm. Finally, the recovered data are sent to an MPEG decoder on the PC and the video is displayed on its monitor. The results are shown in Figure 7.12 for $|s| = 7.5|v|$, where SDPC has more than 3dB performance improvement over the regular THP scheme (with dither) in low SNR

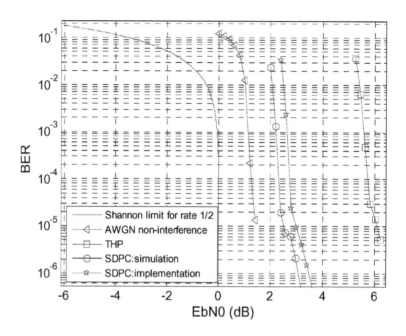

Figure 7.12: SDPC performance.

regime, which is remarkable. Compared to the perfect noninterfering AWGN channel, SDPC has about 1 dB performance gap due to power loss [24]. We also notice that the implementation loss is within 0.5 dB relative to the simulation. It is worth noting that SDPC works at any ratio, and the performance gain is more significant when $|s|$ and $|v|$ are close to each other.

7.6 Conclusion

In this chapter, we have developed a DPC-based broadcast and unicast hybrid system and characterize its achievable rate region via the generalization of the information theoretic capacity of downlink channels. Interference-free unicast is achieved by applying DPC precancellation at the transmitter. We have developed fast power loading algorithms that offer comparable performance with the optimal power allocation. In the multicell scenario, we have presented transmission scheme to achieve interference-free unicast and evaluated their performance. The numerical results show significant performance improvement of the hybrid network over the traditional TDM system. Finally, we developed an efficient and practical SDPC scheme that approaches the promised channel capacity.

Appendix A: Proof of Theorem 6.2

We shall prove Theorem 6.21 with the following Lemma [28, pp. 473].

Lemma 7.1

In a multicarrier multiuser unicast only (downlink) SISO system, OFDMA is the optimal transmission scheme to maximize the sum capacity and the subcarrier is always assigned to the user who has the best channel on this subcarrier.

Let $\pi_n(\cdot)$ denotes the permutation of the $K+1$ user indices on subcarrier n, such that

$$\left|h_{\pi_n(1)}\right| \leq \left|h_{\pi_n(2)}\right| \leq \cdots \leq \left|h_{\pi_n(K+1)}\right|,$$

$$\text{with } h_{\pi_n(k)} = \begin{cases} h_n^u(\pi_n(k)), & \text{for UC: } 1 \leq \pi_n(k) \leq K \\ h_n^T, & \text{for BC: } \pi_n(k) = K+1. \end{cases}$$

To achieve a specific boundary point of the capacity region $(\ R^o \quad R_{sum}^u\ )$, we assume $\left\{g_{\pi_n(\cdot)}\right\}$ are the optimal power allocation on subcarrier n. Then the $K+1$ dimensional hybrid rate vector on subcarrier n is given by Theorem 6.1:

$$\boldsymbol{R}_{\pi_n} = (R_{\pi_n(1)}, \ldots, R_{\pi_n(K+1)}):$$

$$R_{\pi_n(k)} = B_n \log_2 \left(1 + \frac{g_{\pi_n(k)}}{z_{\pi_n(k)} + \sum_{j=k+1}^{K+1} g_{\pi_n(j)}} \right),$$

where $z_{\pi_n(k)} = \frac{N_o B_n}{|h_{\pi_n(k)}|^2}$ is user $\pi_n(k)$'s effective noise. Without loss of generality, we assume that i_n is the permutation index of broadcast user L_A, and $1 < i_n < K+1$. Then the two-dimensional hybrid capacity of the boundary point is:

$$R_n^b = B_n \log_2 \left(1 + \frac{g_{\pi_n(i_n)}}{z_{\pi_n(i_n)} + \sum_{j=i_n+1}^{K+1} g_{\pi_n(j)}} \right) \tag{7.20}$$

$$R_{n,sum}^u = \sum_{k=1}^{i_n-1} B_n \log_2 \left(1 + \frac{g_{\pi_n(k)}}{z_{\pi_n(k)} + \sum_{j=k+1}^{K+1} g_{\pi_n(j)}} \right) \tag{7.21}$$

$$+ \sum_{k=i_n+1}^{K+1} B_n \log_2 \left(1 + \frac{g_{\pi_n(k)}}{z_{\pi_n(k)} + \sum_{j=k+1}^{K+1} g_{\pi_n(j)}} \right).$$

Applying Lemma 6.1 to the first summation term in Equation (7.21): If $g_{\pi_n(k)} \neq 0$ for $k < i_n - 1$, we can increase $R_{n,sum}^u$ by assigning all the power of the first $i_n - 2$ users to user $\pi_n(i_n - 1)$ without affecting the second term and R_n^b. Thus, we have

$$g_{\pi_n(k)} = 0 \quad \text{for } k < i_n - 1.$$

The same analysis applies to the second summation term in (7.21), i.e.,

$$g_{\pi_n(k)} = 0 \quad \text{for } i_n < k < K+1.$$

Therefore, Equations (7.20)–(7.21) become

$$R_n^b = B_n \log_2 \left(1 + \frac{g_{\pi_n(i_n)}}{z_{\pi_n(i_n)} + g_{\pi_n(K+1)}} \right)$$

$$R_{n,sum}^u = B_n \log_2 \left(1 + \frac{g_{\pi_n(i_n-1)}}{z_{\pi_n(i_n-1)} + g_{\pi_n(i_n)} + g_{\pi_n(K+1)}} \right)$$

$$+ B_n \log_2 \left(1 + \frac{g_{\pi_n(K+1)}}{z_{\pi_n(K+1)}} \right).$$

Achieving a boundary point of capacity region $(\ R^o \quad R_{sum}^u\ )$ is equivalent to the following optimization problem:

$$\arg\max_{g} R_{n,sum}^{u} = B_n \log_2 \left(1 + \frac{g_{\pi_n(K+1)}}{z_{\pi_n(K+1)}} \right) \tag{7.22}$$

$$+ B_n \log_2 \left(1 + \frac{g_{\pi_n(i_n-1)}}{z_{\pi_n(i_n-1)} + g_{\pi_n(i_n)} + g_{\pi_n(K+1)}} \right)$$

$$s.t. \ R_n^b = B_n \log_2 \left(1 + \frac{g_{\pi_n(i_n)}}{z_{\pi_n(i_n)} + g_{\pi_n(K+1)}} \right) \geq R_n^{b0}$$

$$g_{\pi_n(i_n-1)} + g_{\pi_n(i_n)} + g_{\pi_n(K+1)} \leq P, \ g_{\pi_n(\cdot)} \geq 0$$

$$\forall \ n, \ 1 \leq n \leq N.$$

It can be shown that (7.22) is a convex optimization problem, and the optimal solution requires $g_{\pi_n(i_n-1)} = 0$, i.e., it requires OFDMA transmission scheme for unicast users, and subcarrier n is used for broadcasting as well as unicasting to the best user.

References

[1] Cisco, Cisco visual networking index: Global mobile data traffic forecast update, 2014–2019, in Cisco, 2014.

[2] F. Hartung and U. "Horn, Delivery of broadcast services in 3g networks", *IEEE Transaction on Broadcasting*, vol. 53, no. 1, pp. 188–199, Mar. 2007.

[3] G. T. 26.346, Multimedia broadcast/multicast service (mbms): Protocols and codecs," vol. 11.6.0, 2013, http://www.3gpp.org/DynaReport/26346.htm.

[4] J. Mitola, "Cognitive radio for flexible mobile multimedia communications," in Mobile Multimedia Communications, 1999. (MoMuC 99) 1999 IEEE International Workshop on, 1999, pp. 3–10.

[5] S. Haykin, "Cognitive radio: Brain-empowered wireless communications," *IEEE Journal on Selected Areas in Communications*, vol. 23, no. 2, pp. 201–220, August 2005.

[6] B. Ray, "How to build a national cellular wireless network for 50m," April 2011, the Register.

[7] T. Dudda and T. Irnich, "Capacity of cellular networks deployed in TV white space," in IEEE International Symposium on Dynamic Spectrum Access Networks (DySPAN12), Oct, 16–19.

[8] L. Shi, K. Sung, and J. Zander, "Secondary spectrum access in TV-bands with combined co-channel and adjacent channel interference constraints," in IEEE International Symposium on Dynamic Spectrum Access Networks (DySPAN12), Oct, 16–19.

[9] J. Erman and K. Ramakrishnan, "Understanding the super-sized traffic of the Super Bowl," in Proceedings of the 2013 Conference on Internet Measurement Conference, ser. IMC 13. New York, NY, USA: ACM, 2013, pp. 353–360.

[10] U. Horn, R., Keller, and N. Niebert, "Interactive mobile streaming services — convergence of broadcast and mobile communication," *EBU Technical Review*, 1999, pp. 14–19.

[11] W. Klingenberg and A. Neutel, "MEMO — A hybrid DAB/GSM communication system for mobile interactive multimedia services," *Proc. ECMAST '98*, Berlin 1998.

[12] R. Keller, T. Lohmar, R. Tonjes, and Thielecke, "Convergence of cellular and broadcast networks from a multi-radio perspective," *IEEE Personal Communications*, vol. 8, no. 2, pp. 51–56, April 2001.

[13] O. Benali et al., "Framework for an evolutionary path toward 4G by means of cooperation of networks," *IEEE Communications Magazine*, pp. 82–89, May 2004.

[14] F. Allamandri et al., "Service platform for converged interactive broadband broadcast and cellular wireless," *IEEE Trans. on Broadcasting*, vol. 53, no. 1, pp 200–211, March 2007.

[15] ETSI Standard: EN 300 744 V1.5.1, Digital Video Broadcasting (DVB): Framing structure, channel coding and modulation for digital terrestrial television, 2004.

[16] H. Ekstrom, et al., "Technical solutions for the 3G long-term evolution," *IEEE Commun. Mag.*, vol. 44, no. 3, pp. 38–45, Mar. 2006.

[17] H. Li, D. W. Matolak, "Phase noise and fading effects on system performance in MT-DS-SS," *IEEE Trans. on Veh. Tech.*, vol. 54, no. 5, pp.1759–1767, Sep. 2005.

[18] IEEE 802.16e-2005, IEEE "standard for local and metropolitan area networks—Part 16: Air interface for fixed broadband wireless access systems," IEEE, 2005.

[19] H. Li and H. Liu, "Capacity of OFDMA-based wirelessMAN", Book chapter in *Mobile WiMAX: Toward Broadband Wireless Metropolitan Area Networks*, edited by Y. Zhang and H. Chen, Auerbach Publications, CRC press, New York, 2007.

[20] J. Bingham, "Multicarrier modulation for data transmission: An idea whose time has come," *IEEE Commun. Mag.*, Vol. 28, No. 5, pp. 5–14, May 1990.

[21] A. Cohen and A. Lapidoth, "The Gaussian watermarking game," *IEEE Trans. Inform. Theory*, vol. 48, pp. 1639–1667, June 2002.

[22] C. Hua, C. Yu, and H. Zhongji, "Dirty paper coding with phase reshaping: New integration scheme for broadcast and unicast," in Personal, Indoor and Mobile Radio Communications, 2009 IEEE 20th International Symposium, 2009, pp. 2355–2359.

[23] S. Gaur, J. Acharya, and G. Long, "Enhancing ZF-DPC performance with receiver processing," *Wireless Communications, IEEE Transactions*, vol. 10, no. 12, pp. 4052–4056, 2011.

[24] M. Sharif and B. Hassibi, "A comparison of time-sharing, DPC, and beamforming for MIMO broadcast channels with many users," Communications, IEEE Transactions, vol. 55, no. 1, pp. 11–15, 2007.

[25] R. Zamir, S. Shamai, and U. Erez, "Nested linear/lattice codes for structured multiterminal binning." *IEEE Trans. Info. Theory*, vol. 48, no. 6, pp. 1250–1276, June 2002.

[26] B. Liu, H. Li, H. Liu, and S. Roy, "DPC-based hierarchical broadcasting: Design and implementation," *Vehicular Technology, IEEE Transactions*, vol. 57, no. 6, pp. 3895–3900, 2008.

[27] A. Goldsmith, *Wireless Communications*. New York: Cambridge University Press, 2005.

[28] L. Li and A. Goldsmith, "Capacity and optimal resource allocation for fading broadcast channels-Part I: Ergodic capacity," *IEEE Trans. Inform. Theory*, vol. 47, no. 3, pp. 1103–1127, Mar. 2001.

[29] D. Tse, "Optimal power allocation over parallel gaussian broadcast channels", unpublished, http://www.eecs.berkeley.edu/~dtse/broadcast2.pdf.

[30] H. Li and H. Liu, "An analysis of uplink OFDMA optimality," *Wireless Communications, IEEE Transactions*, vol. 6, no. 8, pp. 2972–2983, 2007.

[31] D. Hughes-Hartog, The capacity of a degraded spectral Gaussian broadcast channel PhD thesis, Stanford University, July 1995.

[32] S. Shamai, O. Somekh, and B. Zaidel, "Multi-cell communications: An information theoretic perspective." Joint Workshop on Communications and Coding, Donnini(Florence), Italy, October 2004.

[33] M. Eriksson, "Dynamic single frequency networks," *IEEE Journal on Selected Areas in Communications*, vol. 19, no. 10, pp. 1905–1914, October 2001.

[34] H. Li, G. Ru, Y. Kim, and H. Liu, "OFDMA capacity analysis in MIMO channels," *Information Theory, IEEE Transactions*, vol. 56, no. 9, pp. 4438–4446, Sep. 2010.

[35] Y. Seokhyun and K. Donghee, "System level performance of broadcast/unicast service overlay using superposition coding," in Personal, Indoor and Mobile Radio Communications, 2007. PIMRC 2007. IEEE 18th International Symposium, 2007, pp. 1–5.

[36] D. Kim and S. Yoon, "Superposition of broadcast and unicast in wireless cellular systems," *IEEE Communications Magazine*, vol. 46, no. 7, pp. 110–117, 2008.

[37] H. M. Costa, "Writing on dirty paper," *Information Theory, IEEE Trans.*, vol. 29, no. 3, pp. 439–441, 1983.

[38] H. Weingarten, Y. Steinberg, and S. Shamai, "The capacity region of the Gaussian MIMO broadcast channel," in Information Theory, 2004. ISIT 2004, Proceedings International Symposium, pp. 174, 2004.

[39] E. Visotsky and U. Madhow, "Optimum beamforming using transmit antenna arrays," *Proc. IEEE Veh. Tech. Conf.*, vol. 1, pp. 851–856 May 1999.

[40] H. Harashima and H. Miyakawa, "Matched-transmission technique for channels with intersymbol interference," *Communications, IEEE Transactions*, vol. 20, no. 4, pp. 774–780, 1972.

[41] M. Tomlinson, "New automatic equaliser employing modulo arithmetic," *Electronics Letters*, vol. 7, no. 5, pp. 138–139, 1971.

[42] R. Wesel and J. Cioffi, "Achievable rates for tomlinson-harashima precoding," *Information Theory, IEEE Transactions*, vol. 44, no. 2, pp. 824–831, 1998.

[43] W. Yu, D. Varodayan, and J. Cioffi, "Trellis and convolutional precoding for transmitter-based interference presubtraction," *Communications, IEEE Transactions*, vol. 53, no. 7, pp. 1220–1230, 2005.

[44] R. Zamir, S. Shamai, and U. Erez, "Nested linear/lattice codes for structured multiterminal binning," *Information Theory, IEEE Transactions*, vol. 48, no. 6, pp. 1250–1276, 2002.

[45] U. Erez, S. Shamai, and R. Zamir, "Capacity and lattice strategies for canceling known interference," *Information Theory, IEEE Transactions*, vol. 51, no. 11, pp. 3820–3833, 2005.

[46] U. Erez and S. ten Brink, "A close-to-capacity dirty paper coding scheme," *Information Theory, IEEE Transactions*, vol. 51, no. 10, pp. 3417–3432, 2005.

Chapter 8

Cooperation Based Dynamic Spectrum Sharing in CRNs

Ning Zhang, Haibo Zhou, Shaohua Wu, Ying Wang, Jon W. Mark, and Xuemin (Sherman) Shen

CONTENTS

Cognitive radio network is a promising solution to solve the spectrum scarcity problem by allowing dynamic spectrum sharing between unlicensed users and licensed users. In order to avoid interfering with licensed users, unlicensed users need to perform spectrum sensing. However, spectrum sensing might be inaccurate due to multipath fading and shadowing. Therefore, cooperation based dynamic spectrum sharing are introduced. Specifically, when unlicensed users want to transmit, they can be coordinated to cooperatively sense the spectrum bands to maximize the total expected available time. The coordination problem is formulated as a nonlinear integer programming problem, which is proved to be NP-complete. Then, the problem is first transformed to an associated stochastic optimization problem, which is solved by cross-entropy (CE) method of stochastic optimization. When unlicensed users are idle, they can earn credits by acting relays for licensed users to improve the latter's performance such as the secrecy. The earned credits can be utilized for spectrum trading in the future when they have traffic. The procedure of payment negotiation and transmission power allocation is modeled by Stackelberg game. By analyzing the game, the unlicensed users can determine the transmission powers for cooperation, while the licensed user can select the best payment. Finally, simulation results are provided to demonstrate the performance of the cooperation based proposed schemes.

8.1 Introduction

We have witnessed a massive growth in mobile data, which almost doubles every year. Moreover, it is expected to skyrocket in the foreseeable future due to the proliferation of devices and data-hungry applications. On one hand, the number of devices increases exponentially. It is reported that the number of the connected devices is around 25 billion in 2020 [1], due to development of mobile networks, the promising machine-to-machine (M2M) application, Internet of things (IoT), and Internet of Vehicles (IoV) [2, 3]. On the other hand, the wireless data is also boomed by the data-hungry applications, such as video streaming and online gaming. In the future, more and more multimedia-rich applications will emerge, leading to a tremendous increase in mobile data.

Such a dramatic increase in mobile traffic and devices imposes a huge demand on radio spectrum. However, radio spectrum, as a natural resource, is scarce and limited. Currently, the spectrum is managed by government agencies, such as Federal Communications Commission (FCC) in the USA, which assigns spectrum to licensed users for exclusive use on a long-term basis, aiming at avoiding interference among various wireless systems. Unfortunately, this static spectrum management policy has created a severe shortage of spectrum for unlicensed users. Moreover, spectrum scarcity problem is further exacerbated by spectrum underutilization of licensed users. The main cause is that licensed users generally do not fully utilize the assigned bandwidths most of the time, while unlicensed users are being starved for spectrum availability. Cognitive radio network (CRN) is a promising paradigm created in an attempt to provide high bandwidth to the users and improve spectrum utilization [1, 4–7, 9]. It enables dynamic spectrum sharing between unlicensed users and licensed users, in the fashion that unlicensed users can make use of the underutilized spectrum when licensed users are absent [10] [7].

In CRN, unlicensed users and licensed users are refereed to as secondary users (SUs) and primary users (PUs), respectively. In order to identify the unused spectrum (spectrum holes) and avoid the interference to PUs, spectrum sensing should be conducted by SUs. Particularly, the SU scans a certain spectrum range and detects whether the PU is active or not. Then, it selects the available spectrum band for access. The SU has to refrain from transmission in the current band and searches for a new band when the PU reclaims the frequency band. Therefore, spectrum sensing is so critical to both the PUs and SUs, which typically requires a high detection probability and a low false-alarm probability. However, spectrum sensing results might be inaccurate because of multipath fading and shadowing. For instance, when the SU experiences deep fading, it cannot detect the active PU. Then SU will make a wrong decision and access the channel, affecting the operation of PUs.

8.2   Cooperation-Based Dynamic Spectrum Sharing

To overcome the aforementioned issues, user cooperation can be leveraged, mainly in two forms: cooperative spectrum sensing and cooperative cognitive radio networking (CCRN) [12, 13]. For the former, cooperation is performed among SUs, where multiple SUs cooperate with each other to improve the spectrum sensing performance. For the latter, cooperation is carried out between SUs and PUs, where SUs cooperate with PUs to improve the PUs' transmission performance and then gain spectrum access opportunities as a reward.

8.2.1   *Cooperative Spectrum Sensing*

Cooperative spectrum sensing can improve the detection performance, which exploits the spatial diversity and multiuser diversity [14]. Instead of relying on individual decision, multiple SUs perform spectrum sensing and share the sensing results to make a combined decision through cooperation. Particularly, centralized cooperative spectrum sensing is performed in a three-step fashion. First, individual SUs perform local sensing locally. Then, all the cooperating SUs report the sensing results to the fusion center (FC), e.g., a common receiver. Last, the FC combines all the sensing results to make a final decision on whether the spectrum band is idle or busy. For distributed cooperative spectrum sensing, where there is no FC, SUs exchange the detection results among themselves and then converge to a final decision after several iterations. Through cooperation, a combined sensing decision can be derived from the spatially collected observations, which helps to overcome the deficiency of individual observations. It has been demonstrated that cooperative spectrum sensing can deal with multipath fading and shadowing effectively, mitigate the receiver uncertainty problem, and, hence, improve the detection performance significantly [15–18].

8.2.2   *Cooperative cognitive radio networking*

Cooperative communications have the potential to improve the transmission rate, save energy, enhance the reliability, and so on. When the source transmits message to the destination, the nodes in between also overhear it. Those intermediate nodes can process the received signal and retransmit it to the destination. Therefore, at the destination, the multiple copies of the message can be utilized to improve the reception performance by exploring the spatial diversity.

Because of those benefits, cooperative networking can be leveraged by the CRN to deal with challenges in spectrum sensing and to better explore transmission opportunities. In CCRN, SUs can cooperate with PUs to improve the latter's performance in terms of transmission rate, reliability, energy efficiency, and so on, and in return gain transmission opportunities [19–26]. Specifically, SUs can act as relaying nodes to improve transmission performance of PUs. Then, the PUs grant a period of time to the SUs as a reward. By leveraging cooperation between PUs and SUs, a "win-win" situation is created, where the PU's performance is enhanced and SUs can access the channel in the rewarding time. By this emerging cooperative networking, SUs can be relieved from the burden of spectrum sensing.

In this chapter, cooperative spectrum sensing in a multi-channel CRN is studied first, where multiple SUs cooperate with each other to detect unused channels and then share them. Specifically, for spectrum sensing, the objective of the CRN is to maximize the expected available time while keeping the interference to PUs under a predefined level. With the dynamics in the channel usage characteristics and the detection capacities, the coordination problem is formulated as a nonlinear integer programming problem. To find the solution efficiently, the deterministic optimization problem is first transformed to an associated stochastic optimization problem, which is then solved by the CE method of stochastic optimization. Then, we study cooperative cognitive radio networking framework, whereby the PU's security can be enhanced through cooperation with the SUs in present of multiple eavesdroppers when SUs have no traffic requirement. Two partner selection algorithms have been devised, which can select suitable SUs, acting as relays or jammers to maximize the secrecy rate. A game-theoretic incentive mechanism has been proposed to stimulate the SUs to participate into cooperation. With the proposed cooperative scheme, all the cooperative SUs can gain a ceratin amount of credits, which can be used in the future when needed.

The remainder of the chapter is organized as follows. Cooperative spectrum sensing and cooperative cognitive radio networking are studied in Section 7.3 and Section 7.4, respectively. Concluding remarks are provided in Section 7.5.

8.3 Cooperative Spectrum Sensing in Multi Channel Environments

In this section, we study cooperative spectrum sensing in a multi-channel environment. Each SU can choose only one channel in spectrum sensing due to the limitation of hardware. The objective of the CRN is to maximize the expected available time of all the channels, given that the PUs are sufficiently protected. To accomplish this, user scheduling needs to be studied to decide which SU to sense which channel. We consider a more general scenario: i) the various detection performance of individual SUs; and ii) the different usage characteristics of channels, such as average sojourn idle time and the probability of being idle.

8.3.1 System Model

In a CRN, the spectrum is divided into a set of channels with a fixed frequency bandwidth. There are K licensed channels, each of which can be either busy or idle. N SUs ($N \geq K$) seek for transmission opportunities through spectrum sensing. An ON-OFF channel usage model is adopted to model the states of each channel, which alternates between ON (busy) and OFF (idle). To avoid interference to PUs, SUs can access the channel only when it is in the state OFF. Suppose that PU_j operates over channel j and the state of each channel changes independently. Denote by α_j the transition rate for channel j ($1 \leq j \leq K$) from state ON to state OFF and β_j vice versa.

Spectrum sensing is carried out to detect the status of the channels. The popular spectrum sensing techniques contain energy detection, cyclostationary detection, and matched filtering. Energy detection is adopted due to the simplicity and minimal time overhead. Then, the detection probability p_d and the false alarm probability p_f are defined as follows:

$$p_d = Pr(D > \delta|H_1), \ p_f \ = Pr(D > \delta|H_0), \tag{8.1}$$

where H_1 and H_0 are the cases where the PU is present and absent, respectively. δ is the detection threshold and D is the test statistic. In particular, $D = \frac{1}{M}\sum_{n=1}^{M}|y(n)|^2$, where M is the number of samples in an observation period and $y(n)$ is the n-th sample of the received signal.

Without loss of generality, we consider the case of the complex-valued phase shift keying (PSK) signal and circular symmetric complex gaussian (CSCG) noise. According to [27], in this case, the false alarm probability of SU_i for channel j can be given by

$$p_f(i,j) = Q((\frac{\delta}{\sigma^2} - 1)\sqrt{M}), \tag{8.2}$$

where $Q(\cdot)$ is the complementary distribution function of the standard Gaussian. The Neyman-Pearson criterion is considered [28], where the false alarm probability is fixed. In other words, the false alarm probabilities are the same for all SUs, and we denote it by p_f for simplicity.

The detection probability for SU_i to sense channel j is given as follows:

$$p_d(i,j) = Q((\frac{\delta}{\sigma^2} - \overline{\gamma}_{i,j} - 1)\sqrt{\frac{M}{2\overline{\gamma}_{i,j} + 1}}), \tag{8.3}$$

where $\bar{\gamma}_{i,j}$ is the average received signal-to-noise ratio (SNR) from PU_j at SU_i. In particular, $\bar{\gamma}_{i,j} = \frac{P_{PU}h_{i,j}}{\sigma^2}$, where P_{PU} is the transmission power of the PU, $h_{i,j}$ is the average channel gain from PU_j to SU_i, and σ^2 is the noise power.

Given $p_f(i,j)$, based on (8.2) and (8.3), the detection probability $p_d(i,j)$ can be calculated as follows:

$$p_d(i,j) = Q(\frac{1}{\sqrt{2\bar{\gamma}_{i,j}+1}}(Q^{-1}(p_f(i,j)) - \sqrt{M}\gamma_{i,j})). \tag{8.4}$$

8.3.2 Spectrum Sensing in Multi-channel CRNs

8.3.2.1 Cooperative Spectrum Sensing

In cooperative spectrum sensing, the sensing results are combined based on a fusion rule, which can be AND rule, OR rule, the soft combination rule, or the majority rule [6]. When OR rule is adopted, PUs are considered to be present if at least one SU claims the presence of PUs. Suppose that each SU selects a channel for sensing at one time, and let $\mathbf{S}_j$ be the set of SUs selecting channel j. Then, the cooperative detection probability and false alarm probability can be given as follows:

$$F_d(j) = 1 - \prod_{i \in \mathbf{S}_j}(1 - p_d(i,j)) = 1 - \prod_{i \in \mathbf{S}_j} p_m(i,j) \tag{8.5}$$

$$F_f(j) = 1 - \prod_{i \in \mathbf{S}_j}(1 - p_f(i,j)) = 1 - \prod_{i \in \mathbf{S}_j} p_s(i,j), \tag{8.6}$$

where $p_m(i,j) = Pr(D < \delta|H_1) = 1 - p_d(i,j)$ and $p_s(i,j) = Pr(D < \delta|H_0) = 1 - p_f(i,j)$. The cooperative misdetection probability F_m^j is defined as the probability that the presence of PU is not detected, i.e., $F_m^j = 1 - F_d^j$.

When AND rule is adopted, the channel is considered to be busy if all the SUs claim the presence of PUs. Then, the cooperative detection probability and false alarm probability are respectively given by

$$F_d(j) = \prod_{i \in \mathbf{S}_j} p_d(i,j), \quad F_f(j) = \prod_{i \in \mathbf{S}_j} p_f(i,j). \tag{8.7}$$

8.3.2.2 Sensing Coordination

Suppose that AND rule is adopted by the secondary system. The objective of the secondary system is to maximize the expected available time of all the channels, under the constraint that the PUs are sufficiently protected. In the following, the sensing coordination/scheduling problem is formulated first. Then, a CE based approach is proposed to solve the problem.

According to the ON-OFF model, the sojourn times of ON state and OFF state, i.e., T_{ON}^j and T_{OFF}^j, for channel j, follow exponential distributions with means given by

$$\bar{T}_{ON}^j = \frac{1}{\alpha_j}, \quad \bar{T}_{OFF}^j = \frac{1}{\beta_j}. \tag{8.8}$$

Denote by P_{ON}^j and P_{OFF}^j the probabilities that channel j is in the ON state and OFF state, respectively, which can be given by

$$P_{ON}^j = \frac{\beta_j}{\alpha_j + \beta_j}, \quad P_{OFF}^j = \frac{\alpha_j}{\alpha_j + \beta_j}. \tag{8.9}$$

If channel j is detected to be in the OFF state when it is actually idle, the SUs can access the channel for a period of $\overline{T}_{OFF}^j$ on average. We define a channel selection matrix $\mathbf{I} = (I_{i,j})_{N \times K}$, where $I_{i,j} = \{0,1\}$ indicates whether or not SU_i selects channel j in spectrum sensing. Specifically, SU_i selects channel j for sensing when $I_{i,j} = 1$. Based on $\mathbf{I}$, the set of SUs choosing channel j can be determined by $\mathbf{S}_j = \{SU_i, I_{i,j} = 1\}$. The target is to maximize the total average available time, which can be formulated as follows:

$$\max_{\mathbf{I}} \sum_{j=1}^{j=K} \overline{T}_{OFF}^j P_{OFF}^j (1 - F_f(j))$$

$$s.t. \sum_{j=1}^{j=K} I_{i,j} \leq 1, i \in \{1,2,...,N\} \tag{8.10}$$

$$(1 - F_d(j)) P_{ON}^j \leq P_i$$

$$I_{i,j} = \{0,1\},$$

where P_i is the probability of inferencing the PU over channel i.

By using exterior point method, the constraint that $(1 - F_d(j)) P_{ON}^j \leq P_i$ can be removed. Then, the above problem can be transformed into the following format:

$$\max_{\mathbf{I}} \sum_{j=1}^{j=K} [\overline{T}_{OFF}^j P_{OFF}^j (1 - F_f(j)) - A(F_d(j)) U_0 (1 - F_d(j)) P_{ON}^j]$$

$$s.t. \sum_{j=1}^{j=K} I_{i,j} \leq 1, i \in \{1,2,...,N\} \tag{8.11}$$

$$I_{i,j} = \{0,1\},$$

where $U_0 > 0$ is a linear penalty factor when the constraint $(1 - F_d(j)) P_{ON}^j \leq P_i$ is violated. $A(F_d(j))$ is the indicator function, where $A(F_d(j)) = 1$ when $(1 - F_d(j)) P_{ON}^j \geq P_i$, and $A(F_d(j)) = 0$, otherwise.

8.3.2.3 Cross-Entropy-Based Approach

In the following, the CE method of stochastic optimization is employed to solve the sensing scheduling problem. The main idea of CE approach is to generate random samples according to a specified stochastic policy, and update the stochastic policy based on the outcome to produce a "better" sample in the next iteration.

We first define the strategy space $\mathbb{S}$ for SUs as follows:

$$\mathbb{S} := \{ch_1, ch_2, ..., ch_K\}, \tag{8.12}$$

where each SU can choose only one channel from $\mathbb{S}$.

The probability vector associated with the strategy space is defined as follows:

$$\mathbb{P}_t^i := \{p_{1,t}^i, p_{2,t}^i, ..., p_{K,t}^i\}, \sum_{j=1}^{K} p_{j,t}^i = 1, \tag{8.13}$$

where $\mathbb{P}_t^i$ is the stochastic policy of SU_i on the strategy space $\mathbb{S}$ at t-th iteration, and $p_{j,t}^i$ is the probability that SU_i chooses channel j.

The detailed procedure consists of the following five main steps:

1. (Initialization). The iteration counter t is set to 1. The uniform distribution on the strategy space $\mathbb{S}$ is selected as the initial stochastic policy $\mathbb{P}_0^i$ of all SUs. In other words, each SU selects the strategy from $\mathbb{S}$ uniformly, with the probability $1/K$.

2. (Sample Generation). According to the stochastic policy, Z samples are generated:

$$\mathbb{S}^i(z) := \{I_{i,1}(z), I_{i,2}(z), ..., I_{i,K}(z)\}, \tag{8.14}$$

where $\mathbb{S}^i(z)$ is the z-th strategy vector of SU_i with only one element to be "1". The associated probability for $I_{i,j}$ to be "1" is $p_{j,t}^i$.

3. (Performance evaluation). With the generated samples, the utilities $U(z)$ can be obtained according to (8.11). Then, the resulting $U(z)$ is arranged in a non-increasing order based on the values, i.e., $U^1 \geq U^2 \geq ... \geq U^Z$. Let υ be the $(1-\rho)$-th sample. We have $\upsilon = U_{\lceil(1-\rho)Z\rceil}$, where ρ is the percentage of samples obsolete at each iteration and $\lceil \cdot \rceil$ is the ceiling function.

4. (Stochastic Policy Update). Based on the outcomes, $\mathbb{P}_t^i := \{p_{1,t}^i, p_{2,t}^i, ..., p_{K,t}^i\}$ is updated using the following equation:

$$p_{j,t}^i = \frac{\sum_{z=1}^{N} X_{U^z \geq \upsilon} I_{i,j}(z) = 1}{\sum_{z=1}^{N} X_{U^z \geq \upsilon}}, \tag{8.15}$$

where $X_{U^z \geq \upsilon}$ is given by

$$X_{U^z \geq \upsilon} = \begin{cases} 1 & U^z \geq \upsilon \\ 0 & \text{otherwise.} \end{cases} \tag{8.16}$$

5. When the stopping criterion is met, e.g., the maximum number of iterations (i.e., T), then stop; otherwise, increase the iteration counter t by 1, and reiterate from step 2.

8.3.3   Simulation Results

In this section, simulations are carried out to evaluate the performance of the proposed CE based approach. In a 2 km×2 km area, PUs are located inside the circle with 1 km radius, while SUs are randomly distributed outside the circle. The transmission power of PUs and the noise power is set to 10 mw and -80 dB, respectively. The channel gain between a generic SU and a PU is calculated by $h = \frac{k}{d^\mu}$, where $k = 1$ and $\mu = 3.5$. The false alarm p_f is set to 0.1. The average results are obtained through Monte Carlo simulation.

Figure 8.1 shows the convergence speed of the proposed CE algorithm when the number of channel is set to 5. It can be seen that the CE algorithm converges after a few iterations, which means all SUs select a channel for sensing, with probability 1. It can also be seen that a larger number of SUs results in a larger utility because of the user diversity gain.

Figure 8.2 shows the utility of the secondary system, with respect to the number of SUs for the case with 4 channels. We compare the proposed CE algorithm with the greedy algorithm in [29]. Greedy 1 algorithm does not consider the dynamics of channels and detection probabilities of SUs, while Greedy 2 algorithm does. It can be seen that the CE algorithm can achieve higher utility than the Greedy algorithms because of the iterative update mechanism.

Figure 8.3 shows the utility of the secondary system, with respect to the number of channels when we have 10 SUs. It can be seen that the utility increases as the number of channels increases. Moreover, it can also be seen that the Greedy 2 algorithm performs slightly better than the Greedy 1 algorithm, while the proposed CE algorithm can achieve the highest utility among these algorithms.

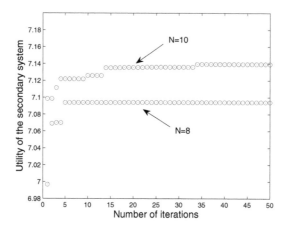

Figure 8.1: Convergence of CE algorithm.

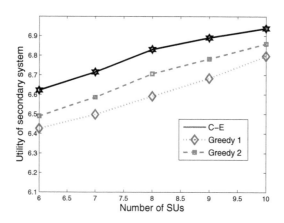

Figure 8.2: Utility of SUs vs. the number of SUs.

8.4   Cooperative Cognitive Radio Networking

In this section, we study the user cooperation to enhance the PU's security when SUs have no traffic, where the PU cooperates with SUs to transmit message securely in presence of multiple eavesdroppers. To stimulate SUs for cooperation, the PUs grant credits to them. The earned credits can be utilized by SUs for spectrum leasing in the future when they have traffic. In other words, the SUs can accumulate credits through cooperation with PUs and consume credits in the spectrum trading market when needed. In this study, we mainly focus on the following issues: i) with whom to cooperate; ii) how to determine and share the credits. To address those issues, a cooperative framework is proposed, whereby the PU selects multiple SUs and stimulates them by granting an amount of rewards. Specifically, multiple SUs acting as cooperative relays and jammers are selected by the PU using greedy-based approach. Then, the PU and the SUs negotiate for the payment and transmission power, which is modeled by a two-layer game. At the top layer, a buyer-seller game is utilized, where the PU pays to buy the service provided by the SUs. At the bottom layer, all the SUs share the reward by determining their transmission powers in a distributed way, which is formulated

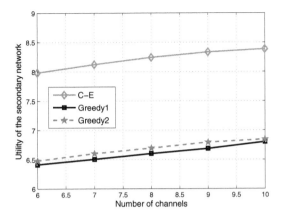

Figure 8.3: Utility of SUs vs. the number of channels.

as a non-cooperative power selection game. By analyzing the game, the SUs can determine the transmission powers for cooperation, while the PU can select the best payment.

8.4.1 System Model

The system consists of a primary transmitter as the source (S), a primary receiver as the destination (D), M intermediate SUs ($i = 1, 2, ..., M$), and multiple eavesdroppers (E). The PU aims to increase the secrecy rate through cooperation with SUs, while SUs request rewards from the PU, which can be utilized in spectrum leasing in the future. To improve the security level, S has to select the suitable cooperative SUs.

We consider a slow, flat, block Rayleigh fading environment, where channels keep fixed in a time slot and vary independently over different time slots. The channel coefficients from S to D and S to a specific E are denoted by h_{sd} and h_{se}, respectively. The channel coefficient from S to SU $i \in \mathcal{M}$ is denoted by h_s^i. Similarly, the channel coefficients from SU $i \in \mathcal{M}$ to D and E are h_d^i and h_e^i, respectively. The channel state information (CSI) is considered to be available, including D-related CSI (D-CSI) and E-related CSI (E-CSI), which is a common assumption in the literature [30–33]. In addition, white Gaussian noise is assumed with zero mean, and the one-side power spectral density is N_0.

8.4.2 Partner Selection

Secrecy rate is utilized as the performance metric for secure communication, which is defined as the difference between the transmission rate at D and that at E. In the following, the secrecy rate through cooperation is analyzed, and then the suitable partners are determined.

At D, the SNR γ_{sd} from the source S is given by

$$\gamma_{sd} = \frac{P_s |h_{sd}|^2}{N_0},$$ (8.17)

where P_s is the transmission power of S.

Suppose that SU i is in the relay set $\mathbb{R}$, then the SNR from relay i using amplify-and-forward

(AF) mode can be given as follows [34]:

$$\gamma_d^i = \frac{1}{N_0} \frac{P_s \left|h_s^i\right|^2 P_i \left|h_d^i\right|^2}{P_s \left|h_s^i\right|^2 + P_i \left|h_d^i\right|^2 + N_0}, i \in \mathbb{R},$$
(8.18)

where P_i is the transmission power of SU i.

Suppose that SU j is in the jammer set $\mathbb{J}$, the interference γ_d^j at the destination caused by jammer j is:

$$\gamma_d^j = \frac{P_j \left|h_d^j\right|^2}{N_0}, j \in \mathbb{J}.$$
(8.19)

Then, the achievable rate at D can be given as follows:

$$R_d = \frac{W}{2} \log_2(1 + \frac{\gamma_{sd} + \sum_{i\in\mathbb{R}} \gamma_d^i}{1 + \sum_{j\in\mathbb{J}} \gamma_d^j}).$$
(8.20)

For a generic eavesdropper, e.g., k-th E, the SNR γ_{se} from S can be given by

$$\gamma_{se} = \frac{P_s \left|h_{se}\right|^2}{N_0}.$$
(8.21)

The SNR γ_e^i from relay i, where $i \in \mathbb{R}$, can be given as follows:

$$\gamma_e^i = \frac{1}{N_0} \frac{P_s \left|h_s^i\right|^2 P_i \left|h_e^i\right|^2}{P_s \left|h_s^i\right|^2 + P_i \left|h_e^i\right|^2 + N_0}, i \in \mathbb{R}.$$
(8.22)

The interference γ_e^j at E caused by jammer j can be given as follow:

$$\gamma_e^j = \frac{P_j \left|h_e^j\right|^2}{N_0}, j \in \mathbb{J}.$$
(8.23)

Similarly, the achievable rate at the k-th E is obtained as below:

$$R_e^k = \frac{W}{2} \log_2(1 + \frac{\gamma_{se} + \sum_{i\in\mathbb{R}} \gamma_e^i}{1 + \sum_{j\in\mathbb{J}} \gamma_e^j}).$$
(8.24)

According to the definition of secrecy rate, the secrecy rate is given by

$$R_{sec}^k = R_d - R_e^k,$$
(8.25)

where R_d and R_e^k are given in (8.20) and (8.24), respectively.

Since we have multiple eavesdroppers, the overall secrecy rate R_{sec} is given by

$$R_{sec} = \max\{0, \min_k\{R_d - R_e^k\}\}.$$
(8.26)

To maximize the secrecy rate, the source first selects the suitable cooperative relays and jammers, given that the transmission power of the potential participants is fixed. This problem can be formulated as follows:

$$\max_{X_{i,j}, \forall i \in \{1,2,...,M\}} R_{sec}$$

$$\text{s.t.} \sum_{j\in\{R,J,N_u\}} X_{i,j} = 1, \forall i \in \{1,2,...,M\}$$

$$X_{i,j} \in \{0,1\}, \forall i \in \{1,2,...,M\} \text{ and } \forall j \in \{R,J,N_u\}.$$

Specifically, the binary variable $X_{i,j}$ indicates the role of node i, where j can be $\{R,J,N_u\}$, which correspond to act as a relay (R), a jammer (J), or keep silent (N_u). For instance, when $X_{i,R} = 1$, node i acts as a relay. The secrecy rate $R_{sec} = \frac{W}{2}\log_2(1 + \frac{\gamma_{sd}+\sum_{i\in\mathbb{R}}\gamma_d^i}{1+\sum_{j\in\mathbb{J}}\gamma_d^j}) - \frac{W}{2}\log_2(1 + \frac{\gamma_{se}+\sum_{i\in\mathbb{R}}\gamma_e^i}{1+\sum_{j\in\mathbb{J}}\gamma_e^j})$, where the relay and jammer set can be determined by $\mathbb{R} = \{i, X_{i,R} = 1\}$ and $\mathbb{J} = \{i, X_{i,J} = 1\}$. Exclusive search can obtain the optimal solution. However, the complexity is high, since the search space is exponential to the number of intermediate nodes. Instead, a greedy partner selection algorithm is developed, as shown in Algorithm 7.1. The main idea is to select the best cooperative SU and its role at each round until the overall secrecy rate cannot be improved.

Algorithm 1 Greedy Parter Selection Algorithm

Require: $\mathcal{M}, h_s^i, h_d^i, h_e^i, \forall i \in \mathcal{M}$.
Ensure: Partner selection results $\mathbb{R}$ and $\mathbb{J}$
  1: **(Initialization)**: Set $R_{sec} = 0, \forall i \in \mathcal{M}$.
  2: **for** $i \leftarrow 1$ to M **do**
  3:     **for** $j \in \{R,J,N_u\}$ **do**
  4:         $X_{i,j} = 1$
  5:         Calculate R'_{sec}
  6:     **end for**
  7:     Find the maximum R'_{sec}
  8:     **if** $R'_{sec} > R_{sec}$ **then**
  9:         $R_{sec} = R'_{sec}$
 10:         $X_{i,j} = \arg\max R'_{sec}$
 11:     **end if**
 12: **end for**
 13: **return** $\mathbb{R} = \{i, X_{i,R} = 1\}$ and $\mathbb{J} = \{i, X_{i,J} = 1\}$

8.4.3   Resource Allocation

After the cooperative SUs are determined, the PU and SUs determine payment and transmission power to maximize their own utilities, respectively. The negotiation procedure is modeled as a two-layer game. At the top layer, a buyer-seller game is utilized, in which the PU pays for the service provided by the SUs. At the bottom layer, the SUs select the transmission power to share the reward in a distributed way, which is formulated as a non-cooperative power selection game. Through analysis, the SUs can choose the best transmission powers for cooperation, while the PU can select the best payment.

8.4.3.1   Utility Functions

The utility of the primary source is given by

$$U_s = \lambda_1 R_{sec} - R_m, \tag{8.27}$$

where λ_1 is the profit per secrecy rate, and $0 \leq R_m \leq R_{max}$ is the payment for the cooperative SUs.

The cooperative relays and jammers share the payment according to their contributions to the achieved secrecy rate. The payment that the SU can obtain is proportional to its contribution during the cooperation. For the relay, it aims to increase the perfect secrecy of the relaying link, compared with that of the eavesdropper link, the contribution can be approximately given by $\frac{P_r^i|h_{rd}^i|}{|h_{re}^i|}$. For jammer, it is leveraged to increase more artificial noise at eavesdropper than at the destination, the contribution can be approximately given by $\frac{P_j^i|h_{je}^i|}{|h_{jd}^i|}$.

Based on the above analysis, the utility of the selected node i is defined as

$$U_i = \frac{P_i r_i}{\sum_{j \subseteq \mathbb{C}} P_j r_j} R_m - \lambda_2 P_i,$$

where $\mathbb{C} := \mathbb{R} \uplus \mathbb{J}$ is the set of selected nodes with the size N, λ_2 is the cost rate for transmission power, and the contribution factor r_i is defined as follows:

$$r_i = \begin{cases} \frac{|h_d^i|}{|h_e^i|}, & i \in \mathbb{R} \\ \frac{|h_e^i|}{|h_d^i|}, & i \in \mathbb{J}. \end{cases} \tag{8.28}$$

The buyer-seller game can be analyzed by the backward induction method. First, the optimal strategies (i.e., the transmission powers) of the SUs are analyzed, given the fixed strategy of the PU (i.e., the payment). Second, based on the outcomes of the first step, the PU its optimal strategy, aware of the effects of its decision on the strategies of the SUs. By doing so, the best strategies of both the PU and the SUs can be obtained, such that the respective utilities can be maximized.

8.4.3.2 Noncooperative Power Selection Game

To stimulate the cooperation of the SUs, the PU pays for their service. Each SU gets a portion of payment according to its contribution. For a given payment, each cooperative SU strives to maximize its own utility by selecting transmission power, which is modeled as a non-cooperative power selection game.

Definition 8.1 Non-cooperative power selection game is defined by $G = \{\mathbb{C}, \{\mathbb{S}_i\}, \{U_i\}\}$, where $\mathbb{C}$ is the set of players, $\mathbb{S}_i$ is the strategy set of SU i, and U_i is the utility function of SU i.

Note that $\mathbb{S}_i$ is the transmission power that SU i can choose, and the utility function of node i is given as follows:

$$U_i = \frac{P_i r_i}{\sum_{j \subseteq \mathbb{C}} P_j r_j} R_m - \lambda_2 P_i.$$

Theorem 8.1
There exists a unique Nash equilibrium in the noncooperative power selection game $G = \{\mathbb{C}, \mathbb{S}_i, \{U_i\}\}$.

Due to the space limitation, the proof is omitted, and interested readers can refer to [35].

Since U_i is concave with respect to P_i, the best response correspondence can be obtained by setting the first derivative of U_i, with respect to P_i to 0, as follows:

$$\frac{\partial U_i}{\partial P_i} = -\frac{-r_i R_m A_i + \lambda_2 A_i^2 + 2\lambda_2 A_i P_i r_i + \lambda_2 P_i^2 r_i^2}{\left(\sum_{j \subseteq \mathbb{C}} P_j r_j\right)^2} = 0, \tag{8.29}$$

where $A_i = \sum_{j \neq i, j \subseteq \mathbb{C}} w_j P_j r_j$. By solving it, the solutions are given by

$$
P_i^* = \begin{cases} 0 & \text{if } \sum_{j \neq i, j \subseteq \mathbb{C}} P_j r_j \geq \frac{R_m P_i r_i}{\lambda_2} \\ \frac{1}{r_i} \left(\sqrt{\frac{R_m P_i r_i A_i}{\lambda_2}} - A_i \right) & \text{if } \sum_{j \neq i, j \subseteq \mathbb{C}} P_j r_j < \frac{R_m P_i r_i}{\lambda_2} \text{ and } \frac{1}{r_i} \left(\sqrt{\frac{R_m P_i r_i A_i}{\lambda_2}} - A_i \right) < P_{max} \\ P_{max} & \text{otherwise.} \end{cases} \tag{8.30}
$$

By solving the equations set (8.29), we can find the unique equilibrium as follows:

$$
P_i^* = [\min\{\frac{R_m r_i B_i}{\lambda_2 (r_i + B_i)^2}, P_{max}\}]^+, \tag{8.31}
$$

where $B_i = \frac{(N-1)r_i}{\sum_{j=1}^{N} \frac{r_i}{r_j} - N + 1}$.

8.4.3.3 Source Node Utility Maximization

Based on the results of the power selection game, the PU selects the optimal payment to maximize its utility, which can be formulated as:

$$
\max_{R_m} \ U_s = \lambda_1 R_{sec} - R_m
$$
$$
s.t. \ 0 \leq R_m \leq R_{max}. \tag{8.32}
$$

Specifically, R_{sec} is obtained when the SUs use the transmission power given by (8.31), which is a function of R_m. Therefore, the objective function of the PU becomes a function with respect to one single parameter R_m. The classic approach can be utilized to find the extremum by setting the first derivative of U_s, with respect to R_m equal to 0, and then compare the extremum with the boundary to find the best payment R_m^*. Finally, we can obtain the best strategy of SUs by substituting R_m^* into (8.31).

8.4.4 Simulation Results

In this section, simulations are carried out to evaluate the performance of the proposed scheme. The simulation is set up as follows. In a 1 km× 1 km area, the source, the destination, and two eavesdroppers are located at the origin, (1 km, 0.5 km), (1 km, −0.5 km), and (0.8 km, −0.4 km), respectively, while a set of nodes are located in between. The maximum transmission power of all nodes are set to 1 W, while the noise power is set to −70 dB. The average power gains between nodes is calculated by the path loss with exponent $\mu = 3.5$. The maximum power is set to 10 W.

To evaluate the performance of the proposed partner selection algorithm, with respect to the number of intermediate nodes, Monte Carlo simulation is carried out, which consists of 500 trials. At each trial, a number of intermediate nodes are uniformly distributed in the area. Figure 8.4 shows the average secrecy rate using the exhaustive search algorithm, the proposed greedy algorithm, and single relay and jammer selection algorithm in [36]. The exhaustive search algorithm has the best performance, which provides a performance benchmark. It can be seen that the proposed algorithm can achieve a higher secrecy rate, compared with the single relay and jammer selection algorithm. This is because they can fully exploit the benefits of cooperation by leveraging multiple relays and jammers.

In the following, we validate the incentive mechanism in the network scenario, as shown in Figure 8.5. The source, destination, and eavesdroppers are fixed at the same location as before, while 15 intermediate nodes are distributed at the locations marked in the figure. The reward that the source can choose ranges from 0 to 100. Figure 8.6 shows the averaged utility of the source, with respect to the amount of reward, for different λ_1 and λ_2. It can be seen that the overall utility

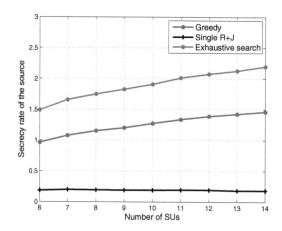

Figure 8.4: Comparison among different partner selection algorithms.

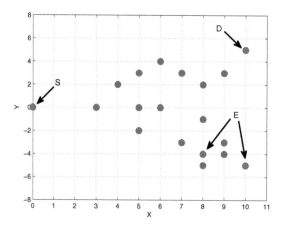

Figure 8.5: The network scenario for simulation.

first increases and then decreases as the reward increases. The reason is that, at the beginning, the partners are willing to devote more transmission power during cooperation with increasing reward, which leads to an increase in the secrecy rate. However, when the reward keeps rising, the cost also increases, which will lower the overall utility. It can also be seen that there exists an optimal reward, such that the utility can be maximized. It can also be seen that a larger λ_1 leads to a greater utility and payment because the source node cares more about the secrecy rate and is willing to pay more reward to increase the secrecy rate. Moreover, a larger λ_2 leads to a lower utility, since the intermediate node cares more about their energy consumption, and it will devote less power to cooperate given the same payment.

Figure 8.7 shows the utilities of SUs, averaged over fading distribution. It can be seen that the SUs, who contribute to increase the secrecy rate of the PU, can receive a certain amount of reward through cooperation. It implies that all the partners have the incentive for cooperation. Moreover, the node located at (0.9 km, −0.4 km) act as a jammer (node 13), while other nodes receiving non-zero rewards act as relays.

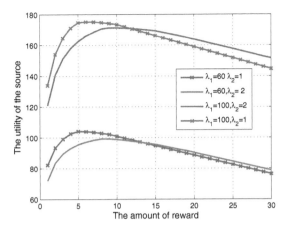

Figure 8.6: Utility of the source vs. the amount of rewards.

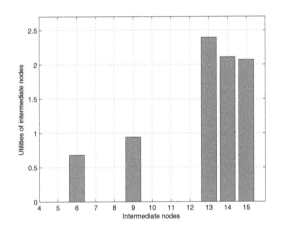

Figure 8.7: Utilities of intermediate nodes averaged over fading when $\lambda_1 = 60$ and $\lambda_2 = 1$.

8.5 Conclusion

In this chapter, we have studied cooperation-based dynamic spectrum sharing in cognitive radio networks. We have investigated cooperative spectrum sensing in a multi-channel scenario to maximize the expected available time when unlicensed users have traffic to send. When unlicensed users are idle, they can earn credits by acting relays or jammers for licensed users to improve the latter's secrecy. The earned credits can be utilized for spectrum trading in the future when they have traffic. Numerical results have demonstrated that, with the proposed cooperative schemes, the SUs can obtain longer average access time and gain a ceratin amount of credits through cooperation.

To better explore the spectrum access opportunities, further research is required in the following aspects: i) an integrated dynamic spectrum sharing framework could be devised; ii) the effect of imperfect CSI on cooperation has to be studied, since the channel estimation cannot be perfect; and iii) security and privacy issues should be considered because cooperation may involve malicious users [37, 38].

References

[1] URL: http://www.qualcomm.com/media/documents/wireless-networks-1000x-more-small-cells.

[2] N. Lu, N. Zhang, N. Cheng, X. Shen, J. W. Mark, and F. Bai, "Vehicles meet infrastructure: Toward capacity–cost tradeoffs for vehicular access networks," *IEEE Transactions on Intelligent Transportation Systems*, vol. 14, pp. 1266–1277, 2013.

[3] N. Lu, N. Cheng, N. Zhang, X. Shen, and J. W. Mark, "Connected vehicles: Solutions and challenges," *IEEE Internet of Things Journal*, vol. 1, no. 4, pp. 289–299, 2014.

[4] I. Akyildiz, W. Lee, M. Vuran, and S. Mohanty, "A survey on spectrum management in cognitive radio networks," *IEEE Communications Magazine*, vol. 46, no. 4, pp. 40–48, 2008.

[5] S. Gong, P. Wang, Y. Liu, and W. Zhuang, "Robust power control with distribution uncertainty in cognitive radio networks," *IEEE Journal on Selected Areas in Communications*, vol. 31, no. 11, pp. 2397–2408, 2013.

[6] N. Zhang, H. Liang, N. Cheng, Y. Tang, J. W. Mark, and X. Shen, "Dynamic spectrum access in multi-channel cognitive radio networks," *IEEE Journal on Selected Areas in Communications*, vol. 32, no. 11, pp. 2053–2064, 2014.

[7] S. Gunawardena and W. Zhuang, "Service response time of elastic data traffic in cognitive radio networks," *IEEE Journal on Selected Areas in Communications*, vol. 31, no. 3, pp. 559–570, 2013.

[8] I. Akyildiz, W. Lee, M. Vuran, and S. Mohanty, "Next generation/dynamic spectrum access/cognitive radio wireless networks: A survey," *Computer Networks*, vol. 50, no. 13, pp. 2127–2159, 2006.

[9] N. Cheng, N. Zhang, N. Lu, X. Shen, J. W. Mark, and F. Liu, "Opportunistic spectrum access for cr-vanets: A game-theoretic approach," *IEEE Transactions on Vehicular Technology*, vol. 63, pp. 237–251, 2014.

[10] J. Mitola III and G. Maguire Jr., "Cognitive radio: Making software radios more personal," *IEEE Personal Communications*, vol. 6, no. 4, pp. 13–18, 1999.

[11] S. Haykin, "Cognitive radio: Brain-empowered wireless communications," *IEEE Journal on Selected Areas in Communications*, vol. 23, no. 2, pp. 201–220, 2005.

[12] N. Zhang, H. Zhou, K. Zheng, N. Cheng, J. W. Mark, and X. Shen, "Cooperative heterogeneous framework for spectrum harvesting in cognitive cellular network," *IEEE Communications Magazine*, vol. 53, no. 5, pp. 60–67, 2015.

[13] P. Li, S. Guo, W. Zhuang, and B. Ye, "On efficient resource allocation for cognitive and cooperative communications," *IEEE Journal on Selected Areas in Communications*, vol. 32, no. 2, pp. 264–273, 2014.

[14] G. Ganesan and Y. Li, "Cooperative spectrum sensing in cognitive radio networks," in *Proceedings of IEEE DySPAN*. IEEE, 2005, pp. 137–143.

[15] D. Cabric, S. Mishra, and R. Brodersen, "Implementation issues in spectrum sensing for cognitive radios," in *Proceedings of the 38th. Asilomar Conference on Signals, Systems, and Computers,*, pp. 772–776, 2004.

[16] A. Ghasemi and E. S. Sousa, "Collaborative spectrum sensing for opportunistic access in fading environments," in *Proceedings of IEEE DySPAN*, 2005.

[17] S. Mishra, A. Sahai, and R. Brodersen, "Cooperative sensing among cognitive radios," *Proceedings of IEEE ICC*, 2006.

[18] W. Lee and I. Akyildiz, "Optimal spectrum sensing framework for cognitive radio networks," *IEEE Transactions on Wireless Communications*, vol. 7, no. 10, pp. 3845–3857, 2008.

[19] J. Zhang and Q. Zhang, "Stackelberg game for utility-based cooperative cognitiveradio networks," in *Proc. of ACM MobiHoc'09*, 2009.

[20] O. Simeone, I. Stanojev, S. Savazzi, Y. Bar-Ness, U. Spagnolini, and R. Pickholtz, "Spectrum leasing to cooperating secondary ad hoc networks," *IEEE Journal on Selected Areas in Communications*, vol. 26, no. 1, pp. 203–213, 2008.

[21] N. Zhang, N. Lu, N. Cheng, J. W. Mark, and X. Shen, "Cooperative spectrum access towards secure information transfer for CRNs," *IEEE Journal on Selected Areas in Communications*, vol. 31, no. 11, pp. 2453–2464, 2013.

[22] S. Hua, H. Liu, M. Wu, and S. Panwar, "Exploiting MIMO antennas in cooperative cognitive radio networks," in *Proceedings IEEE INFOCOM*, 2011.

[23] Y. Han, A. Pandharipande, and S. Ting, "Cooperative decode-and-forward relaying for secondary spectrum access," *IEEE Transactions on Wireless Communications*, vol. 8, no. 10, pp. 4945–4950, 2009.

[24] Y. Yi, J. Zhang, Q. Zhang, T. Jiang, and J. Zhang, "Cooperative communication-aware spectrum leasing in cognitive radio networks," in *Proceeding of IEEE DySPAN'10*, 2010.

[25] N. Zhang, N. Cheng, N. Lu, H. Zhou, J. W. Mark, and X. Shen, "Risk-aware cooperative spectrum access for multi-channel cognitive radio networks," *IEEE Journal on Selected Areas in Communications*, vol. 32, no. 3, pp. 516–527, 2014.

[26] I. Stanojev, O. Simeone, U. Spagnolini, Y. Bar-Ness, and R. Pickholtz, "Cooperative arq via auction-based spectrum leasing," *IEEE Trans. Commun.*, vol. 58, no. 6, pp. 1843–1856, 2010.

[27] Y.-C. Liang, Y. Zeng, E. C. Peh, and A. T. Hoang, "Sensing-throughput tradeoff for cognitive radio networks," *IEEE Transactions on Wireless Communications*, vol. 7, no. 4, pp. 1326–1337, 2008.

[28] G. Ganesan, Y. Li, B. Bing, and S. Li, "Spatiotemporal sensing in cognitive radio networks," *IEEE Journal on Selected Areas in Communications*, vol. 26, no. 1, pp. 5–12, 2008.

[29] H. Yu, W. Tang, and S. Li, "Optimization of cooperative spectrum sensing in multiple-channel cognitive radio networks," in *Proceedings of IEEE GLOBECOM*, 2011.

[30] L. Dong, Z. Han, A. Petropulu, and H. V. Poor, "Improving wireless physical layer security via cooperating relays," *IEEE Transactions on Signal Processing*, vol. 58, no. 3, pp. 1875–1888, 2010.

[31] H. Wang, Q. Yin, and X. G. Xia, "Distributed beamforming for physical-layer security of two-way relay networks," *IEEE Transactions on Signal Processing*, vol. 60, pp. 3532–3545, 2012.

[32] G. Zheng, L. Choo, and K. Wong, "Optimal cooperative jamming to enhance physical layer security using relays," *IEEE Transactions on Signal Processing*, vol. 59, no. 3, pp. 1317–1322, 2011.

[33] J. Li, A. Petropulu, and S. Weber, "On cooperative relaying schemes for wireless physical layer security," *IEEE Transactions on Signal Processing*, vol. 59, no. 99, pp. 4985–4997, 2011.

[34] J. Laneman, D. Tse, and G. Wornell, "Cooperative diversity in wireless networks: Efficient protocols and outage behavior," *IEEE Transactions on Information Theory*, vol. 50, no. 12, pp. 3062–3080, 2004.

[35] N. Zhang, N. Cheng, N. Lu, X. Zhang, J. Mark, and X. Shen, "Partner selection and incentive mechanism for physical layer security," *IEEE Transactions on Wireless Communication*, vol. PP, no. 99, p. 1, 2015.

[36] I. Krikidis, J. Thompson, and S. McLaughlin, "Relay selection for secure cooperative networks with jamming," *IEEE Transactions on Wireless Communications*, vol. 8, no. 10, pp. 5003–5011, 2009.

[37] Q. Shen, X. Liang, X. Shen, X. Lin, and H. Y. Luo, "Exploiting geo-distributed clouds for a e-health monitoring system with minimum service delay and privacy preservation," *IEEE Journal of Biomedical and Health Informatics*, vol. 18, no. 2, pp. 430–439, 2014.

[38] S. Li, H. Zhu, Z. Gao, X. Guan, K. Xing, and X. Shen, "Location privacy preservation in collaborative spectrum sensing," in *Proceedings IEEE INFOCOM*, 2012.

Chapter 9

Cooperation in Cognitive Radio Networks

Junni Zou

CONTENTS

Cooperative communications [1]- [3], a new technique for mitigating path loss and channel fading, has attracted much attention. It enables users to relay data for each other and thus creates a virtual multiple-input-multiple-output (MIMO) system for cooperative diversity. Recently, incorporations of cooperation concept into cognitive radio networks has become a new cognitive radio paradigm. It employs cooperative relay to assist transmission and improve spectrum efficiency. Cooperation in cognitive radio networks is mainly classified into two categories: i) cooperation among secondary users; ii) cooperation between primary users and secondary users. The first category aims at improving the performance of secondary transmission, in which a secondary user acts as a relay and assists transmissions of other secondary users [4] [5]. Generally, the solutions for traditional cooperative communications are valid for cooperation among secondary users. The only difference is that dynamic spectrum access must be considered in the latter. The second category benefits both primary and secondary users in which different rights of primary users and secondary users to the spectrum are taken into account, thus, it is more challenging than the first category. The cooperation between primary and secondary users can be further divided into two classes: i) secondary users coopera-tively relay data for primary users; ii) primary users cooperatively relay data for secondary users. In this chapter, we discuss the pricing and spectrum allocation of these two cooperation classes from the perspective of game theory.

9.1 Secondary Users Relaying for Primary Users

The motivation of secondary users relaying for primary users is that secondary users work as coop-erative relays for primary users and can, on one hand, increase the quality of service (e.g., in terms of rate or probability of outage) of the primary transmission. On the other hand, secondary users can obtain more opportunity to access the channel. For instance, Simeone et al. in [6] proposed a cooperation-based spectrum leasing scheme, where a primary user leases the owned spectrum to a subset of secondary users for a fraction of time in exchange for cooperation from secondary users. Zhang et al. in [7] proposed a cooperative cognitive radio network framework, in which some sec-ondary users are selected by primary users as cooperative relays, and in return, they obtain more spectrum access opportunities. Similar works can also be found in [8]– [10]. In the following, we model the spectrum access problem of secondary users relaying for primary users as a Stackelberg game, and analytically show the existence of a Nash equilibrium.

9.1.1 System Model

Consider a CR system consisting of a primary user (PU) and a set $\mathcal{S}_{\text{tot}}$ of secondary users (SUs). The PU is equipped with a primary transmitter (PT) and a primary receiver (PR). Each SU i, $i \in \mathcal{S}_{\text{tot}}$, is equipped with a secondary transmitter ST i and a secondary receiver SR i. The channels between nodes are modeled as independent proper complex Gaussian random variables, changing slowly but stable within each slot. Let h_p denote the channel gain between PT and PR, $h_{ps,i}$ be the channel gain between PT and ST i, $h_{sp,i}$ be the channel gain between ST i and PR, and $h_{s,i}$ be the channel gain between ST i and SR i, for any $i \in \mathcal{S}_{\text{tot}}$. Further, let P_p and $P_{s,i}$ represent the transmit power of PT and ST i, respectively, and N_0 be the power of additive white Gaussian noise.

As the authorized user, the PU has the exclusive usage right of the licensed spectrum band. The PU may choose a subset $\mathcal{S}$, $\mathcal{S} \subseteq \mathcal{S}_{\text{tot}}$, of the SUs as the cooperative relays, and in return, give these SUs the chance to access the channel. The selected SUs are allowed to use the channel only if they relay for the PU and meanwhile make some payment to the PU.

Each time slot of data transmission is divided into two parts: T_p and T_s, as shown in Figure 9.1. T_p, $0 \leq T_p \leq 1$ is used for primary transmission, and T_s, $T_s = 1 - T_p$ is used for secondary transmission. In the primary transmission period, the first duration of α is used for the PT to transmit its data to the STs, and the second duration of $(1 - \alpha)$ is used for cooperative transmission. In the secondary transmission period, the STs selected by the PT transmit their own data in a time-division multiplexing access (TDMA) mode.

In primary transmission period: In the first fraction, the PT broadcasts its data to all the STs, and the transmission rate is determined by the worst channel $h_{ps,i}$,

$$R_{ps} = W\log_2\left(1 + \frac{\min_{i \in \mathcal{S}}|h_{ps,i}|^2 P_p}{N_0}\right). \tag{9.1}$$

In the second fraction, both the PT and the STs transmit the primary data to the PR. For a maximum cooperative diversity, assume that distributed space-time coding (DSTC) cooperation is

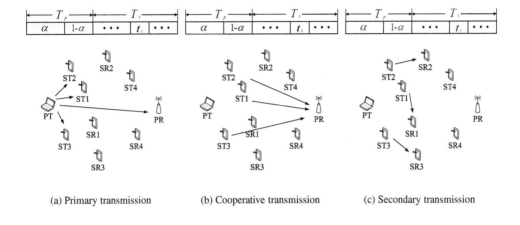

(a) Primary transmission   (b) Cooperative transmission   (c) Secondary transmission

Figure 9.1: Architecture of secondary users relaying for primary users.

deployed among the the STs, then the PU's transmission rate with cooperation is given by

$$R_{sp} = W \log_2 \left(1 + \frac{|h_p|^2 P_p}{N_0} + \sum_{i \in \mathcal{S}} \frac{|h_{sp,i}|^2 P_{s,i}}{N_0} \right). \tag{9.2}$$

Assume that the decode-and-forward (DF) protocol is employed to implement the cooperation between the PU and the SUs. Therefore, the achievable rate of the PU is

$$R_p = \min \{ \alpha R_{ps}, (1 - \alpha) R_{sp} \}. \tag{9.3}$$

It is proved that R_p is maximized when we have

$$\alpha R_{ps} = (1 - \alpha) R_{sp}. \tag{9.4}$$

So we have

$$\alpha = \frac{R_{sp}}{R_{ps} + R_{sp}}. \tag{9.5}$$

Thus, the achievable rate of the PU is

$$R_p = \frac{R_{ps} R_{sp}}{R_{ps} + R_{sp}}. \tag{9.6}$$

In the secondary transmission period: Clearly, the achievable rate of SU i is given by:

$$R_{s,i} = W \log_2 \left(1 + \frac{|h_{s,i}|^2 P_{s,i}}{N_0} \right), \forall i \in \mathcal{S} \tag{9.7}$$

9.1.2   *Stackelberg Game Formulation*

We assume that both primary and secondary users are selfish, and they strategically maximize their individual utility. In particular, the strategy of the PU is to select a subset $\mathcal{S}$, $|\mathcal{S}| = K$, of the SUs for cooperation and determine the spectrum price c per unit time, and the strategy of SU i is to determine the access time t_i it will purchase from the PU. In this way, the joint pricing and access time selection problem is modeled as a Stackelberg game, where the PU plays as the leader, and the SUs play as the follower. We can use backward induction to find the solution of the game. In the first step, given the cooperative relay subset $\mathcal{S}$ and the spectrum price c, each SU i determines its optimal access time t_i. In the second step, based upon the SUs' access time selection, the PU finds out the optimal $\mathcal{S}$ and c.

The utility function of the PU is defined as:

$$U_p = (T_p R_p)^b + c \sum_{i \in \mathcal{S}} t_i, \tag{9.8}$$

where we have $(0 < b < 1)$.

Correspondingly, the utility function of each SU i is given by

$$U_{s,i} = (t_i R_{s,i})^b - c t_i. \tag{9.9}$$

Given the cooperative relay set $\mathcal{S}$ and spectrum price c provided by the PU, the SUs in the set $\mathcal{S}$ compete with each other for maximizing its own utility $U_{s,i}$ by selecting its access time t_i, which forms a noncooperative access time selection game $< \mathcal{S}, t_i, U_{s,i} >$.

Definition 9.1   The optimal access time profile $\mathbf{t} = (t_1, \ldots, t_K)$ is the desirable outcome of an access time selection game, with which every SU i in $\mathcal{S}$ achieves the maximum utility, i.e.,

$$U_{s,i}(t_i, \mathbf{t}_{-i}) \geq U_{s,i}(t_i', \mathbf{t}_{-i}), \ \forall t_i \neq t_i',$$

where $\mathbf{t}_{-i} \triangleq (t_1, \ldots, t_{i-1}, t_{i+1}, \ldots, t_K)$, which we call the supplementary access time profile of t_i. When the optimal access time profile $\mathbf{t}$ occurs, the game reaches a Nash equilibrium (NE).

According to [11], a noncooperative game $< \mathcal{S}, t_i, U_{s,i} >$ has an NE if, for all $i \in \mathcal{S}$, 1) the assigned access time set $\mathbf{t_i}$ of player i is a nonempty compact convex subset of a Euclidian space; 2) the utility function $U_{s,i}$ is continuous and quasiconcave on t_i.

Theorem 9.1
The proposed access time selection game with the optimal access time profile in Definition 8.1, has an NE.

Proof 9.1   It is straightforward to show 1) for the feasible access time set. To show 2), it is clear the utility function $U_{s,i}$ is continuous, so we will just prove the utility function is quasiconcave. We differentiate the utility function $U_{s,i}$, with respect to t_i, and obtain:

$$\frac{\partial U_{s,i}}{\partial t_i} = b (t_i R_{s,i})^{b-1} R_{s,i} - c. \tag{9.10}$$

Furthermore, we have

$$\frac{\partial^2 U_{s,i}}{\partial t_i^2} = b(b-1)(t_i R_{s,i})^{b-2} R_{s,i}^2 < 0. \tag{9.11}$$

Therefore, the utility function $U_{s,i}$ is a concave function of t_i, thus, is also a quasiconcave function, and the existence of the NE is established.   ■

According to Theorem 8.1, we can obtain the optimal strategy t_i^* by solving the following equation,

$$\frac{\partial U_{s,i}}{\partial t_i} = b (t_i R_{s,i})^{b-1} R_{s,i} - c = 0. \tag{9.12}$$

We define the optimal access time function of SU i with respect to price c as:

$$t_i^* (c) = \left(b R_{s,i}{}^b \right)^{\frac{1}{1-b}} c^{\frac{1}{b-1}} . \tag{9.13}$$

Substituting (9.13) into (9.9), the maximum utility of SU i becomes a function of price c:

$$U_{s,i}^* (c) = \left(b^{\frac{b}{1-b}} - b^{\frac{1}{1-b}} \right) R_{s,i}{}^{\frac{b}{1-b}} c^{\frac{b}{b-1}} . \tag{9.14}$$

As we have $0 < b < 1$, the function $U_{s,i}^*(c)$ is monotonic, decreasing with respect to price c, but always positive. The higher spectrum price, the less time the SU would rent, and vice versa.

Then, the total access time required by all the SUs is

$$T_s^* (c) = \sum_{i \in S} t_i^* (c) = A c^{\frac{1}{b-1}} . \tag{9.15}$$

where $A = \sum_{i \in S} \left(b R_{s,i}{}^b \right)^{\frac{1}{1-b}}$.

We now analyze the strategy of the PU. Assume that the SUs ask for the access time from the PU based upon $T_s^* (c)$ in (9.15), then the PU optimizes its strategy (c, S) to maximize the revenue. In particular, the PU needs to solve the following optimization problem,

$$\max_{c,S} U_p = (T_p R_p)^b + c T_s^* (c) . \tag{9.16}$$

According to (9.15) and the relationship between T_p and T_s, we convert the above maximization problem into the following format,

$$\max_{c,S} U_p = \left(\left(1 - A c^{\frac{1}{b-1}} \right) R_p \right)^b + A c^{\frac{b}{b-1}} . \tag{9.17}$$

Therefore, the objective of the PU is to select the optimal c to maximize its utility.

Lemma 9.1
There exists a unique c that maximizes the utility of the PU when S is given, i.e.,

$$c^* = \frac{R_p^b}{\left(1 + A R_p^{\frac{b}{b-1}} \right)^{b-1}} . \tag{9.18}$$

Proof: Since we have

$$\frac{\partial U_p}{\partial c} = -A \frac{b}{b-1} R_p^b \left(1 - A c^{\frac{1}{b-1}} \right)^{b-1} c^{\frac{2-b}{b-1}} + A \frac{b}{b-1} c^{\frac{b}{b-1}-1} \tag{9.19}$$

$$\frac{\partial^2 U_p}{\partial c^2} = b(b-1) \left[R_p^b \left(1 - A c^{\frac{1}{b-1}} \right)^{b-2} + A^{-1} c^{\frac{b-2}{b-1}} \right] \cdot \left(-\frac{1}{b-1} A c^{\frac{1}{b-1}-1} \right)^2 < 0, \tag{9.20}$$

thus, U_p is a concave function with respect to c. Therefore, the optimal c is reached when $\frac{\partial U_p}{\partial T_p} = 0$.

■

Theorem 9.2
For any $i \in S$, the optimal t_i^* and c^*, given by (9.13) and (9.18), is an equilibrium for the proposed Stackelberg game.

Proof 9.2   *Proof:* When the PU gives out its price c, each SU selects its optimal t_i^* according to (9.13). According to Theorem 8.1 and Lemma 8.1, $\{t_i^*\}$ is the unique equilibrium for the game $< S, t_i, U_{s,i} >$. Moreover, as PU's utility U_p is concave in c, which means that the PU can always find an optimal solution c^*. Therefore, the t_i^* in (9.13) and c^* in (9.18) is the Stackelberg equilibrium. ■

9.1.3   Simulation Results

In this subsection, we present some numerical results to show the performance of the proposed cooperation framework. We consider a geometric model, as shown in Figure 9.2, where the PT is located at (0m, 0m), the PR is located at (200m, 0m), and 10 SU pairs are located randomly on a square centered at (100m, 0m). The channel gains are $\frac{0.097}{d^\gamma}$, where d is the distance between two nodes, and the path-loss exponent is $\gamma = 4$. The transmit power of the PT is $0.1W$, and the transmit power of each ST is $0.01W$.

Figure 9.3 shows the evolution of PU's utility with the number of the selected cooperative relays K. Here, U_{P_c} represents PU's utility achieved with the proposed cooperative transmission, and U_{P_d} represents PU's utility achieved with direct transmission, where no cooperation is employed and the PU occupies the channel all the time itself. It is observed that the utility of the PU under cooperative transmission outperforms that under direct transmission. Furthermore, we see that PU's cooperative utility increases with the number of the selected relays K, while PU's direct utility keeps unchanged, regardless of the number of the relays K.

Figure 9.4 shows the utility of secondary network (the total utility of all the SUs) vs. the number of the selected cooperative relays K. It is found that the utility of secondary network increases with

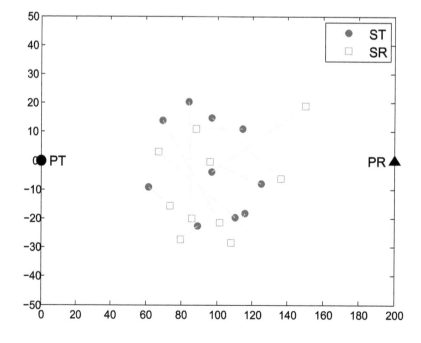

Figure 9.2: Topology of simulation network.

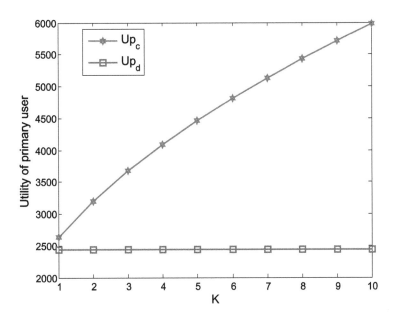

Figure 9.3: Utility of PU vs. the number of the selected cooperative relays.

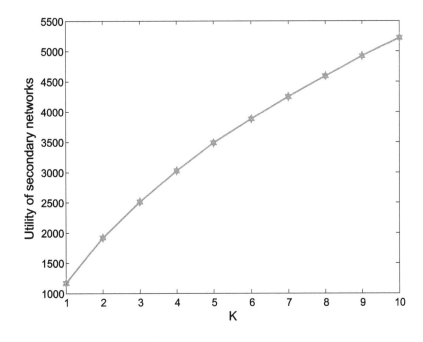

Figure 9.4: Utility of secondary network vs. the number of the selected cooperative relays.

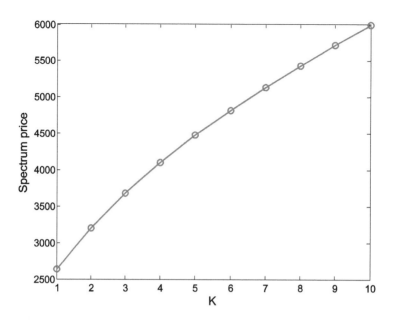

Figure 9.5: Spectrum price vs. the number of the selected cooperative relays.

the number of the relays. The reason is that, with more relays accessing the channel, more SUs would get a positive utility, resulting the increase of the utility of the entire secondary network.

Figure 9.5 illustrates the relationship between the spectrum price c and the number of the selected cooperative relays K. It is seen that the spectrum price c is proportional to the number of the relays. This is due to the fact that the more SUs access the channel, the stronger competition for the limited spectrum access time, and the higher the spectrum price would be.

9.2 Primary Users Relaying for Secondary Users

The secondary users relaying for primary users concentrated on the scenario where the resources owned by primary users are more than capable of supporting their target quality of service, thereby they can lease a certain fraction of the channel access time that is idle temporarily to secondary users, and in return secondary users help relaying for primary transmission. Since primary users can achieve the target throughput by themselves, the rate enhancement or the cooperation from secondary users actually has less attraction. Instead, they might be more interested in the benefits in other formats (e.g., revenue).

This motivates us to consider a new cooperation way, i.e., primary users cooperatively relay data for secondary users on the premise that this would not do harm to the performance of their own transmissions. In exchange, secondary users pay to primary users for the cooperative transmit power as well as the spectrum being used in cooperation. Therefore, primary users earn the revenue by selling the under-utilized resource, and secondary users increase the traffic rates by exploiting cooperative diversity, thus leading to a win-win situation.

In this new cooperative transmission, secondary users require to compete for channel bands and cooperative transmit power of primary users. In this section, we consider an auction-based spectrum access and power allocation scheme. Primary users, i.e. auctioneers, sell a portion of the channel access time and cooperative transmit power to secondary users for economic return. Secondary

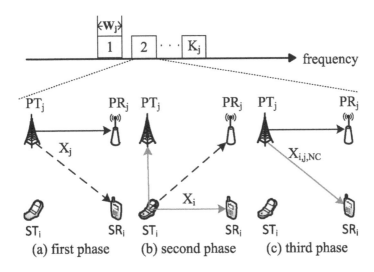

(a) first phase (b) second phase (c) third phase

Figure 9.6: Transmissions in three phases.

users, i.e. bidders, purchase channel and power from primary users for utility maximization. The prices announced by primary users for the channel and power are determined by the ascending clock auction algorithm.

9.2.1 Network Modeling and Notations

Consider a CR system consisting of a primary network composed of M primary links, and a secondary network composed of N secondary links. In the primary network, each PU j is equipped with a PT and a PR, and has K_j non-overlapping narrowband channels. Assume that the channels owned by the same PU have the same carrier frequency, while those of different PUs are different so as to avoid interference of different primary transmissions. In the secondary network, each SU i is equipped with an ST and an SR. The PTs act as relays to assist SUs' transmissions by network coding and the amplify-and-forward relaying protocol. Also, assume that each channel of the PT can be accessed by only one ST at the same time, and the channel occupancy by the STs is maintained by each PT itself. For simplicity, we consider the scenario where the total number of channels in the system equals to the number of SUs, i.e., $N = \sum_{j=1}^{M} K_j$, such that each SU can access one channel.

The structure of our CR frame consists of an auction slot and a data transmission slot. In the auction slot, the ST, which intends to send data to its SR, selects a desired PT and joins the channel and power auction organized by that PT. The data transmission slot is further divided into three phases, as shown in Figure 9.6. At each channel of PU j, in the first phase, PT j sends its data to its receiver. Meanwhile, the data are overheard by SR i, who is designated to that channel by PT j. In the second phase, ST i transmits its data to its receiver, which are also overheard by PR j. In the third phase, PT j combines together its own data sent in the first phase and the data overheard in the second phase and sends the additive data out. Then both PR j and SR i can extract their desired data from the combined data by subtracting the data it overheard. Note that in Figure 9.6, the solid lines indicate the intended communications, and the dotted lines represent the interference.

We now give out some operating assumptions of our system model: 1) Assume that the channels change slowly and the channel gain is stable within each frame. 2) Assume that the channel state information (CSI) can be accurately measured at each receiver, and this information can be sent to other receivers through an error-free control channel. Note that for slow-fading channel, i.e., the channel coherence time is large enough, the CSI can be accurately estimated within a sufficiently long period of observation.

In the first phase: At each channel of PU j, PT j transmits its signal to its destination PR j with power P_j. Assume that the total transmit power P_{U_j} of PT j is equally used at each channel at this phase, i.e., we have $P_j = P_{U_j}/K_j$. The signals received at PR j and SR i are respectively given by

$$Y_{PT_j}^{PR_j} = \sqrt{P_j} G_{PT_j}^{PR_j} X_j + n_{PR_j}, \tag{9.21}$$

$$Y_{PT_j}^{SR_i} = \sqrt{P_j} G_{PT_j}^{SR_i} X_j + n_{SR_i}, \tag{9.22}$$

where Y_A^B represents the signal received at B from A, X_j is the information symbol transmitted by PT j with $E[|X_j|^2] = 1$, and $n_{\{.\}}$ is the additive white Gaussian noise (AWGN) with variance σ^2. G_A^B denotes the channel gains from A to B, which is also the channel gains from B to A. The amplitude $|G_A^B|^2$ is exponentially distributed, with rate parameter $\lambda_A^B = (d_A^B)^\alpha$, where d_A^B denotes the distance between A and B, and α is the path-loss exponent.

The signal-to-noise-ratio (SNR) of X_j at PR j in the first phase is

$$\Gamma_{PR_j}(1) = \frac{P_j |G_{PT_j}^{PR_j}|^2}{\sigma^2}. \tag{9.23}$$

In the second phase: ST i transmits its signal with power P_i. The signals received by SR i, PT j, and PR j are

$$Y_{ST_i}^{SR_i} = \sqrt{P_i} G_{ST_i}^{SR_i} X_i + n_{SR_i}, \tag{9.24}$$

$$Y_{ST_i}^{PT_j} = \sqrt{P_i} G_{ST_i}^{PT_j} X_i + n_{PT_j}, \tag{9.25}$$

$$Y_{ST_i}^{PR_j} = \sqrt{P_i} G_{ST_i}^{PR_j} X_i + n_{PR_j}, \tag{9.26}$$

where X_i is the signal transmitted by ST i in this phase with $E[|X_i|^2] = 1$.

Thus, the SNR of X_i at SR i in the second phase is

$$\Gamma_{SR_i}(2) = \frac{P_i |G_{ST_i}^{SR_i}|^2}{\sigma^2}. \tag{9.27}$$

$$Y_{PT_j,\mathrm{NC}}^{PR_j} = \sqrt{P_{i,j}} G_{PT_j}^{PR_j} X_{i,j,\mathrm{NC}} + n_{PR_j}, \tag{9.28}$$

$$Y_{PT_j,\mathrm{NC}}^{SR_i} = \sqrt{P_{i,j}} G_{PT_j}^{SR_i} X_{i,j,\mathrm{NC}} + n_{SR_i}, \tag{9.29}$$

where $P_{i,j}$ is PT j's cooperative transmit power for ST i, and $X_{i,j,\mathrm{NC}}$ is the normalized energy data symbol defined as

$$X_{i,j,\mathrm{NC}} = \frac{X_j + \sqrt{P_i} G_{ST_i}^{PT_j} X_i + n_{PT_j}}{\sqrt{1 + P_i |G_{ST_i}^{PT_j}|^2 + \sigma^2}}. \tag{9.30}$$

The signal $Y_{PT_j,\mathrm{NC}}^{PR_j}$ received at PR j contains the information of both X_j and X_i, where X_i is the interference signal overheard in the second phase that can be completely removed. This yields

$$\hat{Y}_{PT_j,\mathrm{NC}}^{PR_j} = \frac{\sqrt{P_{i,j}} G_{PT_j}^{PR_j}}{\sqrt{1 + P_i |G_{ST_i}^{PT_j}|^2 + \sigma^2}} \left(X_j + n_{PT_j} \right) + n_{PR_j}. \tag{9.31}$$

In the third phase, the SNR at PR j is then given by

$$\Gamma_{PR_j}(3) = \frac{P_{i,j}|G_{PT_j}^{PR_j}|^2}{\sigma^2(1 + P_i|G_{ST_i}^{PT_j}|^2 + P_{i,j}|G_{PT_j}^{PR_j}|^2 + \sigma^2)}. \tag{9.32}$$

Correspondingly, the achievable rate from PT j to PR j when relaying for ST i is

$$R_{j,i} = \frac{W_j}{3}\log_2(1 + \Gamma_{PR_j}(1) + \Gamma_{PR_j}(3)), \tag{9.33}$$

where W_j is the bandwidth of the channel. The factor $1/3$ comes from the fact that three phases are required to fulfill one cooperative transmission.

Therefore, the achievable rate from PT j to PR j, when relaying for K_j STs over all the channels, is

$$R_{j,C} = \sum_{i=1}^{K_j} R_{j,i} = \sum_{i=1}^{K_j} \frac{W_j}{3}\log_2(1 + \Gamma_{PR_j}(1) + \Gamma_{PR_j}(3)). \tag{9.34}$$

Similarly, the useful signal, the SNR at SR i, and the achievable rate from ST i to SR i are, respectively, given by

$$\hat{Y}_{PT_j,\text{NC}}^{SR_i} = \frac{\sqrt{P_{i,j}}G_{PT_j}^{SR_i}}{\sqrt{1 + P_i|G_{ST_i}^{PT_j}|^2 + \sigma^2}}\left(\sqrt{P_i}G_{ST_i}^{PT_j}X_i \right.$$

$$\left. + n_{PT_j}\right) + n_{SR_i}, \tag{9.35}$$

$$\Gamma_{SR_i}(3) = \frac{P_{i,j}|G_{PT_j}^{SR_i}|^2 P_i|G_{ST_i}^{PT_j}|^2}{\sigma^2(1 + P_i|G_{ST_i}^{PT_j}|^2 + P_{i,j}|G_{PT_j}^{SR_i}|^2 + \sigma^2)}, \tag{9.36}$$

$$R_{i,j} = \frac{W_j}{3}\log_2(1 + \Gamma_{SR_i}(2) + \Gamma_{SR_i}(3)). \tag{9.37}$$

In contrast, if PT j does not cooperate with any secondary user but occupies all the channels all the time itself (we refer to it as direct transmission in this paper), the achievable rate from PT j to PR j over all the channels becomes

$$R_{j,D} = K_j W_j \log_2(1 + \Gamma_{PR_j}(1)). \tag{9.38}$$

9.2.2 *Joint Spectrum and Power Auction*

Each PU $j \in \{1,...,M\}$ sells two heterogeneous commodities (channels and cooperative power) among N SUs. The supply of PU j can be denoted by a vector $\mathbf{S_j} = (K_j, P_{U_j})$, which consists of the number of the licensed channels and the available cooperative power PU j has. The supply of the entire system is then denoted by $\mathbf{S} = (\mathbf{S_1},...,\mathbf{S_M})$. Let λ_j^1 and λ_j^2 be the prices of a channel and a power unit PU j asks for. The price vector of PU j then is denoted by $\lambda_\mathbf{j} = (\lambda_j^1, \lambda_j^2)$. It is noted that the channels owned by the same PU are assumed to be identical, i.e., they have the same bandwidth, carrier frequency, modulating scheme, etc., thus, they sell at the same price.

The utility of PU j is defined as the summation of its achievable rate and the payoff it receives in channel and power auction. That is

$$U_{j,C}(\mathbf{S_j}, \lambda_\mathbf{j}) = gR_j + \lambda_\mathbf{j}\mathbf{S_j}^T$$

$$= g\sum_{i=1}^{K_j} R_{j,i} + \lambda_j^1 K_j + \lambda_j^2 P_{U_j}, \tag{9.39}$$

where g is a positive constant providing conversion of units.

Similarly, we have the utility function of PU j in direct transmission as

$$U_{j,D} = gR_{j,D} = gK_jW_j\log_2(1 + \Gamma_{PR_j}(1)). \tag{9.40}$$

When purchasing the channel and cooperative power, each SU i wishes to maximize its transmission rate with minimum cost. Therefore, SU i requires to address two questions in the auction: 1) How to choose a PU from M PUs for cooperation and 2) How much cooperative power should it request from that PU. Formally, we formulate the solutions of these two questions as the bids of SU i:

$$\mathbf{Q_i} = (\mathbf{Q_{i,1}}, ..., \mathbf{Q_{i,j}}, ..., \mathbf{Q_{i,M}})^T, \tag{9.41}$$

where $\mathbf{Q_{i,j}} = (C_{i,j}, P_{i,j}), \forall i \in \{1,...,N\}, \forall j \in \{1,...,M\}$, is a resource demand vector. $P_{i,j}$ represents the required cooperative power of SU i from PU j. $C_{i,j} \in \{0,1\}$ specifies that whether SU i is willing to buy a channel from PU j. If it is, $C_{i,j} = 1$; otherwise, $C_{i,j} = 0$. It is worth mentioning that the channel and the cooperative power are two different types of commodities, of which the channel is indivisible and the cooperative power is divisible. Therefore, the channel is available in a supply of one and, thus, can be assigned either totally or nothing. The cooperative power can be offered at any quantity of $P_{i,j}$, subject to the constraint that $0 \leq \sum_{i=1}^{N} P_{i,j} \leq P_{U_j}$.

Furthermore, the proposed cooperation architecture requires the SU to buy the channel and the cooperative power from the same PU. It implies that the channel and the power cannot be sold independently at each PU, but should be offered as a bundle [12]. For each SU, it can only purchase the entire bundle or nothings. When SU i does not buy the channel from PU j, i.e., $C_{i,j} = 0$, it will not receive any power from that PU, i.e., we have $P_{i,j} = 0$. Similarly, if PU j does not assign any power to SU i, i.e., $P_{i,j} = 0$, it is not allowed to sell the channel to that SU, i.e., we have $C_{i,j} = 0$. Since each SU can access only one channel, we have $\sum_{j=1}^{M} C_{i,j} \leq 1, \forall i \in \{1,...,N\}$. Correspondingly, the power demand vector $\mathbf{P_i} = (P_{i,1}, P_{i,2}, ..., P_{i,M})$ of each SU is an M-dimensional vector with at most one non-zero element.

To depict an SU's satisfaction with the received channel and power from PU j, we define a utility function of SU i to PU j as:

$$U_{i,j}(\mathbf{Q_{i,j}}, \lambda_j) = gR_{i,j}(P_{i,j}) - \lambda_j^1 C_{i,j} - \lambda_j^2 P_{i,j}, \tag{9.42}$$

where in the right side of the equation, the first term is SU i's gain (achievable rate) achieved in cooperation from PU j, and the second and the third term are the payment to PU j.

Then the utility function of SU i is defined as:

$$U_i = \max_{s \in \{1,...,M\}} U_{i,s}(\mathbf{Q_{i,s}}, \lambda_s). \tag{9.43}$$

If SU i decides to purchase channel and power from PU j, i.e., we have $C_{i,j} = 1$, and $C_{i,k} = 0$ for any $k \neq j$, then the optimal cooperative power demand of SU i can be achieved by solving the following utility maximization problem:

$$\max_{P_{i,j}} \quad U_{i,j}(\mathbf{Q_{i,j}}, \lambda_j) = gR_{i,j}(P_{i,j}) - \lambda_j^1 - \lambda_j^2 P_{i,j}$$
$$\text{s.t.} \quad 0 \leq P_{i,j} \leq P_{U_j}. \tag{9.44}$$

It can be seen that the above maximization problem is a concave optimization problem, as the objective function is strictly concave and the constraint set is convex [13]. We can find the optimal power demand $P_{i,j}^*$ by taking the derivative of $U_{i,j}(Q_{i,j}, \lambda_j)$ with respect to $P_{i,j}$ as

$$\frac{\partial U_{i,j}(\mathbf{Q_{i,j}}, \lambda_j)}{\partial P_{i,j}} = g\frac{\partial R_{i,j}(P_{i,j})}{\partial P_{i,j}} - \lambda_j^2 = 0. \tag{9.45}$$

$$P_{i,j}^{*}(\lambda_j^2) = \max\left[0,\min\left(\frac{\sqrt{\beta_{ij}^2\gamma_{ij}^2 + \frac{4W_j'}{\lambda_j^2}(\alpha_{ij}\beta_{ij}\gamma_{ij} + \beta_{ij}^2\gamma_{ij})} - (2\alpha_{ij}\gamma_{ij} + \beta_{ij}\gamma_{ij})}{2(\alpha_{ij} + \beta_{ij})}, P_{U_j}\right)\right] \tag{9.27}$$

For simplicity, we define

$$W_j' = \frac{gW_j}{3\ln(2)}; \qquad \alpha_{ij} = 1 + \frac{P_i|G_{ST_i}^{SR_i}|^2}{\sigma^2}$$

$$\beta_{ij} = \frac{P_i|G_{ST_i}^{PT_j}|^2}{\sigma^2}; \qquad \gamma_{ij} = \frac{1 + P_i|G_{ST_i}^{PT_j}|^2 + \sigma^2}{|G_{PT_j}^{SR_i}|^2}. \tag{9.46}$$

Substituting (9.46) into (9.45), we have the optimal cooperative power demand $P_{i,j}^{*}(\lambda_j^2)$ shown in (9.27).

We model a multi-auctioneer, multi-bidder, and multi-commodity auction game to efficiently allocate the channels and cooperative power of M PUs among N SUs. Each PU j, i.e. the auctioneer, iteratively announces the prices of its two commodities. Each SU i, i.e. the bidder, responds to each PU j by submitting its demand $\mathbf{Q_{i,j}}$, which reports the quantities of the channel and the power it wishes to purchase at these prices. PU j then calculates the *cumulative clinch* and credits the channel and the power to the SUs at the current prices by ascending clock auction algorithm [14]. Thereafter, PU j adjusts the prices according to the relationship between the total demand and the total supply. This process repeats until the prices converge at which the total demand is less than or equal to the total supply. During this process, several important operations, including PU selection, reserve pricing, resource crediting, and payment calculation, are involved.

1) PU selection

The PU selection occurs on each SU at each auction clock, by which the SU determines from which PU it buys a channel and how much cooperative power it requests from that PU. For example, at the very beginning of the auction, i.e., at the time when the auction clock index τ is set to zero, each PU j makes an initialization and announces its initial prices in a form of $\lambda_\mathbf{j}(0) = (\lambda_j^1(0), \lambda_j^2(0))$ to all the SUs. Based on these prices, SU i selects the desired PU. To do that, SU i sets $C_{i,j} = 1, \forall j \in \{1, ..., M\}$, and separately solve M utility maximization problems defined in (9.44). Then it finds out the desired PU j that incurs the maximum utility, that is

$$j = \arg\max_{k \in \{1,...,M\}} U_{i,k}(\mathbf{Q_{i,k}}, \lambda_\mathbf{k}). \tag{9.28}$$

Thereafter, SU i places its bids to PU j as $C_{i,j}(\lambda_j^1(0)) = 1$, and $P_{i,j}(\lambda_j^2(0)) = P_{i,j}^{*}(\lambda_j^2(0))$ defined in (9.27). For any other PU k, $\forall k \in \{1, ..., M\}$ and $k \neq j$, it sets the bids to $C_{i,k}(\lambda_k^1(0)) = 0$ and $P_{i,k}(\lambda_k^2(0)) = 0$.

2) Reserve pricing

The reserve price refers to the lowest prices at which the PU is willing to sell the channels and power to the SUs. It guarantees a certain amount of the revenue for the PU, even when competition is weak. Generally, the reserve price can be defined as the expense of relaying for the SUs, which reflects the adverse effects of SUs' transmissions on PU's performance, such as device depreciation, channel occupation, and power consumption.

In this work, the PU can choose either to use all the resource itself (i.e. direct transmission) or to sell a fraction of the resource to the SUs (i.e. cooperative transmission). To encourage the PU to share the resource with the SUs, its reserve price vector $\lambda_{j,R}$ can be set in such a way that at which the utility of the PU achieved in cooperative transmission is no less than that in direct transmission. That is to say, the value of $\lambda_{j,R}$ satisfies:

$$U_{j,C}(\mathbf{S_j}, \lambda_{j,R}) \geq U_{j,D}. \tag{9.29}$$

The value of $U_{j,C}(\mathbf{S_j}, \lambda_{j,R})$ varies when different set of K_j SUs access PU j's channels. So, this value is uncertain before the auction, as the PU has no idea which SUs would finally access its channels. However, the PU can estimate its lower bound $\overline{U_{j,C}}(\mathbf{S_j}, \lambda_{j,R})$ by finding out the set $\mathcal{D}$ of K_j SUs, with which PU j achieves the minimum transmission rates $\overline{R_{j,C}}$. That is,

$$\overline{U_{j,C}}(\mathbf{S_j}, \lambda_{j,R}) \approx g\overline{R_{j,C}} + \lambda_{j,R}^1 K_j + \lambda_{j,R}^2 P_{U_j}, \tag{9.30}$$

where we have $\overline{R_{j,C}} = \sum_{i \in \mathcal{D}} R_{j,i}$, in which $R_{j,i}$ is calculated before the auction under the assumption that PU j equally allocates its cooperative power among all the channels, i.e. $P_{i,j} = P_{U_j}/K_j$.

Then we set

$$\overline{U_{j,C}}(\mathbf{S_j}, \lambda_{j,R}) = U_{j,D}. \tag{9.31}$$

Combining (9.34) and (9.40) into (9.31), we have

$$\sum_{i \in \mathcal{D}} \frac{gW_j}{3} \log_2\left(1 + \Gamma_{PR_j}(1) + \Gamma_{PR_j}(3)\right) + \lambda_{j,R}^1 K_j$$
$$+ \lambda_{j,R}^2 P_{U_j} = \frac{gK_j W_j}{3} \log_2(1 + 3\Gamma_{PR_j}(1)). \tag{9.32}$$

We cannot acquire $\lambda_{j,R}^1$ and $\lambda_{j,R}^2$ simultaneously from (9.32). One simplest solution is to set $\lambda_{j,R}^1 = 0$ and get $\lambda_{j,R}^2$ by

$$\lambda_{j,R}^2 = \frac{1}{P_{U_j}} \left[\frac{gK_j W_j}{3} \log_2(1 + 3\Gamma_{PR_j}(1)) \right.$$
$$\left. - \sum_{i \in \mathcal{D}} \frac{gW_j}{3} \log_2(1 + \Gamma_{PR_j}(1) + \Gamma_{PR_j}(3)) \right]. \tag{9.33}$$

The reserve price is very easily implemented in ascending clock auction. For example, the PU can initialize $\lambda_j(0)$ to $\lambda_{j,R}$, such that it can always get a larger utility from choosing cooperation.

3) Resource crediting

At each auction clock $\tau = 0, 1, \ldots$, PU j collects N SUs' bids, and computes the total required channels and power of these SUs. Let $C_j^{tal}(\lambda_j^1(\tau)) = \sum_{i=1}^{N} C_{i,j}(\lambda_j^1(\tau))$ and $P_j^{tal}(\lambda_j^2(\tau)) = \sum_{i=1}^{N} P_{i,j}(\lambda_j^2(\tau))$ represent the total channel and power demand at PU j at clock τ, respectively. Further, let $E_j^1(\lambda_j^1(\tau)) = C_j^{tal}(\lambda_j^1(\tau)) - K_j$ and $E_j^2(\lambda_j^2(\tau)) = P_j^{tal}(\lambda_j^2(\tau)) - P_{U_j}$ represent the excess channel and power demand at PU j, respectively. Then PU j adjusts its price vector according to the excess demand.

Case 1: $E_j^1(\lambda_j^1(\tau)) > 0$ and $E_j^2(\lambda_j^2(\tau)) > 0$. It tells that the total demand for the power as well as for the channel exceeds the supply. Due to the indivisibility, K_j channels cannot be divided and fairly allocated among more than K_j competitors. Therefore, none of the channels would be credited to any competitor. For bundling sale, the power would not be credited to any competitor, either. So we have

$$\hat{C}_{i,j}(\lambda_j^1(\tau)) = 0, \ \hat{P}_{i,j}(\lambda_j^2(\tau)) = 0, \ \forall i \in \{1,2,...,N\}, \tag{9.34}$$

where $\hat{C}_{i,j}(\lambda_j^1(\tau))$ and $\hat{P}_{i,j}(\lambda_j^2(\tau))$ are the cumulative clinch, which are the amounts of the channel and power that are credited to SU i at the price $\lambda_j(\tau)$.

After finishing the computation of the cumulative clinch, PU j updates its price vector with $\lambda_j^1(\tau+1) = \lambda_j^1(\tau) + \mu_j^1$, and $\lambda_j^2(\tau+1) = \lambda_j^2(\tau) + \mu_j^2$, where $\mu_j^1 > 0$ and $\mu_j^2 > 0$ are step sizes, and announces this new price vector to all SUs. Each SU then re-selects a PU based on the new announced prices and starts a new bidding round.

Case 2: $E_j^1(\lambda_j^1(\tau)) > 0$ and $E_j^2(\lambda_j^2(\tau)) \leq 0$. In this case, there are more than K_j SUs competing for K_j channels at PU j, whose total power demand is less than PU j's supply. Similar to Case 1, neither the channel nor the power would be credited to any competitor. Therefore, the cumulative clinch of the channel and power to the SUs are also calculated by (9.34). Finally, the price of the channel is updated by $\lambda_j^1(\tau+1) = \lambda_j^1(\tau) + \mu_j^1$. While the price of per unit power remains unchanged as $\lambda_j^2(\tau+1) = \lambda_j^2(\tau)$ for the sake that the total power demand does not exceed the supply.

Case 3: $E_j^1(\lambda_j^1(\tau)) \leq 0$ and $E_j^2(\lambda_j^2(\tau)) > 0$. In this case, the competition for the power is fierce, while that for the channel is weak. As the supply of the channels is sufficient, the channels can be credited to all the competitors who bid for them. Moreover, the power can be credited to each competitor in terms of their opponents' demands. Thus, for each SU i with $C_{i,j}(\lambda_j^1(\tau)) = 1$, we have

$$\hat{C}_{i,j}(\lambda_j^1(\tau)) = 1,$$
$$\hat{P}_{i,j}(\lambda_j^2(\tau)) = \max\left(0, P_{U_j} - \sum_{k=1, k\neq i}^N P_{k,j}(\lambda_j^2(\tau))\right). \tag{9.35}$$

For each SU i with $C_{i,j}(\lambda_j^1(\tau)) = 0$, we have

$$\hat{C}_{i,j}(\lambda_j^1(\tau)) = 0, \ \hat{P}_{i,j}(\lambda_j^2(\tau)) = 0. \tag{9.36}$$

Thereafter, PU j updates the price vector with $\lambda_j^1(\tau+1) = \lambda_j^1(\tau)$, and $\lambda_j^2(\tau+1) = \lambda_j^2(\tau) + \mu_j^2$.

Case 4: $E_j^1(\lambda_j^1(\tau)) \leq 0$ and $E_j^2(\lambda_j^2(\tau)) \leq 0$. It shows that the supply of both channel and power is sufficient for all the competitors. Therefore, each competitor would be credited according to its demand. Namely,

$$\hat{C}_{i,j}(\lambda_j^1(\tau)) = C_{i,j}(\lambda_j^1(\tau)), \ \hat{P}_{i,j}(\lambda_j^2(\tau)) = P_{i,j}(\lambda_j^2(\tau)). \tag{9.37}$$

Then two prices are kept unchanged, with $\lambda_j^1(\tau+1) = \lambda_j^1(\tau)$, and $\lambda_j^2(\tau+1) = \lambda_j^2(\tau)$.

Additionally, the demand $\mathbf{Q_{i,j}}(\lambda_j(\tau))$ of SU i from PU j is a function of PU j's announced price $\lambda_j(\tau)$. If PU j's price is too high at τ, SU i, which chose PU j at $\tau-1$, might give up it and choose another PU at τ, then all the channel and power clinched to SU i before at PU j become unclinched. Thus, all the credits SU i received before from PU j should be cleared. So, in the above four cases, for SU i with $C_{i,j}(\lambda_j^1(\tau)) = 0$ and $C_{i,j}(\lambda_j^1(\tau-1)) = 1$, we have

$$\hat{C}_{i,j}(\lambda_j^1(\tau')) = 0, \ \hat{P}_{i,j}(\lambda_j^2(\tau')) = 0, \ \forall \tau' \in \{0,1,...,\tau-1\}. \tag{9.38}$$

4) Payment calculation
Assuming that the supply meets the total demand for each PU at clock $\tau = T$, i.e., $E_j^1(\lambda_j^1(T)) =$

0 and $E_j^2(\lambda_j^2(T)) \le 0, \forall j \in \{1,...,M\}$, then the auction converges to an equilibrium price vector $\lambda_{\mathbf{j}}^* = \lambda_{\mathbf{j}}(T)$. Consider that the supply P_{U_j} might not be fully covered at $\lambda_{\mathbf{j}}^*$, i.e., $E_j^2(\lambda_j^2(\tau)) < 0$. For each SU i with $\hat{P}_{i,j}(\lambda_j^2(T)) \ne 0$, its cumulative clinch is re-calculated by [14]:

$$\hat{P}_{i,j}(\lambda_j^2(T)) = P_{i,j}(\lambda_j^2(T)) +$$
$$\frac{P_{i,j}(\lambda_j^2(T-1)) - P_{i,j}(\lambda_j^2(T))}{\sum_{i=1}^{N} P_{i,j}(\lambda_j^2(T-1)) - \sum_{i=1}^{N} P_{i,j}(\lambda_j^2(T))} \left[P_{U_j} - \sum_{i=1}^{N} P_{i,j}(\lambda_j^2(T)) \right].$$
(9.39)

Such that we have $E_j^1(\lambda_j^1(T)) = 0$ and $E_j^2(\lambda_j^2(T)) = 0$.

Finally, the quantities of the channel and power that are assigned to the SU are given by

$$C_{i,j}^* = \hat{C}_{i,j}(\lambda_j^1(T)), \quad P_{i,j}^* = \hat{P}_{i,j}(\lambda_j^2(T)),$$
$$\forall i \in \{0,1,...,N\}, \forall j \in \{0,1,...,M\}.$$
(9.40)

Correspondingly, the payment for the channel from SU i to PU j is

$$V_{i,j}^1 = \lambda_j^1(0)\hat{C}_{i,j}(\lambda_j^1(0))$$
$$+ \sum_{\tau=1}^{T} \lambda_j^1(\tau) \left(\hat{C}_{i,j}(\lambda_j^1(\tau)) - \hat{C}_{i,j}(\lambda_j^1(\tau-1)) \right),$$
(9.41)

and the payment for the power from SU i to PU j is

$$V_{i,j}^2 = \lambda_j^2(0)\hat{P}_{i,j}(\lambda_j^2(0))$$
$$+ \sum_{\tau=1}^{T} \lambda_j^2(\tau) \left(\hat{P}_{i,j}(\lambda_j^2(\tau)) - \hat{P}_{i,j}(\lambda_j^2(\tau-1)) \right).$$
(9.42)

A complete channel and power auction algorithm is shown in Algorithm 8.1. The communication overhead arise from the transmissions of the bids from the SU and the price vector from the PU, which are negligible compared to the main traffic. Since bids from different SUs may arrive at the PU at different time, the proposed auction algorithm can also run in an asynchronous way. At each auction clock, the PU collects new bids until a timeout value $\overline{T}$ has passed. For the SUs whose bids it has received, it uses the new bids; for those slow ones that it has not heard from in this round, it uses the most recent bids from them.

9.2.3 Theoretic Analysis

First, we specify a generic economic model: M auctioneers wish to allocate K types of commodities among N bidders. For each auctioneer j, its available supply is $\mathbf{S_j} = (S_j^1, ..., S_j^K)$, its announced price vector is $\lambda_{\mathbf{j}} = (\lambda_j^1, ..., \lambda_j^K)$, and its allocation to bidder i is $\mathbf{A_{i,j}} = (A_{i,j}^1, ..., A_{i,j}^K)$. For each bidder i, its demand from auctioneer j at price $\lambda_{\mathbf{j}}$ is $\mathbf{Q_{i,j}}(\lambda_{\mathbf{j}}) = (Q_{i,j}^1(\lambda_j^1), ..., Q_{i,j}^K(\lambda_j^K))$, its payment to auctioneer j is $V_{i,j}$, and it has a function $F_{i,j}(\mathbf{Q_{i,j}})$, with respect to $\mathbf{Q_{i,j}}$.

Definition 9.2 [15] A Walrasian Equilibrium is a $M \times K$ price vector $\lambda^* = \begin{pmatrix} \lambda_1^{1*} & \lambda_1^{2*} & \cdots & \lambda_1^{K*} \\ \vdots & \vdots & \vdots & \vdots \\ \lambda_M^{1*} & \lambda_M^{2*} & \cdots & \lambda_M^{K*} \end{pmatrix}$ and a

---

Algorithm 2 The Proposed Channel and Power Auction Algorithm

Initialization

Sets clock index $\tau = 0$;

Each PU $j \in \{1,...,M\}$ announces its initial price vector $\lambda_{\mathbf{j}}(0) = (\lambda_j^1(0), \lambda_j^2(0))$ to all the SUs.

Iteration Step $(\tau = 0,1,2,...)$

At a SU $i \in \{1,...,N\}$:

Input: Receives the price vector $\lambda_{\mathbf{j}}(\tau) = (\lambda_j^1(\tau), \lambda_j^2(\tau))$ from each PU $j \in \{1,...,M\}$.

Finds out the desired PU j according to (9.28);

Calculates the optimal power demand $P_{i,j}^*(\lambda_j^2(\tau))$ in (9.27) from PU j;

Sets the bid $\mathbf{Q_{i,j}}(\tau) = (1, P_{i,j}^*(\lambda_j^2(\tau)))$ to PU j;

For any other PU $k \neq j$, sets the bid $\mathbf{Q_{i,k}}(\tau) = (0,0)$.

Output: The new bid $\mathbf{Q_{i,j}}(\tau) = (C_{i,j}(\lambda_j^1(\tau)), P_{i,j}(\lambda_j^2(\tau)))$ to each PU.

At a PU $j \in \{1,...,M\}$:

Input: Collects the bid $\mathbf{Q_{i,j}}(\tau) = (C_{i,j}(\lambda_j^1(\tau)), P_{i,j}(\lambda_j^2(\tau)))$ from each player $i \in \{1,...,N\}$.

Calculates the excess channel demand by $E_j^1(\lambda_j^1(0)) = \sum_{i=1}^N C_{i,j}(\lambda_j^1(0)) - K_j$;

Calculates the excess power demand by $E_j^2(\lambda_j^2(0)) = \sum_{i=1}^N P_{i,j}(\lambda_j^2(0)) - P_{U_j}$;

If $E_j^1(\lambda_j^1(\tau)) > 0$ and $E_j^2(\lambda_j^2(\tau)) > 0$

$\hat{C}_{i,j}(\lambda_j^1(\tau)) = 0$, $\hat{P}_{i,j}(\lambda_j^2(\tau)) = 0$, $\forall i \in \{1,2,...,N\}$;

$\lambda_j^1(\tau+1) = \lambda_j^1(\tau) + \mu_j^1$, $\lambda_j^2(\tau+1) = \lambda_j^2(\tau) + \mu_j^2$;

Else if $E_j^1(\lambda_j^1(\tau)) > 0$ and $E_j^2(\lambda_j^2(\tau)) \leq 0$

$\hat{C}_{i,j}(\lambda_j^1(\tau)) = 0$, $\hat{P}_{i,j}(\lambda_j^2(\tau)) = 0$, $\forall i \in \{1,2,...,N\}$;

$\lambda_j^1(\tau+1) = \lambda_j^1(\tau) + \mu_j^1$, $\lambda_j^2(\tau+1) = \lambda_j^2(\tau)$;

Else if $E_j^1(\lambda_j^1(\tau)) \leq 0$ and $E_j^2(\lambda_j^2(\tau)) > 0$

$\hat{C}_{i,j}(\lambda_j^1(\tau)) = 0$, $\hat{P}_{i,j}(\lambda_j^2(\tau)) = 0$, for SU i with $C_{i,j}(\lambda_j^1(\tau)) = 1$;

$\lambda_j^1(\tau+1) = \lambda_j^1(\tau)$, $\lambda_j^2(\tau+1) = \lambda_j^2(\tau) + \mu_j^2$;

Else

$\hat{C}_{i,j}(\lambda_j^1(\tau)) = C_{i,j}(\lambda_j^1(\tau))$, $\hat{P}_{i,j}(\lambda_j^2(\tau)) = P_{i,j}(\lambda_j^2(\tau))$, $\forall i \in \{1,2,...,N\}$;

For SU i with $C_{i,j}(\lambda_j^1(\tau)) = 0$ and $C_{i,j}(\lambda_j^1(\tau-1)) = 1$, sets $\hat{C}_{i,j}(\lambda_j^1(\tau')) = 0$, $\hat{P}_{i,j}(\lambda_j^2(\tau')) = 0$, $\forall \tau' \in \{0,1,...,\tau-1\}$;

Output: The new price vector $\lambda_{\mathbf{j}}(\tau+1) = (\lambda_j^1(\tau+1), \lambda_j^2(\tau+1))$ to all SUs.

Iterates until the price vectors of all the PUs converge at $\tau = T$, then proceeds to the final step.

Final Step

For SU i with $\hat{P}_{i,j}(\lambda_j^2(T)) \neq 0$, each PU j updates the cumulative clinch $\hat{P}_{i,j}(\lambda_j^2(T))$ by (9.39);

Calculates the quantities of the channel and power that are assigned to each SU i by (9.40).

$N \times M \times K$ allocation vector $\mathbf{A}^* = \begin{pmatrix} \mathbf{A}_{1,1}^* & \mathbf{A}_{1,2}^* & \cdots & \mathbf{A}_{1,M}^* \\ \vdots & \vdots & & \vdots \\ \mathbf{A}_{N,1}^* & \mathbf{A}_{N,2}^* & \cdots & \mathbf{A}_{N,M}^* \end{pmatrix}$, such that

$$\mathbf{Q}(\lambda^*) = \begin{pmatrix} \mathbf{Q}_{1,1}(\lambda^*) & \mathbf{Q}_{1,2}(\lambda^*) & \cdots & \mathbf{Q}_{1,M}(\lambda^*) \\ \vdots & \vdots & & \vdots \\ \mathbf{Q}_{N,1}(\lambda^*) & \mathbf{Q}_{N,2}(\lambda^*) & \cdots & \mathbf{Q}_{N,M}(\lambda^*) \end{pmatrix} = \mathbf{A}^*, \text{ and } S_j^k = \sum_{i=1}^N A_{i,j}^{k*}, \forall j \in \{1,...,M\}, \forall k \in \{1,...,K\}.$$

According to Definition 8.1, when the auction reaches a Walrasian Equilibrium, the excess demand of each bidder is zero, and the aggregate demand equals to the supply for each commodity.

Lemma 9.2

[14] An auction has a Walrasian Equilibrium if it satisfies:

(1) Pure private values: Bidder i's value, i.e. $F_{i,j}(\mathbf{Q_{i,j}})$, for the demand vector $\mathbf{Q_{i,j}}$ does not change when bidder i learns other bidders' information.

(2) Quasilinearity: Bidder i's utility from receiving the demand vector $\mathbf{Q_{i,j}}$ in return for the payment $V_{i,j}$ is given by $F_{i,j}(\mathbf{Q_{i,j}}) - V_{i,j}$.

(3) Monotonicity: The function $F_{i,j}(\mathbf{Q_{i,j}})$ is increasing, i.e., if $\mathbf{Q'_{i,j}} > \mathbf{Q_{i,j}}$, then $F_{i,j}(\mathbf{Q'_{i,j}}) > F_{i,j}(\mathbf{Q_{i,j}})$.

(4) Concavity: The function $F_{i,j}(\mathbf{Q_{i,j}})$ is concave.

Theorem 9.3

The proposed multi-auctioneer, multi-bidder, and multi-commodity auction has a Walrasian Equilibrium.

Pure private values: In the proposed auction, bidder i's demand $\mathbf{Q_{i,j}}$ is a function of the price vector $\lambda_\mathbf{j}$. That is to say, the value of $R_{i,j}(\mathbf{Q_{i,j}})$ is uniquely determined by the announced price vector $\lambda_\mathbf{j}$. As long as $\lambda_\mathbf{j}$ is fixed, $\mathbf{Q_i}$ would always remain unchanged, regardless of the demands of other bidders.

Quasilinearity: It is obvious that the utility function $U_{i,j}(\mathbf{Q_{i,j}}, \lambda_\mathbf{j}) = gR_{i,j}(P_{i,j}) - \lambda_j^1 C_{i,j} - \lambda_j^2 P_{i,j}$ in (9.41) is a linear function of the price λ_j^1 and λ_j^2.

Monotonicity: It can be easily found that $\partial R_{i,j}(P_{i,j})/\partial P_{i,j} > 0$, therefore, the achievable rate function $R_{i,j}(P_{i,j})$ is increasing.

Concavity: Note that if a function $f(x)$ is twice-differentiable, then $f(x)$ is strictly concave if and only if $f''(x)$ is negative. Since we have $\partial^2 R_{i,j}(P_{i,j})/\partial^2 P_{i,j} < 0$, the achievable rate function $R_{i,j}(P_{i,j})$ is concave with respect to $P_{i,j}$.

Therefore, there exists a Walrasian Equilibrium for the proposed auction.

Theorem 8.1 shows the existence of a Walrasian Equilibrium for the proposed auction algorithm, but it does not tell us how it can efficiently converge to an equilibrium. As previous mentioned, the price adjustment of the proposed auction is directly controlled by the excess demand. If there is excess demand (e.g., for the power) at PU j, i.e., $E_j^2(\lambda_j^2(\tau)) > 0$, the price is increased by μ_j^2. Otherwise, it keeps fixed. In fact, such price adjustment mechanism can be viewed as a discrete version of the Walrasian tâtonnement [16], i.e.,

$$\lambda_j^k(\tau+1) = \lambda_j^k(\tau) + \dot{\lambda}_j^k(\tau), \tag{9.43}$$

where we have

$$\dot{\lambda}_j^k(\tau) = \begin{cases} \mu_j^k, & \text{if } E_j^k(\lambda_j^k(\tau)) > 0; \\ 0, & \text{otherwise.} \end{cases} \tag{9.44}$$

This ascending clock process continually drives the price vector λ to converge to λ^*, at which $\mathbf{E}(\lambda^*) = \mathbf{0}$.

Mathematically, we use the following Lyapunov stability theorem [17] to prove the convergence of the proposed algorithm.

Lemma 9.3

*(**Lyapunov's Theorem**) Consider an autonomous system and its equilibrium point $\dot{x} = 0$. This equilibrium point is globally stable if there exists a Lyapunov function $V(x)$, which is continuously*

differentiable, such that
(1) $V(x) > 0$, $\forall x \neq 0$; (positive definite)
(2) $\dot{V}(x) \leq 0, \forall x$; (seminegative definite)
(3) $V(x) \to \infty$, when $\|x\| \to \infty$.

Theorem 9.4
Starting from any sufficiently small price vector $\lambda(\mathbf{0})$, the proposed auction algorithm converges to a Walrasian Equilibrium price vector λ^* in finite iterations.

According to Lyaounov's Theorem, if we can find a Lyapunov function for the proposed algorithm (which can be considered as a nonlinear autonomous system), such that all three conditions are satisfied, then the equilibrium point of the dynamical system is globally asymptotically stable. Similar to [14], we define the Lyapunov function as:

$$L(\lambda(\tau)) = \lambda(\tau) \cdot \mathbf{S} + \sum_{i=1}^{N} U_i(\mathbf{Q_i}, \lambda(\tau)). \tag{9.45}$$

It's easy to see that $L(\lambda) > 0$ when $\lambda \neq 0$, and $L(\lambda) \to \infty$, when $\|\lambda\| \to \infty$. By taking the derivative of this Lyapunov function, we have

$$
\begin{aligned}
\dot{L}(\lambda(\tau)) &= \frac{\partial L(\lambda(\tau))}{\partial \lambda(\tau)} = \mathbf{S} \cdot \dot{\lambda}(\tau) - \sum_{i=1}^{N} \mathbf{Q_i}(\lambda(\tau)) \cdot \dot{\lambda}(\tau) \\
&= (\mathbf{S} - \sum_{i=1}^{N} \mathbf{Q_i}(\lambda(\tau))) \cdot \dot{\lambda}(\tau) \\
&= -\mathbf{E}(\lambda(\tau)) \cdot \dot{\lambda}(\tau).
\end{aligned}
\tag{9.46}
$$

Clearly, when the demand exceeds the supply, i.e. $\mathbf{E}(\lambda(\tau)) > \mathbf{0}$, the price increases at τ, i.e. $\dot{\lambda}(\tau) > \mathbf{0}$, we then have $\dot{L}(\lambda(\tau)) < 0$. If the supply meets the demand and, thus, $\lambda(\tau) \to \lambda^*$, we have $\dot{\lambda}(\tau) = \mathbf{0}$, and $\dot{L}(\lambda(\tau)) = 0$. Therefore, the seminegative definite condition $\dot{L}(\lambda(\tau)) \leq 0$ is satisfied.

9.2.4 Simulation Results

In this section, we present simulation results to demonstrate the performance of the proposed joint spectrum and power allocation algorithm. We consider a scenario as shown in Figure 9.7, where there are two PUs and six SUs in the network. PU 1 has 2 channels, and PU 2 has 4 channels. The channel gains are $(\frac{0.097}{d^\alpha})^{\frac{1}{2}}$, where d is the distance between two nodes, and the path-loss exponent is $\alpha = 4$. The transmit power of each SU is $0.01W$, and the noise variance is $\sigma^2 = 10^{-13}$. Without a special specification, the transmit power of PU 1 is $2W$, the transmit power of PU 2 is $1W$, the initial power and channel price of PU 1 are 0.8, and the initial power and channel price of PU 2 are 1.

Figure 9.8 shows the convergence performance of the proposed allocation algorithm, where the step sizes of both PUs' channel prices are set to $\mu_1^1 = \mu_2^1 = 0.2$, and the step sizes of both PUs' power prices are set to $\mu_1^2 = \mu_2^2 = 0.5$. It is observed that four prices converge at different speeds, and the entire auction game converges after 71 iterations. Compared to the convergent power prices (i.e., the optimal values) of two PUs, we find that the optimal power price of PU 2 is higher than that of PU 1. It indicates that the power competition at PU 2 is stronger than that at PU 1, thus leading to

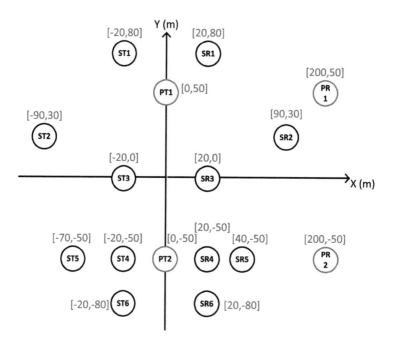

Figure 9.7: Six-SU two-PU simulation network.

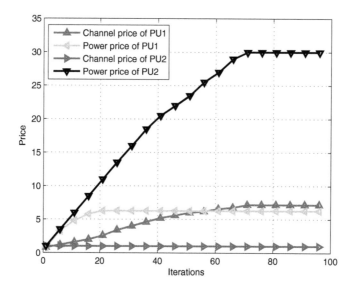

Figure 9.8: Convergence of PUs' prices.

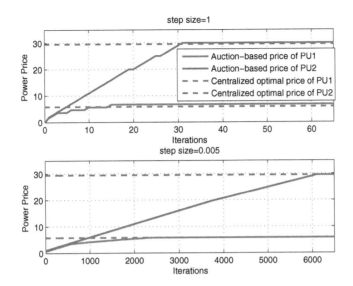

Figure 9.9: Impact of the step size on the convergence performance.

more auction clocks and a higher convergent power price. Also, it is noticed that the channel price of PU 2 remains unchanged throughout the entire auction process. The reason is that the number of SUs that choose PU 2 is always no more than the number of PU 2's available channels. Therefore, the channel price of PU 2 keeps at the initial price during the whole auction.

To evaluate the impact of the step size on the convergence performance, we set the step sizes of two PUs' power prices to 0.005 and 1, respectively. As can be seen in Figure 9.9, the power prices converge much faster with larger step sizes than with smaller step sizes. For example, the power price of PU 1 takes only 15 iterations to reach the convergence when its step size is 1. In contrast, it needs 2329 iterations to converge when the step size is 0.005. However, a smaller step size drives the auction to converge to the same point achieved by the centralized optimization algorithm [13], while a larger step size can only approximate to this optimum.

Figure 9.10 displays the evolution of the PU selection at SU 3, where "1" (Y-axis) represents SU 3 selects PU 1, and "2" represents it chooses PU 2. It is found that SU 3's selection varies between PU 1 and PU 2, and is finally stable at PU 2. According to the proposed algorithm, the SU at each auction clock, chooses the PU that currently incur the maximum utility. Due to different excess demand, two PUs might conduct different price adjustments at each clock. As a result, the PU selected by SU 3 at clock τ might not bring it the maximum utility at $\tau + 1$, then SU 3 leaves that PU and selects another one at $\tau + 1$. For instance, SU 3 selects PU 2 at $\tau = 9$ and changes to PU 1 at $\tau = 10$.

We now adjust the transmit power supplies of two PUs simultaneously within a range of $[1W, 7W]$, and keep other settings unchanged. It is seen in Figure 9.11 that the optimal power price of each PU decreases with the increase of its power supply. This is due to the fact that the less power available at a PU, the smaller possibility that this supply can meet the total power demand of the SUs, the higher optimal power price would be, and vice versa.

Figure 9.12 and Figure 9.13 display the cooperative power and the corresponding utility achieved at each SU, where the transmit power supplies of two PUs are simultaneously increased from $1W$ to $7W$. Note that in these cases, SU 1 and SU 2 always choose PU 1, and SU 3~6 always choose PU 2. It is observed that the cooperative power allocated to each SU increases with the power supply of each PU. On one hand, the more power available at the PU, the more power is assigned

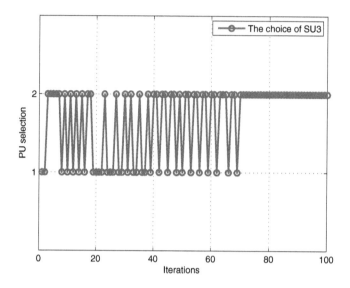

Figure 9.10: PU selection at SU 3.

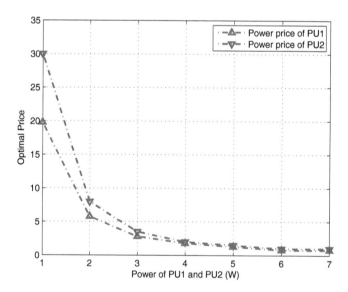

Figure 9.11: Optimal power price of each PU.

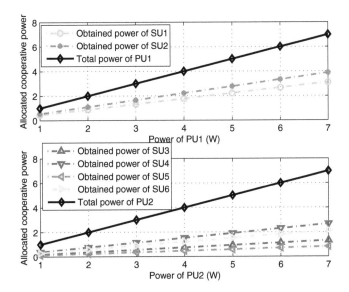

Figure 9.12: Cooperative power allocated to SUs.

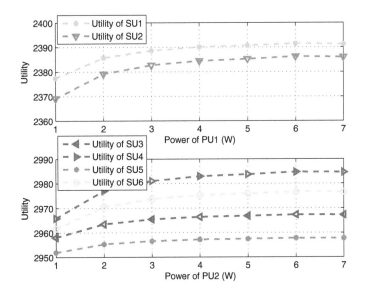

Figure 9.13: Utility achieved at SUs.

to each SU, and the higher achievable rate each SU achieves. On the other hand, the power price decreases as the power supply increases. As a result, the utility of each SU increases with the power supply of the PU.

Figure 9.14 compares the utilities of PUs achieved in cooperative transmission (CT) with those in direct transmission (DT), where the initial channel prices of two PUs are set to 0, and their initial

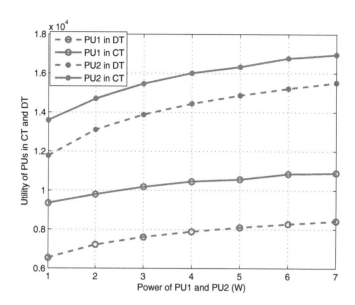

Figure 9.14: Cooperative transmission vs. direct transmission for the PU.

power prices are determined by Eq. (9.33). Such initial prices ensure that the utility of the PU in cooperative transmission is no less than that in direct transmission from the very beginning of the auction. With the ascending clock, the utilities of two PUs in cooperative transmission are always larger than in direct transmission.

Figure 9.15 shows the impacts of the initial power price on PU's utility achieved in CT, where the power supplies of two PUs are simultaneously set to $1W$, $4W$, and $7W$, respectively, and the initial channel prices of two PUs are set to 0. It is observed that each PU's utility in both CT and DT increases with the power supply. When we fix the power supply and adjust PU's initial power price, we find that PU's utility in CT is smaller than that in DT when the initial price is less than the reserved price. As the initial price becomes higher than the reserved price, the utility in CT is larger than that in DT. Once the increasing initial price exceeds a point at which the SUs quit the auction for a non-positive utility, the PU's utility in CT comes to zero.

Figure 9.16 compares the utilities of SUs achieved in CT with those in DT, where the power supply of each PU is $5W$. Note that SU in CT refers to that 6 SUs work in the proposed cooperative way, and SU in DT refers to that these SUs only uses 6 channels of two PUs and sends the data to the receiver by direct transmission. When SU in DT, the achievable rate from ST i to SR i by using PT j's channel is $R_{i,j,\mathrm{D}} = \frac{W_j}{3}\log_2(1 + \frac{P_i|G_{ST_i}^{SR_i}|^2}{\sigma^2})$. It is observed that the SU can always achieve a larger utility when it works in the proposed cooperative way, regardless of its transmit power (Here we only show the utilities of SU2 and SU5, and other SUs' performance can be analyzed in a similar manner). With such a large power supply, the SU can get more power at a lower price, thus receiving larger utility in CT method than in DT method.

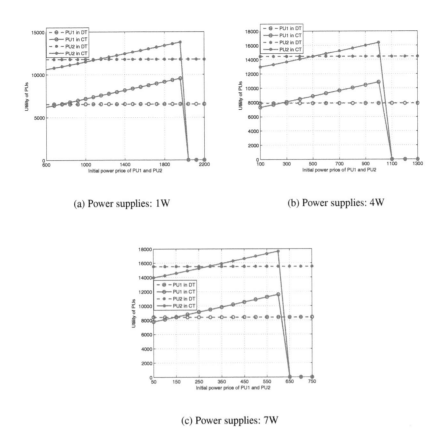

(a) Power supplies: 1W

(b) Power supplies: 4W

(c) Power supplies: 7W

Figure 9.15: Impacts of initial price on the utility achieved in CT.

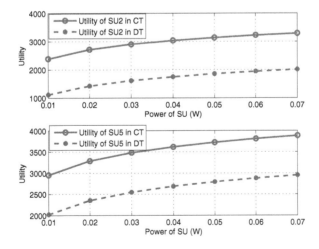

Figure 9.16: Cooperative transmission vs. direct transmission for the SU.

9.3 Summary

In this chapter, we discussed the joint pricing and resource allocation problem under two kinds of cooperative cognitive radio frameworks. In secondary users relaying for primary users, primary users aim at increasing the quality of service from secondary users' relaying help, and the spectrum access problem is modeled as a Stackelberg game. In primary users relaying for secondary users, primary users assist secondary transmissions and earn revenue from selling the spectrum and cooperative power to secondary users. The trade between primary users and secondary users is modeled as an auction with two bundling commodities. The auction algorithm and its convergence performance are investigated.

References

[1] A. Sendonaris, E. Erkip, and B. Aazhang, "User cooperation diversity — Part I: System description," *IEEE Transactions on Communications*, vol. 51, pp. 1927–1938, Nov. 2003.

[2] A. Sendonaris, E. Erkip, and B. Aazhang, "User cooperation diversity — Part II: Implementation aspects and performance analysis," *IEEE Transactions on Communications*, vol. 51, pp. 1939–1948, Nov. 2003.

[3] A. Nosratinia, T. Hunter, and A. Hedayat, "Cooperative communication in wireless networks," *IEEE Communications Magazine*, vol. 42, pp. 74–80, Oct. 2004.

[4] J. Jia, J. Zhang, and Q. Zhang, "Cooperative relay for cognitive radio networks," in *Proc. of IEEE INFOCOM*, pp. 2304–2312, Apr. 2009.

[5] X. Gong, W. Yuan, W. Liu, W. Cheng, and S. Wang, "A cooperative relay scheme for secondary communication in cognitive radio networks," in *Proc. of IEEE GLOBECOM*, pp. 1–6, Nov. 2008.

[6] O. Simeone, I. Stanojev, S. Savazzi, Y. Bar-Ness, U. Spagnolini, and R. Pickholtz, "Spectrum leasing to cooperating secondary ad hoc networks," *IEEE Journal on Selected Areas in Communications*, vol. 26, no. 1, pp. 203–213, Jan. 2008.

[7] J. Zhang and Q. Zhang, "Stackelberg game for utility-based cooperative cognitive radio networks," in *Proc. of ACM MobiHoc*, May 2009.

[8] X. Hao, M. Cheung, V. Wong, and V. Leung, "A Stackelberg game for cooperative transmission and random access in cognitive radio networks," in *Proc. of IEEE International Symposium on Personal, Indoor and Mobile Radio Comunications (PIMRC)*, Sep. 2011.

[9] Y. Yi, J. Zhang, Q. Zhang, T. Jiang, and J. Zhang, "Cooperative communication-aware spectrum leasing in cognitive radio networks," in *Proc. of IEEE Symposium on New Frontiers in Dynamic Spectrum (DySPAN)*, Apr. 2010.

[10] H. Wang, L. Gao, X. Gan, X. Wang, and E. Hossain, "Cooperative Spectrum Sharing in Cognitive Radio Networks: A Game-Theoretic Approach," in *Proc. IEEE ICC*, 2010.

[11] M. Osborne and A. Rubinstein, *A Course in Game Theory*. MIT Press, 1999.

[12] B. Yannis and E. Brynjolfsson, "Bundling information goods: pricing, profits, and efficiency," *Management Science*, vol. 45, no. 12, pp. 1613-1630, Dec. 1999.

[13] S. Boyd and L. Vandenberghe, *Convex Optimization*. Cambridge University. Press, 2004.

[14] L. M. Ausubel, "An efficient dynamic auction for heterogeneous commodities," *The American Economic Review*, vol. 96, no. 3, 2000.

[15] H. R. Varian, *Microeconomic Analysis*, 3rd Edition. W. W. Norton and Company, 1992.

[16] L. Walras, *Elements of Pure Economics*. Allen and Unwin, 1954.

[17] S. Shakkottai and R. Srikant, *Network Optimization and Control*. Now Publishers Inc; Boston – Delft, 2007.

Chapter 10

Cross-Layer Design for Spectrum Efficiency

Dr. Mohammad Robat Mili and Prof. Farokh Marvasti

CONTENTS

Abstract

This chapter describes a cross-layer design for reliable transmission in cognitive radio networks under spectrum sharing approach. In a cognitive radio scenario that consists of the primary users and the secondary users, secondary users opportunistically are allowed to access the existing spectrum without adverse effect on primary users. A cognitive radio network is allowed to detect its communication environment, replace the parameters of its communication scheme to raise the quality-of-service (QoS) for secondary users, and decrease the interference to primary users. Alternatively, the other approach involved in designing a cognitive radio network is to allow simultaneous transmission of primary users and secondary users, which is also termed as spectrum sharing. In this technique, a secondary transmitter can transmit while its maximum interference to the primary receiver is smaller than the predefined threshold. However, a secondary user must control its transmit power to get a reasonable transmission rate. In cognitive radio networks, the main issue is how to guarantee QoS in different applications. High capacity and minimum bit error rate (BER), varying as a function of the channel quality, are two of the major QoS requirements and are interference-limited in mobile communication systems. Therefore, in this chapter, we maximize the ergodic capacity and minimize the average BER under different constraints at the primary users. The effect of reducing channel state information (CSI) at the secondary transmitter is discussed for both optimization problems. Finally, simulation results are presented to support analytical results.

10.1   Introduction

The development of wireless services has recently led to the high demand of limited and valuable radio spectrum. However, the frequency allocation chart shows that most of the frequency bands have been occupied, under which a significant amount of spectrum is under-utilized. In order to balance between these two issues, cognitive radio networks have been proposed. Cognitive radio is a technique to improve the utilization efficiency of the existing radio spectrum. A cognitive radio network consists of primary users (PU) and secondary users (SU) such that the SUs are allowed to opportunistically access the existing spectrum without adverse effect on the PUs [1]–[5].

A cognitive radio network is allowed to detect its communication environment, replace the parameters of its communication scheme to raise the quality of service for SUs and decrease the interference to PUs. Consequently, spectrum sensing must avoid possible collision with PUs. In addition, a medium access control (MAC) layer protocol with sensing ability is required to fairly allocate resources among SUs and avoid collision with PUs [6], [7].

Alternatively, the other approach involved in designing a cognitive radio network is to allow simultaneous transmission of primary users and secondary users, which is also termed as spectrum sharing. In this technique, a secondary transmitter (ST) can transmit, while its maximum interference to the primary receiver (PR) is kept smaller than the predefined threshold. However, an SU must control its transmit power to get a reasonable transmission rate [8]. This paper focuses on the second approach.

The ergodic and outage capacity offered by the dynamic spectrum sharing approach in a single-antenna fading primary network has been investigated in [9]– [11] under average and peak interference power constraints at the PR. These constraints at the PR belong to one of the following two types: The first one is the long-term constraint that regulates the average interference across all the fading state, and the other one is the short-term constraint that limits the instantaneous interference over each fading state. However, in [9]– [11], the interference from the primary transmitter (PT) to the secondary receiver (SR) is ignored and the capacity is evaluated based on the signal-to-noise ratio (SNR).

A similar system model has been applied in recent works. For example, in [12], by employing Jensen's inequality on the objective function, the impact of the interference from PT to SR was assumed as a constant value, which leads only to approximate expression for the ergodic capacity. The authors in [13] also approximated the interference from PT to SR by its average power.

Furthermore, in all aforementioned works, it was assumed that perfect CSI is available at both the receiver and the transmitter. However, providing such side information in practice is not an easy task. The transmitter requires a feedback path between the transmitter and receiver to get the side information [14], [15].

In this chapter, we consider a fading environment with the received power constraint at a third party's receiver on average value. Ergodic capacity can be first obtained by optimal utilization of the transmitted power over time, in which the received power constraint is met. The ergodic capacity is defined as the maximum achievable rate averaged over all the fading blocks. Ergodic capacity is a good performance limit indicator for delay-insensitive services, when the codeword length can be sufficiently long to span over all the fading blocks. The three levels of CSI available at the ST are discussed, namely, CSI between ST-SR, between ST-PR, and between PT-SR. Under average interference power constraint, the power at ST depends on CSI between ST-SR, between ST-PR, and between PT-SR. However, under peak interference power constraint, the power at ST only depends on CSI between ST-PR. Then, we find the minimum BER of the cognitive radio network over different constraints at the PR taking into account the interference from PT. The constraints include either average or peak interference power constraint of secondary users at the primary's receiver. New expressions for computing the minimum BER in the case of Nakagami-m channels are derived. The effects of reducing the CSI at the ST is investigated and expressions for corresponding minimum BERs is derived. [16]– [21].

10.2   System Model

Figure 10.1 shows a spectrum sharing scenario where a cognitive radio link consisting of a transmitter and a receiver uses the same bandwidth for transmission with an existing primary link consisting of a PT and a PR. A flat fading channel with perfect CSI at the receiver and transmitter of the secondary user is considered. The secondary link between ST and SR is characterized by instantaneous channel power gain g_1 and the AWGN n_1. The noise n_1 is an independent random variable with the distribution $CN(0, N_0)$ (circularly symmetric complex Gaussian variable with mean zero and variance N_0).

The channel between ST and PR with instantaneous channel power gain g_0 has also been assumed. We consider the effect of the interference coming from the PT with constant power ρ on

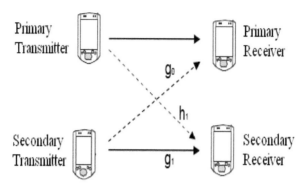

Figure 10.1: System model for spectrum sharing.

the SR. The instantaneous power at ST can be written as P. The instantaneous received signal-to-interference-plus-noise ratio (SINR) at the SR is

$$SINR = \frac{Pg_1}{N_0 + \rho h_1},\qquad(10.1)$$

where h_1 denotes the interference channel power gain between the PT and the SR.

In the following two sections, we derive expressions for evaluating the maximum capacity and minimum BER under different constraints.

10.3   Maximization of the Ergodic Capacity

10.3.1   *Average Interference Power Constraint*

We consider a third party's receiver in a fading environment, with a received average power constraint. Channel capacity can be obtained by optimal utilization of the transmitted power over time, in which the received power constraint is met.

In order to discuss the significance of having g_1, g_0 and h_1 at the ST, ergodic capacity is evaluated under different scenarios. In the first scenario, the optimum power allocation P is a function of g_1, g_0 and h_1. In the second scenario, the CSI h_1 at ST is reduced thus P becomes a function of g_1 and g_0. In the third scenario, we reduce the CSI g_0 at ST, consequently, P will be a function of g_1 and h_1. In next scenario, channel side information g_1 is not made available at ST, so P becomes a function of g_0 and h_1. Finally, all CSI g_1, g_0, and h_1 at ST in the last scenario are reduced, then P simplifies into a constant.

In what follows, we will explain these scenarios in more detail.

10.3.1.1 Full CSI $[P(g_1, g_0, h_1)]$

The ergodic capacity of the secondary link can be found by solving the following optimization problem.

$$\max_{P \geq 0} \quad \mathbb{E}\left[\ln\left(1 + \frac{P(g_1, g_0, h_1)g_1}{N_0 + \rho h_1}\right)\right] \tag{10.2a}$$

$$s.t \quad \mathbb{E}\left[P(g_1, g_0, h_1)g_0\right] \leq Q_{average}, \tag{10.2b}$$

where the transmit power of ST depends on all channel gains g_1, g_0, and h_1. Equation (10.2b) represents the average interference power constraint, which can be used to guarantee a long-term QoS of PU, and $Q_{average}$ is the maximum average received power limit at PR. The above optimization problem is equivalent to solving the following Lagrangian approach,

$$L(P, \lambda) = \mathbb{E}\left[\ln\left(1 + \frac{P(g_1, g_0, h_1)g_1}{N_0 + \rho h_1}\right)\right] - \lambda\left(\mathbb{E}\left[P(g_1, g_0, h_1)g_0\right] - Q_{average}\right), \tag{10.3}$$

where λ is the nonnegative dual variable corresponding to the constraint (10.2b). Taking the derivative of the Lagrangian in (10.3), with respect to $P(g_1, g_0, h_1)$ and letting the derivative equal to zero, yields [22]

$$\frac{L(P, \lambda)}{\partial P} = \mathbb{E}\left[\frac{g_1}{P(g_1, g_0, h_1)g_1 + N_0 + \rho h_1} - \lambda g_0\right] = 0, \tag{10.4}$$

which results in

$$P(g_1, g_0, h_1) = \frac{1}{\lambda g_0} - \frac{N_0 + \rho h_1}{g_1}. \tag{10.5}$$

Note that the optimum power allocation P, the instantaneous power at the ST, is a function of g_1, g_0, and h_1. In (10.5) by considering the constraint $P(g_1, g_0, h_1) \geq 0$, we get

$$\frac{g_0}{g_1} \leq \frac{1}{\lambda(N_0 + \rho h_1)}. \tag{10.6}$$

The parameter λ can be obtained by solving

$$Q_{average} = \mathbb{E}\left[\frac{1}{\lambda} - \frac{g_0}{g_1}(N_0 + \rho h_1)\right], \tag{10.7}$$

which satisfies the complementary slackness conditions [22]. We can get the maximum capacity by substituting (10.5) in (10.2a)

$$C = \mathbb{E}\left[\ln\left(\frac{1}{\lambda}\frac{g_1}{g_0(N_0 + \rho h_1)}\right)\right]. \tag{10.8}$$

We substitute $x = \frac{g_0}{g_1}$ and $y = (N_0 + \rho h_1)$ in (10.7), which yields

$$Q_{average} = \int_0^\infty \int_{x < \frac{1}{\lambda y}} \left(\frac{1}{\lambda} - xy\right) f_x(x) f_y(y) dx dy. \tag{10.9}$$

In the case of Rayleigh fading, the probability density function (PDF) of the ratio between two exponential random variables $\frac{g_0}{g_1}$ can be expressed as [23]

$$f_{\frac{g_0}{g_1}}(x) = \frac{1}{(1 + x)^2}, \tag{10.10}$$

and the PDF of the sum $N_0 + \rho h_1$ becomes

$$f_{(N_0+\rho h_1)}(y) = \frac{1}{\rho} e^{-\frac{y-N_0}{\rho}}. \tag{10.11}$$

By using (10.10) and (10.11), Equation (10.9) becomes

$$Q_{average} = \frac{1}{\rho} \int_{N_0}^{\infty} \int_0^{\frac{1}{\lambda y}} \left(\frac{1}{\lambda} - xy \right) \frac{e^{-\frac{y-N_0}{\rho}}}{(1+x)^2} dx dy. \tag{10.12}$$

Upon invoking [25, eq. (2.113), (4.222.8) and (4.331.2)], Equation (10.12) reduces into the following form:

$$Q_{average} = \frac{1}{\lambda} \left[1 + e^{\frac{1+\lambda N_0}{\lambda P}} (\lambda \rho - 1) E_i \left(-\frac{1+\lambda N_0}{\lambda \rho} \right) \right.$$
$$\left. - e^{\frac{N_0}{\rho}} \lambda \rho E_i \left(-\frac{N_0}{\rho} \right) - \lambda (N_0 + \rho) \ln \left(1 + \frac{1}{\lambda N_0} \right) \right], \tag{10.13}$$

where $E_i(.)$ is the exponential integral function defined as $E_i(x) = \int_{-\infty}^x \frac{e^t}{t} dt$ [25]. We can find λ for a given $Q_{average}$ from Equation (10.13). It is worth noting that determining λ from (10.13) requires the use of numerical integration.

Similarly, we obtain the channel capacity as

$$C = \int_0^{\infty} \int_{x < \frac{1}{\lambda y}} \ln \left(\frac{1}{\lambda} \frac{1}{xy} \right) f_x(x) f_y(y) dx dy$$
$$= \frac{1}{\rho} \int_{N_0}^{\infty} \int_0^{\frac{1}{\lambda y}} \ln(\frac{1}{\lambda} \frac{1}{xy}) \frac{e^{-\frac{y-N_0}{\rho}}}{(1+x)^2} dx dy. \tag{10.14}$$

By changing the variable $t = \frac{1}{x}$ and using [25, eq. (2.727.3), (4.337.1) and (4.331.2)] in (10.14), we obtain the following closed-form expression:

$$C = e^{\frac{N_0}{\rho}} E_i \left(-\frac{N_0}{\rho} \right) - e^{\frac{1+\lambda N_0}{\lambda \rho}} E_i \left(-\frac{1+\lambda N_0}{\lambda \rho} \right) + \ln \left(1 + \frac{1}{\lambda N_0} \right). \tag{10.15}$$

Equation (10.15) is a new closed-form expression for ergodic capacity when the ST knows all instantaneous channel gains g_1, g_0, and h_1.

10.3.1.2 Partial CSI: Reduced Only CSI $h_1 [P(g_1, g_0)]$

Here, we find the maximum capacity with a reduced side information where h_1 is not made available at the ST. Hence, by disregarding h_1, the power of ST depends on g_1 and g_0. The maximum capacity problem (10.2a) subject to (10.2b) changes into the following form:

$$\max_{P \geq 0} \mathbb{E} \left[\ln \left(1 + \frac{P(g_1, g_0) g_1}{N_0 + \mathbb{E}[\rho h_1]} \right) \right] \tag{10.16a}$$

$$s.t \quad \mathbb{E}[P(g_1, g_0) g_0] \leq Q_{average}, \tag{10.16b}$$

where the transmit power is now only a function of (g_1, g_0) and independent of h_1. Following the same procedure used to derive (10.5) in the previous section, the optimal power allocation in the optimization problem (10.16a) subject to (10.16b) is given by

$$P(g_1, g_0) = \frac{1}{\lambda g_0} - \frac{N_0 + \rho}{g_1}. \tag{10.17}$$

In (10.17), by considering the constraint $P(g_1, g_0) \geq 0$, we have

$$\frac{g_1}{g_0} \geq \lambda \left(N_0 + \rho\right). \tag{10.18}$$

Then, by replacing (10.17) into (10.16b) and considering equality, we get

$$Q_{average} = \mathbb{E}\left[\frac{1}{\lambda} - \frac{g_0}{g_1}\left(N_0 + \rho\right)\right]. \tag{10.19}$$

The parameter λ can be obtained in terms of $Q_{average}$ by using the nonlinear Equation (10.19) and, hence, the maximum capacity becomes

$$C = \mathbb{E}\left[\ln\left(\frac{1}{\lambda}\frac{g_1}{g_0\left(N_0 + \rho\right)}\right)\right]. \tag{10.20}$$

By using (10.10), we find $Q_{average}$ as follows:

$$\begin{aligned}
Q_{average} &= \int_0^{\frac{1}{\lambda(N_0+\rho)}} \left(\frac{1}{\lambda} - x(N_0+\rho)\right)\frac{1}{(1+x)^2}dx \\
&= \frac{\frac{1}{\lambda(N_0+\rho)} - (1 + \lambda\left(N_0+\rho\right))\ln(1 + \frac{1}{\lambda(N_0+\rho)}) + 1}{\frac{1}{(N_0+\rho)} + \lambda},
\end{aligned} \tag{10.21}$$

and the maximum capacity becomes

$$\begin{aligned}
C &= \int_0^{\frac{1}{\lambda(N_0+\rho)}} \ln(\frac{1}{\lambda}\frac{1}{x(N_0+\rho)})\frac{1}{(1+x)^2}dx \\
&= \ln\left(1 + \frac{1}{\lambda\left(N_0 + \rho\right)}\right),
\end{aligned} \tag{10.22}$$

which is a closed-form expression for ergodic capacity when ST is only dependent on g_0 and g_1.

10.3.1.3 *Partial CSI: Reduced Only CSI* $g_0[P(g_1, h_1)]$

Here, the maximum capacity with a reduced side information g_0 at the ST is computed. Therefore, we disregard the effect of g_0 from the power allocation, resulting in the following optimization problem:

$$\max_{P \geq 0} \quad \mathbb{E}\left[\ln\left(1 + \frac{P(g_1, h_1)g_1}{N_0 + \rho h_1}\right)\right] \tag{10.23a}$$

$$s.t \quad \mathbb{E}\left[P(g_1, h_1)g_0\right] \leq Q_{average}. \tag{10.23b}$$

Note that the constraint is equivalent to

$$s.t \quad \mathbb{E}\left[P(g_1, h_1)\right] \leq Q_{average}, \tag{10.24a}$$

because g_0 is independent of g_1 and h_1. Similarly, by applying the Lagrangian approach, we get the optimal power allocation as

$$P(g_1, h_1) = \frac{1}{\lambda} - \frac{N_0 + \rho h_1}{g_1}. \tag{10.25}$$

We can find $Q_{average}$ as following

$$Q_{average} = \frac{1}{\rho}\int_{N_0}^{\infty}\int_{\lambda y}^{\infty}\left(\frac{1}{\lambda} - \frac{y}{g_1}\right)e^{-g_1}e^{-\frac{y-N_0}{\rho}}dg_1 dy. \tag{10.26}$$

For $N_0 = 0$, (10.26) can be evaluated in the following closed-form:

$$Q_{average} = \frac{1}{\lambda} + (-1+\gamma)\rho - \rho \ln(1+\lambda\rho).$$

The maximum capacity becomes

$$C = \frac{1}{\rho} \int_{N_0}^{\infty} \int_{\lambda y}^{\infty} \ln(\frac{1}{\lambda}\frac{g_1}{y}) e^{-g_1} e^{-\frac{y-N_0}{\rho}} dg_1 dy$$

$$= -E_i(-\lambda N_0) + e^{\frac{N_0}{\rho}} E_i\left(-\frac{N_0 + \lambda \rho N_0}{\rho}\right). \tag{10.27}$$

Equation (10.27) is an expression for ergodic capacity when the ST knows only channel gains g_1 and h_1.

10.3.1.4 Partial CSI: Reduced Only CSI $g_1[P(g_0, h_1)]$

In order to find the impact of having g_1 at the ST, the maximum capacity with no g_1 at the ST is computed. So, we ignore g_1 from the optimization problem (10.2a) subject to (10.2b), yielding the equation

$$\max_{P\geq 0} \quad \mathbb{E}\left[\int_0^{\infty} \ln\left(1 + \frac{P(g_0,h_1)g_1}{N_0 + \rho h_1}\right) e^{-g_1} dg_1\right] \tag{10.28a}$$

$$s.t \quad \mathbb{E}[P(g_0,h_1)g_0] \leq Q_{average}. \tag{10.28b}$$

Following the same procedure by applying the Lagrangian approach, we find the optimal power allocation as

$$P(g_0, h_1) = \int_{(N_0+\rho h_1)\lambda g_0}^{\infty} \left(\frac{1}{\lambda g_0} - \frac{N_0 + \rho h_1}{g_1}\right) e^{-g_1} dg_1$$

$$= \frac{e^{-(N_0+\rho h_1)\lambda g_0}}{\lambda g_0} - (N_0 + \rho h_1)$$

$$\times \left(\Gamma\left(0, (N_0+\rho h_1)\lambda g_0\right) + \ln\left((N_0+\rho h_1)\lambda g_0\right)\right). \tag{10.29}$$

Accordingly, $Q_{average}$ becomes

$$Q_{average} = \frac{1}{\rho} \int_{N_0}^{\infty} \int_0^{\infty} \frac{1}{\lambda} (e^{-y\lambda g_0} - y\lambda g_0(\Gamma(0, y\lambda g_0))$$

$$+ \ln(y\lambda g_0)))e^{-g_0} e^{-\frac{y-N_0}{\rho}} dg_0 dy. \tag{10.30}$$

For $N_0 = 0$, we can obtain the following closed form result

$$Q_{average} = \frac{1}{\lambda} + (\gamma - 1)\rho - \rho \ln(\lambda\rho) + \rho U(0, -1, \frac{1}{\lambda\rho}), \tag{10.31}$$

where γ is the Euler's constant and $U(a,b,z)$ is the confluent hypergeometric function. The parameter λ can be obtained in terms of $Q_{average}$, and, finally, the maximum capacity is expressed as Equation (10.32),

$$C = \frac{1}{\rho} \int_0^{\infty} \int_{N_0}^{\infty} \int_{y\lambda g_0}^{\infty} \ln(1 + g_1 B) e^{-g_1} e^{-g_0} e^{-\frac{y-N_0}{\rho}} dg_1 dy dg_0$$

$$= \frac{1}{\rho} \int_{N_0}^{\infty} \int_0^{\infty} \left(e^{\frac{1}{B}}\Gamma(0, y\lambda g_0 + \frac{1}{B}) + e^{-y\lambda g_0}\ln(1 + y\lambda g_0 B)\right) e^{-\frac{y-N_0}{\rho} - g_0} dg_0 dy, \tag{10.32}$$

where $B = \frac{e^{-y\lambda g_0}}{y\lambda g_0} - (\Gamma(0, y\lambda g_0) + \ln(y\lambda g_0))$. We observe that closed-form expressions are not obtainable for (10.32) and, hence, we need to solve the equation numerically.

10.3.1.5 Without CSI [Constant P]

Here, all the channel side information that can be available at the ST are reduced and the power at the ST becomes constant. Hence, the maximum capacity is calculated by ignoring h_1, g_0, and g_1 from the optimization problem, yielding the equation

$$\max_{P \geq 0} \quad \mathbb{E}\left[\ln\left(1 + \frac{Pg_1}{N_0 + \rho h_1}\right)\right] \tag{10.33a}$$

$$s.t \quad \mathbb{E}[Pg_0] \leq Q_{average}. \tag{10.33b}$$

Here, we find the maximum capacity with no CSI available at ST, and, therefore, the power becomes a constant and independent of channel gains. We can simplify the above optimization problem into

$$\max_{P \geq 0} \quad \mathbb{E}\left[\ln\left(1 + \frac{Pg_1}{N_0 + \rho h_1}\right)\right] \tag{10.34a}$$

$$s.t \quad \mathbb{P} \leq Q_{average}, \tag{10.34b}$$

which, in the case of Rayleigh fading, gives

$$C = \mathbb{E}\left[\ln\left(1 + \frac{Q_{average}g_1}{N_0 + \rho h_1}\right)\right]$$
$$= \int_0^\infty \int_0^\infty \ln\left(1 + \frac{Q_{average}g_1}{N_0 + \rho h_1}\right) e^{-g_1} e^{-h_1} dg_1 dh_1. \tag{10.35}$$

Therefore, the capacity in this case simplifies into

$$C = \frac{e^{\frac{N_0(1+\rho-Q_{average})}{Q_{average}}} E_i\left(-\frac{N_0(1+\rho)}{Q_{average}}\right) - e^{\frac{N_0}{\rho}} E_i\left(-\frac{N_0(1+\rho)}{\rho}\right)}{\frac{\rho}{Q_{average}} - 1}, \tag{10.36}$$

which is a closed-form expression for capacity with no CSI available at ST. Numerical results that compare the ergodic capacities under different CSI will be given at the end of this section.

10.3.2 Peak Interference Power Constraint

The peak power constraint is more appropriate when the QoS is limited by the instantaneous SINR at the receiver. Therefore, we use the following optimization problem,

$$\max_{P \geq 0} \quad \mathbb{E}\left[\ln\left(1 + \frac{Pg_1}{N_0 + \rho h_1}\right)\right] \tag{10.37a}$$

$$s.t \quad Pg_0 \leq Q_{peak}, \tag{10.37b}$$

where Equation (10.37b) denotes peak interference power constraint, and Q_{peak} is the peak received power limit at the existing PR. We obtain the maximum capacity if the power is replaced by

$$P(g_0) = \frac{Q_{peak}}{g_0}, \tag{10.38}$$

where, in this case, $P(g_0)$ is only a function of g_0 and independent of g_1 and h_1.

In order to investigate the impact of having g_0 at the ST, ergodic capacity is evaluated under two different scenarios. In the first scenario, the optimum power allocation P is a function of g_0, and in the second scenario, the CSI g_0 at ST is reduced and P is a constant.

10.3.2.1 Full CSI $[P(g_0)]$

In this case, we obtain the maximum capacity when the power at ST is a function of g_0 as in the following expression:

$$C = \mathbb{E}\left[\ln\left(1 + \frac{Q_{peak}}{N_0 + \rho h_1}\frac{g_1}{g_0}\right)\right]. \tag{10.39}$$

We substitute $x = \frac{g_1}{g_0}$ and $y = (N_0 + \rho h_1)$ in Equation (10.39), resulting in,

$$C = \frac{1}{\rho}\int_{N_0}^{\infty}\int_0^{\infty} \ln\left(1 + \frac{Q_{peak}x}{y}\right)\frac{e^{-\frac{y-N_0}{\rho}}}{1+x^2}\,dx\,dy. \tag{10.40}$$

The equation (10.40) can be simplified into (10.41)

$$C = \int_0^{\infty}\frac{1}{2(1+x^2)}\left(e^{\frac{N_0}{\rho}}\left(2E_i\left(-\frac{N_0}{\rho}\right) + e^{\frac{Q_{peak}x}{\rho}}\left(-2E_i\left(-\frac{N_0+Q_{peak}x}{\rho}\right)\right.\right.\right.$$
$$\left.\left.\left. -2\ln\left(\frac{Q_{peak}x}{\rho}\right) + \ln\left(\frac{Q_{peak}^2 x^2}{\rho^2}\right)\right)\right) + 2\ln\left(1 + \frac{Q_{peak}x}{N_0}\right)\right)dx, \tag{10.41}$$

which can be calculated numerically.

10.3.2.2 Without CSI [Constant P]

Here, we find the maximum capacity with a reduced side information, where g_0 is not provided at the ST. Thus, the power of ST becomes a constant.

The maximum capacity problem changes into the following form:

$$C = \mathbb{E}\left[\ln\left(1 + \frac{Q_{peak}g_1}{N_0 + \rho h_1}\right)\right]. \tag{10.42}$$

Likewise, in the case of Rayleigh fading, the maximum capacity becomes

$$C = \frac{e^{\frac{N_0(1+\rho-Q_{peak})}{Q_{peak}}}E_i\left(-\frac{N_0(1+\rho)}{Q_{peak}}\right) - e^{\frac{N_0}{\rho}}E_i\left(-\frac{N_0(1+\rho)}{\rho}\right)}{\frac{\rho}{Q_{peak}} - 1}. \tag{10.43}$$

Equation (10.43) is closed-form expression for capacity under peak interference power constraint when P is constant.

10.3.3 Numerical Results

In this section, we present some numerical results for the maximum capacity under average/peak interference power constraints and different CSI levels. We assume that $N_0 = 1$.

10.3.3.1 Average Interference Power Constraint

Figure 10.2 and Figure 10.3 display capacity vs. $Q_{average}$ under average interference power constraint for different values of ρ. Comparing Figure 10.2 with Figure 10.3 indicates that the interference from the PT can have a big impact on the capacity. As we can see, capacity in all cases increases with increasing $Q_{average}$. Further examination of Figure 10.2 and Figure 10.3 reveals that the highest capacity occurs when g_1, g_0, and h_1 at the power of ST are included, while the lowest capacity occurs when g_1, g_0, and h_1 are excluded.

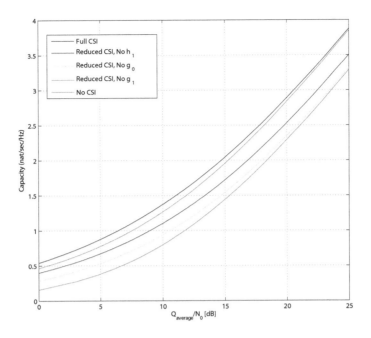

Figure 10.2: Impact of reducing CSI under average interference power constraint with $\rho = 5dB$.

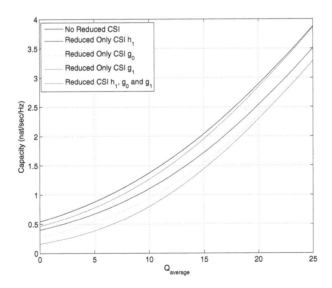

Figure 10.3: Impact of reducing CSI under average interference power constraint with $\rho = 10dB$.

Another important observation is that the capacity difference between no reduced CSI and reducing only CSI g_1 is very small, such that having side information g_1 at ST has negligible effect on the system performance. Again, we observe that when only CSI g_0 is reduced, the secondary link loses all the capacity advantage that can be achieved by having all side information. Therefore, g_0 has the highest impact on the capacity of the system, while having g_1 has very minimal impact. This

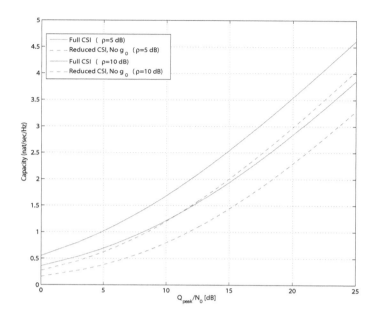

Figure 10.4: Impact of reducing CSI under peak interference power constraint.

is because the CSI g_0 directly affects the optimization problem, but CSI g_1 is inside the logarithmic function and has less impact. Furthermore, the effect of having only h_1 is less than g_0 and bigger than g_1, such that by reducing only CSI h_1, we lose almost half of the obtainable capacity.

10.3.3.2   Peak Interference Power Constraint

Figure 10.4 shows the capacity under peak interference power constraint for two different values of ρ. In this case, the capacity almost linearly increases with increasing Q_{peak}. Comparing Figure 10.4 against Figure 10.2 and Figure 10.3, we observe the performance of the channel under peak interference power constraint is almost the same as under average interference power constraint. As can be seen, the difference between the capacity in both cases when g_0 is either included or ignored is considerable.

10.4   Minimization of the Bit Error Rate (BER)

The BER in most common types of digital modulation schemes in wireless communication takes one of the following forms [24]:

$$BER = \begin{cases} \frac{1}{2}\exp(-SINR) & e.g.\ DPSK \\ aQ(\sqrt{bSINR}) & e.g.\ BPSK, QPSK \end{cases}, \tag{10.44}$$

where $Q(x)$ is the Q-function, and $a,b > 0$. Equation (10.44) applies to a wide class of modulation schemes. For example, exact results follow for quadrature phase-shift keying (QPSK) and binary phase shift keying (BPSK) with $(a,b) = (1,2)$. Furthermore, in the case of multiple phase-shift keying (M-PSK), (a,b) is $\left(\frac{1}{\log_2 M}, \sin^2(\frac{\pi}{M}) \times \log_2 M\right)$, and for quadrature amplitude modulation (QAM), (a,b) becomes $\left(\frac{2}{\log_2 M}, \frac{3\log_2 M}{M-1}\right)$ to approximate the BER.

In the following two sections, we derive expressions for evaluating the minimum average BER under different constraints and different digital modulations.

10.4.1 Minimum BER Under Average Interference Power Constraint

10.4.1.1 Minimization of Exp(-SINR)

The minimum BER under the average interference power constraint can be obtained by solving the following optimization problem:

$$\min_{P \geq 0} \quad \mathbb{E}\left[\frac{1}{2}\exp\left(-\frac{P(g_1,g_0,h_1)g_1}{N_0+\rho h_1}\right)\right] \tag{10.45a}$$

$$s.t \quad \mathbb{E}\left[P(g_1,g_0,h_1)g_0\right] \leq Q_{average}. \tag{10.45b}$$

Equation (10.45b) represents the average interference power constraint, which can be used to guarantee a long-term QoS of PU. $Q_{average}$ is the maximum average received power limit at PR. The optimal power allocation, $P(g_1,g_0,h_1)$, is obtained by forming the Lagrangian

$$L(P,\lambda) = \mathbb{E}\left[\frac{1}{2}\exp\left(-\frac{P(g_1,g_0,h_1)g_1}{N_0+\rho h_1}\right)\right]$$
$$+ \lambda\left(\mathbb{E}\left[P(g_1,g_0,h_1)g_0\right] - Q_{average}\right), \tag{10.46}$$

where λ is the nonnegative dual variable. By applying the karush-kuhn-tucker (KKT) conditions [22], the optimal power allocation satisfies the following equation,

$$\frac{L(P,\lambda)}{\partial P} = \mathbb{E}\left[-\frac{g_1}{N_0+\rho h_1}\frac{1}{2}\exp(-\frac{P(g_1,g_0,h_1)g_1}{N_0+\rho h_1}) + \lambda g_0\right] = 0, \tag{10.47}$$

which results in

$$P(g_1,g_0,h_1) = \frac{N_0+\rho h_1}{g_1}\ln\left(\frac{1}{N_0+\rho h_1}\frac{1}{2\lambda}\frac{g_1}{g_0}\right). \tag{10.48}$$

Note that the optimum power allocation P, the instantaneous power at the ST, is a function of the channel gains g_1, g_0, and h_1. By considering the constraint $P(g_1,g_0,h_1) \geq 0$ in (10.48), we get

$$\frac{1}{2}\frac{1}{N_0+\rho h_1}\frac{g_1}{g_0} \geq \lambda. \tag{10.49}$$

The parameter λ^*, which satisfies the complementary slackness conditions [22], can be obtained by solving

$$Q_{average} = \mathbb{E}\left[\frac{g_0}{g_1}(N_0+\rho h_1)\ln\left(\frac{1}{N_0+\rho h_1}\frac{1}{2\lambda^*}\frac{g_1}{g_0}\right)\right]. \tag{10.50}$$

We can get the minimum BER by substituting (10.48) in (10.45a)

$$BER = \lambda^*\mathbb{E}\left[\frac{g_0}{g_1}(N_0+\rho h_1)\right]. \tag{10.51}$$

In order to find the impact of having g_0 and h_1 at the ST, *BER* is evaluated under two different special cases. In the first case, the interference from the PT is ignored ($\rho = 0$), and then the optimum power allocation P becomes only a function of g_1 and g_0. Thus, we can study the effect of having extra CSI g_0 at the ST. In the second case, the interference from the PT is included, and the optimum power allocation P is a function of g_1,g_0, and h_1, which leads to study the effect of having extra CSI g_0 and h_1 at the ST.

10.4.1.2 Special Case 1: The Effect of Having Extra CSI g_0 at ST

In order to focus on the effect of having g_0 at the ST, we assume that $\rho = 0$ and P is only a function of g_1 and g_0.

In the case of a Nakagami-m fading model, the channel power gain is distributed as a gamma distribution [26]

$$f(x) = \frac{m^m}{\Gamma(m)} x^{m-1} \exp(-mx) \quad x \geq 0, \tag{10.52}$$

where $\Gamma(x)$ is the Gamma function. Note that if g_1 and g_0 are independent gamma random variables with parameters m_1 and m_0, respectively, then the PDF of the ratio g_1/g_0 becomes (e.g., [23, pp 695])

$$f_{g_1/g_0}(x) = \frac{x^{m_1-1}(1+x)^{-m_1-m_0}}{B(m_1, m_0)}, \tag{10.53}$$

where $B(a,b) = \frac{\Gamma(a)\Gamma(b)}{\Gamma(a+b)}$ is the beta function. By using (10.53) and assuming $m_0 = m_1 = m$, (10.50) becomes

$$Q_{average}(\lambda^*) = \frac{N_0}{B(m,m)} \int_{2\lambda^*N_0}^{\infty} \ln\left(\frac{x}{2\lambda^*N_0}\right) \frac{x^{m-1}}{x}(1+x)^{-2m}dx, \tag{10.54}$$

and upon using [25, eq.(1.512.3) and eq.(3.197.2)], (10.54) can be rewritten as

$$Q_{average}(\lambda^*) = \frac{\left(\frac{1}{2\lambda^*N_0}\right)^m}{2\lambda^*(1+m)^2 B(m,m)}$$
$$\times \; {}_3F_2\left(2m, 1+m, 1+m; 2+m, 2+m; -\frac{1}{2\lambda^*N_0}\right), \tag{10.55}$$

in which ${}_3F_2(a_1, a_2, a_3; b_1, b_2; z)$ is the hypergeometric function.

It is worth noting that we can find λ^* for a given $Q_{average}$ from Equation (10.55). Then, the minimum BER (10.51) is determined by

$$BER = \frac{\lambda^*N_0}{B(m,m)} \int_{2\lambda^*N_0}^{\infty} x^{m-2}(1+x)^{-2m}dx, \tag{10.56}$$

which is reduced, upon the change of variable $x = -\frac{1}{t}$, to the simple closed-form

$$BER = -\lambda^*N_0(-1)^{-m} \frac{B_{-\frac{1}{2\lambda^*N_0}}(1+m, 1-2m)}{B(m,m)}, \tag{10.57}$$

where $B_a(c,d)$ is the incomplete Beta function. When $m = 1$ (Rayleigh fading), we obtain the following simplified equations instead of (10.55) and (10.57):

$$Q_{average}(\lambda^*) = -N_0\left(\ln(1 + \frac{1}{2\lambda^*N_0}) + Li_2(-\frac{1}{2\lambda^*N_0})\right) \tag{10.58}$$

and

$$BER = \lambda^*N_0 B_{-\frac{1}{2\lambda^*N_0}}(2, -1), \tag{10.59}$$

where $Li_2(.)$ is the polylogarithm function of order 2 [25].

10.4.1.3 Special Case 2: The Effect of Having Extra CSI g_0 and h_1 at ST

Here, the effect of interference coming from the PT on the SR is considered by studying the impact of providing channel gain g_0 and h_1 at the ST. We substitute $x = \frac{g_1}{g_0}$ and $y = (N_0 + \rho h_1)$ in (10.50), which yields

$$Q_{average}(\lambda^*) = \int_0^\infty \int_{x>2y\lambda^*} \frac{y}{x} \ln(\frac{x}{2\lambda^* y}) f_x(x) f_y(y) du dy, \qquad (10.60)$$

Closed-form result for (10.60) can be obtained in the special case of Rayleigh fading, where $m_0 = m_1 = 1$. In this case, the PDF of $\frac{g_1}{g_0}$ becomes

$$f_{\frac{g_1}{g_0}}(x) = \frac{1}{(1+x)^2}, \qquad (10.61)$$

and the PDF of $N_0 + \rho h_1$ for $m = 1$ in Equation (10.52) can be expressed as

$$f_{(N_0+\rho h_1)}(y) = \frac{1}{\rho} e^{-\left(\frac{y-N_0}{\rho}\right)} \qquad y \geq N_0. \qquad (10.62)$$

By using (10.61) and (10.62) in (10.60), we obtain

$$Q_{average}(\lambda^*) = \frac{1}{\rho} \int_{N_0}^\infty \int_{2y\lambda^*}^\infty \left(\frac{y}{x} \ln\left(\frac{x}{2\lambda^* y}\right)\right)$$
$$\times \frac{1}{(1+x)^2} \left(e^{-\left(\frac{y-N_0}{\rho}\right)}\right) du dy. \qquad (10.63)$$

In the special case of $N_0 = 0$, we can obtain closed-form results with the aid of [25, eq.(1.512.3), eq.(3.197.2), eq.(3.326.2), eq.(4.352.2), eq.(4.337.5), eq.(4.358.2), and eq.(3.351.2)]. This can be expressed as

$$Q_{average}(\lambda^*) = \frac{1}{4\rho\lambda^{*2}} \left[\lambda^*\rho(2e^{\frac{1}{2\lambda^*\rho}}(2\lambda^*\rho - 1)E_i(\frac{-1}{2\lambda^*\rho}) + \lambda^*\rho(\pi^2 + 2\ln^2 2 \right.$$
$$+ 2\gamma(\gamma - 4 - \ln 4) + \ln 256 + 2\ln\rho\ln(4\rho) - 4(\gamma - 2)\ln(\lambda^*\rho)$$
$$\left. + 2\ln\lambda^* \ln(4\lambda^*\rho^2))) - G_{3\,4}^{4\,1}\left(\frac{1}{2\lambda^*\rho}\Big|_{-2,-2,-2,0}^{-2,-1,-1}\right)\right], \qquad (10.64)$$

where $E_i(.)$ is the exponential integral function and γ is the Euler's constant value, and $G_{p\,q}^{m\,n}\left(.\big|_{(b_q)}^{(a_p)}\right)$ is the Meijer function. Similarly, we substitute $u = \frac{g_0}{g_1}$ and $y = (N_0 + \rho h_1)$ in (10.51) to get

$$BER = \lambda^* \int_0^\infty \int_{u<\frac{1}{2y\lambda^*}} (uy) f_u(u) f_y(y) du dy, \qquad (10.65)$$

which results in

$$BER = \frac{\lambda^*}{\rho} \int_{N_0}^\infty \int_0^{\frac{1}{2y\lambda^*}} \frac{uy}{(1+u)^2} e^{-\left(\frac{y-N_0}{\rho}\right)} du dy. \qquad (10.66)$$

By using [25, eq.(2.113.2), eq.(4.222.8), eq.(3.353.5), and eq.(4.352.2)], the double integration in (10.66) can be represented as Equation (10.67)

$$BER = \frac{1}{4\rho\lambda^*} \left[-e^{\frac{1+2\lambda^*N_0}{2\lambda^*\rho}}(1 - 2\lambda^*\rho + 4\rho^2\lambda^{*2})E_i(-\frac{1+2\lambda^*N_0}{2\lambda^*\rho}) \right.$$
$$\left. + 2\lambda^*\rho\left(2e^{\frac{N_0}{\rho}}\rho\lambda^*E_i(-\frac{N_0}{\rho}) - 2\lambda^*(N_0 + \rho)\ln(\frac{2\lambda^*N_0}{1+2\lambda^*N_0}) - 1\right)\right]. \qquad (10.67)$$

10.4.1.4 Minimization of $aQ(\sqrt{bSINR})$

The minimum BER under the average interference power constraint can be obtained by using $Q(x) = \frac{1}{2}\operatorname{erfc}\left(\frac{x}{\sqrt{2}}\right)$ and then solving the following optimization problem

$$\min_{P \geq 0} \quad \mathbb{E}\left[\frac{a}{2}\operatorname{erfc}\sqrt{\frac{b}{2}\frac{P(g_1,g_0,h_1)g_1}{N_0+\rho h_1}}\right] \tag{10.68a}$$

$$s.t. \quad \mathbb{E}\left[P(g_1,g_0,h_1)g_0\right] \leq Q_{average}. \tag{10.68b}$$

Equation (10.68b) represents the average interference power constraint, which can be used to guarantee a long-term QoS of PU, where $Q_{average}$ is the maximum average received power limit at PR. The error function can be written as the following identity, which can be employed to simplify the BER analysis in fading environments [27, eq.(4.2)]

$$\operatorname{erfc}\sqrt{x} = \frac{2}{\pi}\int_0^{\pi/2}\exp\left(-\frac{x}{\cos^2(\theta)}\right)d\theta. \tag{10.69}$$

Therefore, (10.68a) becomes

$$\min_{P \geq 0} \quad \mathbb{E}\left[\frac{a}{\pi}\int_0^{\pi/2}\exp\left(-\frac{b}{2}\frac{P(g_1,g_0,h_1)g_1}{N_0+\rho h_1}\frac{1}{\cos^2(\theta)}\right)d\theta\right], \tag{10.70}$$

where the expectation is with respect to the channel gains g_1, g_0, and h_1. Notice that (10.70) subject to (10.68b) is mathematically equivalent to the following problem:

$$\min_{P \geq 0} \quad \mathbb{E}\left[\frac{a}{2}\exp\left(-\frac{b}{2}\frac{P'(g_1,g_0,h_1,\theta)g_1}{N_0+\rho h_1}\frac{1}{\cos^2(\theta)}\right)\right] \tag{10.71a}$$

$$s.t. \quad \mathbb{E}\left[P'(g_1,g_0,h_1,\theta)g_0\right] \leq Q_{average}, \tag{10.71b}$$

where the expectation is with respect to g_1, g_0, h_1, and θ. Here, we regarded the integration in (10.70), with respect to θ as expectation with respect to a dummy random variable θ, which is uniformly distributed over $(0, \pi/2)$. $P'(g_1, g_0, h_1, \theta)$ also represents a dummy power allocation, which is a function of g_1, g_0, h_1, and θ.

The optimal power, P', is obtained by forming the Lagrangian

$$L(P',\lambda) = \mathbb{E}\left[\frac{a}{2}\exp\left(-\frac{b}{2}\frac{P'(g_1,g_0,h_1,\theta)g_1}{N_0+\rho h_1}\frac{1}{\cos^2(\theta)}\right)\right]$$
$$+ \lambda\left(\mathbb{E}\left[P'(g_1,g_0,h_1,\theta)g_0\right] - Q_{average}\right), \tag{10.72}$$

where λ is the nonnegative dual variable. Taking the derivative of the Lagrangian with respect to $P'(g_1, g_0, h_1, \theta)$ and letting the derivative equal to zero yields [22]

$$\frac{L(P',\lambda)}{\partial P'} = \mathbb{E}\left[-\frac{g_1}{N_0+\rho h_1}\frac{ab}{4\cos^2(\theta)}\exp\left(-\frac{b}{2\cos^2(\theta)}\frac{P'(g_1,g_0,h_1,\theta)g_1}{N_0+\rho h_1}\right) + \lambda g_0\right] = 0, \tag{10.73}$$

which results in

$$P'(g_1,g_0,h_1,\theta) = \frac{2}{b}\frac{N_0+\rho h_1}{g_1}\cos^2(\theta)\ln\left[\frac{ab}{4\lambda g_0}\frac{g_1}{N_0+\rho h_1}\frac{1}{\cos^2(\theta)}\right]. \tag{10.74}$$

Then, we can find $P(g_1, g_0, h_1)$ as the average of $P'(g_1, g_0, h_1, \theta)$ over θ

$$P(g_1, g_0, h_1) = \frac{2 \times 2}{b\pi} \frac{N_0 + \rho h_1}{g_1} \int_0^{\pi/2} \cos^2(\theta) \ln[\frac{ab}{4\lambda g_0} \frac{g_1}{N_0 + \rho h_1} \frac{1}{\cos^2(\theta)}] d\theta, \qquad (10.75)$$

which gives

$$P(g_1, g_0, h_1) = \frac{1}{b} \frac{N_0 + \rho h_1}{g_1} \ln\left(\frac{1}{e} \frac{ab}{\lambda g_0} \frac{g_1}{N_0 + \rho h_1}\right). \qquad (10.76)$$

In (10.76), by considering the constraint $P(g_1, g_0, h_1) \geq 0$, we have

$$\frac{ab}{\lambda e} \frac{1}{N_0 + \rho h_1} > \frac{g_0}{g_1}.$$

The parameter λ^*, which satisfies the following complementary slackness conditions, can be obtained by solving

$$Q_{average} = \mathbb{E}\left[\frac{1}{b} \frac{g_0}{g_1} (N_0 + \rho h_1) \ln\left(\frac{1}{e} \frac{ab}{\lambda^* g_0} \frac{g_1}{N_0 + \rho h_1}\right)\right]. \qquad (10.77)$$

We can get the minimum BER by substituting (10.76) in (10.68a) as

$$BER = \mathbb{E}\left[\frac{a}{2} \mathrm{erfc}\sqrt{\frac{1}{2} \ln\left(\frac{1}{e} \frac{ab}{\lambda^* g_0} \frac{g_1}{N_0 + \rho h_1}\right)}\right]. \qquad (10.78)$$

In order to find the impact of having g_0 and h_1 at the ST, *BER* is evaluated under two different special cases. In the first case, the interference from the PT is ignored ($\rho = 0$), and then the optimum power allocation P becomes only a function of g_1 and g_0. Thus, we can study the effect of having extra CSI g_0 at the ST. In the second case, the interference from the PT is included, and the optimum power allocation P is a function of g_1, g_0, and h_1, which leads to studying the effect of having extra CSI g_0 and h_1 at the ST.

Special Case 1: The Effect of Having Extra CSI g_0 at ST

In order to focus on the effect of having g_0 at the ST, we assume that $\rho = 0$, and then P becomes a function of only g_1 and g_0.

By invoking (10.53) and assuming $m_0 = m_1 = m$, $Q_{average}$ in the case of Nakagami fading becomes

$$Q_{average}(\lambda^*) = \frac{N_0}{B(m,m)} \frac{1}{b} \int_{\frac{\lambda^* e}{ab} N_0}^{\infty} \ln\left(\frac{ab}{\lambda^* e} \frac{x}{N_0}\right) \frac{x^{m-1}}{x} (1+x)^{-2m} dx. \qquad (10.79)$$

The integral in (10.79) can be evaluated into a closed-form expression as follows:

$$Q_{average}(\lambda^*) = \frac{N_0}{bB(m,m)(1+m)^2} \left(\frac{ab}{eN_0\lambda^*}\right)^{1+m}$$

$$\times {}_3F_2\left(2m, 1+m, 1+m; 2+m, 2+m; -\frac{ab}{eN_0\lambda^*}\right), \qquad (10.80)$$

in which ${}_3F_2(a_1, a_2, a_3; b_1, b_2; z)$ is the hypergeometric function. Equation (10.80) can be simplified in the case of Rayleigh fading ($m = 1$) as

$$Q_{average}(\lambda^*) = \frac{N_0}{6b}\left(\pi^2 - 6\ln\left(1 + \frac{ab}{eN_0\lambda^*}\right)\right.$$

$$\left. + 3\ln^2\left(\frac{ab}{eN_0\lambda^*}\right) + 6Li_2\left(-\frac{eN_0\lambda^*}{ab}\right)\right), \qquad (10.81)$$

where $Li_2(.)$ is the polylogarithm function of order 2. It is worth noting that we can find λ^* for a given $Q_{average}$ from Equation (10.81). Likewise, the minimum BER can be expressed as

$$BER = \frac{a}{2B(m,m)} \int_{\frac{\lambda^* e}{ab} N_0}^{\infty} \text{erfc} \sqrt{\frac{1}{2} \ln(\frac{ab}{\lambda^* e} \frac{x}{N_0})} x^{m-1} (1+x)^{-2m} dx. \quad (10.82)$$

(10.82) gives the minimum BER when $\rho = 0$ and P is only a function of g_1 and g_0.

Special Case 2: The Effect of Having Extra CSI g_0 and h_1 at ST

Here, the effect of interference on the SR coming from the PT is considered by studying the impact of providing channel gain g_0 and h_1 at the ST. We substitute $x = \frac{g_1}{g_0}$ and $y = (N_0 + \rho h_1)$ in Equation (10.77) to find

$$Q_{average}(\lambda^*) = \int_{N_0}^{\infty} \int_{\frac{\lambda^* e}{ab} y}^{\infty} \frac{y}{bx} \ln\left(\frac{ab}{\lambda^* e} \frac{x}{y}\right) f_x(x) f_y(y) dx dy. \quad (10.83)$$

By using (10.61) and (10.62) in (10.83), we obtain

$$Q_{average}(\lambda^*) = \int_{N_0}^{\infty} \int_{\frac{\lambda^* e}{ab} y}^{\infty} \frac{y}{bx} \ln\left(\frac{ab}{\lambda^* e} \frac{x}{y}\right) \frac{e^{-\left(\frac{y-N_0}{\rho}\right)}}{(1+x)^2} dx dy. \quad (10.84)$$

In the special case of $N_0 = 0$, we obtain the closed-form result as Equation (10.85)

$$Q_{average}(\lambda^*) = \frac{1}{4be^2\lambda^{*2}} \left[e\lambda^*\rho (4e^{\frac{ab}{e\lambda^*\rho}} (e\lambda^*\rho - ab) E_i(\frac{-ab}{e\lambda^*\rho}) \right.$$
$$+ e\lambda^*\rho (2\gamma^2 - 8\gamma + \pi^2 - 4\ln\frac{ab}{e\lambda^*} + 2\ln^2\rho - 4\ln\rho(\gamma - 2 + \ln\frac{ab}{e\lambda^*})$$
$$\left. + (4\gamma - 4 + 2\ln\frac{ab}{e\lambda^*})\ln\frac{ab}{e\lambda^*})) - 4a^2b^2 G_{3\,4}^{4\,1}(\frac{ab}{e\lambda^*\rho} \Big|_{-2,-2,-2,0}^{-2,-1,-1}) \right], \quad (10.85)$$

where $E_i(.)$ is the exponential integral function and γ is Euler's constant value, and $G_{pq}^{mn}\left(.\Big|_{(b_q)}^{(a_p)}\right)$ is the Meijer function [25]. Likewise, the following equation is obtained for the minimum BER:

$$BER = \int_{N_0}^{\infty} \int_{\frac{\lambda^* e}{ab} y}^{\infty} \frac{a}{2} \text{erfc} \sqrt{\frac{1}{2} \ln\left(\frac{ab}{\lambda^* e} \frac{x}{y}\right)} \frac{e^{-\left(\frac{y-N_0}{\rho}\right)}}{(1+x)^2} dx dy. \quad (10.86)$$

We observe that closed-form expressions are not obtainable for (10.82) and (10.86), and, hence, we need to solve the equations numerically.

10.4.2 Minimum BER Under Peak Interference Power Constraint

The peak power constraint is more appropriate when the QoS is limited by the instantaneous SINR at the receiver. Therefore, in this subsection, we replace the following equation denoting peak interference power constraint with (10.45b) and (10.68b)

$$\text{s.t } Pg_0 \le Q_{peak}, \quad (10.87)$$

where Q_{peak} is the peak received power limit at the existing PR. We obtain the minimum BER if the power is replaced by

$$P(g_0) = \frac{Q_{peak}}{g_0} \quad (10.88)$$

where, in this case, $P(g_0)$ is only a function of g_0 and independent of g_1 and h_1

10.4.2.1 Minimization of Exp(-SINR)

We can find the minimum BER, in this case, by substituting Equation (10.88) in (10.45a)

$$BER = \frac{1}{2}\mathbb{E}\left[\exp\left(-\frac{Q_{peak}}{N_0 + \rho h_1}\frac{g_1}{g_0}\right)\right]. \tag{10.89}$$

We also substitute $x = \frac{g_1}{g_0}$ and $y = (N_0 + \rho h_1)$ in Equation (10.89), resulting in

$$BER = \frac{1}{2}\int_0^\infty \int_0^\infty \exp(-\tfrac{x}{y}Q_{peak})f_x(x)f_y(y)dudy. \tag{10.90}$$

We arrive at the following closed-form expression with the aid of [25, eq.(3.324.1), eq.(9.34.3), and eq.(7.811.5)] in the special case of $m = 1$ and the interference limited scenario ($N_0 = 0$) by using the PDF of x and y in (10.61) and (10.62)

$$BER = \frac{1}{2\rho}\int_{N_0}^\infty \int_0^\infty \exp(-\tfrac{x}{y}Q_{peak})\frac{e^{-\left(\frac{y}{\rho}\right)}}{(1+x)^2}dxdy$$

$$= \frac{1}{2}G_{13}^{31}\left(\frac{Q_{peak}}{\rho}\Big|_{0,1,1}^{0}\right). \tag{10.91}$$

The above closed-form expression gives the minimum BER under exponential function.

10.4.2.2 Minimization of $aQ(\sqrt{bSINR})$

We minimize the BER minimization problem as

$$BER = \frac{1}{2}\mathbb{E}\left[\text{erfc}\sqrt{\frac{Q_{peak}}{N_0 + \rho h_1}\frac{g_1}{g_0}}\right]. \tag{10.92}$$

We substitute $x = \frac{g_1}{g_0}$ and $y = (N_0 + \rho h_1)$ in Equation (10.92), and the BER can be expressed as

$$BER = \frac{1}{\pi\rho}\int_0^{\pi/2}\int_{N_0}^\infty \int_0^\infty \exp\left(-\frac{xQ_{peak}}{y\cos^2(\theta)}\right)f_x(x)f_y(y)dxdyd\theta. \tag{10.93}$$

We arrive at the following closed-form expression in the special case of $m = 1$ and the interference limited scenario ($N_0 = 0$) by using the PDF of x and y in (10.61) and (10.62),

$$BER = \frac{1}{\pi\rho}\int_0^{\pi/2}\int_0^\infty \int_0^\infty \exp(-\frac{uQ_{peak}}{y\cos^2(\theta)})\frac{e^{-\left(\frac{y}{\rho}\right)}}{(1+u)^2}dudyd\theta$$

$$= \frac{\Gamma(\frac{1}{2})}{2\pi}G_{24}^{41}\left(\frac{Q_{peak}}{\rho}\Big|_{\frac{1}{2},0,1,1}^{0,1}\right), \tag{10.94}$$

where $G_{pq}^{mn}\left(.\big|_{(b_q)}^{(a_p)}\right)$ is the Meijer function.

10.4.3 Numerical Results

In this section, we present some numerical results for the minimum BER under different scenarios. We also assume $(a, b) = (1, 2)$ which means that BPSK or QPSK is considered. Throughout this section, lines represent the results obtained from the analytical results, and symbols represent the Monte Carlo simulation results. Both simulation and analytical results are closely matched, which supports the validity of the presented analysis.

10.4.3.1 Average Interference Power Constraint

Special Case 1: The effect of having extra CSI g_0 at ST

Figure 10.5 and Figure 10.6 display BER vs. $Q_{average}/N_0$ under average interference power constraint for different channel models without interference from the PT for DPSK and BPSK, respectively. The Nakagami parameter indicates the severity of fading, such that for Rayleigh fading m=1, and for an AWGN channel without fading $m = \infty$ [15]. As we can see, BER in all cases exponentially decreases with increasing $Q_{average}/N_0$. In order to discuss the significance of having channel gain g_0 at the ST, we also include the minimum BER results with a reduced side information, where g_0 is not made available at the ST. Hence, by disregarding g_0, the power of ST depends only on g_1. The minimum BER problem (10.45a) subject to (10.45b) in DPSK or (10.68a) subject to (10.68b) in BPSK reduce into the simplified form

$$\min_{P \geq 0} \quad \mathbb{E}\left[\frac{1}{2}\exp\left(-\frac{P(g_1)g_1}{N_0}\right)\right]$$

$$s.t \quad \mathbb{E}\left[P(g_1)\right] \leq Q_{average}$$

or

$$\min_{P \geq 0} \quad \mathbb{E}\left[\frac{1}{2}\mathrm{erfc}\sqrt{\frac{P(g_1)g_1}{N_0}}\right]$$

$$s.t \quad \mathbb{E}\left[P(g_1)\right] \leq Q_{average},$$

for which the expectation is only with respect to g_1. Examining Figure 10.5 and Figure 10.6, it can be seen that the BER, when the power is a function of g_1 and g_0, ST-SR, and ST-PR CSI, is always lower than that when the power depends only on g_1, ST-SR CSI. Another important observation is that the difference between the BERs in both cases is very small, such that the side information between the ST and the PR has negligible effect on the system performance. Due to difference between exponential function and complementary error function, we can observe that with the same $Q_{average}/N_0$, the BER under BPSK is always lower than the BER under DPSK.

Special Case 2: The effect of having extra CSI g_0 and h_1 at ST

The behavior of BER vs. $Q_{average}/N_0$, considering the effect of interference coming from the PT on the SR under average interference power constraint in Rayleigh fading for DPSK and BPSK, are shown in Figure 10.7 and Figure 10.8, respectively, with $\rho = 10dB$. Comparing these figures with Figure 10.5 and Figure 10.6 indicates that the interference from the PT can have a big impact on BER. In addition, the minimum BER results with a reduced side information g_0 and h_1 at the ST is also plotted. Therefore, we disregard the effect of g_0 and h_1 from the power allocation, resulting in the simplified optimization problem

$$\min_{P \geq 0} \quad \mathbb{E}\left[\frac{1}{2}\exp\left(-\frac{P(g_1)g_1}{N_0 + E[\rho h_1]}\right)\right]$$

$$s.t \quad \mathbb{E}\left[P(g_1)\right] \leq Q_{average}$$

or

$$\min_{P \geq 0} \quad \mathbb{E}\left[\frac{1}{2}\mathrm{erfc}\sqrt{\frac{P(g_1)g_1}{N_0 + E[\rho h_1]}}\right]$$

$$s.t \quad \mathbb{E}\left[P(g_1)\right] \leq Q_{average},$$

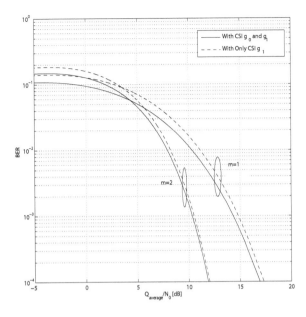

Figure 10.5: Effect of having g_0 at the ST for different fading channel models under average interference power constraint in DPSK.

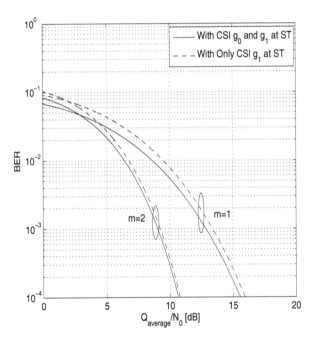

Figure 10.6: Effect of having g_0 at the ST for different fading channel models under average interference power constraint in BPSK.

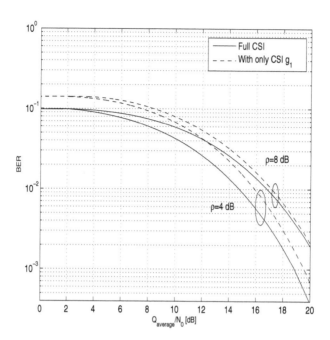

Figure 10.7: Effect of having g_0 and h_1 at the ST under average interference power constraint for DPSK.

where, in this case, the expectation is with respect to g_1 only. This is in contrast to (10.45a) and (10.68a), where the expectation is with respect to g_1, g_0, and h_1. Figure 10.7 and Figure 10.8 show that the BER, when the power is a function of g_1, g_0, and h_1 is always lower than that when the power depends only on g_1. Again we observe that the difference between the BERs in both cases is negligible, such that having g_0 and h_1 at ST has little effect on the system performance.

10.4.3.2 *Peak Interference Power Constraint*

The behavior of BER vs. Q_{peak}/N_0, considering the effect of interference coming from the PT on the SR under peak interference power constraint for Rayleigh fading over DPSK and BPSK, are shown in Figure 10.9 and Figure 10.10, respectively, with $\rho = 10dB$. Likewise, we also plot the minimum BER results with a reduced side information g_0 at the ST. Hence, we can ignore the effect of g_0 from the optimization problem, resulting in the following

$$\min_{P \geq 0} \quad \mathbb{E}\left[\frac{1}{2}\exp\left(-\frac{Pg_1}{N_0 + \rho h_1}\right)\right]$$
$$s.t \quad P \leq Q_{peak}.$$

or

$$\min_{P \geq 0} \quad \mathbb{E}\left[\frac{1}{2}\mathrm{erfc}\sqrt{\frac{Pg_1}{N_0 + \rho h_1}}\right]$$
$$s.t \quad P \leq Q_{peak}.$$

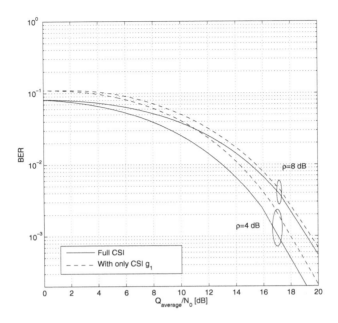

Figure 10.8: Effect of having g_0 and h_1 at the ST under average interference power constraint for BPSK.

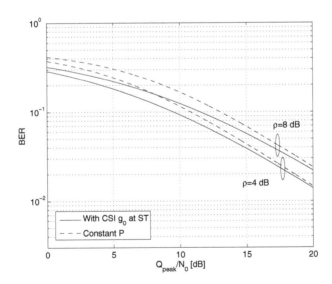

Figure 10.9: Effect of having g_0 at the ST for peak interference power constraint under DPSK, considering the interference from PT.

Similarly, the difference between the BERs in both cases, when g_0 is either included or ignored, is very small.

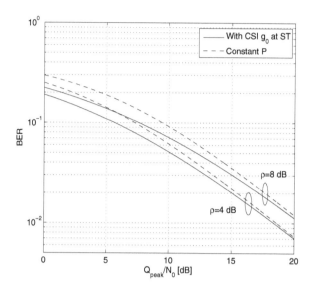

Figure 10.10: Effect of having g_0 at the ST for peak interference power constraint under BPSK, considering the interference from PT.

10.5   Conclusion

In this chapter, we considered a spectrum-sharing system and evaluated the maximum ergodic capacity and minimum BER subject to either average or peak constraint on the interference power. We investigated the effect of different levels of channel side information, which can be provided at the secondary transmitter. In most cases, closed-form results were derived for the capacity and BER. Using some results from the numerical analysis, the maximum capacity of the secondary link highly depends on having side information between secondary transmitter and primary receiver at the secondary transmitter. However the side information between secondary transmitter and secondary receiver at the secondary transmitter has negligible impact to the average capacity.

Evidently, with the same predefined parameters, minimum BER under average interference power constraints is lower than that under peak interference power constraints. One important observation made is that providing the extra side information between the secondary transmitter and primary receiver, and also between primary transmitter and secondary receiver at the secondary transmitter, requiring intersystem message-passing, have negligible effect on the BER. Therefore, the results found in this chapter can be used as a tradeoff between performance and complexity.

References

[1] J. Mitola, "Cognitive radio architecture evolution," Proc. *IEEE*, vol. 97, no. 4, pp. 626–641, April 2009.

[2] FCC, Facilitating opportunities for flexible, efficient, and reliable spectrum use employing cognitive radio technologies, notice of proposed rule making and order, FCC 03-322, Dec. 2003.

[3] L. Cao and H. Zheng, "SPARTA: Stable and efficient spectrum access in next generation dynamic spectrum networks," in *Proc. IEEE INFOCOM*, pp. 870–878, Apr. 2008.

[4] O. Holland, A. Attar, N. Olaziregi, N. Sattari, and A. Aghvami, "A universal resource awareness channel for cognitive radio," in *Proc. IEEE PIMRC*, pp. 1–5, Sep. 2006.

[5] Y. Yuan, P. Bahl, R. Chandra, T. Moscibroda, and Y. Wu, "Allocating dynamic time-spectrum blocks in cognitive radio networks," in *Proc. ACM Int. Symp. MobiHoc*, pp. 130–139, 2007.

[6] S. Haykin, "Cognitive radio: Brain-empowered wireless communications," *IEEE J. Sel Areas Commun*, vol. 23, no. 2, pp. 201–220, Feb. 2005.

[7] Q. Zhao and B. Sadler, "A survey of dynamic spectrum access," *IEEE Signal Process. Mag.*, vol. 24, no. 3, pp. 79–89, May 2007.

[8] M. Gastpar, "On capacity under receive and spatial spectrum-sharing constraints," *IEEE Trans. Inform. Theory*, vol. 53, no. 2, pp. 471–487, Feb. 2007.

[9] A. Ghasemi and E. S. Sousa, "Fundamental limits of spectrum-sharing in fading environments," *IEEE Trans. Wireless Commun.*, vol. 6, no. 2, pp. 649–658, Feb. 2007.

[10] L. Musavian and S. Aissa, "Capacity and power allocation for spectrum-sharing communications in fading channels," *IEEE Trans. Wireless Commun.*, vol. 8, no. 1, pp. 148–156, Jan. 2009.

[11] X. Kang, Y. Liang, A. Nallanathan, H. Krishna, and R. Zhang, "Optimal power allocation for fading channels in cognitive radio networks: Ergodic capacity and outage capacity," *IEEE Trans. Wireless Commun.*, vol. 8, no. 2, pp. 940–951, Feb. 2009.

[12] M. G. Khoshkholgh, K. Navaie, and H. Yanikomeroglu, "Access strategies for spectrum sharing in fading environment: Overlay, underlay, and mixed," *IEEE Trans. Mobile Computing*, vol. 9, no. 12, pp. 1780–1793, Dec. 2010.

[13] A. Attar and V. Krishnamurthy, "Interference diversity gain and its application in multi-channel systems: Capacity maximization and QoS guarantee strategies," *IEEE Trans. Commun.*, vol. 59, no. 1, pp. 248–257, Jan 2011.

[14] A. Jovicic and P. Viswanath "Cognitive radio: An information-theoretic perspective," *IEEE Trans. Inform. Theory* , vol. 55, no. 9, pp. 3945–3958, Sep. 2009.

[15] A. J. Goldsmith and P. P. Varaiya, "Capacity of fading channels with channel side information," *IEEE Trans. Inf. Theory*, vol. 43, no. 6, pp. 1986–1992, Nov. 1997.

[16] M. R. Mili and K. Hamdi, "The effect of different levels of side information on the ergodic capacity in cognitive radio networks,"in *Proc. IEEE Global Communications Conference (GLOBECOM)*, 2014.

[17] M. R. Mili, K. Hamdi, and Farokh Marvasti, "Optimization in cognitive radio networks with multiple secondary links", Submitted to *IEEE Transaction on Vehicular Technology*, 2015.

[18] M. R. Mili and K. Hamdi, "Minimum BER analysis in interference channels,"*IEEE Trans. Wireless Commun.*, vol. 12, no. 7, pp. 3191–3201, July 2013.

[19] M. R. Mili and K. Hamdi, "Minimum BER analysis in cognitive radio,"in *Proc. IEEE Global Communications Conference (GLOBECOM)*, 2012.

[20] M. R. Mili and K. Hamdi, "Minimum BER analysis and the effect of extra channel side information on BER,"in *Proc. IEEE 23rd Personal, Indoor and Mobile Radio Communications (PIMRC)*, 2012.

[21] M. R. Mili and K. Hamdi, "Interference evaluation in ad-hoc cognitive radio networks,"in *Proc. IEEE 76th Vehicular Technology Conference (VTC)*, 2012.

[22] S. Boyd and L. Vandenberghe, *Convex Optimization*. Cambridge, UK: Cambridge University Press, 2004.

[23] E. W. Weisstein, *CRC Concise Encyclopedia of Mathematics*. CRC Press, 1998.

[24] Y. M. Shobowale and K. A. Hamdi, "A unified model for interference analysis in unlicenced frequency bands," *IEEE Trans. Wireless Commun.*, vol.8, no.8, pp. 4004–40013, Aug. 2009.

[25] I. S. Gradshteyn and I. M. Ryzhik, *Table of Integrals, Series, and Products*, 7th Ed. San Diego, CA: Academic Press, 2007.

[26] A. J. Goldsmith, *Wireless Communications*. New York: Cambridge University Press, 2005.

[27] M. K. Simon and M. S. Alouini, *Digital Communication Over Fading Channels*, 2nd Ed., John Wiley & Sons Inc., 2005.

Chapter 11

Spectrum-Aware Routing in Cognitive Radio Vehicular Ad Hoc Networks (CR-VANETs): Challenges and Solutions

Ahmed M. Ahmed, Ala Abu Alkheir, and Hussein T. Mouftah

CONTENTS

The abundance of spectrum resources available for cognitive radio networks (CRNs) and the unpredictable activities of licensed users, i.e., primary users (PUs), make channel allocation an inherently dynamic multi-objective optimization problem. A CRN has to choose an operating channel that is not being used by PUs, meets its Quality of Service (QoS) demands, has no interference from other CR users or networks, have low PU activities, etc. This selection process becomes even more challenging for cognitive radio vehicular ad hoc networks (CR-VANETs) due to the absence of a central network coordinator, and the fast changing network topology and spectral environment [1, 2]. This chapter studies spectrum-aware routing techniques for CR-VANETs. This class of CRNs should accommodate a wide range of QoS classes, including the delay-intolerant safety information, while operating in a very dynamic environment that makes the process of establishing and maintaining reliable routing paths a highly complicated process [3, 4]. To tackle this issue, this chapter starts by identifying the main distinguishing features of CR-VANETs. The process of reliable spectrum acquisition is then studied as a prerequisite to route establishment. The chapter then discusses the various metrics and algorithms used to build routing paths using both end-to-end approaches and opportunistic approaches. Route maintenance is then studied, where we develop a rigorous mathematical model for the losses incurred due to the unpredictable activities of PUs. A comparative study is then conducted using a network simulator. In this study, we compare the performance of the various routing techniques, quantify the losses incurred due to PU activities, and devise some insights into proactive route maintenance.

11.1   Introduction

The increasing number of vehicles on the road made the deployment of intelligent transportation systems (ITS) a high priority target for several national governments around the world to improve transportation safety and efficiency while reducing the environmental and financial cost of road congestions. Several developed nations such as Europe, the United States, Japan, South Korea, and China, are undergoing massive investments in the field of ITS. In an ITS, vehicles act as floating sensors, making use of information and communication technologies (ICT) to monitor, share, and adapt to different road conditions [5]. Cooperation amongst vehicles and between vehicles and the infrastructure both at the system and application levels is key to the realization of ITS. Vehicular ad hoc networks (VANETs) play a crucial role in achieving this cooperation through multihop vehicle-to-vehicle (V2V) and vehicle-to-infrastructure (V2I) communication. A multitude of cooperative vehicular applications have already been implemented and evaluated in realistic vehicular environments. Safety applications include adaptive cruise control, enhanced driver awareness, and collision avoidance. Transport efficiency applications include cooperative traffic information and monitoring systems, as well as cooperative freight and fleet management applications for inter-urban transportation. Moreover, vehicular applications are expected to follow the Internet trend and grow in the direction of common entertainment, including multimedia and gaming [6]. Some of the most developed examples of such systems and testbeds include, among others, the Cooperative Vehicle-Infrastructure Systems (CVIS) project [7], the CO-OPerative SystEms for Intelligent Road Safety (Coopers) project [8], and the Pre-DRIVE C2X project [9].

In terms of standardization for wireless access technology for VANETs, dedicated short-range communication (DSRC) [10–13], a generic name for short-range point-to-point communication, was used for early ITS applications, such as fee collection in toll plazas [2]. DSRC channels were reserved worldwide in the 5.9 GHz band. The IEEE 802.11p standard [14,15], mainly an adaptation of the former IEEE 802.11a, is used for the PHY/MAC aspects of high speed vehicular (V2V and V2I) communications in the 5.9 GHz band, while a family of IEEE 1609 standards build on top of that defining various functionalities related to networking, management, transport, and application layers to form the wireless access in vehicular environments (WAVE) architecture shown in 26.1. WAVE short message protocol (WSMP) is defined for applications with strict latency requirements, such as in safety applications. WAVE also defines a multi-channel operation for IEEE 802.11p, which includes a control channel and six service channels. While the IEEE 1609 WAVE architecture is based only on IEEE 802.11p, the International Standards Organization (ISO) is developing a generic ITS reference architecture aiming at a generic protocol stack that enables convergence of different access technologies, including cellular (GSM/GPRS, UMTS), older DSRC standards, IEEE 802.11 standards, Bluetooth, etc. [16]. This is known as the silo-application-driven approach. In terms of spectrum allocation, in the USA, the FCC has allocated 75 MHz at 5.9 GHz for V2V and V2I; in Europe, due to the unavailability of 75 MHz of continuous spectrum in the DSRC band, the Car2Car consortium proposed to allocate two 10 MHz channels at 5.9 GHz range for safety critical applications. Finally, Japan has allocates 10 MHz in the 700 MHz band, which has very good propagation properties for vehicular communications [2].

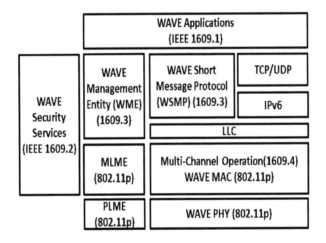

Figure 11.1: IEEE 1609 WAVE architecture.

Demand for vehicular communications is expected to grow due to several reasons. First, the number of wireless-enabled vehicles, while currently very low and with limited bandwidth requirements, is expected to increase considerably. Second, vehicular communication applications is expected to grow, beside road safety, reliability, and transport efficiency, in the direction of common entertainment applications (e.g., gaming and video) and information applications (e.g., driver assistance through real-time feeds on traffic, weather, and visual inputs from external cameras) [17]. Many of these applications need strict bandwidth and/or delay requirements to function properly. This high demand for spectrum is especially foreseen in urban environments during peak traffic hours. Third, federal agencies have been recently subjected to strong pressure from major industry players (Qualcomm, Cisco, etc.) to release a part of the 5.9 GHz band for more general usages, such as WiFi. And, hence, these bands could be jointly used by ITSs and other applications. In fact, some recent studies suggest that the spectrum allocation foreseen by the IEEE 802.11p standard might be inadequate to support strict delay and guaranteed delivery requirements of safety applications in peak hours of traffic [18]. All of these reasons, and others, call for adoption of novel techniques and approaches to address this looming spectrum scarcity problem in VANETs, and to allow for efficient use of radio spectrum resources.

Spectrum scarcity, a growing concern in telecommunications in general and not just VANETs, is mainly a factor of the spectrum allocation policy, which is fixed rather than dynamic and leads to overcrowding of some frequency bands, like the ISM band, while other frequency bands, like the DTV bands, are sporadically used. cognitive radio (CR) technology or dynamic spectrum access is a key solution to enable more efficient spectrum utilization, and to increase the spectrum available to a communication system through opportunistic use of licensed spectrum [19]. CR principles are already being thought of with the aim of increasing available spectrum in a number of communication systems on different scales. And, hence, different IEEE standards groups are working on CR-based spectrum sharing techniques, including: IEEE 802.22 [20] wireless regional area networks (WRANs) that uses CR on TV bands in rural areas, and the IEEE 802.16h [21] (improved coexistence mechanisms for license-exempt operation) that is working on efficient interference management and resource allocation dynamic spectrum access over WiMax [4]. This motivated the interest to analyze how these CR principles can be applied to the vehicular environment with the goal of increasing the available bandwidth to face the looming spectrum scarcity problem and to analyze the challenges that arise in such an environment.

A CR-VANET is composed of communicating vehicles and roadside units (RSUs) with reconfigurable software defined radio (SDR) devices and intelligent CR functionalities, that enable network nodes to reconfigure all the aspects of radio control, including the operating spectrum frequencies, the transmitted power levels, etc. CR-VANETs, however, cannot be considered a mere application of the CR technology to VANETs, as they have some unique characteristics that must be accounted for and assumptions that can be exploited. For example, the vehicular mobility has a significant impact on the spectrum management process where, unlike static CR networks, in CR-VANETs, spectrum availability experienced by each vehicle changes over time as a function of both mobility and PU's activities. However, on the positive side, cooperation among vehicles can be exploited to enhance spectrum sensing and decision process [22]. Also, the constrained and predictable nature of a vehicle's trajectory can be leveraged so that CR-enabled vehicles can collect anticipated spectrum availability information at future locations along their path. This requires multi-hop dissemination of spectrum-related information along the streets [3]. Furthermore, spectrum sensing can take several forms: stand-alone or local, distributed and cooperation based, database (DB) assisted, or hybrid methods. In this introduction, we briefly discuss the main characteristics of CR-VANETs, highlighting main characteristics carried over from VANETs and cognitive radio ad hoc networks (CRAHNs), followed by new characteristics unique to CR-VANETs. Then, we list a number of application areas for CR-VANETs. Finally, we discuss different architectures of CR-VANETs.

11.1.1 Characteristics of CR-VANETs

The main components of any CR-enabled device are an SDR, a sensor, a knowledge DB, optimization tools, and a learning/reasoning engine. A CR-enabled vehicle will have some additional components such as a localization system like a global positioning system (GPS), navigation tools, and possibly additional (non SDR) radios. These components enable the vehicle to engage in a cognition cycle that consists of multiple phases: observe, analyze, reason, and act. The vehicle's sensors are responsible for the observation and information gathering phase (e.g., different signals, their modulation types, noise, transmission power, etc.) through spectrum sensing and exchange of control messages with neighbor vehicles. A CR-vehicle should have a cognitive engine that implements the complex policies and self-configuration needed to meet the QoS requirements of the application at hand. The SDR is responsible for the action phase by reconfiguring its operational parameters, on-the-fly, across the different layers of the protocol stack. The SDR transceiver can transmit on one of the channels dedicated to vehicular communication in the 5.9 GHz or can access other channels of the frequency spectrum, including licensed ones. Therefore, each CR-enabled vehicle in a CR-VANET implements an opportunistic channel access policy to find the channel that best satisfies the QoS requirements of the running application, while guaranteeing that PUs are not affected. CR-VANETs inherit several characteristics from traditional VANETs and CR networks, but also have their own unique characteristics that must be considered during the design and implementation of network protocols. In [4], the authors highlight these characteristics (inherited and novel). In the following, we summarize these characteristics as follows

- Characteristics inherited from CR networks

  1. ***Spectrum management operations:*** A CR-enabled vehicle must implement all spectrum management functionalities that comprise the cognitive cycle: spectrum sensing, spectrum decision, spectrum sharing, and spectrum mobility [19]. These functionalities have been extensively studied in the literature of CR systems, through several protocol proposals at the PHY and MAC layers [23]. Most of these protocols can be integrated into existing vehicular communications protocol stacks, however, they need to be adapted to the specific characteristics of CR-VANETs. For example, the RTS/CTS mechanism needed to counter the hidden terminal problem might be impractical when the primary network is DTV, where primary transmitters are transmit-only TV broadcasting towers and primary receivers are receive-only TV devices.

  2. ***Reconfigurability:*** The SDR allows for the dynamic reconfiguration of transmission parameters across all layers of the protocol stack. The reconfiguration is the output of the node cognitive engine, which, can employ different techniques, such as: genetic algorithms [24], reinforcement-based learning [25], and statistical learning [26]. For example, in reinforcement-based learning, an agent senses the environment, performs an action, and receives a numerical reward. The agent aims to learn the sequence of actions that maximizes the sum of rewards over a finite number of future steps through trial and error. However, these techniques need to be adapted for CR-VANETs due to its highly dynamic nature, which might not allow the CR-enabled vehicle to learn the relationship between the performed actions and the received rewards to improve its behavior over time [4]. Cooperation techniques can be of huge benefit in improving the convergence of the learning process [25].

- Characteristics inherited from VANETs

  1. ***Mobility characteristics:*** Mobility characteristics depend upon the vehicular scenario at hand. For example, in highway scenarios, groups of vehicles move at approximately the same speed, leading to a quasi-stationary relative mobility and formation of clusters of vehicles. This behavior can be exploited to facilitate spectrum management functionalities of CR-VANETs by creating a virtual infrastructure in the VANET. Cluster leaders can act as a fusion center for local spectrum sensing reports of other cluster members to determine the presence or absence of a PU. Cluster leaders can also perform channel selection and channel access coordination for the different members of the cluster [27–29]. And, hence, the election process of a cluster leader is of utter importance in clustering schemes, and should be based on the mobility characteristics of each vehicle as well as on CR-related parameters [30].

  2. ***Fragmented topology:*** The density of vehicles significantly fluctuates over the course

of a day, and across different roads, and, hence, might cause frequent network partitions. Opportunistic protocols employing a store-carry-and-forward approach proposed for delay-tolerant networks (DTN) perform well in such intermittently connected topologies [31]. When a network partition occurs, a vehicle, employing the store-carry-and-forward approach, stores temporarily the message until it encounters another vehicle moving toward the final destination.

3. ***Sufficient space and power supply:*** Traditional CRAHNs and wireless sensor networks (WSNs) employ energy saving schemes at different layers of the protocol stack as network devices usually have limited size energy resources [19]. However, CR-VANETs can assume abundant energy resources due to the recharge of batteries from the vehicles' energy resources. Therefore, energy conservation schemes are not considered a priority in communication protocol design for CR-VANETs. Moreover, vehicles have sufficient space to accommodate more sizeable onboard units (OBU) that can support more advanced CR capabilities.

- Novel characteristics and assumptions

  1. ***Presence of a common control channel (CCC):*** The implementation of a reliable CCC, while challenging in classical CR systems due to the visitor role of the (secondary) CR user [32], is more guaranteed in a CR-VANET by using the control channel (CCH) in the 5.9 GHz band foreseen by the IEEE 802.11p/1609.4 protocol. All vehicles are synchronized, which can be achieved by the GPS, and cyclically switch between the CCH and service channels (SCHs). The presence of a CCC is necessary for exchange of spectrum information between vehicles as well as other control traffic necessary for routing and cooperative techniques. However, the use of the CCH for the CCC raises the concern of the CCH getting saturated in congested scenarios (e.g., congested roads during peak hours) [18].

  2. ***Spatio-temporal activity of PUs:*** A CR device, in a static environment, monitors the ON/OFF variations of a channel attempting to learn its occupation pattern and utilizes this knowledge in the spectrum decision process [33]. In a vehicular network, a CR-enabled vehicle might enter the interference region of multiple PUs, which complicates the process of learning a channel occupation pattern. Moreover, the responsiveness of a sensing scheme, which is the delay before a CR vehicle detects the presence of a PU, is very important.

  3. ***Utilization of spectrum DB:*** The FCC foresees the creation of spectrum DB with detailed PU locations, such as the TV query system in the United States. Integrating spectrum database information into digital maps available at CR vehicles can be used in spectrum-aware routing techniques. However, access to such databases might not always be possible due to intermittent and short-range V2I connectivity or due to lack of frequent updates of PUs' activity, and, hence, the FCC still foresees a role for local sensing as well as cooperation-based approaches [34].

  4. ***Impact of mobility:*** A moving CR vehicle will collect sensing samples at different locations inside the interference region of a PU with different degrees of correlations due to shadowing and multipath fading effects. How a single vehicle merges these observations is still not well explored. Also, the availability of digital road maps and GPS in modern vehicles enables future knowledge of spectrum resources available on a vehicle's path, thus, deciding on a channel schedule to be used before a given area.

  5. ***Role of cooperation:*** Cooperation techniques among CR-enabled vehicles can be exploited at the system and application level [4]. Cooperative sensing schemes exploit the spatial diversity of the sensing samples to increase the accuracy of PU detection [22]. In literature, different fusion schemes have been investigated. A fusion

scheme aims to weigh each sample according to its relevance in the final sensing decision. Furthermore, cooperation in spectrum sharing and decision algorithms is key to distributed channel allocation schemes. Due to the evolving topology of a CR-VANETs, the set of cooperating neighbor vehicles might change dynamically, which has to be considered in the design of cooperative schemes.

11.1.2 Applications of CR-VANETs

CR-VANETs can change existing and emerging vehicular applications. The major factors affecting the applications are the choice of transmission frequency, the available bandwidth for the application, and the interference caused in that range. Some major areas of possible applications of CR-VANETs include:

1. ***Vehicle-to-vehicle communication:*** In high-traffic areas, incidents, such as accidents, road blockages, road repairs, and slow traffic, can cause significant delays. Communicating relevant information from vehicles closer to affected areas to other vehicles following behind can mitigate these delays and their incurred financial and environmental impact by enabling drivers farther away from the area of impact to change course or take other preventive measures. Furthermore, V2V communication can be crucial in boosting safety features, like collision avoidance, through periodic exchange of data, such as average velocity, acceleration, and brake status, among neighbor vehicles. For example, Honda and Volkswagen-led consortium has developed a practical system operating in the 5.9 GHz with a transmission range limited to a few dozen meters to warn drivers of potential hazards at intersections [35]. The idea is to use CR-principles to exploit the lower frequencies of the licensed TV band in the sub-gigahertz band, which has better range, to improve the responsiveness of such collision avoidance mechanism by reducing the number of needed hops to reach a given area. Aside from range extension, opportunistic spectrum access can be used to offload lower priority flows (traffic), such as lower priority video flows and P2P traffic to additional spectrum using CR and save the WAVE CCH for safety applications [2].

2. ***Entertainment and information systems:*** The popularity of wireless applications fuels an ever increasing demand for bandwidth. A number of new applications with strict bandwidth and QoS requirements are emerging, including: video streaming, real-time video driver assistance, weather and road conditions updates, etc. While many car manufacturers have recently started to provide Internet connectivity through cellular networks, cellular networks might suffer from spectrum congestion in urban areas and lower (or absence) of coverage in some rural areas. Thus, CR technology is an attractive solution to face these scalability issues with the increasing number of vehicles and emerging application.

3. ***Public safety communication:*** Large-scale natural disasters, such as Hurricane Katrina, have led to the breakdown of the public safety communications infrastructure. Public safety personnel had to, sometimes, resort to non-electric means of communication. Mobile public safety personnel operating in the field during such outages or beyond the reach of fixed communication infrastructure can benefit from work on CR-VANETs to enable distributed spectrum access in the un-congested licensed frequencies.

4. ***Reprogrammable vehicular telematics:*** New communications standards are regularly being proposed, such as DVB-H, DVB-T2, WiMaX, 802.11p, LTE, and HSDPA, and some of these standards might not succeed commercially. Furthermore, spectrum usage regulations, such as operating frequencies and transmit power limits, differ from one country to another. All of this creates a dilemma for global car manufacturers on which technology to deploy in their vehicles, especially that a vehicle has a lifecycle that exceeds 15 years and onboard

communicating devices should not become obsolete during that period. CR combined with SDR is key to design future-proof onboard communicating devices that are context aware and adaptable [2].

11.1.3 CR-VANETs Network Architecture

CR-VANETs can be classified in terms of network architecture into three categories, shown in 11.2 [17]. The first category is completely ad hoc, where a CR-VANET is formed between vehicles only with no infrastructure. This category has low deployment cost, yet needs more involved coordination among travelling vehicles to exchange sensing results and effectively share spectrum opportunities (SOPs). The second category makes use of RSUs or local BSs, which act as data repository used and updated, with time and location stamped local sensing results, by passing vehicles. RSUs may have an out-of-band connection to each other. This category can assist cooperative techniques as well as techniques that utilize predictable vehicular mobility nature and better detects and protects PUs, communications. The third category relies on a centralized BS, and is rather theoretical than practical, as it creates a traffic bottleneck and a single point of failure. However, it is used as a benchmark to evaluate performance of spectrum management schemes from the first two categories against a scenario with global network knowledge.

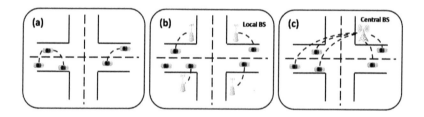

Figure 11.2: Three deployment architectures for CR-VANETs: (a) vehicle to vehicle only, (b) multiple local BSs or RSUs, and (c) centralized BS-serving vehicles.

11.2 Channel Acquisition Techniques

In a CR-VANET, the channel acquisition and access problems becomes more complex (when compared to a traditional vehicular network) due to the time and location-varying spectrum availability, QoS requirements for transmissions by vehicles' OBUs, and the interference constraints for PUs (i.e., the probability of collision between licensed and unlicensed users should be bounded). Channel acquisition can be classified into spectrum sensing and spectrum decision discussed in the next two sections.

11.2.1 Spectrum Sensing

Spectrum sensing is a key component of CR-VANETs to ensure proper protection of PUs traffic and proper detection of SOPs. However, the high mobility of the vehicular environment can deteriorate the performance of the spectrum sensing module in two ways. First, the wireless channel effects including multipath fading, shadowing, Doppler effects, frequency selectivity, etc. These effects can vary according to the vehicular scenario (i.e. urban vs. highway). Second, the fast change in a vehicle's location means that a vehicle can experience different PUs along its path. These PUs can have different characteristics, activity patterns, and protection requirements. Three different sensing techniques have been identified in literature for spectrum sensing in CR-VANETs: per-vehicle sensing, geolocation-based sensing, and cooperative spectrum sensing (CSS), which can be infrastructure-based or completely ad hoc.

11.2.1.1 Per-Vehicle Sensing

Per-vehicle, or stand-alone, sensing uses any of the different CR sensing techniques proposed for CR systems, such as energy detection, matched-filter, feature detection, etc. [19]. Several advanced signal processing techniques have been proposed in literature to enhance the sensing performance of theses techniques in fast time-varying channels and in high-speed environments using MIMO, wavelet transform, neural networks, and iterative solutions [36–40]. The main advantage of per-vehicle sensing is that it requires minimal, if any, network support. However, the performance of per-vehicle sensing deteriorates in realistic outdoor scenarios due to small-scale (multipath) and large-scale (shadowing) fading effects.

11.2.1.2 Geolocation-Based Sensing

A geolocation DB keeps track of PUs' locations, types, and specific protection requirements in its area of coverage. And, hence, a CR-enabled vehicle (CRV) can obtain a list of channels that are available at a specific location, as well as the allowed transmission parameters over these channels to protect vulnerable PUs. Therefore, this requires each CRV to be equipped with a self-localization device (e.g., a GPS), and a secure Internet connection to provide its location to the DB system. Moreover, CRVs can integrate obtained PU information from the spectrum database with digital maps onboard. Geolocation-based sensing is mainly envisioned for opportunistic access over the TV band spectrum, and, hence, the FCC in the US and OFCOM in the UK foresee the utilization of geolocation databases as the main solution to provide accurate spectrum information about TV whitespaces for SUs in CR-VANETs as they provide maximum protection to PUs receivers [34].

While geolocation-based sensing can offer high PU receivers protection, there are concerns on the implementation and update cost, as well as the coverage area of building such a database system. Moreover, in congested urban scenarios, mobile vehicles can generate a significant query overhead, making the geolocation database a network bottleneck. For this reason, in [34], the authors investigate the optimal ratio between cooperation and spectrum database querying to minimize network utilization needed for PUs detection while guaranteeing proper protection for these PUs. A distributed (bio-inspired) protocol for network deployment is then proposed to enable vehicles with multi-interfaces to dynamically decide the detection mode to use. Another joint technique that utilizes geolocation information in conjunction with per-vehicle or CSS is proposed in [41].

11.2.1.3 Cooperative Spectrum Sensing

As mentioned above, the impact of small-scale (multipath) fading, and large-scale fading (shadowing) can deteriorate stand-alone spectrum sensing performance. A spectrum sensing study of the performance and accuracy of spectrum sensing in a vehicular environment using classical energy detection sensing is performed in [42]. Spectrum measurements were taken from a moving vehicle in different locations of the city of Boston in the United States, including an open space, a downtown location, and a straight street with moderately high buildings on one side and an open area on the other. Sensing accuracy significantly decreases in the presence of obstructions due to large-scale fading even at lower speeds. CSS can be used to tackle these problems by increasing the spatial diversity of sensing samples available to each vehicle [22]. CSS can be aided with infrastructure or completely ad hoc.

- *Infrastructure-based sensing*

  This category of CSS makes use of fixed RSU or small base stations to coordinate the spectrum sensing process and, in some cases, act as a fusion center for local sensing observations reported by moving vehicles to perform spectrum decisions. In [43], a sensing coordination framework is proposed, where an RSU coordinates spectrum sensing activities of the surrounding SUs in a CR-VANET. The RSU senses the channels sequentially and repeatedly in a proactive manner through the use of energy detection, which is a fast and effective method of identifying the presence of PUs. An approaching SU with data traffic sends a request to the RSU, which responds to the SU with coordination instructions in the form of an ordered list of available channels. The RSU orders the list of available channels according to their quality as well as QoS requirements in the SU request. The SU, whose location can be different than the RSU, performs stand-alone fine sensing with a sensing method of its choice (e.g., energy detection, cyclo-stationary, etc.). After identifying an available channel, an SU sends a pilot signal on the selected channel to the intended receiving SU, which responds with a pilot signal after sensing the channel to verify its availability at its side. This coordination mechanism reduces the time needed to find an SOP, by reducing and ordering the searching space of available channels.

  In [44] and [18], the authors propose a cognitive network architecture to dynamically extend the CCH of the WAVE protocol, which is used by vehicles to transmit safety-related information. Vehicles detect and report to RSUs available spectrum resources on the 5.9 GHz ISM band along their path together with their own traffic information. RSUs forward gathered data from vehicles to a processing unit that connects several RSUs. This processing unit hosts a fuzzy-logic-based spectrum allocation algorithm that infers the actual CCH contention conditions for each SU, and dynamically extends the CCH bandwidth in network congestion scenarios by using the vacant frequencies detected by the sensing module.

- *Ad hoc cooperative spectrum sensing*

  This category of CSS is completely ad hoc and does not benefit from any fixed infrastructure. And, hence, data fusion and spectrum decision is performed in a distributed fashion rather than at a centralized RSU or fusion center. A collaborative spectrum sensing scheme based on belief propagation is proposed in [45] to combine distributed observations and exploit redundancies in both space and time. The spatial correlation is utilized by message passing among neighboring vehicles, while time correlation due to the limited speed of the vehicles is exploited by defining virtual vehicles. Assuming a single channel, a vehicle i periodically uses energy detection to sense the channel and broadcast its hard decision belief on the presence of a PU, $S_i = 1$ if a PU is detected, and $S_i = 0$ otherwise. Each vehicle then combines received belief vectors from neighbor vehicles with its own belief to generate a new belief. A time-slot is divided into a sensing period followed by a number of iterations for exchange of belief vectors followed by a data transmission period. The network is expected to enter steady state after several iterations. The focus is to derive two functions: a local function ϕ_i operating on the local sensing results, and a compatibility function ($\psi_{i,j} = \eta, if\,S_i = S_j, 0.5 < \eta < 1$) between two vehicles, i and j. Finally, after several iterations, each vehicle calculates the network state to decide on PU presence as follows:

$$P(\mathbf{S}|\mathbf{X}) = \prod_{i=1}^{N} \phi_i(s_i|x_i) \prod_{i \neq j} \psi_{i,j}(s_i, s_j|x_i, x_j). \tag{11.1}$$

Drawbacks of this scheme include: the speed of convergence of the scheme, as well as control overhead, and computational complexity, especially with increasing number of neighbors in a multi-channel scenario.

CSS can enable a vehicle to detect spectrum availability in future locations by exploiting sensing information from other cooperating vehicles. In [42] and [3], the authors introduce the concept of a spectrum horizon, where the network topology is divided into short segments or cells, and a vehicle keeps spectrum information for up to h segments. The authors then introduce a collaborative sensing and decision algorithm, where each vehicle maintains a local spectrum availability database with a spectrum availability entry for every licensed channel. Each vehicle performs energy detection spectrum sensing for one of M licensed channels every T_s seconds that has the lowest number of sensing samples. A correlation-aware fusion scheme is used to merge sensing samples collected from other vehicles every T_d seconds to perform a spectrum decision. Correlation between samples of two vehicles is a function of the distance between the two vehicles. Every T_b seconds, each vehicle broadcasts its spectrum availability database over a CCC.

11.2.2   Spectrum Decision

Spectrum decision is the capability to decide the best spectrum band among the available spectrum bands, identified by the spectrum sensing module, according to the QoS requirements of the applications. This decision consists of two steps: First, each spectrum band needs to be characterized in terms of radio environment and PU activity. Then, the most suitable spectrum band must be chosen based on this characterization. The following are main functionalities required for spectrum decision:

1. **Spectrum characterization:** A spectrum band is characterized in terms of radio environment and PU activity. First, in terms of radio environment, a spectrum band is characterized by several parameters including: operating frequency, bandwidth, time variance, interference power, path loss which is related to distance to next hop, and to operating frequency, MAC layer frame errors, MAC layer delay, etc. Second, characterization in terms of PU activity. In literature, PU activity is mostly modeled as a two-state ON-OFF process [46–51]. The ON (busy) state represents the period where a PU is active, and the OFF (idle) state represents the period where a PU is idle [52]. The ON and OFF periods are exponentially distributed. Other PU activity models are proposed in [53] to characterize PUs in cellular networks, and in [54] for wireless local area networks (LANs).

   In [55], the authors define a set of spectrum characterization metrics with/without spatial-awareness and compare them in terms of the total communication duration and the amount of transmitted data. These metrics are: (i) Rate-based: choose channel with the highest rate regardless of PU activity and, thus, may cause frequent channel switches; (ii) rate and utilization-based: choose a channel with a high rate and a low PU activity; (iii) rate and the expected OFF period-based: choose a channel with a high rate and a long-residence time; (iv), (v), and (vi) are the same as (i), (ii), and (iii), but multiplied with the range or distance to the next hop. Note, that there is a number of next hop nodes, not necessarily on the same channel.

2. **Spectrum Selection:** In CR-VANETs, numerous combinations of route and spectrum exist between a source-destination pair. Hence, the selection rule is closely coupled with the routing protocol, giving rise to joint routing and spectrum decision algorithms. Furthermore, spectrum decision should support transmission in multiple spectrum bands for several reasons including: It allows transmission at a lower power resulting in less interference with PUs; CR users can adopt multi-radio techniques, where each radio interface tunes to different non-contiguous spectrum bands for different users and transmit data simultaneously [50,56].

11.3 The Routing Problem

Network connectivity in VANETs is a crucial matter, which depends upon the density of vehicles, as well as the channel environment. In a CR-VANET, interference and congestion from PUs as well as other SUs (vehicles) becomes an additional factor. Network connectivity also depends upon the network scenario at hand. For example, in an urban environment, especially at peak hours, the high density of vehicles can lead to more robust connectivity due to prevailed multiple paths, while obstacles, such as buildings, structures, and hills, can lead to frequent loss of connectivity due to multipath fading and shadowing. On the other hand, in a highway scenario, while fewer obstacles can lead to better channels and better spectrum sensing, high vehicles mobility causes Doppler spread. Also, while high vehicle density leads, in general, to more robust network connectivity, it also entails increased PU activity and higher SU contention over primary bands. Finally, the sudden appearance of PUs in different locations can lead to more frequent route failures. The routing problem in CR-VANETs, a sub-class of CRAHNs, is to create and maintain wireless multi-hop paths among SUs by deciding both the relay nodes and the spectrum band to be used on each link of the path [57].

Routing protocols proposed for mobile ad hoc networks (MANETs) and wireless mesh networks (WMNs) cannot be directly applied to CR-VANETs due to the unique characteristics and challenges stemming from high vehicles mobility and dynamic PUs activity. These challenges include the following:

1. **Spectrum-awareness:** Efficient routing solutions will require a tight coupling between the routing module and the spectrum management functionalities. In literature, three spectrum-awareness scenarios are studied:

   (a) SUs are provided with spectrum information by an external entity, such as a database of TV towers' whitespaces [58].

   (b) SUs can gather information about the spectral environment through spectrum sensing mechanisms [42].

   (c) Hybrid schemes of the previous two, where a subset of nodes employ cellular 3G transceivers to connect to a central spectrum database. The query overhead can be significant in a mobile congested network. Spectrum sensing techniques can reduce the frequency of querying the spectrum databases, and provide system robustness and resilience in situations where a spectrum database is not available [34].

2. **Setup of routes:** This will heavily rely on the dynamics of the available spectrum, which is affected by mobility of SUs and the activity of PUs. Classical route quality metrics, such as delay, energy efficiency, throughput, fairness, etc., need to be coupled with new measures, such as path stability and spectrum availability. For example, in terms of spectrum availability scenarios, experiencing high PU activity might favor opportunistic routing techniques designed for disconnected networks, rather than end-to-end techniques that are more suited to moderate-to-low PU activity scenarios [59]. In terms of path stability, PUs behavior have to be taken into consideration, in addition to more common ad hoc networks aspects, such as mobility and wireless channel considerations.

3. **Route maintenance/reparation:** Frequent and unpredictable route failures are a possibility in CRAHNs due to the sudden appearance of a PU in a given location, which may require frequent path rerouting through different nodes (relays) or spectrum handoff to other channels.

The routing problem at hand exhibits a lot of similarities with routing in multi-channel, multi-hop ad hoc networks, with the added challenges of coexistence with PUs. More precisely, the CR-VANET should be transparent to coexisting PUs, and, hence, we cannot assume any feedback from or control over PUs [60]. A CR vehicle needs to interrupt its transmission whenever a PU activity is detected, and, hence, the design of routing protocols for CR-VANETs need to estimate the stability of different routes and choose the most stable ones. Therefore, routing protocols proposed for CR-VANETs tend to characterize the non-permanent availability of spectrum bands due to PUs appearance to maximize the likelihood of meeting CRV QoS requirements. Furthermore, routing metrics, a main component of any routing protocol that determine the quality of different routes, tend to be hybrids that combine a number of atomic metrics to obtain a global one [61]. Finally, some routing protocols explore multiple paths between a given source-destination pair as well as operating multiple backup channels over a CR link to further enhance network connectivity. Very few routing protocols have been specifically proposed for CR-VANETs in literature. However, many routing protocols have recently been proposed for the more general CRAHNs, and some of them can be applied to CR-VANETs, or at least provide valuable insight into solving the routing problem in CR-VANETs. In this section, we investigate the design of routing protocols for CR-VANETs. First, we highlight the challenges faced in designing a routing metric for a CR network and the different techniques to combine a number of routing metrics. Then, we discuss some of the routing protocols that have been proposed for CR-VANETs.

11.3.1 Routing Metrics

A good routing metric design for CRAHNs should address several challenges. Some of these challenges are inherited from traditional networks, including: nodes mobility, wireless medium properties, channel scarcity, etc. In addition to these, a number of factors extend classical routing metrics, such as delay to address the constraints imposed by the PUs. A good routing metric for CRAHNs will assign different weights to different channels based on the probability that an ongoing transmission will be interrupted by a returning PU or another interfering SU. Channel switching time, in the event of PU detection, has to be taken into account by the routing metric. Switching delay may involve the time required to inform the next hop neighbor of the switching decision, in addition to the time needed to tune the radio to a new channel, referred to as the tuning time, which is usually a function of the frequency separation between the old and new channels [62]. Another factor that affects the routing metric is the deafness problem common with multi-channel communication, where a node tuned to one channel cannot receive signals transmitted on a different channel. As discussed before, the deafness problem can be tackled through the use of a CCC to disseminate route initialization and maintenance data, or sending the same data over all channels, or the use of channel synchronization schemes. While using a CCC is more bandwidth efficient than sending the same data over all channels, it faces the risk of becoming a bottleneck for communication. Synchronization schemes, on the other hand incur higher delays as well as bandwidth inefficiencies, especially with frequent channel switching. All these challenges, among others, make a cross-layer routing metric that takes into account features from different layers a must and not an option [61].

In [61], the authors present a taxonomy, Figure 11.3, of the different routing metrics that have been used in multi-hop CR networks. The authors categorize routing metrics into metrics for single-path routing algorithms and those for multipath routing algorithms. Here, some of these metrics that are more relevant to CR-VANETs are briefly discussed.

11.3.1.1 Single-Path Routing Metrics

A single routing protocol may combine a number of metrics, some of which are classical, such as delay and hop-count, while others target the CR aspect of CRAHNs, such as spectrum availability. In the following, we briefly discuss these metrics, giving examples from some routing protocols.

1. **Delay:** A classical routing that captures the end-to-end delay, which includes channel switching time, MAC backoff time, queueing delay, and transmission delay. Channel switching delay can be a constant [63] independent of new and old channels, or proportional to the differ-

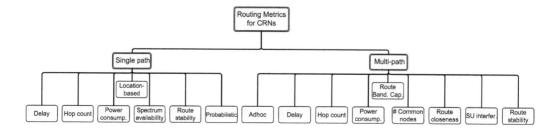

Figure 11.3: Taxonomy of routing metrics used in different CRAHNs routing protocols [61].

ence between the new and old channel, as in [62], where the proportionality constant is taken to be 10 ms/10 MHz. The MAC backoff time caused by the contention between the different nodes and is given in [62] by

$$\frac{1}{(1-p_c)(1-(1-p_c)^{\frac{1}{Num-1}})}W_0, \tag{11.2}$$

where p_c is the probability that a contending node experiences a collision, Num is the number of contending nodes on a given channel, and W_0 represents the minimum contention window size. The queueing delay at node i for flow j is mainly caused by the waiting packets from other flows, and is given in [64] for a round-robin queueing discipline among flows by

$$\sum_{k=1,k\neq j}^{k=Num-1} \frac{P}{B}, \tag{11.3}$$

where P is the packet length and B is the channel bandwidth. The transmission time over a link taking into consideration the expected number of retransmissions is know as the Effective transmission time (ETT) [65] and is given by

$$\frac{L}{r(1-p)}, \tag{11.4}$$

where L is the average packet size, r is the transmission rate, and p is the packet error rate. In [66], an end-to-end routing protocol for mobile CRAHNs called society for environmental awareness and rehabilitation of child and handicapped (SEARCH) sends route request (RREQ) packets on all channels using greedy geographic forwarding within a focus region as shown in Figure 11.4. The destination then knows the end-to-end delay on each channel. An intermediate node on a given channel that cannot find a next hop closer to the destination than itself will go into PU avoidance phase using perimeter routing as in greedy perimeter state routing (GPSR) [67] and will flag this point as a decision point. Finally, the destination starts from the channel with the least delay, and assesses the benefit of channel switching at the decision points to reduce end-to-end delay.

2. **Hop count:** An indicator of other metrics, as it traditionally reflects less delay and less resource consumption, as the packet passes through fewer intermediate nodes. In CRAHNs, hop count can be used as a coarse routing metric. For example, SAMER [68] is a two-tier link state routing approach, where a forwarding mesh is first built based on the hop-count for long-term optimality, followed by opportunistic exploitation of available channels by allowing for deviation from the shortest-path route.

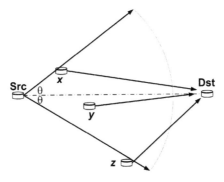

Figure 11.4: Greedy geographic forwarding on a given channel within a focus region in SEARCH [66].

3. **Location-based:** Geographic forwarding is used in several routing protocols in conjunction with other metrics. As mentioned above SEARCH selects the next hop node with the greatest distance advancement toward the destination within a focus region. The route decision made by the destination is, however, based on end-to-end delay. Other protocols, such as MP-JSRCA, [69] also select next hop candidates from a sector region toward the destination. However, the next hop is selected according to a data transmission cost (DTC) that is a weighted sum of the cost of mobility and the cost of interference to PUs, as well as other SUs. In IPSAG, a node selects the closest neighbor to the destination as the next hop, as long as it shares a common available channel with itself and satisfies some channel quality measures, such as signal to noise ratio (SNR) [70]. An any path routing protocol called CoRoute is proposed in [60], where each node forms a forwarding set of neighbors closest to the destination. Nodes in the forwarding set relay received packets according to a priority based on the cumulative ETT toward the destination. Any path routing techniques increase route reliability by providing backup relay nodes in case higher priority nodes cannot relay traffic [71]. In [72], an opportunistic location-based routing protocol is proposed, where the sending node transmits a request-to-forward (RTF) message that includes the location of the destination as well as itself, the set of available channels, and the needed transmission rate on the CCC using maximum transmission power. Relaying neighbors reply to the RTF message after a random period that is determined according to proximity to the destination.

4. **Spectrum availability:** Spectrum availability of a CR link refers to the bandwidth available for communication between sender and receiver, considering both PU's activity and other SU's activity. For example, in SAMER [68], the path from source to destination is selected according to path spectrum availability (PSA) metric, which is the minimum spectrum availability of all the links forming the path. The spectrum availability of any given link is given by

$$T \times B \times (1 - P_{\text{loss}}),$$ (11.5)

where T is the fraction of time during which the link is available for communication between sender and receiver and can be estimated using MAC layer information, B is the available bandwidth, and P_{loss} is the PER or packet loss rate of the CR link, which can be calculated using injected periodic pilot packets or can be estimated probabilistically like in CoRoute [60].

5. **Route stability:** Unstable routes due to high mobility or PU activity will cause frequent rerouting requests, which consume resources and degrade performance, especially in time critical applications. Route stability can be reflected in terms of links available times, as in

the STOD-RP protocol [63], where the routing metric combines route stability with end-to-end delay. The the link cost in STOD-RP is given by

$$[O_{ca} + O_p + \frac{L}{r}] \frac{1}{1-p} \frac{1}{T}, \tag{11.6}$$

where O_{ca} is the channel access overhead in μs, O_p is the protocol overhead in μs, p is the PER, L is the average packet size, r is the data rate, and, finally, T is the predicted available time for the channel, which is estimated from the statistical history of PUs' activities. The routing metric is the summation of all links costs in a route. The Coolest Path protocol in [73] uses a similar link cost, but a different approach, to calculate the overall path cost. Rather than the sum link costs, the Coolest Path protocol uses the *max* operator or a linear combination of the maximum and the sum of links costs to calculate the path cost. The authors show accumulated cost is better in terms of path longevity for high PU activity, while the *max* cost is better at low PU arrival rate and high PU channel occupancy. Finally, route stability can also be implicit, as in SEARCH [66], where route construction avoids routing through PU activity regions and circumvents these regions by information gathering in periodic beacons.

6. **Probabilistic:** In [74], Khalife et al., propose a source-based routing protocol with a probabilistic routing metric to find the most probable path (MPP) to the destination. The MPP path is also statistically the most stable path to the destination that can satisfy a given bandwidth demand D in a network with N nodes and M orthogonal channels. The authors assume that the probability distribution function (PDF) of the primary network interference, as perceived by SUs, follows a log-normal distribution [75]. The link metric is then given by

$$-\log Pr[C \geq D + U], \tag{11.7}$$

where C is the channel's Shannon capacity, and U is a system memory that accounts for the cognitive interference in the vicinity of the sender and receiver nodes. The overall path metric is the summation of link metrics for all links constituting the path.

11.3.1.2 Multipath Routing Metrics

Multipath routing aims to provide redundant routes between source and destination to account for the sudden arrival of a PU. Many of the multipath metrics are similar to single-path routing metrics, including delay, hop-count, and route stability. In the following, we discuss some new routing metrics specific to multipath routing protocols.

1. **Route-bandwidth capacity:** The multipath routing and spectrum access (MRSA) protocol [76] uses the classical dynamic source routing (DSR) algorithm to discover multiple candidate paths and select the path with maximum bandwidth capacity. A path bandwidth capacity is the minimum bandwidth capacity over all nodes constituting the path. A number of nodes sharing a common link each will get a fair share of its bandwidth capacity, and a number of flows passing through a common node each will get a fair share of its bandwidth capacity. Secondary routes are selected based on the same metric, and by preferring routes with a lower number of common nodes with previously selected routes. Other protocols, like SPEctrum-aware routing (SPEAR) [77] protocol, use a similar approach to MRSA, but instead use hop-count to prefer secondary routes.

2. **Route closeness:** The route closeness metric [78] is based on the intuition that multiple routes that are far away from each other are less likely to be subjected to the same active PUs. In other words, if routes are far enough from each other, they cannot all be interrupted by the same PU at the same time. An example of calculating the closeness of two routes is shown

in Figure 11.5, where $PuER(L)$ is the region around link L between two SUs $S1$ and $S2$ that can be interrupted by a PU, which is surrounded by an approximation polygon for more efficient calculations. The second part in the figure shows the definition of closeness between two links, L_1 and L_2, as the intersection $PuER(L_1) \cap PuER(L_2)$. Finally, the closeness of two routes is the sum of the pairwise closeness of their links.

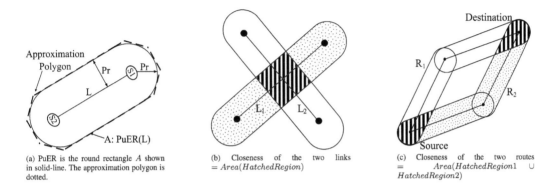

(a) PuER is the round rectangle A shown in solid-line. The approximation polygon is dotted.

(b) Closeness of the two links $= Area(HatchedRegion)$

(c) Closeness of the two routes $= Area(HatchedRegion1 \cup HatchedRegion2)$

Figure 11.5: Definition of routes closeness from [78].

3. **Dead zone penetration:** Instead of changing active route paths to avoid areas of PU activity, the dead zone penetration (DZP) [79] routing protocol and metric tolerates PU activity by using cooperative beamforming by neighboring nodes to null out SU transmission at the PUs. An example is shown in Figure 11.6, where Node 1 maintains constructed route $(1 - 2 - 4)$ in the presence of a PU in this region, instead of choosing the alternate route that goes through Node 5 by allowing Nodes 2 and 3 to cooperatively send data packets to Node 4.

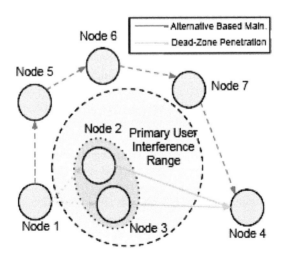

Figure 11.6: Dead zone penetration example from [79].

4. **SU interference:** While most routing metrics tend to select routes that are less likely to be affected by future PU activities, this class of routing metrics takes into consideration the interference among SUs themselves as well. In MRSA [76], a channel that is selected by an SU node in the route is not used for two hops to reduce intra-path interference and contention. The bandwidth footprint product (BFP) metric is defined in [80], which captures both spectrum usage (bandwidth) and spatial occupancy (footprint). The footprint is the interference area of a transmitting node. The authors formulate an optimization problem to minimize the sum BFP in a network. Multipath routing is allowed in their formulation by allowing flow splitting to achieve better performance.

11.3.2 Routing Approaches

Routing approaches for CR-VANETs can be classified into end-to-end approaches and opportunistic approaches. End-to-end routing solutions use ad hoc on-demand distance vector (AODV-like) methodology to collect key parameters, and in route formation through broadcast of RREQ packets. A source node either broadcasts multiple RREQ packets over all available channels if no CCC is used, or a single time in the presence of a CCC. Upon receiving a RREQ packet, an intermediate SU adds to it its spectrum information and link cost. The destination SU receives multiple RREQ packets from multiple paths, and performs joint route and channel selection. The destination SU, then, transmits an RREP packet over the minimum cost route. In some protocols, the destination determines the end-to-end route and delegates channel allocation to intermediate node, where each intermediate node knowing the channel selections of its destination-side part of the route selects its pre-hop channel to minimize cost of path to the destination. In opportunistic approaches, on the other hand, the source SU, as well as each intermediate SU, selects in a distributed manner its next hop relay and channel from its neighbors.

11.3.2.1 Opportunistic Approaches

CoVanet, a cognitive vehicular ad hoc network multi-radio multichannel cross-layer architecture is proposed in [60]. CoVanet allows vehicles opportunistic access to WiFi channels to counter the limited capacity of dedicated DSRC channels. However, WiFi channels in urban areas are already heavily subscribed by residential customers near roadside, as well as various wireless devices in the The industrial, scientific and medical (ISM) bands, which are given priority by CoVanet and considered the PUs. Due to shadowing effects, AP's interference range can differ from one segment of the road to another. Each vehicle is assumed to be equipped with two radios, R1 and R2, for data and a radio for the CCC, and, hence, vehicles are able to transmit and receive packets concurrently. R1 is tuned to the SU's receiving channel and changes slowly as the vehicle travels down the urban grid, while R2 is switched dynamically among current neighbors' receiving channels for data transmission and reception.

Each vehicle performs periodic sensing of its receiving channel and shares local sensing results with its neighbors using HELLO packets. A quiet period is observed by all SUs, during which PU sensing takes place, for example, a 20 ms every second. Time synchronization can be achieved using the vehicles' GPS systems. Monitored channels are described by two-state (ON/OFF) semi-Markov model, where the expected busy (ON) and idle (OFF) times are expressed by exponential distribution with rates λ and μ, and with cumulative distribution functions given by

$$P(T_{idle} < t) = 1 - e^{-\lambda t}, \tag{11.8a}$$

$$P(T_{busy} < t) = 1 - e^{-\mu t}. \tag{11.8b}$$

Each SU estimates its receiving channel workload, ω, as

$$\omega = \frac{T_{busy}}{T_{busy} + T_{idle}}. \tag{11.9}$$

The channel workload is then used to calculate the expected channel capacity according to

$$R_i = R_0 \times (1 - \omega), \tag{11.10}$$

where R_0 is the physical data rate (e.g., 11 Mbps). The authors assume that co-located SUs choosing a channel as their receiving channel, will share it equally, and, hence, the channel capacity per node can be given by

$$R_i' = \frac{R_i}{N(i)}, \tag{11.11}$$

where $N(i)$ is the number of SUs selecting the channel i within radio range. Finally, an SU will select the receiving channel, j, among multiple channel as follows:

$$j = \arg\max_i R_i'. \tag{11.12}$$

An SU then tunes its first radio (R1) to j. HELLO packets are flooded after determining the receiving channels. HELLO packets contents include monitored channel information, selected channel, neighbor information, own activity (whether it has a flows to transmit), and current geo-location information. A random jitter is inserted before each HELLO packet to avoid collision with other neighbors' HELLO packets.

A main contribution of [60] is a cognitive anypath routing protocol called CoRoute that utilizes geographical location and sensed channel information to find an end-to-end route in a rapidly changing channel condition and workload. CoRoute increases network throughput by selecting low interference channels and exploiting alternate paths. CoRoute adopts ETT value as a link metric, and cumulative ETT as a path quality metric. ETT indirectly reflects packet loss and channel bandwidth. To calculate ETT, the authors start by estimating the expected transmission count (ETX) over a link. Naturally, ETX can be measured by broadcasting probing packets at very slow rate during a predefined window. However, this would be inaccurate in CoRoute, due to high mobility, with considerable overhead, since each vehicle must send packets on multiple channels. Instead, the authors derive ETX probabilistically by estimated PU workload and distance between a sender and a receiver. Let p_f and p_r be the packet loss probabilities in forward and reverse transmissions, respectively, then

$$p = 1 - (1 - p_f)(1 - p_r), \tag{11.13a}$$

$$ETX = \sum_{k=1}^{\infty} k p^{k-1}(1-p) = \frac{1}{1-p}. \tag{11.13b}$$

The probability of successfully receiving a packet based on distance d with data rate k is denoted by $P_d(d,k)$. Data rate k is achievable when the signal to noise ratio (SNR) is above a threshold Ψ_k as follows:

$$P_d(d,k) = \Pr\left(\frac{A^2}{\eta} > \Psi_k\right) = 1 - \frac{\gamma(m, \frac{m}{\Omega}\Psi_k)}{\Gamma(m)}, \tag{11.14}$$

where η is thermal noise, $\gamma(\cdot,\cdot)$ is the incomplete gamma function, A is the random variable that represents the amplitude of the received signal in a vehicular environment modeled by the Nakagami distribution, $\Gamma(\cdot)$ is the Gamma function, m is the fading intensity that depends on the environment and the distance between the transmitter and the receiver and can be varied to multi-path fading distributions, such as Ricean and Rayleigh, and, finally, Ω is the received power derived from the two-ray path loss model. The probability of successfully receiving packets also depends on interference from the PU. The packet error rate (PER) is determined by the collision duration between the PU's and the SU's packets and by SNR (i.e. bit error rate) discussed previously. Let $P_c(t,k)$ be the error rate of packets received, with data rate k and collision duration t. To calculate $P_c(t,k)$, we need to calculate the average collision duration between by the PU and the SU given by

$$E[t] = P_0 T_m + \sum_{t=1}^{T_m} \left(1 - \prod_{n=0}^{t-1} P_n\right)(T_m - t), \tag{11.15}$$

where T_m is the maximum overlapped time given by

$$T_m = \frac{1}{T_{bit}} \min\{T_{busy}(PU), T_{busy}(SU)\}, \tag{11.16}$$

where T_{bit} is the bit duration. The error probability of forward and reversed transmission can then be calculated as p_f or $p_r = 1 - [P_d(d,k)(1 - P_c(t,k))]$. From ETX, ETT can be calculated as

$$ETT = ETX \times \frac{S}{R}, \tag{11.17}$$

where S is packet size and R is data rate.

CoRoute uses opportunistic anypath routing following the forwarding set technique introduced in [71]. SUs form neighbor tables using the periodic HELLO packets exchanged over the CCC. From this neighbor table, a source or intermediate SU selects a number of vehicles geographically close to the destination to form a forwarding set J. An SU uses multiple broadcasts to transmit the packet to all vehicles in J, since they may have different receiving channels. Each vehicle in J has a priority to relay the received packet based on its EATT to a pre-defined node. Initially, this node is the node two-hop from the relay and closest to the destination. However, as packets reach the destination node, EATT values start propagating back from the destination to the source through periodic Hello messages in a Bellman-Ford-like manner, and, hence, the pre-defined node becomes the destination. Relaying priority is enforced by making the MAC layer contention window (CW) proportional to EATT, i.e., smaller CW for lower EATT. To avoid superfluous multiple transmissions, a higher priority relay broadcasts a notification message on the CCC to the sender and all forwarders in J upon successful relaying of the packet. Recall, vehicles in the forwarding set are operating on different channels, but they all are continuously listening to the CCC. A lower priority relay drops the packet upon hearing a notification from a higher priority relay on the CCC, otherwise it transmits the packet upon its counter timeout. To calculate EATT at node i, let d_j be the ETT of link (i, j), $d_{i,J}$ be the ETT of node i to forwarding set J given by

$$d_{i,J} = \frac{S}{R_j} \frac{1}{1 - \prod_{j \in J}(1 - p_j)}, \tag{11.18}$$

where $p_j = \frac{1}{ETX}$ is the probability of successful transmission over link to node j, and R_j is the link data rate, then D_i is the EATT at node i, calculated using Bellman equation as

$$D_i = d_{i,J} + D_J, \tag{11.19}$$

where D_J is the weighted sum of each vehicle's contribution across the forwarding set given by

$$D_J = \sum_{j \in J} \omega_j D_j, \quad \text{where} \sum_{j \in J} \omega_j = 1, \tag{11.20}$$

where ω_j is the probability that forwarder j is successfully used given by

$$\omega_j = p_j \frac{\prod_{k=1}^{j-1}(1 - p_k)}{1 - \prod_{j \in J}(1 - p_j)}. \tag{11.21}$$

11.3.2.2 End-to-End Approaches

In [81], the authors proposed the expected path duration maximized routing (EPDM-R) protocol, which is a modification of the routing metric of AODV from minimizing hop-count to maximizing the expected path duration (EPD). The EPD of an end-to-end route is equivalent to the minimum expected link duration (ELD) over all its nodes. To estimate ELD for each link, the authors assume that the quality of each link is high and that co-channel interference is removed by using MIMO techniques. Hence, the network topology is mainly influenced by vehicular mobility and spectrum availability. To evaluate a path's EPD, the authors use the freeway mobility model [82] to describe or approximate vehicular mobility, and the call-based model [53] to describe PU behavior.

In [83], the authors propose a proactive routing protocol for CR-VANETs called Cog-OLSR, which is an advancement of the well known OLSR routing protocol. The authors consider a CR-VANET, where PUs are residential WiFi APs and SUs are vehicles. Like EPDM-R, the authors assume the quality of each link to be high and that co-channel interference is removed using MIMO techniques. Periodic channel sensing is used to maintain updated information on available SOPs. The life time of a link between two SUs over a given SOP, i.e., the link state, is estimated considering PU activity characterization, which is modeled by an ON/OFF process each modeled with the exponential distribution, as well as the velocity, distance, and angle of separation between neighboring nodes. Periodic HELLO packets are exchanged over the CCC among neighbors to advertise estimated link states. Finally, vehicular mobility follows the graph walk mobility model. The operation of Cog-OLSR is then similar to the OLSR protocol.

11.4 Simulating End-to-End Protocols in CRAHNs

The availability of comprehensive simulators capable of handling a large number of dynamically varying spectrum parameters is a crucial factor in CR-VANET, and the more general CRAHN, research. In literature, a CR extension has been proposed to a number of existing network simulators. cognitive radio cognitive network (CRCN) [84] is a simulation framework based on NS-2 that provides multi/single radio and muti-channel support per node. CRCN is mainly geared toward MAC layer research rather than routing protocols research. In [85], another CR extension for NS-2 called CogNS is proposed. However, CogNS does not incorporate multiple radios per node. As for OMNet++ [86], a CR simulation extension is proposed in [87], which also focuses mainly on evaluating CR MAC layer protocols. A simulator written in C++, proposed in [88], offers a modular approach providing a full network layer stack. However, it is not extensively tested by the general networking community, and, hence, it will ensue more overhead to port well-established protocols in different layers, such as different MAC protocols, and different channel models. Also, unlike popular simulators such as OMNET++, NS-2, and NS-3, it does not provide a wider user-base and support community. In this section, we briefly discuss two of the more comprehensive CR extensions proposed in literature. The first extension is based on NS-2, while the second is based on NS-3.

11.4.1 NS2-CRAHN

In [89], the authors introduce NS2-CRAHN, an extension of the NS-2 simulator designed to support realistic simulation of CRAHNs. NS2-CRAHN provides an accurate yet flexible model of the cognitive cycle implemented by each CR user and PUs' activities. The authors then use the NS2-CRAHN tool to evaluate the performance of end-to-end protocols (e.g., routing and transport layer protocols) for CRAHNs. In the following, we discuss the main building blocks of this simulation tool. Consider, SUs equipped with K radio transceivers that can be tuned to any of M primary channels, where $K < M$, i.e., the number of radios per node is smaller than the number of available channels. Different channels may have dissimilar raw channel bandwidth. The link-layer solution

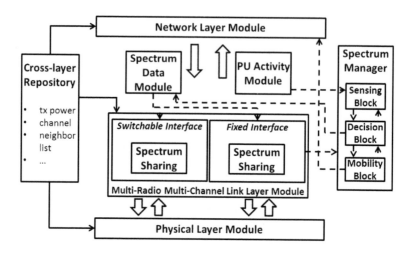

Figure 11.7: NS2-CRAHN architecture.

adopted in [89] is based on the interface assignment strategy proposed in [90], which classifies available radio interfaces into fixed and switchable ones. The fixed interfaces stay on a specified channel for longer time intervals, while the switchable interfaces are more dynamic and switch among the available channels to maintain network connectivity. A distributed protocol is to be used to assign a channel to the fixed interface of each CR user. This channel can change over time for different reasons, e.g., traffic load. The simplest case is $K = 2$, i.e., each CR user is equipped with a fixed interface tuned to a receiving channel and responsible for spectrum sensing on that channel, and a switchable interface is used to transmit data to other CR users by switching to their receiving channels and to transmit and receive control data on/from the CCC.

The PU activity is modeled by the exponential ON/OFF model [50], where each PU has two states: an ON (busy) state representing the period in which a PU occupies the channel, and an OFF (idle) state representing the period in which the channel can be used by a CR user. Switching between the ON/OFF states follows a birth-death Markovian process. The ON/OFF states of channel i follow exponential distributions, with mean $\mathcal{T}_{ON}^i$ and $\mathcal{T}_{OFF}^i$, respectively.

Each CR user alternates between sensing the channel for a sensing time equal to t_s, followed by a transmission time equal to T_0. If the spectrum sensing block detects a PU in a channel, then the CR user vacates the channel immediately and continues its communication in another portion of the spectrum. The spectrum mobility block performs the spectrum handoff and protocol-reconfiguration, while the spectrum decision block is responsible for selecting a new channel based on the CR user's QoS requirements. If the current channel is found free from PU activity, a MAC layer coordination scheme, implemented by the spectrum sharing block, is used to enable the CR user to transmit data on the channel while preventing collisions with other CR users.

The architectural model of the NS2-CRAHN simulator is shown in Figure 11.7, where the authors of [89] have added the following extendible stand-alone C++ modules:

1. **PU activity module:** This module describes (i) PUs locations and characteristics and (ii) PUs activity over time in the different spectrum bands. All the information about PUs are contained in a PU-log file generated offline and composed of two part:

    - The first part contains entries with the format $< id, x, y, channel, tpower >$, where id is the unique PU identifier, x, y are its location, *channel* and *tpower* are the channel and transmitting power used by the PU, respectively.

    - The second part describes PU activity over time with entries of the format $<$

$id, arrival_{time}, departure_{time} >$, where $arrival_{time}$ is the simulation time when the PU enters the ON period and starts transmitting, and $departure_{time}$ is the simulation time when it enters the OFF period and stops transmitting.

2. **Spectrum data module:** This module describes PHY characteristics of each channel. Also utilizing a channel-log file with entries of the format $< id, frequency, bandwidth, noise >$, where id is the channel identifier, $frequency$ is the channel central frequency, $bandwidth$ is the raw bandwidth of the channel (e.g., 11Mb/s), and $noise$ is the average value of the noise on that channel. Using these spectrum data, it is possible to model the average BER experienced by the receiver model for QPSK modulation [91].

3. **Spectrum manager module:** The spectrum manager module implements the cognitive cycle for each CR user. This module is composed of the following three building blocks:

   (a) **Spectrum sensing block:** Channel sensing is implemented as a lookup function on the PU-log file for the current channel. The sensing block can also simulate sensing accuracy. The final outcome of the sensing block is a binary response, whether a PU signal is detected or not on the current channel. More specifically, a CR user C performing sensing on channel i at time t checks the PU-log file if there is an entry P satisfying two conditions:

      i. First, P is transmitting on the same channel, or adjacent channels, for the time interval $[t; t + t_s]$.

      ii. Second, the amount of power injected on channel i by node P and received by node C, P_r^C, is higher than a given sensitivity threshold P_{th}^C.

      Signal propagation is modeled using a generalized free-space model as follows: $P_r^C = \frac{P_t \cdot C_t}{(d^\alpha)} \cdot k$, where P_t is the transmitting power of P, C_t captures different transmission properties, such as the antenna gain and height, α is the attenuation factor, d is the physical distance between P and C, and k is the overlap factor between channel i and the central frequency of P. If both conditions apply, then PU is in the ON state. However, the authors also introduce a probability of successful detection probability, P_d, to simulate PU detection failure situation.

   (b) **Spectrum decision block:** This block is responsible for (i) deciding the spectrum policy and (ii) choosing the next channel to be used by CR users. As for spectrum policy, two policies can be used:

      i. **Switch immediately:** CR user immediately vacates the current channel as a PU is detected.

      ii. **Stay and wait policy:** CR user stops transmitting on current channel but does not vacate it. A notification is sent to the sensing block, and as soon as the PU activity ceases, the CR user re-starts its operation on the current channel.

      In the case of channel switch, three selection and allocation schemes of the next channel can be used:

      i. **Random allocation:** The CR user randomly chooses a new channel among the available N channels.

      ii. **Sequential allocation:** The CR user visits all the N channels using a round-robin algorithm: $next\_channel = (current\_channel + 1)\% N$.

      iii. **PU-aware allocation:** The CR user estimates the amount of interference on each channel by each PU and by neighbor CR users before selecting the next channel.

   (c) **Spectrum mobility block:** After deciding on the next channel to switch to, the mobility block is invoked. This block simulates the channel switching delay by using a timer.

No communication is allowed during the handoff operation. After the completion of the handoff process, the spectrum sensing block is invoked to sense PU activity on the new channel.

4. **Multi-radio multi-channel link layer module:** This module implements the multi-radio multi-channel environment for a CR user consisting of:

   (a) **Link Layer Management:** A periodic HELLO message is broadcasted by each CR user on all available channels to inform its neighbours about the channel used by its fixed interface. When a CR user (A) needs to communicate with another CR user (B), it tunes its switching interface to the channel used by Bs fixed interface and starts transmitting. Each CR user operates two timers on the fixed radio interface: a sensing timer for a t_s time interval to detect the presence or lack thereof a PU, and an operational timer for a T_0 time interval where it can send/receive data on the current channel.

   (b) **Spectrum Sharing:** Each radio interface implements a spectrum sharing scheme based on CSMA-CA MAC scheme.

5. **Network Layer Module:** The routing protocol used.

6. **Cross-Layer Repository Module:** This module enables information sharing among the different layers of the protocol stack. Examples of information it may contain include: PHY layer information (e.g., current transmit power), MAC layer information (e.g., current size of the backoff window), and network layer information (e.g., current neighbors list).

11.4.2   CRE-NS3

Another simulation tool, CRE-NS3, for CRAHNs based on the network simulator 3 (NS-3) [92] is proposed in [93]. It demonstrates execution time and memory usage improvements over earlier tools based on the NS-2 environment. NS-3 provides several advantages over NS-2 [94] including:

1. A new core written in C++.

2. Greater support for wireless communications.

3. Mobility models support for vehicular networks for highway and urban scenarios.

4. Expandable modular architecture.

5. Extensive documentation via HTML Doxygen [95].

6. NS-3 code can be easily adapted to work in real devices.

The authors work realize the first CR extension for NS-3 (CRE-NS3) with the following features:

1. CR capabilities such as sensing, PU detection, channel hand-off and decision making provided at the different network layers.

2. The ability to query a PU activity database.

3. The ability to combine cognitive and non-cognitive wireless nodes in one test scenario.

4. Seamless support for multi-channel and multi-radio node architectures.

5. Ability to include cognitive and non-cognitive interfaces in one node.

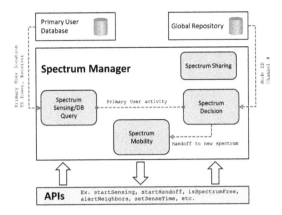

Figure 11.8: Building blocks of NS-3 CR extension [93].

6. New application programming interfaces (APIs) to create and access node-level and network-level features without major code changes.

7. Dynamic configuration of sensing/hand-off times from the command line.

8. Documentation through Doxygen.

9. Release of the full source code, with additional guides on how to compile and run trial examples.

A spectrum manager block, shown in Figure 11.8, masks the inner CR API calls leading to an organized modular approach and serving as a black box to the other modules in NS-3. The different layers in the network simulator keep a reference to the spectrum manager instance and use its functionalities via exposed APIs and hooked listeners such as: startSensing(), startHandoff(channel), is SpectrumFree(channel), alertNeighbors(), etc. The spectrum manager block consists of several sub modules. These sub modules map to the cognitive cycle as follows

1. Spectrum Sensing/Database Query
   This submodule is responsible for checking the existence of a PU in a given channel at a given time period. Similar to NS2-CRAHN, PU activity is inferred from a static PU Database loaded before the simulation starts.

2. Spectrum Decision:
   This submodule is responsible for determining, based on sensing/querying results, wether a hand-off is necessary or not. If hand-off is necessary, this submodule also needs to determine the channel to switch to. The spectrum decision submodule is linked to a global repository which keeps track of current occupied channels by all CR nodes in the simulator, which can be used, for example, to determine the least occupied channel to switch to.

3. Spectrum Mobility:
   Responsible for initiating the handoff protocol in the current node.

4. Spectrum Sharing:
   This submodule employs the built-in carrier sensing MAC 802.11 standards in NS-3 to make sure available spectrum is shared without collisions among CR nodes operating on the same channel.

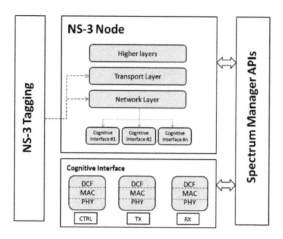

Figure 11.9: CR node layered architecture [93].

The layered architecture of CRE-NS3 is shown in Figure 11.9. Several APIs and listeners are exposed to the different networking layers. For example, network researchers working on a CR application can use an nFode's CR features by calling the respective APIs in the spectrum manager. A new cognitive interface block is shown in Figure 11.9, where a CR node can have any number of these cognitive interfaces. Each interface is composed of three separate MAC-PHY layers. The first, the CTRL interface, is for communicating control packets, such as HELLO, RREQ, and RREP packets of the AODV routing protocols and ARP messages over the CCC. The second, the TX interface, is a switchable interface used to transmit data messages to neighboring nodes on their receiving channels. Switching and transmission times can be defined using the NS-3 attribute system, which is a mechanism to pass parameters on the command line without the need to recompile the core of the simulator to change the value of various parameters. The TX distributed coordination function (DCF) use multiple MAC queues, one for each active channel. The TX interface switches among these queues randomly, or in round robin fashion, or according to other policies. The third and final interface is the receiving (RX) interface, which is also switchable. The RX interface is responsible for PU sensing and initiating hand-offs when PUs are detected. The probability of detection error can be defined using the NS-3 attribute system. The cognitive interface makes all new calls through the spectrum manager block API, providing a cleaner and easier cross-layer referencing, as opposed to having each layer hold references to several other network layers.

A new sensing state is added at the PHY layer of the RX interface. This sensing state is similar to that of the hand-off state, where the PHY layer instructs the DCF to halt de-queueing from the respective MAC queue, while channel sensing or hand-off operation is ongoing. When the RX interface starts sensing, hand-off or transmission, it transitions along the cognitive cycle depicted in the state machine shown in Figure 11.10. The cycle is triggered by the sense state. If no PU activity is detected, the state moves to the transmit state for a predefined period of data transmission time, then moves back to the sense state. Otherwise, a PU is detected, the state machine moves to the decision state. The CR node, based on the spectrum decision policies, either stays on the busy channel, then no transmission will happen and the state immediately returns back to the sense state, or a hand-off is decided, and then the state machine moves to the handoff state after the decision block decides which channel to hand-off to. After the completion of the hand-off process, sensing is triggered again before confirming channel availability and resuming data transmission.

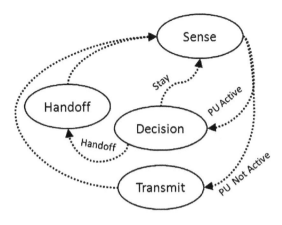

Figure 11.10: RX Interface state machine [93].

11.5 Open Research Issues

In literature, few routing protocols have been proposed for CR-VANETs. Routing protocols and metrics proposed for the general CRAHNs need to be tested and adapted to suit the unique challenges of CR-VANETs. In this section, we highlight some open research issues and future direction in the field of routing for CR-VANETs.

11.5.1 Design of Global Routing Metrics

Routes in multi-hop CR-VANETs, both single-path and multi-path, are rarely characterized by a single performance metric. However, it is usually a combination of different atomic routing metrics in order to achieve some performance tradeoff. These tradeoffs can be scenario specific. Therefore, the design and calculation of these high-level routing metrics, as well as the application of the relevant tools of multi-objective optimization, is an ongoing research issue. Multi-objective optimization aims to simultaneously optimize several conflicting objectives subject to certain constraints. A survey of the theory and application of multi-objective optimization techniques can be found in [96].

11.5.2 QoS Routing and Realtime Applications

Routing protocols with QoS support and differentiated services that can meet the challenges of CR-VANETs is still an open research issue. Multi-path routing techniques that can select backup routes or multiplex data traffic over different routes together with multi-objective metrics that combine a number of QoS metrics, such as delay, jitter, bandwidth, route stability, etc., are crucial tools to achieve this objective.

11.5.3 Security-Based Routing

Secure routing protocols that operate in the presence of malicious nodes in the CR-VANET is another open research area. A survey on communication protocols and security in CRAHNs is given in [97]. Common attacks in CRAHNs include: jamming attacks to trigger denial of service (DoS) to legitimate PUs and SUs, especially if the attack is on the CCC; PU emulation (PUE) attacks launched by a malicious or selfish SU masquerading as a PU to obtain access to a given channel rather than sharing it with other SUs; PUE attacks, which can also occur during channel switching, leading to data throughput degradation; and, finally, cross-layer attacks that target multiple layers of the TCP/IP protocol stack simultaneously to achieve specific attack goals.

11.5.4   More Accurate PU Modeling

Almost all routing protocols for CR-VANETs and CRAHNs in general assume an ON/OFF static PU model. An open research area is the inclusion of more realistic PU traffic models, as well as mobile PUs and the impact of that on CR-VANETs' routing protocols

11.6   Conclusions

In this chapter, we reviewed the design of routing protocols for CR-VANETs, which utilize CR principles to increase the bandwidth available for communication in the vehicular environments. However, CR-VANETs have their unique characteristics that must be accounted for when benefiting from the CR paradigm. To give a broad perspective of the factors that need to be considered in the design of a routing protocol for CR-VANETs, some of the main characteristics of CR-VANETs, both new and those carried over from VANETs and CRAHNs, were highlighted. Then, a brief discussion of the different network architectures of a CR-VANET, as well as a number of possible areas for the application of CR-VANETs, is given. Furthermore, there was also a summary of channel acquisition techniques, which include spectrum sensing and spectrum decision.

The routing problem in CR-VANETs is to create and maintain wireless multi-hop paths among SUs by deciding both the relay nodes and the spectrum band to be used on each link. While few routing protocols have been specifically proposed in literature for CR-VANETs, a multitude of routing protocols has been proposed for the general CRAHNs. However, the majority of these protocols are designed for static environments and do not consider the topological challenges of the vehicular environment. We shed light on some of these protocols and, more precisely, on the design of routing metrics, which can be valuable when addressing routing in CR-VANETs. Routing approaches proposed for CR-VANETs can be classified into opportunistic approaches, where channel and relay selection is performed by each node in a distributed manner, or end-to-end, where a complete route is first formed from source to destination before actual data transfer. We discuss these approaches with example routing protocols proposed in literature.

The evaluation of routing protocols is usually done through computer simulations. This chapter overviews some of the simulation tools or extensions to known network simulators used to simulate network layer protocols in CR-VANETs. Finally, we briefly highlight some of the open areas for further research in the design of routing protocols for CR-VANETs.

References

[1] I. F. Akyildiz, W.-Y. Lee, and K. R. Chowdhury, "CRAHNs: Cognitive radio ad hoc networks," *Ad Hoc Netw.*, vol. 7, no. 5, pp. 810–836, Jul 2009.

[2] K. Singh, P. Rawat, and J.-M. Bonnin, "Cognitive radio for vehicular ad hoc networks (CR-VANETs): Approaches and challenges," *EURASIP Journal on Wireless Communications and Networking*, no. 1, 2014.

[3] M. Di Felice, K. Chowdhury, and L. Bononi, "Analyzing the potential of cooperative cognitive radio technology on inter-vehicle communication," in *IFIP Wireless Days (WD)*, Oct. 2010, pp. 1–6.

[4] M. Di Felice, K. R. Chowdhury, and L. Bononi, *Cognitive Radio Vehicular Ad Hoc Networks: Design, Implementation, and Future Challenges.* John Wiley & Sons, Inc., 2013, pp. 619–644. [Online]. Available: http://dx.doi.org/10.1002/9781118511305.ch18

[5] A. Ezell, "Explaining international it application leadership: Intelligent transportation systems," University of Zurich, Department of Informatics, *Tech. Rep.*, Jan 2010.

[6] T. L. Willke, P. Tientrakool, and N. Maxemchuk, "A survey of inter-vehicle communication protocols and their applications," *IEEE Communications Surveys and Tutorials*, vol. 11, pp. 3–20, 2009.

[7] EU. (2006–2010) Cooperative vehicle-infrastructure systems (cvis) eu fp7 project @ONLINE. [Online]. Available: http://www.cvisproject.org

[8] ——. (2006–2010) Cooperative systems (coopers) for intelligent road safety eu fp6 project @ONLINE. [Online]. Available: http://www.coopers-ip.eu/

[9] ——. (2006–2010) Pre-drivec2x eu fp7 project @ONLINE. [Online]. Available: http://www.pre-drive-c2x.eu/

[10] C. Cseh, "Architecture of the dedicated short-range communications (dsrc) protocol," in *48th IEEE Vehicular Technology Conference, 1998.*, vol. 3, May 1998, pp. 2095–2099 vol.3.

[11] J. Zhu and S. Roy, "MAC for dedicated short range communications in intelligent transport system," *Communications Magazine, IEEE*, vol. 41, no. 12, pp. 60–67, Dec 2003.

[12] L. Cheng, B. Henty, D. Stancil, F. Bai, and P. Mudalige, "Mobile vehicle-to-vehicle narrowband channel measurement and characterization of the 5.9 GHz dedicated short range communication (DSRC) frequency band," *Selected Areas in Communications, IEEE Journal on*, vol. 25, no. 8, pp. 1501–1516, Oct 2007.

[13] J. Kenney, "Dedicated short-range communications (DSRC) standards in the united states," *Proceedings of the IEEE*, vol. 99, no. 7, pp. 1162–1182, July 2011.

[14] "Ieee standard for information technology—local and metropolitan area networks—specific requirements– part 11: Wireless lan medium access control (mac) and physical layer (phy) specifications amendment 6: Wireless access in vehicular environments," *IEEE Std 802.11p-2010 (Amendment to IEEE Std 802.11-2007 as amended by IEEE Std 802.11k-2008, IEEE Std 802.11r-2008, IEEE Std 802.11y-2008, IEEE Std 802.11n-2009, and IEEE Std 802.11w-2009)*, pp. 1–51, July 2010.

[15] D. Jiang and L. Delgrossi, "Ieee 802.11p: Towards an international standard for wireless access in vehicular environments," in *IEEE Vehicular Technology Conference (VTC), Spring 2008.*, May 2008, pp. 2036–2040.

[16] E.-I. Transport. (2014, Dec) Intelligent transport systems @ONLINE. [Online]. Available: http://www.etsi.org/technologies-clusters/technologies/intelligent-transport

[17] M. Di Felice, R. Doost-Mohammady, K. Chowdhury, and L. Bononi, "Smart radios for smart vehicles: Cognitive vehicular networks," *Vehicular Technology Magazine, IEEE*, vol. 7, no. 2, pp. 26–33, June 2012.

[18] K. Fawaz, A. Ghandour, M. Olleik, and H. Artail, "Improving reliability of safety applications in vehicle ad hoc networks through the implementation of a cognitive network," in *Telecommunications (ICT), 2010 IEEE 17th International Conference on*, April 2010, pp. 798–805.

[19] I. F. Akyildiz, W.-Y. Lee, M. C. Vuran, and S. Mohanty, "Next generation/dynamic spectrum access/cognitive radio wireless networks: A survey," *Comput. Netw.*, vol. 50, no. 13, pp. 2127–2159, Sep. 2006.

[20] "Ieee draft standard for information technology—telecommunications and information exchange between systems wireless regional area networks (WRAN)—specific requirements part 22: Cognitive wireless ran medium access control (MAC) and physical layer (PHY) specifications: Policies and procedures for operation in the tv bands amendment: Enhancement for broadband services and monitoring applications," *IEEE P802.22b/D5, April 2015*, pp. 1–312, May 2015.

[21] "Ieee standard for local and metropolitan area networks part 16: Air interface for broadband wireless access systems amendment 2: Improved coexistence mechanisms for license-exempt operation," *IEEE Std 802.16h-2010 (Amendment to IEEE Std 802.16-2009)*, pp. 1–223, July 2010.

[22] I. F. Akyildiz, B. F. Lo, and R. Balakrishnan, "Cooperative spectrum sensing in cognitive radio networks: A survey," *Phys. Commun.*, vol. 4, no. 1, pp. 40–62, March 2011.

[23] C. Cormio and K. R. Chowdhury, "A survey on {MAC} protocols for cognitive radio networks," *Ad Hoc Networks*, vol. 7, no. 7, pp. 1315–1329, 2009.

[24] W. Yang, D. Ban, W. Liang, and W. Dou, "A genetic algorithm for joint resource allocation in cooperative cognitive radio networks," in *Wireless Communications and Mobile Computing Conference (IWCMC), 2011 7th International*, July 2011, pp. 167–172.

[25] K. Chowdhury, R. Doost-Mohammady, W. Meleis, M. Di Felice, and L. Bononi, "Cooperation and communication in cognitive radio networks based on tv spectrum experiments," in *World of Wireless, Mobile and Multimedia Networks (WoWMoM), 2011 IEEE International Symposium on a*, June 2011, pp. 1–9.

[26] C. Clancy, J. Hecker, E. Stuntebeck, and T. O'Shea, "Applications of machine learning to cognitive radio networks," *Wireless Communications, IEEE*, vol. 14, no. 4, pp. 47–52, August 2007.

[27] C. Guo, T. Peng, S. Xu, H. Wang, and W. Wang, "Cooperative spectrum sensing with cluster-based architecture in cognitive radio networks," in *Vehicular Technology Conference, 2009. VTC Spring 2009. IEEE 69th*, April 2009, pp. 1–5.

[28] D. Li and J. Gross, "Robust clustering of ad-hoc cognitive radio networks under opportunistic spectrum access," in *Communications (ICC), 2011 IEEE International Conference on*, June 2011, pp. 1–6.

[29] A. Alsarhan and A. Agarwal, "Cluster-based spectrum management using cognitive radios in wireless mesh network," in *Computer Communications and Networks, 2009. ICCCN 2009. Proceedings of 18th Internatonal Conference on*, Aug 2009, pp. 1–6.

[30] L. BONONI, M. DI FELICE, and S. PIZZI, "Dba-mac: Dynamic backbone-assisted medium access control protocol for efficient broadcast in vanets," *Journal of Interconnection Networks*, vol. 10, no. 04, pp. 321–344, 2009.

[31] Z. Zhang, "Routing in intermittently connected mobile ad hoc networks and delay tolerant networks: Overview and challenges," *Communications Surveys & Tutorials, IEEE*, vol. 8, no. 1, pp. 24–37, 2006.

[32] L. Lazos, S. Liu, and M. Krunz, "Spectrum opportunity-based control channel assignment in cognitive radio networks," in *Sensor, Mesh and Ad Hoc Communications and Networks, 2009. SECON '09. 6th Annual IEEE Communications Society Conference on*, June 2009, pp. 1–9.

[33] A. Anandkumar, N. Michael, and A. Tang, "Opportunistic spectrum access with multiple users: Learning under competition," in *INFOCOM, 2010 Proceedings IEEE*, March 2010, pp. 1–9.

[34] M. Di Felice, A. Ghandhour, H. Artail, and L. Bononi, "Integrating spectrum database and cooperative sensing for cognitive vehicular networks," in *Vehicular Technology Conference (VTC Fall), 2013 IEEE 78th*, Sept 2013, pp. 1–7.

[35] C. C. C. Consortium. (2007) Car 2 car communication consortium manifesto @ONLINE. [Online]. Available: http://www.car-to-car.org

[36] K. Hassan, R. Gautier, I. Dayoub, E. Radoi, and M. Berbineau, "Non-parametric multiple-antenna blind spectrum sensing by predicted eigenvalue threshold," in *Communications (ICC), 2012 IEEE International Conference on*, June 2012, pp. 1634–1629.

[37] S. Kharbech, I. Dayoub, E. Simon, and M. Zwingelstein-Colin, "Blind digital modulation detector for MIMO systems over high-speed railway channels," in *Communication Technologies for Vehicles*, ser. Lecture Notes in Computer Science, M. Berbineau, M. Jonsson, J.-M. Bonnin, S. Cherkaoui, M. Aguado, C. Rico-Garcia, H. Ghannoum, R. Mehmood, and A. Vinel, eds. Springer Berlin Heidelberg, 2013, vol. 7865, pp. 232–241. [Online]. Available: http://dx.doi.org/10.1007/978-3-642-37974-1_19

[38] K. Hassan, I. Dayoub, W. Hamouda, and M. Berbineau, "Automatic modulation recognition using wavelet transform and neural networks in wireless systems," *EURASIP Journal on Advances in Signal Processing*, vol. 2010, p. 42, 2010.

[39] K. Hassan, I. Dayoub, W. Hamouda, C. Nzeza, and M. Berbineau, "Blind digital modulation identification for spatially-correlated MIMO systems," *Wireless Communications, IEEE Transactions on*, vol. 11, no. 2, pp. 683–693, February 2012.

[40] E. Simon and M. Khalighi, "Iterative soft-kalman channel estimation for fast time-varying MIMO-OFDM channels," *Wireless Communications Letters, IEEE*, vol. 2, no. 6, pp. 599–602, December 2013.

[41] S. Pagadarai, A. M. Wyglinski, and R. Vuyyuru, "Characterization of vacant uhf tv channels for vehicular dynamic spectrum access," in *Vehicular Networking Conference (VNC), 2009 IEEE*, Oct 2009, pp. 1–8.

[42] M. Di Felice, K. Chowdhury, and L. Bononi, "Cooperative spectrum management in cognitive vehicular ad hoc networks," in *IEEE Vehicular Networking Conference (VNC)*, Nov. 2011, pp. 47–54.

[43] X. Y. Wang and P.-H. Ho, "A novel sensing coordination framework for cr-vanets," *Vehicular Technology, IEEE Transactions on*, vol. 59, no. 4, pp. 1936–1948, May 2010.

[44] A. J. Ghandour, K. Fawaz, H. Artail, M. Di. Felice, and L. Bononi, "Improving vehicular safety message delivery through the implementation of a cognitive vehicular network," *Ad Hoc Networks*, vol. 11, no. 8, pp. 2408 – 2422, 2013.

[45] H. Li and D. Irick, "Collaborative spectrum sensing in cognitive radio vehicular ad hoc networks: Belief propagation on highway," in *IEEE Vehicular Technology Conference (VTC-Spring)*, May 2010, pp. 1–5.

[46] A. Al Daoud, M. Alanyali, and D. Starobinski, "Secondary pricing of spectrum in cellular cdma networks," in *New Frontiers in Dynamic Spectrum Access Networks, 2007. DySPAN 2007. 2nd IEEE International Symposium on.* IEEE, 2007, pp. 535–542.

[47] C.-T. Chou, N. Sai Shankar, H. Kim, and K. G. Shin, "What and how much to gain by spectrum agility?" *IEEE J. Sel. A. Commun.*, vol. 25, no. 3, pp. 576–588, Apr. 2007. [Online]. Available: http://dx.doi.org/10.1109/JSAC.2007.070408

[48] H. Kim and K. G. Shin, "Efficient discovery of spectrum opportunities with MAC-layer sensing in cognitive radio networks," *IEEE Transactions on Mobile Computing*, vol. 7, no. 5, pp. 533–545, May 2008. [Online]. Available: http://dx.doi.org/10.1109/TMC.2007.70751

[49] H. Kim and K. Shin, "Fast discovery of spectrum opportunities in cognitive radio networks," in *New Frontiers in Dynamic Spectrum Access Networks, 2008. DySPAN 2008. 3rd IEEE Symposium on*, Oct 2008, pp. 1–12.

[50] W.-Y. Lee and I. F. Akyildiz, "Optimal spectrum sensing framework for cognitive radio networks," *Trans. Wireless. Comm.*, vol. 7, no. 10, pp. 3845–3857, Oct. 2008.

[51] Q. Zhao, L. Tong, A. Swami, and Y. Chen, "Decentralized cognitive MAC for opportunistic spectrum access in ad hoc networks: A pomdp framework," *IEEE J. Sel. A. Commun.*, vol. 25, no. 3, pp. 589–600, Apr 2007. [Online]. Available: http://dx.doi.org/10.1109/JSAC.2007.070409

[52] K. Sriram and W. Whitt, "Characterizing superposition arrival processes in packet multiplexers for voice and data," *IEEE J.Sel. A. Commun.*, vol. 4, no. 6, pp. 833–846, Sep. 2006. [Online]. Available: http://dx.doi.org/10.1109/JSAC.1986.1146402

[53] D. Willkomm, S. Machiraju, J. Bolot, and A. Wolisz, "Primary users in cellular networks: A large-scale measurement study," in *New Frontiers in Dynamic Spectrum Access Networks, 2008. DySPAN 2008. 3rd IEEE Symposium on*, Oct 2008, pp. 1–11.

[54] S. Geirhofer, L. Tong, and B. M. Sadler, "Cognitive radios for dynamic spectrum access – dynamic spectrum access in the time domain: Modeling and exploiting white space," *Comm. Mag.*, vol. 45, no. 5, pp. 66–72, May 2007. [Online]. Available: http://dx.doi.org/10.1109/MCOM.2007.358851

[55] K. Tsukamoto, Y. Omori, O. Altintas, M. Tsuru, and Y. Oie, "On spatially-aware channel selection in dynamic spectrum access multi-hop inter-vehicle communications," in *Vehicular Technology Conference Fall (VTC 2009-Fall), 2009 IEEE 70th*, Sept 2009, pp. 1–7.

[56] R. W. Brodersen, A. Wolisz, D. Cabric, and S. M. M. (UC Berkeley, "Corvus: A cognitive radio approach for usage of virtual unlicensed spectrum," 2004.

[57] M. Cesana, F. Cuomo, and E. Ekici, "Routing in cognitive radio networks: Challenges and solutions," *Ad Hoc Networks*, vol. 9, no. 3, pp. 228 – 248, 2011. [Online]. Available: http://www.sciencedirect.com/science/article/pii/S157087051000079X

[58] F. .-. FCC, "Unlicensed Operation in the TV Broadcast Bands in ET Docket No. 04-186," Nov. 2008. [Online]. Available: https://apps.fcc.gov/edocs_public/attachmatch/FCC-08-260A1.pdf

[59] H. Khalif, N. Malouch, and S. Fdida, "Multihop cognitive radio networks: To route or not to route," *IEEE Network*, vol. 23, no. 4, pp. 20–25, 2009.

[60] W. Kim, S. Oh, M. Gerla, and K. Lee, "Coroute: A new cognitive anypath vehicular routing protocol," in *Wireless Communications and Mobile Computing Conference (IWCMC), 2011 7th International*, July 2011, pp. 766–771.

[61] M. Youssef, M. Ibrahim, M. Abdelatif, L. Chen, and A. Vasilakos, "Routing metrics of cognitive radio networks: A survey," *Communications Surveys Tutorials, IEEE*, vol. 16, no. 1, pp. 92–109, First 2014.

[62] H. Ma, L. Zheng, X. Ma, and Y. luo, "Spectrum aware routing for multi-hop cognitive radio networks with a single transceiver," in *Cognitive Radio Oriented Wireless Networks and Communications, 2008. CrownCom 2008. 3rd International Conference on*, May 2008, pp. 1–6.

[63] G.-M. Zhu, I. Akyildiz, and G.-S. Kuo, "Stod-rp: A spectrum-tree based on-demand routing protocol for multi-hop cognitive radio networks," in *Global Telecommunications Conference, 2008. IEEE GLOBECOM 2008. IEEE*, Nov 2008, pp. 1–5.

[64] G. Cheng, W. Liu, Y. Li, and W. Cheng, "Joint on-demand routing and spectrum assignment in cognitive radio networks," in *Communications, 2007. ICC '07. IEEE International Conference on*, June 2007, pp. 6499–6503.

[65] H.-P. Shiang and M. van der Schaar, "Distributed resource management in multihop cognitive radio networks for delay-sensitive transmission," *IEEE Transactions on Vehicular Technology*, vol. 58, no. 2, pp. 941–953, Feb 2009.

[66] K. Chowdhury and M. Di. Felice, "Search: A routing protocol for mobile cognitive radio ad-hoc networks," *Computer Communications*, vol. 32, no. 18, pp. 1983–1997, 2009.

[67] B. Karp and H. T. Kung, "GPSR: Greedy perimeter stateless routing for wireless networks," in *Proceedings of the 6th Annual International Conference on Mobile Computing and Networking*, ser. MobiCom '00, 2000, pp. 243–254.

[68] I. Pefkianakis, S. Wong, and S. Lu, "Samer: Spectrum aware mesh routing in cognitive radio networks," in *New Frontiers in Dynamic Spectrum Access Networks, 2008. DySPAN 2008. 3rd IEEE Symposium on*, Oct 2008, pp. 1–5.

[69] F. Tang, L. Barolli, and J. Li, "A joint design for distributed stable routing and channel assignment over multihop and multiflow mobile ad hoc cognitive networks," *Industrial Informatics, IEEE Transactions on*, vol. 10, no. 2, pp. 1606–1615, May 2014.

[70] C.-I. Badoi, V. Croitoru, and R. Prasad, "Ipsag: An IP spectrum aware geographic routing algorithm proposal for multi-hop cognitive radio networks," in *Communications (COMM), 2010 8th International Conference on*, June 2010, pp. 491–496.

[71] R. Laufer and L. Kleinrock, "Multirate anypath routing in wireless mesh networks," in *in INFOCOM 2009, IEEE.*, 2009, pp. 37–45.

[72] J. Kim and M. Krunz, "Spectrum-aware beaconless geographical routing protocol for mobile cognitive radio networks," in *Global Telecommunications Conference (GLOBECOM 2011), 2011 IEEE*, Dec 2011, pp. 1–5.

[73] X. Huang, D. Lu, P. Li, and Y. Fang, "Coolest path: Spectrum mobility aware routing metrics in cognitive ad hoc networks," in *Distributed Computing Systems (ICDCS), 2011 31st International Conference on*, June 2011, pp. 182–191.

[74] H. Khalife, S. Ahuja, N. Malouch, and M. Krunz, "Probabilistic path selection in opportunistic cognitive radio networks," in *Global Telecommunications Conference, 2008. IEEE GLOBECOM 2008. IEEE*, Nov 2008, pp. 1–5.

[75] H. Bany Salameh, M. Krunz, and O. Younis, "MAC protocol for opportunistic cognitive radio networks with soft guarantees," *IEEE Transactions on Mobile Computing*, vol. 8, no. 10, pp. 1339–1352, Oct. 2009.

[76] X. Wang, T. T. Kwon, and Y. Choi, "A multipath routing and spectrum access (mrsa) framework for cognitive radio systems in multi-radio mesh networks," in *Proceedings of the 2009 ACM Workshop on Cognitive Radio Networks*, ser. CoRoNet '09, 2009, pp. 55–60.

[77] A. Sampath, L. Yang, L. Cao, H. Zheng, and B. Y. Zhao, "High throughput spectrum-aware routing for cognitive radio networks," *Proc. of IEEE Crowncom*, 2008.

[78] I. Beltagy, M. Youssef, and M. El-Derini, "A new routing metric and protocol for multipath routing in cognitive networks," in *Wireless Communications and Networking Conference (WCNC), 2011 IEEE*, March 2011, pp. 974–979.

[79] M. Karmoose, K. Habak, M. El Nainay, and M. Youssef, "Dead zone penetration protocol for cognitive radio networks," in *Wireless and Mobile Computing, Networking and Communications (WiMob), 2013 IEEE 9th International Conference on*, Oct 2013, pp. 529–536.

[80] Y. Shi and Y. Hou, "A distributed optimization algorithm for multi-hop cognitive radio networks," in *INFOCOM 2008. The 27th Conference on Computer Communications. IEEE*, April 2008, pp. –.

[81] J. Liu, P. Ren, S. Xue, and H. Chen, "Expected path duration maximized routing algorithm in CR-VANETS," in *Communications in China (ICCC), 2012 1st IEEE International Conference on*, Aug 2012, pp. 659–663.

[82] F. Bai, N. Sadagopan, and A. Helmy, "Important: A framework to systematically analyze the impact of mobility on performance of routing protocols for adhoc networks," in *INFOCOM 2003. Twenty-Second Annual Joint Conference of the IEEE Computer and Communications. IEEE Societies*, vol. 2, March 2003, pp. 825–835 vol.2.

[83] T. Stephan and K. Karuppanan, "Cognitive inspired optimal routing of olsr in vanet," in *Recent Trends in Information Technology (ICRTIT), 2013 International Conference on*, July 2013, pp. 283–289.

[84] "Cognitive radio cognitive network simulator," http://faculty.uml.edu/Tricia_Chigan/Research/CRCN_Simulator.htm.

[85] V. Esmaeelzadeh, R. Berangi, S. Sebt, E. Hosseini, and M. Parsinia, "Cogns: A simulation framework for cognitive radio networks," *Wireless Personal Communications*, vol. 72, no. 4, pp. 2849–2865, 2013. [Online]. Available: http://dx.doi.org/10.1007/s11277-013-1184-y

[86] "Omnet++ network simulation framework," http://www.omnetpp.org/.

[87] J. Marinho and E. Monteiro, "Cognitive radio simulation based on omnet++/mixim," in *11ł Conferłncia sobre Redes de Computadores, CRC 2011*, 2011.

[88] M. Zhan, P. Ren, and M. Gong, "An open software simulation platform for cognitive radio," in *Wireless Communications Networking and Mobile Computing (WiCOM), 2010 6th International Conference on*, Sept 2010, pp. 1–4.

[89] M. D. Felice, K. R. Chowdhury, W. Kim, A. Kassler, and L. Bononi, "End-to-end protocols for cognitive radio ad hoc networks: An evaluation study," *Performance Evaluation*, vol. 68, no. 9, pp. 859 – 875, 2011, special issue: *Advances in Wireless and Mobile Networks*.

[90] P. Kyasanur and N. H. Vaidya, "Routing and link-layer protocols for multi-channel multi-interface ad hoc wireless networks," *SIGMOBILE Mob. Comput. Commun. Rev.*, vol. 10, no. 1, pp. 31–43, Jan. 2006. [Online]. Available: http://doi.acm.org/10.1145/1119759.1119762

[91] T. Rappaport, *Wireless Communications: Principles and Practice*, 2nd ed. Upper Saddle River, NJ, USA: Prentice Hall PTR, 2001.

[92] T. Henderson, M. Lacage, and G. Riley, "The ns-3 network simulator," *Software package retrieved from http://www. nsnam. org*, 2011.

[93] A. Al-Ali and K. Chowdhury, "Simulating dynamic spectrum access using ns-3 for wireless networks in smart environments," in *Sensing, Communication, and Networking Workshops (SECON Workshops), 2014 Eleventh Annual IEEE International Conference on*, June 2014, pp. 28–33.

[94] "Network simulator version 2," http://www.isi.edu/nsnam/ns/.

[95] "Doxygen," http://www.stack.nl/~dimitri/doxygen/.

[96] R. Marler and J. Arora, "Survey of multi-objective optimization methods for engineering," *Structural and Multidisciplinary Optimization*, vol. 26, no. 6, pp. 369–395, 2004. [Online]. Available: http://dx.doi.org/10.1007/s00158-003-0368-6.

[97] N. Meghanathan, "A survey on the communication protocols and security in cognitive radio networks," *International Journal of Communication Networks and Information Security (IJC-NIS)*, vol. 5, no. 1, 2013.

Chapter 12

The Spectrum Sharing Games

Pu Yuan, Yong Xiao, Guoan Bi, and Shan Luo

CONTENTS

12.1 Introduction

12.1.1 Motivation of Game Theory Based Spectrum Sharing

Spectrum sharing is a promising technology to solve the spectrum under-utilization problem for the next generation wireless networking systems. Currently, the optimization of the spectrum sharing system is mainly limited by the following facts: (1) centralized optimization approach for the spectrum sharing network requires collecting the information throughout the entire network, which is inefficient due to the computational and communication complexity; (2) distributed multi-agent optimization problem is notoriously difficult, and there is still a lack of a simple and general framework that can be applied to solve the distributed optimization problem; (3) the spectrum sharing network can be dynamic and time-varying. Simply applying the empirical model or pre-defined policies cannot always guarantee the optimal performance. Most existing works assume the network topology is fixed and the behavior of the wireless device cannot adapt to the changing radio environment.

Developing simple and distributed optimization framework is very important for the spectrum sharing-based wireless systems. Game theory is a set of mathematical tools that have been introduced to model and analyze the decision-making problem between interactive players who may have conflicts of interest. It has been shown to be an ideal tool for investigating the interactions of the autonomous players in spectrum sharing networks.

More specifically, using the game theoretical tool, mobile nodes in a spectrum sharing network can be modeled as the players in a game. By applying different game models, we can analyze and predict the potential outcome of the players for different spectrum sharing networking systems. In this approach, choosing the most appropriate game theoretical model to analyze each specific spectrum sharing network is an important problem.

12.1.2 Spectrum Sharing Types

By allowing the mobile device to flexibly adapt its operation to the surrounding environment, dynamic spectrum sharing has the potential to improve the spectral efficiency in wireless communication networks. The spectrum are not assigned to a fixed mobile device but are opportunistically reused by multiple mobile devices, hence, the spectrum under-utilization is avoided. The main challenge for this network is that each mobile device needs to intelligently decide how to compete for the limited wireless resources (e.g., time, spectrum, and space) without causing intolerably harmful performance degradation for other devices.

For example, in temporal spectrum sharing (also called spectrum underlay, dynamic/opportunistic spectrum access [25]) networks, mobile devices are allowed to send signals in the spectrum when it is idle. In this network, each mobile device needs to continuously sense the availability of the vacant sub-bands. Based on the sensing result, mobile devices make a binary decision on the presence of the other users. In other words, temporal spectrum sharing allows the same spectrum to be shared by different users during different time.

In contrast, the spatial spectrum sharing [18] [21] (or spectrum overlay, dynamic spectrum sharing [25]) allows the mobile devices to tolerate a small increase of interference power. In this system, the mobile devices can transmit signals at the same time but are required to control their transmit powers to ensure that the resulting interference power at each mobile device is below an acceptable level [7]. As the transmission of mobile devices in spatial spectrum sharing network occurs simultaneously, the game theory can be applied to optimization of the sub-band allocation and power control problems in spatial spectrum sharing networks.

In this chapter, we mainly focus on the spatial spectrum sharing network and, to simplify our description, in the rest of this chapter, we refer to the spatial spectrum sharing network as spectrum sharing network.

---

[0]School of Electrical and Electronic Engineering, Nanyang Technological University, Singapore.

12.1.3 Application of Game Theory in Spectrum Sharing

As described in Chapter 1, most of the game theoretical models can be classified into two types: non-cooperative game and cooperative game. In the non-cooperative game, the players are selfish and only interested in improving its own profit. In the cooperative game, each mobile device can form groups with other devices and cooperate with other group members to further improve their profits. In the rest of this section, we give detailed description for the existing works applying the above two types of games to analyze the spectrum sharing networks.

12.1.3.1 Application of Non-cooperative game

In a non-cooperative game-based spectrum sharing network, one of the most common solution concepts used to analyze users' interaction is the Nash equilibrium (NE).

In [17], the authors point out that the efficiency of the NE can be degraded by the competition among autonomous players. Different approaches that can help to improve the performance of players have been investigated. A variety of non-cooperative game approaches for distributed interference control have been proposed to solve the interference management problems [4, 8].

Using the Stackelberg game model to handle the interference control problem was considered in [1]. It assumes that the operator can charge a price for each user accessing each sub-band and can use the value of the price to regulate the received power at the base station (BS) in a code division multiple access (CDMA) network. The author designed a mechanism that can minimize the information exchange between the BS and the mobile nodes. A similar game theoretical model has been applied in [9] to study a femto-cell network where the licensed subscribers (LS) and unlicensed subscribers (ULS) shared the common spectrum. They formulated a non-cooperative game to analyze the distributed interference control problem and proposed two different pricing schemes to discuss the impact to the pay-offs using different pricing schemes. In [18], the authors considered the setup that the spectrum is divided into sub-bands, and they proposed a non-cooperative game model to enable the ULS join the sub-bands sequentially, while the interference to the LS was controlled by the price charged by the operator.

12.1.3.2 Application of cooperative game

It is known that exploring the benefits gained by cooperation among the players may improve the performance of the wireless communication systems. In a cooperative game, the players can jointly coordinate their behavior to improve the overall performance of the networking system. A widely applied cooperative game model in the study of wireless communications is the coalitional game.

The coalitional game is usually utilized to investigate the cooperative behaviors and interactions among the users in the wireless communication systems, where the mobile subscribers seek to form coalitions in case this can provide mutual benefits, compared to acting alone. In [13], three kinds of coalitional games and their applications in wireless communications have been summarized. They pointed to the potential of these games in modeling the wireless communication problems. Recently, the coalition formation game has been applied to analyze interaction among cooperative users in the spectrum sharing network. In [6], the coalition formation game had been used to model the cooperation between the mobile nodes at different locations in a cell. The authors proposed a mechanism that the mobile nodes located near the BS can help to improve the (QoS) of the mobile nodes at the cell boundary. Coalition was formed between them when the overall performance of the network was improved. In [11], the rate allocation problems for Gaussian multiple access channels was investigated using coalitional game model. In [19], the authors considered a coalition formation game among the secondary users in the cognitive radio (CR) network. The ULSs formed disjoint coalitions to cooperatively utilize the spectrum. Together with the Stackelberg game between the ULs and ULSs, the authors provided a hierarchical game framework toward the solution to jointly optimize the resource allocation problems in CR networks.

Although the coalitional game has been widely used to study the problems in wireless communications, most of the game model only considers forming disjoint coalitions. In other words, denoting C_j as a coalition, then C_1 and C_2 are disjoint coalitions if $C_1 \cap C_2 = \emptyset$. In contrast, C_1 and C_2 are overlapping coalitions if $C_1 \cap C_2 \neq \emptyset$. In practical communication systems, enabling the overlapping of coalitions may further improve the performance. For example, one user D_k forms a coalition with D_j to cooperatively transmit in sub-band m. If D_k still has spare power, it may cooperative with D_i on the sub-band l to support more data rate. However, so far, only limited works have been reported for overlapping coalitional games [20, 23, 24]. [24], the authors studied how small cell BSs coordinate with each other to achieve efficient transmission. By allowing the small cells to form overlapping coalitions to jointly schedule the transmission of their subscribers, they found that the performance of mobile nodes in the cell edge was improved. In [23], the authors developed a hierarchical game theoretical framework to jointly optimize the power and sub-band allocations of mobile devices in a spectrum sharing based heterogeneous networks (HetNets). The hierarchical game is established by integrating an overlapping coalition formation (OCF) game which models the cooperative behaviors of the unlicensed subscribers into a Stackelberg based power control game to protect the licensed subscribers. They proved that the OCF-game are 2^K-finite and subsequently the existence of core, which makes the optimal sub-band allocation possible.

12.2 Non-cooperative Spectrum Sharing Game

The non-cooperative game has been widely adopted to analyze the spectrum sharing in wireless network. In this section, we investigate the power control problem using a non-cooperative game which serves as an example to demonstrate how to apply game theoretical tools to study the spectrum sharing networking systems.

12.2.1 Network Model

Consider a multiple access channel with M sub-bands which can be accessed by a set of N mobile users denoted as $S = \{S_1, S_2, \ldots, S_N\}$. We assume each user can access multiple sub-bands to maximize the data rate. Each sub-band can be reused by multiple users to further improve the overall network capacity using spectrum sharing. We assume users are selfish, and each user is only interested in maximizing its own profit.

The transmit power of each user should satisfy the following power constraint,

$$\sum_{m=1}^{M} p_{S_k}^m \leq \overline{p}, \tag{12.1}$$

where $p_{S_k}^m$ is the transmit power of user S_k on sub-band m, and $\overline{p}$ is the total power. We assume that all users have the same total transmit power constraint.

Each user can access arbitrary sub-bands and simultaneously transmit over multiple sub-bands, it is important for each user to properly allocate the total transmit power among different sub-bands.

We seek a stable solution among all the users, once this solution is reached by all the users, no user has the incentive to unilaterally deviate from it. This solution corresponds to the NE if we formulate the above problem as a non-cooperative game. We formally define the NE for the above model as follows:

Definition 12.1 *The power allocation matrix* $\boldsymbol{P}^* = [\boldsymbol{p}_{S_1}^*, \boldsymbol{p}_{S_2}^*, \ldots, \boldsymbol{p}_{S_K}^*]$ *is an NE if the constraint in (12.1) is satisfied and the optimal transmit power* $\boldsymbol{p}_{S_k}^*$ *of each user* $S_k \in S$ *is given by*

$$\boldsymbol{p}_{S_k}^* = \arg \max_{p_{S_k}^m \geq 0} \pi_{S_k}(\boldsymbol{p}_{S_k}, \boldsymbol{p}_{-S_k}^*), \forall S_k \in S. \tag{12.2}$$

12.2.2 Pay-off definition

We define the pay-off function of each user S_k as

$$\pi_{S_k} = \sum_{m=1}^{M} \log\left(1 + \gamma_{S_k}^m\right). \tag{12.3}$$

where $\gamma_{S_k}^m$ is the signal to interference and noise ratio (SINR) of the multiple access channel consisting of the additive white noise and the interference come from co-channel users, i.e., we can write $\gamma_{S_k}^m$ as,

$$\gamma_{S_k}^m = \frac{g_{S_k}^m p_{S_k}^m}{\sigma^2 + \sum_{l=1, l \neq k}^{K} g_{S_l, k}^m p_{S_l}^m}, \tag{12.4}$$

where $g_{S_l, k}$ is the channel gain between user S_l and S_k.

Each user S_k tries to solve the following optimization problem:

$$\max \ \pi_{S_k}\left(p_{S_k}, P_{-S_k}\right) \tag{12.5}$$

$$\text{s.t.} \sum_{m=1}^{M} p_{S_k}^m \leq \overline{p}, \tag{12.6}$$

$$p_{S_k}^m \geq 0. \tag{12.7}$$

The above problem is a maximization problem with linear inequality constraints that can be solved using the existing optimization tools.

12.2.3 Power Control for Multi-user Spectrum Sharing Networks

Exhaustive searching for all the possible power control strategies among all users is mathematically intractable. Fortunately, low-cost sub-optimal solutions have been proposed in [22] [14] [15]. These solutions can be regarded as variations of the classic iterative water-filling (IWF) algorithm. The IWF algorithm aims to solve the power allocation problem in multiple access channel by performing the water-filling procedure iteratively.

In fact, we can find that the setup in the iterative water-filling algorithm can be one to one correspondent to the elements in a non-cooperative game.

- The wireless users can be modeled as the players.

- The transmit power of each user can be modeled as the action of each player.

- The interference plus noise measured by each user can be regarded as the observation of each player about others' action

- The water-filling solution for the power allocation obtained by each user in each iteration corresponds to the best response function of each player

- The transmit rate achieved by each user corresponds to the pay-off obtained by each player

We can prove the following results about the pay-off optimization problem defined in (12.7).

Theorem 12.1 *In the proposed K users M sub-bands multiple access channel, $p_{S_k}^{m*}$ is an optimal solution to the pay-off sum maximization problem if and only if $p_{S_k}^{m*}$ is the water-filling solution vector given by*

$$p_{S_k}^{m*} = \left(\frac{1}{\beta_{S_k}} - \frac{\sigma^2 + g_{S_0, k}^m p_{S_0}^{m*} + \sum_{l=1, l \neq k}^{K} g_{S_l, k}^m p_{S_l}^{m*}}{h_{S_k}^m}\right)^+. \tag{12.8}$$

Proof 12.1

1) The if part can be obtained directly from the problem.

   The solution given in (12.8) is exactly the same as the single user scenario except an additive interference term is given by $g_{S_0,k}^m p_{S_0}^m + \sum_{l=1,l\neq k}^K g_{S_l,k}^m p_{S_l}^m$. Hence, if we consider the interference as noise, there are no correlation between the optimal power allocation of each user. Then problem is just a linear combination of a series of individual pay-off maximization problem, i.e., $\max_{\boldsymbol{P}} \sum_{k=1}^K \pi_{S_k} = \max_{\boldsymbol{p}_{S_1}} \pi_{S_1} + ,\dots + \max_{\boldsymbol{p}_{S_K}} \pi_{S_K}$. Therefore if each of the $\boldsymbol{p}_{S_k}$ optimizes π_{S_k}, then the collection of $\boldsymbol{p}_{S_1}, \boldsymbol{p}_{S_2}, \dots \boldsymbol{p}_{S_K}$ will optimize $\sum_{k=1}^K \pi_{S_k}$.

2) Let us prove the only if part. Suppose all the users allocate their power using (12.8), there exists a user S_k, and its power allocation $\boldsymbol{p}'_{S_k}$ is equal to the one-time water-filling solution of (12.3). Since it is the optimum point and all other users have fixed their power allocation, then the interference term in (12.3) becomes constant. Subsequently, we take the interference as noise and the optimal solution of problem (12.3) is given by $\boldsymbol{p}_{S_k}^{m*}$, which is the water-filling solution, and will satisfy $\boldsymbol{p}_{S_k}^{m*} > \boldsymbol{p}'_{S_k}$. Hence, it contradicts with our assumption. Therefore the power allocation of S_k should be the water-filling solution.

We rewrite the result of (12.8) as,

$$p_{S_k}^{m*} = \left(\frac{1}{\beta_{S_k}} - \frac{I_{S_k}^m}{g_{S_k}^m} \right)^+, \tag{12.9}$$

where $I_{S_k}^m = \sigma^2 + g_{S_0,k}^m p_{S_0}^m + \sum_{l=1,l\neq k}^K g_{S_l,k}^m p_{S_l}^m$ represents the interference plus noise. ■

12.2.4 Distributed Algorithm

Following the same line as the previous section, we have the following distributed power control algorithm for the spectrum sharing networks.

Algorithm 3 Two-Layer Iterative Water-Filling Algorithm

Consider a K-user system with M sub-bands. We denote the power cap constraint as $\overline{p}$. We denote ε, δ and η as small constants and $\boldsymbol{u}$ as a unit vector of length M.
Initialization:
$P = \overline{p}$.
WHILE $\|\boldsymbol{p}_i(t+1) - \boldsymbol{p}_i(t)\| \leq \varepsilon$
FOR $i = 1$ to N,
FOR $m = 1$ to K,
$N_{-i}^m(t) = \sum_{j=1,j\neq i}^N p_j^m(t-1)H_{ji}^m + \sigma^2$
$p_i^m(t) = WF(N_{-i}^m(t))$,
END.
If P does not converge in a limited iterations.
Set $\overline{p} = \overline{p} - \eta$
END.

The IWF algorithm contains two-layered loops, the inner loop is a is the power allocation algorithm, and the outer loop controls the amount of transmit power per-user. The convergence proof of the IWF algorithm is based on the contract-mapping theorem [2].

Lemma 12.1 *For any given fixed linear pricing function, the water-filling updating function* $WF(\boldsymbol{p}_{-i}, \mu^m)$ *converges to a fixed solution if the channel gain of the interfering channel gain* $g_{S_l,k}^m$, $l \neq k$, *is sufficiently weak compared with the signal channel gain* $g_{S_k,k}^m$. *More specifically, it is given by,*

$$\sum_{k=1, k \neq k\prime} \max_{m \in l_{S_k} \cap l_{k\prime}} \left\{ \frac{g_{S_k,k\prime}^m}{g_{S_k,k}^m} \right\} \leq 1, \forall k\prime \in \mathcal{N}, \tag{12.10}$$

$$\sum_{k\prime=1, k\prime \neq k} \max_{m \in l_{S_k} \cap l_{k\prime}} \left\{ \frac{g_{S_k,k\prime}^m}{g_{S_k,k}^m} \right\} \leq 1., \forall k \in \mathcal{N}. \tag{12.11}$$

Algorithm 3 can be implemented in a distributed manner. The only external information $h_{S_k}^m$ can be estimated using pilot signal or calculated using some empirical path loss model if real-time estimation is invalid.

12.3 Cooperative Spectrum Sharing Game

The cooperative game can be adopted to the wireless communications to overcome the inefficiency caused by competition. The key difference between the non-cooperative game and cooperative game is the distinct behavior of the players. In the non-cooperative game, the player acts selfishly as he only cares about maximizing his own profit. In cooperative game, the player's action follows certain agreement, and the goal of each individual player is to maximize the overall benefit of the player group in the agreement. In this section, we formulate a simple cooperative game to analyze the sub-band allocation problems for spectrum sharing networks.

12.3.1 Network Model

We consider a scenario that the spectrum is divided into M sub-bands and shared by K multiple users. Each user can choose multiple sub-bands to transmit signals simultaneously and multiple users can share the same sub-band. We utilize a game theoretical approach to map the sub-band allocation problem into an overlapping coalition formation game (OCF-game). In this game, each user has an amount of power resource to distribute in different sub-bands, and the achieved data rate depends on the parameters of the sub-bands and the transmit strategies of other users.

We first present formal definitions of the coalition and imputation.

Definition 12.2 ([12], chapter 9) *We denote the set of all players as* $\mathcal{K}$, *and we define a coalition* $\mathcal{C}$ *is a non-empty sub-set of* $\mathcal{K}$, *i.e.,* $\mathcal{C} \subseteq \mathcal{K}$. *Specially,* $\mathcal{K}$ *is referred as the grand coalition. A coalitional game is defined as* $(\mathcal{C}, v)$ *where* v *is the value function mapping a coalition structure* $\mathcal{C}$ *to a real value* $v(\mathcal{C})$. *If for any two disjoint coalitions* $\mathcal{C}_1$ *and* $\mathcal{C}_2$ *in a coalition game,* $\mathcal{C}_1 \cap \mathcal{C}_2 = \emptyset$, $\mathcal{C}_1 \text{and} \mathcal{C}_2 \subset \mathcal{K}$, *we have,*

$$v(\mathcal{C}_1 \cup \mathcal{C}_2) \geq v(\mathcal{C}_1) + v(\mathcal{C}_2), \tag{12.12}$$

then we say this game satisfies the super-additive property. Given two coalitions $\mathcal{C}_1$ *and* $\mathcal{C}_2$, *we say* $\mathcal{C}_1$ *and* $\mathcal{C}_2$ *overlaps if* $\mathcal{C}_1 \cap \mathcal{C}_2 \neq \emptyset$.

Definition 12.3 *We define an imputation as a pay-off vector satisfying both group and individual rationalities. A pay-off vector* $\boldsymbol{\pi}$ *is a division of the value* $v(\mathcal{C})$ *to all the coalition members, i.e.,* $\boldsymbol{\pi} = [\pi_{S_1}, \cdots, \pi_{S_K}]$. *We say* $\boldsymbol{\pi}$ *is group rational if* $\sum_{k=1}^{K} \pi_{S_k} = v(\mathcal{C})$ *and individual rational if* $\pi_{S_k} \geq v(\{S_k\}), \forall S_k \in \mathcal{C}$.

If comparing the above definition to which of the non-cooperative game, we observe that the evaluation about the outcome of the game is different. The outcome of the non-cooperative game is mapped to the value of individual player, while the outcome of the coalitional game are mapped to the value of the coalition.

An important property to judge the quality of the coalition is the stability. Obviously, a coalition that achieves poor value will not be stable, as the member player will seek chances to form other coalitions or at least form a single coalition. Hence, for a coalition that is stable, the value achieved by all the member players of the coalition should satisfy the super-additive condition. Obviously, only under the above condition it is possible for the coalition to provide its member player better pay-off than if it acts alone.

If a coalitional game satisfies the super-additive condition, then the grand coalition is formed, and the game focuses on finding the optimal imputation to form the grand coalition. However if the supper-additive condition does not hold, then the game focuses on finding optimal partition of the grand coalition. In this case, a core of the coalitional game is defined as a set of stable coalition formation structures in which no player can profitably deviate from them. This is different from the one defined in the coalitional games which is a set of imputations stabilizing the grand coalition.

On the other hand, a coalition with unfair pay-off division is also not stable. For example, if a member player in the coalition can not get a pay-off as good as if it acts alone, it would rather leave the coalition and form a single coalition. The pay-off division of coalition $\mathcal{C}$ is specified by the imputation.

We denote the set of sub-bands as $\mathcal{B}$ and the set of users as $\mathcal{K}$. The frequency selective fading is considered here, i.e., channel fading in different sub-bands is interdependent. We assume the channel state is time-invariant in each time block. The additive noise in each sub-band is assumed to be white Gaussian.

Each user can apply multiple sub-bands to transmit, i.e., the same portion of sub-bands can be reused by more than one users. Let g_{kj}^m be the channel gain between source node of S_k and destination node of S_jth. Let $\boldsymbol{p}_{S_k} = [p_{S_k}^1, ..., p_{S_k}^M]$ be the power allocation vector of S_k; note that $p_{S_k}^m = 0$ implies that sub-band m is not used by S_k.

On the other hand, multiple users can apply for the same sub-band at the same time. We denote the set of users utilizing the same sub-band m as $\mathcal{L}_m$, i.e., $\mathcal{L}_m = \{k : p_{S_k}^m > 0\}$. $\mathcal{L}_m = \emptyset$ means no user uses sub-band m, $\mathcal{L}_m = S_k$ means sub-band m is exclusively occupied by S_k, and $|\mathcal{L}_m| \geq 2$ means sub-band m has been shared by two or more users.

Here we consider the scenario that users can cooperatively transmit the signal with co-channel peers to improve their pay-offs. We consider the power cap as the physical limitation of the maximum transmit power of the mobile device

$$\sum_{m=1}^{M} p_{S_k}^m \leq \overline{p}, \tag{12.13}$$

where $p_{S_k}^m$ is the transmit power of S_k on sub-band m, and $\overline{p}$ is the power cap. This constraint specifies the maximum transmit power of each user, due to the hardware limitation and the battery life. Similar setup considering both the total power and per-band power constraints is investigated in [5].

The power constraints limit the number of sub-band accessed by each users. Hence, an important problem is how users can smartly form the overlapping coalitions to maximize their pay-off.

12.3.2 Cooperative Game Formulation

The proposed coalition formation game focuses on two questions: (1) how the coalition members coordinate with each other, and (2) how a coalition formation structure can be established among users.

To answer the first question, we introduce the virtual Multiple Input Multiple Output technique as the cooperation scheme among the users in the same coalition. The users in the same sub-band m form a coalition to transmit and receive signal cooperatively. Using the virtual MIMO technique, we can convert the communication within one coalition into a virtual $\mathcal{L}_m$-input $\mathcal{L}_m$-output channel. We choose the virtual MIMO as the cooperation strategy for two reasons: (1) it is shown to achieve the upper-bound of the rate for a multiple access channel [16], and (2) it is shown to satisfy the proportional fairness [19]. Therefore, we obtain that the capacity sum of all users in the virtual MIMO channel m as

$$\sum_{S_k \in \mathcal{L}_m} r_{S_k} = \sum_{S_k \in \mathcal{L}_m} \log\left(1 + \lambda_{S_k}^m p_{S_k}^m\right), \tag{12.14}$$

where $\lambda_{S_k}^m$ is the k_{th} non-zero eigenvalue of matrix $\boldsymbol{G}^T_{\{S_k \in \mathcal{L}_m\}} \boldsymbol{G}_{\{S_k \in \mathcal{L}_m\}}$, and $\boldsymbol{G}_{\{S_k \in \mathcal{L}_m\}}$ is the channel gain matrix of users in the same sub-band. For example, if $\{S_1, ..., S_n\}$ are in the same sub-band m, then the matrix is given by

$$\boldsymbol{G}_{\{S_k \in \mathcal{L}_m\}} = \begin{bmatrix} g_{11}^m & g_{12}^m & \cdots & g_{1n}^m \\ g_{21}^m & g_{22}^m & \cdots & g_{2n}^m \\ \cdot & \cdot & \cdot & \cdot \\ \cdot & \cdot & \cdot & \cdot \\ g_{n1}^m & g_{n2}^m & \cdots & g_{nn}^m \end{bmatrix}. \tag{12.15}$$

In above matrix, g_{jk}^m is the ratio of the channel gain between source node of user S_j and destination node of user S_k to the received interference power at S_k in sub-band m.

The main objective of proposed scheme is to solve the following problems:

(1) *Sub-band allocation problem*: We investigate the strategies of users for sub-band accessing.

(2) *Coalition formation problem*: We investigate how the users form overlapping coalitions to improve the data rate.

We formulate the cooperative spectrum sharing problem as an OCF-game, while the member players in the same coalition aim to maximize the sum rate of all members in the coalition:

$$\pi_{S_k}(\boldsymbol{p}_{S_k}) = r_{S_k}(\boldsymbol{p}_{S_k}). \tag{12.16}$$

Furthermore, since S_k can transmit in multiple sub-bands at the same time, it aims to maximize the sum of the pay-offs obtained from all the active sub-band, under the constraints given in (12.13).

The member users in the same coalition coordinate their transmission to improve the sum pay-off. Assuming that the overlapping coalition formation structure is fixed, i.e., each S_k already obtained a fixed λ_{S_k}, then each user S_k will obtain a pay-off defined by

$$\pi_{S_k}^m(p_{S_k}^m, \lambda_{S_k}^m) = \log\left(1 + \lambda_{S_k}^m p_{S_k}\right). \tag{12.17}$$

The optimal power allocation of S_k is obtained by solving the following maximization problem:

Problem 12.1

$$\max_{\boldsymbol{p}} \pi_{S_k}(\boldsymbol{p}_{S_k}, \boldsymbol{\lambda}_{S_k})$$

$$= \sum_{m=1}^{M} \left(\log\left(1 + \lambda_{S_k}^m p_{S_k}\right)\right), \tag{12.18}$$

$$S.t. \quad \sum_{m=1}^{M} p_{S_k}^m \leq \bar{p}.$$

Problem 12.1 can be easily solved using convex optimization method. Hence the optimal power is obtained as

$$p_{S_k}^{m*} = \arg\max_p \pi_{S_k}^m (p_{S_k}^m, \lambda_{S_k}^m), \tag{12.19}$$

$$= \left(\frac{1}{\beta_{S_k}} - \frac{1}{\lambda_{S_k}^m} \right)^+, \tag{12.20}$$

where β_{S_k} is the coefficient chosen to satisfy the constraint.

The corresponding sub-band allocation indicator is

$$\boldsymbol{l}_{S_k}^* = \{l_{S_k}^{m*}, m = 1, 2, ..., M.\},$$

$$l_{S_k}^{m*} = \begin{cases} 1, \text{if } \boldsymbol{p}_{S_k}^* > 0, \\ 0, \text{otherwise.} \end{cases} \tag{12.21}$$

From the results above, we see that the optimal solution of the transmit power in $\boldsymbol{p}_i$ only depends on $\lambda_{S_k}^m$, which are the key outcomes of the OCF-game. The pay-off division factor $\lambda_{S_k}^m$ is obtained from the coalitional game among users. We focus on finding optimal coalition formation structure, i.e., we investigate the optimal coalition partitioning of the grand coalition.

When overlapping is enabled among coalitions, the coalitions are no longer disjoint sub sets of the player set, as defined in a non-overlapping coalitional game. In the OCF-game, the concept *partial coalition* is utilized:

Definition 12.4 *The partial coalition is defined on a vector $\boldsymbol{p}^m = (p_{S_1}^m, ..., p_{S_K}^m)$, where $p_{S_k}^m$ is the fractional resource of S_k dedicated to coalition m. Note that $p_{S_k}^m = 0$ means S_k not being in this coalition. A coalition structure is a collection $\boldsymbol{P} = (\boldsymbol{p}^1, ..., \boldsymbol{p}^M)$ of partial coalitions.*

Remark 12.1 *In a non-overlapping coalition formation game, a coalition is just a subset of the player set. For a player set of size N, the number of coalition formation structures is given by the Bell number B_N, where $B_N = \sum_{k=0}^{N-1} \binom{N-1}{k} B_k$ is the possible number of coalition structures and B_k is the number of ways to partition the set into k items.*

We use the following example for illustration. Considering a player set $\{S_1, S_2\}$, the set of possible coalitions formed by this player set is given by $\{\{S_1\}, \{S_2\}, \{S_1, S_2\}\}$. Hence, the resulting coalition structure is $\{S_1, S_2\}$ or $\{\{S_1\}, \{S_2\}\}$. However, in the OCF-game, the concept of partial coalition not only shows who joins the coalition, but also indicates how many resources each player contributes to this coalition. If the resource is continuous, there are generally infinite an number of partial coalitions. For example, for players set $\{S_1, S_2\}$, the set of partial coalitions may be $\{\{0, 1\}, \{0.2, 0.3\}, \{1, 1\}, \{0.5, 0\}, ...\}$. It means that the concept of coalition is considered as a special case of the partial coalition, where each player joins only one coalition with all its resources.

Definition 12.5 *An OCF-game is denoted by $G = (\mathcal{K}, \mathcal{M}, \boldsymbol{P}, \boldsymbol{v})$, where*

- $\mathcal{K} = \{1, 2, ..., K\}$ *is the set of players.*

- $\mathcal{M} = \{1, 2, ..., M\}$ *is the set of sub-bands.*

- $\boldsymbol{P}$ *is the power allocation matrix, where the row $\boldsymbol{p}_{S_k} = (p_{S_k}^1, p_{S_k}^2, ..., p_{S_k}^M)$ represents how player S_k assigns its power on different sub-bands, and the column $\boldsymbol{p}^m = (p_{S_1}^m, p_{S_2}^m, ..., p_{S_K}^m)$ represents the power each player spends on sub-band m. $\boldsymbol{p}^m = (p_{S_1}^m, p_{S_2}^m, ..., p_{S_K}^m)$ also corresponds to a partial coalition.*

- $\boldsymbol{v}(\boldsymbol{C}^m) : \mathbb{R}^n \longrightarrow \mathbb{R}^+$ *is the value function, which represents the total pay-off of a partial coalition $\boldsymbol{C}^m$.*

Definition 12.6 *We define a game to be U-finite if for any coalition structure that arises in this game, the number of all possible partial coalitions is bounded by U.*

The sum rate achieved by forming a coalition is given by (12.14), hence, the value function of the partial coalition $\boldsymbol{p}^m$ is defined as the pay-off sum on sub-band m. The value function of the partial coalition $\boldsymbol{p}^m$ is given by

$$v(\boldsymbol{p}^m, \boldsymbol{\lambda}^m) = \sum_{S_k \in \mathcal{L}_m} r_{S_k}. \tag{12.22}$$

It is proved in [19] that the pay-off division among coalition members satisfies the proportional fairness [10], if the benefit allocated to each member equals to its contribution to the overall rate in sub-band m, i.e.,

$$r_{S_k}^m = \log(1 + \lambda_{S_k}^m p_{S_k}^m). \tag{12.23}$$

The solution of the optimal power vector $p_{S_k}^m$ of S_k is given in (12.20). Hence, the users can optimize the pay-off sum by choosing $\lambda_{S_k}^m$. Furthermore, since $\lambda_{S_k}^m$ depends on the coalition structure, then finding optimal $\lambda_{S_k}^m$ is equivalent to choosing an optimal coalition structure. Therefore, the power allocation of the user side can be equivalently achieved by the coalition formation game.

12.3.3 Coalitional Behavior Analysis

There are two types of actions for each player in an OCF-game, which are the coalitional action and the overlapping action. The former defines how the resources are being allocated among the member players in one coalition, and the later defines how resources are being allocated between players in the overlapping parts of multiple coalitions. These are the key features to differentiate the OCF-game from the non-overlapping coalition formation game.

As mentioned previously, the users accessing the same sub-band form a coalition. The cooperation among the member players is achieved by forming a virtual MIMO channel. The pay-off division relies on assigning $\boldsymbol{\lambda}$ to the players, which can be considered as the contribution of each coalition member to the sum rate. Since the users can join multiple coalitions, the proposed game becomes an OCF-game. As the resources of a user is the total transmit power, the profit is the pay-off sum obtained from all coalitions. The users need to distribute the resources in each sub-band properly to maximize the pay-off. For the proposed OCF-game, we have the following definition.

Definition 12.7 *For a set of users $\mathcal{S}$, a coalition structure on $\mathcal{S}$ is a finite list of vectors (partial coalitions) $\boldsymbol{P} = (\boldsymbol{p}^1, ..., \boldsymbol{p}^M)$ that satisfies (i) $\sum_{k=1}^{K} h_{S_k}^m p_{S_k}^m \leq \overline{Q}$, (ii) $\sup \boldsymbol{p}^m \subseteq \mathcal{S}$ for all $m = 1, ..., M$, and (iii) $\sum_{m=1}^{M} p_{S_k}^m \leq \overline{p}$ for all $j \in \mathcal{S}$.*

The power allocation matrix also indicates the utilization status of sub-bands. The constraint (i) states that the transmit power of user in each sub-band is bounded, (ii) states that the overlapping coalition is a sub set of the grand coalition, and (iii) states that the sum of transmit power is upper-bounded.

Proposition 12.1 *The proposed OCF-game is 2^n-finite.*

Proof 12.1 *Suppose a partial coalition $\boldsymbol{p}^{m*} = \{p_{S_k}^{m*} : k = 1, 2, ...K\}$ is formed on sub-band m, in which the positive power $p_{S_k}^{m*}$ is given by (12.20), i.e.,*

$$\boldsymbol{p}^{m*} = \arg\max_{\boldsymbol{p}^m} \pi(\boldsymbol{p}^m). \tag{12.24}$$

We define the support of $\boldsymbol{p}^{m*}$ as,

$$supp(\boldsymbol{p}^{m*}) = \{S_k : p_{S_k}^{m*} > 0, k = 1, 2, ...K\}^m, \tag{12.25}$$

which defines a coalition of users regardless of the resource distribution. Hence, for any other partial coalition $\boldsymbol{p}^{m'}$ with the support $supp(\boldsymbol{p}^{m*})$, we have

$$\pi(\boldsymbol{p}^{m*}) \geq \pi(\boldsymbol{p}^{m'}), \tag{12.26}$$

i.e., the partial coalition $\boldsymbol{p}^{m*}$ blocks all other partial coalitions formed on sub-band m, which involves with $supp(\boldsymbol{p}^{m*})$.

Therefore, we can say that the partial coalition $\boldsymbol{p}^{m*}$ in our proposed game is one-to-one correspondent to the coalition $\{S_k : p_{S_k}^{m*} > 0, k = 1, 2, ...K\}^m$ formed on sub-band m. Since $\{S_k\}^m \subseteq \mathcal{K}$, i.e., $\{S_k\}^m$ is a subset of $\mathcal{K}$, the number of all possible partial coalitions is equal to the number of the subset of $\mathcal{K}$, which is given by,

$$\sum_{n=1}^{K} \binom{K}{n} = 2^n - 1. \tag{12.27}$$

Hence, the proposed game is 2^n-finite. ■

This result above indicates that the coalition structure is reduced to a finite set, which enables us to find the core of the proposed game. In a traditional coalitional game, which studies the grand coalition, which is a finite set of all players, the core is a set of imputations, i.e., efficient pay-off division vectors that satisfy individual rationality, which stabilizes the grand coalition. However, many practice problems are naturally inefficient with the cooperation of all players. We are interested in investigating a stable coalition structure that optimizes the pay-off sum, i.e., to find an optimal partitioning of the grand coalition. Following the same line in [3], we define the core of the OCF-game for the sub-bands allocation.

Definition 12.8 *For a set of players $\mathcal{I} \subseteq \mathcal{K}$, a tuple $(\boldsymbol{P}_{\mathcal{I}}, \boldsymbol{\pi}_{\mathcal{I}})$ is the core of an OCF-game $G = (\mathcal{K}, \boldsymbol{v})$. If for any other set of player $\mathcal{J} \subseteq \mathcal{K}$, any coalition structure $\boldsymbol{P}_{\mathcal{J}}$ on $\mathcal{J}$, and any imputation $\boldsymbol{y}_{\mathcal{J}}$, we have $p_j(\mathcal{C}_{\mathcal{J}}, \boldsymbol{y}_{\mathcal{J}}) \leq p_i(\mathcal{C}_{\mathcal{I}}, \boldsymbol{\pi}_{\mathcal{I}})$ for some player $j \in J$.*

Theorem 12.2 *[3] Given an OCF-game $G = (\mathcal{K}, \boldsymbol{v})$, if $\boldsymbol{v}$ is continuous bounded, monotone and U-finite for some $U \in \mathbb{N}$, then an outcome $(\mathcal{C}_{\mathcal{S}}, \boldsymbol{\pi})$ is in the core of $(\mathcal{K}, \boldsymbol{v})$ iff $\forall S \in N$,*

$$\sum_{j \in S} p_j(\mathcal{C}_{\mathcal{S}}, \boldsymbol{\pi}) \leq v^*(S), \tag{12.28}$$

where $v^*(S)$ is the least upper bound on the value that the members of S can achieve by forming the coalition.

Proposition 12.2 *The proposed OCF-game of sub-band allocation has non-empty core.*

Proof 12.2 *(1) Continuous: The value function in (12.22) is the difference between a log function and a linear function, which is obviously continuous.*

(2) Monotone: The interference power constraint in (12.13) limits the total transmit power in sub-band m indirectly by pricing in the Stackelberg game. Hence, for S_k, the power it allocate in sub-band m is bounded by $p_{S_k}^{m*}$. Since the pay-off function of S_k $\pi(p_{S_k}^m)$ is concave, then for any $\pi(p_{S_k}^{m'}) \in [0, p_{S_k}^{m*}]$, we have $\pi(p_{S_k}^{m'}) \leq \pi(p_{S_k}^{m*})$. Therefore, for any $\boldsymbol{p}^{m'}$, $\boldsymbol{p}^{m*}$, such that $p_{S_k}^{m'} \leq p_{S_k}^{m*}$, we have $v(\boldsymbol{p}^{m'}) \leq v(\boldsymbol{p}^{m*})$, i.e., the value function is monotone.

(3) *Bounded: According to the proof of in (2), the value function is bounded by* $\nu(\boldsymbol{p}^{m*})$, *where*
$\boldsymbol{p}^{m*} = (p_{S_1}^m, p_{S_2}^m, ..., p_{S_K}^m)$ *satisfies* $\sum_{k=1}^K h_{S_k}^m p_{S_k}^m = \overline{Q}$.

(4) *U-finite: The proof can be referred to proposition 12.1.*

(5) *The inequality: The equality of (12.28) is always taken in the proposed game, since the value
function is the summation of individual pay-off of the member players.*

■

12.3.4 Distributed Algorithm

In this section, we discuss the algorithms that can reach the coalition structure in the core of the
coalition formation game and the SE of the hierarchical game.

Algorithm 4 OCF Algorithm for Sub-band Allocation

---

(1) *Initialization:*

(a) The users sequentially send the pilot signal to obtain the channel information. They can
estimate their pay-offs in each of the sub-bands when the sub-bands are exclusively used by
S_k.

(b) Each user S_k broadcasts the sub-band combination $\boldsymbol{l}_{S_k}^*$ that maximizes its pay-off sum,

$$\boldsymbol{l}_{S_k}^* = [l_{S_k}^{(1)}, l_{S_k}^{(2)}, ..., l_{S_k}^{(n)}]. \tag{12.29}$$

Let $\mathcal{R}^* = \{\boldsymbol{l}_{S_k}^* : k \in \{1, ..., K\}\}$.

(2) *Negotiating:*

- (a) The active users in the same sub-bands, i.e., all the active users in $\mathcal{R}^*$, must negotiate
with each other about the pay-off division factor $\lambda_{S_k}^m$.

- (b) After negotiation, each user S_k obtains a set of $\lambda_{S_k}^m$ corresponding to each sub-bands. S_k
solves problem (12.1) and obtains a new sub-band allocation, which maximizes its pay-off.
All the S_k update and broadcast their optimal sub-band allocation. Step 2) is repeated until
no user wants to change its occupied sub-bands.

12.4 Spectrum Sharing with Priorities

In spectrum sharing network, the users may have different priorities to access the spectrum. One the
instance is the cognitive wireless network. In the cognitive wireless network, there are primary users
(PU) and secondary users (SU). The spectrum is licensed to the PU for its utilization at any time,
while the SU can only access the spectrum opportunistically. Hence, the spectrum sharing policy in
network should consider the priority order of the users, i.e., it should first fulfill the transmit demand
of the PUs, and then allow the SU to transmit without violating the PU.

12.4.1 Network Model

Suppose we have a spectrum-sharing-based network consisting of M PUs and K SUs. The mth PU and kth SU are labeled as P_m and S_k, respectively. The spectrum belongs to P_m is labeled as sub-band m. The secondary user can lease the spectrum from the primary users without violating their transmission. To enable the transmission of SU while give the PU motivation to share its spectrum, we assume that the SU will make payment to the PU for interfering. Each SU tries to maximize its profit by trade-off between the achieved transmit rate and the price paid to PUs. The PU tries to maximize its revenue by leasing the spectrum while avoiding the licensed spectrum to be overcrowded.

The Stackelberg game in which the players are categorized into leaders and followers can be adopted to model such a scenario. The PU, who has higher priority to utilize the spectrum, acts as leader who moves first to take advantage in the game. The action is to adjust the price for accessing the spectrum. The SU acts as follower to move after the PU and maximize its own profit under the control of PU. The action is to choose suitable sub-band to access and optimize the transmit power accordingly.

12.4.2 The Stackelberg Game Formulation

The Stackelberg game is defined between the PU and the SUs.

Formally, we define the Stackelberg game between PUs and SUs as follows.

Definition 12.9 *A Stackelberg game is denoted by $G = (\mathcal{K}, \mathcal{M}, \boldsymbol{P}, \boldsymbol{v})$, where*

- $\mathcal{N} = \{1, 2, ..., N\}$ *is the set of leaders that are the PUs;*
- $\mathcal{C} = \{C_1, C_2, ..., C_M\}$ *is the set of followers that are the SUs;*
- μ^m *is the price charged in each sub-band, which is controlled by the leader;*
- p_{S_k} *is the power allocation of S_k;*
- π_{P_m} *is the pay-off of the PUs;*
- π_{S_k} *is the pay-off of the SUs.*

Hence, we define the pay-off of the PU as the payment sum collected from the SUs:

$$\pi_{p_{S_k}} = (\mu - \gamma) \sum_{K}^{k=1} p_{S_k} h_{S_k, P_m}, \tag{12.30}$$

where γ is the cost factor of PU for providing access to the SUs.

The pay-off of the SU is the defined as the achieved data rate minus the cost function:

$$\pi_{S_k} = \log_2 \left(1 + \frac{p_{S_k} g_{k,k}}{\sigma^2 + p_0 h_{m,k}} \right) - \mu p_{S_k} h_{S_k, P_m}, \tag{12.31}$$

where $I_{S_k} = \sigma^2 + p h_{m,k}$ is the interference plus noise. Note that the following assumptions are required:

- PU is able to measure its overall received interference power.

- SU needs to estimate the channel gains of its connected channels. Note that SUs do not need to communicate with each other or to know the interference power constraint $\overline{Q}$.

- A dedicated channel is required for broadcasting and receiving pricing coefficient μ by PU and SU, respectively.

The game worked in this flow. In each iteration, the PU will broadcast suitable interference prices for each sub-band based on the current measure of interference. Then each of the SUs will choose a sub-band to access and optimize its transmit power on the choosing band.

Hence, for any given current interference price μ, the SU solves (12.31) to obtain the optimal power allocation

$$p_{S_k}^{m*} = \left(\frac{1}{\mu h_{S_k}} - \frac{I_{S_k}}{g_{S_k}} \right)^+. \tag{12.32}$$

Substituting (12.32) into (12.30), we have the following problem:

$$\max_{\mu} \mu^m \left(\frac{1}{\mu^m} - \frac{I_{S_k}^m}{g_{S_k}^m} \right)^+ g_{S_k}^m. \tag{12.33}$$

Solving above problem, we obtain the optimal pricing coefficient

$$\mu^{m*} = \sqrt{\frac{\gamma K}{\sum_{k=1}^{K} \frac{g_{S_k}}{h_{S_k} I_{S_k}}}}. \tag{12.34}$$

Note that there are two boundary values defining the feasible region of the interference price μ, which we labeled as $\overline{\mu}$ and $\underline{\mu}$, respectively. The upper-bound is derived from the fact $p^* \geq 0$. Hence, for any SU k in any sub-band m, we have

$$\frac{1}{\mu h_{S_k}} - \frac{I_{S_k}}{g_{S_k}} \geq 0. \tag{12.35}$$

Then the upper-bound $\overline{\mu}$ is given by $\max_{S_k, P_m} \frac{g_{S_k}}{I_{S_k} h_{S_k}}$.

The following theorem shows the SE is guaranteed by the proposed scheme.

Theorem 12.3 *The pair of p_{S_k} for $i \in [1,k]$ defined in (12.32) and μ^m* defined in (12.34) is the unique SE for the proposed Stackelberg game if both p_{S_k} and μ^m* are feasible, i.e., $\underline{\mu} < \mu^m* \leq \overline{\mu}$.*

Proof 12.3 *Here, we only give a brief outline of the proof due to limit of space. First, note that payoff π_{S_k} for SU i is a concave function if μ is fixed, and, hence, there is a unique maximum point given in (5). It is also observed from (5), for a given μ, the optimal value of p_{S_k} is unique. Finally, if we substitute the optimal transmit powers of SUs into the optimization problem of PU P_m, it is found that the resulting π_{P_m} is a concave function of μ. We, hence, can claim that p_{S_k} and μ achieve the SE of the game if both values are feasible.*

12.4.3 Distributed Algorithm

We provide the distributed algorithm to achieve the SE of the game.

- Step (1): PU first chooses a large initial pricing coefficient $\mu(0)$ that cannot be afforded by all SUs,

- Step (2): PU gradually decreases the price, i.e., $\mu(t) = \mu(t-1) - \varepsilon$, to make each SU sequentially join the licensed spectrum, i.e., assume SU S_k is the first SU to find that the pricing coefficient μ is affordable when it accesses the best available sub-band. SU S_k will occupy mth sub-band of the licensed spectrum, and the price of m is labeled as $\mu^m = \mu(t)$. Then, PU will raise the price of sub-band m-, or the occupation of SU S_k will "scare way" other SUs who try to access the mth sub-band (i.e., other SUs will find high interference when they try to access the sub-band). If the price continues to decrease, assume SU S_j will be the second SU to join the licensed spectrum. S_j will occupy the nth sub-band in the licensed spectrum.

- Step (3): With the pricing coefficient decrease, SUs will sequentially join the licensed spectrum, as discussed in Step (2). PU will stop decreasing the price if it finds that all the available sub-band has been occupied.

References

[1] Tansu Alpcan, Tamer Başar, Rayadurgam Srikant, and Eitan Altman. "CDMA uplink power control as a noncooperative game." *Wireless Networks*, 8(6):659–670, 2002.

[2] D.P. Bertsekas and J.N. Tsitsiklis. *Parallel and Distributed Computation*. Old Tappan, NJ (USA), Prentice Hall Inc., 1989.

[3] Georgios Chalkiadakis, Edith Elkind, Evangelos Markakis, and Nicholas R Jennings. "Overlapping coalition formation." In *Internet and Network Economics*, pages 307–321, Springer, 2008.

[4] V. Chandrasekhar, J. G. Andrews, T. Muharemovic, Z. Shen, and A. Gatherer. "Power control in two-tier femtocell networks." *IEEE Transactions on Wireless Communications*, 8(8):4316–4328, 2009.

[5] Jie Gao, Sergiy A. Vorobyov, and Hai Jiang. "Cooperative resource allocation games under spectral mask and total power constraints." *IEEE Transactions on Signal Processing*, 58(8):4379–4395, 2010.

[6] Zhu Han and H. Vincent Poor. "Coalition games with cooperative transmission: A cure for the curse of boundary nodes in selfish packet-forwarding wireless networks." *IEEE Transactions on Communications*, 57(1):203–213, 2009.

[7] Simon Haykin. "Cognitive radio: Brain-empowered wireless communications." *Journal on Selected Areas in Communications*, 23(2):201–220, 2005.

[8] Sudharman K. Jayaweera and Tianming Li. "Dynamic spectrum leasing in cognitive radio networks via primary-secondary user power control games." *IEEE Transactions on Wireless Communications*, 8(6):3300–3310, 2009.

[9] Xin Kang, Rui Zhang, and Mehul Motani. "Price-based resource allocation for spectrum-sharing femtocell networks: A stackelberg game approach." *IEEE Journal on Selected Areas in Communications*, 30(3):538–549, 2012.

[10] Frank P. Kelly, Aman K. Maulloo, and David K.H. Tan. "Rate control for communication networks: Shadow prices, proportional fairness and stability." *Journal of the Operational Research Society*, pages 237–252, 1998.

[11] Richard J. La and Venkat Anantharam. "A game-theoretic look at the Gaussian multiaccess channel." *DIMACS Series in Discrete Mathematics and Theoretical Computer Science*, 66:87–106, 2004.

[12] Roger B. Myerson. *Game Theory*. Harvard University Press, 2013.

[13] Walid Saad, Zhu Han, Mérouane Debbah, Are Hjorungnes, and Tamer Basar. "Coalitional game theory for communication networks." *IEEE Signal Processing Magazine*, 26(5):77–97, 2009.

[14] G. Scutari, D.P. Palomar, and S. Barbarossa. "Asynchronous iterative water-filling for gaussian frequency-selective interference channels." *IEEE Transactions on Information Theory*, 54(7):2868–2878, 2008.

[15] K.W. Shum, K.K. Leung, and C.W. Sung. "Convergence of iterative waterfilling algorithm for gaussian interference channels." *IEEE Journal on Selected Areas in Communications*, 25(6):1091–1100, 2007.

[16] Emre Telatar. "Capacity of multi-antenna gaussian channels." *European Transactions on Telecommunications*, 10(6):585–595, 1999.

[17] Beibei Wang, Yongle Wu, and KJ Liu. "Game theory for cognitive radio networks: An overview." *Computer Networks*, 54(14):2537–2561, 2010.

[18] Yong Xiao, Guoan Bi, and Dusit Niyato. "Distributed optimization for cognitive radio networks using stackelberg game." In *International Conference on Communication Systems (ICCS)*, pages 77–81, IEEE, 2010.

[19] Yong Xiao, Guoan Bi, Dusit Niyato, and Luiz A. DaSilva. "A hierarchical game theoretic framework for cognitive radio networks." *IEEE Journal on Selected Areas in Communications*, 30(10):2053–2069, 2012.

[20] Yong Xiao, Kwang-Cheng Chen, Chau Yuen, Zhu Han, and Luiz A. DaSilva. "A Bayesian overlapping coalition formation game for device-to-device spectrum sharing in cellular networks." *To appear in IEEE Trans. Wireless Commun., 2015*. Avaliable at https://sites.google.com/site/xyong2007/publication.

[21] Yong Xiao, Chau Yuen, Paolo Di Francesco, and Luiz A. DaSilva. "Dynamic spectrum scheduling for carrier aggregation: A game theoretic approach." In *International Conference on Communications (ICC)*, pages 2672–2676, IEEE, 2013.

[22] W. Yu, G. Ginis, and J.M. Cioffi. "Distributed multiuser power control for digital subscriber lines." *IEEE Journal on Selected Areas in Communications*, 20(5):1105–1115, 2002.

[23] Pu Yuan, Yong Xiao, and Guoan Bi. "Towards cooperation by carrier aggregation in heterogeneous networks: A hierarchical game approach." *Under review at IEEE Transactions on Vehicular Technology*, 2015.

[24] Zengfeng Zhang, Lingyang Song, Zhu Han, and Walid Saad. "Coalitional games with overlapping coalitions for interference management in small cell networks." *IEEE Transactions on Wireless Communications*, 2014.

[25] Qing Zhao and Brian M Sadler. "A survey of dynamic spectrum access." *IEEE Signal Processing Magazine*, 24(3):79–89, 2007.

Chapter 13

Game-Theoretic Opportunistic Spectrum Sharing

Yuhua Xu and Alagan Anpalagan

CONTENTS

13.1   Introduction

With the explosive increase in wireless transmission demands, traditional static, and pre-determined spectrum allocation approaches can not meet the requirements of flexibility, adaptability and intelligence for the terminals (users). Cognitive radio has been established as an innovative framework for intelligent opportunistic spectrum sharing (OSS) which could observe the environment, make intelligent decisions and reconfigure the hardware [1]. In OSS systems, the devices are required to be autonomous and smart. Basically, mutual interactions in the wireless environment, mainly including competition, interference and coordination, should be well addressed when all the devices are autonomous and smart. Game theory [2] is a powerful tool to study the interactions among multiple autonomous decision-makers and has been extensively applied in wireless communication networks. This chapter presents novel game models and discusses distributed learning techniques for OSS systems.

In game-theoretic solutions, equilibria are the desirable and stable outcomes, in which no decision-maker is willing to change its strategy unilaterally. For game-theoretic solutions, there are two key steps [3]: designing the game model and finding equilibrium-based stable solutions. Since game theory is a branch of applied mathematics and originally studied in the field of economics, some new challenges with regard to information constraints should be addressed when applied in wireless communication networks. The purpose of this chapter is to present some examples for the application of game theory and distributed learning solutions for OSS systems, and show the general methodology of designing game-theoretic solutions for wireless networks.

The rest of this chapter is organized as follows. Section 12.2 presents preliminary of game theory and distributed learning. In Section 12.3, two graphical game models for spatial OSS systems are proposed. In Section 12.4, a robust game for dynamic OSS systems is presented. Finally, some future directions are presented in Section 12.5.

13.2   Game Theory and Distributed Learning

13.2.1   Motivation of Using Game Theory

Game theory is an important branch of applied mathematics that was originally studied in economics and has been extensively applied into biology [4], social activities [5] and engineering [6] in recent years. In about 2000, game theory was successfully applied to wireless optimization problems [7,8]. Through about 15 years' development, it has become an important optimization approach for wireless networks. On one hand, game theory is very suitable for solving wireless resource management problems that can directly be formulated as economic events and activities, e.g., spectrum auction [9–11] and incentive mechanism [12]. On the other hand, it can also be applied to solve other wireless optimization problems involving multiple interactive users, e.g., power control [13], self-organizing networking [14], multiple access control [15], and heterogeneous network selection [16].

The motivations of using game models in OSS systems are as follows:

- In the absence of a centralized controller, the users usually access the spectrum in a distributed and autonomous manner. In game models, the players also act distributively and autonomously. Thus, game theory provides a good decision framework for OSS.

- Due to the open nature of wireless channels, the decisions of cognitive users are interactive, i.e., a user is affected by the actions of others. Traditional optimization approaches mainly focused on the system throughput but ignoring the interactions. In contrast, game models provide efficient approaches for analyzing the interactions among multiple decision-makers, which mainly include interference, competition, and cooperation in OSS systems.

- A distinct feature of OSS is that the users are able to make intelligent decisions. From the perspective of network control and optimization, multiple intelligent decision-makers may cause disorder and chaos. Furthermore, the users may belong to different systems and holders, and hence, access the spectrum with self-interested utilities. The selfish behavior of the users can be well captured by (non-cooperative) game models, in which the individual payoffs, such as achievable throughput, delay, and experienced interference, can be mapped into utility functions. More importantly, through analyzing the stable points (equilibria), the effects of interactive behaviors on the system performance can be predicted, and, hence, it is possible to improve the system designing and optimization.

13.2.2 Preliminary of Game Theory and Distributed Learning

13.2.2.1 Basic Game Models

Generally, a non-cooperative game is denoted as $\mathcal{G} = \{\mathcal{N}, \mathcal{A}_n, u_n\}$, where $\mathcal{N}$ is the set of players, $\mathcal{A}_n$ is the available action set of player n, and u_n is the utility function of n. For presentation, denote $a_n \in \mathcal{A}_n$ as the chosen action of n, and a_{-n} as the action profile of all players except player n. Due to the interactions among the players, the utility function is generally expressed as $u_n(a_n, a_{-n})$.

In addition to pure strategy, the players can also choose mixed strategies over the available action set. Specifically, a mixed strategy is denoted as $\sigma_n(a_n)$, which represents the probability of player n choosing action a_n. The mixed strategy action space can be expressed as $\Sigma_n = \{\sigma_n : \sum_{a_n \in \mathcal{A}_n} \sigma_n(a_n) = 1, 0 \leq \sigma_n(a_n) \leq 1\}$. Denote the mixed strategy profile of the players as $\sigma = \{\sigma_1, \ldots, \sigma_N\}$. Similarly, denote σ_{-n} as the mixed strategy profile of the players except n, and then the expected utility function of player n is given by $u_n(\sigma_n, \sigma_{-n}) = \sum_{a \in \mathcal{A}} \left(\prod_{n \in \mathcal{N}} \sigma_n(a_n) \right) u_n(a)$.

In non-cooperative games, each player maximizes its individual utility. In order to analyze the interactions among the players, it needs to study the stable solutions of the game. In the following, definitions of the well-known Nash equilibrium (NE) and exact potential game (EPG), which admits promising properties in terms of NE existence, were presented respectively.

Definition 12.1 (Nash equilibrium [17]). An action profile $a^* = (a_1^*, \ldots, a_{-n}^*)$ is a pure strategy NE if and only if no player can improve its utility by deviating unilaterally, i.e.,

$$u_n(a_n^*, a_{-n}^*) \geq u_n(a_n, a_{-n}^*), \forall n \in \mathcal{N}, \forall a_n \in \mathcal{A}_n, a_n \neq a_n^*. \tag{13.1}$$

Definition 12.2 (Exact potential game [18]). A game is an EPG if there exists an exact potential function $\phi_e : A_1 \times \cdots \times A_N \to R$, such that for all $n \in \mathcal{N}$, all $a_n \in \mathcal{A}_n$, and $a_n' \in \mathcal{A}_n$,

$$u_n(a_n, a_{-n}) - u_n(a_n', a_{-n}) = \phi_e(a_n, a_{-n}) - \phi_e(a_n', a_{-n}). \tag{13.2}$$

In other words, the change in the utility function caused by an arbitrary unilateral action change of a user is the same with that in the exact potential function. EPGs have been widely applied to wireless

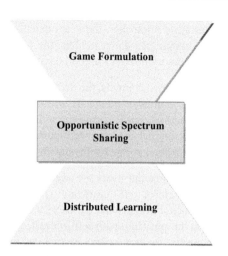

Figure 13.1: Framework of game-theoretic solutions for OSS systems.

communication systems. Potential game exhibits several nice properties, and the most important two are as follows: (i) every potential game has at least one pure strategy NE, (ii) any global or local maxima of the potential function constitutes a pure strategy NE. In this chapter, several potential games for OSS systems are presented.

Besides NE, there are also some other useful concepts of equilibria in game models, e.g., correlated equilibrium [19], evolutionary stable strategy [20], and conjectural equilibrium [21].

13.2.2.2 Distributed Learning for Achieving Equilibria

For OSS systems, a framework of game-theoretic optimization is proposed, as shown in Figure 13.1, which consists of two key steps [3]: (i) game formulation and (ii) distributed learning.

1. **Game formulation.** With regard to game formulation, the first task is to identify the players and their available actions, and define suitable utility functions. The available action set can be a singular optimization variable or a combination of multiple variables. Defining utility function is key to game formulation, since it eventually determines the properties and performance of the game-theoretic models. In practice, there are three rules for designing utility function [3]: (i) making the NE become or approach to the optimal solution, (ii) capturing the inherent features of wireless communications, and (iii) having clear physical meanings, e.g., it should explicitly be related to throughput, delay, interference and energy-efficiency.

   Potential game [18] has been regarded as an efficient game-theoretic model for wireless communication networks. In potential games, there is a potential function, such that the change in the utility function caused by the unilateral action change of an arbitrary player has the same trend with that in the potential function, i.e., both increasing or decreasing. Potential game has at least one pure strategy NE, and all its NE points are located in the global or local maxima of the potential function. Thus, the NE solutions would be desirable if the potential function is directly related to the original optimization objective. Furthermore, to achieve the goal that the stable game solutions are optimal (near-optimal), another efficient method is including pricing of using the resources, i.e., defining the utility function as the received payoff minus the cost of using the amount of a particular resource.

2. **Distributed learning.** Identifying equilibria of game models is a challenge, while finding them is another challenge. In pure game theory, players are assumed to perfectly monitor

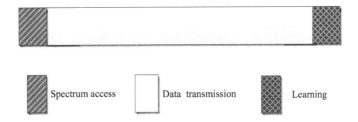

Figure 13.2: Learning procedure in opportunistic spectrum sharing.

the environment and the actions chosen by other players. With this assumption, the players can use some efficient algorithms, e.g., best response [2] and fictitious play [22], to adjust their strategies toward Nash equilibria. However, the assumption of perfect monitoring does not hold in wireless communication networks. In fact, the users always encounter with uncertain, dynamic, and incomplete information constraints in wireless communications. Thus, the players need to observe the results of mutual interactions, e.g., interference, collision, or competition, learn useful information, and then adjust their strategies toward some desirable solutions.

Typically, the learning procedure in OSS systems is shown in Figure 13.2. Denote $a_n(k)$ as the action of user n and $a_{-n}(k)$ as the action profile of all other players except n at the kth iteration, respectively. Due to the mutual interactions (interference, congestion, or competition) among users, the received instantaneous payoff $r_n(k)$ of each player is jointly determined by the action profile of all players, and it may be deterministic or random. Generally, the players update their actions based on the current action-payoff information, i.e.,

$$a_n(k+1) = G(a_n(k), a_{-n}(k); r_n(k)),$$ (13.3)

where $G(x,y;z)$ specifies the update rule. The above rule is called *coupled* learning, as it needs to know $a_{-n}(k)$. In the presence of an imperfect monitor, it is desirable to develop the following *uncoupled* learning algorithms:

$$a_n(k+1) = F(a_n(k); r_n(k)).$$ (13.4)

With the learning algorithms, the system evolves as: $\{a_n(k), a_{-n}(k)\} \rightarrow \{r_n(k), r_{-n}(k)\} \rightarrow \{a_n(k+1), a_{-n}(k+1)\}$, and the objective is to converge to a stable action profile that maximizes the system utility.

13.3 Graphical Game for Spatial Opportunistic Spectrum Sharing

In this section, two graphical game models for OSS systems are presented, in which the transmission of a user only interferes with its neighboring users directly. The games are proved to be potential games, with the network throughput and network interference level serving as the potential function, respectively. Finally, a distributed learning algorithm that only requires information exchange among neighboring users is proposed to achieve the global solution. Note that the main analysis and results in this section were presented in [23].

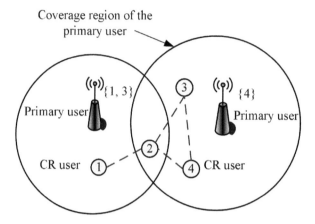

{1 3} Occupied channel set   — —   Interference relationship

Figure 13.3: An example of the considered opportunistic spectrum sharing system. The dashed lines represent the interference relationship between the users, which is inherently determined by the physical distance.

13.3.1 System Model and Problem Formulation

Consider an OSS system involving N cognitive users and M licensed channels. Without loss of generality, it is assumed that the number of users is larger than that of channels, i.e., $N > M$. Denote the user set as $\mathcal{N}$, i.e., $\mathcal{N} = \{1, 2, \ldots, N\}$, and the licensed channel set as $\mathcal{M}$, i.e., $\mathcal{M} = \{1, \ldots, M\}$. For simplicity of analysis, it is assumed that all the channels support the same transmission rate for all users[1]. Note that this represents the case that all channels have the same bandwidth and yield the same transmission rate to all the users, although the users may experience different channel conditions. This assumption holds in some practical systems, e.g., IEEE 802.16d/e standard [24]. Moreover, it is assumed that the spectrum opportunities are static or vary slowly in time.

To characterize the heterogeneous spectrum opportunities, a channel availability vector is defined as $\mathbf{C}_n = \{C_{n1}, C_{n2}, \ldots, C_{nM}\}$, where $C_{nm} = 1$ indicates that channel m is available for user n, and $C_{nm} = 0$ means it is unavailable. An example of the considered OSS system is shown in Figure 13.3, which involves four CR users, two primary users and four licensed channels (1,2,3,4). For the presented example, the channel availability vector for the users are $\mathbf{C}_1 = \{0\,1\,0\,1\}$, $\mathbf{C}_2 = \{0\,1\,0\,0\}$, and $\mathbf{C}_3 = \mathbf{C}_4 = \{1\,1\,1\,0\}$, respectively.

The transmission of a user usually only interferes with the nearby users, due to the limited transmitting power. Thus, the interference relationship between the users can be characterized using an interference graph, which is eventually determined by the distance between users. The interference graph is constructed as follows: Each CR user corresponds to a node on the interference graph; furthermore, if the distance between two CR users m and n is less than a predefined interference distance D_I, then they are connected by an edge. Denote $\mathcal{E}$, $\mathcal{E} \subset N^2$, as the edge set, and J_n as the set of connected (neighboring) users for user n, i.e.,

$$J_n = \{m \in \mathcal{N} : (n, m) \in \mathcal{E}\}. \tag{13.5}$$

Due to the hardware limitation, it is assumed that the CR users can sense all channels simultaneously

---

[1] However, the proposed game-theoretic solution can be easily extended to scenarios with heterogeneous channel rates.

but transmit on one channel at a time [25]; moreover, the spectrum sensing is perfect. Collision channel model is considered, i.e., a collision occurs when more than one neighboring user transmits simultaneously on the same channel. The slotted Aloha mechanism is used to solve the contention among the users. Specifically, a user transmits with probability p in each slot, while being silent with probability $(1-p)$.

Denote A_n as the available channel set of user n, i.e., $A_n = \{m \in \mathcal{M} : C_{nm} = 1\}$, and $a_n \in A_n$ as the channel chosen by user n. When there is no channel available for user n, i.e., $A_n = \emptyset$, the user n does not choose any channel and keep silent, i.e., $a_n = \emptyset$; otherwise, $a_n \neq \emptyset$. Then, for an arbitrary user with non-empty action, the achievable throughput is determined by

$$g_n(a_1, \ldots, a_N) = p \prod_{k \in J_n} (1-p)^{f(a_n, a_k)}, \tag{13.6}$$

where J_n is the neighboring user set of n, as specified by (13.5), and $f(a_n, a_k)$ is the following indication function:

$$f(a_n, a_k) = \begin{cases} 1, & a_n = a_k \\ 0, & a_n \neq a_k. \end{cases} \tag{13.7}$$

According to (13.6), the network throughput is expressed as

$$U_0 = \sum_{n \in \mathcal{N}} g_n = p \sum_{n \in \mathcal{N}} \prod_{k \in J_n} (1-p)^{f(a_n, a_k)}. \tag{13.8}$$

Then, the first optimization objective is to maximize the network throughput, i.e.,

$$(P1): \quad \max U_0. \tag{13.9}$$

Besides the approaches of maximizing the network throughput explicitly, there are other alternative and efficient methods that implicitly maximize the network throughput via interference reduction [26, 27]. Motivated by such ideas, the problem is also considered from the perspective of network collision minimization. For a user n with non-empty action a_n, the individual collision level is defined as the number of neighboring CR users that compete for the same channel, i.e.,

$$s_n = \sum_{k \in J_n} f(a_n, a_k). \tag{13.10}$$

Based on (13.10), (13.6) can be re-written as $g_n(a_1, \ldots, a_N) = p(1-p)^{s_n}$. It is observed that lower value of s_n is desirable from the user-side as higher throughput can be achieved. Also, lower aggregate collision level is more preferable for improving the overall network performance. Formally, the network collision level I_0 is defined as the total number of competing neighboring pairs selecting the same channel, i.e.,

$$I_0 = \frac{1}{2} \sum_{n \in \mathcal{N}} s_n = \frac{1}{2} \sum_{n \in \mathcal{N}} \sum_{k \in J_n} f(a_n, a_k). \tag{13.11}$$

Thus, the second optimization objective is as follows:

$$(P2): \quad \min I_0. \tag{13.12}$$

Both problems $P1$ and $P2$ are combinatorial optimization problems and NP-hard [28]. Heuristic methods can be applied, but they cannot obtain the optimal solutions. In addition, one may use exhaustive search in a centralized manner to obtain the global solutions, but it has intractable computational complexity. Thus, a distributed approach with low complexity that can achieve the optimal solutions is desirable. In the following, two kinds of graphical games are proposed to solve the problem. The first one is a local altruistic game, which maximizes the network throughput, and the second one is a local congestion game, which minimizes the network collision level.

13.3.2 Local Altruistic Game for Network Throughput Maximization

In traditional non-cooperative game models, players always act selfishly. Generally, such approaches cannot guarantee to obtain the global optimization. To improve the efficiency of the games, local altruistic behaviors among neighboring users is considered, which is motivated by local cooperative behavior in biographical systems [29, 30]. Specifically, the utility function of each player n in the local altruistic game is defined as follows:

$$U1_n(a_n, a_{J_n}) = g_n(a_n, a_{J_n}) + \sum_{k \in J_n} g_k(a_k, a_{J_k}), \tag{13.13}$$

where $g_k(a_k, a_{J_k})$ is the individual achievable throughput of player k, as characterized by (13.6). The defined utility function consists of two parts: the individual throughput of player n and the aggregate throughput of all the neighbors. That is, a player not only considers itself but also considers its neighbors. This is why it is called local altruistic game. Then, the local altruistic game is expressed as follows:

$$(\mathcal{G}1): \max_{a_n \in A_n} U1_n(a_n, a_{J_n}) \quad \forall n \in \mathcal{N}. \tag{13.14}$$

Theorem 13.1

The local altruistic game $\mathcal{G}1$ is an exact potential game that has at least one pure strategy NE, and the optimal solution of the network throughput maximization problem P1 constitutes a pure strategy NE of $\mathcal{G}1$.

Proof 13.1   Refer to [23].

According to Theorem 13.1, an important result can be observed that the global solution constitutes a pure strategy NE of the local altruistic game. With this promising result, the global solution of $P1$ can be obtained in a distributed manner. ■

13.3.3 Local Congestion Game for Network Collision Minimization

In this part, another graphical game from the perspective of minimizing network collision level is proposed. From (13.6), it is seen that the individual achievable throughput of a user is a decreasing function of the individual experienced level $s_n = \sum_{k \in J_n} f(a_n, a_k)$, which implies that minimizing s_n is equivalent to maximizing the individual achievable throughput. This motivates us to define the utility function as follows:

$$U2_n(a_n, a_{J_n}) = -\sum_{k \in J_n} f(a_n, a_k), \tag{13.15}$$

where J_n is the neighboring user set of player n, and $f(a_n, a_k)$ is the indication delta function, as specified by (13.7). Then, the local congestion game is expressed as follows:

$$(\mathcal{G}2): \max_{a_n \in A_n} U2_n(a_n, a_{J_n}) \quad \forall \in \mathcal{N}. \tag{13.16}$$

The above defined utility function is motivated by the congestion games [31], in which the utility function is defined as a function of the number of the players selecting the same action. Note that the utility function of the proposed local congestion game is only dependent on its neighbors, whereas that of congestion game is dependent on all other players. This differentiates the proposed local congestion game from the traditional congestion game significantly.

Theorem 13.2

The local congestion game $\mathcal{G}2$ is an exact potential game that has at least one pure strategy NE, and *the optimal solution of the network collision minimization problem P2 constitutes a pure strategy NE of* $\mathcal{G}2$.

Proof 13.2 Refer to [23].

Also, according to Theorem 13.2, an important result can be observed that the global solution constitutes a pure strategy NE of the local congestion game. Thus, the global solution of *P2* can be obtained in a distributed manner. ■

13.3.4 Spatial Adaptive Play for Achieving Global Optimization

Based on Theorems 13.1 and 13.2, if there exists an algorithm that can achieve the optimal NE points of the two games, the global optimum (i.e., network throughput maximization or network collision minimization) can be obtained through distributed implementation. However, normally multiple NE points exist in $\mathcal{G}_1$ ($\mathcal{G}_2$), and most of them are suboptimal [2]. Furthermore, the tasks of identifying and finding the optimal NE points of games are different, and the latter is generally much harder than the former [2]. In the following, a learning algorithm is proposed to achieve the optimal NE points of the games. The learning algorithm is based on an existing algorithm, called *spatial adaptive play (SAP)*, which is originally designed for investigating the stochastic stability of social networks [32]. In any potential games, the SAP algorithm converges to a pure NE, which maximizes the potential function with arbitrarily higher probability [32,33].

In order to implement the SAP algorithm, the game is extended to a mixed strategy form. Specifically, the mixed strategy for player n at iteration k is denoted as the probability distribution $q_n(k) \in \Delta(A_n)$, where $\Delta(A_n)$ denotes the set of probability distributions over action set A_n. In SAP, only one player is randomly selected to update its selection, while all other players repeat their selections. This process is repeated until some stop criterions are met (see Algorithm 1).

Note that in Step 2 of SAP, the exchanged information of different games is different. Specifically, in local altruistic game, the neighbors exchange the current channel selection $a_n(k)$ and the current individual achievable throughput $g_n(k)$. In the local congestion game, it only requires to exchange the current channel selection $a_n(k)$ among the neighbors. In Step 5, the stop criterion is dependent on specific applications, and the following are some examples: (i) for $\forall n \in \mathcal{N}$, the individual throughput g_n remains unchanged for a certain number of iterations, and (ii) for $\forall n \in \{k : A_k \neq \emptyset\}$, there exists a component of $p_n(k)$, which is sufficiently approaching one, say 0.99.

13.3.4.1 Convergence and Optimality

Denote the set of available selection profiles of all the CR users as $\mathcal{A}$, i.e., $\mathcal{A} = A_1 \otimes \cdots \otimes A_N$, then the asymptotic behavior of SAP is determined by the following theorems.

Theorem 13.3

In an exact potential game in which all players adhere to SAP, the unique stationary distribution $\mu(a) \in \Delta(\mathcal{A})$ *of the joint action profiles,* $\forall \beta > 0$, *is given as:*

$$\mu(a) = \frac{\exp\{\beta\Phi(a)\}}{\sum_{s\in\mathcal{A}} \exp\{\beta\Phi(s)\}}, \tag{13.18}$$

where $\Phi()$ *is the potential function.*

Proof 13.3 Refer to [23]. ■

Algorithm 1: Spatial adaptive play (SAP) for graphical games

Step 1: Initially, set $k = 0$ and let each CR user $n \in \mathcal{N}$ randomly selects an available channel $a_n(0)$ from its available channel set A_n with equal probability.
Step 2: All the CR users exchange information with their neighbors.
Step 3: A CR users is randomly selected, say i.
Step 4: All other CR users repeat their selections, i.e., $a_{-i}(k+1) = a_{-i}(k)$. Meanwhile, with the information received from the neighbors, user i calculates the utility functions over its all available actions, i.e., $U_i(\bar{a}_i, a_{J_i}(k)), \forall \bar{a}_i \in A_i$. Then, it randomly chooses an action according to the mixed strategy $q_n(k+1) \in \Delta(A_i)$, where the a_nth component $q_i^{a_i}(k+1)$ of the mixed strategy is given as

$$q_i^{a_i}(k+1) = \frac{\exp\{\beta U_i(a_i, a_{J_i}(k))\}}{\sum_{\bar{a}_i \in A_i} \exp\{\beta U_i(\bar{a}_i, a_{J_i}(k))}} \tag{13.17}$$

for some learning parameter $\beta > 0$. The utility function $U_i(a_i, a_{J_i})$ in the above equation for local altruistic game is $U1_i(a_i, a_{J_i})$, which is specified by (13.13), and that of local congestion game is $U2_i(a_i, a_{J_i})$, which is specified by (13.15).
Step 5: If the predefined maximum number of iteration steps is reached, stop; else go to Step 2.

Theorem 13.4
With a sufficiently large β, SAP achieves the global optimum of problems P1 or P2 with an arbitrarily high probability.

Proof 13.4   According to the distribution as characterized by (13.18), SAP asymptotically converges to action profiles that maximize the potential function as β goes sufficiently large. Thus, according to the relationship between the potential functions and optimization problems $P1$ and $P2$, it can be concluded that SAP converges to their global optimum with an arbitrarily high probability. For detailed analysis, refer to [23].

In Theorem 13.4, arbitrarily high probability means that the converging probability sufficiently approaches one. For example, suppose that there are four NE points in the local altruistic game, leading to the following network throughput, $U = \{8\ 8.5\ 8.5\ 9\}$. Denote P_c as the probability of converging to the global maximum $U_0 = \max\{U\} = 9$. According to Theorem 13.3, it is known $P_c = 0.9867$ for $\beta = 10$. In addition, P_c sufficiently approaches one when β increases, e.g., $P_c = 0.9999$ for $\beta = 20$ and $P_c = 1 - 6.1 \times 10^{-7}$ for $\beta = 30$.

Theorem 13.4 validates the optimality of SAP in the proposed graphical games. It is a desired learning algorithm because the optimal solutions for the network throughput maximization problem $P1$ and the network collision minimization problem $P2$ are achieved via just local information exchange between neighbors.   ■

13.3.4.2   Implementation Issues

For practical implementation, the empirical frequency of the channel selection profile $a(k)$ asymptotically converges to the stationary distribution $\mu(a)$ given by (13.18), as the iteration number goes sufficiently large. That is, the algorithm asymptotically converges to a global optimum as the iteration number goes sufficiently large, but may converge to a global or local optimum in finite iterations. Thus, there is a tradeoff between learning iteration and optimality, and, hence, the selection of iteration number should be application-dependent in practice.

The learning parameter β in Step 4 of SAP balances the tradeoff between exploration and exploitation. Smaller β implies that the users are more willing to choose a suboptimal action to explore, whereas higher β implies that they are prone to choose the best response action. In particular, when $\beta = 0$, user i selects any action $a_i \in A_i$ with equal probability, while $\beta \to \infty$ means that it selects an action from its best response set, i.e., $a_i(k+1) \in \arg\max_{a_i \in A_i} U_i(a_i, a_{J_i}(k))$. Therefore, it is advisable that the value of β is set to small value at the beginning phase, and while increasing as the learning algorithm iterates [34].

Although SAP achieves the global optimum, there are still two drawbacks: (i) a mechanism is needed to coordinate the learning procedure, such that only one player is scheduled to update its action in each iteration, e.g., the random token mechanism given in [34], and (ii) the convergence speed is slow. To overcome the above drawbacks, concurrent spatial adaptive play (C-SAP) can be applied. In C-SAP, multiple players are selected in an autonomous fashion in each iteration, and then they concurrently update their actions. For detailed analysis on C-SAP, refer to [23].

13.3.5 Simulation Results

In the simulation study, the users are randomly located in a square region. The licensed channels are independently occupied by the primary users and are idle with the same probability θ, $0 < \theta < 1$. However, note that static spectrum opportunities are considered, i.e., the channel states remain unchanged during the convergence of the learning algorithm. The transmission rate of the channels is set to $R = 1$Mbps. In order to balance the tradeoff between exploration and exploitation of the learning algorithm, the learning parameter is set to $\beta = k$, where k is the iteration number.

13.3.5.1 Convergence of the SAP-Based Learning Algorithm

Consider a small network with four primary users and 20 CR users, as shown in Figure 13.4. There are three licensed channels, i.e., $\mathcal{N} = \{1, 2, \ldots, 20\}$ and $\mathcal{M} = \{1, 2, 3\}$. The interference distance is set to $D_I = 250$. In this scenario, the maximum number of possible channel selection profiles of all the users is about 3×10^9.

For an arbitrary realization of the heterogeneous spectrum opportunities (the spectrum opportunities are randomly generated by the channel idle probability $1 - \theta$), the convergence behavior of the local altruistic game is shown in Figure 13.5, in which the global optimum is obtained by using the exhaustive search method. It is noted from the figure that SAP catches up with the global optimum. Also, for the same realization of the heterogeneous spectrum opportunities, the convergence behavior of the local congestion game is shown in Figure 13.6. It is noted that SAP converges to the global optimum. The results presented in Figure 13.5 and Figure 13.6 validate the optimality of the proposed local altruistic game and local congestion game, in terms of maximizing network throughput or minimizing network collision level.

13.3.5.2 Throughput Performance

From the above analysis, it is known that the neighboring relationship is eventually determined by the interference range D_I. To be more specific, larger D_I means that a user has more neighbors, and, as a result, the number of competing neighbors increases accordingly. In the following, the impact of the interference range D_I on the throughput performance of the proposed game theoretic solutions is studied. Consider a network consisting of 20 CR users and four primary users. There are three licensed channels. The SAP learning algorithm is applied to achieve the optimal NE. The maximum number of learning iterations is set to 200, and the results are obtained by simulating 10^4 trials for different spectrum opportunities and different initial channel selection profiles.

a) *Scenario of small interfering range:* In the first scenario, consider a small interfering range $D_I = 250$m, and the network deployment is shown in Figure 13.4. The expected network

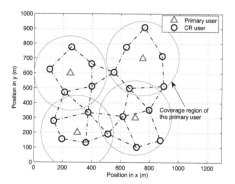

Figure 13.4: Interference graph for a small cognitive radio network (CRN) with four primary users and 20 CR users (The interference range is set to $D_I = 250$m).

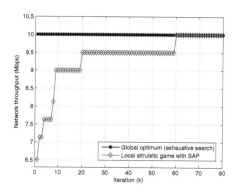

Figure 13.5: Convergence behavior of the local altruistic game for arbitrary spectrum opportunities (20 CR users).

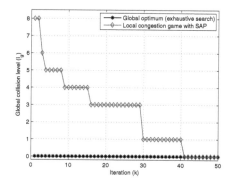

Figure 13.6: Convergence behavior of the local congestion game for arbitrary spectrum opportunities (20 CR users).

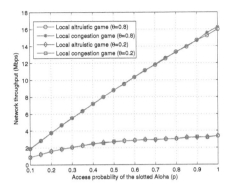

Figure 13.7: Expected network throughput for small interference range ($D_I = 250$m).

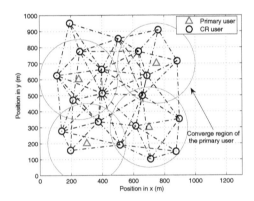

Figure 13.8: Interference graph for a CRN with large interference range ($D_I = 400$m).

throughput of all the users when varying the access probability in the Aloha mechanism, p, is shown in Figure 13.7. It can be seen that the expected network throughput increases almost linearly with p. The reason is that the users are located sparsely with small interference range, and, hence, there are sufficient spectrum opportunities. As a result, neighboring users are spread over different channels, and the collision between neighbors becomes trivial. Approximately, the expected achievable throughput of a user can be expressed as $g_n = \theta R p$, and the network throughput as $U_0 = N\theta R p$. Thus, the results in the figure hold.

Moreover, it is observed that the gap between the expected network throughput of the two game theoretic solutions is trivial. The reason is as follows: as there are sufficient spectrum opportunities, the optimal NE solutions for the two games are that neighbors are spread over different channels. Therefore, the optimal solutions for the network throughput maximization problem $P1$, and the network collision minimization problem $P2$ are the same in most cases.

b) *Scenario of large interference range:* In the second scenario, consider a relatively large interference range $D_I = 400$m. The deployment of the simulated cognitive radio network (CRN) is shown in Figure 13.8. The expected network throughput when varying the access probability, p, is as shown in Figure 13.9.

For the larger channel idle probability, i.e., $\theta = 0.8$, it is noted that the achievable throughput

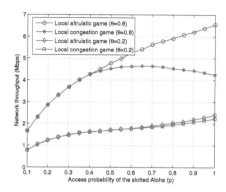

Figure 13.9: Expected network throughput for large interference range ($D_I = 400$m).

for both local altruistic game and local congestion game increases non-linearly with p. The reason is that in this case, the spectrum opportunities become limited, and then the number of competing neighbors could not be ignored anymore. In particular, it is noted that the achievable network throughput for the local altruistic game keeps increasing when p increases. On the other hand, there is a peak in the achievable expected network throughput for the local congestion game (i.e., $p_{max} \approx 0.65$), as can be expected in any Aloha transmission mechanism.

It is also observed that when the access probability is less than a value, i.e., $p \leq 0.4$, the obtained network throughput of local congestion game is close to that of local altruistic game. However, as the access probability increases, i.e., $p > 0.4$, there is an increasing throughput gap. The reason is as follows: $(1-p)^{s_n}$ decreases significantly when p increases, which makes the connection between the network collision minimization problem $P2$ and the network throughput maximization problem $P1$ weak. In other words, there exists a channel selection profile that minimizes the network collision, whereas it does not maximize the network throughput.

For a smaller channel idle probability, i.e., $\theta = 0.2$, the achievable network throughput for local altruistic game is slightly greater than that of the local congestion game.

13.3.6 Discussion

The most important characteristics of the discussed local altruistic game and local congestion game are that global optimum is achieved via local information exchange. The key point is that the game models are carefully designed, such that the utility functions are properly aligned with the global objectives. It is believed that the methodology and results presented provide a good understanding for distributed decision problems. In particular, the rationale behind local altruistic game have been applied to solve other active research problems in wireless networks, e.g., heterogeneous network selection [35], and joint user scheduling and power allocation in Long Term Evolution-Advanced (LTE-A) multi-cell networks [36].

13.4   Robust Game for Dynamic Opportunistic Spectrum Sharing

In most existing studies, the spectrum opportunities are assumed to be static during the learning procedure. Such an assumption leads to tractability but may not be true in practice. In this section, a robust game model for opportunistic spectrum sharing network with time-varying spectrum opportunities is presented. Note that the main analysis and results were presented in [37].

13.4.1   System Model and Problem Formulation

Consider an opportunistic spectrum sharing system involving N users and M licensed channels with transmission rate R_m, $1 \le m \le M$, $N > M > 1$. It is assumed that the primary users occupy the licensed channels in a slotted fashion, and each channel m is idle with probability θ_m in each slot. To make it more practical, the following dynamic and incomplete information constraints are considered: (i) dynamic: the occupation states of the channels are randomly and independently changing from slot to slot, and (ii) incomplete: the channel availability probabilities θ_m, $1 \le m \le M$, are fixed but unknown to the users. There is no centralized controller and no information exchange between the users.

Due to hardware limitation, the users can only select one channel for transmission in a slot [25]. For simplicity, it is assumed that the channel sensing is perfect[2]. If the selected channel is sensed to be idle, the carrier sense multiple access (CSMA) mechanism is used to resolve collision among the users sharing the channel [25]; otherwise, the users keep silent. At the end of the slot, each user receives a random payoff, which is determined by the channel state and the contention among the users, and updates its channel selection strategy using a learning procedure. The basic CSMA mechanism is considered, in which time is divided into mini-slots with equal length, and each user contends for the channel with the same probability p_a in each mini-slot. A channel contention is successful if no other users contend in the same mini-slot. After a successful contention, the successful user transmits data in the residual time slot, and all other competing users keep silent until the next slot. For presentation, denote the useful time after sensing in a slot as T_e, and the mini-slot length as τ.

Denote a_n as the channel selection of user n, c_m as the set of users selecting channel m for transmission, i.e., $c_m = \{n \in \{1,\dots,N\} : a_n = m\}$, and s_m as the number of users in set c_m, i.e., $s_m = |c_m|$. Define the random reward received of each user n in the jth slot as the normalized effective transmission time, which is given by:

$$r_n(j) = \left[(T_e - N_c(s_m)\tau)/T_e\right]\beta_n(s_m)I_m R_m, \tag{13.19}$$

where T_e is the useful time after channel sensing, τ is the length of a mini-slot in the contention period, $N_c(s_m)$ is the number of mini-slots for achieving a successful channel contention when there are s_m users contending for the channel, $\beta_n(s_m)$ indicates whether user n successfully contends or not, and I_m indicates whether channel m is idle or occupied. Note that this rewarding strategy captures the dynamics of the random spectrum opportunities and the interactions among multiple competing users. $N_c(s_m)$ is a geometric random variable [38] with the following probability mass function (PMF):

$$\Pr\{N_c(s_m) = i\} = p_s(1 - p_s)^{i-1}, i \ge 1, \tag{13.20}$$

where $p_s = s_m p_a (1 - p_a)^{s_m - 1}$ represents the overall successful channel contention probability in a mini-slot. $\beta_n(s_m)$ and I_m are Bernoulli random variables with the following PMFs:

$$\Pr\{\beta_n(s_m) = x\} = \begin{cases} \frac{1}{s_m}, & x = 1 \\ 1 - \frac{1}{s_m}, & x = 0 \end{cases}, \tag{13.21}$$

---

[2]However, it would be pointed out that the presented analysis can easily be extended to scenarios with imperfect spectrum sensing.

$$\Pr\{I_m = y\} = \begin{cases} \theta_m, & y = 1 \\ 1 - \theta_m, & y = 0 \end{cases}. \tag{13.22}$$

For presentation, define a *throughput loss function* as follows:

$$f(s_m) = \mathbf{E}[T_e - N_c(s_m)\tau]/T_e, \tag{13.23}$$

where $\mathbf{E}[\cdot]$ takes the expectation over $N_c(s_m)$. Based on (13.19)–(13.23), the expected achievable throughput of user n in a slot can be expressed as:

$$\bar{r}_n(j) = \frac{\theta_m f(s_m) R_m}{s_m}, \tag{13.24}$$

and the system throughput (the aggregate achievable throughput of all the users), is given by:

$$U_s(a) = \sum_{n=1}^{N} \frac{\theta_{a_n} f(s_{a_n}) R_{a_n}}{s_{a_n}} = \sum_{m=1}^{M} \theta_m f(s_m) R_m \delta(s_m), \tag{13.25}$$

where $a = (a_1, \dots, a_N)$ is the channel selection profile for the users, and $\delta(s_m)$ is the following indicator function:

$$\delta_m = \begin{cases} 1, s_m \geq 1 \\ 0, s_m = 0 \end{cases}. \tag{13.26}$$

Then the optimization objective is to maximize the system throughput. However, this task is challenging, since (i) there is no control center and no information exchange among the users, and (ii) the key parameters θ_m and N are unknown. To solve this problem, a game-theoretic distributed learning solution is proposed.

13.4.2 Robust Game Model

The problem of OSS in a dynamic and unknown environment is formulated as a robust game. Formally, the game is denoted by $\mathcal{G}_c = [\mathcal{N}, \{A_n\}_{n\in\mathcal{N}}, \{u_n\}_{n\in\mathcal{N}}]$, where $\mathcal{N} = \{1, \dots, N\}$ is the player (user) set, $A_n = \{1, \dots, M\}$ is the available action (channel) set of player n, and u_n is the utility function, which is defined as the expected achievable throughput, i.e.,

$$u_n(a_n, a_{-n}) \triangleq \mathbf{E}[r_n|(a_n, a_{-n})] = \frac{\theta_{a_n} f(s_{a_n}) R_{a_n}}{s_{a_n}}, \tag{13.27}$$

where r_n is the random reward received by player n, as specified by (13.19), $a_n \in A_n$ represents the channel selection of player n, and $a_{-n} \in A_1 \times \cdots \times A_{n-1} \times A_{n+1} \times \cdots \times A_N$ represents a channel selection profile of all the players, excluding n, where $\times$ denotes the Cartesian product. Note that different from previous games with deterministic utility functions, the utility function in the game involves random components. This is why it is called robust game.

According to Definition 12.1, NE in the game can be expressed in the following form. A channel selection profile $a^* = (a_1^*, \dots, a_N^*)$ is a pure strategy NE point of $\mathcal{G}_c$ if and only if no user can improve its utility function by deviating unilaterally, i.e.,

$$\frac{\theta_{a_n^*} f(s_{a_n^*}) R_{a_n^*}}{s_{a_n^*}} \geq \frac{\theta_{a_n} f(s_{a_n} + 1) R_{a_n}}{s_{a_n} + 1}, \forall n \in \mathcal{N}, \forall a_n \in A_n \backslash \{a_n^*\}, \tag{13.28}$$

where $f()$ is the throughput loss function specified by (13.23), $s_{a_n^*}$ is the number of users selecting channel a_n^*, and $A_n \backslash a_n$ means that a_n is excluded from A_n.

Theorem 13.5

The robust spectrum access game is an exact potential game that has at least one pure strategy NE point.

Algorithm 2: Genie-aided algorithm for achieving Nash equilibria

---

Step 1: Initially, set $\mathcal{N}_1 = \mathcal{N}$, $\mathcal{N}_2 = \emptyset$, $k = 1$, and $s_m(k) = 0$, $1 \leq m \leq M$, where $\emptyset$ is the null set.

Step 2: Randomly select a user $n \in \mathcal{N}_1$ and let it select channel m^*, i.e., $a_n = m^*$, where m^* is determined by:

$$m^* \in \underset{1 \leq m \leq M}{\arg\ \max} \left[\frac{\theta_m f(s_m(k) + 1) R_m}{s_m(k) + 1} \right]. \tag{13.29}$$

That is, m^* is the one that leads to the maximum individual throughput.

Step 3: Exclude n from $\mathcal{N}_1$ and include it in $\mathcal{N}_2$, i.e., $\mathcal{N}_1 = \mathcal{N}_1 \setminus n$ and $\mathcal{N}_2 = \mathcal{N}_2 \cup n$. Then, update $\{s_1(k), \ldots, s_M(k)\}$ according to the following rules:

$$\begin{aligned} s_m(k+1) &= s_m(k) + 1, \quad m = m^* \\ s_m(k+1) &= s_m(k), \qquad m \neq m^* \end{aligned} . \tag{13.30}$$

Step 4: If $\mathcal{N}_1 = \emptyset$, stop; else go to Step 2.

Proof 13.5   Refer to [37].

Although Theorem 13.5 indicates that the proposed robust spectrum access game has at least one pure strategy NE, the total number of NEs is still unknown [2]. In the following, it is assumed that there is an omnipotent genie who knows the channel idle probabilities and can perfectly monitor all selections made by the users in each iteration. Based on this assumption, a genie-aided spectrum access algorithm is proposed, as described by Algorithm 2, to find all NE points of $\mathcal{G}_c$.   ■

Theorem 13.6

The proposed genie-aided selection algorithm converges to a pure strategy NE point of $\mathcal{G}_c$.

Proof 13.6   Refer to [37].

In Step (2) of the proposed genie-aided algorithm, there may be multiple channels simultaneously resulting in the maximum individual throughput for the selected user. That is, these channels are indistinguishable for the user. For general scenarios, it is hard to investigate the achievable system throughput of NE solutions. However, under the assumption (it is tagged as **A1**) that the channels are distinguishable in each iteration, the achievable throughput of NE solutions can be investigated.   ■

Theorem 13.7

Under assumption A1, the robust spectrum access has multiple pure strategy NE points, and all of them lead to the same system throughput.

Proof 13.7   Refer to [37].

Besides throughput, fairness is another key concern in wireless communication systems. In the following, the fairness of the game using Jain's fairness index (JFI) [39] is investigated, which is defined as:

$$
J_{\mathcal{G}_c} = \frac{\left(\sum_{n=1}^{N} u_n\right)^2}{N \sum_{n=1}^{N} u_n^2}, \tag{13.31}
$$

where $u_n \geq 0$, $\forall n$, denotes the expected achievable throughput of user n. JFI translates a resource allocation vector $\{u_1, \ldots, u_N\}$ into a score in the interval of $[1/N, 1]$ and higher JFI implies that the resource allocation is more fair. In particular, $J_{\mathcal{G}_c} = 1$ corresponds to an absolute fair scenario, in which all the users get the same amount of resource, i.e., $u_n = u_0, \forall n$, and $J_{\mathcal{G}_c} = 1/N$ corresponds to an absolute unfair scenario, in which there exists a user, such that $u_n > 0$ and $u_k = 0, \forall k \neq n$. The following theorem characterizes the achieved fairness of the game. ▪

Theorem 13.8
If the channel contention is negligible, $J_{\mathcal{G}_c}$ is no less than $8/9$.

Proof 13.8 Refer to [37].
Although the channel contention overhead cannot be ignored in general scenarios, it is believed that the game-theoretic solution can still achieve good fairness, which will be verified by simulation results. ▪

13.4.3 Stochastic Learning for Achieving Nash Equilibria

There are some learning algorithms converging toward pure strategy NE points for exact potential games, such as regret learning [26], best (better) response dynamic [18], spatial adaptive play [32, 33], and fictitious play [22]. However, these algorithms are only suitable for static scenarios and can not be applied for the considered dynamic OSS system. To solve this problem, in the following, a stochastic learning automata (SLA) [40] based distributed spectrum access algorithm is proposed, with which the users learn from their past action-reward information and gradually adjust their behaviors toward an NE.

To implement the SLA-based distributed learning algorithm, the robust game $\mathcal{G}_c$ is also extended to a mixed strategy. Denote $\mathbf{P} = (\mathbf{p}_1, \ldots, \mathbf{p}_N)$ as a mixed strategy profile of all the users, where $\mathbf{p}_n = (p_{n1}, \ldots, p_{nM})$, and $\forall n \in \mathcal{N}$ is the channel selection probability vector of user n. In particular, p_{nm} is the probability that user n selects channel m. Denote $h_{nm}(\mathbf{P})$ as the expected reward of user n if it employs pure strategy m (i.e., $a_n = m$) but all other users $k, \forall k \in \mathcal{N}$, and $k \neq n$, employ mixed strategy $\mathbf{p}_k$. Formally,

$$
h_{nm}(\mathbf{P}) = \sum_{a_k, k \neq n} u_n(a_1, \ldots, a_{n-1}, m, a_{n+1} \ldots, a_N) \prod_{k \neq n} p_{ka_k}. \tag{13.32}
$$

The proposed SLA-based distributed spectrum access algorithm is described in Algorithm 3. In the strategy update rule of (13.33), $r_n(j)$ serves as a reinforcement signal. Specifically, if a channel is selected and it feeds back a positive reward, i.e., $r_n(j) > 0$, the probability of selecting the channel in the next slot increases. In contrast, if the fed reward is zero, i.e., $r_n(j) = 0$, the probability of selecting the channel in the next slot remains unchanged. Moreover, the proposed learning algorithm is completely distributed, as it is only dependent on the action-reward information of each user. Furthermore, it neither needs any information exchange, nor monitors the actions taken by other users.

In the following, the convergence of the SLA-based algorithm is proved. First, using the ordinary

---

Algorithm 3: SLA-based distributed spectrum access algorithm

Step 1: Initially, set $j = 0$ and the initial channel selection probability vector $p_{nm}(j) = 1/M, \forall n \in \mathcal{N}, m \in \{1, \ldots, M\}$.

Step 2: At the beginning of the jth slot, each user n selects a channel $a_n(j)$ according to its current channel selection probability vector $\mathbf{p}_n(j)$.

Step 3: In each slot, the users perform channel sensing and channel contention. At the end of the jth slot, each user n receives the random reward $r_n(j)$, specified by (13.19).

Step 4: All the users update their channel selection probability vectors according to the following rule:

$$\begin{aligned} p_{nm}(j+1) &= p_{nm}(j) + b\tilde{r}_n(j)(1 - p_{nm}(j)), m = a_n(j) \\ p_{nm}(j+1) &= p_{nm}(j) - b\tilde{r}_n(j)p_{nm}(j), \qquad m \neq a_n(j) \end{aligned}, \tag{13.33}$$

where $0 < b < 1$ is the step size, and $\tilde{r}_n(j)$ is the normalized reward defined as follows:

$$\tilde{r}_n(j) = r_n(j)/(\max_m R_m) = r_n(j)/R_{\max}. \tag{13.34}$$

Step 5: If $\forall n \in \mathcal{N}$, there exists a component of $p_n(j)$, which is approaching one, e.g., larger than 0.99, stop; otherwise, go to Step 2.

---

differential equation (ODE), the long-term behavior of the sequence $\{\mathbf{P}(j)\}$ can be analyzed, and the relationship between the stable points of the ODE and the Nash equilibria of $\mathcal{G}_c$ can be characterized. Secondly, a sufficient condition to achieve NE points for the SLA-based algorithm is established, and it is proved that this condition is satisfied in the proposed robust spectrum access game.

Proposition 13.1
With a sufficiently small step size b, the sequence $\{\mathbf{P}(j)\}$ will converge to $\mathbf{P}^*$, which is the solution of the following ODE:

$$\frac{d\mathbf{P}}{dt} = F(\mathbf{P}), \mathbf{P}_0 = \mathbf{P}(0), \tag{13.35}$$

where $\mathbf{P}_0$ is the initial channel selection probability matrix, and $F(\mathbf{P})$ is the conditional expected function defined as:

$$F(\mathbf{P}) = \mathbf{E}[G(\mathbf{P}(j), \mathbf{a}(j), \mathbf{r}(j))|\mathbf{P}(j)]. \tag{13.36}$$

In (13.36), $G(\mathbf{P}(j), \mathbf{a}(j), \mathbf{r}(j)) = \mathbf{P}(j+1)$ represents the updating rules specified by (13.33).

Proof 13.9   Refer to Theorem 3.1 in [40].   ■

Proposition 13.2
The following are true of the SLA-based algorithm:

1. *All the stable stationary points of (13.35) are the Nash equilibria of $\mathcal{G}_c$.*

2. *All the Nash equilibria of $\mathcal{G}_c$ are the stable stationary points of (13.35).*

Proof 13.10   Refer to Theorem 3.2 in [40].   ■

Theorem 13.9

Suppose that there is a non-negative function $H(\mathbf{P}) : \mathbf{P} \rightarrow R$ for some positive constant c, such that:

$$H(m_1, \mathbf{P}_{-n}) - H(m_2, \mathbf{P}_{-n})$$
$$= c[h_{nm_1}(\mathbf{P}) - h_{nm_2}(\mathbf{P})], \forall n, m_1, m_2, \mathbf{P}, \tag{13.37}$$

where $H(m, \mathbf{P}_{-n})$ is the value of H on the condition that $\mathbf{p}_n$ is a unit vector with the mth component unity, and $h_{nk}(\mathbf{P})$ is specified by (13.32). Then, the SLA-based algorithm converges to a pure strategy NE point of a game.

Proof 13.11   Refer to [37].

Theorem 13.9 establishes a sufficient condition that can guarantee the convergence toward NE. Next, it is proved that the proposed robust spectrum access $\mathcal{G}_c$ satisfies this condition and, hence, it converges to a pure strategy NE point by using the SLA-based learning algorithm.   ■

Theorem 13.10

With a sufficiently small step size b, the proposed SLA-based distributed spectrum access algorithm converges to a pure NE point of $\mathcal{G}_c$.

Proof 13.12   Refer to [37].   ■

13.4.4   Simulation Results

In this subsection, the convergence and the throughput performance of the SLA-based distributed spectrum access algorithm is studied. The slot length is set to $T = 100 \times 10^{-3}$s and the length for spectrum sensing is $T_s = 5 \times 10^{-3}$s. As a result, the effective time spectrum sensing is $T_e = 95 \times 10^{-3}$s. The mini-slot length is set to $\tau = 2 \times 10^{-3}$s, and the access probabilities of all the users is set to $p_a = 0.3$. In addition, the step size of the learning algorithm is set to $b = 0.15$.

The simulation results mainly include two parts: (1) convergence, and (2) throughput and fairness evaluation. Specifically, the achievable system throughput and fairness of the following three schemes are compared: (i) the SLA-based learning algorithm, (ii) the exhaustive search, and (iii) the random selection approach. In the exhaustive search, it is assumed that there is an omnipotent controller that knows all the system parameters, including the channel idle probabilities, θ_m, and the number of users, N, and solves the problem in a centralized manner. In the random selection scheme, the users select the channels with equal probability in each slot.

13.4.4.1   Convergence

Consider an OSS system with six users and three licensed channels. The channel rates and idle probabilities are set to: $R_1 = 2$, $R_2 = 1.5$, $R_3 = 1$, and $\theta_1 = 0.6$, $\theta_2 = 0.7$, $\theta_3 = 0.6$. Using the genie-aided algorithm, the channel selection profile of NE in this case is given by: $s_1 = 3, s_2 = 2, s_3 = 1$. That is, there are three users selecting channel-1, two users selecting channel-2 and one user selecting channel-3 in the NE solution.

The evolution of the channel selection probabilities of an arbitrarily chosen user is plotted in Figure 13.10. It is observed that the selection probability vector evolves from $\{1/3, 1/3, 1/3\}$ to $\{0, 1, 0\}$ in about 250 iterations, which implies that the user finally selects channel-2 for transmission. Moreover, the evolution of the number of users selecting each channel is shown in Figure 13.11. It is seen that the algorithm finally converges to a channel selection result of $s_1^* = 3, s_2^* = 2, s_3^* = 1$, which is exactly the NE solution.

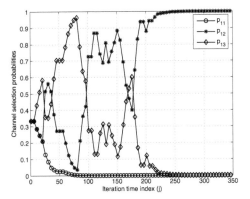

Figure 13.10: Evolution of the channel selection probability of an arbitrarily user ($N = 6$, $R_1 = 2$, $R_2 = 1.5$, $R_3 = 1$, $\theta_1 = 0.6$, $\theta_2 = 0.7$, $\theta_3 = 0.6$).

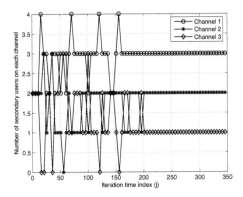

Figure 13.11: Evolution of the number of users selecting each channel ($N = 6$, $R_1 = 2$, $R_2 = 1.5$, $R_3 = 1$, $\theta_1 = 0.6$, $\theta_2 = 0.7$, $\theta_3 = 0.6$).

13.4.4.2 Throughput Performance of Homogeneous OSS Systems

Consider a homogeneous open systems approach (OSA) system, where the licensed channels have the same transmission rates and the same idle probabilities. The system parameters are set to: $R_1 = R_2 = R_3 = 1$, and $\theta_1 = \theta_2 = \theta_3 = 0.6$. Figure 13.12 shows the comparison results of the achievable system throughput when increasing the number of users. For the proposed learning algorithm and random selection approach, the simulation results are obtained by independently simulating 10^5 trials and then taking the average results. It is seen that the proposed learning algorithm significantly outperforms the random selection approach. Furthermore, it is also noted that when the number of users becomes large, e.g., $N \geq 4$, the achievable system throughput of the learning algorithm is close to that of the exhaustive search.

Moreover, the JFI of the three schemes are shown in Figure 13.13, where J_e, $J_{\mathcal{G}_c}$, and J_r denote the JFI of the exhaustive search, the proposed learning algorithm, and the random selection approach, respectively. It is noted that the random selection approach achieves perfect fairness ($J_r \approx 1$), while the proposed learning algorithm and the exhaustive search can also achieve good fairness (both $J_{\mathcal{G}_c}$ and J_e are greater than 0.90).

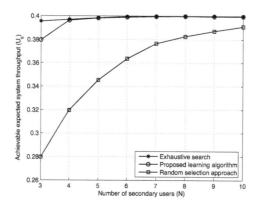

Figure 13.12: Comparison of the achievable system throughput of three channel selection schemes in the homogeneous OSS system ($\theta_1 = \theta_2 = \theta_3 = 0.6$).

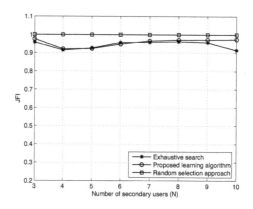

Figure 13.13: Comparison results of the JFI of three channel selection schemes in the homogeneous OSS system ($\theta_1 = \theta_2 = \theta_3 = 0.6$).

13.4.4.3 Throughput Performance of Heterogeneous OSA Systems

Consider a heterogeneous OSA system, where the variation of the licensed channels is large. The system parameters are set as follows: $R_1 = R_2 = R_3 = 1$, and $\theta_1 = 0.2, \theta_2 = 0.4, \theta_3 = 0.8$. Figure 13.14 shows the comparison results of the achievable system throughput when increasing the number of users. It is noted that the achievable system throughput of the proposed learning algorithm is greater than that of the random selection approach. Moreover, as the number of the users N increases, the proposed learning algorithm and exhaustive search algorithms perform closely. This phenomena is referred to as the *price of anarchy*, which is the inherent limitation of the NE solution, and has been well discussed in [41].

The JFI of the three channel selection schemes is shown in Figure 13.15. It is noted in this case that, the random selection approach still achieves perfect fairness, as expected. However, the exhaustive search achieves poor fairness (J_e is about 0.8), whereas the proposed learning algorithm still achieves good fairness (J_{g_c} is greater than 0.95).

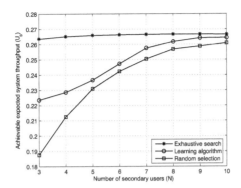

Figure 13.14: Comparison of the achievable system throughput of three channel selection schemes in the heterogeneous system ($\theta_1 = 0.2, \theta_2 = 0.4, \theta_3 = 0.8$).

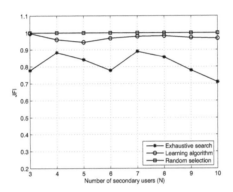

Figure 13.15: Comparison results of the JFI of three channel selection schemes in the heterogeneous system ($\theta_1 = 0.2, \theta_2 = 0.4, \theta_3 = 0.8$).

13.4.5   *Discussion*

From the simulation results, it can be realized that the proposed SLA-based learning algorithm can achieve high system throughput while guaranteeing good fairness. Furthermore, the algorithm is robust with increasing the number of users. Thus, it is claimd that the SLA-based algorithm is desirable in OSS systems where the channel idle probabilities are prior unknown and there is no information exchange among the users.

It is observed from Theorem 13.10 that when b approaches zero, the proposed SLA-based learning algorithm finally converges to an NE point. However, smaller step size b implies a slower convergence speed. Hence, the choice of the step size b involves a tradeoff between accuracy and speed, and is application-dependent. This can be done through practical experiments or training [42, 43].

According to the analysis, it is known that the SLA-based algorithm converges to a pure strategy NE point for any exact potential game. Actually, the SLA-based distributed learning algorithms have been successfully applied in our other recent studies, e.g., distributed interference mitigation in time-varying environment [44], and media access control (MAC)-layer interference mitigation in spatial OSS networks [45].

13.5   Future Directions and Challenges

In this section, some future directions and challenges for the application of game-theoretic solutions OSS are presented.

- **QoE-oriental game-theoretic optimization.** Basically, the purpose of wireless communications is to serve users. Thus, the user perception, i.e., quality of experience (QoE) [46,47], should be included in the game design. However, most game-theoretic models and solutions for resource optimization problems in the existing literature cared more about the achievable throughput instead of QoE. Therefore, it is important and timely to design new QoE-oriental optimization models.

  In methodology, the QoE-oriental optimization differs from previous throughput-oriental optimization significantly. Some new challenges are: (i) how to evaluate the individual user QoE in a complicated wireless environment, and design an efficient model for system QoE evaluation, (ii) as the user perception is generally subjective and discrete, e.g., a user may feel "excellent," "good," "fair," "poor," and "bad" by the method of the mean opinion score [47], it is interesting but challenging to design discrete-QoE-aware solutions. Some preliminary results on QoE-oriental game-theoretic solutions were reported in [48,49].

- **Dynamic spectrum access with carrier aggregation.** Most existing game-theoretic OSS only considered simple scenarios with single channel, which may not achieve high data rates. To achieve the peak data rates required by IMT-Advanced, 3GPP long term evolution (LTE) Release 10 has introduced carrier aggregation (CA) [50]. To further expand LTE capacity, integration of unlicensed carriers (unlicensed spectrum) into the LTE system (called LTE-U) has been proposed [51–53], in which CA is also the key enabling technology. With CA technology, it allows scalable expansion of effective bandwidth provided to a user terminal through simultaneous utilization of radio resources across multiple carriers. It increases the usable spectrum by aggregating resource blocks (RBs), either within a given band or in different frequency bands.

  The new challenges caused by dynamic spectrum access with CA are summarized as follows: (i) generally, there are three different CA types: intra-band contiguous CA, intra-band non-contiguous CA, and inter-band non-contiguous CA. Then, the first task is to formulate the game models properly, taking into account the simultaneous transmission of multiple channels and aggregation cost caused by non-continuous channels [54]; (ii) as the devices in LTE-U belong to different systems and holders, it is important to design efficient spectrum access mechanisms to combat with cheat and malicious behaviors. A few preliminary results on game-theoretic CA were reported in [54,55].

- **Distributed optimization in ultra-dense small cell networks.** The explosive proliferation of the smart devices and wireless services has led to a tremendous increase of the wireless communication traffic. It is reported that the wireless traffic will grow at an annual growth rate of 78% from 2011 to 2016 [56]. Unfortunately, the conventional macro-cell can no longer meet the requirements for higher capacity and throughput. The ultra-dense deployment of small cells with low power access points is considered as an efficient solution to meet a higher data rate in wireless networks [57], because small cells can significantly improve the efficiency of the frequency reuse and spectrum sharing.

  However, a number of key technical challenges have emerged in the small cell wireless networks, such as the diversity distribution of the wireless traffic and the interference management. First, the distribution of mobile traffic varies significantly in time and space, as most of the data requirements come from hot time period and hot spots [58]. Second, the large-scale and self-organized deployment of small cells will inevitably result in the cross-tier interference between small cell and macro-cell and the co-tier interference among small cells. Considering these challenges, the traditional approaches are not applicable in the ultra-dense small cell networks. The combination of game theory and distributed learning would also provide efficient solutions to this problem. We have done some preliminary work on this topic [59, 60].

References

[1] S. Haykin, "Cognitive radio: Brain-empowered wireless communications," *IEEE Journal on Selected Areas in Communications*, vol. 23, no. 2, pp. 201–220, 2005.

[2] R. B. Myerson, Game Theory: Analysis of Conflict, 1991.

[3] Y. Xu, "Research on theory and approaches of game learning in opportunistic spectrum access," PhD dissertation, PLA University of Science and Technology, June, 2014.

[4] R. Axelrod and W. D. Hamilton, "The evolution of cooperation," *Science*, vol. 211, no. 4489, pp. 1390–1396, 1981.

[5] E. Fehr, and U. Fischbacher, "The nature of human altruism," *Nature*, vol. 425, no. 6960, pp. 785–791, 2003.

[6] C. S. Yeung, A. S. Poon, and F. F. Wu, "Game theoretical multi-agent modelling of coalition formation for multilateral trades," *IEEE Transactions on Power Systems,* vol. 14, no. 3, pp. 929–934, 1999.

[7] A. B. MacKenzie and S. B. Wicker, "Game theory and the design of self-configuring, adaptive wireless networks," *IEEE Communications Magazine,* vol. 39, no. 11, pp. 126–131, 2001.

[8] H. Yaïche, R. R. Mazumdar, and C. Rosenberg, "A game theoretic framework for bandwidth allocation and pricing in broadband networks," *IEEE/ACM Transactions on Networking (TON),* vol. 8, no. 5, pp. 667–678, 2000.

[9] D. Niyato, E. Hossain, and Z. Han, "Dynamics of multiple-seller and multiple-buyer spectrum trading in cognitive radio networks: A game-theoretic modeling approach," *IEEE Transactions on Mobile Computing,* vol. 8, no. 8, pp. 1009–1022, 2009.

[10] L. Gao, X. Wang, Y. Xu, and Q. Zhang, "Spectrum trading in cognitive radio networks: A contract-theoretic modeling approach,"*IEEE Journal on Selected Areas in Communications*, vol. 29, no. 4, pp. 843–855, 2011.

[11] D. Niyato and E. Hossain, "Spectrum trading in cognitive radio networks: A market-equilibrium-based approach," *IEEE Wireless Communications,* vol. 15, no. 6, pp. 71–80, 2008.

[12] D. Yang, G. Xue, X. Fang, and J. Tang, "Crowdsourcing to smartphones: incentive mechanism design for mobile phone sensing," in *Proceedings of the 18th Annual International Conference on Mobile Computing and Networking*, pp. 173–184, 2012.

[13] C. U. Saraydar, N. B. Mandayam, and D. Goodman, "Efficient power control via pricing in wireless data networks," *IEEE Transactions on Communications*, vol. 50, no. 2, pp. 291–303, 2002.

[14] V. Srivastava, J. O. Neel, A. B. MacKenzie, R. Menon, L. A. DaSilva, J. E. Hicks, J. H. Reed, and R. P. Gilles, "Using game theory to analyze wireless ad hoc networks," *IEEE Communications Surveys and Tutorials,* vol. 7, no. 1-4, pp. 46–56, 2005.

[15] K. Akkarajitsakul, E. Hossain, D. Niyato, and D. I. Kim, "Game theoretic approaches for multiple access in wireless networks: A survey," *IEEE Communications Surveys and Tutorials,* vol. 13, no. 3, pp. 372–395, 2011.

[16] R. Trestian, O. Ormond, and G.-M. Muntean, "Game theory-based network selection: Solutions and challenges," *IEEE Communications Surveys and Tutorials,* vol. 14, no. 4, pp. 1212–1231, 2012.

[17] J. Nash, "Non-cooperative games,"*Annals of mathematics,* pp. 286–295, 1951.

[18] D. Monderer. and L. S. Shapley, "Potential games," *Games and Economic Behavior*, vol. 14, pp. 124–143, 1996.

[19] R. J. Aumann, "Correlated equilibrium as an expression of Bayesian rationality," *Econometrica: Journal of the Econometric Society,* pp. 1–18, 1987.

[20] P. D. Taylor and L. B. Jonker, "Evolutionary stable strategies and game dynamics," *Mathematical Biosciences,* vol. 40, no. 1, pp. 145–156, 1978.

[21] M. P. Wellman and J. Hu, "Conjectural equilibrium in multiagent learning," *Machine Learning,* vol. 33, no. 2-3, pp. 179–200, 1998.

[22] J. Marden, G. Arslan, and J. Shamma, "Joint strategy fictitious play with inertia for potential games," *IEEE Transactions on Automatic Control*, vol. 54, no. 2, pp. 208 –220, 2009.

[23] Y. Xu, J. Wang, Q. Wu, et al., "Opportunistic spectrum access in cognitive radio networks: Global optimization using local interaction games," *IEEE Journal on Selected Topics in Signal Processing*, vol. 6, no. 2, pp. 180–194, 2012.

[24] "IEEE 802.16e-2005 and IEEE Std 802.16-2004/Cor1-2005," available: http://www.ieee802.org/16/.

[25] Q. Zhao, L. Tong, A. Swami, et al., "Decentralized cognitive MAC for opportunistic spectrum access in ad hoc networks: A POMDP framework," *IEEE Journal on Selected Areas in Communications*, vol. 25, no. 3, pp. 589–600, 2007.

[26] N. Nie and C. Comaniciu, "Adaptive channel allocation spectrum etiquette for cognitive radio networks," *Mobile Networks and Applications*, vol. 11, no. 6, pp. 779–797, 2006.

[27] J. Neel, "Analysis and design of cognitive radio and distributed radio resource management algorithms," PhD dissertation, Virginia Tech, September 2006.

[28] S. Arnborg, "Efficient algorithms for combinatorial problems on graphs with bounded decomposability—A survey," *BIT Numerical Mathematics*, vol. 25, no. 1, pp. 1–23, 1985.

[29] M. A. Nowak, "Five rules for the evolution of cooperation," *Science,* vol. 314, pp.1560–1563, 2006.

[30] F. C. Santos, M. D. Santos, and J. M. Pacheco, "Social diversity promotes the emergence of cooperation in public goods games," *Nature,* vol. 454, pp. 213–49, 2008.

[31] B. Vcking and R. Aachen, "Congestion games: Optimization in competition," in *Proceedings of the 2nd Algorithms and Complexity in Durham Workshop.* Kings College Publications, pp. 9–20, 2006.

[32] H. P. Young, *Individual Strategy and Social Structure.* Princeton University Press, 199

[33] J. Marden, G. Arslan, and J. Shamma, "Cooperative control and potential games," *IEEE Transactions on Systems, Man, and Cybernetics, Part B*, vol. 39, no. 6, pp. 1393–1407, 2009.

[34] Y. Song, C. Zhang, and Y. Fang, "Joint channel and power allocation in wireless mesh networks: A game theoretical perspective," *IEEE Journal on Selected Areas in Communications*, vol. 26, no. 7, pp. 1149–1159, 2008.

[35] Z. Du, Q. Wu, P. Yang, Y. Xu, and Y-D. Yao, "User demand aware wireless network selection: A localized cooperation approach," *IEEE Transactions on Vehicular Technology*, vol. 63, no. 9, pp. 4492–4507, 2014.

[36] J. Zheng, Y. Cai, Y. Liu, Y. Xu, B. Duan, and X. Shen, "Optimal power allocation and user scheduling in multicell networks: Base station cooperation using a game-theoretic approach," *IEEE Transactions on Wireless Communications*, vol. 13, no. 12, pp. 6928–6942, 2014.

[37] Y. Xu, J. Wang, Q. Wu, et al., "Opportunistic spectrum access in unknown dynamic environment: A game-theoretic stochastic learning solution," *IEEE Trans. Wireless Commun.*, vol. 11, no. 4, pp. 1380–1391, 2012.

[38] D. Zheng, W. Ge, and J. Zhang, "Distributed opportunistic scheduling for ad hoc networks with random access: An optimal stopping approach," *IEEE Transactions on Information Theory*, vol. 55, no. 1, pp. 205–222, 2009.

[39] R. Jain, D. Chiu, and W. Haws, "A quantitative measure of fairness and discrimination for resource allocation in shared computer system," *Technical Report*, 1984.

[40] P. Sastry, V. Phansalkar, and M. Thathachar, "Decentralized learning of nash equilibria in multi-person stochastic games with incomplete information," *IEEE Transactions on Systems, Man, and Cybernetics, Part B*, vol. 24, no. 5, pp. 769–777,

[41] L. Law, J. Huang, M. Liu, and S. R. Li, "Price of anarchy for cognitive MAC games," in *Pro. IEEE GLOBECOM 2009,* pp. 1–6, Dec., 2009.

[42] Y. Xing and R. Chandramouli, "Stochastic learning solution for distributed discrete power control game in wireless data networks," *IEEE/ACM Transactions on Networking*, vol. 16, no. 4, pp. 932–944, 2008.

[43] W. Zhong, Y. Xu, M. Tao, et al., "Game theoretic multimode precoding strategy selection for MIMO multiple access channels," *IEEE Signal Processing Letters*, vol. 17, no. 6, pp. 563–566, 2010.

[44] Q. Wu, Y. Xu, J. Wang, et al., "Distributed channel selection in time-varying radio environment: Interference mitigation game with uncoupled stochastic learning," *IEEE Transactions on Vehicular Technology*, vol. 62, no. 9, pp. 4524–4538, 2013.

[45] Y. Xu, Q. Wu, L. Shen, et al., "Opportunistic spectrum access with spatial reuse: Graphical game and uncoupled learning solutions," *IEEE Transactions on Wireless Communications*, vol. 12, no. 10, pp. 4814–4826, 2013.

[46] P. Brooks and B. Hestnes, "User measures of quality of experience: Why being objective and quantitative is important," *IEEE Network*, vol. 24, no. 2, pp. 8–13, 2010.

[47] J. A. Hassan, M. Hassan, S. K. Das, et al., "Managing quality of experience for wireless VOIP using noncooperative games," *IEEE Journal on Selected Areas in Communications*, vol. 30, no. 7, pp. 1193–1204, July 2012.

[48] Y. Xu, J. Wang, Q. Wu, Z. Du, L. Shen, and A. Anpalagan, "A game theoretic perspective on self-organizing optimization for cognitive small cells," *IEEE Communications Magazine*, to appear.

[49] Z. Du, Q. Wu, P. Yang, Y. Xu, J. Wang, and Y-D. Yao, "Exploiting user demand diversity in heterogeneous wireless networks," *IEEE Transactions on Wireless Communicstions*, to appear.

[50] S. Parkvall, A. Furuskar, and E. Dahlman, "Evolution of LTE toward IMT-Advanced," *IEEE Communications Magazine*, vol. 49, no. 2, Feb. 2011.

[51] A. Al-Dulaimi and S. Al-Rubaye, "5G communications race: Pursuit of more capacity triggers LTE in unlicensed band," *IEEE Vehicular Technology Magazine*, vol. 10, no. 1, pp. 43–51, 2015.

[52] F. Liu and E. Bala, "Small cell traffic balancing over licensed and unlicensed bands," *IEEE Transactions on Vehicular Technology*, 2015.

[53] M. Bennis and M. Simsek, "When cellular meets WiFi in wireless small cell networks," *IEEE Communications Magazine*, vol. 51, no. 6, pp. 44–50, 2013.

[54] H. Ahmadi, I. Macaluso, and L. A. DaSilva, "Carrier aggregation as a repeated game: Learning algorithms for convergence to a Nash equilibrium," *IEEE Global Communications Conference (GLOBECOM)*, 2013.

[55] Y. Zhang, C. Kan, and Y. Xu, "Distributed carrier aggregation in small cell networks: A game-theoretic approach," *KSII Transactions on Internet and Information Systems*, submitted.

[56] Cisco visual networking index: Global mobile data traffic forecast update, 2011–2016. Feb. 2012, white paper.

[57] J. G. Andrews, H. Claussen, M. Dohler, S. Rangan, and M. C. Reed, "Femtocells: Past, present, and future," *IEEE Journal on Selected Areas in Communications,* vol. 30, no. 3, pp. 497–508, 2012.

[58] G. Auer, V. Giannini, C. Desset, et al., "How much energy is needed to run a wireless network?" *IEEE Wireless Communication,* vol. 18, no. 5, pp. 40–49, Oct. 2011.

[59] Y. Xu, J. Wang, Y. Xu, et al., "Centralized-distributed spectrum access for small cell networks: A cloud-based game solution," *IEEE Transactions on Vehicular Technology,* under revision.

[60] Y. Xu, Y. Zhang, J. Chen, et al., "Distributed spectrum access for dynamic cognitive small cell networks: A robust graphical game approach," *IEEE Access,* submitted.

Chapter 14

An Adaptive Game Theoretic Framework for Self-coexistence among Cognitive Radio Enabled Smart Grid Networks

Deepak K Tosh and Shamik Sengupta

CONTENTS

14.1   Introduction

Wireless spectrum, the most valuable commodity of the current communication era, has been inefficiently utilized in different point of time and space. This sporadic use of licensed spectrum and overuse of unlicensed bands may not fulfill the data demands of ever-increasing future wireless applications and services, unless the "white spaces" (unused bands) are managed appropriately toward spectrum sharing and usage. A large portion of the spectrum has been allocated statically for TV, government, defenses, public safety etc., and these are used very intermittently, resulting in spectrum under-utilization. Additionally, this policy does not permit access to the unused spectrum resources to meet high network demand by which spectrum owners can generate additional revenue. The suboptimality of such static spectrum allocation policy has led the Federal Communication Commission (FCC) to propose a dynamic spectrum access (DSA) policy, which is expected to overcome the issues caused by static policy via adopting cognitive radio (CR) technology [1]. The CRs (a.k.a. secondary users) are intelligent radios that can perform periodic sensing to detect the absence of licensed or primary user (PU) in a band so it can access the chunk for data communication in an non-interfering manner. Once PU resumes the transmission in its allocated frequency, the secondary user (SU) must sense and switch to a different band promptly [2]. This way, the PUs and SUs can coexist in the same spectrum space, and SUs ensure that PU's ongoing communication will not be disrupted.

Cognitive radios are envisioned as generic smart radios that can take advantage of the unused spectrum bands by suitably adjusting their tranmission/reception parameters based on the sensed wireless environment. The infrequent use of sub-900 MHz TV bands, as found by FCC, led to build IEEE 802.22 wireless regional area network (WRAN) [3] standard, where the cognitive radio network (base station (BS) and consumer premise equipments (CPE)) [4] [5] opportunistically access the TV channels to maximize spectrum utilization and quality of service (QoS). The primary tasks of BS/CPEs are periodic sensing of the spectral space to avoid licensed incumbents' transmission and transmit/receive own data. However, BS additionally manages the on-air activities in its cell that include allocating network resources to CPEs, maintaining appropriate QoS, secure network admission mechanisms, etc. The impact of CR technology is not only limited to cellular/data services in rural/urban areas, but also to the recently emerging pervasive networks such as smart grid (SG) and cyber-physical systems (CPS).

Smart grid [6] has been recognized as an intelligent electrical network working in coherence with every component connected via reliable communication infrastructure to enable electricity demand management, resource efficiency, reliability gains, customer participation and cleaner energy across the electrical grids. SG requires a sophisticated communication backbone for serving applications like Advanced Metering Infrastructure (AMI), Automatic Demand Response (ADR), Feeder Automation (FA), Electric Vehicle (EV) charging, and Mobile Workflow Management (MWM) [7] [8] [9] [10].

The SG communication backbone is envisioned to operate in a multi-tier fashion extending from generating station to consumers' home as shown, in Figure 14.1. Such infrastructure comprises of the following tiers: (1) home area network (HAN), where home devices and appliances including smart meters are connected with each other to provide services for energy management and demand response. (2) neighborhood area network (NAN), which collects electrical usage/demand data from HANs and delivers to the remote grid center; and (3) wide area network (WAN) acts as a electrical backbone network between grid and core utility systems that is used to transport metering data collected from HANs and generation/transmission related data to the remote server for further analysis. For this hierarchical communication infrastructure, there are several options, such as cellular technologies, home broadband solutions, typical wired network via fiber-optical connections, etc. The diversity of communication technology hints that SG communication will most likely be using a mixture of a wide range of technologies. For example, short-range wireless technologies,

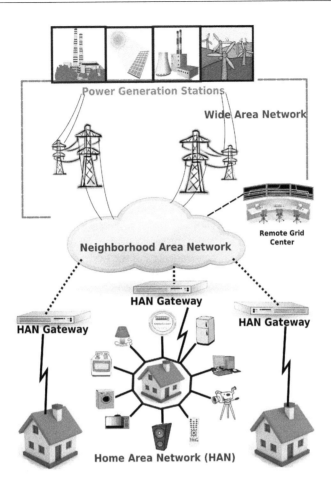

Power Generation Stations

Wide Area Network

Remote Grid Center

Neighborhood Area Network

HAN Gateway

HAN Gateway

HAN Gateway

Home Area Network (HAN)

Figure 14.1: Hierarchical Smart Grid Network Architecture.

like Bluetooth, ZigBee, and IEEE 802.11 could best suit to build the HAN. Cellular networks, like general packet radio service (GPRS), universal mobile telecommunications system (UMTS), 4G, long term evolution (LTE), worldwide interoperability for microwave access (WiMAX), could be used for building the WAN-side of the SG network. As wireless technologies have reached to a level of sophistication, they can be relied upon for providing high quality of service and also can be deployed at a low investment too. For providing such high range communication in SG WAN, it is suitable to adopt the wireless medium, and, most importantly, the DSA enabled cognitive radio technology that best exploits the unused spectrum bands in an efficient way. As the SG core networks and backhaul/distribution networks mostly run through a less-populated area, IEEE 802.22-based cognitive radio network can be a potential candidate for WAN communication infrastructure, which will definitely benefit SG operations in terms of demand management and cost reduction.

As different energy vendors might have their own smart grid structure and their wide area network backbone might share a common space, the SG nodes can have a different level of competition for the unused spectrum. Assuming SG nodes are DSA enabled cognitive radios, they will have to fight among each other to access the free spectrum resources and must be mechanized with a self-coexistence scheme that will find the unused spectrum at a minimum switching cost, thereby improving the spectral efficiency. However, devising such coordination mechanism in such bazaar like environment is indeed a challenging problem, hence it requires the SG networks to strategically access the spectrum. The self-coexistence problem becomes even more difficult when the spectrum chunks are heterogeneous in nature, as the SG networks always look forward to find the contention-free spectrum band of higher quality. Hence it is of utmost requirement to have a self-coexistence mechanism for SG networks and their communicating devices to operate in an intelligent and adaptive manner.

In this chapter, we have considered a game-theoretic and learning model to analyze the self-coexistence problem of smart grid networks (SGN) by adopting tools from behavioral game theory. We first model the competitive wireless environment as a distributed Modified Minority Game (MMG) [11], where several overlapping SG networks compete with their neighbors for an unused spectrum band to have data communication. If the acquired chunk has multiple occupants then each of them should opt one of the binary choices: {stay in the same band or switch to another}. The networks do not have any information about the choices taken by other competing entities, hence is played under incomplete information. This underscores the following questions: how each SG entity decide whether to cooperate or not in this minority game? Is there any equilibrium strategy existing in this game? How can the SG networks learn from their past actions to find a free spectrum? Using the proposed MMG model, it is anticipated that the SG networks will be able to take right decision at appropriate game stage without any direct knowledge about their rivals, such that an unused spectrum chunk can be discovered at minimum number of channel switches. This MMG approach to the self-coexistence problem has several benefits: (1) The SG networks can independently act to find an accessible spectrum without any help from a central authority or centralized allocation mechanism, thus the distributed approach makes the system scalable. (2) The non-cooperation among the SG networks prohibits any direct communication among each other, thereby reducing the communication overhead and resource wastage. (3) Using this robust mechanism the SG networks can maximize their own payoffs or minimize the channel switching cost, which will help to quickly find a free spectrum without wandering around. As a part of this chapter, we investigate the pure and mixed strategy of the modified minority game from each SG network's perspective and most importantly analyze the existence of any equilibrium strategies which will minimize the switching costs of each player[1] if played.

Though MMG mechanism proves and provides the Nash equilibrium strategy for a particular instance of self-coexistence game, it requires every SG network to know how many other SG networks are competing and how many spectrum bands are available. Hence CRNs need to use their potential and learn the effectiveness of their actions from each stage to figure out a contention-free band. This eventually maximizes their net reward over the game stages. This competition model mimics the situation of famous optimal foraging model [12] where a group of birds scavenge over different islands to find food sources of sufficient amount but they are constrained by the energy used to fly over the places and amount of food discovered. Hence the trade-off is to balance the total energy gain from the food sources and scavenging period such that they can survive for longer duration. If many birds start consuming the same food source, they might not gain enough energy to forage. In our situation the SG networks (birds) forage for unused spectrum bands (islands) in a distributed manner to maximize the reward and minimize the contentions (foraging duration). If the spectrum bands are assumed to be homogeneous (similar in transmission characteristics, e.g., bandwidth, data rate, modulation etc.) in nature, the SG networks will more likely fight to access a spectrum band that is free from interference; whereas if the bands are of heterogeneous (distinct transmission characteristics) nature, the SG networks not only fight to occupy a contention-free band but also look for a high utility band at the same time. To adapt in such a chaotic environment, we propose a perception-based learning model to find a homogeneous spectrum chunk void of in-

---

[1]Throughout this chapter, the words "SG networks", "players", "SG nodes", "SG entities" are used interchangeably unless explicitly mentioned otherwise.

terference. A regret minimization based heuristic is presented later to address the self-coexistence issues when the spectrum resources are heterogeneous in nature.

14.2 Challenges and Related Works

Since data transportation of many future real-time technologies is relying on the wireless access medium, it is very important to efficiently manage the spectrum space so that many of such heterogeneous communication networks can coexist. SG [6] is one such technology that will be shaping the future energy sector by enabling efficient demand management, resource utilization, customer satisfaction, grid reliability, etc. To achieve these, SG requires a promising communications technology, like cellular networks, 4G/long term evolution (LTE) equivalent networks, which provide high performance and reliable services. Instead of adopting the costliest communication technologies in a vast SG network, it will be wise to choose dynamic spectrum access triggered communication as the spectrum space is found to be underutilized in spatial and temporal manner. As SG NAN and WAN contain multiple heterogeneous communication networks, proper spectrum management mechanisms need to be devised to effectively utilize the spectrum holes toward successful communication. Hence, self-coexistence among these DSA-enabled SG networks is a research challenge that must be addressed so that every network in the SG environment can serve better. Since the devices in SG networks are equipped with CRs, an intelligent breed of software defined radio (SDR) technology, that allows radio frequency (RF) operating parameters, such as frequency range, modulation type, output power, etc., to be set or altered by software, self-coexistence mechanisms can be well adapted by the SG networks and their devices without any drastic modification in radio architecture.

Research advancements in SDR technology have produced novel algorithms, architectures, and protocols for dynamic spectrum access enabled CRs. A great body of research has been conducted in the field of dynamic spectrum sensing and access, which deals with different decision-making aspects and challenges in CR network settings. [13] provided a proactive spectrum access approach that aimed to build predictive models on spectrum availability for the SUs using their past observations and minimizing interference to primary users. To monitor and detect PU activity in the spectrum space, energy detection and interference temperature measurement techniques have been largely used in [14]–[16]. Spectral correlation-based signal detection for primary spectrum sensing in IEEE 802.22 WRAN systems is presented in [17]. [18] used signature-based spectrum sensing algorithms to analyze the presence of advanced television systems committee (ATSC) Digital television (DTV) signals. To detect the primary user activity in IEEE 802.22 CR networks, [19] presented another mechanism called sequential pilot sensing of advanced television systems committee (ATSC) DTV signals. [20] proposed a novel dynamic frequency hopping (DFH) method that requires cooperation from neighboring WRAN cells to coordinate the DFH operations so that WRAN data transmission can be carried out in parallel with spectrum sensing without any interruptions. This technique helps to minimize the interrupts due to quiet sensing and increase the quality of service. Most of the above-mentioned works focus on PU detection, spectrum sensing, and primary-secondary etiquettes. However, the issues of self-coexistence among secondaries in a (un)coordinated DSA environment are not addressed specifically.

To address the self-coexistence issues among IEEE 802.22 base stations (BS), [5] proposed a utility graph coloring technique to allocate unused spectrum to the BSs. [21] proposed an uplink soft frequency reuse (USFR) technique to address the co-channel self-coexistence issue, where the uplink resource allocation is solved by decoupling to two subproblems: subchannel allocation and transmit power allocation. A round-robin based resource allocation algorithm was proposed for IEEE 802.22 WRAN in [22], which fairly allocates spectrum resources among the WRANs to improve spectrum utilization. Several novel approaches on dynamic spectrum allocation techniques for competing secondary users are discussed in [23–26] that all consider a spectrum broker who has knowledge on the spectrum availability. The spectrum broker acts as the centralized decision-making agent via auction and pricing mechanisms. "Economic behavior" of network nodes is studied in [27] under a constrained channel capacity scenario and assuming different purchasing strategies for accessing relay nodes, access points, and clients in mesh networks; however, dynamic spectrum access technology was not considered in this research work.

The auction mechanisms and pricing models have potential to generate revenue through efficient commercialized secondary spectrum usage. However, there exist several issues and challenges in terms of implementing pricing in the DSA scenario, such as payment transaction method, best-effort-service nature of opportunistic spectrum access, trustworthiness, authentication, and many more. As SG networks will opportunistically share the spectrum under the presence of licensed incumbents, the self-coexistence among the SG networks and their devices in an overlapping region is very significant. Maintaining self-coexistence among SG entities is challenging, because the SG networks do not have any information about which bands other secondary SG networks will choose. In the same scenario, when the networks overlap with each other and permit to share spectrum bands, it is highly probable that greedy networks will try to access the entire bandwidth and, similarly, all other networks act in the same way, resulting in interference among each other. Hence, efficient spectrum access methods need to be devised so that the interference can be minimized and, thus, SG networks can self-coexist.

14.3   Self-Coexistence among DSA-Enabled Smart Grid Networks

In this section, the self-coexistence game is formulated as a dynamic channel[2] switching game, where we consider N SG networks (players) are competing for M separate orthogonal spectrum bands that are unused by the primary incumbents. Multiple SG networks might share a common geographical area, hence, creating interference relationship with a set of nodes. If two SG networks are in the interference range of each other, a common spectrum cannot be used by both, since the quality of service of both SG entities will be affected. The players in this non-cooperative channel switching game aim to grab a spectrum band not accessed by any other nodes at the same time. It is assumed that the players need to know the number of overlapping competitors to successfully participate in the game, which can be found from the other nodes' broadcasting beacons in foreign beacon period (FBP) [28].

14.3.1   Decision problem of SG networks

It is assumed that each SG network (SGN) looks for one of M available spectrum bands for data communication, and an SGN's transmission will succeed if it has the exclusive access to the spectrum band, hence, free from possible interference. If multiple SGNs access the same frequency band simultaneously, then they will have to decide whether to stay or switch in the next game stage, as the current time slot is wasted due to collision. The game terminates when every competing SG network successfully finds a contention-free spectrum chunk. Thus, the underlying, optimization problem is to find a mechanism that enables the SG networks to find a clear spectrum in minimum number of failed transmission stages. As far as the feasible action set of the SG networks is concerned on the occurrence of interference, network i has to decide an action from the binary strategy set of switch

---

[2]Throughout this chapter, we use the words "channel," "band," and "chunk" interchangeably unless explicitly mentioned otherwise.

to another band (looking for a free channel) or stay on the current band (expecting the interfering networks will move away).

The strategies of SG networks to choose either *"switch"* or *"stay,"* involve costs in terms of time units. Upon choosing "switch" strategy, the SG networks have to pay the price of finding another clear spectrum in the game. The cost of discovering a new unused band might take 1 switching, or more than that, since there are N networks competing for exclusive access to one of the M spectrum bands. Therefore, the current strategy (*"switch"*) taken by the network leads to a subgame, and the average cost of finding a clear band will be dependent on the number of networks competing (N) and available spectrum resources (M). We define the expected cost (C_i) of finding a clear channel, if the network chooses the strategy of switching, as

$$E[C_i(s_i, \mathbf{s_{-i}})] = c^{f(N,M)} \tag{14.1}$$

over all possible resulting subgames, where s_i and $\mathbf{s_{-i}}$ denote the strategies chosen by network i and the rest of the networks, respectively. c denotes the cost of switching, and $f(\cdot)$ represents the function that portrays the nature of cost at different N and M. The general intuitions behind modeling this function are as follows: The expected cost of finding an unused spectrum increases as the number of competitors (N) increases, but M remains fixed; while the cost decreases as the number of available spectrum (M) increases, N remains fixed; vary when both N and M simultaneously, then the cost depends on $M:N$ ratio and difference between them. We opt for a simple closed form of $f(N,M) = \frac{NM}{M-N}$, which satisfies the above-mentioned requirements.

The *"stay"* strategy for a network i might lead it to one of the following three scenarios: (i) all of the contended networks, which tried to operate in the same as network i, might choose to switch, thus, leaving the spectrum free for network i; (ii) all of the contended network might choose to *"stay,"* thus, wasting the game current stage and repeating the same game G as the game configuration remains unchanged; (iii) some networks might move away (*"switch"*), while the remaining networks choose to *"stay"* in the same band that creates a subgame G' of the original game G. A detailed explanation of the originated subgame G' will be given later. The cost functions can be mathematically presented as the following:

$$C_i(s_i, \mathbf{s_{-i}}) = \begin{cases} 0 & \text{Case (i)} \\ 1 + C_i(G) & \text{Case (ii)} \\ 1 + C_i(G') & \text{Case (iii)} \end{cases}. \tag{14.2}$$

14.3.2 Self-coexistence game analysis

With the defined strategy set and cost function, we now require finding a mechanism for choosing strategies (*"switch"* or *"stay"*) that minimize the cost incurred, and achieve equilibrium. The players in the game are assumed to be rational in nature and choose their strategies non-cooperatively at each stage that optimize their individual cost only. Analyzing this game, we aim to find the set of strategies such that no competitor benefit more by changing its strategy unilaterally while others' strategies remain unchanged, which is called the Nash equilibrium (NE) [29] strategy profile.

To analyze the NE of the game, we first assign probabilities to each of the strategies in the binary strategy space. Assuming the mixed strategy space of player i as $S_i^{mixed} = \{(\text{switch} = p), (\text{stay} = (1-p))\}$, where network i selects the *"switch"* strategy with probability p and *"stay"* strategy with probability $(1-p)$. Since all the players in this modied minority game (MMG) game are assumed to be homogeneous in nature, the similar mixed strategy space will also be applicable for all of them too. Now, the question that needs to be answered is what values of $(p, 1-p)$ tuple represent the equilibrium solution that will prove the existence of a non-zero probability of *"switch"* or *"stay."*

The game starts with a case where all $(N-1)$ other networks coexist on one band with network i and select a strategy from their mixed strategy space. Irrespective of what strategy chosen by network i, the possible subgames will have one of the following configurations: all $N-1$ players choose to "*switch*," or $N-2$ players choose to "*switch*," i.e., 1 player decides to "*stay*," or , $\cdots$, or 0 players choose to "*switch*," i.e., everyone decides to "*stay*". To derive the NE strategy profile, we find the expected cost of network i under the consideration of the "*switch*" or "*stay*" strategy. If a network choses to "*switch*," the expected switching cost to find a contention-free spectrum over all possible resulting subgames for the network i can be

$$E[C_i^{switch}] = \sum_{j=0}^{N-1} Q_j \times c^{f(N,M)}, \tag{14.3}$$

where, j is the number of other networks that decided to "*switch*," and Q_j denotes the probability of j networks switching out of other $N-1$ networks, which is given by $Q_j = \binom{N-1}{j} p^j (1-p)^{(N-1-j)}$. On the other hand, the expected cost of choosing strategy "*stay*" by network i, can be given as

$$E[C_i^{stay}] = \sum_{j=0}^{N-2} Q_j (1 + E[C_i(G'_{(N-j)})]) + Q_{(N-1)} \times 0, \tag{14.4}$$

where, $E[C_i(G'_{(N-j)})]$ is the expected cost incurred in the subgame $G'_{(N-j)}$. Since the SG networks are rational entities, they would not choose the "*stay*" strategy if the cost of switching is less than the cost of staying, thereby going back to the pure strategy scenario, where NE cannot be achieved [30]. If the case is the other way around, i.e., expected cost of staying is less than switching, a similar reasoning can be applied for the "*stay*" strategy, where NE is difficult to achieve. Therefore, to prove the existence of mixed strategy NE, the network i must be indifferent about choosing strategies "*switch*" or "*stay*" irrespective of actions taken by other networks. The NE mixed strategy profile $(p, 1-p)$ ensures that the strategy chosen by network i is never dominated by strategies of the competing networks. Hence, network i will not deviate unilaterally from the equilibrium strategy space $(p, 1-p)$ to lower the cost. Now, to find the NE strategy profile, we use the principle of indifference by equating Equations (14.3) and (14.4);

$$\sum_{j=0}^{N-2} Q_j (1 + E[C_i(G'_{(N-j)})]) = c^{f(N,M)}. \tag{14.5}$$

Though it is evident from Equation (14.5) that the expected cost of the game at NE is dependent on j, i.e., number of networks that are switching, the cost actually depends on the total number of networks (N) and the number of bands available (M). Therefore, the expected cost for network i in the subgame $G'_{(N-j)}$ is nothing but same as the original game.

Using binomial expansion and detailed mathematical derivations, we obtain the closed form for p as

$$p = \left(\frac{1}{1+c^{f(N,M)}}\right)^{\frac{1}{N-1}}. \tag{14.6}$$

It is now clear from the above equation that for any value of N and M, p has a finite non-zero value, proving the existence of mixed strategy NE profile. In other words, the mixed strategy tuple, $(p, 1-p)$, given in Equation (14.6), constitutes the best response strategy of every network in the game.

14.4 Self-Coexistence Using Multi-Stage Interaction Game

14.4.1 System Description

Considering the same game configuration, we now model this modified minority game as a dynamic multi-stage interaction game [31], where every rational SG network aims to find an unused spectrum by adopting a learning heuristic. The spectrum bands can be assumed as homogeneous (similar in characteristics) or heterogeneous in terms of bandwidth, data rate, operating frequency range, etc. Hence, the typical behavior of the networks, when the bands are homogeneous, will be to find a free chunk irrespective of its quality. On the other hand, if the spectrum qualities are distinct and unique, then the rational SG networks might always look for the best quality channel to maximize their utility. But this might lead to a colliding situation, where no networks can be benefited, thus, it will be interesting to devise a self-coexistence mechanism that captures the optimal foraging behavior of the networks in such a way that their average benefit will be maximized over the period of playing.

14.4.2 Game Settings for Homogeneous Band-Based Self-Coexistence Problem

For the homogeneous spectrum scenario, we use a different utility function for spectrum accesses, which is not the expected cost, as defined in the earlier section. Here, the successful access to a band, if not contended by any other networks, rewards the network i a constant utility α and zero if at least one network chooses the same channel. For the sake of simplicity, we neglect the switching cost or the cost of collisions. In this scenario, each network i has a mixed strategy profile for each of the M spectrum bands, which is represented as $p_i = (p_1^{(i)}, p_2^{(i)}, ..., p_M^{(i)})$, where $0 \leq p_j^{(i)} \leq 1$ is the probability of network i choosing band $j \in M$ to operate and $\sum_{j=1}^{M} p_j^{(i)} = 1$. The utility function for network i can be mathematically presented as:

$$U_i(a_i, a_{-i}) = \begin{cases} \alpha & \text{if } a \neq a_i, \forall a \in a_{-i} \\ 0 & \text{otherwise} \end{cases}, \tag{14.7}$$

where α is a constant utility for all spectrum bands. While exploring the available spectrum bands, it is assumed that the networks do not have any information about the actions chosen by the other SG networks in the competing environment. The networks maintain perceptions about the spectrum bands based on the success/failure feedbacks obtained after accessing them after each game stage. Based on the perception parameter, the networks decide the next step action of whether to stick to the currently acquired band or choose another to explore more. Next, we describe a distributed perception-based learning mechanism that helps to strategize the action of an SG network, which can also be used by other networks simultaneously to find a spectrum void of contention.

14.4.3 Perception-Based Learning Model

The SG networks use the knowledge of utilities received by accessing a spectrum chunk to define the belief/perception about them, which is nothing but a metric to classify the spectrum bands based on how they succeed in accessing the bands. Each SG network i maintains a perception vector $P^{(i)} = (P_1^{(i)}, P_2^{(i)}, P_3^{(i)}, ..., P_M^{(i)})$ corresponding to each spectrum band. If the network i access j^{th} band at stage t, then the perception $P_j^{(i)}$ gets updated based on success/failure to access it. At the starting of every game stage, the strategy to be played is derived from player i's mixed strategy, which is mapped from the perception vector $P_j^{(i)}$. This process is repeated several times until every network possesses a contention-free spectrum, where self-coexistence is achieved in a distributed manner.

At the beginning of the dynamic learning game, $t = 0$, and network i selects a random band j to operate. The utility observed for the action taken at stage t can be defined as $U_{i,a_i(t)}(t)$. As the game does not have any history prior to the starting of game, we initialize the perception $(P^{(i)}(0))$ of all bands with a constant value to avoid biases toward any particular spectrum. To decide the next stage strategy, we need to map the perception vector to an equivalent mixed strategy, and in order to do that we consider the Q-value mapping based on softmax method in reinforcement learning [32] [33]. The perception values of each band are mapped to an equivalent probability expression using the Boltzmann distribution, which helps in exploring a large search space via a control parameter name temperature ($\gamma > 0$). In other words, the perception vector $(P^{(i)}(t))$ of network i is mapped to the corresponding mixed strategy, $p_i(t) = (p_1^{(i)}(t), p_2^{(i)}(t), ..., p_M^{(i)}(t))$, according to the following equation:

$$p_j^{(i)}(t) = \frac{e^{\gamma P_j^{(i)}(t)}}{\sum_{i=1}^{M} e^{\gamma P_j^{(i)}(t)}}, \forall j \in M, \tag{14.8}$$

where γ is the temperature parameter in Boltzmann's distribution, which controls the exploration of the network's strategy space. The value of this dynamic parameter can be increased from a low to high value so that the networks will explore more in the beginning and gradually settle down in exploiting spectrum that has higher perception value.

After finding the mixed strategy at the end of the game stage, the next stage action is decided stochastically by each network i that refers to the operating band for the next period. But prior to that, the networks update their perception vector for the action taken previously. The update rule for network i that accessed spectrum j at stage t is presented in Equation 14.9. The update rule uses the past belief/perception about band j and current period reward to update its belief for the next game stage. If network i has successfully grabbed a band j by taking action $a_i(t) = j$ in game stage t, then the perception value $(P_j^{(i)}(t+1))$ for j^{th} band should increase proportionally to the utility received in the current stage. This increment is relative to the prior established belief about the spectrum band j, hence, successful access will increase the perception value, whereas collision leads to reduction in perception. The perceptions of the un-accessed bands remains unchanged. Algorithm 1 summarizes the distributed procedure for self-coexistence in homogeneous bands scenario using perception-based learning model:

$$P_j^{(i)}(t+1) = \begin{cases} (1 - \mu_t)P_j^{(i)}(t) + \mu_t U_{i,j}(t) & \textbf{if } a_i(t) = j \\ P_j^{(i)}(t) & \textbf{otherwise} \end{cases}, \tag{14.9}$$

where $\mu_t \in (0,1)$ is the smoothing variable factor that balances the choices to explore or not by putting dynamic weight either on the past experience or the current reward. If the weight is high on the past perception value, then exploitation is given more importance, whereas high weight on current reward hints for more exploration.

14.4.4 Game Settings for Heterogeneous Band Based Self-Coexistence Problem

In a heterogeneous spectrum-bands scenario, the networks receive distinct utilities from different bands upon mutual exclusive access, because the characteristics of the spectrum chunks are unique. If we consider M spectrum bands available in the system that can potentially deliver unique utilities $u_1, u_2, ..., u_M$ to the corresponding SG network on exclusive access, the utility function for network i can be presented as:

Algorithm 5 Perception-Based Learning

---

1 Initialize γ and $P_j^{(i)}(0) = \frac{1}{M}$ for all networks $i \in N$ and $j \in M$;
2 **while** $staget \le MaxT$ **do**
3   **for** *all network* $i \in N$ **do**
4   Select a band $j \in M$ based on its mixed strategy equation (3);
5   Observe the utility reward for the stage t, $U_{i,a_i(t)}(t)$;
6   Update the perception $(P_j^{(i)}(t+1))$ for all bands $j \in M$ according to equation (4);
7       $t \leftarrow t+1$
8   **end**
9 **end**

$$U_i(a_i, a_{-i}) = \begin{cases} u_j & \text{if } a \ne a_i, \ \forall a \in a_{-i} \\ 0 & \text{otherwise} \end{cases}. \tag{14.10}$$

Though this scenario seems equivalent to the homogeneous case we discussed before, here the rationality of SG networks makes it even more challenging to maintain self-coexistence among themselves. The greedy networks are always inclined to choose the best spectrum that has the highest reward, and the equivalent decision by all other networks will lead them to collision. Thus, no network will get the exclusive access to the highest utility band but rather incur cost for collision, and eventually the expected payoff reduces. Therefore, SG networks must strategize their actions to acquire a contention-free band of relatively better utility but not necessarily the best channel so that their expected system utility will be maximized over the period of operation. The networks require an experience-based heuristic to explore and learn about the quality of spectrum bands, which will eventually lead to exploit the bands that have high availability and better utility. Here, we propose a heuristic that uses the perception-based learning among with a regret minimization technique to conduct strategic analysis about spectrum choices.

14.4.5 Regret minimization model

Assuming number of bands (M) is higher than number of SG networks, the simultaneous access to the most valuable band never helps the networks, so they must look for an unused spectrum chunk that is free from interference. If all the networks find an exclusive band, then we can say that the system is sub-optimally stable, where the best set of bands are not necessarily exploited. However, the goal of the system is to use the best valued spectrum bands tactfully so that the system's gain will be maximum, and this condition can be the optimal convergence scenario. Though it is hard to achieve the optimal case, we present a regret minimization [34] heuristic that helps to achieve near-optimal system utility.

We assume that all networks know about the number of available bands and their utilities on exclusive access. A regret matching technique is used by the SG networks to maintain the regret differences of actions that would have given higher reward than the current action taken. Hence, the strategy of selecting an action $\bar{\alpha}$ at stage $(t+1)$ should be a function of the average accumulated regret so that the networks will ignore the lowest utility bands and will be more inclined to choose high-rewarding bands. However, the strategy function is missing another factor, which should take care of the possible collisions while choosing the high-valued spectrum chunks. Here we use our previously proposed perception vector to prohibit the networks from choosing the bands that are highly susceptible to contentions, and rather stick to the best-performed band so that the reward over the stages will be positive instead of only the cost of collision. Thus, the strategy of choosing an action $\bar{\alpha}$ for stage $(t+1)$ should be a function of the regret difference $(R_i^{\bar{\alpha}}(t))$ and perception vector $(P_{\bar{\alpha}}^{(i)}(t))$ of network i up to game stage t, which is presented in Equation 14.11. The average regret $(R_i^{\bar{\alpha}})$ accumulated for network i, for all actions, $\bar{\alpha} \in A_i$ up to stage t is given by

$$R_i^{\bar{\alpha}}(t) = \left(\frac{1}{t}\right) \sum_{t'=1}^{t} [U_{i,\bar{\alpha}}(t') - U_{i,\alpha}(t')],$$

where $U_{i,\alpha}(t)$ is the utility reward to network i by choosing action $\alpha \in A_i$ at stage t. Now, the probability of choosing an action $\bar{\alpha}$ at stage $t+1$ can be evaluated based on the following equation, where the normalized regret difference contributes in leading the networks to choose higher utility bands, and the normalized perception value helps in deciding the successfully accessed bands instead of choosing the bands that have high probability of contention:

$$p_i^{\bar{\alpha}}(t+1) = \left[\frac{R_i^{\bar{\alpha},+}(t)}{\sum_{\bar{\alpha} \in A_i}\left[R_i^{\bar{\alpha},+}(t)\right]}\right]\left[\frac{P_{\bar{\alpha}}^{(i)}(t)}{\sum_{\bar{\alpha} \in A_i}\left[P_{\bar{\alpha}}^{(i)}(t)\right]}\right], \tag{14.11}$$

where $R_i^{\bar{\alpha},+}(t) = max(R_i^{\bar{\alpha}}(t),0)$.

14.5 Simulation Model and Results

We conduct simulation experiments for the modified minority game to evaluate the efficiency of derived mixed strategy. Source code for the experiment has been written in C under Linux environment. For multi-stage dynamic game, we simulate various instances of self-coexistence under presence of homogeneous and heterogeneous bands using Matlab version 8.1.

14.5.1 Self-Coexistence Strategy Evaluation

Assuming N SG networks are competing for one of N available spectrum bands, the networks can choose either to *"switch"* and *"stay"* at each decision stage. The system itself converges when all the competing networks capture a spectrum band void of interference from other SG networks. The value of number of SG networks (N) and spectrum bands (M) are provided as inputs to the system.

We show the average system convergence cost in Figure 14.2, when 20 SG networks are competing with each other. The probability of choosing "switch" strategy is varied in the simulation experiment. It is observed that the convergence cost increases with increase in number of available bands. The existence of mixed strategy NE is proven in this experiment, which is evident from the convex nature of the curve in the Figure 14.2 that a minima exists for each simulation instance. The minima is nothing but the NE strategy (p) of the game.

Depending on various network:band ratio ($50\% - 90\%$), we present the corresponding system convergence cost when the game is played using the theoretical Nash mixed strategy. It is found from the Figure 14.3 that the system convergence cost increases almost exponentially with increase in network:band ratio, thereby justifying the proposed cost function.

14.5.2 Multi-Stage Learning Evaluation

In the dynamic multi-stage self-coexistence game of homogeneous spectrum bands, we simulate the proposed perception based learning model to verify the convergence of SG networks. We assume a unity reward from the bands when accessed exclusively and experimented by fixing the number of bands (M) to 150. For different number of competing networks (N), we run our algorithm for 1000 times, and the average results are reported in Figure 14.4. It is found that our perception learning

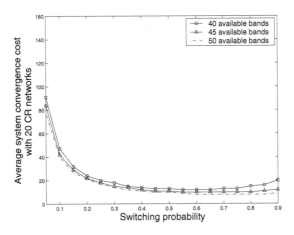

Figure 14.2: Average system convergence cost with 20 SG networks and varying number of bands.

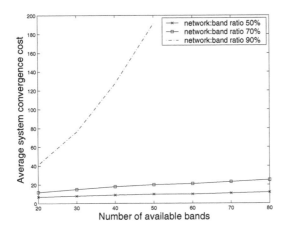

Figure 14.3: Average system convergence cost with varying network:band ratio.

helps the SG networks to quickly find a contention-free band within a few game stages. When the number of competing networks (N) is quite less than the available bands (M), the convergence is achieved sooner compared to the case when N is close to M. This occurs because the competition for free spectrum increases as there are more networks in the system, which is why some networks might have to explore longer to find an unused spectrum.

For self-coexistence in heterogeneous bands scenario, we simulated the regret minimization-based heuristic to achieve optimal system utility where all networks aim to occupy a band with fairly high utility reward. The simulations are repeated for 100 times, and the average results are reported. Using the regret-minimization heuristic, we analyze the convergence nature of SG networks at two different values of network (N):band (M) ratio, i.e., 0.5 and 0.75. From Figure 14.5, we can observe that the number of stages required to converge to optimal solution for 50% ratio mix is less than for 75% due to reduced intensity of competition. Similarly, the number of stages to reach sub-optimal convergence is less for 50% $N : M$ ratio mix. Comparing the number of stages required for optimal and sub-optimal convergence, we found it is quick enough to achieve sub-optimal convergence where every SG network has at least a band to start communication. However, greediness to

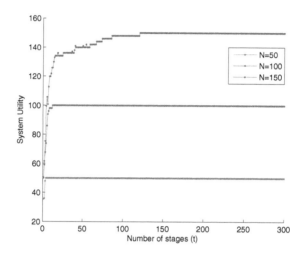

Figure 14.4: System utility vs. number of stages for varying N.

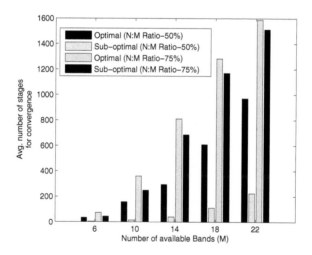

Figure 14.5: Average number of stages for convergence vs. number of bands for 50% and 75% N:M ratio.

reach optimal allocation takes longer duration because the networks collide when they start acting rationally to grab the best valued bands.

14.6   Conclusions

In this chapter, we studied the importances and necessities of self-coexistence mechanism in a distributed wireless environment, which is also motivated from a practical smart grid network point-of-view. We presented a game theoretic model and analysis of self-coexistence among smart grid networks, where each SG network competes for a contention-free spectrum chunk to access. Additionally, we proposed a multi-stage interaction game among SG networks, where networks dynamically select the bands as their strategy, and this is repeated until convergence is achieved. First, we discussed about the solution concepts of the distributed non-cooperative MMG game to analyze the mixed strategy NE for the SG networks. Then, the self-coexistence problem is addressed via a multi-stage interaction game between N networks, competing for M homogeneous or heterogeneous bands. We presented a perception-based learning mechanism for the homogeneous spectrum scenario, and a regret-minimization based heuristic for heterogeneous spectrum scenario to help the networks learn and act quickly to find contention-free spectrum bands and maximize the system utility at the same time.

14.7   Acknowledgment

This research is based upon work supported by the National Science Foundation under NSF CAREER Grant CNS #1346600.

References

[1] I. F. Akyildiz, W.-Y. Lee, M. C. Vuran, and S. Mohanty, "Next generation/dynamic spectrum access/cognitive radio wireless networks: a survey," *Computer Networks*, vol. 50, no. 13, pp. 2127–2159, 2006.

[2] S. Mangold, Z. Zhong, K. Challapali, and C.-T. Chou, "Spectrum agile radio: Radio resource measurements for opportunistic spectrum usage," in *Global Telecommunications Conference, 2004. GLOBECOM'04. IEEE*, vol. 6, 2004, pp. 3467–3471.

[3] Z. Lei and S. J. Shellhammer, "Ieee 802.22: The first cognitive radio wireless regional area network standard," *IEEE communications magazine*, vol. 47, no. 1, pp. 130–138, 2009.

[4] C. Cordeiro, K. Challapali, D. Birru, and S. Shankar, "Ieee 802.22: The first worldwide wireless standard based on cognitive radios," *IEEE International Symposium on Dynamic Spectrum Access Networks*, pp. 328–337, Nov. 2005.

[5] S. Sengupta, S. Brahma, M. Chatterjee, and N. Sai Shankar, "Enhancements to cognitive radio based IEEE 802.22 air-interface," *In proceedings of IEEE International Conference on Communications (ICC)*, pp. 5155–5160, June 2007.

[6] X. Fang, S. Misra, G. Xue, and D. Yang, "Smart grid the new and improved power grid: A survey," *Communications Surveys & Tutorials, IEEE*, vol. 14, no. 4, pp. 944–980, 2012.

[7] V. C. Gungor, D. Sahin, T. Kocak, S. Ergut, C. Buccella, C. Cecati, and G. P. Hancke, "Smart grid technologies: Communication technologies and standards," *Industrial informatics, IEEE Trans. on*, vol. 7, no. 4, pp. 529–539, 2011.

[8] Z. Fan, P. Kulkarni, S. Gormus, C. Efthymiou, G. Kalogridis, M. Sooriyabandara, Z. Zhu, S. Lambotharan, and W. H. Chin, "Smart grid communications: Overview of research challenges, solutions, and standardization activities," *Communications Surveys & Tutorials, IEEE*, vol. 15, no. 1, pp. 21–38, 2013.

[9] R. Ma, H.-H. Chen, Y.-R. Huang, and W. Meng, "Smart grid communication: Its challenges and opportunities," *Smart Grid, IEEE Transactions on*, vol. 4, no. 1, pp. 36–46, 2013.

[10] Y. Yan, Y. Qian, H. Sharif, and D. Tipper, "A survey on smart grid communication infrastructures: Motivations, requirements and challenges," *Communications Surveys & Tutorials, IEEE*, vol. 15, no. 1, pp. 5–20, 2013.

[11] D. Challet and Y.-C. Zhang, "Emergence of cooperation and organization in an evolutionary game," *Physica A: Statistical Mechanics and its Applications*, vol. 246, no. 3, pp. 407–418, 1997.

[12] C. Gass and J. Garrison, "Energy regulation by traplining hummingbirds," *Functional Ecology*, vol. 13, no. 4, pp. 483–492, 1999.

[13] P. Acharya, S. Singh, and H. Zheng, "Reliable open spectrum communications through proactive spectrum access," *ACM International Conference Proceeding Series*, vol. 222, 2006.

[14] G. Ganesan and Y. Li, "Cooperative spectrum sensing in cognitive radio networks," *in Proc. of IEEE DySPAN*, 2005.

[15] A. Ghasemi and E. Sousa, "Collaborative spectrum sensing for opportunistic access in fading environments," *First IEEE International Symposium on New Frontiers in Dynamic Spectrum Access Networks (DySPAN)*, pp. 131–136, Nov. 2005.

[16] A. Sahai, N. Hoven, and R. Tandra, "Some fundamental limits on cognitive radio," *in Forty-second Allerton Conference on Communication, Control and Computing*, 2004.

[17] N. Han, S. Shon, J. Chung, and J. Kim, "Spectral correlation based signal detection method for spectrum sensing in IEEE 802.22 WRAN systems," *In Proceedings of 8th International Conference on Advanced Communication Technology (ICACT)*, vol. 3, Feb. 2006.

[18] S. Chen, "Inequalities for $\mathcal{M}-$matrices and inverse $\mathcal{M}-$matrices," *Linear Algebra and Its Applications*, vol. 426, pp. 610–618, June 2007.

[19] N. Kundargi and A. Tewfik, "Sequential pilot sensing of atsc signals in IEEE 802.22 cognitive radio networks," *IEEE International Conference on Acoustics, Speech and Signal Processing (ICASSP)*, pp. 2789–2792, 2008.

[20] W. Hu, D. Willkomm, M. Abusubaih, J. Gross, G. Vlantis, M. Gerla, and A. Wolisz, "Cognitive radios for dynamic spectrum access—Dynamic frequency hopping communities for efficient IEEE 802.22 operation," *IEEE Communications Magazine*, vol. 45, no. 5, pp. 80–87, May 2007.

[21] B. Gao, J.-M. Park, and Y. Yang, "Uplink soft frequency reuse for self-coexistence of cognitive radio networks," *Mobile Computing, IEEE Transactions on*, vol. 13, no. 6, pp. 1366–1378, 2014.

[22] M. Yoo and S. Hwang, "A self-coexistence method for the IEEE 802.22 cognitive WRAN," *Recent Researches in Automatic Control, Systems, Science and Communications*, pp. 169–172, 2012.

[23] M. Buddhikot, P. Kolodzy, S. Miller, K. Ryan, and J. Evans, "DIMSUMnet: New directions in wireless networking using coordinated dynamic spectrum access," *IEEE Intlernational Symposium on a World of Wireless, Mobile and Multimedia Networks*, pp. 78–85, 2005.

[24] L. Cao and H. Zheng, "Spectrum allocation in ad hoc networks via local bargaining," *In Proc. of SECON*, 2005.

[25] O. Ileri, D. Samardzija, and N. Mandayam, "Demand responsive pricing and competitive spectrum allocation via spectrum server," *In Proc. of IEEE DySPAN*, Nov. 2005.

[26] S. Sengupta, M. Chatterjee, and S. Ganguly, "An economic framework for spectrum allocation and service pricing with competitive wireless service providers," in *New Frontiers in Dynamic Spectrum Access Networks, 2007. DySPAN 2007. 2nd IEEE International Symposium on.* IEEE, 2007, pp. 89–98.

[27] R. Lam, D. Chiu, and J. Lui, "On the access pricing and network scaling issues of wireless mesh networks," *IEEE Transactions on Computers*, vol. 56, no. 11, pp. 1456–1469, Nov. 2007.

[28] "IEEE P802.22/D0.1 Draft Standard for Wireless Regional Area Networks Part 22: Cognitive Wireless RAN Medium Access Control (MAC) and Physical Layer (PHY) specifications: Policies and procedures for operation in the TV Bands," 2006. [Online]. Available: http://www.ieee802.org/22/.

[29] J. Nash, "Equilibrium points in N-person games," In *Proceedings of the National Academy of Sciences*, vol. 36, pp. 48–49, 1950.

[30] S. Sengupta, R. Chandramouli, S. Brahma, and M. Chatterjee, "A game theoretic framework for distributed self-coexistence among IEEE 802.22 networks," In *proc. of IEEE Global Communications Conference (GLOBECOM)*, pp. 1–6, Dec. 2008.

[31] D. Tosh and S. Sengupta, "Self-coexistence in cognitive radio networks using multi-stage perception learning," in *Vehicular Technology Conference (VTC Fall), 2013 IEEE 78th*, pp. 1–5, Sep. 2013.

[32] L. P. Kaelbling, M. L. Littman, and A. W. Moore, "Reinforcement learning: A survey," *Journal of Artificial Intelligence Research*, pp. 237–285, 1996.

[33] A. Bab and R. I. Brafman, "Multi-agent reinforcement learning in common interest and fixed sum stochastic games: An experimental study," *Journal of Machine Learning Research*, vol. 9, no. 12, 2008.

[34] J. R. Marden, G. Arslan, and J. S. Shamma, "Regret based dynamics: Convergence in weakly acyclic games," in *Proceedings of the 6th International Joint Conference on Autonomous Agents and Multiagent Systems.* ACM, p. 42, 2007.

MODELING
ISSUES

Chapter 15

Performance Modeling of Opportunistic Spectrum Sharing

Shensheng Tang

CONTENTS

Abstract

Current radio spectrum usage has been shown to be highly underutilized. Opportunistic spectrum sharing (OSS) is a promising solution for efficient spectrum utilization. In this chapter, we analytically model an opportunistic spectrum sharing system under the conditions of perfect spectrum sensing, unreliable sensing, and tolerable service degradation through queueing theoretic frameworks. The considered OSS system consists of primary users and secondary users that share a set of channels over a coverage area. Both initiating secondary users and ongoing secondary users sense the channels and perform appropriate activities. Either a buffer or infinite queues are incorporated into the proposed models for performance analysis. Sensing errors from either initiating or ongoing secondary users are considered in appropriate models, which may cause false alarm and misdetection events and impose various impacts on both types of users. We solve the steady-state probabilities of the considered systems under perfect sensing, unreliable sensing, and tolerable service degradation. We also derive a set of performance metrics of interest. Different problem Solving methods, e.g., matrix analytic method and generating function technique, are used for solving the system equations. The proposed modeling methods are expected to be used for design and evaluation of future opportunistic spectrum sharing networks.

Keywords: Opportunistic spectrum sharing, primary users, secondary users, perfect sensing, unreliable sensing, tolerable service degradation, false alarm, misdetection, markov process.

15.1   Introduction

Current radio spectrum usage has been shown to be highly underutilized [1][2]. Opportunistic spectrum sharing (OSS) is a promising solution for efficient spectrum utilization. In such an OSS scenario, there are two types of wireless networks. The one that owns the license for spectrum usage is referred to as the *primary system*; its users are referred to as the *primary users* (PUs). The calls generated from primary users constitute the primary traffic (PT) stream. The other network in the same service area is referred to as the *secondary system*; its users are referred to as the *secondary users* (SUs). The calls generated from the SUs constitute the secondary traffic (ST) stream. The SUs equipped with cognitive radios (CRs) are capable of sensing idle frequency channels and opportunistically make use of them without causing harmful interference to the PUs [3]. Thus, the secondary system is also called a cognitive radio (CR) network. The system consisting of the primary and secondary systems is called an OSS system. By allowing SUs to reclaim idle channels, much higher spectrum efficiency can be achieved, even under unreliable spectrum sensing [4][5].

In the OSS wireless system, PUs operate as if there are no SUs in the service area. SUs include initiating SUs (who are searching for channels and try to access the system) and ongoing SUs (who are occupying channels in the system). In general, an initiating SU senses when a channel is idle and then makes use of such a channel. Similarly, an ongoing SU also senses the spectrum and vacates its channel for a PU if one presents on the channel, and then either switches to another idle channel or moves to a buffer. The call waiting in the buffer can reconnect back when a channel becomes available or drop out from the buffer when a predefined maximum waiting time expires.

Much research about OSS or dynamic spectrum access has been developed in the past a few years. In [6], collaborative spectrum sensing was studied as a means to combat the shadowing or fading effects that a user experiences. In [7], a multi-channel medium access control (MAC) protocol was developed to enable the interoperation of the primary-secondary overlay network. In [8], a sensing-based approach was studied for channel selection in spectrum-agile communication systems. In [9], an admission control algorithm, in conjunction with a power control scheme, was proposed for cognitive wireless networks such that quality-of-service (QoS) requirements of all admitted SUs are satisfied. In [10], an online flow control, scheduling and resource allocation algorithm was designed for a CR network, with static PUs and potentially mobile secondary users by the Lyapunov optimization technique that meets the desired objectives and provides explicit performance guarantees. In [11], three opportunistic spectrum access schemes were proposed to analyze the SU performance under given primary constraints, by introducing the two metrics, namely, collision probability and overlapping time. In [12], a collaborative scheme was developed for a group of frequency agile radios to estimate the maximum interference-free transmit power (MIFTP) without causing harmful interference to the primary receivers. In [13], a hard decision-combining-based cooperative spectrum sensing scheme was proposed for CR networks in the presence of a feedback error caused by imperfect channel condition. In [14], a single spectrum sensing scheme with only one cognitive user performing sensing was proposed in both network-centric and user-centric ways for CR networks, and the proposed scheme was further generalized to a multiple spectrum sensing scenario. In [15], a genetic algorithm (GA)-based suboptimal scheduling method was proposed to address throughput and delay issues in CR networks under interference temperature constraints.

In this chapter, we summarize the analytic modeling methods of an OSS system under perfect and unreliable spectrum sensing. We also analyze the system performance by considering tolerable service degradation for the PT and ST calls. The contents are mainly referenced from our prior research works [4][16][17]. The remainder of the Chapter is organized as follows. Section 14.2 describes the system models under the conditions of perfect sensing, unreliable sensing, and tolerable service degradation, respectively. Section 14.3 develops two-dimensional Markovian models for analyzing the system performance under different conditions. Section 14.4 derives some performance metrics of interest. Finally, the chapter is concluded in Section 14.5.

15.2   System Models

In the OSS system, the primary system and secondary system operate independently. When a PT call arrives to the system, its base station (BS) will assign a channel to it if there is one available; otherwise, the PT call will be blocked. Note that a channel being used by an ST call is still seen as an idle channel by the primary system, since here the primary system and secondary system are supposed not to exchange information. SUs perform some periodic sensing to detect the presence or absence of signals from PUs and maintain records of the channel occupancy status. The detection mechanism may involve collaboration with other SUs or an exchange with an associated BS of the secondary system, if any.

Much research on sensing or detection method has been done in the literature, such as [18][19], which is not the focus in this chapter. However, the sensing errors resulting from any selected sensing method will be modeled by false alarm and misdetection probabilities and incorporated in the proposed framework.

SUs opportunistically access the channels that are in idle status. If an initiating ST call (user) finds an idle channel, it makes use of the channel. If the initiating ST call senses a channel being busy, it attempts to sense another channel as a totally new call. If all channels are busy, the ST call is blocked and considered lost from the system. When an ongoing SU detects or is informed (by its BS or collaborative SUs) of an arrival of PT call in its current channel, it immediately leaves the channel and switches to an idle channel, if one is available, to continue the call. If at that time all the channels are occupied, the ST call is placed into a buffer located at its BS (for an infrastructured network) or a virtual queue (for an infrastructureless network). The head-of-line (HOL) ST call in the buffer can reconnect to the system as soon as a channel becomes available before a predefined maximum waiting time expires. In principle, the maximum waiting time of an ST call should be equal to its residence time in the given service area, if the effect of impatience of the queued ST calls is not considered.

In perfect sensing case, the ST calls are able to move in and out of channels without causing any harmful interference with PT calls. However, in practice, unreliable sensing often exists. An initiating SU may incorrectly determine that a channel is busy (and, thus, stops accessing) when, in fact, the channel is idle. In addition, an ongoing SU may also incorrectly determine the presence of a PU on its channel (and, thus, vacates its channel) when, in fact, no PU enters the channel. We refer to the former type of error as type-I false alarm event, and the latter as type-II false alarm event. On the other hand, a PT call that is actively using a given channel may experience disruption if an initiating ST call searching for a free channel incorrectly determines that the channel is idle. We refer to this class of sensing errors as class-A misdetection event. A second class of disruption events to a PT call may occur when an ongoing ST call (user) on a given channel fails to detect the presence of an arriving PT call on that channel. This is referred to as class-B misdetection events. When a misdetection event occurs, both the ST call and the PT call are using the same channel, causing large interference (large "noise") to each other.

Spectrum sensing errors on either an initiating or an ongoing SU may cause false alarm and misdetection events and impose various impacts on both PUs and SUs. When a false alarm event occurs, the initiating SU will not enter the channel, and the ongoing SU will leave its current channel. When a misdetection event occurs, it may cause the following three types of results under intolerable service degradation condition:

- The first call collision result (CCR1): Both the involved calls drop from the system due to the collision (large "noise" incurred).

- The second call collision result (CCR2): Only the PT call drops from the system, but the ST call remains on the channel. In this case, the PU has no patience for the large "noise".

- The third call collision result (CCR3): Only the SU call drops from the system, but the PT call remains on the channel. In this case, the SU has no patience for the large "noise."

We denote the class-A and class-B misdetection probabilities by p_a and p_b, and the type-I and type-II false alarm probabilities by p_{f1} and p_{f2}, respectively. We assume that, when a collision happens, CCR1, CCR2, and CCR3 occur with probability q_1, q_2, and q_3, respectively, and $q_1 + q_2 + q_3 = 1$. One may argue that the sensing errors of type-II and class-B should be identical to that of type-I and class-A, respectively, since they are all due to the SUs. However, in realistic situations, the sensing errors of ongoing SUs may be different from that of initiating SUs. The ongoing users may make larger sensing errors, since they are being in communications and their decision-making time and resource may be less than that of the initiating users. An alternative analysis on the impact of different "sensing errors" by SUs can be found in [20]. Hence, different parameters of sensing errors should be applied to the modeling process. Moreover, for modeling purpose, the simple consideration of identical probability of sensing errors is just a special case of the proposed framework.

Note that there is some relationship between a false alarm probability (p_{f1}/p_{f2}) and a misdetection probability (p_a/p_b) through the signal-to-noise ratio (SNR) [21], which, however, does not impact the formulation of the proposed model. Instead, the proposed model makes this relation as a special case by a setting of $p_{f1} = p_{f2}$ and $p_a = p_b$.

Therefore, the key of design and deployment of an OSS system is to keep the unreliable sensing probabilities as small as possible so that the caused interference (particularly to the PUs) is restricted in a predefined threshold.

The above three CCRs occurs under the assumption of intolerable service degradation condition. Once a misdetection occurs, one of the two calls or both calls on the same channel will drop from the system. On the other hand, if the PT and ST calls can tolerate service degradation to some extent, misdetection events may also provide a chance for the PT and ST calls to both stay on the same channel. When this happens, both the PT and ST calls will experience degraded service. For example, each call may occupy part of the channel bandwidth via sub-rating (cf. [22]); hence, each call will have a reduced service rate. The analysis of this scenario may be applicable to a special secondary system, e.g., tolerable CR network.

In a tolerable CR network, an initiating SU has to sense the channel availability before accessing a channel. An ongoing SU must continue to sense the channel periodically, in case a PT call attempts to use the channel. An ST call occupying a channel, say, channel i, may encounter the following three situations:

- The ST call completes without interruption and leaves the system. This occurs with probability r_{i0}.

- The ST call senses the arrival of a PT call to the channel and switches the call to another channel j ($j \neq i$) with probability r_{ij}.

- The ST call fails to sense the arrival of a primary call to the channel (i.e., class-B misdetection occurs) and remains on the channel. This happens with probability $r_{ii'}$, where $r_{ii'} = 1 - \sum_{j \neq i} r_{ij}$.

In this case, both calls receive degraded service.

The tolerable CR network is modeled by using a queueing network, which affords greater flexibility in modeling channel assignment strategies than the single buffer queueing model used previously. In the single buffer queueing model, the buffer is used only to store the ongoing ST calls that vacate their channels due to the PT calls. In the queueing network model developed here, the ST calls are never blocked or dropped; rather, each ST call enters a queue associated with a particular channel, where it waits (possibly for zero time) until the channel becomes available. Thus, the queueing network model is appropriate for delay-tolerant data traffic.

In the queueing network model, there is no feedback link at a given channel, i.e., an ST call that vacates its channel does not join the queue associated with the same channel. Rather, the ST call attempts to join an alternative idle channel, say, channel j. If channel j is being used at that time, the ST call will join the queue associated with channel j. The queued ST calls access the channel in FCFS (first-come first-served) order as the channel becomes available.

15.3   Performance Analysis

Suppose the spectrum in the service area is divided into N traffic channels serving the PT and ST calls. As described previously, the buffer in the proposed model is used to store the ongoing ST calls that vacate from their current channels but cannot find idle channels in the presence of PT calls on their channels. The maximum number of ongoing ST calls is N. Hence, we set the buffer size as N. Note that the buffer is introduced exclusively for the ongoing ST calls that actively vacate their channels due to their decision of the presence of primary calls; the ST calls that passively drop from their channels due to collisions (large "noise" incurred) do not enter the buffer (they directly drop from the system).

Arrivals of the PT and ST calls are assumed to form independent Poisson processes, with rates λ_1 and λ_2, respectively. The channel occupancy times of the PT and ST calls are assumed to be exponentially distributed, with means $1/\mu_1$ and $1/\mu_2$, respectively. The residence time for the ST calls in the service area is assumed to be exponentially distributed with mean $1/r_2$. These assumptions have been found to be reasonable, as long as the number of users is much more than that of the channels in a service area, and have been widely used in the literature [23]–[26]. We further assume that both types of traffic occupy one channel per call for simplicity. However, the analysis method used here can be extended to handle variable bandwidth requests (cf. [27]).

Let $X_1(t)$ denote the number of PT calls in the OSS system at time t. Similarly, let $X_2(t)$ be the number of ST calls in the system at time t, including the ST calls being served and those waiting in the buffer. The process $(X_1(t), X_2(t))$ is a two-dimensional Markov process, with state space $S = \{(i,j) \mid 0 \leq i, j \leq N\}$. We classify the channel occupancy of the system in state (i, j) as *pre-full* if $i + j < N$, *just-full* if $i + j = N$, and *post-full* if $i + j > N$.

Let $\pi(i, j)$ denote the steady-state probability that the OSS system is in state (i, j). The steady-state system probability vector, with states ordered lexicographically, can be represented as $\pi = (\pi_0, \pi_1, \ldots, \pi_N)$, where $\pi_n = (\pi(n,0), \pi(n,1), \ldots, \pi(n,N)), 0 \leq n \leq N$. The vector π is the solution of equations

$$\pi Q = 0 \text{ and } \pi e = 1, \tag{15.1}$$

where the matrix Q is the infinitesimal generator of the two-dimensional Markov process, and $\mathbf{e}$ and $\mathbf{0}$ are vectors of all ones and zeros, respectively.

In the following, we perform analysis under the conditions of perfect spectrum sensing, unreliable spectrum sensing, and tolerable service degradation, respectively.

A. Perfect Spectrum Sensing

In perfect spectrum sensing, no sensing errors are involved during the sensing process. The transition rate diagrams in pre-full, just-full, and post-full conditions are shown in Figure 14.1 (a), (b), and (c), respectively. In pre-full case, the N channels are not fully used by the PT/ST calls. In just-full case, the number of occupied channels by the PT/ST calls is just N. In post-full case, the number of occupied channels by the PT/ST calls is greater than N. State $(i, j+1)$ moves to state (i, j), with rate $(j+1)\mu_2$ in Figure 14.1 (a) (where $i + j$ ¡ N), $(N-i)\mu_2 + r_2$ in Figure 14.1 (b) (where $i + j = N$), and $(N-i)\mu_2 + (j + 1 - N + i)r_2$ in Figure 14.1 (c) (where $i + j$ ¿ N). State (i, j) cannot move to state $(i, j+1)$ in Figure 14.1 (b) and (c), since all the N channels are occupied, but can move to state $(i+1, j)$, since an ST call has to vacate its channel for the incoming PT call, though all the channels are busy.

From the transition rate diagrams in Figure 14.1 (a), (b), and (c), the matrix Q of the two-dimensional Markov process is obtained as

$$Q = \begin{bmatrix} E_0 & B_0 & 0 & \cdots & 0 & 0 & 0 \\ D_1 & E_1 & B_1 & \cdots & 0 & 0 & 0 \\ \vdots & \vdots & \vdots & \vdots & \vdots & \vdots & \vdots \\ 0 & 0 & 0 & \cdots & D_{N-1} & E_{N-1} & B_{N-1} \\ 0 & 0 & 0 & \cdots & 0 & D_N & E_N \end{bmatrix}, \tag{15.2}$$

where each sub-matrix has size $(N+1)$ by $(N+1)$ and defined by

$$B_i = \lambda_1 I_{N+1}, \quad 0 \le i < N, \tag{15.3}$$
$$D_i = i\mu_1 I_{N+1}, \quad 1 \le i \le N, \tag{15.4}$$
$$E_i = A_i - \bar{\delta}(i)D_i - \bar{\delta}(N-i)B_i, \quad 0 \le i \le N, \tag{15.5}$$

where I_n denotes an n-by-n identity matrix, and $\bar{\delta}(i)$ is 0 when $i = 0$ and 1 otherwise. The matrix A_i has the same size as E_i. The j−th row and k−th column element of matrix A_i, denoted by $A_i(j, k)$, is defined as

$$A_i(j, k) =$$

$$\begin{cases} \lambda_2, & 0 \le i < N, 0 \le j < N-i, k = j+1, \\ j\mu_2, & 0 \le i < N, 1 \le j \le N-i, k = j-1, \\ (N-i)\mu_2 + (j-N+i)r_2, & 1 \le i \le N, N-i < j \le N, k = j-1, \\ -[A_i(j, j-1) + A_i(j, j+1)], & 0 \le i \le N, 0 \le j \le N, k = j, \\ 0, & \text{otherwise,} \end{cases} \tag{15.6}$$

where $A_i(j,k) \hat{=} 0$ for j, k ¡ 0 or j, k ¿ N. Applying the method developed in [25], the steady state probability vector can be determined as

$$\pi_n = \pi_0 \prod_{i=0}^{n} [B_{i-1}(-C_i)^{-1}], \quad 1 \le n \le N, \tag{15.7}$$

where π_0 satisfies $\pi_0 C_0 = \mathbf{0}$ and

$$\pi_0 \left[I + \sum_{n=1}^{N} \prod_{i=1}^{n} [B_{i-1}(-C_i)^{-1}] \right] e = 1. \tag{15.8}$$

The C_i can be recursively determined by $C_N = E_N$ and

$$C_i = E_i + B_i(-C_{i+1})^{-1}D_{i+1}, 0 \le i \le N - 1. \tag{15.9}$$

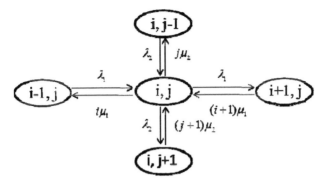

Figure 15.1: (a). The state diagram at (i, j) with pre-full channel occupancy under perfect sensing.

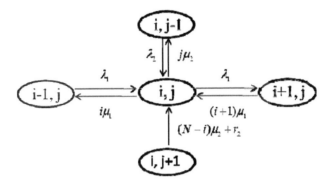

Figure 15.1: (b). The state diagram at (i, j) with just-full channel occupancy under perfect sensing.

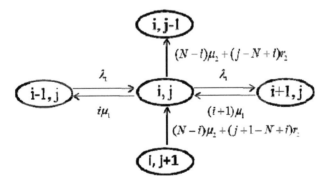

Figure 15.1: (c). The state diagram at (i, j) with post-full channel occupancy under perfect sensing.

B. Unreliable Spectrum Sensing

In this case, sensing errors include type-I and type-II false alarm events and class-A and class-B misdetection events. False alarm events decrease the spectrum utilization. Misdetection events cause harmful interference to PUs/SUs; they may cause the three types of results: CCR1 (with probability q_1), CCR2 (with probability q_2), and CCR3 (with probability q_3).

All of these sensing errors can cause different channel occupancy behavior and lead to different system state transitions. To make the presentation more clearly, however, we only consider the results of CCR1 and CCR2 in the analysis (i.e., $q_3 = 0$). The transition rate diagrams shown in Figure 14.2 (a), (b), and (c) refer to the situation of the pre-full, just-full, and post-full channel occupancy, respectively.

In Figure 14.2 (a), state (i, j) moves to $(i, j-1)$ with rate $j\mu_2 + q_1 p_b \lambda_1$, where $j\mu_2$ is the normal transition due to service completion, and $q_1 p_b \lambda_1$ is the additional transition due to a class-B misdetection with CCR1. State (i, j) moves to $(i-1, j)$, with rate $i\mu_1 + q_1 p_a \lambda_2$, where $i\mu_1$ is the normal transition due to service completion and $q_1 p_a \lambda_2$ is the additional transition due to a class-A misdetection with CCR1. State (i, j) moves to $(i-1, j+1)$, with rate $q_2 p_a \lambda_2$ due to a class-A misdetection with CCR2. State (i, j) moves to $(i, j+1)$, with rate $[1 - p_{f1} - \bar{\delta}(i)p_a]\lambda_2$, where $\bar{\delta}(i) \triangleq 0$ if $i = 0$ and 1 if $i \neq 0$. State (i, j) does not change at the condition of the class-B misdetection with CCR2.

Note that in Figure 14.2 (a), when $i = 0$, we have $p_a = 0$. A type-II false alarm may occur but does not affect the state transition, since the ongoing SU making type-II false alarm can either find another idle channel to continue its service or reconnect back to the system as soon as it enters the buffer, according to the proposed model.

In Figure 14.2 (b), state (i, j) moves to $(i, j-1)$ with rate $\theta_1(j)$, where

$$\theta_1(j) = (1 - p_{f2})j\mu_2 + p_{f2}[(j-1)\mu_2 + r_2] + q_1 p_b \lambda_1, \tag{15.10}$$

where $(1-p_{f2})j\mu_2$ is the normal transition without the occurrence of a type-II false alarm; $p_{f2}[(j-1)\mu_2 + r_2]$ is the transition due to the occurrence of a type-II false alarm (the corresponding ST call goes to the buffer, since at this time it cannot find another idle channel); and $q_1 p_b \lambda_1$ is the additional transition due to a class-B misdetection with CCR1. Similarly, state $(i, j+1)$ moves to (i, j), with rate $\theta_2(i, j)$, where

$$\theta_2(i, j) = [1 - p_{f2}\bar{\delta}(N - i)][(N - i)\mu_2 + r_2]$$
$$+ p_{f2}\bar{\delta}(N - i)[(N - i - 1)\mu_2 + 2r_2] + q_1 p_b \bar{\delta}(N - i)\lambda_1. \tag{15.11}$$

In Figure 14.2 (c), the system is in the post-full channel occupancy status, no initiating ST call enters the system. However, it is possible for an ongoing ST call to leave the channel due to service completion, type-II false alarm, or class-B misdetection with CCR1, and for a waiting ST call in the buffer to reconnect back due to a completion of a PT or ST call. State (i, j) means that there are i PT calls, $N - i$ ongoing ST calls being served and $j-(N-i)$ queued ST calls in the buffer. State (i, j) moves to $(i, j-1)$, with rate $\theta_3(i, j)$, where

$$\theta_3(i, j) = [1 - p_{f2}\bar{\delta}(N - i)][(N - i)\mu_2 + (j - N + i)r_2] +$$
$$p_{f2}\bar{\delta}(N - i)[(N - i - 1)\mu_2 + (j - N + i + 1)r_2] + q_1 p_b \bar{\delta}(N - i)\lambda_1, \tag{15.12}$$

where the first two terms contribute to the cases without and with the occurrence of a type-II false alarm, respectively, and the third term contributes to the case due to the occurrence of a class-B misdetection with CCR1.

Note that in the post-full condition, no type-I false alarm and class-A misdetection events happen. When $i = N$, we have $p_{f2} = 0$ and $p_b = 0$; and no PT calls can enter the system. Note also that state (i, j) does not exist when $i < 0$, $j < 0$, $i > N$, or $j > N$.

From the transition rate diagrams in Figure 14.2 (a), (b), and (c), the matrix Q of the two-dimensional Markov process is obtained with the same form as (14.14) but different sub-

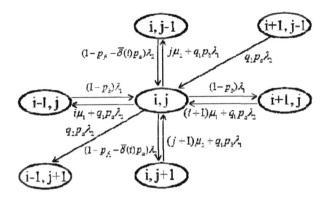

Figure 15.2: (a). The state diagram at (i, j) with pre-full channel occupancy under unreliable sensing.

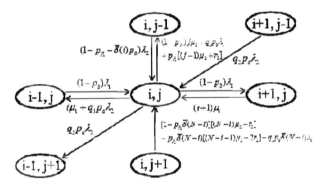

Figure 15.2: (b). The state diagram at (i, j) with just-full channel occupancy under unreliable sensing.

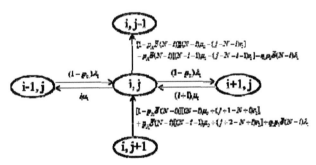

Figure 15.2: (c). The state diagram at (i, j) with post-full channel occupancy under unreliable sensing.

matrices B_i, D_i, and E_i. Each sub-matrix has size $(N+1)$ by $(N+1)$ and defined by

$$B_i(j,k) = (1 - p_b)\lambda_1 I_{N+1} 1_{\{0 \leq i < N\}} \tag{15.13}$$

$$D_i(j,k) = i\mu_1 1_{\{1 \leq i \leq N, 0 \leq j \leq N, k=j\}} + q_1 p_a \lambda_2 1_{\{1 \leq i \leq N, 0 \leq j \leq N-i, k=j\}} \tag{15.14}$$
$$+ q_2 p_a \lambda_2 1_{\{1 \leq i \leq N, 0 \leq j \leq N-i, k=j+1\}},$$

$$E_i = [A_i - \bar{\delta}(i)D_i - \bar{\delta}(N-i)B_i] 1_{\{0 \leq i \leq N\}}, \tag{15.15}$$

where I_n denotes an n-by-n identity matrix, $1_{\{\Phi\}}$ is the indicator function of set Φ defined by

$$1_{\{\Phi\}} \triangleq \begin{cases} 1, & \text{if } \Phi \text{ is true,} \\ 0, & \text{otherwise,} \end{cases}$$

and A_i is a matrix with the same size as E_i and with its element $A_i(j,k)$ given by
$A_i(j,k) =$

$$\begin{cases} [1 - p_{f1} - \bar{\delta}(i)p_a]\lambda_2, & 0 \leq i < N, 0 \leq j < N-i, k = j+1, \\ j\mu_2 + q_1 p_b \lambda_1, & 0 \leq i < N, 1 \leq j < N-i, k = j-1, \\ \theta_1(j), & 0 \leq i < N, j = N-i, k = j-1, \\ \theta_3(i,j), & 1 \leq i \leq N, N-i < j \leq N, k = j-1, \\ -[A_i(j,j-1) + A_i(j,j+1)], & 0 \leq i \leq N, 0 \leq j \leq N, k = j, \\ 0, & \text{otherwise.} \end{cases} \quad (15.16)$$

Applying the same procedure of (14.76), (14.77), and (14.78) to solve the steady state probability vector π_n, $0 \leq n \leq N$, of the system. There are some special cases for the proposed model.

- *Special Case 0: Single Primary System*

  If there are no SUs in the system, that is, $n_2 = 0$, $\lambda_2 = 0$, and $\mu_2 = 0$, the OSS system reduces to a single primary system. In this case, the performance model simplifies as follows:

$$B_i = \lambda_1, \quad 0 \leq i < N; \quad (15.17)$$
$$D_i = i\mu_1, \quad 1 \leq i \leq N; \quad (15.18)$$
$$E_i = i\mu_1 - \bar{\delta}(N-i)\lambda_1, \quad 0 \leq i \leq N; \quad (15.19)$$
$$A_i = 0, \quad 0 \leq i \leq N; \quad (15.20)$$
$$C_i = i\mu_1, \quad 0 \leq i \leq N. \quad (15.21)$$

  Substituting the above equations into (14.76), we obtain

$$\pi_n = \frac{1}{n!}\left(\frac{\lambda_1}{\mu_1}\right)^n \pi_0, \quad 1 \leq n \leq N, \quad (15.22)$$

  and

$$\pi_0^{-1} = \sum_{i=0}^{N} \frac{1}{i!}\left(\frac{\lambda_1}{\mu_1}\right)^i, \quad (15.23)$$

  which is the well-known Erlang loss model [28].

- *Special Case 1: An OSS System with Perfect Sensing*

  In this scenario, both initial and ongoing SUs make perfect spectrum detection, i.e., $p_{f1} = p_{f2} = 0$ and $p_a = p_b = 0$. Thus, the model of Part A is a special case of that of Part B.

- *Special Case 2: An OSS System with Only Initiating SUs Making Sensing Errors*

  In this scenario, only initiating SUs make detection errors, i.e., $p_{f2} = 0$ and $p_b = 0$. If we further assume that the class-A misdetection causes only CCR1, i.e., $q_1 = 1$, the proposed OSS model will become the same as that in [5].

- *Special Case 3: An OSS System with Only Ongoing SUs Making Sensing Errors*

  In this scenario, only ongoing SUs make detection errors, i.e., $p_{f1} = 0$ and $p_a = 0$.

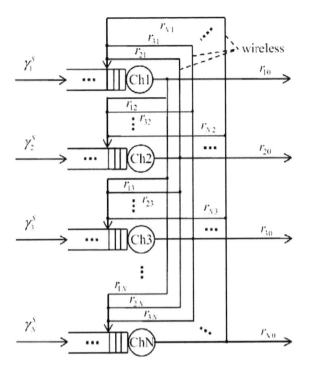

Figure 15.3: Queueing network model of the tolerable CR network.

- *Special Case 4: An OSS System with Only False Alarm Events Occurring*

  This scenario overemphasizes the performance of the primary system. It requires zero interference to the PUs, i.e., $p_a = p_b = 0$.

- *Special Case 5: An OSS System with Only Misdetection Events Occurring*

  In contrast to Special Case 4, this case gives excessive emphasis on the performance of the secondary system. It never considers the false alarm events for the SUs, i.e., $p_{f1} = p_{f2} = 0$.

C. Tolerable Service Degradation

We consider a tolerable CR data network operating over a given service area with the radio spectrum divided into N channels. PT and ST calls arriving to different channels form independent Poisson processes.

Figure 14.3 illustrates the proposed queueing network model of the tolerable CR data network, where each channel is effectively a single server queueing system. The external arriving ST traffic to channel i is assumed to form a Poisson process, with rate γ_i^S, $1 \leq i \leq N$. The service time of ST traffic at channel i is assumed to be exponentially distributed, with parameter μ_i^S, $1 \leq i \leq N$. The dynamics of the PT traffic are not shown explicitly in Figure 14.3, however, the constraints imposed by the PT traffic on the ST traffic are embedded in the model. For example, routing probabilities depend on the activities of the PUs.

From Figure 14.3, the total ST call arrival rate to channel i, λ_i^S, can be obtained from the external arrival rate γ_i^S and the internal arrival rates from other channels.

When the system is in steady-state, the output rate of channel i is equal to its arrival rate.

Hence, we have the following traffic equation [29]:

$$\lambda_i^S = \gamma_i^S + \sum_{j=1}^{N} \lambda_j^S r_{ji}, \quad j \neq i, \ i = 1, 2, \cdots, N, \tag{15.24}$$

where r_{ji} is the routing probability from channel j to channel i. The traffic equation (15.24) can be expressed in vector form:

$$\Lambda_S = \Gamma_S + P' \Lambda_S, \tag{15.25}$$

where $\Lambda_S = \left(\lambda_1^S, \lambda_2^S, \cdots, \lambda_N^S\right), \Gamma_S = \left(\gamma_1^S, \gamma_2^S, \cdots, \gamma_N^S\right)$, and the routing matrix P is

$$P = \begin{bmatrix} 0 & r_{12} & \cdots & r_{1N} \\ r_{21} & 0 & \cdots & r_{2N} \\ \vdots & \vdots & \ddots & \vdots \\ r_{N1} & r_{N2} & \cdots & 0 \end{bmatrix}. \tag{15.26}$$

Since the matrix $(I - P')$ is invertible, (15.25) has a unique solution given by

$$\Lambda_S = (I - P')^{-1} \Gamma_S. \tag{15.27}$$

Next, we analyze the performance of the queue associated with a target channel i under unreliable spectrum sensing, i.e., an ongoing ST call on a given channel may fail to sense the presence of an arriving PT call on the channel. According to the previous definition, such a sensing error is a class-B misdetection, and it occurs with probability p_b. For simplicity, we assume that sensing errors are committed only by ongoing SUs occupying a channel; an initiating SU accessing a new channel for a call is assumed to have enough time to identify the channel status (i.e., $p_a = 0$).

Let the arrival rate and service rate of the PT calls to channel i be λ_i^P and μ_i^P, respectively. When a class-B misdetection occurs, both PT and ST calls will use the same channel and experience degraded service. Hence, each type of call will have a reduced rate, say, $\alpha_P \mu_i^P$ for the PT call and $\alpha_S \mu_i^S$ for the ST call, where $0 \leq \alpha_P, \alpha_S \leq 1$, $\alpha_P + \alpha_S = 1$. We assume that the reduced rate of a PT call satisfies the predefined threshold for the QoS of PUs. Thus, the system will provide higher supportability to SUs at the expense of some tolerable service degradation for PUs.

Let $\{X(t), Y(t), t \geq 0\}$ represent the state of the queue associated with channel i at time t, where $X(t)$ is set to 1 if the system is in the primary mode serving a PT call, 2 if the system is in the secondary mode serving an ST call, and 3 if the system is in the degraded mode serving both calls with reduced rates, and $Y(t)$ denotes the number of ST calls in the system, including the one in service (if any). The process $(X(t), Y(t))$ is a two-dimensional Markov process, with state space $\{(k, j) \mid k = 1, 2, 3; j = 0, 1, 2, \cdots \}$ and state transition diagram shown in Figure 14.4.

Let $\pi_i(k, j)$ denote the steady-state probability that the channel i system is in state (k, j). The

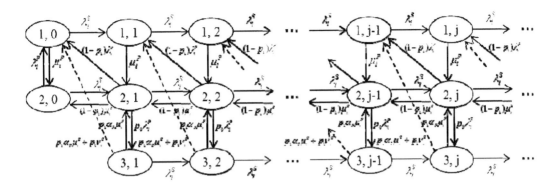

Figure 15.4: State transition diagram of the channel i system in tolerable service degradation.

global balance equations of the system are given as follows:

$$\pi_i(1,0)\left(\lambda_i^S + \mu_i^P\right) = \pi_i(2,0)\lambda_i^P$$
$$+\pi_i(2,1)(1-p_b)\lambda_i^P + \pi_i(3,1)p_b\alpha_S\mu_i^S, \tag{15.28}$$

$$\pi_i(2,0)\left(\lambda_i^P + \lambda_i^S\right) = \pi_i(2,1)(1-p_b)\mu_i^S$$
$$+\pi_i(1,0)\mu_i^P, \tag{15.29}$$

$$\pi_i(1,j)\left(\lambda_i^S + \mu_i^P\right) = \pi_i(2,j+1)(1-p_b)\lambda_i^P$$
$$+\pi_i(1,j-1)\lambda_i^S + \pi_i(3,j+1)p_b\alpha_S\mu_i^S, \tag{15.30}$$

$$\pi_i(2,j)\left(\lambda_i^P + \lambda_i^S + (1-p_b)\mu_i^S\right) = \pi_i(3,j)p_b\alpha_P\mu_i^P$$
$$+\pi_i(2,j-1)\lambda_i^S + \pi_i(1,j)\mu_i^P + \pi_i(2,j+1)(1-p_b)\mu_i^S, \tag{15.31}$$

$$\pi_i(3,j)\left(\lambda_i^S + p_b\alpha_P\mu_i^P + p_b\alpha_S\mu_i^S\right)$$
$$= \pi_i(3,j-1)\lambda_i^S + \pi_i(2,j)p_b\lambda_i^P, \tag{15.32}$$

where $j \geq 1$ and $\pi_i(3,0)\triangleq 0$. It is worth noting that state $(2, 0)$ corresponds to the empty system, $(1, 0)$ corresponds to a PT call in service and no ST call in the system, and $(3, 1)$ corresponds to both a PT call and an ST call being served on the same channel with no other ST call in the system.

In states $(2,j), j \geq 1$, an ongoing ST call contributes to the transition to $(3,j)$, with rate $p_b\lambda_i^P$, when it fails to detect the presence of an arriving PT call, and contributes to the transition to $(1, j-1)$ with rate $(1-p_b)\lambda_i^P$ when it detects correctly. From Figure 14.4, it is easy to find the routing probabilities as

$$r_{i0} = \frac{(1-p_b+p_b\alpha_S)\mu_i^S}{(1-p_b)\lambda_i^P + (1-p_b+p_b\alpha_S)\mu_i^S}, \tag{15.33}$$

$$r_{ij} = \frac{(1-p_b)\lambda_i^P \cdot \beta_{ij}}{(1-p_b)\lambda_i^P + (1-p_b+p_b\alpha_S)\mu_i^S}, \quad j \neq i, \tag{15.34}$$

where β_{ij} denotes the conditional probability that the ST call switches from channel i to channel j, given that it has to switch out from channel i. The value of β_{ij} depends on the channel assignment strategy and real-time measurements. For simplicity, we choose the equal-probability channel assignment strategy, i.e., $\beta_{ij}=1/(N-1)$, which can be substituted into (15.34) to evaluate Λ_S in (15.27).

We shall use generating function techniques (cf. [30], [31]) to solve the above system equations. Define the partial generating functions of the channel i system as

$$G_{k,i}(z) = \sum_{j=0}^{\infty} \pi_i(k,j) z^j, \quad |z| \le 1, k = 1, 2, 3. \tag{15.35}$$

Applying $G_{k,i}(z)$ to the system equations, we obtain the following equations:

$$[\lambda_i^S z^2 - (\lambda_i^S + \mu_i^P) z] G_{1,i}(z) + (1 - p_b) \lambda_i^P G_{2,i}(z)$$
$$+ p_b \alpha_S \mu_i^S G_{3,i}(z) = \pi_i(2,0) \lambda_i^P (1 - p_b - z), \tag{15.36}$$

$$\mu_i^P z G_{1,i}(z) + p_b \alpha_P \mu_i^P z G_{3,i}(z) + [\lambda_i^S z^2 -$$
$$(\lambda_i^P + \lambda_i^S + (1 - p_b) \mu_i^S) z + (1 - p_b) \mu_i^S] G_{2,i}(z)$$
$$= \pi_i(2,0) \lambda_i^P (1 - p_b) \mu_i^S (1 - z), \tag{15.37}$$

$$(\lambda_i^S z - \lambda_i^S - p_b \alpha_P \mu_i^P - p_b \alpha_S \mu_i^S) G_{3,i}(z) +$$
$$+ p_b \lambda_i^P G_{2,i}(z) = \pi_i(2,0) p_b \lambda_i^P. \tag{15.38}$$

Solving (15.36), (15.37), and (15.38), we have

$$G_{k,i}(z) = \frac{\pi_i(2,0) g_{k,i}(z)}{g_{0,i}(z)}, \quad k = 1, 2, 3, \tag{15.39}$$

where

$$g_{0,i}(z) = (\lambda_i^S z - \lambda_i^S - \mu_i^P)[\lambda_i^S z - (1 - p_b) \mu_i^S](\lambda_i^S z - C_0)$$
$$- \lambda_i^S z [\lambda_i^P (\lambda_i^S z - C_0 - \mu_i^P) + C_4] - C_6, \tag{15.40}$$

$$g_{1,i}(z) = -\lambda_i^P (\lambda_i^S z)^2 + \lambda_i^P \lambda_i^S (C_0 + C_1) z$$
$$- \lambda_i^P (C_0 C_1 - C_2 + C_5), \tag{15.41}$$

$$g_{2,i}(z) = -(1 - p_b) \mu_i^S (\lambda_i^S z)^2 +$$
$$+ \lambda_i^S [(1 - p_b) \mu_i^S C_0 + C_3 - C_4] z - C_0 C_3 + \mu_i^P C_2, \tag{15.42}$$

$$g_{3,i}(z) = p_b \lambda_i^P \lambda_i^S (\lambda_i^S z - \lambda_i^P - \lambda_i^S - \mu_i^P) z, \tag{15.43}$$

and

$$C_0 \triangleq \lambda_i^S + p_b \alpha_P \mu_i^P + p_b \alpha_S \mu_i^S, \tag{15.44}$$

$$C_1 \triangleq \lambda_i^P + (1 - p_b)(\lambda_i^S + \mu_i^S), \tag{15.45}$$

$$C_2 \triangleq p_b \lambda_i^P (\lambda_i^S + p_b \alpha_P \mu_i^P), \tag{15.46}$$

$$C_3 \triangleq \lambda_i^P \mu_i^P + (1 - p_b) \mu_i^S (\lambda_i^S + \mu_i^P), \tag{15.47}$$

$$C_4 \triangleq p_b^2 \lambda_i^P \alpha_P \mu_i^P, \tag{15.48}$$

$$C_5 \triangleq p_b^2 \lambda_i^S \alpha_S \mu_i^S, \tag{15.49}$$

$$C_6 \triangleq \lambda_i^P \mu_i^P [(1 - p_b) C_0 + p_b^2 \alpha_S \mu_i^S]. \tag{15.50}$$

Using the normalization condition $\sum_{k=1}^{3} G_{k,i}(1) = 1$, we obtain

$$\pi_i(2,0) = \frac{g_{0,i}(1)}{g_{1,i}(1) + g_{2,i}(1) + g_{3i}(1)}. \tag{15.51}$$

The generating functions $G_{k,i}(z)$, $k = 1, 2, 3$, are then determined by substituting (15.51) into (15.39).

15.4 Performance Metrics

15.4.1 *Selected Metrics for Perfect or Unreliable Spectrum Sensing*

- **Blocking probability of the PT**

  The PT call blocking probability, denoted by P_1, is defined as the probability that upon an arrival of a PT call in a service area all the channels are occupied by PT calls and the arrival request has to be blocked. Thus, we have

  $$P_1 = \sum_{j=0}^{N} \pi(N, j) = \pi_0 \prod_{i=1}^{N} [B_{i-1}(-C_i)^{-1}] \mathbf{e}. \qquad (15.52)$$

- **Blocking probability of the ST**

  The ST call blocking probability, denoted by P_2, is defined as the probability when all the channels in a service area are occupied by either PT calls or ST calls, and no channel is available for a new ST call request. Thus, we have

  $$P_2 = \sum_{i=0}^{N} \sum_{j=N-i}^{N} \pi(i, j). \qquad (15.53)$$

- **Total channel utilization**

  The total channel utilization η is defined as the ratio of the mean number of occupied channels to the total number of channels. We find that

  $$\eta = \frac{1}{N} \left\{ \sum_{i=0}^{N} \sum_{j=0}^{N-i} (i+j)\pi(i,j) + \sum_{i=1}^{N} \sum_{j=N-i+1}^{N} N\pi(i,j) \right\}. \qquad (15.54)$$

- **Mean reconnection probability of queued ST calls**

  As mentioned earlier, an ST call that waits in the buffer due to unavailability of a channel could reconnect back to the system if a channel becomes available before its maximum waiting time expires. The reconnection probability of a given ST call is defined as the probability that the ST call eventually reconnects back to the system before its maximum waiting time expires.

  Suppose that an ST call (referred to as the *test call*) arriving at the buffer finds that there are $j(0 \le j \le N\text{-}1)$ queued calls in the buffer, i, $1 \le i \le N$, PT calls and $(N-i)$ ST calls are being served. It is easily determined that the test call can reconnect back to the system only if $(j+1)$ calls ahead of it leave the system (either leave the channels or leave the buffer) before its maximum waiting time expires. To capture the queueing behavior of the queued ST calls, we introduce a 3-dimensional Markov process $(Z_1(t), Z_2(t), J(t))$ under the condition of post-full channel occupancy, where $Z_1(t)$, $Z_2(t)$, and $J(t)$ represent the number of PT calls, ongoing ST calls, and queued ST calls in the system at time t. The state space of the Markov process is

  $S^* = \{(n_1, n_2, j)|n_1 + n_2 = N, 0 \le j \le N\}$.

  Actually, the above 3-dimensional Markov process $(Z_1(t), Z_2(t), J(t))$ is equivalent to the previous 2-dimensional Markov process $(X_1(t), X_2(t))$ under the *post-full* condition, where $Z_1(t)$, $Z_2(t)$, and $J(t)$ can be determined by $X_1(t)$ and $X_2(t)$, and vice versa.

  Suppose that the test call arriving at the buffer finds j queued ST calls in the buffer, $0 \le j \le N\text{-}1$. The system state can be represented as $(i, N-i, j+1)$ with $0 \le j \le N\text{-}1$. Let γ denote the mean reconnection probability of the queued ST calls in FCFS discipline. Let $\beta(i,j)$ denote the probability that the test call arriving at the buffer eventually reconnects back to the system before its maximum waiting time expires, given that the test call arrives to find i PT calls in service and j ST calls in the buffer ($0 \leq j \leq N\text{-}1$). The mean reconnection probability γ can be expressed as

$$\gamma = \frac{\sum_{i=1}^{N} \sum_{j=0}^{i-1} \pi(i, N-i+j+1)\beta(i,j)}{\sum_{i=1}^{N} \sum_{j=0}^{i-1} \pi(i, N-i+j+1)}. \tag{15.55}$$

The conditional probability $\beta(i,j)$ can be derived under perfect spectrum sensing and unreliable spectrum sensing, respectively. Under perfect spectrum sensing, $\beta(i,j)$ is solved as [4]:

$$\beta(i,j) = \frac{i\mu_1 + (N-i)\mu_2}{i\mu_1 + (N-i)\mu_2 + (j+1)r_2}. \tag{15.56}$$

Under unreliable spectrum sensing, $\beta(i,j)$ is solved as [16]:

$$\beta(i,j) = \frac{i\mu_1 + (N-i)\mu_2 + q_1 p_b \lambda_1}{i\mu_1 + (N-i)\mu_2 + q_1 p_b \lambda_1 + (j+1)r_2}. \tag{15.57}$$

The mean reconnection probability γ can then be calculated by substituting (15.56) or (15.57) into (15.55).

15.4.2 Selected Metrics for Tolerable Service Degradation

- **Mean number of ST calls**

The mean number of secondary calls in the channel i system, denoted by L_i, can be expressed as

$$L_i = \sum_{k=1}^{3} \sum_{j=1}^{\infty} j\pi_i(k,j) = \sum_{k=1}^{3} G'_{k,i}(1). \tag{15.58}$$

- **System supportability**

The supportability of the channel i system for ST calls, denoted by S_i, is defined as the sum of the steady-state probabilities that the channel supports ST calls with both normal service (e.g., full-rate) and degraded service (e.g., sub-rate). Thus,

$$S_i = \sum_{k=2}^{3} \sum_{j=1}^{\infty} \pi_i(k,j) = \sum_{k=2}^{3} G_{k,i}(1) - \pi_i(2, 0). \tag{15.59}$$

- **Interference factor**

The system supportability of ST calls is achieved at the expense of some service degradation for PT calls. We introduce a metric, *interference factor* of the secondary system to the primary system, denoted by I_{sp}, to characterize the impact of service degradation to PT calls. The interference factor is defined as the sum of the probabilities of the states in which PT calls receive degraded service:

$$I_{sp} = \sum_{j=1}^{\infty} j\pi_i(3,j) = G_{3,i}(1). \qquad (15.60)$$

By varying the input parameters, e.g., arrival rates, service rates, detection error, etc., one can maximize the carried load of ST calls subject to limiting the interference factor to a predefined value I_0.

Due to length limitation, the numerical results and simulations for evaluating the above systems are omitted here. However, readers can refer to [4][16][17] for details, where it was shown that, by using OSS, much higher spectrum efficiency can be achieved, even under unreliable spectrum sensing.

15.5 Conclusions

We analytically modeled an OSS system under the conditions of perfect spectrum sensing, unreliable sensing, and tolerable service degradation through queueing theoretic frameworks. The OSS system consists of the PUs that are licensed a set of spectrum resources, and the SUs that opportunistically utilize the spectrum resources. The initiating SUs sense the channels that are unused by the PUs and then make use of them. An ongoing SU vacates its current channel when it detects the presence of a PT call, and enters another idle channel. If no channel is available for the ST call, the call waits in a buffer until either a channel becomes available or a maximum waiting time is reached. The condition of perfect sensing is a special case of unreliable sensing. Unreliable sensing is modeled by type-I and type-II false alarm events and class-A and class-B misdetection events, which brings more flexibility for the purpose of performance evaluation.

In the condition of tolerable service degradation, when a SU detects the presence of a PT call, it switches out from its current channel to another channel to continue its service; however, when the SU fails to detect the PT call, it will remain on the channel with the PT call, and both calls will receive degraded service.

We solve the steady-state probabilities of the considered systems and derive a set of performance metrics of interest. Matrix analytic method and generating function technique are respectively used for solving the system equations of different models. The proposed modeling methods are expected to be used for design and performance evaluation of future opportunistic spectrum sharing networks.

References

[1] M. McHenry, "Frequency agile spectrum access technologies," in Proc. FCC Workshop on Cognitive Radio, May 2003.

[2] G. Staple and K. Werbach, "The end of spectrum scarcity," *IEEE* Spectrum, vol. 41, pp. 48–52, March 2004.

[3] S. Haykin, "Cognitive radio: brain-empowered wireless communications," *IEEE J. Selected Areas in Comm.*, vol. 23, pp. 201–220, Feb. 2005.

[4] S. Tang and B. L. Mark, "Performance analysis of a wireless network with opportunistic spectrum sharing," in *Proc. IEEE Globecom*, pp. 4636–4640, Nov. 2007.

[5] S. Tang and B. L. Mark, "Modeling and analysis of opportunistic spectrum sharing with unreliable spectrum sensing," *IEEE Trans. on Wireless Communications*, vol. 8, pp. 1934–1943, Apr. 2009.

[6] A. Ghasemi and E. Sousa, "Collaborative spectrum sensing for opportunistic access in fading environments," in *Proc. of the frst IEEE Symposium on New Frontiers in Dynamic Spectrum Access Networks*, pp. 131–136, Nov. 2005.

[7] A. Mishra, "A Multi-channel MAC for Opportunistic Spectrum Sharing in Cognitive Networks," in *Military Communications Conference (MILCOM 2006)*, pp. 1–6, Oct. 2006.

[8] X. Liu and S. Shankar, "Sensing-based opportunistic channel access," *Mobile Networks and Applications*, vol. 11, pp. 577–591, Aug. 2006.

[9] L. Le and E. Hossain, "QoS-aware spectrum sharing in cognitive wireless networks," in *Proc. IEEE Globecom'07*, pp. 3563–3567, Nov. 2007.

[10] R. Urgaonkar and M. J. Neely, "Opportunistic scheduling with reliability guarantees in cognitive radio networks," in *Proc. IEEE Infocom*, (Phoenix, AZ), pp. 1301–1309, Apr. 2008.

[11] S. Huang, X. Liu, and Z. Ding, "Opportunistic spectrum access in cognitive radio networks," in *Proc. IEEE Infocom*, (Phoenix, AZ), pp. 1427–1435, Apr. 2008.

[12] B. L. Mark and A. O. Nasif, "Estimation of maximum interference-free transmit power level for opportunistic spectrum access," *IEEE Trans. on Wireless Communications*, vol. 8, pp. 2505–2513, May 2009.

[13] D.-C. Oh and Y.-H. Lee, "Cooperative spectrum sensing with imperfect feedback channel in the cognitive radio systems," *Int. J. of Communication Systems* (IJCS), vol. 23, pp. 763–779, June–July 2010.

[14] H. Chen and H.-H. Chen, "Spectrum sensing scheduling for group spectrum sharing in cognitive radio networks," *Int. J. of Communication Systems* (IJCS), vol. 24, pp. 62–74, January 2011.

[15] D. Gozupek and F. Alagoz, "Genetic algorithm-based scheduling in cognitive radio networks under interference temperature constraints," *Int. J. of Communication Systems* (IJCS), vol. 24, pp. 239–257, February 2011.

[16] S. Tang, "A General Model of Opportunistic Spectrum Sharing with Unreliable Sensing", International Journal of Communication Systems (IJCS), 27:31–44, 2014.

[17] S. Tang and B. L. Mark, "Performance of a Cognitive Radio Network with Tolerable Service Degradation", 7th Int'l Workshop on the Design of Reliable Communication Networks (DRCN 2009), Oct. 25-28, 2009, Washington, D.C., USA.

[18] H. Kim and K. G. Shin, "In-band spectrum sensing in cognitive radio networks: energy detection or feature detection," in *Proc. 14th ACM international conference on Mobile computing and networking*, (San Francisco, CA), pp. 14–25, 2008.

[19] S. Tang and B. L. Mark, "An adaptive spectrum detection mechanism for cognitive radio networks in dynamic traffic environments," in *Proc. IEEE Globecom*, pp. 1–5, Dec. 2008.

[20] S. Huang, X. Liu, and Z. Ding, "Optimal sensing-transmission structure for dynamic spectrum access," in *Proc. IEEE Infocom*, (Rio de Janeiro, Brazil), pp. 2295–2303, Apr. 2009.

[21] H. L. Van Trees, Detection, *Estimation and Modulation Theory: Part I*. New York: John Wiley & Sons, Inc., 2001.

[22] Y.-B. Lin, A. Noerpel, and D. Harasty, "The sub-rating channel assignment strategy for PCS hand-offs," *IEEE Trans. on Vehic. Tech.*, vol. 45, pp. 122–130, Feb. 1996.

[23] Y. Fang, Y.-B. Lin, and I. Chlamtac, "Channel occupancy times and handoff rate for mobile computing and PCS networks," *IEEE Trans. on Computers*, vol. 47, pp. 679–692, June 1998.

[24] Y.-R. Huang, Y.-B. Lin, and J. M. Ho, "Performance analysis for voice/data integration on a finite mobile systems," *IEEE Trans. on Vehic. Tech.*, vol. 49, pp. 367–378, Feb. 2000.

[25] B. Li, L.-Z. Li, B. Li, and X.-R. Cao, "On handoff performance for an integrated voice/data cellular system," *Wireless Networks*, vol. 9, pp. 393–402, Mar.-Apr. 2003.

[26] W. Li and X. Chao, "Modeling and performance evaluation of a cellular mobile network," *IEEE/ACM Trans. on Networking*, vol. 12, pp. 131–145, Feb. 2004.

[27] S. Tang and W. Li, "An adaptive bandwidth allocation scheme with preemptive priority for integrated voice/data mobile networks," *IEEE Trans. on Wireless Communications*, vol. 5, pp. 2874–2886, Oct. 2006.

[28] H. Kobayashi and B. L. Mark, *System Modeling and Analysis: Foundations of System Performance Evaluation*. Upper Saddle River, NJ: Pearson Education, Inc., 2009.

[29] P. G. Harrison and N. M. Patel, *Performance Modelling of Communication Networks and Computer Architectures*. Boston, MA, USA: Addison-Wesley, Jan. 1993.

[30] H. White and L. S. Christie, "Queueing with preemptive priorities or with breakdown," *Operations Research*, vol. 6, pp. 79–95, 1958.

[31] U. Yechiali and P. Naor, "Queuing problems with heterogeneous arrivals and service," *Operations Research*, vol. 19, pp. 722–734, 1971.

Chapter 16

Calculation Methods for Spectrum Availability

Andreas Achtzehn and Petri Mähönen

CONTENTS

Regulatory policies for secondary spectrum access enable more efficient spectrum exploitation and, by defining the rules for locally sharing incumbent spectrum, have the potential to open new markets to interested parties. The design space for such policy frameworks is vast, and comprises a plethora of possibilities to determine which and how frequency bands may be opened up for secondary operations. With regard to this, the reuse of unoccupied parts of the UHF-TV broadcasting bands, the TV whitespaces (TVWS), has gained most attention in academia and industry, paving the way towards the development of similar sharing frameworks for radar and other incumbent technologies. As a global forerunner, the US Federal Communications Commission (FCC) has enabled unlicensed secondary access to the TVWS in their 2008 Notice and Order. In Europe, the UK regulator Ofcom is currently preparing its own technical framework.

A comparison of the two practical examples for regulation-enabled secondary access reveals significant technical differences in how regulators approach the design of spectrum sharing policies. The FCC regulations are based on a static interference power protection concept, which demands incumbent receivers to withstand a certain man-made increase in the local noise floor. In its practical implementation, this "interference temperature" system culminates in a set of exclusion rules that require secondary transmitters to maintain a minimum separation distance to any potential primary device location. On the contrary, the Ofcom proposal implies a regulatory interpretation of interference as a reduction in incumbent service quality, originating from the deployment of other radio transmitters. Since receivers close to the main transmitter sites are less susceptible to service quality reduction, they may withstand higher amounts of absolute interference power, *ceteris paribus*. Building upon this general radio principle, the UK regulator aims at implementing a service-level protection concept that provides secondary access to the TVWS at any location for any channel. In order to guarantee a fixed maximum reduction in the primary service level, secondary devices are required to (sometimes substantially) limit their power emissions.

Whereas the FCC policy outcome can be properly assessed by spatial statistics on locally available spectrum resources, for Ofcom-type of proposals also permissible transmit powers and primary interference levels need to be taken into account. Further statistics on spectrum fragmentation can provide meaningful insight into the practical exploitability of secondary spectrum resources. In this chapter, we therefore compare the two spectrum sharing policy design principles from a conceptual perspective. In order to support the discussion on selecting the "right" protection concept for a particular technology/band, we give an overview of the respective parameter design space. In our assessment of the impact of certain parameter choices, we explore the achievable benefits of spectrum sharing applying the concepts side-by-side for the TVWS of Germany, for which we present key metrics on spectrum availability and quality.

16.1   Incumbent-Newcomer Relationship

At first glance, spectrum sharing models focus exclusively on the technical aspects of how incumbent (primary) systems and new (secondary) systems may coexist. While the incumbent needs to be protected against unacceptable interference, the access policy procures secondary systems with enough spectrum to be economically viable. Beyond this technical side, multilateral negotiations between regulators, incumbents, and newcomers are necessary to find solutions that adequately mediate between economic and public interests, and which can be codified into a technical solution. In this chapter, we focus exclusively on the technical side of the discussion. In the following, we present the framework within which solutions are developed, and enumerate the parameters that constitute the necessary negotiation space. We find this analysis important for highlighting that spectrum availability is, contrary to a physical truth, a direct result of this negotiation process.

Regardless of the interference framework used, the technical discussion is generally focused on the impact that secondary transmissions have on the performance of primary links. In Figure 16.1, we show the typical geometry that is used as a baseline for this relationship analysis. In order to keep the analysis tractable, the primary system is often reduced to a single primary transmitter/receiver pair, which is usually selected to be the one that is most susceptible to secondary interference. The ERP at the output of the primary transmitter device is specified by $\text{ERP}_{\text{TX},p}$[1], and frequency-dependent signal parameters, such as feeder cable losses ($L_{c,pt}$), antenna directionality in azimuth, and polar direction ($A_{TX,p}(\alpha, \theta)$), are taken into account. The geometry is strictly three-dimensional, which permits differentiating between (relatively) exposed and obstructed links. The primary signal traverses through space and gets attenuated. Naturally, the extent of attenuation thereby depends on frequency, the medium, obstacles, multipath propagation, and fading. As scenario-dependent measurement studies are generally impracticable, the mean distance and obstruction-dependent pathloss is approximated as $\text{PL}_{p \to p}$ through generic propagation models,

---

[1]Furthermore, in this chapter, we exclusively use logarithmic notation, i.e. power values are defined relative to a reference power, while attenuations and gains are given by fixed decibel offsets.

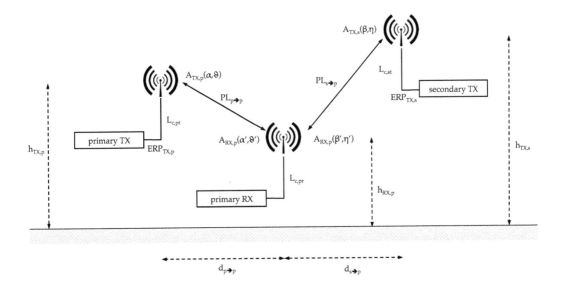

Figure 16.1: System model for secondary spectrum access.

for which the uncertainty due to scenario-specific particularities is modeled through an additive Gaussian component. While in reality propagation is considerably more complex, the Gaussianity assumption establishes a mathematically tractable baseline for subsequent steps. At the primary receiver side, losses and gains from the physical configuration of the setup are incorporated to estimate the geometry-dependent received power $P_{RX,p}(\alpha, \theta)$. In the absence of secondary transmissions, the signal-to-noise ratio (SNR) at the primary receiver is, hence,

$$\text{SNR}_p = P_{RX,p}(\alpha, \theta) - N_0 - \text{NF}, \tag{16.1}$$

where NF is the noise figure of the device. N_0 is the thermal noise floor [11]. The primary receiver can decode the transmitted information if $\text{SNR}_p \geq \text{SNR}_{\text{min,p}}$. $\text{SNR}_{\text{min,p}}$ is a system-specific value that depends on the applied modulation scheme, coding, and other performance aspects of the receiver chain. For large-scale broadcasting systems, the SNR threshold allows us to derive a closed shaped spatial area $\mathcal{C}$, commonly denoted as the *coverage area*, within which a primary receiver is able to decode the signal.

The secondary spectrum access policy specifies the permissible power of a secondary transmitter, $P_{TX,s}(\beta, \eta)$, in the frequency band of primary operations, whereby the link model is aligned with that of the primary link. At the primary receiver, the secondary transmission is observed as a certain interference power $I_{RX,p}$. The signal-to-interference-and-noise ratio (SINR) after introduction of the secondary transmitter into the system becomes

$$\text{SINR}_p = P_{RX,p}(\alpha, \theta) - 10\log_{10}\left(10^{\frac{N_0}{10}} + 10^{\frac{I_{RX,p}}{10}}\right) - \text{NF}, \tag{16.2}$$

where for $P_{TX,s} > -\infty$ it holds that $\text{SINR}_p < \text{SNR}_p$, i.e., a reduction compared to the SNR is inevitable. The secondary access policy subsequently determines where, when, and how much the SNR may be reduced by secondary operations.

16.2   Interference Temperature-Based Primary Protection Model

We will now describe a first practical model for defining a policy that enables secondary transmitters to operate at acceptable power levels. This interference temperature-based system is used in the FCC regulation for the TVWS, and is first mentioned in a report of the Spectrum Efficiency Working Group of the FCC Spectrum Policy Task Force from November 2002 [5]. The interference temperature $T_I(f, \Delta f)$ [14], as introduced herein, is defined (in the notation we adapt from Clancy [1]) as

$$T_I(f, \Delta f) = \frac{I_{\text{RX,p}}}{k_B \Delta f \times 10^6}. \tag{16.3}$$

The unit of $T_I(f, \Delta f)$ is kelvin, whereby the notation is normalized to unit bandwidth by division through Δf, the system bandwidth. An intuitive understanding of the interference temperature is that the received power from the interfering secondary transmitter matches the noise-only power if the receiving antenna is operated at this temperature. Likewise, interference as experienced at the receiver would be equivalent if the temperature was raised to $T_I(f, \Delta f)$ kelvin.

In the derivation of the policy parameters, the weakest primary link is considered, which assumes that the primary receiver is located at the edge of the intended coverage area; see Figure 16.2. For this receiver, it holds that

$$\text{SNR}_p = \text{SNR}_{\text{min,p}} + \gamma, \tag{16.4}$$

where γ is a safety margin set by the primary operator to account for planning uncertainties and other temporal variations. Due to this safety measure, the area where the signal is decodable extends beyond the *coverage contour* and establishes a (potentially large) safety area. Assuming a regular propagation model, the boundary of this is generally referred to as the *noise-limited contour*. Interference from a secondary source reduces the safety area by lowering SNR_p at the coverage contour to

$$\text{SINR}_p = \text{SNR}_{\text{min,p}} + \gamma - 10\log_{10}\left(1 + \frac{T_I(f, \Delta f)}{T}\right), \tag{16.5}$$

whereby no regular service degradation within the coverage area is experienced if γ is larger than $10\log_{10}\left(1 + \frac{T_I(f,\Delta f)}{T}\right)$. If the latter does not hold, the coverage area is reduced. For reasons of geometry, the secondary transmitter is always assumed to operate outside the primary coverage area.

The interference temperature assumption yields a set of four inter-related parameters to be specified in the protection model, namely,

- the permissible secondary transmit power $P_{\text{TX,s}}(\beta, \eta)$, i.e., the maximum acceptable emission from the secondary device in direction of the worst interfered primary receiver that causes interference at that side;

- the minimum secondary separation distance $d_{\text{min},s \to p}$, i.e., the distance to the worst respective interfered primary receiver that must be maintained in order to ensure that the received interference power $I_{\text{RX,p}}$ stays below a reasonable threshold;

- the acceptable erosion margin ε, i.e., the degradation in SNR the primary operator needs to accept from the introduction of the secondary system, whereby $\text{SNR}_p - \text{SINR}_p \leq \varepsilon$ for all locations inside $\mathcal{C}$; and

- the protected coverage contour, defined by SNR_p before introduction of the secondary.

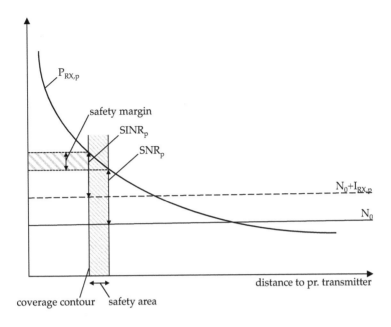

Figure 16.2: Interference temperature model. Effect on coverage and safety margin.

$P_{\text{TX,s}}(\beta, \eta)$ and $d_{\text{min},s \to p}$ are with respect to ε interchangeably. The acceptable erosion margin ε and the coverage-determining SNR_p on hand are subject to negotiation between the primary license holder and the regulator, because they affect the potential operations area size. Without loss of coverage, the erosion margin can be selected up to the safety margin of γ. $P_{\text{TX,s}}(\beta, \eta)$, and $d_{\text{min},s \to p}$ may be selected freely so to match the requirements of one or more secondary users, e.g., by allowing higher transmission powers with decreased spatial spread of the secondary system.

The interference power that the primary receiver must withstand can be calculated from the erosion margin as follows. At the worst-interfered point, it holds that

$$\text{SNR}_p - \text{SINR}_p \overset{!}{=} \varepsilon. \tag{16.6}$$

By expanding (16.6) for (16.1) and (16.2), we find the permissible secondary interference power to be

$$10\log_{10}\left(10^{\frac{N_0}{10}} + 10^{\frac{I_{\text{RX,p}}}{10}}\right) - N_0 = \varepsilon$$

$$\Leftrightarrow I_{\text{RX,p}} = 10\log_{10}\left(10^{\frac{\varepsilon}{10}} - 1\right) + N_0. \tag{16.7}$$

It is worth to note that beside interference from co-channel operations of secondary systems, the secondary transmitter spectral purity in terms of its adjacent-channel-leakage ratio (ACLR) and the quality of primary system receive filters, i.e., their ability to reject out-of-band emissions, need to be accounted for. Particularly, the latter adjacent-channel selectivity (ACS) parameter poses potential challenges, as a posteriori introduced secondary spectrum access may imply a certain quality of primary receivers. If those quality constraints are not met, the secondary will cause unavoidable interference. For a more extensive treatment of this parameter space subject, see, e.g., [12].

Practical interference temperature model implementations make a number of simplifications. In current regulatory practice, the coverage areas are only approximated through applying chartered propagation curves [3, 4]. Furthermore, only very generic assumptions are made on the antenna gains at the primary transmitter and receiver side. The intention of this definition is that through the simplification, the coverage area of the primary system can be easily derived through finding a fixed distance boundary with regard to the primary transmitter location. In reality, this assumption may be too simple. As indicated before, after introduction of the secondary system, the SINR at the primary coverage contour becomes

$$\text{SINR}_p(d_{p\rightarrow p}) = \text{SINR}_{\text{min},p} + \gamma - \varepsilon, \tag{16.8}$$

whereby the pathloss between any two radios is assumed to be monotonically increasing with respective distance. For reasons of geometry, and since secondary operations are not permitted within the circular coverage area, it holds that any reduction in distance to the primary transmitter leads to an increase in distance toward the secondary transmitter. Consequently, the protection model assumes that

$$\text{SINR}_p(d_{p\rightarrow p} - \Delta d) \overset{!}{>} \text{SINR}_p(d_{p\rightarrow p}), 0 < \Delta d < d_{p\rightarrow p}, \tag{16.9}$$

because at the same time the received signal strength from the primary is increased, the interference power is reduced. Expanding this expression we find that some approximations need to be made for the assumption to hold. We list them here briefly, but refer to our extensive study on shortcomings of the plane protection model in [16].

- $\text{PL}_{p\rightarrow p}(d_{p\rightarrow p}) - \text{PL}_{p\rightarrow p}(d_{p\rightarrow p} - \Delta d) > 0$, which applies only for a regular propagation environment. Shadowing from obstacles, such as mountains, can result in lower pathlosses at higher distances.

- $A_{TX,p}(\alpha, \theta) + A_{RX,p}(\alpha', \theta') \geq A_{TX,p}(\alpha, \theta^\dagger) + A_{RX,p}(\alpha', \theta^*)$, where $\theta^\dagger$ and θ^* are the polar angles at distance $d_{p\rightarrow p} - \Delta d$. If antenna gains differ depending on the inclination angle, which is the case for any real physical antenna, then lower antenna gains will result in a different received signal level closer to the transmitter. However, this is only a problem if no proper angular alignment is applied.

- $A_{TX,s}(\beta, \eta) + A_{RX,p}(\beta', \eta') \geq A_{TX,s}(\beta, \eta^\dagger) + A_{RX,p}(\beta', \eta^*)$, where $\theta^\dagger$ and θ^* are the polar angles at distance $d_{p\rightarrow p} - \Delta d$. If antenna gains differ depending on the inclination angle, which is generally the case for any practical antenna, then lower antenna gains will result in a different received interference level. As the primary receiver antenna is generally directed toward the primary transmitter, this antenna gain effect cannot be controlled.

Each of these simplifications is, unfortunately, necessary in order to keep the system model for secondary spectrum access tractable. In order to ensure that these approximations are not affecting primary operations, policymakers currently do need to incorporate sufficient protection margins, e.g., by increasing protection distances or lowering permissible transmit powers. A more detailed scenario analysis could in the future allow reduced safety margins.

16.3   Service-Level-Based Primary Protection Model

While comparably simple to implement, the interference temperature protection model remains a conservative choice. Particularly, the fixed interference power threshold is very protective, and excludes secondary operations, even if some primary receivers were principally capable of withstanding higher interference levels. For this reason, the regulatory community has adopted a service-level-oriented interference definition [9]; only unwanted signal power that actually degrades the incumbent service is considered relevant in the light of this specification. A purely signal-level-based protection model is fundamentally inefficient, because it protects any primary receiver against the *same amount* of interference power, which is selected low enough to cause service degradation only at the weakest of them. Given regular propagation characteristics, however, primary receivers closer to a transmitter are less vulnerable due to their higher SNR, thus, in principle, they could operate even under the impression of higher interference powers. This is the main motivation for a service-level-based primary protection model.

A service protection model integrates service degradation benchmarks into the decision-making. Acceptable interference levels are defined so that

$$Pr\left\{SINR_p \geq SNR_{min,p}\right\} \geq q_{SINR}, \tag{16.10}$$

where $Pr\{\cdot\}$ is the probability of an event, in this case, the probability that the SNR is above the decodability threshold, i.e., that the service is possible. q_{SINR} is the service probability, i.e., the probability that the primary service can be provided, given the experienced SINR. The policy defines the target service probability reduction Δq, with respect to the service probability before deployment of the secondary, q_{SNR}. At any location inside coverage area $\mathcal{C}$, the policy guarantees that

$$q_{SINR} \geq q_{SNR} - \Delta q, \tag{16.11}$$

i.e., that the service probability is not degraded beyond the target level [13]. For secondary transmitters closer to high SNR receivers, the transmit powers may, thus, be increased due to the lower susceptibility of service level reduction. Most importantly, a service-level-based primary protection model does not require defining protection contours around transmitters, thus, all channels are in principle accessible to the secondary (albeit with potentially very low transmit powers).

Contrary to the interference temperature model, where a trade-off between secondary transmit power and minimum separation distance (conditioned on the erosion margin at the coverage edge) exists, the service-level model unifies this to just two probabilistic parameters, namely the permissible service level degradation Δq and the likelihood L that this value is exceeded in a given location within the coverage area. The calculation model is, due to its stochastic nature and in the absence of strict spatial demarcation lines, considerably more complex to compute. Contrary to a fixed signal-level value to find the coverage contour, the coverage area is defined in terms of the initial service probability, q_{SNR}, which can be derived from the signal-level distribution at a given distance. Assuming a generic pathloss model with Gaussian error term, the coverage probability can be derived from the properties of the Gaussian distribution as

$$\begin{aligned} q_{SNR} &= Pr\{SNR_p \geq SNR_{min,p}\} \\ &\quad Pr\{P_{TX,p}(\alpha,\theta) - PL_{p \to p} + A_{RX,p}(\alpha',\theta') - L_{c,pr} - N_0 - NF \dots \\ &\quad -SNR_{min,p} \geq 0\} \\ &= 1 - \frac{1}{2}\text{erfc}\left\{\frac{1}{\sqrt{2}}\frac{m_S}{\sigma_p}\right\}, \end{aligned} \tag{16.12}$$

where $m_S = P_{TX,p}(\alpha,\theta) - \mathbb{E}[PL_{p \to p}] + A_{RX,p}(\alpha',\theta') - L_{c,pr} - N_0 - NF - SNR_{min,p}$ is the mean deviation of the receiver SNR from the minimum operational SNR, and σ_p is the error term standard deviation. Based upon negotiations with the primary user, the policymaker may consider a location as covered (and thereby subject to protection) if $q_{SNR} \geq q_{min}$. While diverging in describing the coverage thresholds, both coverage assumptions for interference and service level-based protection models are, to a large extent, interchangeable; for a fixed value of q_{min}, an equivalent minimum mean SNR can be calculated through inversion of (16.13).

The deployment of a secondary system lowers the service probability of the primary system, i.e., the coverage probability is reduced and the covered user service level degrades. For the new service probability, q_{SINR}, the following inequality holds:

$$q_{SNR} > q_{SINR} \geq q_{SNR} - \Delta q, \tag{16.13}$$

whereby

$$q_{SINR} = \Pr\{SINR_p \geq SNR_{min,p}\} \tag{16.14}$$

$$\Pr\{\underbrace{P_{TX,p}(\alpha, \theta) - PL_{p \to p} + A_{RX,p}(\alpha', \theta') - L_{c,pr}}_{P_{RX,p}} - NF \ldots$$

$$-10\log_{10}\left(10^{\frac{N_0}{10}} + 10^{\frac{I_{RX,p}}{10}}\right) - SNR_{min,p} \geq 0\}. \tag{16.15}$$

The complexity of the service-level protection model originates from the fact that, for a given location of the secondary transmitter, *all* covered locations within a reasonable distance need to be tested in order to find the maximum permissible secondary transmit power. That means we need to find for any given combination of secondary location and secondary transmit channel the distances (cf. pathloss), channel leakage and primary selectivity ratios (due to ACS and ACLR), and angular antenna discriminations to all potentially interfered primary coverage locations. Additionally, some of these constraints are for any real receiver a function of the primary received power. Currently, no relaxations or acceptable heuristics have been proposed, thus, the calculation is highly time-consuming. The practical applicability of the service level protection model is, thus, constrained to non-real time applications or limited area sizes, and requires substantial research not only into the performance of secondary transmitters, but also in the interference susceptibility of primary receivers for the particular primary service under different geometry assumptions. Consequently, service-level-based primary protection models have not yet reached mainstream regulations. Due to their benign features, however, they are highly interesting for scenarios where pure spatial exclusion rules provide only very limited spectrum capacity.

16.4 Spectrum Availability Metrics

We will now introduce three relevant spectrum availability metrics, namely, *channel availability*, *conditional channel availability*, and *secondary transmit power to primary interference and noise ratio*, to quantify the comparative benefit of particular protection models and their parameters. We have chosen the secondary use case scenario in the German TV bands as an example, due to the high system diversity of originally two broadcasting system being merged, the terrain differences between the flat North and the mountainous South, and the large number of neighboring countries. It will allow us to showcase some of the major differences in how spectrum availability metrics are affected by the selection of protection models.

We are aiming at providing quantitative figures that enable a fair comparison between interference temperature and service-level protection concepts. As the two major reference TVWS frameworks proposed by FCC and Ofcom apply to slightly different TV system technologies, DVB-T and ATSC, and as the aim of regulators inside their respective domain slightly differs, we have applied certain unification steps. In particular, we assume that the primary system serves TV sets up to the noise limited contour level, whereby we make a conservative assumption on the signal level this contour is represented by. We adopt the FCC's view of the contour level being based on the F(50,90) propagation curves, i.e., the signal level is used which is exceeded in 50% of the locations for 90% of time. The minimum operational SNR, $SNR_{min,p}$, is selected as 20.3 dB, which is the (simulation-based) decodability threshold of a 64-QAM, code rate 2/3 DVB-T signal for a Ricean fading channel [2]. The selected modulation and coding scheme is a very common configuration in Germany, and the channel assumption is the most conservative possible choice, i.e., the actual likelihood of coverage is very high inside the modelled coverage area. We fix the permissible service reduction to $\Delta q = 0.07$, which is the value also used by Ofcom for the UK. This equals to an increased interference and noise level of 3 dB at the coverage contour. Given that this interference-to-noise ratio represents the erosion margin in the interference temperature model, we can thus set ε to the equivalent value. Antenna gains and cable feeder losses were chosen according to standard values provided in literature. The maximum secondary transmit power was set to 36 dBm.

16.4.1 Coverage Calculation and Channel Availability

The elevation profile between transmitter and receiver in our example scenario is regionally highly diverse, therefore, in the first step of our analysis, we quantify the difference between using the terrain-aware Longley–Rice propagation model [7, 8, 15] for the primary protection area calculation with curve-based modeling applying the ITU P.1546 [10] propagation model. For sake of clarity, we focus here on the interference temperature protection model, because only this model allows defining a channel availability metric in terms of a vector of locally accessible channels. Secondary access schemes that are based on a protection contour approach derive, as a first approximation, the inverse of the spectrum occupancy of the primary system. However, the requirement of protecting adjacent channels for reasons of weak receiver filters (low ACS) or a liberal secondary transmitter spectral mask (high ACLR) will exclude additional channels from spectrum access. For example, in the FCC regulations also adjacent channels next to the broadcasting system operating channels need to be protected. Unless otherwise stated, we follow this modeling assumption.

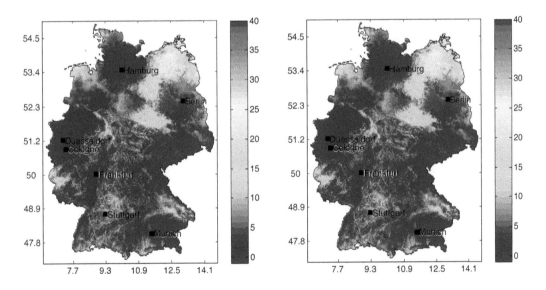

(a) Number of TVWS channels, Longley–Rice propagation model.

(b) Number of TVWS channels, ITU-R P.1546 propagation model.

Figure 16.3: Number of TVWS channels in Germany, as calculated by the interference temperature protection model.

We show in Figure 16.3 maps of Germany, where we color-code the number of available TVWS channels according to the interference temperature protection model, whereby we modify only the underlying primary propagation model. At first visual inspection, we see that the Longley–Rice dataset in Figure 16.3(a) exhibits significantly smaller channel availabilities throughout the entire country, compared to the ITU model in Figure 16.3(b). For the latter, the coverage regions of the individual transmitters become readily visible by the circular low-availability zones that surround the transmitter installations in and nearby larger urban areas, e.g., in Duesseldorf/Cologne, Berlin, or near Munich. Only in the northeastern region of the country on both maps exhibit larger channel availabilities. In both propagation scenarios, the main cities of the country are subject to low secondary spectrum availability.

The interference protection model has a significant shortcoming when used in combination with terrain-aware protection models. We find that the local elevation profile of a transmitter may practically allow reception of a broadcasting signal even far away from the intended coverage region, and, thus, practically outside a realistic protection area. One may consider the example of a mountain installation of a TV tower to establish coverage for a single valley at the foot of that mountain. This is a very common scenario, which can be found, e.g., in the major cities. The tower serves the valley region, but, due to its exposed position, has an obstacle-free propagation path also toward other elevated points outside the intended coverage area. While the broadcasters do not intentionally radiate toward those points (and, thus, would unlikely declare protection requirements for those locations), the coarse protection model assumption treats these locations as equally in need of protection. The effects of exposed positions in the studied terrain are partially counter-acted by the effects that larger mountain ranges do have on the protection assumption. If regions nearby a mountain range are served by a single transmitter, a terrain-agnostic model may indicate these regions to extend beyond the mountain range. This, naturally, is physically inconsistent. Examples of these can be found in the south of the country, at the borders to Switzerland and Austria. However, in practice, one can see in the transmitter antenna patterns that network planners have often taken this case into account, i.e., we find a decreased gain toward the mountain range.

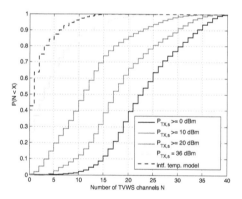

Figure 16.4: Empirical distribution function of the number of TVWS channels in Germany, as calculated by the interference temperature protection-based spectrum availability model.

Following our initial spatial assessment, we will now study quantitative metrics for Germany's TVWS spectrum availability. We have plotted the empirical distribution function (EDF) of the number of available channels for the ITU-R P.1546 propagation model in Figure 16.4. To gather a first approximative understanding of the value of secondary access, we also show the channel availability with respect to the population density, where instead of the fraction of the countries area for which a certain number of TVWS channels is available, the fraction of the population is used as a reference. A surprising observation is made with regard to the population-weighted sampling effects on spectrum availability. The population distribution within the country to a large extent averages out the disparity between TVWS channel counts in urban and rural areas. In the respective distribution function, we see almost no effect of the weighting on the metric for the number of channels a random user would be offered. We consider this to be a peculiarity of the country's population distribution, where only 7 percent of the landmass is uninhabited. Only in few spot regions the population density increases, thus, the area spectrum availability is a reasonable approximation to the population spectrum availability. This is an interesting finding in comparison to earlier whitespace studies for the US, where the sampling on population density provided significantly different findings on the feasibility of spectrum operations [6]. There, a higher population clustering can be assumed.

16.4.2 Minimum Separation Distance and Local Channel Availability

The necessary trade-off between secondary transmit power and required separation distances to the coverage contour changes the overall availability of TVWS channels in the interference temperature protection model. For a fixed erosion margin, the stakeholders may decide whether to favor high power transmissions at the costs of lower channel availability, or whether to open the spectrum for more secondary access of smaller devices.

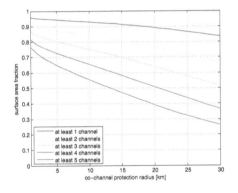

(a) Longley–Rice propagation model.

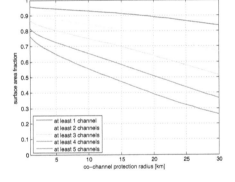

(b) ITU P.1546 propagation model.

Figure 16.5: Availability of TVWS channels in the interference temperature protection model for different co-channel protection distances.

For the case of Germany, we can show how this trade-off significantly affects the the spectrum availability; see Figure 16.5. For the sake of clarity, we have fixed the adjacent channel separation distance, as imposed by the FCC implementation of the interference temperature model to 0 km, i.e., secondary transmissions in directly adjacent channels are allowed once the secondary is outside the coverage area. For the Longley–Rice propagation model, we find that the separation distance has only little effect on the general TVWS availability. By varying the co-channel separation distance between 0 km and 30 km, the fractional area with at least one available channel changes by only 10 percent. Lower separation distances are, thus, of minor use for opening up TVWS. Our analysis shows that the adjacent channel exclusion rule causes harsher constraints on the spectrum availability. Looking at those figures for higher channel availabilities, e.g., the case of more than 5 channels per location, we find that a modification of the separation distances mostly benefits areas where already some spectrum is available. We can conclude that the separation distance is a reasonable design parameter for increasing secondary spectrum capacity, but that it is not sufficient for removing areas that are underserved by secondary access systems.

16.4.3 Conditional Channel Availability

As we can see from the reduced sensitivity of the channel availability on the separation distance, the actual distribution of channel usage dominates the results for the interference temperature-based protection model. TVWS assignment, according to an interference temperature protection model, must necessarily show a certain local fragmentation in the TVWS channel set, because broadcast planners aim at maximizing the separation in the channel assignment of neighboring broadcast towers to minimize adjacent channel interference issues in cheap receivers. Unused intermediate channels thereby act as guard bands. To quantify this effect, we depict in Figure 16.6 the conditional probability of TVWS channel availability. The probability to find two adjacent TVWS channels is high in the studied environment, however, when considering also larger channel aggregations this figure is quickly decaying. For example, if we find an arbitrary channel to be available for secondary use, there is a one in three chance that also the adjacent channel can be used by the secondary. When comparing the availability for four adjacent channels (32 MHz of contiguous whitespaces), the conditional probability of the furthest channel is only approximately 18 percent. Given a free TVWS channel, the probability of finding a free adjacent channel according to Figure 16.6 is only 10 percent. It is very unlikely that a secondary user can find larger numbers of adjacent TVWS channels, thus, if the secondary technology cannot exploit also small channel bandwidths, this may hamper exploitation capabilities. We can conclude that a spectrum availability metric should not only include figures on the total bandwidth of spectrum freed for secondary use, but also on the relative local configuration. This is particularly important for technologies that rely on continuous spectrum access.

16.4.4 Channel Availability In Service Level Protection Models

The spectrum availability is severely constrained in the interference protection model for the German TVWS due to the way the broadcasters have assigned channels to broadcasting regions. Hence, the service-level protection model seems an interesting alternative for enabling interference-free

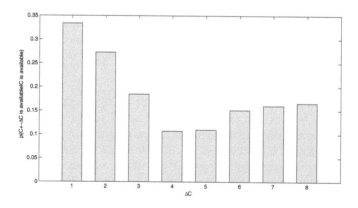

Figure 16.6: Probability that channel $C + \Delta C$ is a TVWS channel if channel C is a TVWS channel.

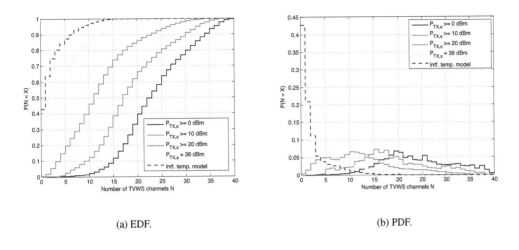

(a) EDF.

(b) PDF.

Figure 16.7: Empirical distribution and relative frequency functions of the number of TVWS channels in North Rhine–Westphalia, conditioned on the permissible transmit power $P_{TX,s}$.

coexistence. To evaluate this, we have chosen one particular populous example regions, the state of North Rhine–Westphalia (NRW), for our further study. NRW spans an area of approximately $34\,000\,\text{km}^2$, and has a population of approximately 18 million citizens. It suffers from low TVWS channel availability, according to the interference temperature model, with on average only 2.02 TVWS channels, compared to the country-wide 5.45 channels.

In the absence of strict exclusion rules, we need to modify the TVWS channel availability metric. In order to acquire a comparable vector of locally available channels, we deem a channel "available" if the permissible secondary transmit power $P_{TX,s}$ exceeds a fixed threshold. In Figure 16.7, we show the relative frequency function and the EDF for this metric side-by-side. The dashed lines show for comparison the unmodified channel availability metric for the interference temperature model. The maximum permissible transmit power plot of Figure 16.7(a) , i.e., $P_{TX,s} = 36\,\text{dBm}$, already shows an advantage of the service level protection model over the interference temperature model. In the NRW case, the EDF universally stays below the respective interference temperature case, i.e., that locally on average more high-power channels become available than before. The quantitative channel distribution shows that the transition to the service-level protection model does not generally change the composition of high-power $36\,\text{dBm}$ channel availabilities. A shift from no-availability to the availability of at least one channel is seen in most cases, as can be observed from the general form of the function, however, the service-level protection model does not resolve the overall channel availability problem for high-power scenarios. This outcome originates from two systems aspects. On the one hand, the service-level protection model allows high power access closer to the main transmitter due to the lower interference susceptibility. Furthermore, the secondary system is allowed to penetrate deeper into adjacent channel regions at high powers.

16.4.5   *Primary System Interference and Spectrum Quality*

While attractive in terms of increasing channel availability, the service-level protection model does not guarantee high-quality spectrum to become available. The interference temperature protection model implicates that the interference from the primary system toward the primary system remains low, because it is determined by the received primary signal strength at the primary receiver plus the minimum separation distance. We can safely infer that the "pollution" of spectrum by the primary

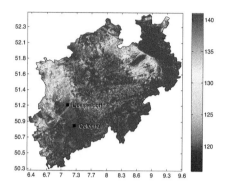

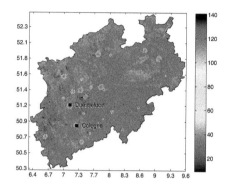

(a) Maximum local primary interference exceedance ratio.

(b) Minimum local primary interference exceedance ratio.

Figure 16.8: Heatmap of the maximum and minimum local primary interference exceedance ratio for North Rhine–Westphalia.

system is negligible in secondary system planning. However, if the service-level protection is applied, secondary systems may operate within the primary coverage region, thus, they are subject to a highly diverse noise plus interference figure. The better availability of spectrum may thus come at the price of high interference values, and the actual benefit of high transmit powers could be reduced due to low SNR at the secondary receivers. In order to illustrate this effect, we define an additional metric, the secondary transmit power to primary interference and noise ratio, which yields a first order approximation of the interference environment in the short range of the transmitter. We calculate it as

$$\text{PINR} = P_{\text{TX,s}} - 10\log_{10}\left(10^{\frac{N_0}{10}} + 10^{\frac{P_{\text{RX,p}}}{10}}\right). \tag{16.16}$$

For the best case channel, we show the PINR distribution in Figure 16.8(a). Locally, up to 15 dB difference can be found, i.e., that the capacity potential of a high-power transmitter reduces to the equivalent of a short-range device, e.g., a Wi-Fi access point. This effect becomes more expressed in the western part of the region. As this figure shows, only the best case, worse performance figures can be expected in the other TVWS channels. To define a lower bound on the achievable theoretical capacity, we show in Figure 16.8(b) the lowest expected PINR for those channels where the permissible secondary transmit power exceeds 20 dBm. We find that the reduction in performance is locally centered at locations of primary transmitters, i.e., that the increase in permissible transmit power is not compensating the increase in interference. This is an expected result, given the capping of transmit powers to 36 dBm, which was adopted in our system model from the Ofcom regulations. Large fractions of the studied regions show a rather static PINR value, i.e., it is possible to approximate the interference in these cases by a fixed SINR reduction.

16.5   Conclusion

Secondary spectrum access is an important cornerstone of efficient spectrum exploitation. In this chapter, we have presented the motivation and architecture of two protection models, which allow interference-free coexistence between primary and secondary systems. Where applied, they will improve wireless capacity and enable new wireless services. We have demonstrated how these models build on a fundamentally different understanding of interference, and how, as a result, the implementation of a coexistence scheme largely differs. In the context of TVWS policies, these models are currently evaluated for their practical applicability in various countries.

In order to demonstrate the strong relationship between protection models and quantitative spectrum availability metrics, we have carried out extensive simulations for the example of the German TVWS spaces. Three different spectrum availability metrics have been evaluated, which show the benefits and disadvantages of simpler interference temperature protection models over a more comprehensive service-level protection. Our simulations show that more comprehensive spectrum occupancy calculations (through terrain-aware propagation modeling) may unintentionally lead to severe secondary capacity constraints. Furthermore, our metric-driven analysis reveals that higher-quality primary receivers can open up more spectrum reuse opportunities. Our proposed spectrum availability metrics, the conditional channel availability, and the power to interference and noise ratio give an initial indication of how well a particular frequency band is suited for secondary operations. Nevertheless, more systematic exploration is necessary, directly taking the particularities of the planned secondary deployment into account, e.g., their susceptibility to interference from the incumbent and their capability to aggregate discontinuous spectrum resources.

References

[1] T. C. Clancy. "Formalizing the interference temperature model." *Wireless Communications and Mobile Computing*, 7(9):1077–1086, 2007.

[2] European Telecommunications Standards Institute. Digital video broadcasting (DVB); framing structure, channel coding and modulation for digital terrestrial television. ETSI EN 300 744 V1.6.1, January 2009.

[3] Federal Communications Commission. Second Memorandum Opinion and Order—In the Matter of Unlicensed Operation in the TV Broadcast Bands (ET Docket No. 04-186) / Additional Spectrum for Unlicensed Devices Below 900 MHz and in the 3 GHz Band (ET Docket No. 02-380), September 2010. 25 FCC Rcd 18661 (2010).

[4] Federal Communications Commission. Third Memorandum Opinion and Order—In the Matter of Unlicensed Operation in the TV Broadcast Bands (ET Docket No. 04-186) / Additional Spectrum for Unlicensed Devices Below 900 MHz and in the 3 GHz Band (ET Docket No. 02-380), April 2012. 27 FCC Rcd 3692 (2012).

[5] Federal Communications Commission Spectrum Policy Task Force. Report of the Spectrum Efficiency Working Group. Federal Communications Commission, November 2002.

[6] K. Harrison, S. M. Mishra, and A. Sahai. "How much white-space capacity is there?" In *Proceedings of the 5th IEEE Symposium on New Frontiers in Dynamic Spectrum Access Networks (DySPAN)*, pages 1–10, April 2010.

[7] G. A. Hufford. "The ITS irregular terrain model, the algorithm." Technical Report 1.2.2, National Telecommunications and Information Administration, 1984.

[8] G. A. Hufford, A. G. Longley, and W. A. Kissick. "A guide to the use of the ITS irregular terrmain model in the area prediction mode." Technical Report 82-100, National Telecommunications and Information Administration, 1982.

[9] International Telecommunication Union. Radio regulations articles, 2012.

[10] International Telecommunication Union Radiocommunication Sector (ITU-R). "P.1546: Method for point-to-area predictions for terrestrial services in the frequency range 30 MHz to 3000 MHz." Technical Report 5, International Telecommunication Union, September 2013.

[11] International Telecommunication Union Radiocommunication Sector (ITU-R). "P.372: Radio noise." Technical Report 11, International Telecommunication Union, September 2013.

[12] K.-M. Kang, J. C. Park, S.-I. Cho, B. J. Jeong, Y.-J. Kim, H.-J. Lim, and G.-H. Im. "Deployment and coverage of cognitive radio networks in TV white space." *IEEE Communications Magazine*, 50(12):88–94, December 2012.

[13] H. R. Karimi. "Geolocation databases for white space devices in the UHF TV bands: Specification of maximum permitted emission levels." In *Proceedings of th 6th IEEE Symposium on New Frontiers in Dynamic Spectrum Access Networks (DySPAN)*, pages 443–454, May 2011.

[14] P. J. Kolodzy. "Interference temperature: A metric for dynamic spectrum utilization." *International Journal on Network Management*, 16(2):103–113, March 2006.

[15] A. G. Longley and P. L. Rice. "Prediction of tropospheric radio transmission over irregular terrain, a computer method—1968." Technical Report ERL 79-ITS 67, Environmental Science Services Administration, July 1968.

[16] J. Nasreddine, J. Riihijärvi, A. Achtzehn, and P. Mähönen. "The world is not flat: Wireless communications in 3D environments." In *Proceedings of the 14th IEEE International Symposium and Workshop on a World of Wireless, Mobile and Multimedia Networks (WoWMoM)*, June 2013.

[17] Ofcom. "Implementing TV white spaces," February 2015.

[18] Ofcom. "Implementing TV white spaces: Annex 1 to 12," February 2015.

Chapter 17

How To Use Novel Methods For Improving The Performance Of Wireless Cognitive Networks

Barbaros Preveze

CONTENTS

Abstract

There are so many attempts in the literature for the performance improvement of different network types or protocols. For network systems, better results for all the performance parameters, such as delay, packet loss, hop count, efficient spectral usage, etc., will cause better throughput results. So, it can obviously be said that the throughput is the main performance parameter directly sensed by the end user.

In this chapter, the system performance improvement of 802.16.j mobile multi-hop ad-hoc network will be discussed by investigating the evaluated results of works done on this protocol before.

Finally, how to investigate the effects of proposing and implementing some novel methods on a system will be explained for further possible performance improvement attempts on the throughput of a system.

17.1   Introduction

In cognitive networks, all the nodes work for improving the overall performance of the whole network rather than improving its own performance, but finally this behavior also causes an improvement of the performances of each node [1].

In a work done before [2], some novel spectrum sharing methods are proposed to decrease the packet loss rate and to improve the system throughput performance by better management of the spectral access and buffer management of the nodes in the network. Then later in another study [3], a novel cognitive buffer management algorithm is proposed that probabilistically arranges the packets stored in the buffers of the nodes and selects correct packet of the correct node to transmit first for providing fewer packet losses and also fewer average hop-counts. It is shown in the same work that a novel proposed buffer management algorithm provides more throughput improvement by replacing the buffer management algorithm of the system.

In these works, WIMAX (worldwide interoperability for microwave access) protocol is selected and simulated in a simulation program. So, we will also work on WIMAX protocol here.

In the next sections of this chapter, the structure of the simulation program that is used in those works and on which the performance tests can be done will be explained. Then the functions running on the simulation will be mentioned, and the types of the resources shared by the nodes and the ways of sharing these resources in the most efficient way will be discussed. Finally, some alternative ways and how they can be used to improve the overall throughput performance of the overall network will be explained.

17.2   Simulation Program

The networks can be simulated by computers to be able to test the effects of some changes on the protocol functions or on network traffic conditions, which are difficult to control and measure in real world networks. Some available network simulators, such as OPNET, Cisco Packet Tracer, or NS2, designed for this purpose can be used for developing the network simulation, and, on the other hand, the network protocols can also be simulated writing the protocol functions in details using C++ or MATLAB too. Here, the simulation program is developed using MATLAB to be able to easily write, modify, edit, or add all desired details of all the functions in the protocol.

In the simulation program, we have a defined the number of nodes moving in the predefined area with random speeds toward random directions using random way point mobility model [4]. The nodes also produce different types of packets, such as data packet, voice packet, or video packet, while they are moving, and each packet with different data type is stored in the corresponding buffer in the node [1, 2, 3, 5, and 6].

At this point, some of the separated buffers will have higher priority than others according to their data type, and some of the packets in the same buffer will also have different priorities with respect to other packets. So, the buffer from which the packets will be retrieved must be selected carefully each time the node starts transmission. In this selection, the buffer fullness rates and the buffer packet types must be taken into consideration.

Then, the packets that will be transmitted over the transmission line must be selected carefully from the selected buffer. In this selection, the route of each packet and the buffer states of all the nodes on their routes will be taken into consideration. These 2 points can be thought in the buffer Management (BM) issue.

Finally, the spectrum access rights of all the nodes will be adjusted for the benefit of overall network performance. This will be done by spectral access (SA) control in the system. In adjustment of the SA of the nodes, again, the buffer fullness rates and the hop-counts of the packets in the buffers of the nodes will be taken into consideration.

By SA algorithm used in the simulation, each node gives the SA right to other nodes when it loses its packet or when some other node needs the SA right to free its buffer for preventing possible packet losses. For this purpose, the congested nodes with their full buffers will also be able to have SA by the permission of all other nodes in the network. This is called most congested access first (MCAF) in the simulation. The term adaptive rate algorithm (AR) in Figure 16.1 adjusts the packet generation and transmission rates on the nodes, such that the packet generation/transmission rate is decreased by the system in a rate proportional to the packet loss rate of the nodes in the system. The overall working structure of the simulation is given in Figure 16.1 for better understanding.

The algorithm shown in Figure 16.1 summarizes the working principle of a node in the cognitive network simulation system for which the working principle of the overall system is explained in [5] in details. That means the overall system will work in coordination and fairly when this working principle is applied on all the nodes.

17.3 How to Improve Network Performance by Modifications Done on the Algorithms of the System

Before trying to improve the performance of a system, the steps given below must be achieved.

a. **The system for which its performance will be improved must be selected, learned, and simulated by its every detail:** The system parameters affecting the system performance and also these system performance parameters must be understood in details, and the system must be simulated perfectly using a network simulator or a programming language. During the preparation of the simulation system, the system parameters must be used carefully as they are defined by the standards.

In this study, we work on 802.16.j multi-hop WIMAX orthogonal frequency division multiple access (OFDMA) system, and the working principle for 802.16.j is given in Figure 16.1 as a flow chart that is implemented for all the nodes running in the simulation program.

b. **Evaluated results must be confirmed by using the results of other works in the literature on the same subject:** After finishing development of the simulation system, the evaluated results must be confirmed by the results of some other works on the same subject in the literature by use of exactly the same parameter values. That is why the literature review is required before setting up the simulation. If the results evaluated by our simulation match with the other results in the literature for the same parameter values, this will confirm the correctness of our simulation program. After this point, we will have a ticket for an attempt of trying to improve the system performance. The simulation results when BM_{OLD} is used as the buffer management algorithm are given in [2].

c. **Some novel methods should be proposed in order to improve the system performance:** After understanding the working principle and setting up the simulation of the system in details, the parameters and the functions affecting the system performance should be selected. Then, some novel methods can be proposed for substitution with the old one in order to improve the system performance. These functions can also be selected from the flow chart given in Figure 16.1.

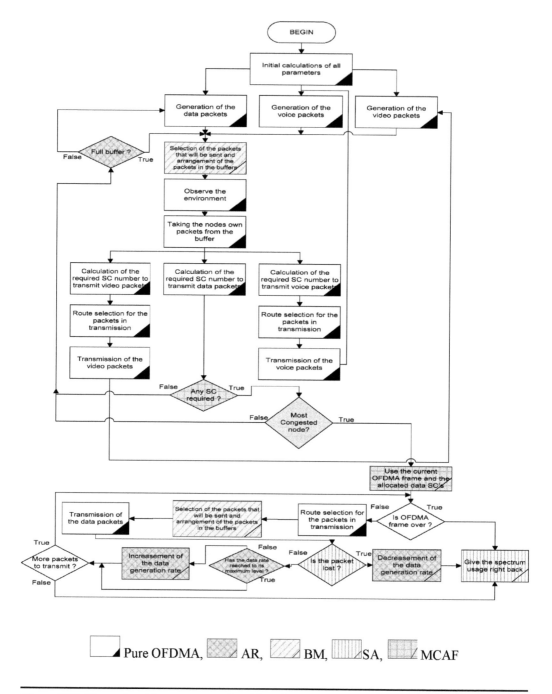

Figure 17.1: Throughput improvement algorithm that can be used in WIMAX OFDMA system [2, 3].

At the end, it must be shown that the evaluated results give better or worse performance than the original simulation results evaluated by the unmodified algorithm.

During all these works, the network must be simulated using as small functions as possible. This makes it possible also to change one of these small functions with a novel one for further improvement attempts in the future. In [3], the buffer management algorithm used in Figure

16.1 is replaced by a novel one called BM_{NOVEL} for the performance improvement. The results of this work have shown that the performance of the system is improved by use of this novel proposed buffer management algorithm.

17.4 Network Performance Improvement by Using Novel Buffer Management Algorithm

The performance of a network can be thought as:

- the average delay amount of the system;

- the average hop count of the system;

- the average packet loss rate of the system;

- the jitter variance of the system; and

- and the average throughput of the system.

But it must be noted that the one that can finally be sensed by the end users is obviously the throughput. Therefore, the throughput is widely selected as the main performance measurement parameter of the networks [1, 2, and 3], and the others are some other performance parameters that already have direct effects on the overall system throughput.

In [3] improvement of the system throughput performance is attempted by increasing the effectiveness of the buffer management of the system. Therefore, a novel BM algorithm shown in Figure 16.2 [3] called BM_{NOVEL} is proposed.

According to the algorithm given in Figure 16.2, the buffer management algorithm which was implemented as explained in [1, 2, and 5], is replaced with a novel one [3] that runs on each node as follows.

a. Each node in the system recognize, the packets in the buffers of the nodes with their calculated routes up to its destination point.

b. The number of packets at each node and the hop-counts of each packet up to its destination point is kept.

c. The number of empty slots at the routes of each packet is calculated.

d. Then the node that has the maximum difference of number of empty slots in next buffers of each node and the number of packets to be transmitted in the same node, and which has minimum average number of hops for transmitting its packets, is selected to make transmissions of their packets.

After applying the novel proposed BM algorithm on the system, some novel algorithms will also be evaluated. In order to be able to show the correctness of the results, the new results must be plotted on the same graph, and they must be compared with the previous ones.

If the new results of the performances is higher than the results of previous ones, then the difference between them gives us the amount of improvement on the system performance.

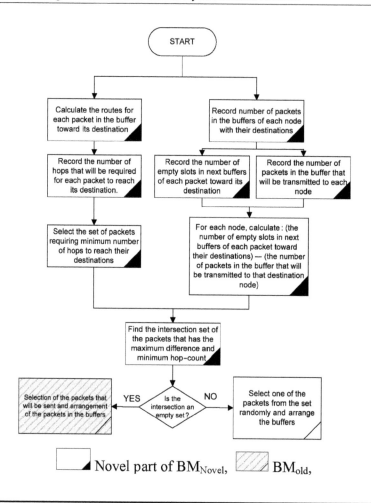

Figure 17.2: BM$_{Novel}$ algorithm (replaced by BM$_{Old}$ in the simulation) [3].

17.5 Theoritical Calculation of the Effects of Novel Algortihm on the Overall System Performance

The probability of having a packet with a specified hop-count (hc) up to a random destination point can be calculated as in Equation 16.1 and Equation 16.2. [1,2 and 3]:

$$P(hc) = \frac{hc \times 8}{\sum_{n=1}^{N} 8 \times n} = \frac{hc}{\sum_{n=1}^{N} n} = \frac{hc}{\frac{N \times (N+1)}{2}} = \frac{2 \times hc}{N \times (N+1)}, \tag{17.1}$$

where, "N" is the considered maximum hc to be used in the simulations/calculations.

After calculating the probability of having a specified number of hop-counts from a random node to another, the average hop count (AHC) value can also be calculated as in Equation 16.2 [1,2, and 3] using each of the P(hc) values calculated in Equation 16.1.

$$AHC(bfr) = \frac{\sum_{bfr=ppf+1}^{buffsize} \sum_{hc=1}^{N} \left(\left[total + \left\langle \sigma > ppf \Rightarrow \left\{ ppf - \sum_{total=1}^{hc} total \right\} \neg \{\sigma\} \right\rangle \right] \times hc \right)}{ppf}$$

$$where \quad \sigma = bfr \times P(hc), total = \sum_{total=1}^{hc} (\sigma \times 10) - (\sigma \bmod 10)/10$$

$$and \quad ppf \; is \; the \; number \; of \; packets.$$

(17.2)

In the simulation program, the first "packets per frame" (ppf) packets (calculated as 11 in [6] and also used as 11 in this work), which have the minimum hop-counts up to their destination points using the fastest path routing algorithm, are selected to be transmitted. The confirmed simulation and calculation results, which evaluated for an average of 1000 simulation runs, are also given in Figure 16.3. [3]. Of course, any kind of modification can be done on any part of the system and can be observed by the simulation.

It is observed on Figure 16.3 that the BM$_{Novel}$ algorithm also decreases the AHC, especially for an increasing number of full slots (X axis on Figure 16.6) in the buffers, while AHC is not affected by the buffer fullness rate for the same case in BM$_{Old}$ algorithm.

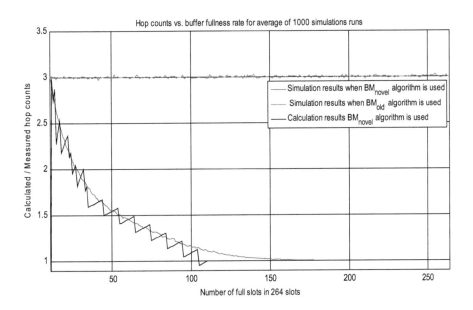

Figure 17.3: Simulation/calculation results of reduced AHC replacing the BM$_{Novel}$ algorithm with the BM$_{Old}$ algorithm [3].

In the AHC calculations and simulations of the BM$_{Novel}$ algorithm, it is seen that the buffer fullness rates of the nodes have a dominant effect on the average hop-count of the network. This is because the nodes with more packets have greater probabilities of having packets with a lower number of hop-counts, which yields a greater probability of being transmitted.

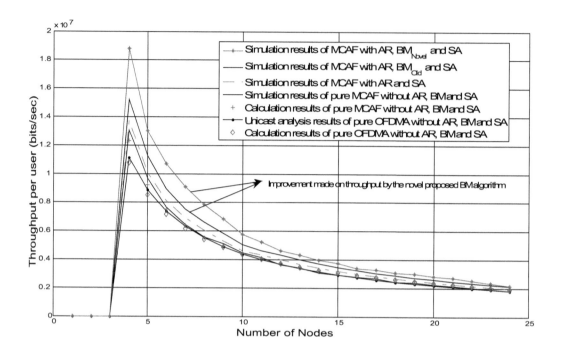

Figure 17.4: Results of throughput rates when BM$_{Novel}$ is used instead of BM$_{old}$, with MCAF, AR, and SA in the system [3].

17.6   Results and Discussion

The simulation results evaluated by use of BM$_{OLD}$ algorithm [1, 2] and the throughput performance improvement amount with respect to using BM$_{NOVEL}$ [3] is shown in Figure 16.4.

In the subject area of computer networks, the main purpose is generally improving the service quality of the network, which is possible by improving the network performance in terms of delay, packet losses, and the throughput as a result of all these parameters.

17.7   Conclussion

In this chapter, some methods for improving the throughput of a network by decreasing the delay and packet loss rate are summarized. The following can be concluded:

- The network simulations must be designed in details with small functions that can be modified in the future.

- The simulation results must be confirmed by some other results evaluated in the literature to show that the evaluated results are correct.

- The expected results after the modification must be first evaluated by mathematical calculations to show that the novel algorithm that we are attempting to embed to the system may generate better results on the system, then the simulation can be modified.

- Any part of an algorithm running on a network can be replaced by a novel one to improve the performance of that part. Then, improving the performance of a part of the system will finally cause a general performance improvement on the overall system performance.

The improvement amount of the investigated work on the overall system performance can then be highlighted as in Figure 16.4 as the success rate of the work.

References

[1] Preveze B. and Safak A. "Throughput improvement of mobile multi-hop wireless networks," Int. Journal of Wireless & Mobile Networks 2010, 2;120–140.

[2] Preveze B. and A. Safak. "Effects of routing algorithms on novel throughput improvement of mobile ad hoc networks." Turk J Elec Eng & Comp Sci 2012, 20:507–522.

[3] Preveze B. "A novel cognitive method for throughput improvement of mobile multi-hop ad hoc networks," *Wireless Personal Communications*, May 2015, Volume 82, Issue 1, pp 229–243

[4] Deborah E., Daniel Z., Li T., Yakov R., and Kannan V., "Source: Demand routing: Packet format and forwarding specification," *Internet RFCs archive*, (version 1)., pp. 1–7, 1995.

[5] *Multimedia over Cognitive Radio Networks: Algorithms, Protocols, and Experiments*, December 4, 2014, Fei Hu, Sunil Kumar. ISBN-13: 978-1482214857.

[6] Preveze B. and Safak A. "Throughput maximization of different signal shapes working on 802.16e mobile multi-hop network using novel cognitive methods." In *WIMO 2010 The Second International Conference on Wireless & Mobile Networks*; 26–28 June 2010; Ankara, Turkey: Springer AIRCC. pp. 71–86.

Chapter 18

Low-Complexity and High-Efficient Scheduling Schemes For Spectrum-Sharing-Based Secondary Transmissions

Haiyang Ding, Tangwen Xu, Daniel B. da Costa, Jianhua Ge, Yinfa Zhang, Wulin Liu, and Ya-Ni Zhang

CONTENTS

423

Abstract

In this chapter, we introduce high-efficient and low-complexity scheduling schemes for several spectrum-sharing-based transmission scenarios. The scenarios are categorized into two groups. The first group focuses on the design of cognitive user scheduling for non-cooperative secondary transmissions. For such, besides guaranteeing the reliable operation of primary user, three energy-efficient and low-complexity schemes are presented that can achieve exactly the same or even superior transmission robustness over that of existing solutions while consuming much less transmit power. In addition, the average rate and bit error rate (BER) performance of the secondary systems are investigated.

The second group concentrates on the spectrum-sharing based cognitive cooperative systems, where the secondary systems resort to multiple relays to improve the reliabilities of the end-to-end dual-hop transmissions. For multi-relay cooperative scenarios, high-efficiency and low-complexity relay selection strategies are designed to improve the transmission robustness of the secondary systems. By analyzing the high-signal-to-noise ratio (SNR) scaling law of the system outage probability, the impact of the number of relays on the diversity and coding gains are characterized. Moreover, the effects of imperfect channel state information (CSI) pertaining to the interference/transmission channels on the diversity gain of secondary systems are evaluated. Under the scenario of single-relay secondary transmissions, high-efficient link selection schemes are proposed which selects an appropriate secondary link for transmission in a distributed manner. The distributed link scheduling schemes can achieve competitive transmission robustness with that of the centralized/optimal link scheduling schemes, while it only needs an extremely low signaling overhead to perform link selection. Under all the foregoing scenarios, both theoretical deductions and numerical plots are employed to highlight the merits of the proposed schemes.

18.1   Energy-Efficient And Low-Complexity Schemes For Non-Cooperative Uplink Cognitive Cellular Networks

In this section, both cognitive users (CUs) and primary user (PU) share the same base station (BS). Under an outage probability protection criterion for the PU, we first propose a round-robin CU scheduling scheme and analytically show its achieved signal-to-interference ratio statistics at the BS, from which it is observed that the round-robin scheduling is much more energy-efficient than opportunistic scheduling to achieve the same mean capacity and BER. Inspired by this interesting observation, a statistics-based CU scheduling scheme is also presented, whose energy-efficiency is further improved. After that, an improved opportunistic scheduling strategy is introduced. Different from the foregoing proposals, the proposed strategy chooses the CU with the best instantaneous channel quality in addition to guaranteeing the reliable operation of PU. For such a strategy, it is analytically shown that in comparison with previous proposals, a greater average rate and a lower average bit error rate (BER) can be achieved. Numerical results are shown to validate the above remarks.

18.1.1 Three Low-Complexity Schemes for Uplink Cognitive Cellular Networks

18.1.1.1 System model

Consider an interference-limited uplink cognitive cellular network, where N CUs ($CU_1, \ldots, CU_N$) and one PU communicate with one BS sharing the same frequency band. For the CUs, a time-division multiple-access (TDMA) scheme is employed for orthogonal channel access. In one time-slot, one CU is selected among all potential ones to access the channel and the detailed selection schemes will be addressed later. Concerning the channel fading characteristic, we consider a general fading scenario, where independent but not necessarily identically distributed (inid) Rayleigh fading channels are assumed. Then, the channel gains from the PU and the k-th CU to BS are represented by G_0 and G_k ($k=1, 2, \ldots, N$), respectively. Accordingly, G_0 and G_k are independent exponential distributed random variables (RVs) with mean $\frac{1}{\lambda_0}$ and $\frac{1}{\lambda_k}$, respectively. In addition, the transmit powers of each CU and the PU are denoted by P and P_0, respectively. In order to guarantee the transmission quality of PU, similar to [1], we adopt an outage probability protection criterion for the PU, which can be formulated as

$$\Pr\left(\ln\left(1 + \gamma_0\right) \leq R_0\right) \leq \zeta_0, \tag{18.1}$$

where $\Pr(\bullet)$ denotes probability, γ_0 is the received signal-to-interference ratio (SIR) at BS from PU, R_0 indicates the target transmission rate in nats/s/Hz, and ζ_0 represents the threshold of the outage probability for the PU.

Regarding the CU selection schemes, we first consider two types of them, namely, opportunistic scheduling proposed by [1] and the proposed round-robin scheduling. According to [1], in each time-slot, opportunistic scheduling scheme selects the CU with the minimum instantaneous channel gain (i.e., $\min_{k=1,\ldots,N}[G_k]$), causing therefore the minimum interference to the PU under the same transmit power P. In contrast, the proposed round-robin scheduling assigns a time-slot to each CU in a circular order without any knowledge of instantaneous CSI. Under the PU protection criterion (17.1), it will be shown that the round-robin scheduling can achieve the same mean capacity and BER with those of opportunistic scheduling but consuming much less power, therefore significantly prolonging the network lifetime and causing less interference to other devices nearby. In the next subsection, assuming a general inid fading scenario, we first re-study the performance of opportunistic scheduling in terms of mean capacity and BER. Then, the same performance metric of the proposed round-robin CU scheduling is investigated and is compared with that of opportunistic scheduling. Finally, a statistics-based CU scheduling is presented, and its performance is studied as well.

18.1.1.2 Opportunistic Scheduling Schemes

For opportunistic scheduling, the received SIR at BS from the PU can be expressed as [1]

$$\gamma_0 = \frac{P_0 G_0}{P \min_k [G_k]}. \tag{18.2}$$

To alleviate notation, let $\gamma_0 = \frac{X}{Y}$, where $X = P_0 G_0$ and $Y = P \min_k [G_k]$. Then, according to order statistics, it is easy to show that the probability density function (PDF) of X and Y can be written, respectively, as $p_X(x) = \frac{\lambda_0}{P_0} e^{-\frac{\lambda_0}{P_0} x}$ and $p_Y(y) = \frac{\sum_{k=1}^{N} \lambda_k}{P} e^{-y \frac{\sum_{k=1}^{N} \lambda_k}{P}}$. Therefore, the cumulative distribution function (CDF) of γ_0 can be calculated as

$$F_{\gamma_0}(z) = \Pr\left(\frac{X}{Y} < z\right) = \int_0^\infty \left(\int_0^{zy} p_X(x) dx\right) p_Y(y) dy = \frac{\lambda_0 P z}{\lambda_0 P z + P_0 \sum_{k=1}^{N} \lambda_k}. \tag{18.3}$$

To satisfy the outage probability protection criterion (17.1) for the PU, we have the following proposition.

Proposition 18.1

For opportunistic scheduling scheme, the transmit power P of the scheduled CU should satisfy

$$P \le \frac{\zeta_0 P_0 \sum_{k=1}^{N} \lambda_k}{(1 - \zeta_0) \lambda_0 (e^{R_0} - 1)} \triangleq P_{\max}. \tag{18.4}$$

It is noteworthy that for the independent and identically distributed (iid) fading scenarios as considered in [1], by setting $\lambda_k = \lambda_0$ for $k \in \{1, ..., N\}$, (17.4) reduces to [1].

Proof 18.1 By Combining (17.1) and (17.3), and knowing that $F_{\gamma_0}(z)$ is a monotonic increasing function with respect to P, (17.4) is attained. ■

In order to evaluate the mean capacity and BER of opportunistic scheduling, it is necessary to analyze the PDF of the received SIR at BS from the scheduled CU. To maximize the system performance, $P = P_{\max}$ is assumed and the received SIR for the scheduled CU is defined as $\gamma_{\max} \triangleq \frac{P_{\max} \min_k [G_k]}{P_0 G_0}$. For $P = P_{\max}$ in (17.2), we have $\gamma_{\max} = \frac{1}{\gamma_0}$. Thus, using the fundamental theorem given by [2], we can attain the PDF of $\gamma_{\max}$ as

$$p_{\gamma_{\max}}(\gamma) = \frac{\lambda_0 P_0 P_{\max} \sum_{k=1}^{N} \lambda_k}{\left(\lambda_0 P_{\max} + \gamma P_0 \sum_{k=1}^{N} \lambda_k\right)^2} = \frac{\frac{\zeta_0}{(1-\zeta_0)(e^{R_0}-1)}}{\left[\frac{\zeta_0}{(1-\zeta_0)(e^{R_0}-1)} + \gamma\right]^2}. \tag{18.5}$$

Note that (17.5) is exactly the same with [1] after substituting [1] into [1] and simplifying. From (17.5), following the same procedure as utilized in [1], we can attain the mean capacity, namely, C, and BER for the cognitive systems, which are listed below, and interested readers can refer to [1] for details.

$$C = \begin{cases} \frac{\ln \delta}{\delta - 1}, & \delta \ne 1 \\ 1, & \delta = 1 \end{cases}, \tag{18.6}$$

$$P_b \le \frac{1}{2\delta} e^{\frac{1}{\delta}} \mathrm{Ei}\left(-\frac{1}{\delta}\right) + \frac{1}{2}, \tag{18.7}$$

where $\mathrm{Ei}(\bullet)$ denotes the exponential integral function [3] and $\delta = \left(e^{R_0} - 1\right) / \left(\frac{1}{1 - \zeta_0} - 1\right)$.

Now, it can be concluded that the performance of opportunistic scheduling does not vary with the fading characteristic within the cognitive systems. Nevertheless, it is worth noting that for opportunistic scheduling, the maximum allowable transmit power changes with the fading statistics, as indicated in Proposition 17.1.

18.1.1.3 Round-Robin CU Scheduling Scheme

For round-robin scheduling, at each time-slot, one predetermined CU is scheduled to access the channel, and each CU is scheduled to a different time-slot in a circular order. Without loss of generality, we assume, at the n-th time-slot, CU_n is scheduled to access the channel. Consequently, the received SIR at BS from the PU can be written as

$$\gamma_0^{(n)} = \frac{P_0 G_0}{P G_n}. \tag{18.8}$$

Then, following a similar procedure as employed to calculate (17.3), one can derive the CDF of $\gamma_0^{(n)}$ as

$$F_{\gamma_0^{(n)}}(z) = \frac{\lambda_0 P z}{\lambda_0 P z + \lambda_n P_0}. \tag{18.9}$$

Now, to satisfy the outage probability protection criterion (17.1) for the PU, the following proposition is made.

Proposition 18.2
For round-robin CU scheduling scheme, the transmit power P of the scheduled CU_n should satisfy

$$P \leq \frac{\zeta_0 \lambda_n P_0}{(1 - \zeta_0) \lambda_0 (e^{R_0} - 1)} \triangleq P_{\max}^{(n)}. \tag{18.10}$$

In particular, for iid fading scenarios, we have $P_{\max}^{(n)} = \zeta_0 P_0 / [(1 - \zeta_0)(e^{R_0} - 1)]$.

Proof 18.2 Combining (17.1) and (17.9), and knowing that $F_{\gamma_0^{(n)}}(z)$ is a monotonic increasing function with respect to P, (17.10) is achieved. ■

Next, we analyze the PDF of the received SIR for CU_n. To maximize the system performance, $P = P_{\max}^{(n)}$ is adopted. Thus, the received SIR for CU_n is defined as $\gamma_{\max}^{(n)} \triangleq \frac{P_{\max}^{(n)} G_n}{P_0 G_0}$. Invoking the fundamental theorem [2] again, the PDF of $\gamma_{\max}^{(n)}$ is derived as

$$p_{\gamma_{\max}^{(n)}}(\gamma) = \frac{\lambda_n \lambda_0 P_0 P_{\max}^{(n)}}{\left(\lambda_0 P_{\max}^{(n)} + \lambda_n P_0 \gamma\right)^2} = \frac{\frac{\zeta_0}{(1-\zeta_0)(e^{R_0}-1)}}{\left[\frac{\zeta_0}{(1-\zeta_0)(e^{R_0}-1)} + \gamma\right]^2}. \tag{18.11}$$

From (17.11), it is observed that the PDF of $\gamma_{\max}^{(n)}$ is exactly the same with (17.5). Also, for $\forall n \in \{1, 2, ...N\}$, the PDF of $\gamma_{\max}^{(n)}$ remains the same, yielding, therefore, the same mean capacity and BER performance for different CU. This means that the achieved performance (in terms of mean capacity and BER) of round-robin scheduling is the same with that of opportunistic scheduling. Furthermore, the ratio of average power-consumption of round-robin scheduling to that of opportunistic scheduling is

$$\rho = \frac{\frac{1}{N} \sum_{n=1}^{N} P_{\max}^{(n)}}{P_{\max}} = \frac{1}{N}. \tag{18.12}$$

Note that (17.12) tells us that, to achieve the same mean capacity or BER, round-robin scheduling only needs to allocate $(1/N)$-th transmit power of the counterpart for opportunistic scheduling, which is a tremendous improvement in energy-efficiency. Besides, we should also note that, for opportunistic scheduling, not only the instantaneous CSI is needed to select CU, but also channel statistics are required to allocate transmit power (according to Proposition 17.1), whereas for round-robin scheduling, only channel statistics are needed to determine the transmit power, as shown in Proposition 17.2.

18.1.1.4 A Statistics-Based CU Scheduling Scheme

Although round-robin scheduling is energy-efficient, its energy-efficiency can be further improved by introducing opportunistic mechanism. By scheduling the CU with the best channel statistics (rather than instantaneous CSI) in each time-slot, the energy-efficiency can be further improved. Specifically, in each time-slot, the statistics-based scheduling scheme selects the CU, satisfying $k* = \arg\min_k [\lambda_k]$, and the corresponding received SIR at BS from the PU can be written as

$$\gamma_0^{(k*)} = \frac{P_0 G_0}{P G_{k*}}. \tag{18.13}$$

Then, using the same method as employed to calculate (17.3), the CDF of $\gamma_0^{(k*)}$ can be expressed as

$$F_{\gamma_0^{(k*)}}(z) = \frac{\lambda_0 P z}{\lambda_0 P z + P_0 \min_k [\lambda_k]}. \tag{18.14}$$

In order to satisfy the outage probability protection criterion (17.1) for the PU, we present the following proposition.

Proposition 18.3
For the statistics-based CU scheduling scheme, the transmit power P of the scheduled CU should satisfy

$$P \leq \frac{\zeta_0 P_0 \min_k [\lambda_k]}{(1 - \zeta_0) \lambda_0 (e^{R_0} - 1)} \triangleq P_{\max}^{(k*)}. \tag{18.15}$$

Particularly, for iid fading scenarios, one can attain $P_{\max}^{(k*)} = \zeta_0 P_0 / \left[(1 - \zeta_0) (e^{R_0} - 1) \right]$.

Proof 18.3   Similar to the proof of Proposition 17.2. ■

Now, by combing (17.4) and (17.3), we have

$$\rho = \frac{P_{\max}^{(k*)}}{P_{\max}} = \frac{\min_k [\lambda_k]}{\sum_{k=1}^{N} \lambda_k} \leq \frac{\min_k [\lambda_k]}{N \cdot \min_k [\lambda_k]} = \frac{1}{N}, \tag{18.16}$$

where the equality holds for iid fading scenarios. Next, denoting the received SIR for CU_{k*} as $\gamma_{\max}^{(k*)} \triangleq \frac{P_{\max}^{(k*)} G_{k*}}{P_0 G_0}$, we can attain the PDF of it, which is the same with (17.11), yielding, therefore, the same mean capacity and BER with the two schemes aforementioned.

Remark 17.1   In summary, the merits of the proposed two types of CU scheduling schemes are twofold:

(a) **Low-complexity:** They do not need the instantaneous CSI while achieving the same performance (in terms of mean capacity and BER) with that of opportunistic scheduling, which can practically reduce the complexity and cost of the cognitive radio devices.

(b) **Energy-efficient:** Round-robin CU scheduling scheme only need to consume $(1/N)$-th transmit power of the counterpart for opportunistic scheduling scheme. Moreover, when the channel statistics are used for CU selection, the statistics-based CU scheduling scheme can be employed, and it further increases the energy-efficiency, i.e., ρ will be less than $1/N$ for inid fading scenarios, as shown in (17.4). Hence, these two CU scheduling schemes can efficiently prolong the network lifetime.

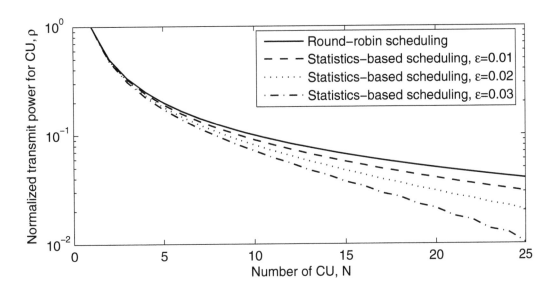

Figure 18.1: Normalized transmit vs. number of CU.

18.1.1.5 Numerical Examples and Discussions

Figure 17.1 plots the normalized transmit power of the CU for round-robin scheduling and statistics-based scheduling in inid fading scenarios. The normalization is done according to (17.12) and (17.4), respectively. Herein, $\lambda_k = \lambda_0 + (-1)^k \cdot k \cdot \varepsilon$, where $\lambda_0 = 1$, $k \in \{1, 2, ..., N\}$, and ε indicates the difference among the channels. It is observed that, with an increase of N, the power-consumption of the proposed schemes relative to that of opportunistic scheduling decreases significantly, and the statistics-based scheduling performs better than round-robin scheduling, as expected. In particular, for N=25 and $\varepsilon = 0.03$, the power consumption of statistics-based scheduling is only 1% of that for opportunistic scheduling, which further demonstrates the superiority of the proposed scheme.

18.1.1.6 Conclusions

In this section, we presented two types of energy-efficient and low-complexity CU scheduling schemes, namely, round-robin CU scheduling and statistics-based CU scheduling were studied in a non-cooperative spectrum-sharing scenario, where both CUs and PU share the same BS. It was shown that round-robin scheduling is much more energy-efficient than opportunistic scheduling to achieve the same mean capacity and BER. In addition, it was manifested that the statistics-based scheduling can achieve a superior energy-efficiency than that of the foregoing schemes. Representative numerical examples were shown to validate the above remarks.

18.2 An Improved Scheduling Scheme for Uplink Cognitive Cellular Networks

This section advocates a novel opportunistic scheduling strategy for the aforementioned uplink cognitive cellular network. Different from previous proposals, the proposed strategy chooses the CU with the best instantaneous channel quality in addition to guaranteeing the reliable operation of PU. For such a strategy, it is analytically shown that in comparison with previous proposals, a greater mean capacity and a lower BER can be achieved. Interestingly, it is manifested that with an increase in the number of CUs, both the mean capacity and the average BER of the proposed strategy are ameliorated. Moreover, it is indicated that under the same outage protection criterion for the PU, a much lower transmit power is sufficient for the proposed strategy to optimize the system performance, especially for a large number of CUs, as compared with previous strategies.

18.2.1 System Model

We focus on the same interference-limited uplink cognitive cellular network, as considered in the last section, where N CUs share one BS with a PU over the same frequency band. We assume that all the channels suffer from inid Rayleigh fading. As a result, denoting G_k ($k=1, \ldots, N$) and G_0 as the channel gains from the k-th CU to BS and from the PU to BS, respectively, it follows that G_k ($k=0, 1, \ldots, N$) conform to inid exponential distributions with mean $1/\lambda_k$. It is also assumed that each CU has the same transmit power constraint P, and P_0 indicates the transmit power of the PU. In contrast to [1] and [4], which scheduled the CU with the minimum channel gain to access the channel, the proposed strategy selects the CU with the best instantaneous channel quality, in addition to satisfying the following outage protection regulation for the PU:

$$P_r \left\{ \ln \left(1 + \gamma_0 \right) \leq R_0 \right\} \leq \zeta_0, \tag{18.17}$$

where $\gamma_0 = \frac{P_0 G_0}{P \max_k [G_k]}$ ($k=1,2,\cdots N$) is the received SIR at BS from the PU, ζ_0 denotes the outage probability protection threshold for the PU, and R_0 represents the target transmission rate. Hereafter, we use $F_X()$ and $p_X()$ to denote the CDF and PDF of a random variable (RV) X, respectively.

18.2.2 An Improved Scheduling Scheme

In what follows, we investigate the mean capacity and the average BER of secondary system for the proposed strategy. With this aim, we first calculate the transmit power range of the scheduled CU in order to satisfy the outage probability protection regulation at the PU, which is formulated in the proposition as below.

Proposition 18.4
Under the outage probability protection criterion (17.1) for the PU, the transmit power of the selected CU P should satisfy the following expression:

$$\sum_{l=1}^{N} \sum_{\substack{S_l \subseteq \{1,2,\ldots,N\} \\ |S_l| = l}} (-1)^{l+1} \left(1 - \frac{\sum\limits_{i \in S_l} \lambda_i P_0}{\lambda_0 P \left(e^{R_0} - 1 \right) + \sum\limits_{i \in S_l} \lambda_i P_0} \right) \leq \zeta_0. \tag{18.18}$$

Proof 18.4   Let $X = P_0 G_0$ and $Y = P \max_k [G_k]$, where G_k ($k=0,1,\ldots N$) are inid exponential distributed RVs, with mean $1/\lambda_k$ ($k=0,1,\ldots,N$). The PDF of X and Y can be expressed as

$$p_X(x) = \frac{\lambda_0}{P_0} e^{-\frac{\lambda_0}{P_0}x}, x > 0, \tag{18.19}$$

$$p_Y(y) = \frac{1}{P} \sum_{l=1}^{N} \sum_{\substack{S_l \subseteq \{1,2,\dots N\} \\ |S_l| = l}} (-1)^{l+1} \sum_{i \in S_l} \lambda_i \exp\left(-\sum_{i \in S_l} \lambda_i y \middle/ P\right), y > 0, \tag{18.20}$$

which in turn leads to the PDF of $\gamma_0 = \frac{P_0 G_0}{P \max_k[G_k]} = \frac{X}{Y}$ as

$$p_{\gamma_0}(z) = \int_0^\infty y p_X(yz) p_Y(y) dy$$

$$= \frac{\lambda_0}{P_0 P} \sum_{l=1}^{N} \sum_{\substack{S_l \subseteq \{1,2,\dots N\} \\ |S_l| = l}} (-1)^{l+1} \sum_{i \in S_l} \lambda_i \int_0^\infty y \exp\left(-\left(\frac{\lambda_0 z}{P_0} + \frac{\sum_{i \in S_l} \lambda_i}{P}\right) y\right) dy$$

$$= \lambda_0 P_0 P \sum_{l=1}^{N} \sum_{\substack{S_l \subseteq \{1,2,\dots N\} \\ |S_l| = l}} (-1)^{l+1} \sum_{i \in S_l} \lambda_i \middle/ \left(\lambda_0 P z + \sum_{i \in S_l} \lambda_i P_0\right)^2. \tag{18.21}$$

Thus, the CDF of γ_0 can be expressed as

$$F_{\gamma_0}(x) = \int_0^x p_{\gamma_0}(z) dz = \sum_{l=1}^{N} \sum_{\substack{S_l \subseteq \{1,2,\dots,N\} \\ |S_l| = l}} (-1)^{l+1} \left[1 - \sum_{i \in S_l} \lambda_i P_0 \middle/ \left(\lambda_0 P x + \sum_{i \in S_l} \lambda_i P_0\right)\right]. \tag{18.22}$$

Summarizing the preceding results, one can arrive at (17.18), which completes the proof. ■

To enhance the secondary system performance, the selected CU allocates its maximum allowable transmit power, namely, P_{CU}, as long as (17.1) is satisfied. Thus, P_{CU} should satisfy the following equation:

$$\sum_{l=1}^{N} \sum_{\substack{S_l \subseteq \{1,2,\dots,N\} \\ |S_l| = l}} (-1)^{l+1} \left(1 - \frac{\sum_{i \in S_l} \lambda_i P_0}{\lambda_0 P_{CU}(e^{R_0} - 1) + \sum_{i \in S_l} \lambda_i P_0}\right) = \zeta_0. \tag{18.23}$$

Next, we derive the PDF of the received SIR at BS from the selected CU, which can be expressed as $\gamma_{CU} = \frac{P_{CU} \max_k[G_k]}{P_0 G_0}$ $(k = 1, 2, \cdots N)$. Substituting P with P_{CU} in $\gamma_0 = \frac{P_0 G_0}{P \max_k[G_k]}$ $(k = 1, 2, \cdots N)$, we notice that $\gamma_{CU} = 1/\gamma_0$. Therefore, using the fundamental theorem in [2], the PDF of the received SIR at BS for the scheduled CU can be written as

$$p_{\gamma_{CU}}(x) = \lambda_0 P_0 P_{CU} \sum_{l=1}^{N} \sum_{\substack{S_l \subseteq \{1,2,\dots,N\} \\ |S_l| = l}} (-1)^{l+1} \sum_{i \in S_l} \lambda_i \middle/ \left(\sum_{i \in S_l} \lambda_i P_0 x + \lambda_0 P_{CU}\right)^2. \tag{18.24}$$

Consequently, the mean capacity of the selected CU can be given by

$$C = E\left[\ln\left(1 + \gamma_{CU}\right)\right] = \int_0^\infty \ln\left(1 + x\right) p_{\gamma_{CU}}(x) \, dx$$

$$\overset{(a)}{=} \lambda_0 P_{CU} \sum_{l=1}^{N} \sum_{\substack{S_l = \{1,2,\ldots,N\} \\ |S_l| = l}} (-1)^{l+1} \frac{\ln\left(\sum_{i \in S_l} \lambda_i P_0 \Big/ \lambda_0 P_{CU}\right)}{\sum_{i \in S_l} \lambda_i P_0 - \lambda_0 P_{CU}}, \tag{18.25}$$

where step (a) is due to [3]. As before, herein we consider the binary phase-shift keying (BPSK) modulation. Thus, the average BER of the selected CU can be evaluated as

$$P_b = \int_0^\infty Q\left(\sqrt{2x}\right) p_{\gamma_{CU}}(x) \, dx$$

$$= \frac{1}{\sqrt{2\pi}} \int_0^\infty e^{-\frac{y^2}{2}} \int_0^{\frac{y^2}{2}} \lambda_0 P_0 P_{CU} \sum_{l=1}^{N} \sum_{\substack{S_l \subseteq \{1,2,\ldots,N\} \\ |S_l| = l}} (-1)^{l+1} \sum_{i \in S_l} \lambda_i \Big/ \left(\sum_{i \in S_l} \lambda_i P_0 x + \lambda_0 P_{CU}\right)^2 dx \, dy$$

$$= \sum_{l=1}^{N} \sum_{\substack{S_l \subseteq \{1,2,\ldots,N\} \\ |S_l| = l}} (-1)^{l+1} \left(\frac{1}{2} - \frac{\lambda_0 P_{CU}}{\sqrt{\pi}} \int_0^\infty \frac{e^{-y^2}}{\sum_{i \in S_l} \lambda_i P_0 y^2 + \lambda_0 P_{CU}} \, dy\right)$$

$$\overset{(b)}{=} \frac{1}{2} \sum_{l=1}^{N} \sum_{\substack{S_l \subseteq \{1,2,\ldots,N\} \\ |S_l| = l}} (-1)^{l+1} \left(1 - \sqrt{\frac{\lambda_0 P_{CU} \pi}{\sum_{i \in S_l} \lambda_i P_0}} \exp\left(\frac{\lambda_0 P_{CU}}{\sum_{i \in S_l} \lambda_i P_0}\right) erfc\left(\sqrt{\frac{\lambda_0 P_{CU}}{\sum_{i \in S_l} \lambda_i P_0}}\right)\right), \tag{18.26}$$

where $Q(x) = \frac{1}{\sqrt{2\pi}} \int_x^\infty e^{-t^2/2} \, dt$ and step (b) is due to [5].

18.2.3   Numerical Examples and Discussions

In this section, our analytical results are validated through simulations. Herein, we assume $\lambda_i = \lambda_0 + (-1)^i i\varepsilon$, where $\lambda_0 = 1$, $i \in \{1,2,\ldots,N\}$, and ε denotes the difference among the involved channels. To ensure a fair comparison of two strategies, a same value of ε is assumed, i.e., $\varepsilon = 0.02$. In all the plots, we can observe that the analytical curves match perfectly with the simulated curves.

Figure 17.2 and Figure 17.3 draw a performance comparison between the proposed strategy and previous proposals in terms of the mean capacity and the average BER of the selected CU, respectively. It is shown that, compared with previous strategies, the proposed strategy achieves a higher mean capacity and a lower average BER for the same ζ_0. In addition, with an increase of ζ_0 (or with a decrease of R_0), the mean capacity and the average BER of the secondary system improve considerably.

It is worthwhile to mention that the mean capacity and the average BER of the previous proposals [1], [4], and [6] do not vary with the number of CUs. However, for the proposed strategy in this Chapter, the case is quite different, as shown in Figure 17.4 and Figure 17.5. In particular, both the mean capacity and the average BER of the proposed strategy are ameliorated with an increase in the number of CUs.

Figure 17.6 illustrates the transmit power ratio of the proposed strategy to the previous proposals in [1], where the impacts of the number of CUs and the value of ζ_0 are considered. It is shown that with an increase in ζ_0 or in the number of CUs, the transmit power ratio significantly decreases. For

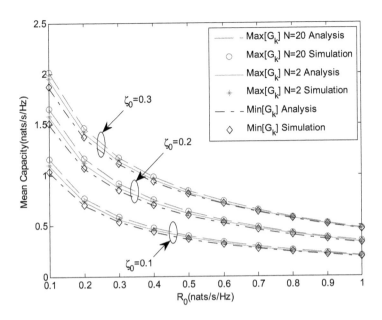

Figure 18.2: Mean capacity of the selected CU vs. the target transmission rate R_o for different scheduling schemes.

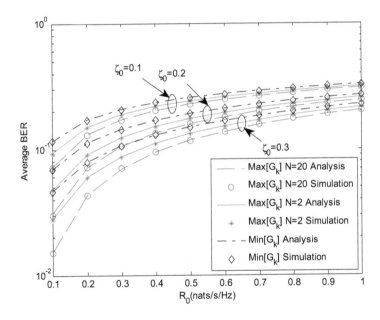

Figure 18.3: Average BER of the selected CU vs. the target transmission rate R_o for different scheduling schemes.

instance, when N=20 and $\zeta_0 = 0.3$, the proposed strategy merely consumes 1.18% transmit power of the counterpart in [1]. Moreover, a higher N or ζ_0 yields superior performance for the secondary system.

An intuitive explanation for the foregoing phenomenon is as follows. Under the outage protec-

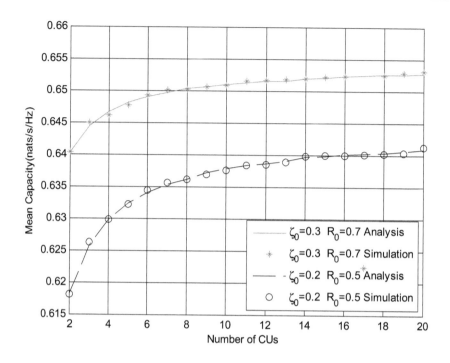

Figure 18.4: Mean capacity of the scheduled CU of the proposed strategy vs. the number of CUs.

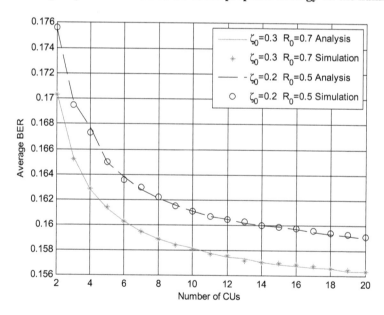

Figure 18.5: Average BER of the scheduled CU of the proposed strategy vs. the number of CUs.

tion criterion (17.1) for the PU, the "weakest-channel-quality selection rule" of [1] leads to a higher transmit power of the secondary system, whereas our proposed strategy, which selects the secondary user with the strongest channel quality, incurs a lower secondary transmit power. Nonetheless, the mean capacity and the average BER of the secondary system are determined by the received SIR at

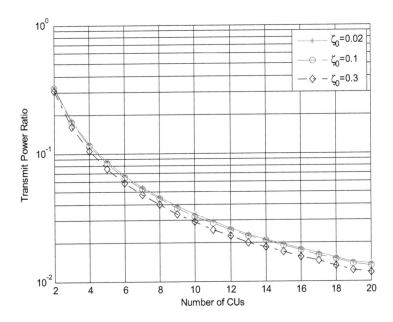

Figure 18.6: Transmit power ratio of the proposed strategy to the previous proposal vs. the number of CUs.

BS from the selected CU, which rely not only on the secondary transmit power, but also on the channel power gain of the scheduled secondary user. This makes it possible for the proposed strategy to attain superior performance to the proposal in [1]. The above deductions are actually the underlying motivations of this Chapter, which are also validated by the comprehensive numerical results in this section. Additionally, the proposed strategy has the need of a much lower secondary transmit power in comparison with that in [1], which is very attractive in practice.

18.2.3.1 Conclusions

In this section, we proposed a novel opportunistic scheduling strategy for an uplink cognitive cellular network. The proposed strategy selects the CU with the best instantaneous channel quality to access the shared spectrum while protecting the PU from excess interference. For such a strategy, we analyzed the mean capacity, the average BER, and the transmit power of the secondary system, and compared then with previous proposals. Numerical results indicated that compared with previous solutions, the proposed strategy consumed much less transmit power and achieved superior performance in terms of mean capacity and BER, especially when the number of CUs is large.

18.3   High-Efficient and Low-Complexity Relay Selection Strategies for Cooperative Cognitive Relaying Systems

Three low-complexity relay selection strategies, namely, selective amplify-and-forward (S-AF), selective decode-and-forward (S-DF), and amplify-and-forward with partial relay selection (PRS-AF), in spectrum-sharing scenarios are studied. For such, the respective asymptotic outage behaviors of the secondary system are analyzed, from which the diversity and coding gains are derived and compared. Unlike the coding gain, which is shown to be very sensitive with the position of the primary receiver, the diversity gain of the secondary system is the same as the non-spectrum-sharing system.

In addition, depending on the cooperative strategy employed, an increase of the number of relays may lead to severe loss of the coding gain.

18.4 Efficient Relay Selection Strategies for Multi-Relay Cognitive Cooperative Transmissions

18.4.1 System model

We consider a cooperative spectrum-sharing system where primary and secondary users coexist in a given geographical area and share the same frequency band. Our analysis will focus on the secondary communication. In this case, one source node S communicates with one destination node D using one out of N relay nodes R_k, k=1, 2, ..., N. More specifically, before the source transmission, one relay is selected among N available ones to cooperate with S, which was named as "*proactive selection*" in [7]. Due to the presence of obstacles, there is no direct link between S and D. Similar to [8] and [9], it is assumed that the N relays are clustered relatively close together (location-based clustering) and have been selected by a long-term routing process for establishing a link between S and D [10].

18.4.2 Three half-duplex cognitive relaying protocols

Concerning the relaying procedure, three *proactive* half-duplex protocols are considered as relaying strategies. All these strategies select only one relay to cooperate with S depending on the CSI of a single or both hops. In what follows, we introduce the three protocols one by one.

(1) *S-AF:* The relay that provides the highest end-to-end ratio SNR is selected among all N participating relays. Then, the end-to-end SNR can be written as

$$\gamma_{end}^{S\text{-}AF} = \max_{k=1,\ldots,N} \left[\frac{\frac{P_S|f_k|^2}{N_0} \frac{P_{R_k}|g_k|^2}{N_0}}{\frac{P_S|f_k|^2}{N_0} + \frac{P_{R_k}|g_k|^2}{N_0} + 1} \right], \quad (18.27)$$

where f_k and g_k denote the channel coefficients of the links $S \to R_k$ and $R_k \to D$, respectively, P_S and P_{Rk} are the transmit powers of S and R_k, respectively, and N_0 is the mean power of the additive white Gaussian noise at R_k and D. For convenience, we define $\gamma_{f_k} \triangleq P_S|f_k|^2/N_0$ as the SNR of the link $S \to R_k$ and $\gamma_{g_k} \triangleq P_{R_k}|g_k|^2/N_0$ as the SNR of the link $R_k \to D$.

(2) *S-DF:* The "best" relay that maximizes the minimum SNR related to the links $S \to R_k$ and $R_k \to D$ is chosen, so that the end-to-end SNR can be expressed as

$$\gamma_{end}^{S\text{-}DF} = \max_{k=1,\ldots,N} \left[\min \left[\gamma_{f_k}, \gamma_{g_k} \right] \right]. \quad (18.28)$$

(3) *PRS-AF:* The relay selection process is based on the quality of the links pertaining to the first-hop only. In this case, the end-to-end SNR can be written as

$$\gamma_{end}^{PRS\text{-}AF} = \frac{\gamma_{f_M} \gamma_{g_M}}{\gamma_{f_M} + \gamma_{g_M} + 1}, \quad (18.29)$$

where $\gamma_{f_M} \triangleq \max_{k=1,\ldots,N} \left[\gamma_{f_k} \right]$ and γ_{g_M} represents the SNR of the link $R_M \to D$, with R_M being the selected relay.

In this section, the three aforementioned relaying protocols will be investigated in a spectrum-sharing scenario, where a secondary transmitter uses its maximum allowable transmit power within its peak power constraint while satisfying the interference temperature requirement perceived at the PU. The source node and the selected relay adaptively adjust their transmit powers P_S and P_{Rk} so that the interference temperature constraint Q at PU is satisfied. Consequently, P_S and P_{Rk} should satisfy the following:

$$P_S = \begin{cases} P, & |h_1|^2 \leq Q/P \\ Q/|h_1|^2, & |h_1|^2 > Q/P \end{cases}, \quad P_{Rk} = \begin{cases} P, & |h_{2,k}|^2 \leq Q/P \\ Q/|h_{2,k}|^2, & |h_{2,k}|^2 > Q/P \end{cases}, \tag{18.30}$$

in which P is the peak transmit power of each secondary node, h_1 and $h_{2,k}$ are the fading channel coefficients from S to PU and from R_k to PU, respectively. Herein, we assume that S and R_k have perfect CSI of the links $S \rightarrow$PU and $R_k \rightarrow$PU (i.e., h_1 and $h_{2,k}$), respectively, and this can be realized by direct feedback from PU or indirect feedback from a third-party, such as a band manager in charge of the shared spectrum resources [11].

It is worthwhile to mention that, as in [12], the detailed protocol between the primary transmitter and primary receiver is ignored, and the interference from the primary transmitters can be translated into the noise term of the secondary system under the assumption that many primary transmitters exist and the interference from them follows a white Gaussian distribution, which can be justified by the Central Limit Theorem.

Concerning the channel model, we assume that the channels pertaining to each link undergo independent Rayleigh flat fading. Therefore, we have $|h_1|^2 \sim \mathrm{E}\left(d_1^{-\alpha}\right)$, $|h_{2,k}|^2 \sim \mathrm{E}\left(d_{2,k}^{-\alpha}\right)$, $|f_k|^2 \sim \mathrm{E}\left(d_{f,k}^{-\alpha}\right)$, and $|g_k|^2 \sim \mathrm{E}\left(d_{g,k}^{-\alpha}\right)$, k=1, 2, ..., N, where α is the path loss exponent, and d_{index} stands for the distance between the respective transmitters and receivers. Due to the location-based clustering of the relays, the relays are assumed to be very close to each other so that we have $d_{f,k} = d_f$, $d_{g,k} = d_g$, and $d_{2,k} = d_2$.

18.4.3 Asymptotic Analysis of System Performance

In order to evaluate the asymptotic outage behavior of the three relay selection strategies under study, we use $\gamma \overset{\Delta}{=} 1/N_0$ to represent the system SNR in the subsequent discussions [13] [14]. Accordingly, the high SNR regime arises when $\gamma \rightarrow \infty$. Based on the outage probability, which is defined as the probability that the instantaneous capacity per unit bandwidth of the secondary system is below a predefined end-to-end spectral efficiency $\Re_s$ bps/Hz, the respective diversity and coding gains will be derived. With this aim, note that in high SNR regimes, the outage probability P_{out} can be expressed in terms of these two measures as $P_{out} \simeq (G_c\gamma)^{-G_d}$, where G_d and G_c represent the diversity and coding gains, respectively.

(1) *S-AF strategy*

The outage probability of the secondary system can be formulated as

$$P_{out}^{\text{S - AF}} = \Pr\left[\frac{1}{2}\log_2\left(1 + \max_{k=1,\ldots,N}\frac{\gamma_{f_k}\gamma_{g_k}}{\gamma_{f_k} + \gamma_{g_k} + 1}\right) < \Re_s\right], \tag{18.31}$$

in which the factor 1/2 accounts for the fact that the overall transmission is split into two phases (half-duplex). When P_S and P_{Rk} (k=1, ..., N) are known, the independence among the dual-hop links allows us to rewrite (17.4) as

$$P_{out}^{\text{S - AF}}\bigg|_{P_S,P_{R_k}} = \prod_{k=1}^{N}\Pr\left[\frac{\gamma_{f_k}\gamma_{g_k}}{\gamma_{f_k} + \gamma_{g_k} + 1} < \rho \overset{\Delta}{=} 2^{2\Re_s} - 1\right], \tag{18.32}$$

where $P_{\text{out}}^{\text{S - AF}}\Big|_{P_S,P_{R_k}}$ represents the outage probability conditioned on P_S and P_{Rk} so that

$P_{\text{out}}^{\text{S - AF}} = E\left[P_{\text{out}}^{\text{S - AF}}\Big|_{P_S,P_{R_k}}\right]$. Now, let $\phi_k = P_S|f_k|^2$ and $\psi_k = P_{R_k}|g_k|^2$. For arbitrary P_S and

P_{Rk}, it follows that $\phi_k \sim E\left(P_S d_f^{-\alpha}\right)$ and $\psi_k \sim E\left(P_{R_k}d_g^{-\alpha}\right)$. Then, making use of [14], we arrive at

$$\lim_{\gamma\to\infty} \frac{\Pr\left[\frac{\gamma\phi_k\psi_k}{\gamma\phi_k+\gamma\psi_k+1} < \frac{\rho}{\gamma}\right]}{\rho/\gamma} = \frac{1}{P_S d_f^{-\alpha}} + \frac{1}{P_{R_k}d_g^{-\alpha}}. \tag{18.33}$$

Substituting (17.6) into (17.5), we have

$$\begin{aligned}
P_{\text{out}}^{\text{S - AF}}\Big|_{P_S,P_{R_k}} &= \left(\frac{\rho}{\gamma}\right)^N \prod_{k=1}^N \left(\frac{1}{P_S d_f^{-\alpha}} + \frac{1}{P_{R_k}d_g^{-\alpha}}\right) + o\left(\left(\frac{1}{\gamma}\right)^N\right) \\
&\simeq \left(\frac{\rho}{\gamma}\right)^N \underbrace{\prod_{k=1}^N \left(\frac{1}{P_S d_f^{-\alpha}} + \frac{1}{P_{R_k}d_g^{-\alpha}}\right)}_{\Xi},
\end{aligned} \tag{18.34}$$

where $o(\cdot)$ satisfies $\lim_{x\to 0} o(x)/x = 0$. Note that Ξ in (17.1) can be expanded as [15]

$$\Xi = \sum_{l=0}^N \left(\frac{1}{d_f^{-\alpha}P_S}\right)^l \sum_{m=1}^{\binom{N}{l}} \prod_{n=1}^{N-l} \frac{1}{d_g^{-\alpha}P_{R_n^{(m)}}}, \tag{18.35}$$

in which the term $\left(\frac{1}{d_f^{-\alpha}P_S}\right)^l \sum_{m=1}^{\binom{N}{l}} \prod_{n=1}^{N-l} \frac{1}{d_g^{-\alpha}P_{R_n^{(m)}}}$ stands for the sum of all $\binom{N}{l}$ possible products containing $(N-l)$ distinct P_{Rk} in the expansion of Ξ, and the index m symbolizes the m-th possible product of all $\binom{N}{l}$ possible ones. Now, plugging (17.2) into (17.1) and taking the expectation of the latter, it yields

$$P_{\text{out}}^{\text{S - AF}} = E\left[P_{\text{out}}^{\text{S - AF}}\Big|_{P_S,P_{R_k}}\right] \simeq \left(\frac{\rho}{\gamma}\right)^N \sum_{l=0}^N (d_f^\alpha)^l E\left[\left(\frac{1}{P_S}\right)^l\right] \sum_{m=1}^{\binom{N}{l}} \prod_{n=1}^{N-l} \left(d_g^\alpha E\left[\frac{1}{P_{R_n^{(m)}}}\right]\right). \tag{18.36}$$

Before giving continuity to our analysis, the following lemma will play a crucial role in this regard.

Lemma 18.1

The l-th moment of $1/P_S$ and $1/P_{Rk}$ can be expressed, respectively, as

$$E\left[\left(\frac{1}{P_S}\right)^l\right] = \left(\frac{1}{P}\right)^l \left(1 - e^{-d_1^\alpha \frac{Q}{P}}\right) + \frac{\Gamma\left(l+1,d_1^\alpha Q/P\right)}{\left(d_1^\alpha Q\right)^l}, \tag{18.37}$$

$$E\left[\left(\frac{1}{P_{R_k}}\right)^l\right] = \left(\frac{1}{P}\right)^l \left(1 - e^{-d_2^\alpha \frac{Q}{P}}\right) + \frac{\Gamma\left(l+1,d_2^\alpha Q/P\right)}{\left(d_2^\alpha Q\right)^l}, \tag{18.38}$$

where $\Gamma\left(\cdot,\cdot\right)$ denotes the incomplete gamma function [3].

Proof 18.5   Based on the definition of P_S given in (17.30), $E\left[\left(1/P_S\right)^l\right]$ can be written as

$$E\left[\left(\frac{1}{P_S}\right)^l\right]=E\left[\left(\frac{1}{\min\left[P,Q\big/|h_1|^2\right]}\right)^l\right].\tag{18.39}$$

From the probability statistical theory for the calculation of expectation, (17.1) can be rewritten as

$$E\left[\left(\frac{1}{P_S}\right)^l\right]=\left(\frac{1}{P}\right)^l\Pr\left[|h_1|^2\le\frac{Q}{P}\right]+\int_{Q/P}^{\infty}(x/Q)^l\,d_1^{\alpha}e^{-d_1^{\alpha}x}dx.\tag{18.40}$$

Making use of [3], (17.11) is attained. Using the same rationale, (17.38) can be readily obtained.

Now, from the lemma above and after some arrangements, (17.36) can be rewritten as

$$P_{\text{out}}^{\text{S - AF}}\simeq\left(\frac{\rho}{\gamma}\right)^N\sum_{l=0}^{N}\left(d_f^{\alpha}\right)^l\left[\left(\frac{1}{P}\right)^l\left(1-e^{-d_1^{\alpha}Q/P}\right)+\frac{\Gamma\left(l+1,d_1^{\alpha}Q/P\right)}{\left(d_1^{\alpha}Q\right)^l}\right]$$
$$\times\left(\begin{array}{c}N\\l\end{array}\right)\left(d_g^{\alpha}\left[\frac{1}{P}\left(1-e^{-d_2^{\alpha}Q/P}\right)+\frac{\Gamma\left(2,d_2^{\alpha}Q/P\right)}{d_2^{\alpha}Q}\right]\right)^{N-l}.\tag{18.41}$$

A lower bound for $P_{\text{out}}^{\text{S - AF}}$ can also be attained. For such, first note that the outage probability can be written as

$$P_{\text{out}}^{\text{S - AF}}=\Pr\left[\max_{k=1,\dots,N}\frac{\gamma_{f_k}\gamma_{g_k}}{\gamma_{f_k}+\gamma_{g_k}+1}<\rho\right]$$

$$\ge\Pr\left[\bigcap_{k=1}^{N}\left(\min\left[\gamma_{f_k},\gamma_{g_k}\right]<\rho\right)\right]$$

$$=\Pr\left[\bigcap_{k=1}^{N}\left(\min\left[\frac{\min\left[P,Q\big/|h_1|^2\right]|f_k|^2}{N_0},\frac{\min\left[P,Q\big/|h_{2,k}|^2\right]|g_k|^2}{N_0}\right]<\rho\right)\right]\overset{\Delta}{=}F_{\gamma_{\text{up}}}^{\text{S - AF}}(\rho).\tag{18.42}$$

Now, let $X\overset{\Delta}{=}|h_1|^2$, knowing that the PDF of X is $p_X(x)=d_1^{\alpha}e^{-d_1^{\alpha}x}$, (17.4) can be expressed as

$$F_{\gamma_{\text{up}}}^{\text{S - AF}}(\rho)=\int_0^{\infty}\Pr\left[\bigcap_{k=1}^{N}\left(\min\left[\frac{\min[P,Q/x]|f_k|^2}{N_0},\frac{\min\left[P,Q\big/|h_{2,k}|^2\right]|g_k|^2}{N_0}\right]<\rho\right)\right]$$
$$\times p_X(x)\,dx$$

$$=\int_0^{\infty}\prod_{k=1}^{N}\underbrace{\left\{\Pr\left[\min\left[\frac{\min\left[P,Q/x\right]|f_k|^2}{N_0},\frac{\min\left[P,Q\big/|h_{2,k}|^2\right]|g_k|^2}{N_0}\right]<\rho\right]\right.}_{I_k}\tag{18.43}$$
$$\times d_1^{\alpha}e^{-d_1^{\alpha}x}dx,$$

where I_k can be expressed as

$$
I_k = 1 - \Pr\left[\frac{\min[P,Q/x]|f_k|^2}{N_0} > \rho\right] \Pr\left[\frac{\min\left[P,Q\big/|h_{2,k}|^2\right]|g_k|^2}{N_0} > \rho\right]
$$
$$
= 1 - \Pr\left[|f_k|^2 > \frac{\rho N_0}{\min[P,Q/x]}\right]\left(1 - F_{\gamma_{g_k}}(\rho)\right).
$$
(18.44)

By substituting (17.2) into (17.1), the following is written

$$
F_{\gamma_{up}}^{S\text{-}AF}(\rho) = \underbrace{\int_0^{\frac{Q}{P}} d_1^\alpha e^{-d_1^\alpha x} \prod_{k=1}^N \left[1 - e^{-d_f^\alpha \frac{\rho N_0}{P}}\left(1 - F_{\gamma_{g_k}}(\rho)\right)\right] dx}_{\Omega}
$$
$$
+ \underbrace{\int_{\frac{Q}{P}}^\infty d_1^\alpha e^{-d_1^\alpha x} \prod_{k=1}^N \left[1 - e^{-d_f^\alpha \frac{\rho N_0 x}{Q}}\left(1 - F_{\gamma_{g_k}}(\rho)\right)\right] dx}_{\Theta}.
$$
(18.45)

To proceed forward, we need the CDFs of γ_{g_k}, which is addressed in the following lemma. ■

Lemma 18.2

The CDFs of γ_{f_k} and γ_{g_k} can be formulated as

$$
F_{\gamma_{f_k}}(\rho) = \left(1 - e^{-d_1^\alpha Q/P}\right)\left(1 - e^{-N_0 \rho d_f^\alpha/P}\right) + e^{-d_1^\alpha Q/P}\left(1 - \frac{Qd_1^\alpha}{d_f^\alpha N_0 \rho + Qd_1^\alpha}e^{-N_0 \rho d_f^\alpha/P}\right),
$$
(18.46)

$$
F_{\gamma_{g_k}}(\rho) = \left(1 - e^{-d_2^\alpha Q/P}\right)\left(1 - e^{-N_0 \rho d_g^\alpha/P}\right) + e^{-d_2^\alpha Q/P}\left(1 - \frac{Qd_2^\alpha}{d_g^\alpha N_0 \rho + Qd_2^\alpha}e^{-N_0 \rho d_g^\alpha/P}\right).
$$
(18.47)

As $\gamma \to \infty$, (17.46) and (17.47) can be asymptotically expressed as

$$
F_{\gamma_{f_k}}(\rho) \simeq \frac{\rho d_f^\alpha}{\gamma P} + e^{-d_1^\alpha Q/P}\frac{\rho d_f^\alpha}{\gamma Q d_1^\alpha},
$$
(18.48)

$$
F_{\gamma_{g_k}}(\rho) \simeq \frac{\rho d_g^\alpha}{\gamma P} + e^{-d_2^\alpha Q/P}\frac{\rho d_g^\alpha}{\gamma Q d_2^\alpha}.
$$
(18.49)

Proof 18.6   From the total probability theorem, $F_{\gamma_{f_k}}(\rho)$ can be written as

$$
F_{\gamma_{f_k}}(\rho) = \Pr\left[|h_1|^2 \leq Q/P\right]\Pr\left[\gamma_{f_k} \leq \rho \,\big|\, |h_1|^2 \leq Q/P\right] + \Pr\left[|h_1|^2 > Q/P, \gamma_{f_k} \leq \rho\right]
$$
$$
= \Pr\left[|h_1|^2 \leq Q/P\right]\Pr\left[P|f_k|^2\big/N_0 \leq \rho\right] + \Pr\left[|h_1|^2 > Q/P, \frac{Q|f_k|^2}{|h_1|^2 N_0} \leq \rho\right].
$$
(18.50)

Making use of [2], it is easy to arrive at (17.46) from (17.3). Using a similar procedure, (17.47) is also attained. Now, applying the Taylor series expansion in (17.46) and (17.47), as $N_0 \to 0$, (17.1) and (17.2) are obtained, respectively. ■

Therefore, by substituting (17.47) into Ω of (17.45), it yields

$$\Omega = \left(1 - e^{-d_1^\alpha \frac{Q}{P}}\right)\left[1 - e^{-d_f^\alpha \frac{\rho N_0}{P}}\left(1 - F_{\gamma_{g_k}}(\rho)\right)\right]^N. \tag{18.51}$$

In addition, by substituting (17.47) into Θ and making use of the binomial theorem, it follows that

$$
\begin{aligned}
\Theta &= \int_{\frac{Q}{P}}^{\infty} d_1^\alpha e^{-d_1^\alpha x}\left[1 - e^{-d_f^\alpha \frac{\rho N_0 x}{Q}}\left(1 - F_{\gamma_{g_k}}(\rho)\right)\right]^N dx \\
&= \int_{\frac{Q}{P}}^{\infty} d_1^\alpha e^{-d_1^\alpha x}\sum_{l=0}^{N}\binom{N}{l}(-1)^l e^{-d_f^\alpha \frac{\rho N_0 x}{Q}l}\left(1 - F_{\gamma_{g_k}}(\rho)\right)^l dx \\
&= \sum_{l=0}^{N}\binom{N}{l}(-1)^l\left(1 - F_{\gamma_{g_k}}(\rho)\right)^l \frac{d_1^\alpha}{d_1^\alpha + d_f^\alpha \rho N_0 l/Q}e^{-\left(d_1^\alpha + d_f^\alpha \frac{\rho N_0 l}{Q}\right)\frac{Q}{P}}.
\end{aligned} \tag{18.52}
$$

Finally, by plugging (17.4) and (17.5) into (17.45), a closed-form expression for $F_{\gamma_{up}}^{\text{S - AF}}(\rho)$ can be attained. As $\gamma \to \infty$ (or, equivalently, $N_0 \to 0$), making use of the Taylor series expansion for the exponential functions in (17.4) and from (17.2), Ω can be asymptotically written as

$$\Omega \simeq \left(1 - e^{-d_1^\alpha \frac{Q}{P}}\right)\left[\frac{\rho\left(d_f^\alpha + d_g^\alpha\right)}{P} + \frac{\rho d_g^\alpha e^{-d_2^\alpha \frac{Q}{P}}}{Q d_2^\alpha}\right]^N \left(\frac{1}{\gamma}\right)^N. \tag{18.53}$$

To determine the asymptotic expression of Θ, first we rewrite Θ as

$$\Theta = \int_{\frac{Q}{P}}^{\infty} d_1^\alpha e^{-d_1^\alpha x}\underbrace{\left[1 - e^{-d_f^\alpha \frac{\rho N_0 x}{Q}}\left(1 - F_{\gamma_{g_k}}(\rho)\right)\right]^N}_{\chi} dx, \tag{18.54}$$

in which, as $N_0 \to 0$, χ can be asymptotically expressed as

$$
\begin{aligned}
\chi &\simeq \left[1 - \left(1 - d_f^\alpha \frac{\rho N_0 x}{Q}\right)\left(1 - N_0\left(\frac{\rho d_g^\alpha}{P} + e^{-d_2^\alpha Q/P}\frac{\rho d_g^\alpha}{Q d_2^\alpha}\right)\right)\right]^N \\
&\simeq N_0^N\left[\left(\frac{\rho d_g^\alpha}{P} + e^{-d_2^\alpha Q/P}\frac{\rho d_g^\alpha}{Q d_2^\alpha}\right) + \frac{d_f^\alpha \rho x}{Q}\right]^N \\
&= N_0^N\sum_{n=0}^{N}\binom{N}{n}\left(\frac{d_f^\alpha \rho}{Q}\right)^n\left(\frac{\rho d_g^\alpha}{P} + e^{-d_2^\alpha Q/P}\frac{\rho d_g^\alpha}{Q d_2^\alpha}\right)^{N-n} x^n.
\end{aligned} \tag{18.55}
$$

Then, by substituting (17.8) into (17.7) and relying on [3], (17.7) can be asymptotically expressed as

$$\Theta \simeq N_0^N\sum_{n=0}^{N}\binom{N}{n}\left(\frac{d_f^\alpha \rho}{Q}\right)^n\left(\frac{\rho d_g^\alpha}{P} + e^{-d_2^\alpha Q/P}\frac{\rho d_g^\alpha}{Q d_2^\alpha}\right)^{N-n} d_1^{-n\alpha}\Gamma\left(n+1, \frac{d_1^\alpha Q}{P}\right). \tag{18.56}$$

Now, combining (17.45), (17.6), and (17.9), a lower-bound for $P_{\text{out}}^{\text{S - AF}}$ at high SNR can be obtained as

$$
\begin{aligned}
P_{\text{out}}^{\text{S - AF}} &\geq \left(1 - e^{-d_1^\alpha \frac{Q}{P}}\right)\left[\frac{d_f^\alpha + d_g^\alpha}{P} + \frac{d_g^\alpha e^{-d_2^\alpha \frac{Q}{P}}}{Q d_2^\alpha}\right]^N \left(\frac{\rho}{\gamma}\right)^N \\
&\quad + \left(\frac{\rho}{\gamma}\right)^N\sum_{n=0}^{N}\binom{N}{n}\left(\frac{d_f^\alpha}{Q}\right)^n\left(\frac{d_g^\alpha}{P} + \frac{d_g^\alpha e^{-d_2^\alpha Q/P}}{Q d_2^\alpha}\right)^{N-n} d_1^{-n\alpha}\Gamma\left(n+1, \frac{d_1^\alpha Q}{P}\right) \\
&\propto \left(\frac{1}{\gamma}\right)^N.
\end{aligned} \tag{18.57}
$$

Using the relation given by $\gamma_{f_k}\gamma_{g_k}/(\gamma_{f_k}+\gamma_{g_k}+1) \geq \frac{1}{2}\min\left[\gamma_{f_k},\gamma_{g_k}\right]$ [16], an upper-bound for $P_{\text{out}}^{\text{S - AF}}$ can also be derived as

$$P_{\text{out}}^{\text{S - AF}} \leq \left(1-e^{-d_1^{\alpha}\frac{Q}{P}}\right)\left[\frac{d_f^{\alpha}+d_g^{\alpha}}{P}+\frac{d_g^{\alpha}e^{-d_2^{\alpha}\frac{Q}{P}}}{Qd_2^{\alpha}}\right]^N\left(\frac{2\rho}{\gamma}\right)^N$$
$$+\left(\frac{2\rho}{\gamma}\right)^N\sum_{n=0}^{N}\binom{N}{n}\left(\frac{d_f^{\alpha}}{Q}\right)^n\left(\frac{d_g^{\alpha}}{P}+\frac{d_g^{\alpha}e^{-d_2^{\alpha}Q/P}}{Qd_2^{\alpha}}\right)^{N-n}d_1^{-n\alpha}\Gamma\left(n+1,\frac{d_1^{\alpha}Q}{P}\right) \quad (18.58)$$
$$\propto \left(\frac{1}{\gamma}\right)^N.$$

Remark 17.1 (G_d and G_c for S-AF): From (17.34), it can be observed that under non-spectrum sharing scenario, $G_d = N$ and $G_c = \dfrac{P}{\rho\left(d_f^{\alpha}+d_g^{\alpha}\right)}$, whereas from (17.1) and (17.2), it can be seen that under the spectrum-sharing scenario, $G_d = N$ and

$$\frac{1}{2\rho}\left[\left(1-e^{-d_1^{\alpha}\frac{Q}{P}}\right)\left(\frac{d_f^{\alpha}+d_g^{\alpha}}{P}+\frac{d_g^{\alpha}e^{-d_2^{\alpha}\frac{Q}{P}}}{Qd_2^{\alpha}}\right)^N\right.$$
$$\left.+\sum_{n=0}^{N}\binom{N}{n}\left(\frac{d_f^{\alpha}}{Q}\right)^n\left(\frac{d_g^{\alpha}}{P}+\frac{d_g^{\alpha}e^{-d_2^{\alpha}Q/P}}{Qd_2^{\alpha}}\right)^{N-n}d_1^{-n\alpha}\Gamma\left(n+1,\frac{d_1^{\alpha}Q}{P}\right)\right]^{-\frac{1}{N}} \leq G_c \leq$$
$$\frac{1}{\rho}\left[\left(1-e^{-d_1^{\alpha}\frac{Q}{P}}\right)\left(\frac{d_f^{\alpha}+d_g^{\alpha}}{P}+\frac{d_g^{\alpha}e^{-d_2^{\alpha}\frac{Q}{P}}}{Qd_2^{\alpha}}\right)^N\right.$$
$$\left.+\sum_{n=0}^{N}\binom{N}{n}\left(\frac{d_f^{\alpha}}{Q}\right)^n\left(\frac{d_g^{\alpha}}{P}+\frac{d_g^{\alpha}e^{-d_2^{\alpha}Q/P}}{Qd_2^{\alpha}}\right)^{N-n}d_1^{-n\alpha}\Gamma\left(n+1,\frac{d_1^{\alpha}Q}{P}\right)\right]^{-\frac{1}{N}}. \quad (18.59)$$

In addition, from (17.41) we arrive at an exact expression for G_c as

$$G_c = \frac{1}{\rho}\left[\sum_{l=0}^{N}\left(d_f^{\alpha}\right)^l\left[\left(\frac{1}{P}\right)^l\left(1-e^{\frac{-Qd_1^{\alpha}}{P}}\right)+\frac{\Gamma\left(l+1,\frac{Qd_1^{\alpha}}{P}\right)}{(d_1^{\alpha}Q)^l}\right]\binom{N}{l}\right.$$
$$\left.\times\left(d_g^{\alpha}\left[\frac{1}{P}\left(1-e^{\frac{-Qd_2^{\alpha}}{P}}\right)+\frac{\Gamma\left(2,\frac{Qd_2^{\alpha}}{P}\right)}{d_2^{\alpha}Q}\right]\right)^{N-l}\right]^{-\frac{1}{N}}. \quad (18.60)$$

(2) *S-DF strategy*

For S-DF strategy, the outage probability can be formulated as

$$P_{\text{out}}^{\text{S - DF}} = \Pr\left[\max_{k=1,\dots,N}\left[\min\left[\gamma_{f_k},\gamma_{g_k}\right]\right] < \rho\right] = \Pr\left[\bigcap_{k=1}^{N}\left(\min\left[\gamma_{f_k},\gamma_{g_k}\right] < \rho\right)\right]. \quad (18.61)$$

It is noteworthy that (17.5) and $F_{\gamma_{\text{up}}}^{\text{S - AF}}(\rho)$ coincide with each other, which means the outage probability of the S-DF strategy is a lower-bound for that of the S-AF strategy.

Remark 17.2 (G_d and G_c for S-DF): From (17.1) and (17.5), it is shown that $G_d = N$ and

$$G_c = \frac{1}{\rho}\left[\left(1-e^{-d_1^{\alpha}\frac{Q}{P}}\right)\left(\frac{d_f^{\alpha}+d_g^{\alpha}}{P}+\frac{d_g^{\alpha}e^{-d_2^{\alpha}\frac{Q}{P}}}{Qd_2^{\alpha}}\right)^N\right.$$
$$\left.+\sum_{n=0}^{N}\binom{N}{n}\left(\frac{d_f^{\alpha}}{Q}\right)^n\left(\frac{d_g^{\alpha}}{P}+\frac{d_g^{\alpha}e^{-d_2^{\alpha}Q/P}}{Qd_2^{\alpha}}\right)^{N-n}d_1^{-n\alpha}\Gamma\left(n+1,\frac{d_1^{\alpha}Q}{P}\right)\right]^{-\frac{1}{N}}. \quad (18.62)$$

When d_1 and d_2 tend to infinity, it is easy to arrive at $G_c = \frac{P}{\rho\left(d_f^\alpha + d_g^\alpha\right)}$, which is the coding gain for the non-spectrum-sharing scenario, as obtained previously for the S-AF strategy.

(3) *PRS-AF strategy*

For PRS-AF strategy, the outage probability can be expressed as

$$P_{out}^{PRS\text{-}AF} = \Pr\left\{\gamma_{end}^{PRS\text{-}AF} = \frac{\gamma_{f_M}\gamma_{g_M}}{\gamma_{f_M}+\gamma_{g_M}+1} < \rho\right\}$$
$$\geq \Pr\left\{\min\left[\gamma_{f_M},\gamma_{g_M}\right] < \rho\right\}$$

$$= F_{\gamma_{f_M}}(\rho) + F_{\gamma_{g_M}}(\rho) - F_{\gamma_{f_M}}(\rho)F_{\gamma_{g_M}}(\rho) \triangleq F_{\gamma_{up}}^{PRS\text{-}AF}(\rho), \qquad (18.63)$$

where

$$F_{\gamma_{g_M}}(\rho) = F_{\gamma_{g_k}}(\rho). \qquad (18.64)$$

In what follows, $F_{\gamma_{f_M}}(\rho)$ will be derived. First, from the definition of γ_{f_M}, we have

$$F_{\gamma_{f_M}}(\rho) = \Pr\left[\max_{k=1,\ldots,N}\left[\frac{\min\left[P,Q/|h_1|^2\right]|f_k|^2}{N_0}\right] < \rho\right]$$
$$= \int_0^\infty \left[\prod_{k=1}^N \Pr\left[\frac{\min[P,Q/x]|f_k|^2}{N_0} < \rho\right]\right]P_{|h_1|^2}(x)\,dx \qquad (18.65)$$
$$= \int_0^\infty \left(1 - e^{-\frac{N_0\rho d_f^\alpha}{\min[P,Q/x]}}\right)^N d_1^\alpha e^{-d_1^\alpha x}dx.$$

Now, relying on the relation between P and Q/x, (17.9) can be further written as

$$F_{\gamma_{f_M}}(\rho) = \left(1 - e^{-d_1^\alpha \frac{Q}{P}}\right)\left(1 - e^{-\frac{N_0\rho d_f^\alpha}{P}}\right)^N + \underbrace{\int_{\frac{Q}{P}}^\infty \left(1 - e^{-\frac{N_0\rho d_f^\alpha x}{Q}}\right)^N d_1^\alpha e^{-d_1^\alpha x}dx}_{\eta}. \quad (18.66)$$

Invoking the binomial theorem and performing the required integral above, η can be expressed in closed-form so that $F_{\gamma_{f_M}}(\rho)$ is given by

$$F_{\gamma_{f_M}}(\rho) = \left(1 - e^{-\frac{d_1^\alpha Q}{P}}\right)\left(1 - e^{-\frac{N_0\rho d_f^\alpha}{P}}\right)^N + \underbrace{\sum_{n=0}^N \binom{N}{n}(-1)^n \frac{d_1^\alpha}{d_1^\alpha + d_f^\alpha \frac{N_0\rho n}{Q}} e^{-\left(d_1^\alpha + d_f^\alpha \frac{N_0\rho n}{Q}\right)\frac{Q}{P}}}_{\eta}.$$

$$(18.67)$$

By substituting (17.47) and (17.11) into (17.7), a closed-form expression is attained for $F_{\gamma_{up}}^{PRS\text{-}AF}(\rho)$. Note that as $N_0 \to 0$, $F_{\gamma_{g_k}}(\rho) \propto N_0$ according to *Lemma 17.2*. Hence, in order to determine the diversity order of $F_{\gamma_{up}}^{PRS\text{-}AF}(\rho)$, the power terms in N_0 smaller than or equal to 1 are required. From (17.11), it can be seen that $\left(1 - e^{-\frac{d_1^\alpha Q}{P}}\right)\left(1 - e^{-\frac{N_0\rho d_f^\alpha}{P}}\right)^N \propto N_0^N$ as $N_0 \to 0$. Now, knowing that $\frac{1}{d_1^\alpha + d_f^\alpha \frac{N_0\rho n}{Q}} \simeq d_1^{-\alpha} - \frac{nd_f^\alpha \rho}{Qd_1^{2\alpha}}N_0$ and $e^{-d_f^\alpha \frac{N_0\rho n}{P}} \simeq 1 - d_f^\alpha \frac{N_0\rho n}{P}$, η can be further simplified as

$$\eta \simeq e^{-d_1^\alpha \frac{Q}{P}}\sum_{n=0}^N \binom{N}{n}(-1)^n + N_0\rho d_f^\alpha e^{-d_1^\alpha \frac{Q}{P}}\left(\frac{1}{P} + \frac{d_1^{-\alpha}}{Q}\right)\sum_{n=0}^N \binom{N}{n}(-1)^{n+1}n. \quad (18.68)$$

Note that the first term in (17.12) is equal to zero, according to binomial theorem, and the second term is also equal to zero, based on [15]. Therefore, as $\gamma \to \infty$, $F_{\gamma_{up}}^{PRS \text{-} AF}(\rho)$ can be asymptotically written as

$$F_{\gamma_{up}}^{PRS \text{-} AF}(\rho) \simeq \frac{\rho}{\gamma P}\left(d_g^\alpha + \frac{P}{Qd_2^\alpha}e^{-Qd_2^\alpha/P}\right). \tag{18.69}$$

Similar to (17.2), an upper-bound for $P_{out}^{PRS \text{-} AF}$ can be derived as

$$P_{out}^{PRS \text{-} AF} \leq \frac{2\rho}{\gamma P}\left(d_g^\alpha + \frac{P}{Qd_2^\alpha}e^{-Qd_2^\alpha/P}\right). \tag{18.70}$$

Remark 17.3 (G_d and G_c for PRS-AF): From (17.13) and (17.14), it is shown that $G_d=1$ and

$$\frac{P}{2\rho\left[d_g^\alpha + \frac{P}{Qd_2^\alpha}e^{-Qd_2^\alpha/P}\right]} \leq G_c \leq \frac{P}{\rho\left[d_g^\alpha + \frac{P}{Qd_2^\alpha}e^{-Qd_2^\alpha/P}\right]}. \tag{18.71}$$

As $d_2 \to \infty$, one can arrive at the limits regarding the non-spectrum-sharing scenario, which is $\frac{P}{2\rho d_g^\alpha} \leq G_c \leq \frac{P}{\rho d_g^\alpha}$.

18.4.4 Numerical Examples and Discussions

In this section, under the considered spectrum-sharing scenario, the three cooperative strategies are compared in terms of the outage probability and coding gain. Our analytical results have been validated through simulations. For illustrative purposes and without loss of generality, we normalize to unity the distance between S and the center of the relays-cluster and distance between the center of the relays-cluster and D, yielding, therefore, $d_f = d_g = 1$. In addition, we set the pathloss exponent α to 4. To ensure a fair comparison of the three cooperative strategies, unit transmit power and equal power division among cooperating nodes are assumed, i.e., $P=1/2$.

Figure 17.7 depicts the outage probability vs. system SNR. The primary receiver is placed to be in line with the source and the center of the relays-cluster. Note that for S-AF strategy, (17.41) coincides with (17.57), although they are in different forms to each other. For S-DF strategy, the exact analytical results match well with simulations, and the asymptotic curves are tight bounds in the medium and high SNR regions. For PRS-AF strategy, both the lower-bound and the asymptotic curves are very tight at high SNR regime. Regarding the performance comparisons of the three cooperative strategies, the performance of S-DF is superior to that of S-AF, and S-AF is higher than PRS-AF, as expected. Note that in high SNR regions, the performances of S-DF and S-AF are very close to each other.

Figures 17.8, 17.9, and 17.10 show the variation of the outage probability with an increase in N for the three cooperative strategies, respectively. Note that the cross point of the asymptotes in SNR axis corresponds to the inverse of coding gain. From Figures 17.8, 17.9, and 17.10, it is demonstrated that for S-AF and S-DF, an increase of N leads to a decrease of the outage behavior, as well as a decrease of coding gain, whereas for PRS-AF the outage probability and coding gain do not change with N.

Figure 17.11 compares the coding gains for the three cooperative strategies. The coding gains for non-spectrum-sharing scenario are also plotted for comparison purposes. For all three cooperative strategies, it can be observed that the shorter the distance between the secondary system and PU, the greater the loss of the coding gain. And the coding gains for S-AF and S-DF are much smaller

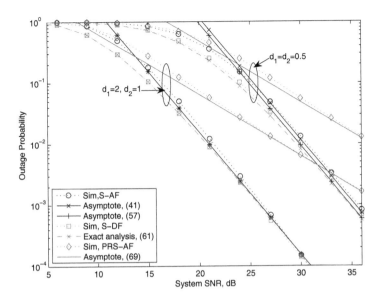

Figure 18.7: Outage probability vs. system SNR $(N = 2, R_s = 1\,\text{bps/Hz and } Q = 0\,\text{dB})$.

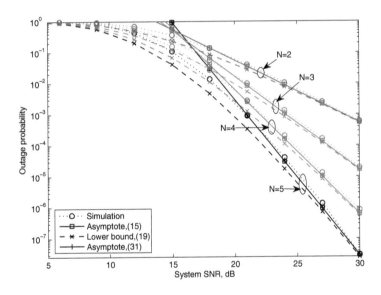

Figure 18.8: Outage probability vs. system SNR for S-AF strategy $(d1 = d2 = 0.75, R_s = 1\,\text{bps/Hz and } Q = 0\,\text{dB})$.

than that for PRS-AF. Interestingly, the coding gain of S-AF is the same with that of S-DF. Also, it is shown that for S-AF and S-DF, an increase in N is detrimental to the coding gain of the secondary system, especially when the value of d is small.

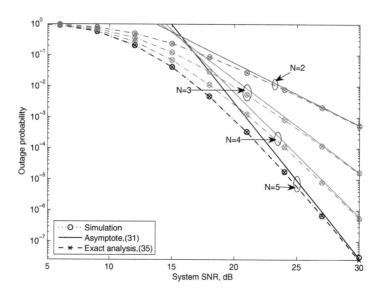

Figure 18.9: Outage probability vs. system SNR for S-DF strategy $(d1 = d2 = 0.75, R_s = 1\text{bps}/\text{Hz} \text{ and } Q = 0\text{dB})$.

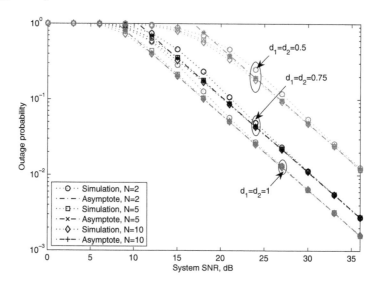

Figure 18.10: Outage probability vs. system SNR for PRS-AF strategy $(R_s = 1\text{bps}/\text{Hz} \text{ and } Q = 0\text{dB})$.

18.4.5 Conclusions

In this section, considering a spectrum-sharing scenario, the asymptotic outage behaviors of three relay selection strategies were investigated. In this case, a comprehensive comparison among these relay selection strategies was performed in terms of outage probability, and the impacts of the number of relays and the distance between the primary and secondary systems on the coding gain were also examined. These results are of great importance for the wireless communications field and find applicability in the design of cooperative networks under spectrum-sharing scenarios.

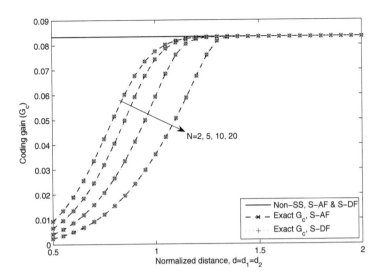

Figure 18.11: Coding gain vs. normalized distance d $(R_s = 1\text{bps}/\text{Hz} \text{ and } Q = 0\text{dB})$. **"SS" denotes "Spectrum Sharing."**

18.5 Distributed Link Scheduling for Single-Relay Secondary Relaying Transmission

In this section, we look at a distributed link selection scheme for cognitive selection AF relaying networks. For such a scheme, by approximating the instantaneous SNR ratio of dual-hop relaying link by its tight bound, a closed-form lower bound expression for the considered system is derived. Both analytical and numerical results show that in comparison with the centralized selection scheme, the proposed scheme incurs extremely low signaling overhead to achieve almost the same system outage performance. In addition, for both schemes, the influence of relay placement on the average amount of feedback overhead is investigated, and some useful conclusions are drawn as well.

18.5.0.1 System Model

As in [17], consider an interference-limited cognitive radio networks (CRN) with a variable-gain AF relay, where a CU S transmits information to a CU-receiver D with the aid of a half-duplex AF relay R. All the secondary terminals occupy the same licensed spectrum band allocated to the PU P. Also, it is assumed that all the channels undergo inid Rayleigh flat fading. As a result, by denoting $|h_{AB}|^2$ as the channel power gain associated with the link $A \rightarrow B$ ($A \in \{S,R\}$, $B \in \{P,R,D\}$), it follows that $|h_{AB}|^2$ conforms to exponential distribution, with mean Ω_{AB}. In addition, $P_S = I_p / |h_{SP}|^2$ and $P_R = I_p / |h_{RP}|^2$ indicate the transmit powers of S and R, respectively, where I_p represents the maximum tolerable interference power of P, and N_0 is the noise variance of the additive white Gaussian noise (AWGN) at B. Hereafter, $F_X(.)$ and $f_X(.)$ are used to denote the CDF and PDF of a random variable X, respectively.

18.5.0.2 Distributed Link Selection Scheme

According to [17], the received SNRs at CU receiver via the direct link and the relaying link can be, respectively, expressed as

$$\gamma_{DT} = \frac{I_p \left| h_{SD} \right|^2}{N_0 \left| h_{SP} \right|^2}, \tag{18.72}$$

$$\gamma_{AF} = \frac{\gamma_{SR} \gamma_{RD}}{\gamma_{SR} + \gamma_{RD} + 1}, \tag{18.73}$$

where $\gamma_{SR} = \frac{I_p |h_{SR}|^2}{N_0 |h_{SP}|^2}$ and $\gamma_{RD} = \frac{I_p |h_{RD}|^2}{N_0 |h_{RP}|^2}$ denote the instantaneous link SNRs pertaining to the first-hop and the second-hop relaying links, respectively. For the case of centralized link scheduling, the link with the higher received SNR at the destination will be chosen, which means that the end-to-end instantaneous SNR of the cognitive system can be formulated as

$$\gamma_D = \max \left\{ \gamma_{AF}, \gamma_{DT} \right\}. \tag{18.74}$$

In contrast, by modifying the link selection criterion (17.4) by its upper-bound, i.e., $\gamma_D = \max \left\{ \gamma_{AF}, \gamma_{DT} \right\} \leq \max \left\{ \min \left[\gamma_{SR}, \gamma_{RD} \right], \gamma_{DT} \right\}$, we propose an efficient link decision mechanism that can be implemented in a distributed manner. To be more specific, the link selection begins with the comparison between γ_{SR} and γ_{DT} at S. If $\gamma_{DT} \geq \gamma_{SR}$, it follows that $\gamma_{DT} \geq \min \left[\gamma_{SR}, \gamma_{RD} \right]$. In this case, the direct link will be selected according to the proposed link selection criterion. Otherwise, S sends a "fail" message to D. Upon receiving the "fail" message, D compares γ_{RD} with γ_{DT}. In the case of $\gamma_{DT} \geq \gamma_{RD}$, we have $\gamma_{DT} \geq \min \left[\gamma_{SR}, \gamma_{RD} \right]$, such that the direct link will be chosen. For such a case, D will broadcast a "success" message to inform S that the direct link should be selected. Otherwise, D forwards a "fail" message to S to indicate that the relaying link should be chosen.

Based on the preceding distributed selection criterion, in what follows, we derive the outage lower-bound of the cognitive system. As mentioned in [17], both γ_{AF} and γ_{DT} rely on the item $|h_{SP}|^2$, which indicates the statistical dependence between γ_{AF} and γ_{DT}. Consequently, we adopt the conditional statistics with respect to $|h_{SP}|^2$ for γ_{AF} and γ_{DT}. Let $X = |h_{SP}|^2$, $Y = \frac{I_p |h_{SR}|^2}{N_0}$, by defining $\gamma_{SR} = \frac{Y}{X}$, the CDF of γ_{SR} conditioned on $|h_{SP}|^2$ is given by

$$F_{\gamma_{SR} || h_{SP} |^2} (\gamma | X) = \Pr \left(\frac{Y}{X} < \gamma | X \right) = 1 - \exp \left\{ -\frac{N_0}{\Omega_{SR} I_p} x \gamma \right\}, \quad \begin{matrix} \gamma > 0 \\ x > 0 \end{matrix}. \tag{18.75}$$

In addition, applying the total probability theorem, the CDF of γ_{RD} can be written as

$$F_{\gamma_{RD}} (\gamma) = 1 - \frac{\Omega_{RD} I_p}{N_0 \Omega_{RP} \gamma + \Omega_{RD} I_p}. \tag{18.76}$$

By approximating γ_{AF} by $\gamma'_{AF} = \min[\gamma_{SR}, \gamma_{RD}]$, one can readily derive the CDF of γ'_{AF} conditioned on $|h_{SP}|^2$ as

$$F_{\gamma'_{AF} || h_{SP} |^2} \left(\gamma | |h_{SP}|^2 \right) = 1 - \exp \left\{ -\frac{N_0}{\Omega_{SR} I_p} x \gamma \right\} \frac{\Omega_{RD} I_p}{N_0 \Omega_{RP} \gamma + \Omega_{RD} I_p}, \quad \begin{matrix} \gamma > 0 \\ x > 0 \end{matrix}. \tag{18.77}$$

Furthermore, the CDF of γ_{DT} conditioned on $|h_{SP}|^2$ can be calculated as

$$F_{\gamma_{DT} || h_{SP} |^2} \left(\gamma | |h_{SP}|^2 \right) = 1 - \exp \left\{ -\frac{N_0}{\Omega_{SD} I_p} x \gamma \right\}, \quad \begin{matrix} \gamma > 0 \\ x > 0 \end{matrix}. \tag{18.78}$$

As a consequence, the lower-bound for the CDF of γ_D conditioned on $|h_{SP}|^2$ is given by

$$\begin{aligned} F^{LB}_{\gamma_D || h_{SP} |^2} \left(\gamma | |h_{SP}|^2 \right) &= F_{\gamma_{DT} || h_{SP} |^2} \left(\gamma | |h_{SP}|^2 \right) \cdot F_{\gamma'_{AF} || h_{SP} |^2} \left(\gamma | |h_{SP}|^2 \right) \\ &= 1 - \exp \left\{ -\frac{N_0}{\Omega_{SD} I_p} x \gamma \right\} - \exp \left\{ -\frac{N_0}{\Omega_{SR} I_p} x \gamma \right\} \frac{\Omega_{RD} I_p}{N_0 \Omega_{RP} \gamma + \Omega_{RD} I_p} \\ &\quad + \exp \left\{ -\left(\frac{1}{\Omega_{SD}} + \frac{1}{\Omega_{SR}} \right) \frac{N_0}{I_p} x \gamma \right\} \frac{\Omega_{RD} I_p}{N_0 \Omega_{RP} \gamma + \Omega_{RD} I_p}, \quad \begin{matrix} \gamma > 0 \\ x > 0 \end{matrix}. \end{aligned} \tag{18.79}$$

Applying the total probability theorem to (17.9), the CDF of γ_D can be expressed as

$$F_{\gamma_D}^{LB}(\gamma) = \int_0^\infty F_{\gamma_D||h_{SP}|^2}^{LB}\left(\gamma||h_{SP}|^2\right) f_{|h_{SP}|^2}(x)\,dx$$

$$= 1 - \frac{1}{\lambda_{SD}\gamma+1} - \frac{1}{(\lambda_{SR}\gamma+1)(\lambda_{RD}\gamma+1)} + \frac{1}{(\lambda_{RD}\gamma+1)(\lambda_{SD}\gamma+\lambda_{SR}\gamma+1)}, \; \gamma>0, \; (18.80)$$

where $\lambda_{SD} = \frac{N_0\Omega_{SP}}{I_p\Omega_{SD}}$, $\lambda_{SR} = \frac{N_0\Omega_{SP}}{I_p\Omega_{SR}}$ and $\lambda_{RD} = \frac{N_0\Omega_{SP}}{I_p\Omega_{RD}}$, The outage event occurs when the end-to-end SNR falls below a predetermined threshold γ_{th}. As a result, replacing γ with γ_{th} in (17.10), the lower-bound of the outage probability for the distributed link selection scheme can be obtained as $P_{out}^{LB} = F_{\gamma_D}^{LB}(\gamma_{th})$.

Furthermore, when the instantaneous interference channel information is available at cognitive transmitters, the centralized scheme always needs to feedback γ_{RD} from D to S, whereas the proposed distributed scheme merely requires the success/fail decision feedback[1]. The worst case occurs when S cannot make a link selection based on its local CSI γ_{DT} and γ_{SR}, yielding thus, 2-bit feedback overhead, where 1-bit is consumed to inform D that $\gamma_{DT} < \gamma_{SR}$, and the other 1-bit is used to feedback the local decision from D to S. On the other hand, the best scenario occurs when $\gamma_{DT} \geq \gamma_{SR}$. In this case, no signaling feedback is required for the whole distributed decision process.

To summarize, the average signaling overhead of the distributed selection scheme can be calculated as

$$
\begin{aligned}
\Phi &= 2 \times \Pr(\gamma_{DT} < \gamma_{SR}) + 0 \times \Pr(\gamma_{DT} \geq \gamma_{SR}) \\
&= 2 \times \int_0^\infty \int_0^\infty f_{\gamma_{SR}||h_{SP}|^2}(\gamma|X) F_{\gamma_{DT}||h_{SP}|^2}(\gamma|X) f_{|h_{SP}|^2}(x)\,d\gamma dx \\
&= 2 \times \left(1 - \frac{\Omega_{SD}}{\Omega_{SD}+\Omega_{SR}}\right).
\end{aligned}
\tag{18.81}
$$

It is obvious that the signaling overhead of the distributed protocol reduces with a decrease in Ω_{SR} (i.e., an increase in distance between S and R) and amounts to 2-bit at most, as shown by our subsequent numerical results.

18.5.1   Numerical Examples and Discussions

In this section, we consider a linear network topology and model the average channel power gains of the link $A \to B$ as $\Omega_{AB} = l_{AB}^{-\partial}$. Herein, l_{AB} represents the distance between $A \in \{S,R\}$ and $B \in \{P,R,D\}$, and the pathloss exponent is set to $\partial = 4$. For illustration purposes and without loss of generality, we assume $\gamma_{th} = 1dB$ in the following discussion.

Figure 17.12 compares the outage probability of the distributed and the centralized link selection schemes, where instantaneous CSI knowledge of the interference channels is taken into account. We consider a co-linear network topology, where the coordinates of S, D, and P are, respectively, set to $(0,0)$, $(1,0)$, and $(0.55,0.55)$, and R is located at $(0.3,0) / (0.5,0)$. For the purpose of comparison, the outage probability of the centralized scheme with 4-bit quantization feedback of γ_{RD} is also plotted. It is noticed that for both cases of l_{SR}=0.3 and l_{SR}=0.5, the distributed and the centralized schemes achieve almost the same outage performance. However, as shown by the Monte Carlo simulation curves in Figure 17.12, 4-bit quantization feedback of γ_{RD} is required for the centralized scheme to achieve comparable outage performance, whereas our proposal has the need of no more than 2-bit feedback overhead for the overall link selection process. A detailed explanation for the advantage of the proposed scheme is as follows. When the link selection is implemented in a centralized manner, D always has to feedback the CSI γ_{RD} to S to perform the link selection at S. This incurs at least 4-bit

---

[1] As mentioned in [18], 1-bit binary symbol "1" or "0" is sufficient to denote the "success" or "fail" message.

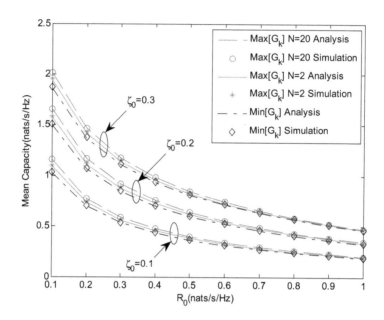

Figure 18.12: Comparisons of the outage probability between the distributed scheme and the centralized scheme 1.

quantization feedback overhead, as manifested by our numerical results. In contrast, our proposed scheme merely requires the feedback of local decisions ("success/fail") rather than the feedback of γ_{RD}, leading, thus, to 2-bit overhead even in the worst case. On the other hand, relying on the channel conditions, the best case appears when S can make link selection based on its local CSI γ_{DT} and γ_{SR}. In this case, the direct link will be chosen without the need of any signaling feedback. The above discussions are validated by the numerical plots in Figure 17.12.

Figure 17.13 (a) and (b) draw the average feedback overhead of both distributed and centralized schemes vs. the distance between S and R, in which the exact analytical results of the distributed schemes are obtained from (17.11). Herein, we consider a linear network topology and assume $l_{SD}=1$. By employing the same quantization rule as [18] and [19], our Monte-Carlo simulations show that at least 4-bit quantization feedback of γ_{RD} is required for the centralized scheme. From Figure 17.13 (a), it is shown that when $0 < l_{SR} < 1$, the average feedback overhead of the distributed scheme decreases with an increase in l_{SR}. The foregoing phenomenon can be explained as follows. Placing R closer to D can improve the average channel quality of the $R \rightarrow D$ link while deteriorating that of the $S \rightarrow R$ link. For the distributed scheme, a lower quality of the $S \rightarrow R$ link makes the event $\{\gamma_{DT} \geq \gamma_{SR}\}$ happens with a high probability, which by its turn leads to a lower feedback overhead. On the other hand, when $l_{SR} < 1$, the average signaling overhead of distributed scheme decreases with an increase in l_{SR}, as shown in Figure 17.13(b). In particular, on average, no more than 1-bit feedback overhead is adequate for our proposed scheme to make an efficient link selection.

In summary, since the proposed distributed link selection rule can achieve excellent system outage performance at the expense of extremely low signaling overhead, it seems to be an attractive solution for practical cognitive relaying scenarios.

18.5.1.1 Conclusion

In this section, we presented a distributed link selection scheme for a cognitive AF relaying network. Then, assuming perfect CSI knowledge of the interference links at the secondary nodes, we derived

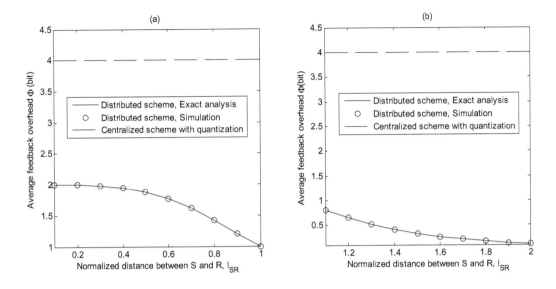

Figure 18.13: Average feedback overhead of different schemes vs. the distance between S and R. Herein, (a) shows the case of lSR = lSD - lRD, lSD=1, while (b) describes the case of lSR = lSD + lRD, lSD=1.

closed-form expressions for the outage lower-bound, as well as the average feedback overhead of the proposed scheme. In addition, numerical results manifested that the proposed scheme is more efficient than previous proposals to perform link decision.

References

[1] D. Li, "Performance analysis of uplink cognitive cellular networks with opportunistic scheduling," *IEEE Commun. Lett.,* vol. 14, no. 9, pp. 827–829, Sep. 2010.

[2] A. Papoulis, *Probability, Random Variables, and Stochastic Processes*, 4th edition. McGraw-Hill, 2002.

[3] I. S. Gradshteyn and I. M. Ryzhik, *Table of Integrals, Series, and Products*, 7th edition. San Diego, CA: Academic, 2007.

[4] L. Fan, X. Lei, A Comment on "Performance analysis of uplink cognitive cellular networks with opportunistic scheduling", *IEEE Commun. Lett.,* vol. 14, no. 4, pp. 361–361, Apr. 2010

[5] M. Abramowitz and I. A. Stegun, *Handbook of Mathematical Functions with Formulas ,Graphs, and Mathematical Tables*, 10th edition. Dover, 1972.

[6] H. Ding, J. Ge, D. B. da Costa and Z. Jiang, "Energy-efficient and low-complexity schemes for uplink cognitive cellular networks," *IEEE Commun. Lett.,* vol. 14, no.12, pp. 1101–1103, Dec. 2010.

[7] A. Bletsas, H. Shin, and M. Z. Win, "Cooperative communications with outage-optimal opportunistic relaying," *IEEE Trans. Wireless Commun.,* vol. 6, no. 9, pp. 3450–3460, Sep. 2007.

[8] D. B. da Costa and S. Aïssa, "Capacity analysis of cooperative systems with relay selection in Nakagami-*m* fading," *IEEE Commun. Lett.,* vol. 13, no. 9, pp. 637–639, Sep. 2009.

[9] I. Krikidis, J. Thompson, S. McLaughlin, and N. Goertz, "Amplify-and-forward with partial relay selection," *IEEE Commun. Lett.,* vol. 12, no. 4, pp. 235–237, Apr. 2008.

[10] J. N. Al-Karaki and A. E. Kamal, "Routing techniques in wireless sensor networks: a survey," *IEEE Wireless Commun.,* vol. 11, no. 6, pp. 6-28, Dec. 2004.

[11] J. M. Peha, "Approaches to spectrum sharing," *IEEE Commun. Mag.,* vol. 43, no. 2, pp. 10-12, Feb. 2005.

[12] T. W. Ban, W. Choi, B. C. Jung, and D. K. Sung, "Multi-user diversity in a spectrum sharing system," *IEEE Trans. Wireless Commun.,* vol. 8, no. 1, pp. 102–106, Jan. 2009.

[13] Y. Zhao, R. Adve, and T. J. Lim, "Symbol error rate of selection amplify-and-forward relay systems," *IEEE Commun. Lett.,* vol. 10, no. 11, pp. 757–759, Nov. 2006.

[14] Y. Zhao, R. Adve, and T. J. Lim, "Improving amplify-and-forward relay networks: optimal power allocation versus selection," *IEEE Trans. Wireless Commun.*, vol. 6, no. 8, pp. 3114–3123, Aug. 2007.

[15] F. Xu, F. C. M. Lau, Q. F. Zhou, and D.-W. Yue, "Outage performance of cooperative communication systems using opportunistic relaying and selection combining receiver," *IEEE Signal Process. Lett.*, vol. 16, no. 4, pp. 237–240, Apr. 2009.

[16] P. A. Anghel and M. Kaveh, "Exact symbol error probability of a cooperative network in a Rayleigh-fading environment," *IEEE Trans. Wireless Commun.*, vol. 3, no. 5, Sep. 2004.

[17] T. Q. Duong, V. N. Q. Bao, G. C. Alexandropoulos, and H. -J. Zepernick, "Cooperative spectrum sharing networks with AF relay and selection diversity," *Electron. Lett.*, vol. 47, no.20, pp. 1149–1151,Sep. 2011.

[18] H. Ding, J. Ge, D. B. da Costa, and Z. Jiang, "Link selection schemes for selection relaying systems with transmit beamforming: New and efficient proposals from a distributed concept," *IEEE Trans. Veh. Technol.*, vol.61, no. 2, pp. 533–552, Feb. 2012.

[19] J. Jiang, R. M. Buehrer, and W. H. Tranter, "Antenna diversity in multiuser data networks," *IEEE Trans. Commun.*, vol. 52, no. 3, pp. 490–497, Mar. 2004.

MIMO-ORIENTED DESIGN

Chapter 19

Capacity Scaling of MIMO Broadcast Channels with Finite-rate Feedback

Ali Tajer and Xiaodong Wang

CONTENTS

Abstract

Multiple-antenna downlink transmission offers significant capacity improvement when the transmit-side channel state information (CSI) is available. The sum-rate capacity with *infinite*-rate feedback (full or partial CSI) scales linearly with the number of transmit antennas (multiplexing gain) and double logarithmically with the number of users (multiuser diversity gain). This chapter presents a new scheduling scheme that requires only *finite*-rate feedback and yet retains the optimal multiplexing and multiuser diversity gains achievable by dirty-paper coding and show that its sum-rate throughput scales similar to that achievable by dirty-paper coding. While the proposed scheduling schemes are asymptotically optimal, they also exhibit a good performance for practical network sizes.

19.1 Introduction

Multiple-antenna communication offers considerable improvement in spectral efficiency and link reliability. The capacity of a single-user multiple-input multiple-output (MIMO) channels increases linearly with the minimum number of transmit/receive antennas *with* or *without* channel state information (CSI) being known to the transmitter or the receiver [1–5]. More recent results show that deploying N_t transmit antennas in a multi-user downlink system allows for simultaneous transmissions to multiple users, resulting in an N_t-fold increase in the channel capacity (multiplexing gain) [6–11]. This promising result for MIMO broadcast channels, however, relies on the availability of perfect CSI at the transmitter (CSIT). The need for perfect CSIT is more highlighted when considering that without CSIT, the sum-rate capacity grows only logarithmically with N_t [12].

Moreover, it has been shown that for networks with large number of users $K \gg N_t$, by opportunistic scheduling, the sum-rate capacity exhibits a double-logarithmic growth in K, which reflects the inherent multiuser diversity of the network [12–14]. Specifically, by using random beamforming with partial CSIT [12], or dirty-paper coding with perfect CSIT [13], the sum-rate capacity scales as $K \log \log K N_r$, where N_r is the number of receive antennas for each user[1]. Based on dirty-paper coding and beamforming, several other schemes that achieve the optimal capacity scaling have been developed [15].

All the aforementioned multiplexing and multiuser diversity gains of different schemes, however, rely on the availability of *infinite*-rate feedback, which is difficult to sustain in practice. A few lines of works have been extended to address the effect of channel quantization and *finite*-rate feedback on the capacity of MIMO broadcast channels in networks with limited number of users [16, 17] and large number of users [18–25], which we briefly go over next.

19.1.1 Related Work

Throughput degradation caused by limited-rate feedback for networks where the number of users is equal to the number of transmit antennas is considered in [16]. For zero-forcing beamforming, it has been shown that in order to achieve full multiplexing gain, the number of feedback bits per user must linearly increase with the signal-to-noise ratio (SNR) and the sum-rate capacity is shown to scale as $N_t \log \text{SNR}$ in the large SNR regime [16]. Sum-rate capacity bounds for zero-forcing dirty-paper coding with quantized channel vector feedback and normalized channel vector feedback are provided in [17], showing that both feedback schemes achieve the same bound at the asymptote of high SNRs.

For networks with large number of users, studies in [19, 20] consider feeding back *unquantized* channel magnitudes, which, in theory, requires infinite-rate feedback, and quantized channel direction vectors. In [19] a scheme based on zero-forcing beamforming precoder and semi-orthogonal user selection [26] is developed (ZFBF-SUS). The analyses reveal an interesting interplay between the required number of feedback bits (for quantizing channel direction vector), the number of users,

---

[1]It is noteworthy that the content of feedback required for random beamforming consists of one real number and $\log M$ bits per mobile user, as opposed to $2M$ real numbers for dirty-paper coding.

and SNR. It is also shown that when the number of feedback bits tends to infinity, the optimal scaling for the sum-rate throughput is achievable. In [20], a joint beamforming and user selection protocol is proposed (OSDMA-TF). It considers orthogonal beamforming precoders and designs a feedback mechanism constrained by the aggregate amount of feedback. This schemes is shown to achieve the optimal sum-rate capacity scaling if the aggregate feedback ideally approaches infinity.

Orthogonal random beamforming (ORB) and opportunistic user selection with partial CSI feedback, which consists of one scalar value signal to interference and noise ratio (SINR) per user, has been proposed in [12]. Also, feedback of perfect SINR, in conjunction with quantized channel direction vector, is considered in [27] (PU2RC). Both schemes are shown to achieve the optimal capacity scaling, where perfect SINR is available at the transmitter. A modification of ORB with opportunistic user selection has been presented in [21], where formulations followed by empirical results show the merits of the method even for networks with small number of users (OSDMA-BS).

The effect of finite-rate feedback on the capacity of single input and single output (SISO) and multiple input and single output (MISO) multiuser channels is also looked into in [22,23], where it is proposed to opportunistically select one user at a time for transmission. Although these schemes can achieve multiuser diversity gain, since they do not allow for simultaneous transmissions to multiple users, they cannot capture the multiplexing gain and, hence, their capacity scales as $\log \log K$.

Most recently, Diaz *et al.* [18] have proposed a scheduling scheme with limited feedback and have shown that, interestingly, it asymptotically achieves the optimal scaling. In [18] a *homogeneous* network, where all users are *equally* distant from the base station, is considered. The users are arranged into N_t subgroups, and the users within each subgroup only measure the SNR of one beam assigned to them *a priori*. Then each user feeds back one information bit to the base station, indicating whether its measured SNR is above a certain threshold set by the base station. In the case that for a specific beam there are more than one eligible users, one of them is selected randomly, and if there is not an eligible user, one user is picked at random. The amount of feedback per user is only one bit, and the aggregate amount of feedback for the scheme is also shown to scale like $\log K$. While the achievable sum-rate throughput of this scheduling algorithm is asymptotically optimal, for practical network sizes it has a considerable gap with the throughput achievable by dirty-paper coding.

19.1.2 Contributions

In this chapter, we propose a scheduling algorithm based on ORB [12, 14] for *heterogeneous* networks. It considers a general geographical distribution of the users, as they are randomly distributed and experience different path losses. In this scheme, N_t randomly generated beams are used to transmit to a set of users, which are selected based on the information conveyed by finite-rate feedback discussed in details in Section 19.3. According to our proposed feedback mechanism, the users are assigned quantization thresholds, which are essentially distinct and vary by the distance of the users from the base station. Each user quantizes the SINR of its most favorable beam using its designated quantization threshold. Our user selection and scheduling scheme, unlike [19, 20, 22], does not require any information about channel direction vectors, which will significantly alleviate the burden of feedback. Also as opposed to [12, 19, 21] for asymptotic optimality, we do not require either the channel gain or SINR be perfectly known to the base station.

A note is also warranted regarding the difference between our 1-bit quantization and that of [18]. In [18], the users are grouped into smaller subgroups, each pre-assigned to a beam. This means that a user pre-assigned to a specific beam can only be scheduled along that beam. Since such pre-assignment ignores the instantaneous channel states, it might so happen that the assignments are suboptimal, and some perturbation in the assignment would enhance the communication quality. This translates to a throughput loss specially for the networks with practical sizes.

We show that for a multiple-antenna downlink transmission with N_t transmit antennas and K users each equipped with N_r receive antennas, our 1-bit quantization of the best SINR of each user is sufficient to achieve the optimal sum-rate capacity scaling, i.e., $N_t \log\log KN_r$. Therefore, as far as achieving the optimal capacity scaling is concerned, SINR quantization with higher resolutions, while desirable, is not necessary. Also, we demonstrate that this scheme has a good achievable sum-rate throughput for networks with limited number of users.

The remainder of the chapter is organized as follows. Section 19.2 describes the system. The proposed scheduling algorithm is discussed in Section 19.3. Section 22.4 provides analysis on the throughput scaling. Section 19.5 provides simulation results, and Section 22.6 concludes the chapter.

19.2   System Model

We consider a multiple-antenna downlink transmission with N_t transmit antennas and K users, each equipped with N_r receive antennas. Let $\boldsymbol{x}(t) \in \mathbb{C}^{N_t \times 1}$ be the transmitted signal vector, and $\boldsymbol{H}_k \in \mathbb{C}^{N_r \times N_t}$ be the channel matrix of the k-th receiver, where $H_k(i,j)$ represents the channel gain from transmit antenna j to receive antenna i. The additive white Gaussian noise at the k-th receiver is denoted by $\boldsymbol{z}_k \in \mathbb{C}^{N_r \times 1}$, where $\boldsymbol{z}_k \sim \mathcal{CN}(\boldsymbol{0}, \boldsymbol{I}_{N_r})$. Also, let $\boldsymbol{y}_k(t) \in \mathbb{C}^{N_r \times 1}$ be the received signal vector at the k-th user. Therefore,

$$\boldsymbol{y}_k(t) = \sqrt{\gamma_k}\boldsymbol{H}_k\boldsymbol{x}(t) + \boldsymbol{z}_k(t), \quad k = 1, \ldots, K. \tag{19.1}$$

The transmitter satisfies an average power constraint P, i.e., $\mathbb{E}[x^\dagger x] \le P$. In non-homogenous networks, as different users undergo different path-loss and shadowing, we include the terms γ_k to account for these effects. We consider block-fading channels with independent fading and assume the entries of $\boldsymbol{H}_k$ are distributed as iid complex Gaussian $\mathcal{CN}(0,1)$.

Considering no information feedback about channel direction vectors motivates exploiting orthogonal random beamforming, as also used in [12, 14]. For this purpose, at the beginning of each block transmission, N_t orthonormal vectors $\{\boldsymbol{\phi}_m(t)\}_{m=1}^{N_t}$, where $\boldsymbol{\phi}_m(t) \in \mathbb{C}^{N_t \times 1}$, are generated randomly according to an isotropic distribution [4]. Then at each time instant, the m-th information stream, denoted by $x_m(t)$, is multiplied by $\boldsymbol{\phi}_m(t)$, and the transmitted signal is constructed as

$$\boldsymbol{x}(t) = \sum_{m=1}^{N_t} x_m(t)\boldsymbol{\phi}_m(t). \tag{19.2}$$

The information streams $\{x_m(t)\}_{m=1}^{N_t}$ are assumed to be statistically independent. Also, we consider equal average transmission power per antenna, i.e., $\mathbb{E}[|x_m(t)|^2] = \frac{P}{N_t}$.

We assume that the k-th user knows $\boldsymbol{H}_k\boldsymbol{\phi}_m(t)$, $m = 1, \ldots, N_t$, perfectly and instantaneously. We denote the transmission SNR per information stream by $\rho \triangleq \frac{P}{N_t}$. Also, throughout the chapter, we use the notation $a_K \doteq b_K$ to denote that $\lim_{K \to \infty} \frac{a_K}{b_K} = 1$. Operators $\overset{.}{\le}$ and $\overset{.}{\ge}$ are defined accordingly. All the logarithms, unless otherwise mentioned, are in base 2, and throughput are in bits/sec/Hz.

19.3   Scheduling Algorithm ($N_r = 1$)

Each user is capable of decoding any of the N_t information streams by treating others as noise. Therefore, the m-th information stream can be decoded by the k-th user with SINR:

$$
\begin{aligned}
\mathrm{SINR}_{k,m} &\triangleq \frac{\gamma_k \mathbb{E}\left[|\boldsymbol{H}_k \boldsymbol{\phi}_m x_m|^2\right]}{\mathbb{E}\left[|z_k + \sqrt{\gamma_k}\sum_{l\neq m}\boldsymbol{H}_k\boldsymbol{\phi}_l x_l|^2\right]} \\
&= \frac{|\boldsymbol{H}_k\boldsymbol{\phi}_m|^2}{\frac{1}{\rho\gamma_k} + \sum_{l\neq m}|\boldsymbol{H}_k\boldsymbol{\phi}_l|^2}, \qquad k=1,\dots,K, \ m=1,\dots,N_t. \quad (19.3)
\end{aligned}
$$

We define the most favorable beam of the k-th user as

$$
m_k^* \triangleq \arg\max_{1\le m\le N_t} \mathrm{SINR}_{k,m}. \qquad (19.4)
$$

Now, we consider one-bit quantization of SINR_{k,m_k^*}. For this purpose, as discussed in details in Section 22.4, for each user k, a threshold level $\beta(\gamma_k, K) \ge 1$, which is a function of the number of users K, as well as γ_k, is set by the base station. Then, each user k compares SINR_{k,m_k^*} with this threshold. If $\mathrm{SINR}_{k,m_k^*} \ge \beta(\gamma_k, K)$, user k feeds back the index of its most favorable beam, m_k^*, which requires $\log N_t$ information bits. Such feedback also implicitly declares that the user has satisfied the threshold constraint. On the other hand, if $\mathrm{SINR}_{k,m_k^*} < \beta(\gamma_k, K)$, user k refrains from feeding back any information and will not be considered by the base station as a candidate for scheduling.

Corresponding to each beam m, we define a set consisting of the indices of the users who have fed back m, i.e.,

$$
\mathcal{B}_m \triangleq \left\{ k \,\bigg|\, m = \arg\max_{1\le l\le N_t} \mathrm{SINR}_{k,l} \text{ and } \mathrm{SINR}_{k,m} \ge \beta(\gamma_k, K), \ k=1,\dots,K \right\}. \qquad (19.5)
$$

Next, the base station randomly selects one user from each set $\mathcal{B}_m$ for $m=1,\dots,N_t$ and schedules this user to receive the information stream x_m. Note that the sets $\mathcal{B}_m$ are mutually exclusive, and, as a result, it is guaranteed that no user will be receiving information from than one beam. Although finding no eligible user for the m-th beam ($|\mathcal{B}_m| = 0$) implies a loss in the sum-rate, as we will show, this loss will be vanishing by increasing the number of users.

Corresponding to each beam m, we also define the set

$$
\mathcal{A}_m \triangleq \left\{ k \,\big|\, \mathrm{SINR}_{k,m} \ge \beta(\gamma_k, K), \ k=1,\dots,K \right\}, \qquad (19.6)
$$

which contains the indices of *all* the users for which the SINR of decoding x_m is above the threshold. Although in general $\mathcal{B}_m \subset \mathcal{A}_m$, for certain choices of $\beta(\gamma_k, K)$, we have certain relationship between $\mathcal{B}_m$ and $\mathcal{A}_m$.

Lemma 19.1

For any choice of $\beta(\gamma_k, K) > 1$, we have $\mathcal{B}_m = \mathcal{A}_m$ for all $m=1,\dots,N_t$.

Proof 19.1   See Appendix 19.7.   ■

19.4 Throughput Scaling Analysis

The capacity of a MIMO broadcast channel with full CSI feedback is achieved by dirty-paper coding and is given by [6, 8, 9]

$$R^{\text{DP}} = \max_{\text{tr}(\Sigma_x)=P} \log \det \left(1 + \sum_{k=1}^{K} H_k^\dagger P_k H_k \right),$$

where P_i is the transmit power at the i-th transmit antenna. This capacity for fixed P and N_t scales as [12, 13]

$$\mathbb{E}_H[R^{\text{DP}}] \doteq N_t \log\log K N_r. \tag{19.7}$$

19.4.1 Throughput Scaling for $N_r = 1$

First, we consider the case of single-antenna receivers ($N_r = 1$) and analyze how the sum-rate throughput scales with the number of users, K. The aggregate throughput, denoted by R^Q, is the summation of the data-rates that individual beams support. Conditioned on any given set of users indices $\{\mathcal{B}_m\}_{m=1}^{N_t}$, that the base station identifies after receiving the feedback information from the users, for the sum-rate throughput we have

$$\mathbb{E}_H[R^Q \mid \{\mathcal{B}_m\}] = \sum_{i=1}^{N_t} \mathbb{E}_H[R_i^Q \mid \{\mathcal{B}_m\}], \tag{19.8}$$

where R_i^Q is the throughput of the i-th beam. Furthermore, the rate R_i^Q for each beam depends on the number of users expressed interest in the i-th beam by feeding back its index. According to our scheduling scheme, the user scheduled to receive the information stream x_i is randomly picked from the set $\mathcal{B}_i$. Therefore, conditioned on a given set $\mathcal{B}_m$, we have

$$\mathbb{E}_H[R_i^Q \mid \{\mathcal{B}_m\}] = \frac{1}{|\mathcal{B}_i|} \sum_{j \in \mathcal{B}_i} \mathbb{E}_H\left[\log\left(1 + \text{SINR}_{j,i}\right) \,\Big|\, \mathcal{B}_i \right]. \tag{19.9}$$

For any choice of the threshold level $\beta(\gamma_k, K) > 1$ from Lemma 19.1, we have $\mathcal{A}_m = \mathcal{B}_m$. As a result, by invoking (19.8) and (19.9), we find that

$$\mathbb{E}_H[R^Q] = \sum_{\{\mathcal{A}_m\}_m} P(\{\mathcal{A}_m\}_m) \sum_{i=1}^{N_t} \frac{1}{|\mathcal{A}_i|} \sum_{j \in \mathcal{A}_i} \mathbb{E}_H\left[\log\left(1 + \text{SINR}_{j,i}\right) \,\Big|\, \mathcal{A}_i \right]. \tag{19.10}$$

By utilizing the result of the following lemma, we can simplify the expansion of $\mathbb{E}_H[R^Q]$ in (19.10).

Lemma 19.2

For a continuous random variable X, increasing function $g(\cdot)$ and real values $b \geq a$

$$\mathbb{E}\left[g(X) \mid X \geq b\right] \geq \mathbb{E}\left[g(X) \mid X \geq a\right].$$

Proof 19.2   See Appendix 19.8.   ■

For each $m = 1, \ldots, N_t$, let us define $\text{SINR}_m^{(j)}$ as the j-th largest element of the set $\{\text{SINR}_{k,m}\}_{k=1}^{K}$.

Therefore, for the inner summation in (19.10), we get

$$\sum_{j\in\mathcal{A}_i}\mathbb{E}_{\boldsymbol{H}}\left[\log\left(1+\mathrm{SINR}_{j,i}\right)\,\Big|\,\mathcal{A}_i\right]$$

$$=\sum_{j\in\mathcal{A}_i}\mathbb{E}_{\boldsymbol{H}}\left[\log\left(1+\mathrm{SINR}_{j,i}\right)\,\Big|\,\mathrm{SINR}_{j,i}\geq\beta(\gamma_j,K);\forall l\notin\mathcal{A}_i:\mathrm{SINR}_{l,i}<\beta(\gamma_l,K)\right]$$

$$\geq\sum_{j\in\mathcal{A}_i}\mathbb{E}_{\boldsymbol{H}}\left[\log\left(1+\mathrm{SINR}_{j,i}\right)\,\Big|\,\mathrm{SINR}_{j,i}\geq\min_{l}\beta(\gamma_l,K);\forall l\notin\mathcal{A}_i:\mathrm{SINR}_{l,i}<\beta(\gamma_l,K)\right]\quad(19.11)$$

$$=\sum_{j\in\mathcal{A}_i}\mathbb{E}_{\boldsymbol{H}}\left[\log\left(1+\mathrm{SINR}_{j,i}\right)\,\Big|\,\mathrm{SINR}_{j,i}\geq\min_{l}\beta(\gamma_l,K);\forall l\notin\mathcal{A}_i:\mathrm{SINR}_{l,i}<\min_{l}\beta(\gamma_l,K)\right]$$

$$(19.12)$$

$$=\sum_{\ell=1}^{|\mathcal{A}_i|}\mathbb{E}_{\boldsymbol{H}}\left[\log\left(1+\mathrm{SINR}_i^{(\ell)}\right)\,\Big|\,\mathrm{SINR}_m^{(\ell)}\geq\min_{j}\beta(\gamma_j,K)\right]\quad(19.13)$$

$$\geq\sum_{\ell=1}^{|\mathcal{A}_i|}\mathbb{E}_{\boldsymbol{H}}\left[\log\left(1+\mathrm{SINR}_i^{(\ell)}\right)\right].\quad(19.14)$$

Inequality in (19.11) holds according to Lemma 19.2. Transition from (19.11) to (19.12) is justified by taking into account that SINR of different users are statistically independent, which is due to statistical independence of channel matrices $\boldsymbol{H}_k$. Therefore, changing the condition enforced on $\mathrm{SINR}_{l,m}$ does not affect the statistics of $\mathrm{SINR}_{i,m}$. Equation (19.12) implies that $\mathcal{A}_m$ contains the $|\mathcal{A}_m|$ largest elements of the set $\{\mathrm{SINR}_{k,m}\}_{k=1}^{K}$, which is mathematically expressed in (19.13). The last step is driven by again using Lemma 19.2. Therefore, from (19.10)–(19.14) we get

$$\mathbb{E}_{\boldsymbol{H}}[R^{\mathrm{Q}}]\geq\sum_{\{\mathcal{A}_m\}_m}P(\{\mathcal{A}_m\}_m)\sum_{i=1}^{N_t}\frac{1}{|\mathcal{A}_i|}\sum_{\ell=1}^{|\mathcal{A}_i|}\mathbb{E}_{\boldsymbol{H}}\left[\log\left(1+\mathrm{SINR}_i^{(\ell)}\right)\right]$$

$$=\sum_{\sum K_i=K}P(|\mathcal{A}_i|=K_i,\,\forall i)\sum_{i=1}^{N_t}\frac{1}{K_i}\sum_{\ell=1}^{K_i}\mathbb{E}_{\boldsymbol{H}}\left[\log\left(1+\mathrm{SINR}_i^{(\ell)}\right)\right]$$

$$=\sum_{i=1}^{N_t}\sum_{K_i=1}^{K}P(|\mathcal{A}_i|=K_i)\frac{1}{K_i}\sum_{\ell=1}^{K_i}\mathbb{E}_{\boldsymbol{H}}\left[\log\left(1+\mathrm{SINR}_i^{(\ell)}\right)\right]$$

$$\times\underbrace{\sum_{\sum j\neq i K_j=K-K_i}P\left(|\mathcal{A}_j|=K_j\,\forall j\neq i\,\big|\,|\mathcal{A}_i|=K_i\right)}_{=1}$$

$$=\sum_{i=1}^{N_t}\sum_{\ell=1}^{K}\mathbb{E}_{\boldsymbol{H}}\left[\log\left(1+\mathrm{SINR}_i^{(\ell)}\right)\right]\underbrace{\sum_{K_i=\ell}^{K}\frac{P(|\mathcal{A}_i|=K_i)}{K_i}}_{\overset{\triangle}{=}q_\ell^i}\quad(19.15)$$

$$=\sum_{i=1}^{N_t}\sum_{\ell=1}^{K}q_\ell^i\,\mathbb{E}_{\boldsymbol{H}}\left[\log\left(1+\mathrm{SINR}_i^{(\ell)}\right)\right]\overset{\triangle}{=}\bar{R}_L^{\mathrm{Q}}.\quad(19.16)$$

For all $i=1,\ldots,N_t$, we also set $q_0^i\overset{\triangle}{=}P(|\mathcal{A}_i|=0)$. Note that for any i, the sequence $\{q_\ell^i\}_{\ell=1}^{K}$ is a valid probability mass function (PMF) as

$$\sum_{\ell=0}^{K} q_{\ell}^{i} = q_{0}^{i} + \sum_{\ell=1}^{K}\sum_{k=\ell}^{K}\left[\frac{1}{k}P\Big(|\mathcal{A}_i| = k\Big)\right] = q_{0}^{i} + \sum_{k=1}^{K}\frac{1}{k}\sum_{\ell=1}^{k}P(|\mathcal{A}_i| = k) = \sum_{k=0}^{K}P\Big(|\mathcal{A}_i| = k\Big) = 1.$$

Now, we concentrate on analyzing how $\bar{R}_L^{Q}$ scales, which provides a lower-bound on the throughput scaling law of our protocol. It can be readily verified that for $m = 1,\ldots,N_t$, $\{\text{SINR}_{k,m}\}$ are independently and identically distributed with the cumulative distribution function (CDF) [12]

$$F_k(x) = 1 - \frac{e^{-x/\rho\gamma_k}}{(x+1)^{N_t-1}}. \tag{19.17}$$

Note that because of the statistical independence of the channel matrices $\boldsymbol{H}_k$, the SINRs of different users are independent; but, due to experiencing different path-loss and shadowing (γ_k), they do not have identical distributions. Since the elements of $\{\text{SINR}_{k,m}\}_{k=1}^{K}$ are not iid, for more mathematical tractability, we construct two other sets, which consist of lower-bounds and upper-bounds on the elements of $\{\text{SINR}_{k,m}\}_{k=1}^{n}$. Specifically, we define $\gamma_{\min} = \min_k \gamma_k$ and $\gamma_{\max} = \max_k \gamma_k$ and

$$\mathcal{S}_L(m) \triangleq \left\{ \mathcal{S}_L(k,m) \,\Big|\, \mathcal{S}_L(k,m) \triangleq \frac{|\boldsymbol{H}_k\boldsymbol{\phi}_m|^2}{1/\rho\gamma_{\min} + \sum_{l\neq m}|\boldsymbol{H}_k\boldsymbol{\phi}_l|^2},\; k = 1,\ldots,K \right\}, \tag{19.18}$$

$$\text{and}\quad \mathcal{S}_U(m) \triangleq \left\{ \mathcal{S}_U(k,m) \,\Big|\, \mathcal{S}_U(k,m) \triangleq \frac{|\boldsymbol{H}_k\boldsymbol{\phi}_m|^2}{1/\rho\gamma_{\max} + \sum_{l\neq m}|\boldsymbol{H}_k\boldsymbol{\phi}_l|^2},\; k = 1,\ldots,K \right\}, \tag{19.19}$$

where it can be easily shown that $\mathcal{S}_L(k,m) \leq \text{SINR}_{k,m} \leq \mathcal{S}_U(k,m)$. Moreover, the elements in $\mathcal{S}_L(m)$ are iid with CDF

$$F_{\min}(x) = 1 - \frac{e^{-x/\rho\gamma_{\min}}}{(x+1)^{N_t-1}}, \tag{19.20}$$

and those in $\mathcal{S}_U(m)$ are also iid with CDF

$$F_{\max}(x) = 1 - \frac{e^{-x/\rho\gamma_{\max}}}{(x+1)^{N_t-1}}, \tag{19.21}$$

Next, we provide the following lemma, which is instrumental for further finding lower-bounds and upper-bounds on $\bar{R}_L^{Q}$ given in (19.16). We denote the j-th largest elements of $\mathcal{S}_L(m)$ and $\mathcal{S}_U(m)$ by $\mathcal{S}_L^{(j)}(m)$ and $\mathcal{S}_U^{(j)}(m)$, respectively.

Lemma 19.3
For any beam $m = 1,\ldots,N_t$ and any $\ell = 1,\ldots,K$ we have

$$\mathcal{S}_L^{(\ell)}(m) \leq \text{SINR}_m^{(\ell)} \leq \mathcal{S}_U^{(\ell)}(m).$$

Proof 19.3   See Appendix 19.9. ■

By using the lemma above, for the lower-bound on the throughput given in (19.16), we get

$$\sum_{i=1}^{N_t}\sum_{\ell=1}^{K} q_{\ell}^{i}\,\mathbb{E}_{\boldsymbol{H}}\left[\log\Big(1 + \mathcal{S}_L^{(\ell)}(m)\Big)\right] \leq \bar{R}_L^{Q} \leq \sum_{i=1}^{N_t}\sum_{\ell=1}^{K} q_{\ell}^{i}\,\mathbb{E}_{\boldsymbol{H}}\left[\log\Big(1 + \mathcal{S}_U^{(\ell)}(m)\Big)\right], \tag{19.22}$$

where $\mathcal{S}_L^{(\ell)}(m)$ and $\mathcal{S}_U^{(\ell)}(m)$ are the ℓ-th order statistics of the statistical samples $\mathcal{S}_L(m)$ and $\mathcal{S}_L(m)$, respectively, with parent distributions given in (19.20) and (19.21). Therefore, CDFs of $\mathcal{S}_L^{(j)}(m)$ and $\mathcal{S}_U^{(j)}(m)$ are given by [28]

$$\mathcal{S}_L^{(\ell)}(m) \sim F_{\min}^{(\ell)}(x) \triangleq \sum_{i=0}^{\ell-1} \binom{K}{i} \left(F_{\min}(x)\right)^{K-i} \left(1-F_{\min}(x)\right)^i, \quad \ell = 1,\ldots,K, \quad (19.23)$$

$$\text{and} \quad \mathcal{S}_U^{(\ell)}(m) \sim F_{\max}^{(\ell)}(x) \triangleq \sum_{i=0}^{\ell-1} \binom{K}{i} \left(F_{\max}(x)\right)^{K-i} \left(1-F_{\max}(x)\right)^i, \quad \ell = 1,\ldots,K. \quad (19.24)$$

By using the above distributions and defining

$$F_{\min}^K(x;i) \triangleq \sum_{\ell=1}^K q_\ell^i F_{\min}^{(\ell)}(x), \quad \text{and} \quad F_{\max}^K(x;i) \triangleq \sum_{\ell=1}^K q_\ell^i F_{\max}^{(\ell)}(x),$$

we can rewrite the inequalities in (19.22) as

$$\sum_{i=1}^{N_t} \sum_{\ell=1}^K q_\ell^i \int_0^\infty \log(1+x) dF_{\min}^{(\ell)}(x) \leq \quad \bar{R}_L^Q \quad \leq \sum_{i=1}^{N_t} \sum_{\ell=1}^K q_\ell^i \int_0^\infty \log(1+x) dF_{\max}^{(\ell)}(x),$$

$$\text{or} \quad \sum_{i=1}^{N_t} \int_0^\infty \log(1+x) \, dF_{\min}^K(x;i) \leq \quad \bar{R}_L^Q \quad \leq \sum_{i=1}^{N_t} \int_0^\infty \log(1+x) \, dF_{\max}^K(x;i). \quad (19.25)$$

Therefore, in summary, if for each user k, the threshold level $\beta(\gamma_k, K)$ is set, such that $\beta(\gamma_k, K) > 1$, then $\bar{R}_L^Q$ bounded as in (19.25). In the next step we assess how these bounds scale with increasing K. For this purpose we analyze them individually, and, to start off, we provide the following definitions and lemmas.

Lemma 19.4
For a real variable $x \in [0,1]$ and integer variables K and ℓ, $0 \leq \ell \leq K-1$, the function

$$f(x,\ell) \triangleq \sum_{i=0}^\ell \binom{K}{i} x^{K-i}(1-x)^i$$

is increasing in x.

Proof 19.4   See Appendix 19.10.   ■

Now we define
$$G(x) \triangleq 1 - e^{-x},$$

and let also $G^{(\ell)}(x)$ denote the CDF of the ℓ-th order statistic of statistical samples, with K members and parent distribution $G(x)$. Therefore,

$$G^{(\ell)}(x) \triangleq \sum_{i=0}^{\ell-1} \binom{K}{i} \left(G(x)\right)^{K-i} \left(1-G(x)\right)^i. \quad (19.26)$$

Also, define $G^K(x;i) \triangleq \sum_{\ell=1}^K q_\ell^i G^{(\ell)}(x)$. We now use the result of Lemma 19.4 to prove the following lemma.

Lemma 19.5
For any choice of $\beta(\gamma_k, K) > 1$ for the k-th user, $\bar{R}_L^Q$ is lower-bounded and upper-bounded as

$$\sum_{i=1}^{N_t} \int_0^\infty \log(1 + \rho\gamma_{\min}x)\, dG^K(x;i) - N_t \log\left(\rho\gamma_{\min}(N_t - 1) + 1\right)$$

$$\leq \bar{R}_L^Q \leq \sum_{i=1}^{N_t} \int_0^\infty \log(1 + \rho\gamma_{\max}x)\, dG^K(x;i).$$

Proof 19.5   For any choice of $\beta(\gamma_k, K) > 1$ for the k-th user, as we showed earlier, the bounds on $\bar{R}_L^Q$ provided in (19.25) hold valid. Based on the definition of $G(x)$, we find

$$G\left((N_t - 1)(1 + x) + \frac{x}{\rho\gamma_{\min}}\right) = 1 - \exp\left[-\frac{x}{\rho\gamma_{\min}} - (N_t - 1)\underbrace{(x + 1)}_{\geq \ln(x+1)}\right]$$

$$\geq 1 - \frac{e^{-x/\rho\gamma_{\min}}}{(x+1)^{N_t - 1}} = F_{\min}(x).$$

Then, according to Lemma 19.4, for $\ell = 1, \ldots, K$, we have

$$G^{(\ell)}\left((N_t - 1)(1 + x) + \frac{x}{\rho\gamma_{\min}}\right) = f\left(G\left((N_t - 1)(1 + x) + \frac{x}{\rho\gamma_{\min}}\right), \ell - 1\right)$$

$$\geq f\left(F_{\min}(x), \ell - 1\right)$$

$$= F_{\min}^{(\ell)}(x),$$

and, consequently,

$$G^K\left((N_t - 1)(1 + x) + \frac{x}{\rho\gamma_{\min}}; i\right) = \sum_{\ell=1}^K q_\ell^i\, G^{(\ell)}\left((N_t - 1)(1 + x) + \frac{x}{\rho\gamma_{\min}}\right)$$

$$\geq \sum_{\ell=1}^K q_\ell^i\, F_{\min}^{(\ell)}(x) = F_{\min}^K(x;i). \qquad (19.27)$$

Since $F_{\min}^{(\ell)}(x)$, $\ell = 1, \ldots, K$, are increasing functions in x, so is $F_{\min}^K(x;i)$. By defining $u_i \triangleq F_{\min}^K(x;i)$, (19.27) can be rewritten as $x = (F_{\min}^K)^{-1}(u_i)$. Hence,

$$G^K\left((N_t - 1)\left(1 + (F_{\min}^K)^{-1}(u)\right) + \frac{(F_{\min}^K)^{-1}(u)}{\rho\gamma_{\min}}; i\right) \geq u_i,$$

or, by taking into account that $G^K(x)$ is also invertible, we obtain

$$\left(\rho\gamma_{\min}(N_t - 1) + 1\right)\left(1 + (F_{\min}^K)^{-1}(u_i)\right) \geq 1 + \rho\gamma_{\min}(G^K)^{-1}(u_i). \qquad (19.28)$$

Now, consider the lower-bound in (19.25). Using (19.28), we get

$$\sum_{i=1}^{N_t} \int_0^\infty \log(1+x)\, dF_{\min}^K(x;i) + \sum_{i=1}^{N_t} \log\left(\rho\gamma_{\min}(N_t-1)+1\right)$$

$$= \sum_{i=1}^{N_t} \int_0^1 \log\left(1+(F_{\min}^K)^{-1}(u_i)\right) du_i + \sum_{i=1}^{N_t} \log\left(\rho\gamma_{\min}(N_t-1)+1\right) \int_0^1 du_i$$

$$= \sum_{i=1}^{N_t} \int_0^1 \log\left[\left(\rho\gamma_{\min}(N_t-1)+1\right)\left(1+(F_{\min}^K)^{-1}(u_i)\right)\right] du_i$$

$$\geq \sum_{i=1}^{N_t} \int_0^1 \log\left(1+\rho\gamma_{\min}(G^K)^{-1}(u_i)\right) du_i$$

$$= \sum_{i=1}^{N_t} \int_0^\infty \log(1+\rho\gamma_{\min}x)\, dG^K(x;i). \tag{19.29}$$

By substituting (19.29) into (19.25), we get the following lower-bound on R_L^Q

$$\sum_{i=1}^{N_t} \int_0^\infty \log(1+\rho\gamma_{\min}x)\, dG^K(x;i) - N_t \log\left(\rho\gamma_{\min}(N_t-1)+1\right) \leq \bar{R}_L^Q. \tag{19.30}$$

Also, it can be easily verified that $\forall x \in \mathbb{R}$, $F_{\max}(x) \geq G\left(\frac{x}{\rho\gamma_{\max}}\right)$. Hence,

$$F_{\max}^K(x;i) = \sum_{\ell=1}^K q_\ell^i F_{\max}^{(\ell)}(x) = \sum_{\ell=1}^K q_\ell^i f\left(F_{\max}(x),\ell-1\right) \geq$$

$$\sum_{\ell=1}^K q_\ell^i f\left(G\left(\frac{x}{\rho\gamma_{\max}}\right),\ell-1\right) = \sum_{\ell=1}^K q_\ell^i G^{(\ell)}\left(\frac{x}{\rho\gamma_{\max}}\right) = G^K\left(\frac{x}{\rho\gamma_{\max}};i\right).$$

By defining $u_i \triangleq \frac{F_{\max}^K(x;i)}{\rho\gamma_{\max}}$ and following the same lines as above, we get

$$\sum_{i=1}^{N_t} \int_0^\infty \log(1+x)\, dF_{\max}^K(x;i) \leq \sum_{i=1}^{N_t} \int_0^\infty \log(1+\rho\gamma_{\max}x)\, dG^K(x;i). \tag{19.31}$$

Inequalities in (19.25) and (19.31) together give rise to the following upper-bound on $\bar{R}_L^Q$

$$\bar{R}_L^Q \leq \sum_{i=1}^{N_t} \int_0^\infty \log(1+\rho\gamma_{\max}x)\, dG^K(x;i). \tag{19.32}$$

Combining the lower- and upper-bounds given in (19.30) and (19.32) establishes the desired result.
■

Now we analyze how the sum-rate capacity bounds provided by Lemma 19.5 scale when the number of users, K, increases. We use the result of the following theorem provided in [22] in our analysis.

Theorem 19.1
Let $\{X_k\}_{k=1}^K$ be a family of positive random variables, with finite mean μ_K and variance σ_K^2, also,

$\mu_K \to \infty$ *and* $\frac{\sigma_K}{\mu_K} \to 0$, *as* $K \to \infty$. *Then, for all* $\alpha > 0$, *we have*

$$\mathbb{E}\Big[\log(1 + \alpha X_K)\Big] \doteq \log\Big(1 + \alpha\mathbb{E}[X_K]\Big).$$

Proof 19.6 See [22, Theorem 4]. ■

By providing the following lemma, we shed light on the scaling behavior of the sum-rate capacity bounds given in Lemma 19.5.

Lemma 19.6
If for each user k we set the threshold level $\beta(\gamma_k, K) > 1$, *such that*

$$F_k\Big(\beta(\gamma_k, K)\Big) = 1 - \frac{1}{K}, \tag{19.33}$$

then, for $i = 1, \dots, N_t$ *we have*

$$\int_0^\infty \log(1 + \rho\gamma_{\min}x)\, dG^K(x;i) \quad \doteq \quad \log\log K + \log(\rho\gamma_{\min}),$$

$$and \quad \int_0^\infty \log(1 + \rho\gamma_{\max}x)\, dG^K(x;i) \quad \doteq \quad \log\log K + \log(\rho\gamma_{\max}).$$

Proof 19.7 Since the CDF $F_k(\cdot)$ is a monotonic function, and $1 - \frac{1}{K} \in (0,1)$, for any $N_t, K \in \mathbb{N}$, there is always a unique solution $\beta(\gamma_k, K)$ to (19.33). Also, it can be readily verified that for any given N_t, ρ, and γ_k, for sufficiently large K, we have $\beta(\gamma_k, K) > 1$. For the given threshold above, the probability that the k-th user expresses interest in a beam by feeding back its index is

$$p \triangleq P\Big(\max_m \mathrm{SINR}_{k,m} \geq \beta(\gamma_k, K)\Big) = 1 - \Big[F_k\Big(\beta(\gamma_k, K)\Big)\Big]^{N_t} = 1 - \Big(1 - \frac{1}{K}\Big)^{N_t}, \tag{19.34}$$

which is the same for all the users and beams. Therefore the number of elements in the set $\mathcal{B}_m$ has a binomial distribution, with parameter $p = 1 - (1 - 1/K)^{N_t}$. Hence,

$$q_\ell^i = \sum_{k=\ell}^K \frac{1}{k} P\Big(|\mathcal{A}_i| = k\Big) = \sum_{k=\ell}^K \frac{1}{k} P\Big(|\mathcal{B}_i| = k\Big) = \sum_{k=\ell}^K \frac{p_k}{k}, \text{ where } p_k \triangleq \binom{K}{k} p^k (1-p)^{K-k}. \tag{19.35}$$

For a given number of users K, we define a random variable X_K, distributed as $X_K \sim G^K(x)$, and set $\alpha \triangleq \rho\gamma_{\min}$. Also define

$$\mu_{(\ell)} \quad \triangleq \quad \int_0^\infty x\, dG^{(\ell)}(x),$$

$$\sigma_{(\ell)}^2 \quad \triangleq \quad \int_0^\infty (x - \mu_{(i)})^2\, dG^{(\ell)}(x),$$

$$and \quad \mu_K \quad \triangleq \quad \mathbb{E}[X_K] = \int_0^\infty x\, dG^K(x;i) = \sum_{\ell=1}^K q_\ell^i \int_0^\infty x\, dG^{(\ell)}(x) = \sum_{\ell=1}^K q_\ell^i \mu_{(\ell)}.$$

As given in [28, Sec. 4.6] and discussed in details in [23], for ordered exponentially distributed random variables X_K, we have

$$\sigma_K^2 < 2 + 2\mu_{(1)}\Big(\mu_{(1)} - \mu_{(K)}\Big), \tag{19.36}$$

$$\text{and } \log K + \zeta + \frac{1}{2(K+1)} < \mu_{(1)} < \log K + \zeta + \frac{1}{2K} \quad \Rightarrow \quad \mu_{(1)} \doteq \log K, \quad (19.37)$$

where $\zeta \approx 0.577$ is the Euler-Mascheroni constant. Also,

$$\mu_{(1)} - \log\left(\sum_{\ell=1}^{K} \ell q_{\ell}^{i}\right) - \zeta - 0.5 < \mu_K < \mu_{(1)}.$$

On the other hand, based on the definition of p_k in (19.35) and taking into account that $p = 1 - (1 - \frac{1}{K})^{N_t}$, we have

$$\sum_{\ell=1}^{K} \ell q_{\ell}^{i} = \sum_{\ell=1}^{K} \ell \sum_{k=\ell}^{K} \frac{p_k}{k} = \sum_{k=1}^{K} \frac{p_k}{k} \sum_{\ell=1}^{k} \ell = \sum_{k=1}^{K} \frac{(k+1)}{2} p_k = \frac{1}{2} + \frac{1}{2} \sum_{k=1}^{K} k \binom{K}{k} p^k (1-p)^{K-k} = \frac{Kp+1}{2},$$

$$(19.38)$$

where

$$\lim_{K \to \infty} \frac{Kp+1}{2} = \frac{1}{2} + \frac{1}{2} \lim_{K \to \infty} \frac{1 - \left(1 - \frac{1}{K}\right)^{N_t}}{\frac{1}{K}} = \frac{N_t + 1}{2}.$$

Therefore, as $K \to \infty$

$$\mu_{(1)} - \log\left(\frac{N_t + 1}{2}\right) - \zeta - 0.5 < \mu_K < \mu_{(1)}. \quad (19.39)$$

Equations (19.37) and (19.39) together show that

$$\mu_{(1)} \doteq \mu_K \doteq \log K, \quad (19.40)$$

which also implies that $\mu_K \to \infty$. Taking into account (19.36) and (19.40), we conclude that $\lim_{K \to \infty} \frac{\sigma_K}{\mu_K} = 0$ and, therefore, the conditions of Theorem 19.1 are met. Hence, from Theorem 19.1 for $i = 1, \dots, N_t$

$$
\begin{aligned}
\int_0^\infty \log(1 + \rho \gamma_{\min} x)\, dG^K(x; i) &= \mathbb{E}\left[\log(1 + \rho \gamma_{\min} X_K)\right] \\
&\doteq \log\left(1 + \rho \gamma_{\min} \mathbb{E}[X_K]\right) \\
&= \log\left(1 + \rho \gamma_{\min} \mu_K\right) \\
&\doteq \log\log K + \log(\rho \gamma_{\min}). \quad (19.41)
\end{aligned}
$$

By following the same lines, we can also show that

$$\int_0^\infty \log(1 + \rho \gamma_{\max} x)\, dG^K(x; i) \doteq \log\log K + \log(\rho \gamma_{\max}), \quad (19.42)$$

which completes the proof. ■

Theorem 19.2

For a MISO broadcast channel with N_t transmit antennas, fixed ρ, and the proposed finite-rate feedback mechanism, the sum-rate throughput scales as

$$\mathbb{E}_{\boldsymbol{H}}[R^Q] \doteq N_t \log\log K.$$

Proof 19.8 By applying the result of Lemma 19.6 on the lower- and upper-bounds on $\bar{R}_L^Q$ provided in Lemma 19.5, we get

$$
\lim_{K\to\infty} \frac{\bar{R}_L^Q}{N_t \log\log K} \geq \lim_{K\to\infty} \frac{\sum_{i=1}^{N_t} \int_0^\infty \log(1+\rho\gamma_{\min}x)\, dG^K(x;i) - \sum_{i=1}^{N_t} \log\left(\rho\gamma_{\min}(N_t-1)+1\right)}{N_t \log\log K}
$$

$$
= N_t \lim_{K\to\infty} \frac{\log\log K - \log\left(N_t - 1 + 1/\rho\gamma_{\min}\right)}{N_t \log\log K} = 1, \tag{19.43}
$$

and

$$
\lim_{K\to\infty} \frac{\bar{R}_L^Q}{N_t \log\log K} \leq \lim_{K\to\infty} \frac{\sum_{i=1}^{N_t} \int_0^\infty \log(1+\rho\gamma_{\max}x)\, dG^K(x)}{N_t \log\log K}
$$

$$
= N_t \lim_{K\to\infty} \frac{\log\log K + \log(\rho\gamma_{\max})}{N_t \log\log K} = 1. \tag{19.44}
$$

Inequalities in (19.43) and (19.44) together conclude that $\bar{R}_L^Q \doteq N_t \log\log K$. Now, considering that $\mathbb{E}_{\boldsymbol{H}}[R^Q] \geq \bar{R}_L^Q$ from (19.16), we get $\mathbb{E}_{\boldsymbol{H}}[R^Q] \geq N_t \log\log K$. On the other hand, we know that the optimal scaling throughput scaling for MIMO broadcast channels is $N_t \log\log K$ [13], which completes the proof of the theorem. ■

19.4.2 Throughput Scaling for $N_r > 1$

Thus far, we have considered single-antenna users. As stated earlier in (19.7), by deploying dirty-paper coding and facilitating full CSI feedback, the sum-rate throughput exhibits a double logarithmic growth with the number of receive antennas, N_r. We show that with slight modification to the scheduling algorithm provided in Section 19.3, the same gain can be retained.

We modify the scheduling algorithm to allow each user to receive more than one information stream x_m via its distinct receive antennas. In other words, we preclude the receive antennas to jointly decode the information streams and consider different receive antennas as separate users. As a result, this translates to having effectively KN_r users in the network, and all the analyses provided earlier for K users can be extended for the network with KN_r users. Consequently, for each user, we set the threshold level as

$$
F_k\left(\beta(\gamma_k, K)\right) = 1 - \frac{1}{KN_r}, \tag{19.45}
$$

and as a direct result of Theorem 19.2, we have the following scaling law for MIMO broadcast channels.

Theorem 19.3

For a MIMO broadcast channel with fixed N_t and ρ and finite-rate feedback, the sum-rate throughput scales as

$$
\mathbb{E}_{\boldsymbol{H}}[R^Q] \doteq N_t \log\log KN_r.
$$

Note that according to the modified scheduling algorithm on one hand, each user might be required to feed back the indices of more than one beam, which invokes an increase in the amount of feedback amount per user. On the other hand, increasing the threshold level makes it more stringent for the user to satisfy the threshold and send feedback.

It is noteworthy that in our scheduling protocol, we have considered that each user may receive more than one information stream. On the other hand, if we restrict each user to receive no more than one information stream, the performance will degrade. Indeed, as shown in [12], even with

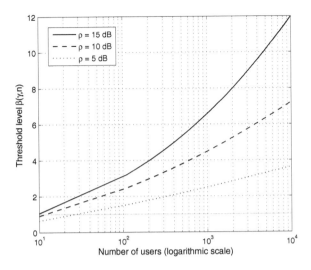

Figure 19.1: Threshold level vs. the number of users for $N_t = 4$ **and** $\gamma_k = 1$.

perfect SINR feedback, the sum-rate throughput scales as $N_t \log\log K$, which does not depend on the number of receive antennas. Therefore, with such restriction, adding more receiver antennas offers no asymptotic gain.

19.5 Simulation Results

In this section, we provide simulation results and numerical evaluations which support our asymptotic analyses and results. First of all, as stated earlier in Lemmas 19.1, 19.5, and 19.6, the asymptotic results hold for the choice of thresholds $\beta(\gamma_k, K) > 1$. As observed analytically, for sufficiently large number of users we expect this constraint be satisfied. By numerical evaluation of (19.33), Figure 19.1 shows how the threshold level changes with the number of users and SNRs. It is seen that when the base station is equipped with $N_t = 4$ transmit antennas and 5dB SNR, with $K \geq 25$ users we have $\beta(\gamma_k, K) > 1$.

Simulation results in Figure 19.2 depict the sum-rate throughput achieved by the proposed scheduling scheme versus those achieved by dirty-paper coding [6], optimal zero-forcing beamforming [15], orthogonal random beamforming with infinite-rate feedback [12], and the 1-bit feedback scheme of [18]. The simulation results in Figure 19.2 show that when the base station has $N_t = 4$ antennas the sum-rate throughput of our scheme, referred to by one-bit ORB, is about 1 bit lower than that of ORB. Note that the one-bit quantization scheme proposed in this chapter is mainly intended to show that the optimal capacity scaling law is attained by finite-rate feedback and has acceptable performance even for small size networks. Therefore, although quantization with as low as 1 bit serves our purpose, it is more desirable to have a higher resolution quantization. As depicted in the figures, by adding only 1 more bit for quantization, it is possible to get fairly close to the sum-rate throughput of ORB.

Figure 19.3 illustrates how the sum-rate throughput changes with varying SNR. We have considered three cases of $K = 10, 25, 100$ in a network with $N_t = 4$ and $N_r = 1$. It is seen that for any fixed SNR, increasing the number of users leads to an increase in the sum-rate throughput. As shown in the figure, the sum-rate throughput saturates beyond certain SNR values. This is justified by con-

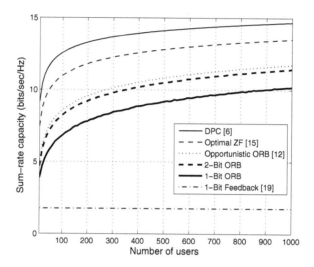

Figure 19.2: Sum-rate capacity vs. the number of users for $N_t = 4$, $N_r = 1$ and $\rho = 10$ dB.

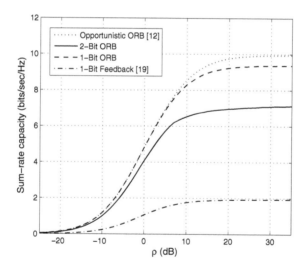

Figure 19.3: Sum-rate capacity vs. SNR for $N_t = 4$ and $N_r = 1$.

sidering that increasing the SNR has a double effect as it boosts the power of the desired signals (x_m) and interference signals ($x_n, n \neq m$), simultaneously. Therefore, as also expected from (19.3), the gain of increasing SNR diminishes as $\rho \to \infty$.

A note is warranted on the degraded performance of the 1-bit feedback scheme of [18] for the setup in the simulations above. Although this scheme is asymptotically optimal, it can be verified that the quantization threshold design is not always guaranteed to be positive (including the case of $N_t = 4$ and $K = 1000$). In such cases, all the users satisfy the threshold condition, and the scheduling scheme will be equivalent to random selection of users (or no CSI available) and, as analytically

Table 19.1   Comparison of Our Proposed Scheme (1-bit ORB) with Other Major Scheduling Schemes

	Throughput scaling	*Feedback per user*	*Complexity*
1-bit ORB	$N_t \log\log K$	$\log N_t$ or 0 bits[a]	Maximum search
Opportunistic ORB [12]	$N_t \log\log K$	$\log N_t$ bits + 1 real number	Maximum search[b]
DPC [13]	$N_t \log\log K$	$2N_t$ real numbers	Very complex
ZFBF [15]	$N_t \log\log K$	$2N_t$ real numbers	Channel inversion
1-bit Feedback [18]	$N_t \log\log K$	1 bit	-
ZFBF-SUS [19]	$< N_t \log\log K^c$	B^d bits + 1 real number	Channel inversion
OSDMA-TF [20]	$< N_t \log\log K^e$	B bits + 1 real number	Maximum search
OSDMA-BS [21]	-	$Q^f + \log N_t$ bits	Maximum search
Opportunistic BF [22]	$\log\log K$	L^g bits	Maximum search

[a] $\log N_t$ when the threshold constraint is satisfied and 0 otherwise.
[b] A search among real numbers for finding the maximum.
[c] This holds only when $B \to \infty$, or when perfect CSI feedback is possible.
[d] B is the number of bits required for channel vector quantization.
[e] This holds when the aggregate amount of feedback $\to \infty$.
[f] For quantizing SINR.
[g] For beam selection.

studied in [12], under such conditions, the sum-rate throughput will be a constant independent of the number of transmit antennas and the number of users.

Finally, we remark that the orthogonality constraint among the beams is to facilitate the analysis, in particular, to obtain (19.17). In practice, our simulations show that by simply employing (non-orthogonal) random beams incur little attendant performance loss.

In Table 19.1, we briefly summarize and compare the merits of the major scheduling schemes available in the literature. As observed in the table, there exist an interplay between the achievable sum-rate throughput, the amount of feedback, and processing complexity at the base station. While schemes with more feedback and more computationally complex can perform closer to the capacity, they are more difficult to implement in practice. This motivates the development of low-complexity techniques that require very little amount of feedback.

19.6   Conclusion

In this chapter, we have proposed a scheduling scheme for multiple-antenna downlink transmission, which requires finite-rate feedback. We have demonstrated that this scheduling scheme is capable of exploiting the full multiplexing gain, as well as the inherent multiuser diversity gain, and attains the optimal sum-rate capacity scaling achievable by dirty-paper coding and perfect channel state information feedback. Furthermore, this scheme has a good performance for practical network sizes.

Appendix
19.7   Proof of Lemma 1

It is easy to see that $\mathcal{B}_m \subset \mathcal{A}_m$ as $\forall k \in \mathcal{B}_m$, $\text{SINR}_{k,m} \geq \beta(\gamma_k, K)$ and, therefore, by the definition of $\mathcal{A}_m$, $k \in \mathcal{A}_m$. Conversely, we also show that $\mathcal{A}_m \subset \mathcal{B}_m$ as follows.

By using the definition of $\mathcal{A}_m$, for any $k \in \mathcal{A}_m$ we have $\text{SINR}_{k,m} \geq \beta(\gamma_k, K) > 1$, or,

$$\forall k \in \mathcal{A}_m, \quad |\boldsymbol{H}_k \boldsymbol{\phi}_m|^2 > \frac{1}{\rho \gamma_k} + \sum_{l \neq m} |\boldsymbol{H}_k \boldsymbol{\phi}_l|^2 > |\boldsymbol{H}_k \boldsymbol{\phi}_l|^2, \quad \text{for } \forall l \neq m,$$

which implies that

$$\forall l \neq m, \quad \text{SINR}_{k,l} = \frac{|\boldsymbol{H}_k \boldsymbol{\phi}_l|^2}{\frac{1}{\rho \gamma_k} + \sum_{i \neq l} |\boldsymbol{H}_k \boldsymbol{\phi}_l|^2} < \frac{|\boldsymbol{H}_k \boldsymbol{\phi}_l|^2}{|\boldsymbol{H}_k \boldsymbol{\phi}_m|^2} < 1 < \text{SINR}_{k,m}.$$

Therefore, $\text{SINR}_{k,m} = \max_l \text{SINR}_{k,l}$. Also, since $\text{SINR}_{k,m} \geq \beta(\gamma_k, K)$ it is concluded that $k \in \mathcal{B}_m$. Hence, for any $k \in \mathcal{A}_m$, we have $k \in \mathcal{B}_m$, or $\mathcal{A}_m \subset \mathcal{B}_m$. This establishes the proof of the lemma.

19.8 Proof of Lemma 2

$$
\begin{aligned}
\mathbb{E}\Big[g(X) \mid X \geq b\Big] &= \int_b^\infty g(x) f_{X|X \geq b}(x) \, dx \\
&= \frac{1}{P(X \geq b)} \int_b^\infty g(x) f_X(x) \, dx \\
&= \left[\frac{1}{P(X \geq b)} - \frac{1}{P(X \geq a)} \right] \int_b^\infty \underbrace{g(x)}_{\geq g(b)} f_X(x) dx + \frac{1}{P(X \geq a)} \int_b^\infty g(x) f_X(x) dx \\
&\geq \left[1 - \frac{P(X \geq b)}{P(X \geq a)} \right] g(b) + \frac{1}{P(X \geq a)} \int_b^\infty g(x) f_X(x) dx \\
&= \frac{g(b)}{P(X \geq a)} P(a \leq X \leq b) + \frac{1}{P(X \geq a)} \int_b^\infty g(x) f_X(x) dx \\
&\geq \frac{1}{P(X \geq a)} \int_a^b g(x) f_X(x) \, dx + \frac{1}{P(X \geq a)} \int_b^\infty g(x) f_X(x) dx \\
&= \int_a^\infty g(x) f_{X|X \geq a}(x) \, dx = \mathbb{E}\Big[g(X) \mid X \geq a\Big].
\end{aligned}
$$

19.9 Proof of Lemma 3

By induction, we show that for any $j = 1, \ldots, K$, $\mathcal{S}_L^{(\ell)}(m) \leq \text{SINR}_m^{(\ell)}$.
(1) For $\ell = 1$, we have

$$\text{SINR}_m^{(1)} = \max_k \text{SINR}_{k,m} \geq \max_k \mathcal{S}_L(k,m) = \mathcal{S}_L^{(1)}(m).$$

(2) Assumption: For some $\ell = l$, we have $\mathcal{S}_L^{(l)}(m) \leq \text{SINR}_m^{(l)}$.
(3) Claim: For $\ell = l + 1$, we show that $\mathcal{S}_L^{(l+1)}(m) \leq \text{SINR}_m^{(l+1)}$.
Each of the $(K - l)$ terms $\text{SINR}_m^{(l+1)}, \ldots, \text{SINR}_m^{(K)}$ is greater than one corresponding element in the set $\mathcal{S}_L(m)$. Therefore, there cannot be more than l elements in $\mathcal{S}_L(m)$ that are all greater than $\text{SINR}_m^{(l+1)}, \ldots, \text{SINR}_m^{(K)}$.

Now, if $\mathcal{S}_L^{(l+1)}(m) > \text{SINR}_m^{(l+1)}$, then, by using the assumption, all the $(l+1)$ terms $\mathcal{S}_L^{(1)}(m)$, $\mathcal{S}_L^{(2)}(m), \ldots, \mathcal{S}_L^{(l+1)}(m)$ are greater than all the $(K - l)$ terms $\text{SINR}_m^{(l+1)}, \ldots, \text{SINR}_m^{(K)}$. Therefore, we have found $(l+1)$ elements in $\mathcal{S}_L(m)$ that are all greater than $\mathsf{SINR}_m^{(l+1)}, \ldots, \mathsf{SINR}_m^{(K)}$, and this contradicts with what we found earlier. Hence, we should have $\mathcal{S}_L^{(l+1)}(m) \leq \mathsf{SINR}_m^{(l+1)}$.

By following the same lines, we can show that also for $j = 1, \ldots, K$, we always have $\mathsf{SINR}_m^{(\ell)} \leq \mathcal{S}_U^{(\ell)}(m)$, which concludes the proof of the lemma.

19.10   Proof of Lemma 4

By the expansion of $\left(x + (1-x)\right)^K$, we have

$$
\begin{aligned}
f(x, \ell) &= 1 - \sum_{i=\ell+1}^{K} \binom{K}{i} x^{K-i}(1-x)^i \\
&= 1 - \sum_{i=\ell+1}^{K} \binom{K}{K-i} x^{K-i}(1-x)^i \\
&= 1 - \sum_{k=0}^{K-(\ell+1)} \binom{K}{k} (1-x)^{K-k} x^k \\
&= 1 - f(1-x, K-\ell-1),
\end{aligned}
$$

where it can be concluded that $f'(u, \ell)\big|_{u=x} = f'(u, K-\ell-1)\big|_{u=1-x}$. So it is sufficient to show that $f'(x, \ell) \geq 0$ for $x \leq \frac{1}{2}$, and for all $\ell = 1, \ldots, K-1$. For this purpose, we consider two cases of $\ell \leq \lfloor \frac{K}{2} \rfloor$ and $\ell > \lfloor \frac{K}{2} \rfloor$.

Case 1: $\ell \leq \lfloor \frac{K}{2} \rfloor$

$$
\begin{aligned}
f'(x, \ell) &= \sum_{i=0}^{\ell} \binom{K}{i}(K-i)x^{K-i-1}(1-x)^i \\
&\quad - \binom{K}{i} i x^{K-i}(1-x)^{i-1} \\
&= \sum_{i=0}^{\ell} \binom{K}{i} x^{K-i-1}(1-x)^{i-1}\left[K(1-x) - i\right],
\end{aligned}
$$

where, since $0 \leq i \leq \ell$, it can be shown that for $x \leq \frac{1}{2}$

$$
\begin{aligned}
K(1-x) - i &\geq K(1-x) - \ell \\
&\geq K(1-x) - \frac{K}{2} \\
&= \frac{K}{2}(1-2x) \\
&\geq 0.
\end{aligned}
\tag{19.46}
$$

Case 2: $\ell > \lfloor \frac{K}{2} \rfloor$

Define $a_\ell = 1 - \frac{1}{2}\delta(\lfloor \frac{K}{2} \rfloor - i)$, where $\delta(\cdot)$ is the Dirac delta function. Therefore, we get

$$
f(x, \ell) = f(x, K-\ell-1)
$$

$$
+ \sum_{i=K-\ell}^{\lfloor \frac{\ell}{2} \rfloor} a_i \binom{\ell}{i}\left[x^{K-i}(1-x)^i + x^i(1-x)^{K-i}\right].
$$

For $x \le \frac{1}{2}$, we get

$$f'(x, \ell) = f'(x, K - \ell - 1)$$

$$+ \sum_{i=K-i}^{\lfloor \frac{K}{2} \rfloor} a_i \binom{K}{i} \left\{ x^{K-i-1}(1-x)^{i-1} \left[K - i - Kx \right] \right.$$

$$+ \underbrace{x^{i-1}(1-x)^{K-i-1}}_{\ge x^{K-i-1}(1-x)^{i-1}} \left. \left[i - Kx \right] \right\}$$

$$\ge f'(x, \underbrace{K - \ell - 1}_{\le \lfloor \frac{K}{2} \rfloor})$$

$$+ \sum_{i=K-\ell}^{\lfloor \frac{K}{2} \rfloor} a_i \binom{\ell}{j} x^{K-i-1}(1-x)^{i-1} \underbrace{\left[K - 2Kx \right]}_{\ge 0}$$

$$\ge 0. \tag{19.47}$$

From (19.46) and (19.47), it is concluded that for $x \le \frac{1}{2}$, $f(x, \ell)$ is an increasing function of x, which completes the proof.

References

[1] E. Telatar, "Capacity of multi-antenna Gaussian channel," *Europ. Trans. Telecommun.*, vol. 10, no. 6, pp. 585–595, Nov. 1999.

[2] G. J. Foschini and M. J. Gans, "On limits of wireless communications in a fading environment when using multiple antennas," *Wireless Personal Commun.*, vol. 6, no. 3, pp. 311–335, Mar. 1998.

[3] L. Zheng and D. N. C. Tse, "Communication on the Grassmann manifold: A geometric approach to the noncoherent multiple-antenna channel," vol. 48, no. 2, pp. 359–383, Feb. 2002.

[4] B. Hassibi and T. L. Marzetta, "Multiple-antennas and isotropically random unitary inputs: the received signal density in closed form," vol. 48, no. 6, pp. 1473–1484, Jun. 2002.

[5] A. Goldsmith, S. A. Jafar, N. Jindal, and S. Vishwanath, "Capacity limits of MIMO channels," vol. 21, no. 5, pp. 684–702, Jun. 2003.

[6] G. Caire and S. Shamai, "On the achievable throughput of a multiantenna Gaussian broadcast channel," vol. 49, no. 7, pp. 1691–1706, Jul. 2003.

[7] H. Weingarten, Y. Steinberg, and S. Shamai, "The capacity region of the Gaussian multiple-input multiple-output broadcast channel," vol. 52, no. 9, pp. 3936–3964, Sep. 2006.

[8] P. Viswanath and D. N. C. Tse, "Sum capacity of the vector Gaussian broadcast channel and uplink-downlink duality," vol. 49, no. 8, pp. 1912–1921, Aug. 2003.

[9] S. Vishwanath, N. Jindal, and A. Goldsmith, "Duality, achievable rates, and sum-rate capacity of gaussian MIMO broadcast channels," vol. 49, no. 10, pp. 2658–2668, Oct. 2003.

[10] N. Jindal and A. Goldsmith, "Dirty-paper coding versus TDMA for MIMO broadcast channels," vol. 51, no. 5, pp. 1783–1794, May 2005.

[11] W. Yu and J. Cioffi, "Sum capacity of vector Gaussian broadcast channels," vol. 50, no. 9, pp. 1875–1892, Sep. 2004.

[12] M. Sharif and B. Hassibi, "On the capacity of MIMO broadcast channel with partial side information," vol. 51, no. 2, pp. 506–522, Feb. 2005.

[13] ——, "A comparison of time-sharing, DPC, and beamforming for MIMO broadcast channels with many users," vol. 55, no. 1, pp. 11–15, Jan. 2007.

[14] P. Viswanath, D. N. C. Tse, and R. Laroia, "Opportunistic beamforming using dumb antennas," vol. 48, no. 6, pp. 1277–1294, Jun. 2002.

[15] T. Yoo and A. Goldsmith, "On the optimality of multiantenna broadcast scheduling using zero-forcing beamforming," vol. 24, no. 3, pp. 528–541, Mar. 2006.

[16] N. Jindal, "MIMO broadcast channels with finite-rate feedback," vol. 52, no. 11, pp. 5045–5060, Nov. 2006.

[17] P. Ding, D. J. Love, and M. D. Zoltowski, "Multiple antenna broadcast channels with shape feedback and limited feedback," vol. 55, no. 7, pp. 3417–3428, Jul. 2007.

[18] J. Diaz, O. Simeone, and Y. Bar-Ness, "Asymptotic analysis of reduced-feedback strategies for MIMO Gaussian broadcast channels," vol. 54, no. 3, pp. 1308–1316, March 2008.

[19] T. Yoo, N. Jindal, and A. Goldsmith, "Multi-antenna downlink channels with limited feedback and user selection," vol. 27, no. 7, pp. 1478–1491, Sep. 2007.

[20] K. Huang, R. W. Heath, and J. G. Andrews, "Space division multiple access with a sum feedback rate constraint," vol. 55, no. 7, pp. 3879–3891, Jul. 2007.

[21] W. Choi, A. Forenza, J. G. Andrews, and R. W. H. Jr, "Opportunistic space division multiple access with beam selection," vol. 55, no. 12, pp. 2371–2380, 2007.

[22] S. Sanayei and A. Nosratinia, "Opportunistic beamforming with limited feedback," vol. 6, no. 8, pp. 2765–2771, Aug. 2007.

[23] ——, "Opportunistic downlink transmission with limited feedback," vol. 53, no. 11, pp. 4363–4372, Nov. 2007.

[24] A. Bayesteh and A. K. Khandani, "How much feedback is required in MIMO broadcast channels?" 2007, submitted, available at http://arxiv.org/pdf/cs/0703143v1.

[25] C. Swannack, E. Uysal-Biyikoglu, and G. W. Wornell, "MIMO broadcast scheduling with limited channel state information," in *Proc. Allerton Conf. Commun. Control & Comput.*, Monticello, IL, Sep. 2005.

[26] ——, "Finding NEMO: near mutually orthogonal sets and applications to MIMO broadcast scheduling," in *Proc. IEEE Int. Conf. Commun. (ICC)*, Seoul, Korea, Jun. 2005.

[27] K. Huang, J. G. Andrews, and R. W. Heath, "Performance of orthogonal beamforming for SDMA with limited feedback," 2007, submitted, available at http://users.ece.utexas.edu/~rheath/papers/2007/TVT1/paper.pdf.

[28] B. C. Arnold, N. Balakrishnan, and H. N. Nagaraja, *A First Course in Order Statistics.* New York: Wiley, 1992.

Chapter 20

On the Achievable Sum-Rate of MIMO Bidirectional Underlay Cognitive Cooperative Networks

Ahmad Alsharoa, Hakim Ghazzai, and Mohamed-Slim Alouini

CONTENTS

Abstract

In this chapter, we consider a multi-input multi-output (MIMO) cooperative cognitive radio (CR) system model under a spectrum sharing set-up, where primary users and secondary users operate on the same frequency band. In the CR underlay mode, secondary users are allowed to exploit the spectrum allocated by primary users in an opportunistic manner by respecting a tolerated temperature limit. The secondary networks employ an amplify-and-forward two-way relaying technique in order to maximize the sum-rate under power budget and interference constraints. Indeed, combined CR, tow-way relaying, and MIMO antennas provide a smart solution for a more efficient usage of the frequency band. Furthermore, we investigate two models of power distributions; discrete power distribution and continuous power distribution. In this context, we formulate an optimization problem that is solved using joint optimization algorithms. For discrete power distribution, we employ heuristic algorithms as iterative and genetic algorithms to find a solution. While for continuous power distribution, first, we derive a closed-form expression of the optimal power allocated to antenna terminals. Then, we employ a heuristic algorithm based on practical swarm optimization algorithm to find the power allocated to secondary relays. In our numerical results, we demonstrate the performance of the proposed schemes for both power distribution types and analyze the impact of several system parameters on the achieved performance. Finally, we compare our proposed scheme with traditional one-way relaying scheme.

Index Terms- Multiple-input multiple-output, cooperative cognitive networks, amplify-and-forward.

20.1   Introduction

Recently, improving both the spectrum usage and the data rate has been widely investigated by wireless communication researchers. Several schemes including cognitive radio (CR), cooperative communication, and multi-input multi-output (MIMO) antennas have been proposed and discussed. The ideas have centered around combining two or more of these solutions together to solve the spectral scarcity and high data rate demand challenges.

20.1.1   Related works

Relay techniques were proposed in modern communication networks to enhance the overall system throughput and extend the network coverage area. With relays, there is a considerable reduction in transmission powers that can lead to the decrease of the interference to neighboring networks. Also, in the case of absence of a direct link between terminals, relays can ensure connectivity and maintain the communication link between the terminals [1]. In the traditional unidirectional transmission, which is also known as one-way relaying (OWR), four time slots are required to accomplish the transmission of different messages between two terminals [2]; see Fig 20.1(a). In order to improve the spectral efficiency, bidirectional transmission, which is also known as two-way relaying (TWR), has attracted significant attention during the last few years [3]. TWR exchanges different messages between two terminals via relays during two time slots only. In the first time slot, the terminals transmit their signals simultaneously to the relays. Subsequently, in the second time slot, the relays broadcast the signal to the terminals [3], as shows Figure 20.1(b). At the terminals side, TWR performs a self-interference cancelation to extract the desired message [4]. The authors in [5]

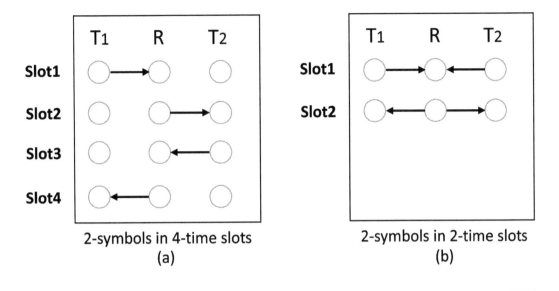

Figure 20.1: (a) OWR, (b) TWR.

investigated the performance of the TWR with an optimal power allocation scheme and showed that TWR provides an improvement of spectral efficiency compared to OWR transmission.

Recently, there has been a great attention to combine the principle of the TWR using amplify-and-forward (AF) strategy, where the relay amplifies the received signal before broadcasting it to the destination, with underlay cognitive radio (CR) in which secondary users utilize the frequency band allocated to primary users under some interference constraints to maintain a certain primary quality-of-service (QoS) [6–12]. All the aforementioned work focused on OWR equipped with a single antenna. The work presented by Li *et. al.* in [7] investigated the joint single relay selection problem to find the optimal power allocation. They also proposed a low complexity approach to maximize the system throughput. In [10–12], heuristic multiple relay selection algorithms for OWR-CR network are investigated.

However, the performance of the network can be further enhanced by employing MIMO antennas that provide extra spatial dimensions. Various previous work have studied the OWR transmission under the MIMO scenario [13, 14] in order to incorporate the benefits of cooperative with MIMO techniques. Some other studies have employed MIMO system with TWR [15, 16]. The authors in [17] have studied the best transmit and receive antennas selection at the two terminals as well as the relays based on minimizing the overall outage probability and maximizing the sum-rate for TWR using the AF protocol. A suboptimal multiple antenna selection scheme with single relay using AF protocol for TWR networks has been investigated in [18]. However, to the best knowledge of the authors, the multiple relay problem in TWR-CR networks with multiple antennas has not been discussed, so far as it is the case in the non-cognitive case.

20.1.2   Contributions

Two models of power distributions, discrete power distribution and continuous power distribution, are investigated for multiple MIMO TWR-CR scheme employing AF strategy. The proposed scheme aims to maximize the secondary sum-rate without affecting the primary QoS by respecting a certain primary interference level in addition to power budget constraints. The main contributions of this chapter can be summarized as follows:

- Formulate optimization problems for discrete and continuous power distribution for MIMO multiple relay TWR-CR networks with AF protocol that maximize the sum-rate of the secondary network by taking into account the power budget of the system in addition to the primary interference level constraints.

- Design practical algorithms to solve the formulated discrete and continuous power distribution optimization problems (i.e., iterative algorithm (IA) and genetic algorithm (GA) for the former distribution and practical swarm optimization (PSO) algorithm for the later distribution).

- Compare the performance of the proposed algorithms with the performance of the optimal or exhaustive search (ES) solution.

- Investigate the impact of the system power budget and primary interference level on the system performance.

- Compare the performance of the TWR-CR scheme with the traditional OWR-CR scheme.

In our discrete power distribution approaches, we assume that each antenna at each cognitive relay can operate with one of the available power levels (i.e., from zero to the maximum peak power) instead of the classic ON-OFF modes only (i.e., the antenna either cooperates with its maximum power or does not cooperate at all), and this will contribute in the maximization of the rate by offering more degrees of freedom to the system.

20.2   System Model and Problem Formulation

We consider a cogniting scheme with one primary user and a cognitive network consisted of two cognitive terminal transceivers, T_1 and T_2, exchanging their messages via L cognitive half-duplex relays using TWR technique, as shows Figure 20.2. We assume that all nodes are equipped with the same number of antennas M. We employ the CR underlay mode in which the secondary users share the spectrum with the primary user by respecting a primary user interference limit denoted I_{th} [6]. Without loss of generality, all the noise variances are assumed to be equal to N_0.

During the first phase, T_1 and T_2 transmit their signals, $\tilde{x}_1$ and $\tilde{x}_2$, to the relays at the same time, with a power denoted $\boldsymbol{P}_1 = [P_1^1, ..., P_1^M]$ and $\boldsymbol{P}_2 = [P_2^1, ..., P_2^M]$, respectively. In the second phase, the relays transmit the amplified signals to the terminals, with a power denoted $\boldsymbol{P}_{r_i} = [P_{r_i}^1, ..., P_{r_i}^M]$, where $i = 1, ..., L$. Let us define $\bar{P}_t$ and $\bar{P}_r$ as the peak powers at the cognitive terminals and at each relay, respectively.

We denote by $\boldsymbol{H}_{1r_i} \in \mathbb{C}^{M \times M}, \boldsymbol{H}_{2r_i} \in \mathbb{C}^{M \times M}, \boldsymbol{H}_{r_ip} \in \mathbb{C}^{M \times M}, \boldsymbol{H}_{1p} \in \mathbb{C}^{M \times M}$, and $\boldsymbol{H}_{2p} \in \mathbb{C}^{M \times M}$ the complex channel mapping matrix between T_1 and the ith relay, the complex channel mapping matrix between T_2 and the ith relay, the complex channel mapping matrix between the ith relay and the primary user, the complex channel mapping matrix between T_1 and the primary user, and the complex channel mapping matrix between T_2 and the primary user, respectively. All the channel gains adopted in our framework are assumed to be constant during the coherence time, with elements h_{ab}^{xy} representing the fading coefficients between transmit antenna y at node a and receive antenna x at node b. In addition to that, channel reciprocity and perfect channel state information at transmitters and receivers are considered. This is not a very benign assumption, as the feedback channel state information from primary users to secondary users is adopted in many cognitive studies [19].

Let V_m and U_m, where $m = 1, 2$, two unitary precoder and decoder matrices, respectively, employed by terminals. In the first phase, T_m employs the precoder matrix, such as $\boldsymbol{x}_m = \boldsymbol{V}_m \tilde{\boldsymbol{x}}_m$, where $\boldsymbol{x}_m$ is the transmitted signal after being precoded by T_m. Subsequently, during the second phase, T_1 and T_2 employ the decoder matrices $\boldsymbol{U}_2$ and $\boldsymbol{U}_1$, respectively, such as $\boldsymbol{r}_1 = \boldsymbol{U}_2^H \boldsymbol{y}_1$ and $\boldsymbol{r}_2 = \boldsymbol{U}_1^H \boldsymbol{y}_2$,

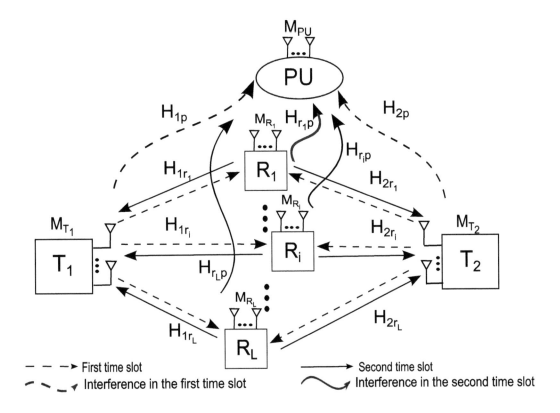

Figure 20.2: MIMO TWR-CR system model.

where $\boldsymbol{y}_1$ and $\boldsymbol{r}_1$ are the received signals at T_1 before and after decoding, respectively, while $\boldsymbol{y}_2$ and $\boldsymbol{r}_2$ are the received signals at T_2 before and after decoding, respectively. The choice of $\boldsymbol{V}_m$ and $\boldsymbol{U}_m$ will be defined later. It is assumed that $\mathbb{E}\left(||\tilde{\boldsymbol{x}}_m||^2\right) = \mathbb{E}\left(||\boldsymbol{x}_m||^2\right) = \mathrm{Tr}\left(\boldsymbol{x}_m\boldsymbol{x}_m^H\right) \leq \bar{P}_t$.

In the first phase, the baseband received signal at the ith relay is given as follows

$$\boldsymbol{y}_{r_i} = \boldsymbol{H}_{1r_i}\boldsymbol{x}_1 + \boldsymbol{H}_{2r_i}\boldsymbol{x}_2 + \boldsymbol{n}_{r_i}, \tag{20.1}$$

where $\boldsymbol{n}_{r_i}$ is the additive Gaussian noise vector at the ith relay. During the second phase, each relay amplifies $\boldsymbol{y}_{r_i}$ by multiplying it by a diagonal matrix $\boldsymbol{W}_i \in \mathbb{R}^{M \times M}$ (containing the amplification factor w_i^k at each antenna k of the ith relay) and broadcasts it to the terminals T_1 and T_2. The amplification factor at the kth antenna of the ith relay can be expressed as

$$|w_i^k|^2 = \frac{P_{r_i}^k}{\sum\limits_{z=1}^{M} P_1^z|h_{1r_i}^{kz}|^2 + \sum\limits_{z=1}^{M} P_2^z|h_{2r_i}^{kz}|^2 + N_0}, \tag{20.2}$$

where $P_{r_i}^k$ denotes as the power at the kth antenna of the ith relay. Finally, the received signals at terminals are given as

$$\boldsymbol{y}_1 = \boldsymbol{A}_2\boldsymbol{x}_2 + \underbrace{\boldsymbol{B}_2\boldsymbol{x}_1}_{\text{Self Interference}} + \boldsymbol{z}_1, \tag{20.3}$$

$$\boldsymbol{y}_2 = \boldsymbol{A}_1\boldsymbol{x}_1 + \underbrace{\boldsymbol{B}_1\boldsymbol{x}_2}_{\text{Self Interference}} + \boldsymbol{z}_2, \tag{20.4}$$

respectively, where

$$A_1 = \sum_{i=1}^{L} H_{2r_i}^T W_i H_{1r_i}, \ B_1 = \sum_{i=1}^{L} H_{2r_i}^T W_i H_{2r_i},$$

$$A_2 = \sum_{i=1}^{L} H_{1r_i}^T W_i H_{2r_i}, \ B_2 = \sum_{i=1}^{L} H_{1r_i}^T W_i H_{1r_i}, \ \text{and} \ z_m = \sum_{i=1}^{L} \left(H_{mr_i}^T W_i n_{r_i} \right) + n_m.$$

n_m is the additive Gaussian noise vectors at T_m, where $m = 1, 2$. By using the knowledge of the side information and channel reciprocity, the terminals can remove the self interference by eliminating their own signals (i.e., x_1 for T_1 and x_2 for T_2). The covariance matrix of the noise z_m can be given as

$$C_{z_m} = \mathbb{E}[z_m z_m^H] = N_0 \sum_{i=1}^{L} H_{mr_i}^T W_i (H_{mr_i}^T W_i)^H + N_0 I_M, \tag{20.5}$$

where I_M denotes the $M \times M$ identity matrix. We then propose to define the precoding and decoding matrices using the singular value decomposition (SVD) of the matrices A_m, where $m = 1, 2$. As such, the sum-rate of the MIMO TWR after SVD can be written as

$$R = \frac{1}{2} \sum_{v=1}^{M} \log_2 \left(1 + \frac{\sigma_{1v}^2 P_1^v}{C_{z_1}(v, v)} \right) + \frac{1}{2} \sum_{u=1}^{M} \log_2 \left(1 + \frac{\sigma_{2u}^2 P_2^u}{C_{z_2}(u, u)} \right), \tag{20.6}$$

where σ_{mq}^2 is the qth eigenvalue of A_m. Finally, the sum-rate maximization problem of MIMO TWR-CR with multiple relays can be formulated as

$$\underset{P_1, P_2, W}{\text{maximize}} \quad R(P_1, P_2, W) \tag{20.7}$$

subject to

$$0 \leq \sum_{v=1}^{M} P_1^v \leq \bar{P}_t, \quad 0 \leq \sum_{u=1}^{M} P_2^u \leq \bar{P}_t, \tag{20.8}$$

$$0 \leq \sum_{k=1}^{M_R} \left(\sum_{v=1}^{M} P_1^v |h_{1r_i}^{kv}|^2 + \sum_{u=1}^{M} P_2^u |h_{2r_i}^{ku}|^2 + N_0 \right) |w_i^k|^2 \leq \bar{P}_r, \quad \forall i = 1, ..., L, \tag{20.9}$$

$$\sum_{v=1}^{M} \sum_{j=1}^{M_{primaryuser}} P_1^v |h_{1p}^{jv}|^2 + \sum_{u=1}^{M} \sum_{j=1}^{M} P_2^u |h_{2p}^{ju}|^2 \leq I_{th}, \tag{20.10}$$

$$\sum_{i=1}^{L} \sum_{j=1}^{M} \sum_{k=1}^{M_R} \left(\sum_{v=1}^{M} P_1^v |h_{1r_i}^{kv}|^2 + \sum_{u=1}^{M} P_2^u |h_{2r_i}^{ku}|^2 + N_0 \right) |w_i^k|^2 |h_{r_ip}^{jk}|^2 \leq I_{th}. \tag{20.11}$$

The constraints (20.8) and (20.9) represent the peak power constraints at the cognitive transceivers, and at each cognitive relay, respectively, while the constraints (20.10) and (20.11) represent the interference constraints in the first and second phase, respectively.

20.3 Problem Solutions

The formulated optimization problem formulated in Section 20.2 is a non-convex problem, and its optimal solution remains unsolved. We investigate the solution of the two power distribution models: discrete power distribution and continuous power distribution. For the discrete power distribution models, we employ two heuristic approaches (i.e, IA and GA) to reach suboptimal solutions to the problem and compare them with the optimal solution using ES. For the continuous power distribution model, we propose to solve the problem into two steps. We first start with the closed-form expression of the optimal terminal power allocation solution by assuming fixed amplification factors at all relay antennas (i.e., maximizing the secondary sum-rate without any control on relay parameters). Then, we employ the PSO algorithm to optimize the relay amplification factors.

20.3.1 Discrete Power Distribution

Before proposing the solution of the discrete optimization problem that will be solved using IA and GA heuristic algorithms, we quantize the power distribution of both the cognitive terminals and relays. Afterwards, we propose two approaches to deal with these maximization problems: IA and GA. A comparison between both approaches are given in Section 20.4.

20.3.1.1 Quantization and Power Distributions

we propose to use a quantization set with discrete number of power levels from zero to the peak antenna power (i.e., it is assumed that the peak power budget allocated at the relays is uniformly distributed at each antenna, $\bar{P}_m^q = \frac{\bar{P}_t}{M}$ and $\bar{P}_r^q = \frac{\bar{P}_r}{M}$). In fact, the qth antenna at each terminal and relay can transmit the amplified signal using one of the power level between 0 and its maximum power budget as $\left(P_m^q \in S = \left\{ 0, \frac{\bar{P}_m^q}{N-1}, \frac{2\bar{P}_m^q}{N-1}, ..., \frac{(N-2)\bar{P}_m^q}{N-1}, \bar{P}_m^q \right\} \right)$ and $\left(P_{r_i}^q \in S = \left\{ 0, \frac{\bar{P}_r^q}{N-1}, \frac{2\bar{P}_r^q}{N-1}, ..., \frac{(N-2)\bar{P}_r^q}{N-1}, \bar{P}_r^q \right\} \right)$, respectively, where N is the number of quantization levels. We assume that the terminal powers at each antenna are equal. By this way, cognitive nodes have more flexibility to allocate their powers in the case where continuous power distribution is not available. This method is considered as a generalization of the ON-OFF mode, where antennas can either transmit or keep silent. Therefore, our goal is to find the optimal power allocation at the cognitive nodes.

20.3.1.2 Iteration Algorithm (IA)

We assume that each antenna has N power levels from zero to the maximum power, i.e., an antenna cooperates with one of the quantized power in S without interfering with the primary user. In the proposed algorithm, we aim to maximize the sum-rate by transmitting the signals with the maximum number of antennas powered with the maximum possible power without affecting the primary users QoS. At the beginning, the transmit powers of all antennas at all cognitive nodes are fixed to their maximum power (i.e., the highest power level in the discrete quantization set S). The algorithm selects the antenna that offers the highest R and satisfies the interference constraint at the same time. Then, it tries to add the maximum number of antennas that can contribute in maximizing the sum-rate. If, during this process, the interference constraint is not satisfied, then the new active antennas have to be powered with the next lower power existing in the discrete quantized power set. At the end, the algorithm converges when power reaches 0 (i.e., no more antenna can be selected, even with the lowest non-zero power). The proposed algorithm is summarized in Algorithm 6.

20.3.1.3 Genetic Algorithm (GA)

In order to employ the GA, we propose to encode the power levels into binary words $b_j^{(k)}, \forall j = 1, \cdots, L+2$ and $\forall k = 1, \cdots, M$, such that each power levels is designed by a binary word. The length of the binary words $b_j^{(k)}$ depends on N (i.e., the number of quantization levels) as follows.

Algorithm 6 Proposed Iterative Algorithm for MIMO TWR-CR Networks with Discrete Power Distribution

---

1: **Initialization:** $R_{max} = 0, P_m^q = \bar{P}_m^q, P_r^q = \bar{P}_r^q, \boldsymbol{L}_{opt}^V = \varnothing$.

2: **while** $P_m^q = 0$ && $P_r^q = 0$ **do**

3:      $l = 1$.

4:      **while** $l \leq (L+2)M$ and $l \notin \boldsymbol{L}_{opt}^V$ **do**

5:          Compute the sum-rate $R^{(l)}$ using (20.6).

6:          $l = l + 1$.

7:      **end while**

8:      Find l_{opt} s.t $R_{opt} = \max\limits_{l} R_l$.

9:      **if** $R_{opt} > R_{max}$ **then**

10:          $\boldsymbol{\varepsilon}(l_{opt}) = 1$.

11:          $R_{max} = R_{opt}$.

12:          $\boldsymbol{L}_{opt}^V = \boldsymbol{L}_{opt}^V \cup \{l_{opt}\}$.

13:      **else**

14:          $P_m^q = P_m^q - \frac{\bar{P}_m^q}{N-1}$.

15:          $P_r^q = P_r^q - \frac{\bar{P}_r^q}{N-1}$.

16:      **end if**

17: **end while**

$\text{length}(b_j^{(k)}) = \lceil \log_2 N \rceil$, where $\lceil . \rceil$ denotes the integer round towards $+\infty$. For instance, if $N = 4$, two bits are sufficient to encode these levels. If $N = 11$, four bits are used to encode the code levels. In the last case, the number of required words is not a power of 2, some binary words are redundant, and they correspond to any valid word. Several solutions were proposed to solve this problem by discarding these words as illegal, assigning them a low utility, or mapping them to a valid word with fixed, random, or probabilistic remapping [20].

In the GA-based approach, we generate randomly G binary strings to form the initial population set, where G denotes the population length. Each string $S_g, \forall g = 1, \cdots, G$, is built by concatenating $(L+2)M$ binary words $b_j^{(k)}$ corresponding to a power level of each cognitive node antenna. Thus, the length of a string is equal to $(L+2)M \log_2 N$. Once the power level of each cognitive node in a string S_g is known, the algorithm verifies whether the interference constraint is satisfied or not. If it is the case, the GA computes the corresponding data rate $R^{(g)}$, which plays the role of the fitness of the string S_g. Otherwise, $R^{(g)} = 0$. Then, the algorithm selects τ, where $(1 \leq \tau \leq G)$, strings that provide the highest data rates and keeps them to the next population, while the $G - \tau$ remaining strings are generated by applying crossovers and mutations to the τ survived parents, as it is shown in Figure 20.3 and Figure 20.4.

Cross-overs consist in cutting two selected random parent strings at a corresponding point that is chosen randomly between 1 and $(L+2)M \lceil \log_2(N) \rceil$. The obtained fragments are then swapped and recombined to produce two new strings. After that, mutation (i.e., changing a bit value of the string randomly) is applied with a probability p. This procedure is repeated until reaching convergence or reaching the maximum iteration number denoted I. The proposed GA with discrete power levels is detailed in Algorithm 7.

Algorithm 7 Proposed Genetic Algorithm for MIMO TWR-CR Networks with Discrete Power Distribution

---

1: **Initialization:** $R_{max} = 0$.
2: Generate a random initial population containing all S_g, $\forall g = 1, \cdots, G$.
3: $itr = 1$.
4: **while** ($itr \leq I$ or not converge) **do**
5:      **for** $g = 1 : G$ **do**
6:          Find P_m^q, $\forall m = \{1,2\}$ and $P_{r_i}^q$, $\forall i = 1, \cdots, L$, $q = 1, \cdots, M$ corresponding to the string S_g.
7:          **if** interference constraint is satisfied **then**
8:              Compute the sum-rate $R^{(g)}$ using (20.6).
9:          **else**
10:              Set $R^{(g)}$ to 0.
11:          **end if**
12:      **end for**
13:      Save R_{max} such that $R_{max} = \max_g \mathbf{R}^{(g)}$.
14:      Keep the best τ strings providing the highest data rates to the next population.
15:      From the survived τ strings, generate $G - \tau$ new strings by applying cross-overs and mutations to generate a new population set.
16:      $itr = itr + 1$.
17: **end while**

20.3.1.4   Complexity Analysis for IA and GA

The formulated problems in Section 20.2 can be, of course, solved via an ES by investigating all possible combinations. This depends on L (i.e., the number of relays in the secondary network), M (i.e., the number of terminals and relays antennas), and N (i.e., the number of quantization levels). Therefore, the ES algorithm needs to perform $\sum_{i=0}^{L+2} \binom{(L+2)M}{i}(N-1)^i = O(N^{(L+2)M})$ tests (compute the achieved rate for different power combinations) to find the optimal solution [21] while our proposed IA and GA require $(N-1)((L+2)M)^2$ and GI times at most to compute the possible achievable rate until reaching a suboptimal solution, respectively. Also, it can be seen that the ES algorithm is not a practical choice due to its high complexity, especially for a large number of relays L, a large number of terminal and relays antenna M, or a high quantization level N. Hence, our proposed algorithms are able to reach a suboptimal solution with a considerable saving in terms of computational complexity. In addition to that, as it will be shown in the sequel, our numerical results show that our proposed algorithms achieve close performance to the ES method. Concerning the convergence of the algorithms, by experiments and for a large number of channel realizations, the proposed algorithms always converge successfully to their suboptimal solutions.

20.3.2   *Continuous Power Distribution*

In this context, we propose to solve our optimization problem formulated in (20.7–20.11) into two steps. first, we start with the closed-form expression of the optimal terminal power allocation solution by assuming fixed amplification factors at all relay antennas (i.e., converts the non-convex

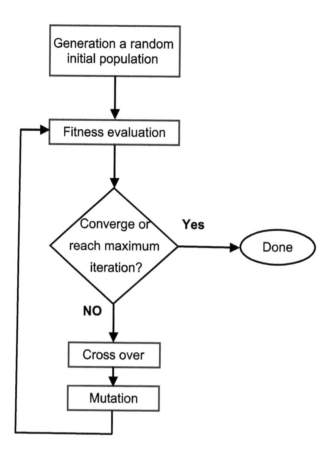

Figure 20.3: GA flow chart.

optimization problem to a convex one). In other words, we maximize the secondary sum-rate without any control on relay parameters. We then employ the PSO algorithm to optimize the relay amplification factors.

20.3.2.1   *Optimal Terminal Power Allocation*

We can solve our convex optimization problem for fixed amplification factors by exploiting its strong duality [22] and finding the Lagrangian multipliers that minimize the dual problem as follows:

$$\min_{\boldsymbol{\lambda} \geq 0} \quad \max_{\boldsymbol{P}_1 \geq 0, \boldsymbol{P}_2 \geq 0} \quad \mathcal{L}(\boldsymbol{\lambda}, \boldsymbol{P}_1, \boldsymbol{P}_2), \tag{20.12}$$

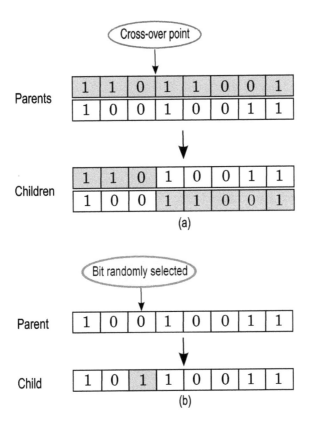

Figure 20.4: Genetic operators (a) cross-over technique and (b) mutation technique.

where $\mathcal{L}$ is the Lagrangian function [22], which is derived as

$$
\begin{aligned}
\mathcal{L}(\boldsymbol{\lambda}, \boldsymbol{P}_1, \boldsymbol{P}_2) =& \frac{1}{2} \sum_{v=1}^{M} \log_2 \left(1 + \frac{\sigma_{1v}^2 P_1^v}{C_{z_1}(v,v)} \right) + \frac{1}{2} \sum_{u=1}^{M} \log_2 \left(1 + \frac{\sigma_{2u}^2 P_2^u}{C_{z_2}(u,u)} \right) - \\
& \lambda_1 \left(\sum_{v=1}^{M} P_1^v - \bar{P}_t \right) - \lambda_2 \left(\sum_{u=1}^{M} P_2^u - \bar{P}_t \right) - \\
& \sum_{i=1}^{L} \lambda_{r_i} \left(\sum_{k=1}^{M} \left(\sum_{v=1}^{M} P_1^v |h_{1r_i}^{kv}|^2 + \sum_{u=1}^{M} P_2^u |h_{2r_i}^{ku}|^2 + N_0 \right) |w_i^k|^2 - \bar{P}_r \right) - \\
& \lambda_{th_1} \left(\sum_{v=1}^{M} \sum_{j=1}^{M} P_1^v |h_{1p}^{jv}|^2 + \sum_{u=1}^{M} \sum_{j=1}^{M} P_2^u |h_{2p}^{ju}|^2 - I_{th} \right) - \\
& \lambda_{th_2} \left(\sum_{i=1}^{L} \sum_{j=1}^{M} \sum_{k=1}^{M} \left(\sum_{v=1}^{M} P_1^v |h_{1r_i}^{kv}|^2 + \sum_{u=1}^{M} P_2^u |h_{2r_i}^{ku}|^2 + N_0 \right) |w_i^k|^2 |h_{r_ip}^{jk}|^2 - I_{th} \right),
\end{aligned}
\tag{20.13}
$$

where $\boldsymbol{\lambda}$ is a vector that contains all the Lagrangian multipliers of the system. λ_1, λ_2, and λ_{r_i}, represent the Lagrangian multipliers related to the peak power budget at T_1, T_2, and i^{th} relay, respectively, while λ_{th_1} and λ_{th_2} represent the Lagrangian multipliers related to the first and second phase interference constraints, respectively.

By taking the derivative of the Lagrangian with respect to the P_m^q, where $q = 1, ..., M$, we find the optimal transmit power allocated to the qth antenna at the mth terminals that maximizes the Lagrangian function, and, consequently, the sum-rate given the amplification factors. Its expression is given as the following:

$$P_m^q = \left(\frac{1}{\ln 2 \left[\lambda_m + \sum_{i=1}^{L} \sum_{k=1}^{M} \lambda_{r_i} |h_{mr_i}^{kq} w_i^k|^2 + \lambda_{th_1} \sum_{j=1}^{M} |h_{mp}^{jq}|^2 + \lambda_{th_2} \sum_{i=1}^{L} \sum_{j=1}^{M} \sum_{k=1}^{M} |h_{mr_i}^{kq} w_i^k h_{r_ip}^{jk}|^2 \right]} - \frac{C_{z_m}(q,q)}{\sigma_{mq}^2} \right)^+ . \tag{20.14}$$

We can employ the subgradient method, ellipsoid method, or other heuristic approaches to find the optimal Lagrangian multipliers of this problem [23]. Hence, to obtain the solution, we can start with any initial values for the different Lagrangian multipliers and evaluate the optimal powers. We then update the Lagrangian multipliers at the next iteration with a step size updated according to the nonsummable diminishing step length policy (see [23] for more details). The updated values of the optimal powers and the Lagrangian multipliers are repeated until convergence.

20.3.2.2 Particle Swarm Optimization (PSO)

In the second step, we employ the PSO algorithm to optimize the terminal powers and amplification factors at each relay antenna, simultaneously. The PSO idea was introduced by Kennedy and Eberhart in 1995 [24]. This idea is inspired by swarm intelligence, social behavior, and food searching of birds flocking and fish schooling. This approach is widely used in several wireless communication fields due to its simplicity (i.e., few parameters to adjust) [25, 26].

First, the PSO generates B random particles (i.e., random amplification factor matrices $\mathbf{W}^{(b)}$, $b = 1, \cdots, B$) of length $M \times L$ (i.e., L and M are the number of relays and number of antennas per relay, respectively) to form an initial population set $\mathcal{S}$. The algorithm computes the achieved sum-rate (20.6) of all particles by computing the optimal terminal powers (20.14) for this fixed amplification factor matrix $\mathbf{W}^{(b)}$. It then finds the particle that provides the global optimal sum-rate for this iteration, denoted $\mathbf{W}^{(b,\text{global})}$. In addition, for each particle b, it memorizes the position of its previous best performance, denoted $\mathbf{W}^{(b,\text{local})}$. After finding these two best values, PSO updates its velocity $V_j^{(b)}$ and its particle positions $\mathbf{W}_j^{(b)}$, respectively, at each iteration t as follows:

$$\mathbf{V}_j^{(b)}(t+1) = \omega \mathbf{V}_j^{(b)}(t) + r_1 \phi_1 \left(\mathbf{W}_j^{(b,\text{local})}(t) - \mathbf{W}_j^{(b)}(t) \right) + r_2 \phi_2 \left(\mathbf{W}_j^{(b,\text{global})}(t) - \mathbf{W}_j^{(b)}(t) \right), \tag{20.15}$$

$$\mathbf{W}_j^{(b)}(t+1) = \left(\mathbf{W}_j^{(b)}(t) + \mathbf{V}_j^{(b)}(t+1) \right)^+, \tag{20.16}$$

where ϕ_1 and ϕ_2 are two random positive numbers generated for each element j, and r_1 and r_2 are the step sizes that a particle takes toward the best individual candidate solution and the global best solution, respectively. In (20.15), ω is the inertia wieght used to control the speed of convergence ($0.8 \leq \omega \leq 1.2$). This procedure is repeated until convergence (i.e., sum-rate remains constant for a several number of iterations or reaching maximum number of iterations).

Details of the PSO algorithm as applied to our optimization problem of interest are given in Algorithm 8.

Algorithm 8 Particle Swarm Optimization Algorithm for MIMO TWR-CR Networks with Continuous Power Levels

---

1: Generate an initial population $\mathcal{S}$ composed of B random particles $\mathbf{W}^{(b)}$, $b = 1 \cdots B$.
2: **while** Not converged **do**
3:     **for** $b = 1, \cdots, B$ **do**
4:         Find the optimal terminal power by computing (20.14) corresponding to the particle $\mathbf{W}^{(b)} \in \mathcal{S}$.
5:         Compute the achieved sum-rate R_b using (20.6).
6:     **end for**
7:     Find $(b_g, t_g) = \arg\max\limits_{b,t} R_b(t)$ (i.e., b_g and t_g indicate the index and the position of the particle that results in the highest sum-rate).
8:     Set $R_{(\text{b,global})} = R_{b_g}(t_g)$ and $\mathbf{W}^{(\text{b,global})} = \mathbf{W}^{b_g}(t_g)$.
9:     Find $t_l = \arg\max\limits_{t} R_b(t)$ for each particle b (i.e., t_l indicates the position of the particle b that results in the highest local sum-rate).
10:    Set $R_{(\text{b,local})} = R_b(t_l)$ and $\mathbf{W}^{(\text{b,local})} = \mathbf{W}^b(t_l)$.
11:    Adjust the velocities and positions of all particles using Equations (20.15) and (20.16), respectively.
12:    Move to the new iteration $t = t + 1$.
13: **end while**

20.3.2.3 Complexity Analysis for PSO Algorithm

PSO is a meta-heuristic algorithm where the exact number of iterations needed to reach the solution is arbitrary and depends on the studied scenario. However, the computational complexity per iteration can be determined. According to (20.15) and (20.16), PSO needs to calculate 5 multiplications and 5 additions for every element of $\mathbf{W}^{(n)}$. Hence, $5(LM^2 + (L+1))N$ multiplications and $5(LM^2 + (L+1))N$ additions are calculated every iteration for the total B particles.

20.4 Results and Discussion

In this section, we provide selected simulation results for identically distributed Rayleigh fading channels to study the performance of the proposed scheme given in Figure 20.2. first, we study the performance of the proposed algorithms for TWR-CR networks with both discrete and continuous power distribution. Then, we compare the performanc of TWR scheme with OWR. The noise variance N_0 is assumed to be equal to 10^{-4}. Without loss of generality, it is assumed that $\bar{P}_t = \bar{P}_r = \bar{P}$.

20.4.1 Performance of the Proposed Algorithms for TWR-CR Networks

20.4.1.1 Simulation Results of Discrete Power Distribution

The GA is executed using these parameters: The mutation probability p is set to 0.5, $\tau = 0.25G$, and the maximum iteration number is $I = 35$.

    The merits of MIMO system over single antenna system are investigated in Figure 20.5. We

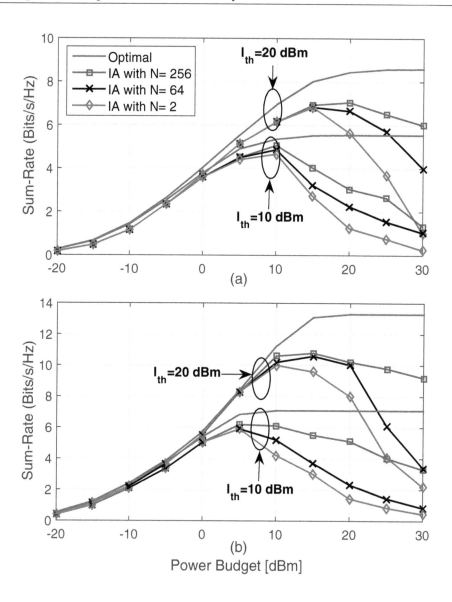

Figure 20.5: Achieved secondary sum-rate vs. power budget for the optimal and IA, with $L = 4$, $I_{th} = \{20, 10\}$ dBm, $N = \{256, 64, 2\}$, (a) $M = 1$, and (a) $M = 2$.

plot the secondary sum-rate for different values of $M = \{1, 2\}$, different values of $N = \{256, 64, 2\}$, and different values of $I_{th} = \{20, 10\}$ dBm. It is noticed that we can improve the performance significantly using the multi-antenna scheme than by using the single antenna scheme. The benefits of using the MIMO system appears clearly, with a considerable data rate improvement when M increases. When $N = 64$, $\bar{P} = 10$ dBm, $I_{th} = 20$ dBm using IA, with $M = 2$ instead of $M = 1$, our proposed algorithm improves the rate by around 70%, since the sum-rate increases from 6 bits/s/Hz to about 10.2 bits/s/Hz.

In low power budget region, IA and the optimal solution have almost the same sum-rate, while in the power budget region, a gap between both methods is obtained. This gap is increasing with higher $\bar{P}$ values. This is justified by the fact that starting from a certain value of $\bar{P}$, the system cannot supply the relays with the whole power budget. Hence, more relays are deactivated. In fact, at high values of

$\bar{P}$, the interference constraint can be affected. For this reason, we have introduced the discretization set to get more degrees of freedom by increasing N; as such, we enhance the secondary sum-rate. It is noted that with the proposed algorithm, when $N \to \infty$, we achieve the performance of the optimal solution.

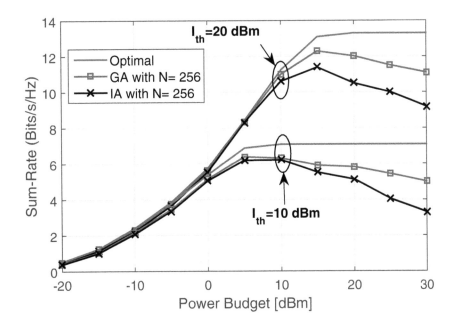

Figure 20.6: Achieved secondary sum-rate vs. the power budget for the optimal and the proposed algorithms, with $M = 2$, $L = 4$, and different values of $I_{th} = \{20, 10\}$ dBm.

To further improve the performance of the system, we proposed to employ the GA (with $G = 32$ random initial strings) to achieve better sum-rate than IA but with more complexity (central processing unit (CPU) time). In the low power budget region, we can notice in Figure 20.6 that both algorithms and the optimal solution have almost the same sum-rate, while in the high power budget region, the benefit of using GA is clearly observed. Indeed, IA is a deterministic approach that reaches always the same suboptimal solution for the same channel realization, while thanks to its random behavior, the GA achieves different suboptimal solutions, even for the same channel realization: It explores several additional options than IA.

The effect of varying I_{th} for GA and different number of antennas is shown in Figure 20.7, where we plot the secondary sum-rate vs. $\bar{P}$ for different values of $I_{th} = \{20, 10\}$, dBm, and different values of equipped antennas $M = \{1, 2\}$.

20.4.1.2 Simulation Results of Continuous Power Distribution

Figure 20.8 depicts the secondary sum-rate obtained using the PSO algorithm described in Algorithm 8 vs. $\bar{P}$. We compare its result to the optimal solution obtained using simulations. In this figure, we plot the achieved secondary sum-rate for different values of $I_{th} = \{20, 10\}$ dBm and $M = \{1, 2\}$ with fixed $L = 4$ vs. $\bar{P}$. Thanks to PSO, the proposed algorithm achieves almost the same performance of the optimal solution. It is shown that the gap for high power budget is reduced compared to discrete power distribution algorithms, since PSO does not depends on N.

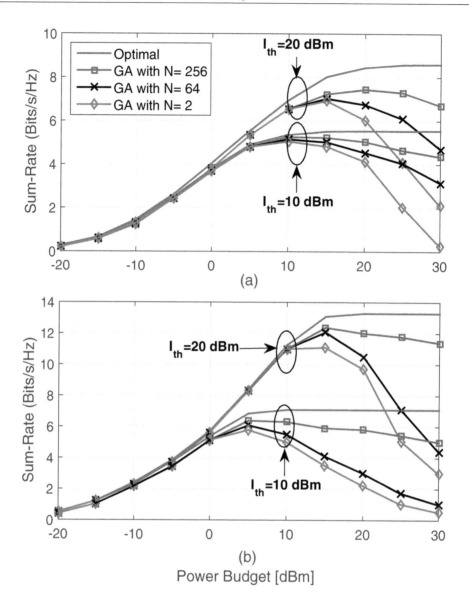

Figure 20.7: Achieved secondary sum-rate vs. power budget for the optimal and GA, with $L = 4$, $I_{th} = \{20, 10\}$ dBm, $N = \{256, 64, 2\}$, (a) $M = 1$, and (a) $M = 2$.

20.4.2   TWR Transmission vs. OWR Transmission

Figure 20.9 depicts the achieved secondary sum-rate of the optimal and proposed algorithms vs. the power budget, $\bar{P}$ with $L = 4$, $I_{th} = 20$ dBm, and different values of $M = \{1, 2\}$ for both OWR and TWR transmissions for discrete algorithms (i.e., IA and GA) and continuous algorithms (i.e., PSO). The secondary sum-rate of both OWR and TWR schemes is compared to the case when only one constraint is applied (either power budget constraint or interference constraint). It can be shown that the optimal solution with interference constraint only is an upper-bound for the case when both constraints are considered. It can be seen that we can almost double the secondary sum-rate by

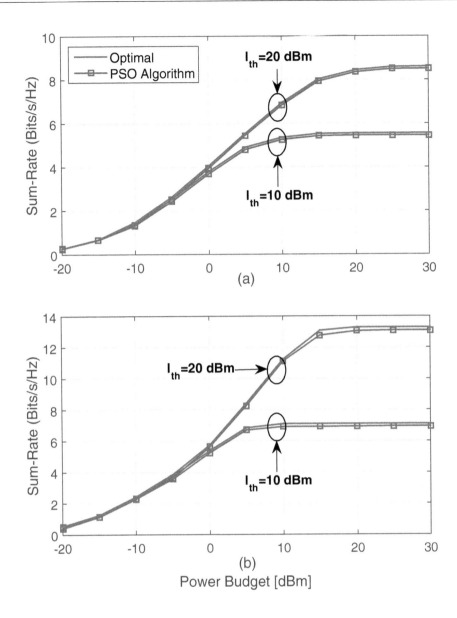

Figure 20.8: Achieved secondary sum-rate vs. power budget for the optimal and PSO algorithm with, $L = 4$, $I_{th} = \{20, 10\}$ **dBm, (a)** $M = 1$**, and (a)** $M = 2$**.**

using TWR transmission instead of using OWR transmission. In addition to that, OWR transmission requires more rate computational analysis than TWR transmission. Indeed, it requires the double number of operations to solve the optimization problem, since it has to execute the algorithm twice (i.e., every two time slots).

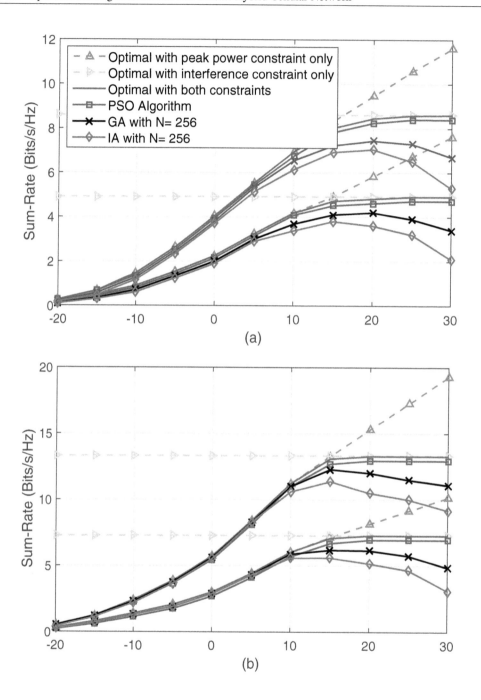

Figure 20.9: Performance comparison between TWR and OWR, with $L = 4$, $I_{th} = 20$ **dBm, (a)** $M = 1$, **and (a)** $M = 2$.

20.5 Summary

In this chapter, we have introduced and solved multiple-input multiple-output underlay cognitive radio two-way relaying optimization problem. More specifically, multiple amplify-and-forward relays and optimized the relay amplification factors adaptively with the terminal transmit powers have been considered. The goal is to maximize the secondary sum-rate while satisfying a power budget and primary quality-of-service constraints. We have investigated two models of power distributions; discrete power distribution and continuous power distribution. Heuristic algorithms for both discrete and continuous power distributions (i.e., iterative and genetic algorithms for the former distribution and practical swarm optimization algorithm for the later distribution) have been proposed and designed to solve the formulated optimization problem. Also, we have investigated the impact of some parameters on the system performance. Furthermore, we have compared the performance of our proposed scheme with one-way relaying scheme under the same set-up.

References

[1] M. Hasna and M.-S. Alouini, "Optimal power allocation for relayed transmissions over Rayleigh-fading channels," *IEEE Transactions on Wireless Communications*, vol. 3, no. 6, pp. 1999–2004, Nov. 2004.

[2] M. Hasna and M.-S. Alouini, "End-to-end performance of transmission systems with relays over Rayleigh-fading channels," *IEEE Transactions on Wireless Communications*, vol. 2, no. 6, pp. 1126–1131, Nov. 2003.

[3] B. Rankov and A. Wittneben, "Spectral efficient protocols for half-duplex fading relay channels," *IEEE Journal on Selected Areas in Communications*, vol. 25, no. 2, pp. 379–389, Feb. 2007.

[4] R. Zhang, Y.-C. Liang, C. C. Chai, and S. Cui, "Optimal beamforming for two-way multi-antenna relay channel with analogue network coding," *IEEE Journal on Selected Areas in Communications*, vol. 27, no. 5, pp. 699–712, Jun. 2009.

[5] K. Jitvanichphaibool, R. Zhang, and Y.-C. Liang, "Optimal resource allocation for two-way relay-assisted OFDMA," *IEEE Transactions on Vehicular Technology*, vol. 58, no. 7, no. 7, pp. 3311–3321, Sep. 2009.

[6] X. Kang, Y.-C. Liang, A. Nallanathan, H. Garg, and R. Zhang, "Optimal power allocation for fading channels in cognitive radio networks: Ergodic capacity and outage capacity," *IEEE Transactions on Wireless Communications*, vol. 8, no. 2, pp. 940–950, Feb. 2009.

[7] L. Li, X. Zhou, H. Xu, G. Li, D. Wang, and A. Soong, "Simplified relay selection and power allocation in cooperative cognitive radio systems," *IEEE Transactions on Wireless Communications*, vol. 10, no. 1, pp. 33–36, Jan. 2011.

[8] C. Luo, F. Yu, H. Ji, and V. Leung, "Distributed relay selection and power control in cognitive radio networks with cooperative transmission," in *Proc. of the IEEE International Conference on Communications (ICC'2010), Cape Town, South Africa*, May 2010.

[9] S. Bayat, R. H. Louie, B. Vucetic, and Y. Li, "Dynamic decentralised algorithms for cognitive radio relay networks with multiple primary and secondary users utilising matching theory," *Transactions on Emerging Telecommunications Technologies*, vol. 24, no. 5, pp. 486–502, May 2013.

[10] J. Xu, H. Zhang, D. Yuan, Q. Jin, and C.-X. Wang, "Novel multiple relay selection schemes in two-hop cognitive relay networks," in *Proc. IEEE 3rd International Conference on Communications and Mobile Computing (CMC'2011)*, Qingdao, China, Apr. 2011.

[11] M. Choi, J. Park, and S. Choi, "Low complexity multiple relay selection scheme for cognitive relay networks," in *Proc. IEEE Vehicular Technology Conference (VTC'2011)*, San Francisco, USA, Sep. 2011.

[12] M. Naeem, D. Lee, and U. Pareek, "An efficient multiple relay selection scheme for cognitive radio systems," in *Proc. IEEE International Conference on Communications Workshops (ICC'2010), Cape Town, South Africa*, May 2010.

[13] B. Wang, J. Zhang, and A. Host-Madsen, "On the capacity of MIMO relay channels," *IEEE Transactions on Information Theory*, vol. 51, no. 1, pp. 29–43, Jan. 2005.

[14] Y. Fan and J. Thompson, "MIMO configurations for relay channels: Theory and practice," *IEEE Transactions on Wireless Communications*, vol. 6, no. 5, pp. 1774–1786, May 2007.

[15] K.-J. Lee, H. Sung, E. Park, and I. Lee, "Joint optimization for one and two-way MIMO AF multiple-relay systems," *IEEE Transactions on Wireless Communications*, vol. 9, no. 12, pp. 3671–3681, Dec. 2010.

[16] Y. Rong, "Joint source and relay optimization for two-way MIMO multi-relay networks," *IEEE Communications Letters*, vol. 15, no. 12, pp. 1329–1331, Dec. 2011.

[17] G. Amarasuriya, C. Tellambura, and M. Ardakani, "Two-way amplify-and-forward multiple-input multiple-output relay networks with antenna selection," *IEEE Journal on Selected Areas in Communications*, vol. 30, no. 8, pp. 1513–1529, Aug. 2012.

[18] H. Park, J. Chun, and R. Adve, "Computationally efficient relay antenna selection for AF MIMO two-way relay channels," *IEEE Transactions on Signal Processing*, vol. 60, no. 11, pp. 6091–6097, Aug. 2012.

[19] R. Sarvendranath and N. Mehta, "Antenna selection in interference-constrained underlay cognitive radios: Sep-optimal rule and performance benchmarking," *IEEE Transactions on Communications*, vol. 61, no. 2, Feb. 2013.

[20] D. Beasley, D. R. Bull, and R. R. Martin, "An overview of genetic algorithms: Part 2, Research topics," *University Computing*, vol. 15, no. 4, pp. 170–181, 1993.

[21] K. H. Rosen, *Discrete Mathematics and its Applications (6th ed.)*. New York, NY: McGraw-Hill, 2007.

[22] S. Boyd and L. Vandenberghe, *Convex Optimization*. New York, NY, USA: Cambridge University Press, 2004.

[23] S. Boyd and A. Mutapcic, "Stochastic Subgradient Methods." *Notes for EE364*, Stanford University, Winter 2006-07.

[24] J. Kennedy and R. Eberhart, "Particle swarm optimization," in *Proc. of the IEEE International Conference on Neural Networks, Perth, Australia*, Nov/Dec. 1995.

[25] H. Chen, C. Tse, and J. Feng, "Minimizing effective energy consumption in multi-cluster sensor networks for source extraction," *IEEE Transactions on Wireless Communications*, vol. 8, no. 3, pp. 1480–1489, Mar. 2009.

[26] S. Efazati and P. Azmi, "Effective capacity maximization in multirelay networks with a novel cross-layer transmission framework and power-allocation scheme," *IEEE Transactions on Vehicular Technology*, vol. 63, no. 4, pp. 1691–1702, May 2014.

Chapter 21

Robust Beamforming Optimization for the Secondary Transmission in a Spectrum Sharing Cognitive Radio Network

Yongwei Huang

CONTENTS

21.1   Introduction

In the spectrum sharing cognitive radio (CR) networks, downlink beamforming design for the secondary transmission has been an intensive research topic in the past decade. Spectrum sharing allows secondary and primary users to access the same channel concurrently, and beamforming techniques can be applied in order to avoid excessive interference caused to the primary users while steering power towards the secondary receivers, and by equipping the secondary transmitter (e.g., a base station or access point) with antenna arrays. For a comprehensive coverage of the recent advances on CR communications and networking, readers are referred to the survey paper [1] and the magazine paper [2]. In particular, [3]–[5] provided readers some recent specific works on optimal CR transmit beamforming.

In this chapter, we are interested in a robust multicast transmit beamforming problem in the secondary transmission of a multiple-input multiple-output (MIMO) spectrum sharing CR network, under the assumption of imperfect channel state information (CSI). A basic and meaningful problem formulation is to minimize transmit power subject to quality-of-service (QoS) constraints on the secondary receivers and interference temperature constraints (or termed as CR interference limiting constraints) on the primary receivers. To proceed, let us first discuss some related works. The multicast transmit beamforming framework for a cellular communication system (i.e. that without CR) under a perfect CSI assumption, was originally developed in [6] (see also the survey paper [7]). In particular, that paper advocated to use semidefinite programming (SDP) relaxation to handle the multicast transmit optimization problems, an idea that has received growing attention in the last two decades. Its robust version under imperfect CSI was later studied in [8]. More recently, the framework has been extended to the CR scenario [4]. Therein, the robust CR multicast beamforming problem (our problem of interest) was also considered for a multiple-input single-output (MISO) spectrum sharing network; the idea was to apply a conservative bound on both the QoS constraints and the interference suppressing constraints, thereby obtaining a quadratically constrained quadratic program (QCQP) formulation which was subsequently approximated by SDP relaxation. As for a non-robust or robust uni-cast beamforming problem, we refer to [7], [9]–[12], and references therein; and for a robust optimization application in radar signal processing, one refers to [13] and references therein.

In an MIMO spectrum sharing CR network, we herein formulate the robust secondary multicast downlink beamforming problem into a robust QCQP problem, and propose two randomized approximation algorithms for it, which can provide better approximation accuracies than the previous method in [4] in a context of a MISO system. Specifically, in one algorithm, we take into consideration an equivalent QCQP reformulation of the robust QCQP problem of multicast beamforming. It is highlighted that the key step to recast the robust problem into the new QCQP is thanks to a closed-form expression for the optimal value of a norm-constrained quadratic optimization, which corresponds to a robust QoS constraint or a robust CR interference limiting constraint. Having the QCQP reformulation, we show that the robust beamforming problem is NP-hard, which means that it is believed that there is no polynomial-time algorithm to find a globally optimal solution. As a compromise, we then obtain a parameterized SDP relaxation problem of the QCQP reformulation, and the parameterized SDP can be solved by searching a one-dimensional parameter over an interval, and a feasibility checking routine using SDP, Then, capitalizing on the optimal solution of the parameterized SDP, we present a Gaussian randomization procedure to generate approximate solutions of the robust beamforming problem in a neat way. In addition, we identify several particularly interesting scenarios, in which the global optimum of the robust problem can be found efficiently.

Those particular cases include the scenarios when there are some (not a lot) primary and secondary receivers involved.

In the other algorithm, we consider a convex SDP relaxation of the robust optimal beamforming problem resorting to *S*-lemma (see, e.g., [14]), an important robust optimization tool. It turns out that the resulting SDP relaxation is looser than the previous parameterized SDP, giving rise to the possibility of returning lower transmit power at a small cost of problem feasibility rate. In addition, we herein present an alternative and new proof for the complex-valued *S*-lemma, and some extensions of the *S*-lemma. At the end, we show the outperformance of the proposed beamformers over the robust design in [4] by numerical simulation results.

The chapter is organized as follows. In Section 21.2, we introduce the system model and formulate the robust optimal beamforming problem. In Sections 21.3 and 21.4 , we propose the randomized approximation algorithms, point out one solvable scenario of the robust beamforming problem, and present a proof of the complex *S*-lemma. In Section 21.5, we demonstrate numerical examples showing the performance of three different algorithms. Finally, the chapter is concluded in Section 21.6.

Notation: We adopt the notation of using boldface for vectors a (lower case), and matrices A (upper case). The transpose operator and the conjugate transpose operator are denoted by the symbols $(\cdot)^T$ and $(\cdot)^H$, respectively. $\mathrm{tr}(\cdot)$ is the trace of the square matrix argument, I and $\mathbf{0}$ denote, respectively, the identity matrix and the matrix (or the row vector or the column vector) with zero entries (their size is determined from the context). The letter j represents the imaginary unit (i.e., $j = \sqrt{-1}$), while the letter i often serves as index throughout this chapter. For any complex number x, we use $\Re(x)$ and $\Im(x)$ to denote, respectively, the real and the imaginary part of x; $|x|$ and $\arg(x)$, represent the modulus and the argument of x respectively, and x^* ($\boldsymbol{x}^*$ or X^*) stands for the (component-wise) conjugate of x ($\boldsymbol{x}$ or X). We employ $A \bullet B$ to stand for the inner product $\mathrm{tr}(AB)$ of Hermitian or symmetric matrices A and B. The Euclidean norm (the Frobenius norm) of the vector $\boldsymbol{x}$ (the matrix X) is denoted by $\|\boldsymbol{x}\|$ ($\|X\|$). The curled inequality symbol $\succeq$ (and its strict form $\succ$) is used to denote generalized inequality: $A \succeq B$ means that $A - B$ is an Hermitian positive semidefinite matrix ($A \succ B$ for positive definiteness). We denote by $\mathcal{H}^N$ ($\mathcal{S}^N$) the space of Hermitian $N \times N$ matrices (the space of real-valued symmetric $N \times N$ matrices), and by $\mathcal{H}^N_+$ ($\mathcal{S}^N_+$) the set of all positive semidefinite matrices in $\mathcal{H}^N$ ($\mathcal{S}^N$). $\mathrm{E}[\cdot]$ represents the statistical expectation. The notation $\mathrm{vec}(X)$ stands for the vector stacked by the columns of X. Denote, respectively, by $\lambda_{\max}(\cdot)$ and $\lambda_{\min}(\cdot)$ the largest the eigenvalue and the least eigenvalue of the argument. Finally, for any optimization problem, $\mathcal{P}$, $v^\star(\mathcal{P})$ represents its optimal value.

21.2 System Model and Problem Formulation

Consider a spectrum sharing CR network that has a secondary N-antenna transmitter sending common information to M secondary receivers, and that there are K primary users (receivers) coexisting in the same spectrum (see Figure 21.1 for a pictorial show). Let $\boldsymbol{H}_m \in \mathbb{C}^{N \times N_m}$ be the MIMO channel from the secondary transmitter to the mth secondary receiver, where the number of receive antennas is denoted by N_m. Let $G_k \in \mathbb{C}^{N \times N_k'}$ be the MIMO channel from the secondary transmitter to the kth primary user, which is equipped with N_k' receive antennas. The signal received by secondary receiver m is given by

$$\boldsymbol{x}_m(t) = \boldsymbol{H}_m^H \boldsymbol{y}(t) + n_m(t), \tag{21.1}$$

where $\boldsymbol{y}(t) \in \mathbb{C}^N$ is the secondary transmit signal vector, and $n_m(t) \in \mathbb{C}^{N_m}$ is Gaussian noise vector, assumed to have zero mean and covariance $\sigma_m^2 I$. There are cases where interference from primary users to the secondary users contributes part of the noise terms $n_m(t)$. Although we may not physically model $n_m(t)$ as being white in those cases (except for $N_m = 1$), we can transform the received model to an equivalent noise-white model by some existing pre-whitening techniques. The sec-

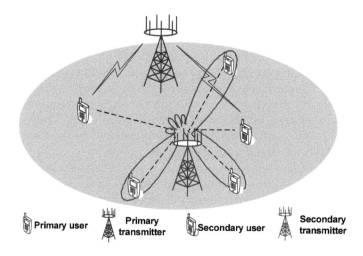

Figure 21.1: Scenario of single-group multicast transmission between the secondary transmitter and receivers in a spectrum sharing CR network.

ondary transmitter employs the multicast transmit beamforming scheme. In words, the secondary transmit signal is expressed as

$$\mathbf{y}(t) = s(t)w,$$

where $w \in \mathbb{C}^N$ is the beamformer weight, and $s(t) \in \mathbb{C}$ is the information signal. We assume $s(t)$ to be with zero mean and unit variance, without loss of generality.

Moreover, assuming maximum ratio combining receive beamforming for all the secondary receivers, the received signal-to-noise ratio (SNR) of secondary user m is

$$\text{SNR}_m = \frac{\|\mathbf{H}_m^H w\|^2}{\sigma_m^2}.$$

Accordingly, the secondary multicast beamformer design of interest may be formulated into the following QCQP (cf. [4,7,15]):

$$
\begin{aligned}
\underset{w}{\text{minimize}} \quad & w^H w \\
\text{subject to} \quad & \|\mathbf{H}_m^H w\|^2 \geq \sigma_m^2 \tau_m, \quad m = 1,\ldots,M, \\
& \|G_k^H w\|^2 \leq \eta_k, \quad k = 1,\ldots,K,
\end{aligned}
\tag{21.2}
$$

where τ_m specifies the minimal QoS of the secondary user m, in terms of SNR, and η_k specifies the maximal allowable interference level from the secondary transmitter to primary user k. Problem (21.2) has been considered in [4], where an effective approximation method via SDP relaxation has been applied. Herein, we further consider the imperfect CSI case.

In practice, one may not have perfect knowledge of the CSI, especially for the links of primary users. The CSI errors may be caused by inaccurate channel estimation, quantization in channel feedback, and outdated CSI effects. Let $\Delta_m \in \mathbb{C}^{N \times N_m}$ and $\Delta_k' \in \mathbb{C}^{N \times N_k'}$ denote the CSI errors associated with $\mathbf{H}_m$ and G_k, respectively. By assuming that the errors Δ_m and Δ_k' are deterministic norm-bounded, a worst-case robust version of (21.2) is given by (cf. problem (8) of [4]):

$$\underset{w}{\text{minimize}} \quad w^H w \tag{21.3a}$$

$$\text{subject to} \quad \underset{\|\Delta_m\| \le \varepsilon_m}{\text{minimize}} \ \|(\boldsymbol{H}_m + \Delta_m)^H w\|^2 \ge \sigma_m^2 \tau_m, \ m = 1, \dots, M, \tag{21.3b}$$

$$\underset{\|\Delta_k'\| \le \varepsilon_k'}{\text{maximize}} \ \|(G_k + \Delta_k')^H w\|^2 \le \eta_k, \ k = 1, \dots, K, \tag{21.3c}$$

where ε_m and ε_k' specify the bounds, or the worst-case magnitudes, of the CSI errors Δ_m and Δ_k', respectively. The robust beamforming problem (21.3) guarantees that for all admissible channel errors, all the secondary users must be served with QoSs no less than the specification $\{\tau_m\}$, and the interferences to all the primary users must be kept below $\{\eta_k\}$.

Robust beamforming problem (21.3) is hard to solve (achieving a globally optimal solution) within polynomial-time computational complexity, due to its nature of non-convexity of problem. Nevertheless, we aim to find an excellent suboptimal (approximate) solution with a lower computational cost, as will be done in next sections.

21.3 Polynomial-Time Approximation Algorithm for the Robust Beamforming Problem

In this section, we will propose a randomized approximation algorithm for the robust beamforming problem (21.3) and, particularly, identify some interesting subclasses, which possess a tight SDP relaxation. Thus, those subclasses of problem (21.3) can be efficiently solved up to global optimality, meaning that they are hiddenly convex programs indeed.

21.3.1 An equivalent QCQP Reformulation of the Robust Optimal Beamforming Problem

Let us start with an equivalent non-convex QCQP reformulation of (21.3). Consider the first robust QoS constraint of (21.3b) in a slightly more general form, and set

$$f_1(w) = \underset{\|E_1^{1/2}\Delta_1\| \le \varepsilon_1}{\text{minimize}} \ \|(\boldsymbol{H}_1 + \Delta_1)^H w\|^2 \ , \tag{21.4}$$

where $E_1 \succ 0$ governs the ellipsoid shape of the error set (or termed as the perturbation set in some robust optimization literature, e.g., [14]) and $w \ne 0$. We claim that the optimal value $f_1(w)$ has a closed-form expression, as stated in the following lemma (see related results in [16]–[18]).

Lemma 21.1

The optimal value for the minimization problem (21.4) (particularly formulating the first robust SNR constraint in the robust optimal beamforming problem (21.3)) has the closed-form expression:

$$f_1(w) = \left(\max\left\{ \|\boldsymbol{H}_1^H w\| - \varepsilon_1 \|E_1^{-1/2} w\|, 0 \right\} \right)^2 . \tag{21.5}$$

Proof 21.1 Suppose $\|H_1^H w\| > \varepsilon_1 \|E_1^{-1/2} w\|$. It then follows that

$$
\begin{aligned}
\|(H_1 + \Delta_1)^H w\| &\geq \|H_1^H w\| - \|(E_1^{1/2}\Delta_1)^H E_1^{-1/2} w\| \\
&\geq \|H_1^H w\| - \|E_1^{1/2}\Delta_1\| \|E_1^{-1/2} w\| \\
&\geq \|H_1^H w\| - \varepsilon_1 \|E_1^{-1/2} w\| \quad (> 0),
\end{aligned}
$$

and the inequality chain becomes equality chain when $\Delta_1 = -\varepsilon_1 \dfrac{E_1^{-1} w w^H H_1}{\|H_1^H w\| \|E_1^{-1/2} w\|}$ (it is seen also that $\|E^{1/2}\Delta_1\| = \varepsilon_1$).

Now assume $\|H_1^H w\| \leq \varepsilon_1 \|E_1^{-1/2} w\|$. By selecting $\Delta_1 = -\dfrac{E_1^{-1} w w^H H_1}{\|E_1^{-1/2} w\|^2}$, one verifies that $\|E_1^{1/2}\Delta_1\| = \|H_1^H w\| / \|E_1^{-1/2} w\| \leq \varepsilon_1$ and $\|(H_1 + \Delta_1)^H w\| = 0$. ■

Note that the closed-form optimal value $f_1(w)$ in (21.5) indeed can be rewritten equivalently as:

$$
f_1(w) = \begin{cases} (\|H_1^H w\| - \varepsilon_1 \|E_1^{-1/2} w\|)^2, & \text{if} \quad \|H_1^H w\| > \varepsilon_1 \|E_1^{-1/2} w\| \\ 0, & \text{if} \quad \|H_1^H w\| \leq \varepsilon_1 \|E_1^{-1/2} w\| \end{cases} . \tag{21.6}
$$

This indicates that those w satisfying $\|H_1^H w\| \leq \varepsilon_1 \|E_1^{-1/2} w\|$ can never be feasible for the robust beamforming problem (since $f_1(w) = 0 > \sigma_1^2 \tau_1$ is never true, in the case).

In particular, if $N_1 = 1$ (hence, H_1 becomes h_1), we conclude that

$$
\underset{\|E_1^{1/2}\delta_1\| \leq \varepsilon_1}{\text{minimize}} \quad |(h_1 + \delta_1)^H w|
$$

has the optimal value $\max\{|h_1^H w| - \varepsilon_1 \|E_1^{-1/2} w\|, 0\}$ (cf. problem (4) in [7] and problem (25) in [16]).

Interestingly, we remark that the optimal value $f_1(w)$ in Lemma 21.5 remains unchanged if $\|E_1^{1/2}\Delta_1\|$ is changed to the spectral norm (the maximal singular value) from the Frobenius norm, due to the fact that the two norms coincide when the argument is of rank one.

Like the proof in Lemma 21.1, it is easy to verify that the maximization problem in the first interference limiting constraint of (21.3c) has the optimal value $(\|G_1^H w\| + \varepsilon_1' \|w\|)^2$.

Therefore, it follows that the robust beamforming problem (21.3) can be recast into

$$
\underset{w}{\text{minimize}} \quad w^H w \tag{21.7a}
$$

$$
\text{subject to} \quad \|H_m^H w\| \geq \sigma_m \sqrt{\tau_m} + \varepsilon_m \|w\|, \, m = 1, \ldots, M, \tag{21.7b}
$$

$$
\|G_k^H w\| \leq \sqrt{\eta_k} - \varepsilon_k' \|w\|, \, k = 1, \ldots, K. \tag{21.7c}
$$

It is known that problem (21.7) is NP-hard [6] (in fact, problem (21.7) has been proved NP-hard, when $\varepsilon_m = 0$, $\forall m$, and $G_k = 0$, $\varepsilon_k' = 0$, $\forall k$). Instead of searching a globally optimal solution for (21.7), one can only resort to efficiently finding a suboptimal (or approximate) solution (e.g., see [4, 6, 8, 19]), as a compromise.

In the following, we will propose randomized, SDP-based, methods for generating an approximate solution of the robust beamforming problem (21.3) (or equivalently (21.7)), as well as presenting some efficiently solvable scenarios of problem (21.3).

21.3.2 Randomized Approximation Algorithm for the Robust Beamforming Problem

Evidently, problem (21.7) is tantamount to the following QCQP problem

$$
\begin{aligned}
\underset{w,\,t}{\text{minimize}} \quad & t^2 \\
\text{subject to} \quad & \|H_m^H w\| \geq \sigma_m \sqrt{\tau_m} + \varepsilon_m t,\ m = 1,\ldots,M, \\
& \|G_k^H w\| \leq \sqrt{\eta_k} - \varepsilon_k' t,\ k = 1,\ldots,K, \\
& \|w\| = t.
\end{aligned}
\tag{21.8}
$$

Note that any feasible point (w,t) of (21.8) must satisfy

$$
\sqrt{\lambda_{\max}(H_m H_m^H)} \geq \frac{\|H_m^H w\|}{\|w\|} \geq \frac{\sigma_m \sqrt{\tau_m}}{t} + \varepsilon_m,\ \forall m
\tag{21.9}
$$

and

$$
\sqrt{\lambda_{\min}(G_k G_k^H)} \leq \frac{\|G_k^H w\|}{\|w\|} \leq \frac{\sqrt{\eta_k}}{t} - \varepsilon_k',\ \forall k.
\tag{21.10}
$$

In fact, the first inequality in (21.9) follows from the basic property $\|w\|^2 \lambda_{\min}(A) \leq w^H A w \leq \|w\|^2 \lambda_{\max}(A)$ for a Hermitian matrix A, and the second inequality in (21.9) is due to the feasibility in (21.8). Likewise, (21.10) is derived.

Observe that $\sqrt{\lambda_{\max}(H_m H_m^H)} - \varepsilon_m > 0$ for all m (otherwise problem (21.8) would be infeasible). It follows from (21.9) and (21.10) that a necessary condition for t to be feasible for (21.8) is

$$
t_0 \leq t \leq t_1,
\tag{21.11}
$$

where the lower-bound t_0 and the upper-bound t_1 are, respectively, given by

$$
t_0 = \max_{1 \leq m \leq M} \left\{ \frac{\sigma_m \sqrt{\tau_m}}{\sqrt{\lambda_{\max}(H_m H_m^H)} - \varepsilon_m} \right\}
\tag{21.12}
$$

and

$$
t_1 = \min_{1 \leq k \leq K} \left\{ \frac{\sqrt{\eta_k}}{\sqrt{\lambda_{\min}(G_k G_k^H)} + \varepsilon_k'} \right\}.
\tag{21.13}
$$

Then problem (21.8), indeed, amounts to the following problem:

$$
\begin{aligned}
\underset{W,\,t}{\text{minimize}} \quad & t & \text{(21.14a)} \\
\text{subject to} \quad & H_m H_m^H \bullet W \geq (\sigma_m \sqrt{\tau_m} + \varepsilon_m t)^2,\ m = 1,\ldots,M, & \text{(21.14b)} \\
& G_k G_k^H \bullet W \leq (\sqrt{\eta_k} - \varepsilon_k' t)^2,\ k = 1,\ldots,K, & \text{(21.14c)} \\
& I \bullet W = t^2, & \text{(21.14d)} \\
& t_0 \leq t \leq t_1, & \text{(21.14e)} \\
& W \succeq 0,\ \text{Rank}(W) = 1, & \text{(21.14f)}
\end{aligned}
$$

where t_0 and t_1 are specified in (21.12) and (21.13), respectively. The SDP relaxation problem of it is (dropping the rank-one constraint):

$$
\begin{aligned}
\underset{W,\,t}{\text{minimize}} \quad & t \\
\text{subject to} \quad & H_m H_m^H \bullet W \geq (\sigma_m \sqrt{\tau_m} + \varepsilon_m t)^2,\ m = 1,\ldots,M, \\
& G_k G_k^H \bullet W \leq (\sqrt{\eta_k} - \varepsilon_k' t)^2,\ k = 1,\ldots,K, \\
& I \bullet W = t^2, \\
& W \succeq 0,\ t_0 \leq t \leq t_1.
\end{aligned}
\tag{21.15}
$$

Note that whenever t is frozen, problem (21.15) is an SDP feasibility problem. Now, let $g(t)$ be the optimal value of such a feasibility problem. In other words, we have

$$g(t) = \begin{cases} t, & \text{if it is feasible} \\ +\infty, & \text{if infeasible} \end{cases}.$$

Note that for $g(t) = t$ (i.e., (21.15) is feasible at t), any feasible W is optimal for the feasibility problem. Therefore, (21.15) amounts to the one-dimensional optimization problem

$$\begin{aligned} \underset{t}{\text{minimize}} \quad & g(t) \\ \text{subject to} \quad & t_0 \le t \le t_1. \end{aligned} \tag{21.16}$$

With this reformulation, problem (21.15) can be solved via (21.16): fixing t, solving the SDP feasibility problem (obtaining $g(t)$), and reducing t iteratively. In the optimization literature, there are some derivative-free methods for solving the one-dimensional optimization problem (21.16). One of these methods is called compass or coordinate search (cf. [20, Algorithm 3.1 and Section 8.1], [21, Algorithm 7.1]). Alternatively, we adopt either the uniform sampling or the Matlab function `fminbnd`, in order to output an optimal solution in our numerical simulations.

Once such a solution $(W^\star, t^\star)$ of (21.15) is obtained, we retrieve a rank-one approximate solution for (21.15) by making use of $W^\star$. Specifically, a randomization procedure is proposed as follows: Take random vectors w_i, $i = 1, \ldots, I$, from the complex normal distribution $\mathcal{N}_\mathbb{C}(0, W^\star)$ and compute

$$\lambda(w_i) = \min \left\{ \frac{\|\boldsymbol{H}_m^H w_i\| - \sigma_m \sqrt{\tau_m}}{\varepsilon_m}, \frac{\sqrt{\eta_k} - \|G_k^H w_i\|}{\varepsilon_k'}, m = 1, \ldots, M, k = 1, \ldots, K \right\}. \tag{21.17}$$

Clearly, if

$$\|w_i\| \le \lambda(w_i), \tag{21.18}$$

then $(w_i, \|w_i\|)$ is feasible for (21.8).

We summarize a randomized approximate solution for problem (21.7) (or equivalently (21.8)) in Algorithm 9.

Algorithm 9 Gaussian randomization procedure for robust beamforming problem (21.7)

Require: H_m, G_k, σ_m, τ_m, ε_m, η_k, ε_k', I;
Ensure: a randomized approximate solution w for (21.7);
1: solve (21.16), and find an optimal solution $(W^\star, t^\star)$;
2: if $\text{Rank}(W^\star) = 1$, then output $w^\star$ with $w^\star w^{\star H} = W^\star$ and terminate;
3: draw random vectors $w_i \in \mathcal{N}_\mathbb{C}(0, W^\star)$, $i = 1, \ldots, I$, and compute $\lambda(w_i)$ by (21.17);
4: pick up w_{i_0}, such that $i_0 = \arg\min\{\|w_i\| : \|w_i\| \le \lambda(w_i), i = 1, \ldots, I\}$.

21.3.3 Solvable Subclasses of the Robust Beamforming Problem via SDP Relaxation

In this subsection, we shall identify two polynomially solvable subclasses of the robust beamforming problem (21.3) (i.e., (21.7) or ((21.8)). Namely, they are: (1) $M + K = 3$ and $N \ge 3$ (i.e., the number of primary and secondary receivers equal three and the number of the transmit antennas is not less than three); (2) $M + K = 2$ (i.e., one primary and one secondary receiver). The results presented here are related to the classes of polynomially solvable QCQPs with "not many" constraints (e.g., see [22] in a context of radar, and references therein).

To proceed the elaboration, let $t^\star$ be a numerical minimizer for $g(t)$ over the interval $[t_0, t_1]$, as in problem (21.16), and $W^\star$ be a corresponding feasible solution, namely, $(W^\star, t^\star)$ complies with the constraints of (21.15). Without loss of generality, assume $M = 2$ and $K = 1$. It then follows that

$$H_m H_m^H \bullet W^\star \geq (\delta_m \sqrt{\tau_m} + \varepsilon_m t^\star)^2, \quad m = 1, 2, \tag{21.19}$$

$$G_1 G_1^H \bullet W^\star \leq (\eta_1 - \varepsilon_1' t^\star)^2, \tag{21.20}$$

$$I \bullet W^\star = (t^\star)^2. \tag{21.21}$$

In order to construct a rank-one matrix fulfilling the above four conditions, we leverage on the rank-one matrix decomposition theorem [23], which is cited as the following lemma.

Lemma 21.2 [23]

Let X be a non-zero $N \times N$ ($N \geq 3$) complex Hermitian positive semidefinite matrix and A_i be Hermitian matrix, $i = 1, 2, 3, 4$, and suppose that $(A_1 \bullet Y, A_2 \bullet Y, A_3 \bullet Y, A_4 \bullet Y) \neq (0, 0, 0, 0)$ for any non-zero complex Hermitian positive semidefinite matrix Y. Then one can find, in polynomial time, a rank-one matrix xx^H, such that x satisfies

$$x^H A_i x = A_i \bullet X, \quad i = 1, 2, 3, 4.$$

We observe that if there exists $(\mu_1, \mu_2, \mu_3, \mu_4) \in \mathbb{R}^4$, such that

$$\mu_1 A_1 + \mu_2 A_2 + \mu_3 A_3 + \mu_4 Q_4 \succ 0, \tag{21.22}$$

then the condition in Lemma 21.2

$$(A_1 \bullet Y, A_2 \bullet Y, A_3 \bullet Y, A_4 \bullet Y) \neq (0, 0, 0, 0) \text{ for any non-zero } Y \succeq 0$$

always holds. It is verified immediately that the condition (21.22) is satisfied for $A_4 = I$ and any other A_i's.

Thus, one can polynomially construct a matrix $w^\star w^{\star H}$ according to the above rank-one decomposition lemma, such that

$$w^{\star H} H_m H_m^H w^\star = H_m H_m^H \bullet W^\star, \quad m = 1, 2,$$

$$w^{\star H} G_1 G_1^H w^\star = G_1 G_1^H \bullet W^\star,$$

$$\|w^\star\|^2 = I \bullet W^\star.$$

This implies that $(w^\star w^{\star H}, t^\star)$ is feasible for (21.15); thus, $(w^\star, t^\star)$ is feasible for (21.8). Therefore, we conclude that $w^\star$ is optimal for (21.8), since the problem shares the same optimal value $t^\star$ with its SDP relaxation problem (21.15).

For the scenario with parameters fulfilling $M + K = 2$, we instead apply another rank-one decomposition theorem (see [24]–[26]), which is cited as follows.

Lemma 21.3 [25]

Suppose that X is a $N \times N$ complex Hermitian positive semidefinite matrix of rank R, and A, B are two $N \times N$ given Hermitian matrices. Then, there is a rank-one decomposition $X = \sum_{r=1}^{R} x_r x_r^H$, such that

$$x_r^H A x_r = \frac{X \bullet A}{R} \quad \text{and} \quad x_r^H B x_r = \frac{X \bullet B}{R}, \quad r = 1, \dots, R.$$

It follows from the lemma that a rank-one matrix ww^H can be obtained efficiently from the general-rank optimal solution $W^\star$ (e.g., see (21.19)–(21.21) with $M = 1$), such that

$$(\boldsymbol{H}_1\boldsymbol{H}_1^H - \frac{(\delta_1\sqrt{\tau_1} + \varepsilon_1 t^\star)^2}{(t^\star)^2}\boldsymbol{I}) \bullet ww^H \leq 0,$$

$$(G_1 G_1^H - \frac{(\eta_1 - \varepsilon_1' t^\star)^2}{(t^\star)^2}\boldsymbol{I}) \bullet ww^H \geq 0.$$

Then, it is seen that $w^\star = \frac{t^\star}{\|w\|}w$ is feasible for (21.8), and the objective function value is the same as the optimal value $t^\star$ of its relaxation problem (21.15). In the same vein, we reach the conclusion that $w^\star$ is optimal for (21.8).

21.4 Another Randomized Approximation Algorithm via Complex-Valued S-Lemma and Convex Relaxation

In this section, we present a convex SDP relaxation of the robust secondary transmit beamforming problem (21.3), via S-lemma (a known robust optimization tool). By capitalizing on the new SDP relaxation, we propose another randomized approximation algorithm for the robust beamforming problem.

To proceed further, let $\delta_m = \text{vec}(\Delta_m)$, $h_m = \text{vec}(\boldsymbol{H}_m)$, and, similarly, δ_k' and g_k are defined. Noting the fact that $\text{tr}(A^H BC) = \text{vec}(A)^H(\boldsymbol{I} \otimes B)\text{vec}(C)$, we can reformulate problem (21.3) equivalently into the following problem:

$$\begin{aligned}
\underset{w}{\text{minimize}} \quad & \boldsymbol{I} \bullet W \\
\text{subject to} \quad & \delta_m^H(\boldsymbol{I} \otimes W)\delta_m + 2\Re(h_m^H(\boldsymbol{I} \otimes W)\delta_m) + h_m^H(\boldsymbol{I} \otimes W)h_m \geq \sigma_m^2\tau_m, \\
& \qquad\qquad\qquad\qquad\qquad \forall\|\delta_m\|^2 \leq \varepsilon_m^2, \, m = 1,\dots,M, \\
& (\delta_k')^H(\boldsymbol{I} \otimes W)(\delta_k') + 2\Re(g_k^H(\boldsymbol{I} \otimes W)\delta_k') + g_k^H(\boldsymbol{I} \otimes W)g_k \leq \eta_k, \\
& \qquad\qquad\qquad\qquad\qquad \forall\|\delta_k'\|^2 \leq (\varepsilon_k')^2, \, k = 1,\dots,K, \\
& W = ww^H.
\end{aligned} \qquad (21.23)$$

In order to further convert (21.23) (with infinitely many constraints) into an equivalent optimization problem with some constraints, we employ the complex-valued S-lemma. Let us start with the complex S-lemma and its new proof.

21.4.1 Complex-Valued S-Lemmas

It is known that S-lemma is a useful tool in optimal control and robust optimization [14]. To proceed the discussion, we cite the well-known S-lemma in real-valued version as follows.

Lemma 21.4 [14]

Let A, B be symmetric matrices of the same size, and let the quadratic form $x^T Ax + 2a^T x + a$ be strictly positive at certain point x_0. Then the condition

$$x^T Bx + 2b^T x + b \geq 0, \forall x : x^T Ax + 2a^T x + a \geq 0 \qquad (21.24)$$

holds true if and only if

$$\exists \lambda \geq 0: \begin{bmatrix} B & b \\ b^T & b \end{bmatrix} - \lambda \begin{bmatrix} A & a \\ a^T & a \end{bmatrix} \succeq 0.$$

For complex-valued parameters, the S-lemma can be extended to include one more quadratic constraint in the set of x as in (21.24) (e.g., see [25, Theorem 3.4] and [27, Lemma 4.1]). Herein, we wish to present an alternative and new proof for the complex-valued S-lemma. Let us begin with a statement of the homogenous case of the complex S-lemma (i.e., all quadratic functions are homogenous).

Lemma 21.5 Homogeneous S-lemma

Let

$$F_0 = \begin{bmatrix} B & b \\ b^H & b \end{bmatrix} \in \mathcal{H}^{N+1}, F_i = \begin{bmatrix} A_i & a_i \\ a_i^H & a_i \end{bmatrix} \in \mathcal{H}^{N+1}, i = 1, 2, \mathbf{y} = \begin{bmatrix} \mathbf{x} \\ t \end{bmatrix} \in \mathbb{C}^{N+1}. \quad (21.25)$$

Let the quadratic forms $\mathbf{y}^H F_1 \mathbf{y} \geq 0$ and $\mathbf{y}^H F_2 \mathbf{y} \geq 0$ be strictly positive at point $\mathbf{y}_0$. Then the condition

$$\mathbf{y}^H F_0 \mathbf{y} \geq 0, \forall \mathbf{y} : \mathbf{y}^H F_i \mathbf{y} \geq 0, i = 1, 2 \quad (21.26)$$

holds true if and only if

$$\exists \lambda \geq 0, \mu \geq 0 : F_0 - \lambda F_1 - \mu F_2 \succeq 0. \quad (21.27)$$

Proof 21.2   The implication from (21.27) to (21.26) is evident, and, therefore, we need to only focus on the proof of reverse direction, namely, the implication from (21.26) to (21.27). Consider the following SDP:

$$\begin{aligned} \underset{X}{\text{minimize}} \quad & \text{tr}(F_0 X) \\ \text{subject to} \quad & \text{tr}(F_i X) \geq 0, i = 1, 2, \\ & \text{tr} X = 1, \\ & X \succeq 0, \end{aligned} \quad (21.28)$$

and its dual:

$$\begin{aligned} \underset{\lambda, \mu, \nu}{\text{maximize}} \quad & \nu \\ \text{subject to} \quad & F_0 - \lambda F_1 - \mu F_2 - \nu I \succeq 0, \\ & \lambda \geq 0, \mu \geq 0, \nu \in \mathbb{R}. \end{aligned} \quad (21.29)$$

Evidently the dual SDP is strictly feasible; now, let us verify the strict feasibility of the primal SDP (21.28). Consider

$$X(\varepsilon) = (1 - \varepsilon) \frac{1}{\|\mathbf{y}_0\|^2} \mathbf{y}_0 \mathbf{y}_0^H + \frac{\varepsilon}{\text{tr} I} I, \varepsilon \in (0, 1). \quad (21.30)$$

Clearly, for a sufficiently small $\varepsilon > 0$, it holds true that $\text{tr}(F_1 X(\varepsilon)) > 0$, $\text{tr}(F_2 X(\varepsilon)) > 0$, $\text{tr} X(\varepsilon) = 1$ and $X(\varepsilon) \succ 0$. Therefore, (21.28) is strictly feasible. Hence, the strong duality theorem (e.g., see [29, Theorem 1.4.2]) holds for (21.28) and (21.29), which means that both the primal and dual SDPs are solvable[1], and they share the same optimal value.

Suppose that $(\{X^\star\}; \{\lambda^\star, \mu^\star, \nu^\star\})$ is an optimal primal-dual pair, and $p^\star$ is the optimal value for (21.28) and (21.29). Since problem (21.28) has three constraints only, hence, by some SDP rank reduction procedure (see, e.g., [9] and [30])[2], there exists a rank-one solution for (21.28), say $\mathbf{y}^\star \mathbf{y}^{\star H}$. Thus it follows that $p^\star := \mathbf{y}^{\star H} F_0 \mathbf{y}^\star$ (the optimal value), $\mathbf{y}^{\star H} F_1 \mathbf{y}^\star \geq 0$ and $\mathbf{y}^{\star H} F_2 \mathbf{y}^\star \geq 0$. From (21.26), we have $p^\star \geq 0$. Then, we have $\nu^\star = p^\star \geq 0$ and $F_0 - \lambda^\star F_1 - \mu^\star F_2 \succeq \nu^\star I \succeq 0$, which implies (21.27). Thus, the proof is complete. ■

---

[1] By "solvable," we mean that the minimization (maximization) problem is feasible, bounded below (above), and the optimal valued is attained [29, page 13].

[2] Alternatively, a specific rank-one decomposition theorem (cf. [24], [25]) or some randomized postprocessing procedure in [15] can be applied.

We note that the proof presented herein is quite different from that in [25]. In fact, the proof therein is based on an local management interface (LMI) description of new matrix co-positive cones, and some results like the bipolar theorem in convex analysis [28] and specific rank-one matrix decomposition theorems [24]. In contrast, we here apply the strong duality result for constructed SDPs (21.28) and (21.29), and a rank reduction procedure [9].

We remark that the equivalence between (21.26) and (21.27) can be characterized as the statement that the two sets are equal to each other:

$$\mathcal{A}_1 = \{F_0 | \boldsymbol{y}^H F_0 \boldsymbol{y} \geq 0, \forall \boldsymbol{y} : \boldsymbol{y}^H F_i \boldsymbol{y} \geq 0, i = 1,2\}$$

and

$$\mathcal{A}_2 = \{F_0 | \exists \lambda \geq 0, \mu \geq 0 \text{ such that } F_0 - \lambda F_1 - \mu F_2 \succeq 0\}.$$

In other words, the convex cone $\mathcal{A}_1$ can be represented by the set $\mathcal{A}_2$ consisting of LMIs (under the Slater condition), and, thus, we say that $\mathcal{A}_1$ is a nice cone in the sense that it is computationally tractable.

By capitalizing on a limiting argument, we can generalize the S-lemma to the inhomogeneous case, stated as follows.

Lemma 21.6 Inhomogeneous S-lemma
Let $f_0(\boldsymbol{x}) = \boldsymbol{x}^H B \boldsymbol{x} + 2\Re(b^H \boldsymbol{x}) + b$, $f_i(\boldsymbol{x}) = \boldsymbol{x}^H A_i \boldsymbol{x} + 2\Re(a_i^H \boldsymbol{x}) + a_i$, $i = 1,2$, and let

$$F_0 = \begin{bmatrix} B & b \\ b^H & b \end{bmatrix} \in \mathcal{H}^{N+1}, F_i = \begin{bmatrix} A_i & a_i \\ a_i^H & a_i \end{bmatrix} \in \mathcal{H}^{N+1}, i = 1,2. \tag{21.31}$$

Let the quadratic functions $f_1(\boldsymbol{x})$ and $f_2(\boldsymbol{x})$ be strictly positive at certain vector $\boldsymbol{x}_0$. Then, the condition

$$f_0(\boldsymbol{x}) \geq 0, \forall \boldsymbol{x} : f_i(\boldsymbol{x}) \geq 0, i = 1,2 \tag{21.32}$$

holds true if and only if

$$\exists \lambda \geq 0, \mu \geq 0 : F_0 - \lambda F_1 - \mu F_2 \succeq 0. \tag{21.33}$$

Note that the constraint set in either (21.32) or (21.26) comprises two inequality quadratic conditions. Recently, the S-lemmas have been extended to the case where the constraint set consists of one inequality and one equality quadratic constraints in [31], due to the key fact that the phase of a complex number provides an additional freedom of degree. To facilitate reading the extension, we include it as the following corollary:

Corollary 21.1
Let $f_0(\boldsymbol{x}) = \boldsymbol{x}^H B \boldsymbol{x} + 2\Re(b^H \boldsymbol{x}) + b$, $f_i(\boldsymbol{x}) = \boldsymbol{x}^H A_i \boldsymbol{x} + 2\Re(a_i^H \boldsymbol{x}) + a_i$, $i = 1,2$, and let

$$F_0 = \begin{bmatrix} B & b \\ b^H & b \end{bmatrix} \in \mathcal{H}^{N+1}, F_i = \begin{bmatrix} A_i & a_i \\ a_i^H & a_i \end{bmatrix} \in \mathcal{H}^{N+1}, i = 1,2. \tag{21.34}$$

Suppose that F_2 is indefinite and that there is $\boldsymbol{x}_0$, such that $f_1(\boldsymbol{x}_0) < 0$ and $f_2(\boldsymbol{x}_0) = 0$. Then, the following two statements are equivalent to each other:

1. $f_0(\boldsymbol{x}) \geq 0$ for all $\boldsymbol{x}$ satisfying $f_1(\boldsymbol{x}) \leq 0$ and $f_2(\boldsymbol{x}) = 0$.

2. There are $\lambda \geq 0$ and $\mu \in \mathbb{R}$ such that

$$F_0 + \lambda F_1 + \mu F_2 \succeq 0.$$

21.4.2 Convex Relaxation for the Robust Beamforming Problem

Clearly, from Lemma 21.6, it follows that the condition

$$x^H Bx + 2\Re(b^H x) + \beta \geq 0, \forall x^H Ax + \alpha \geq 0$$

is equivalent to the linear matrix inequality

$$\exists \lambda \geq 0 : \begin{bmatrix} B - \lambda A & b \\ b^H & \beta - \lambda \alpha \end{bmatrix} \succeq 0 \qquad (21.35)$$

(we set $\mu = 0$ in Lemma 21.6), provided that the Slater condition, $\exists x_0$, such that $x_0^H Ax_0 + \alpha > 0$, holds. Having this LMI reformulation in hand, we can convert the constraints in the beamforming problem (21.23) into equivalent matrix inequality forms.

By (21.35), we can recast each group of (infinitely many) constraints in (21.23) into a matrix inequality constraint. For example, we look at the first group of constraints:

$$\delta_1^H (I \otimes W)\delta_1 + 2\Re(h_1^H (I \otimes W)\delta_1) + h_1^H (I \otimes W)h_1 \geq \sigma_1^2 \tau_1, \forall \|\delta_1\|^2 \leq \varepsilon_1^2.$$

Setting $B = (I \otimes W)$, $b^H = h_1^H (I \otimes W)$, $\beta = h_1^H (I \otimes W)h_1 - \sigma_1^2 \tau_1$, $A = -I$, $\alpha = \varepsilon_1^2$, and capitalizing on (21.35), we can reexpress it into the matrix inequality constraint: $\exists \mu_1 \geq 0$, such that

$$\begin{bmatrix} I \otimes W + \mu_1 I & (I \otimes W)h_1 \\ h_1^H (I \otimes W) & h_1^H (I \otimes W)h_1 - \sigma_1^2 \tau_1 - \mu_1 \varepsilon_1^2 \end{bmatrix} \succeq 0.$$

Note that the above matrix inequality is quadratic with respect to the design beamforming vector w (noting $W = ww^H$). Therefore, in the same vein, the robust beamforming problem (21.23) can be reformulated to the following problem with quadratic matrix inequality (QMI) constraints:

$$\underset{w, \{\mu_m\}, \{\lambda_k\}}{\text{minimize}} \quad I \bullet ww^H \qquad (21.36a)$$

$$\text{subject to} \quad \begin{bmatrix} I \otimes ww^H + \mu_m I & (I \otimes ww^H)h_m \\ h_m^H (I \otimes ww^H) & h_m^H (I \otimes ww^H)h_m - \sigma_m^2 \tau_m - \mu_m \varepsilon_m^2 \end{bmatrix} \succeq 0, \quad (21.36b)$$

$$m = 1, \ldots, M,$$

$$\begin{bmatrix} I \otimes ww^H + \lambda_k I & (I \otimes ww^H)g_k \\ g_k^H (I \otimes ww^H) & g_k^H (I \otimes ww^H)g_k - \eta_k - \lambda_k (\varepsilon_k')^2 \end{bmatrix} \preceq 0, \quad (21.36c)$$

$$k = 1, \ldots, K,$$

$$\mu_m \geq 0, \forall m, \lambda_k \leq 0, \forall k. \qquad (21.36d)$$

Thus, the conventional SDP (or termed as LMI) relaxation problem is formulated:

$$\underset{W, \{\mu_m\}, \{\lambda_k\}}{\text{minimize}} \quad I \bullet W \qquad (21.37a)$$

$$\text{subject to} \quad \begin{bmatrix} I \otimes W + \mu_m I & (I \otimes W)h_m \\ h_m^H (I \otimes W) & h_m^H (I \otimes W)h_m - \sigma_m^2 \tau_m - \mu_m \varepsilon_m^2 \end{bmatrix} \succeq 0, \quad (21.37b)$$

$$m = 1, \ldots, M,$$

$$\begin{bmatrix} I \otimes W + \lambda_k I & (I \otimes W)g_k \\ g_k^H (I \otimes W) & g_k^H (I \otimes W)g_k - \eta_k - \lambda_k (\varepsilon_k')^2 \end{bmatrix} \preceq 0, \quad (21.37c)$$

$$k = 1, \ldots, K,$$

$$W \succeq 0, \mu_m \geq 0, \forall m, \lambda_k \leq 0, \forall k. \qquad (21.37d)$$

This SDP can be solved within polynomial time via an interior-point method, although the size of each linear matrix inequality constraint appears a bit large.

21.4.3 Relations between the Two Relaxation Problems

Note that (21.37) is a convex relaxation of the original robust beamforming problem (21.3), while problem (21.15) is a non-convex relaxation of problem (21.3) (or, equivalently, (21.7)). In what follows, we will discuss some relations between two relaxation problems.

Proposition 21.1
It holds that

1. *if $(W, \sqrt{I \bullet W})$ is feasible for (21.15), then W, together with some $\mu_m \geq 0$ and $\lambda_k \leq 0$, is feasible for (21.37);*

2. *if $(ww^H, \{\mu_m\}, \{\lambda_k\})$ is feasible for (21.37), then $(ww^H, \|w\|)$ is feasible for (21.15).*

Proof 21.3 (1) Since $(W, \sqrt{I \bullet W})$ is feasible for (21.15), hence, $H_m H_m^H \bullet W \geq (\sigma_m \sqrt{\tau_m} + \varepsilon_m \sqrt{I \bullet W})^2, \forall m$, which means that

$$\|W^{1/2} H_m\| - \varepsilon_m \|W^{1/2}\| \geq \sigma_m \sqrt{\tau_m}, \forall m. \tag{21.38}$$

Likewise, we have $\|W^{1/2} G_k\| + \varepsilon_k' \|W^{1/2}\| \leq \sqrt{\eta_k}, \forall k$. Observe that $\|W^{1/2}(H_m + \Delta_m)\| \geq \|W^{1/2} H_m\| - \|W^{1/2} \Delta_m\| \geq \|W^{1/2} H_m\| - \varepsilon_m \|W^{1/2}\|$, for $\Delta_m : \|\Delta_m\| \leq \varepsilon_m$. Therefore, it follows from (21.38) that $\sigma_m^2 \tau_m \leq \min_{\|\Delta_m\| \leq \varepsilon_m} \mathrm{tr}((H_m + \Delta_m)^H W(H_m + \Delta_m))$. Similarly, it has $\eta_k \geq \max_{\|\Delta_k'\| \leq \varepsilon_k'} \mathrm{tr}(G_k + \Delta_k')^H W(G_k + \Delta_k')$. By S-lemma, we conclude that $(W, \{\mu_m\}, \{\lambda_k\})$ is feasible for (21.37) for some $\mu_m \geq 0$ and some $\lambda_k \leq 0$, $m = 1, \ldots, M$ and $k = 1, \ldots, K$.

(2) Let us re-denote $\delta_m = \mathrm{vec}(\Delta_m)$, $h_m = \mathrm{vec}(H_m)$; δ_k' and g_k are defined analogously. Suppose that $(ww^H, \{\mu_m\}, \{\lambda_k\})$ is feasible for (21.37). Let us look into the first constraint. Suppose that $\mu_1 > 0$. It follows from the first constraint of (21.37) and Schur complement lemma that

$$h_1^H (I \otimes ww^H) h_1 - \sigma_1^2 \tau_1 - \mu_1 \varepsilon_1^2 \geq h_1^H (I \otimes ww^H)(I \otimes ww^H + \mu_1 I)^{-1}(I \otimes ww^H) h_1. \tag{21.39}$$

It is straightforward to verify that $(I \otimes ww^H + \mu_1 I)^{-1} = \frac{1}{\mu_1}(I - \frac{I \otimes ww^H}{\mu_1 + \|w\|^2})$ by noting that $(I \otimes ww^H)(I \otimes ww^H) = \|w\|^2(I \otimes ww^H)$, and to check that the right-hand side of (21.39) is equal to $\frac{\|w\|^2}{\mu_1 + \|w\|^2}\|(I \otimes w^H) h_1\|^2$. It follows from (21.39) that

$$
\begin{aligned}
\|(I \otimes w^H) h_1\|^2 &\geq (1 + \frac{\|w\|^2}{\mu_1})(\sigma_1^2 \tau_1 + \mu_1 \varepsilon_1^2) \\
&= \varepsilon_1^2 \|w\|^2 + \sigma_1^2 \tau_1 + \frac{\|w\|^2 \sigma_1^2 \tau_1}{\mu_1} + \mu_1 \varepsilon_1^2 \\
&\geq \varepsilon_1^2 \|w\|^2 + \sigma_1^2 \tau_1 + 2\|w\| \sigma_1 \varepsilon_1 \sqrt{\tau_1},
\end{aligned}
$$

which is equivalent to $H_1 H_1^H \bullet ww^H \geq (\sigma_1 \sqrt{\tau_1} + \varepsilon_1 \sqrt{I \bullet ww^H})^2$. In words, for any $\mu_1 > 0$, we have

$$
\left\{ w : \begin{bmatrix} I \otimes ww^H + \mu_1 I & (I \otimes ww^H) h_1 \\ h_1^H (I \otimes ww^H) & h_1^H (I \otimes ww^H) h_1 - \sigma_1^2 \tau_1 - \mu_1 \varepsilon_1^2 \end{bmatrix} \succeq 0 \right\}
$$
$$
\subseteq \{ w : H_1 H_1^H \bullet ww^H \geq (\sigma_1 \sqrt{\tau_1} + \varepsilon_1 \sqrt{I \bullet ww^H})^2 \}.
$$

By a limiting argument, the inclusion relation still holds for $\mu_1 = 0$. This means that ww^H fulfills the first constraint of (21.15). Similarly, we can show that ww^H also fulfills the second to the M-th constraints.

Now let us deal with the $(M + 1)$-th constraints of (21.37). Due to the feasibility of

$(ww^H, \{\mu_m\}, \{\lambda_k\})$, we see that $I \otimes ww^H + \lambda_1 I \preceq 0$, which means $ww^H + \lambda_1 I \preceq 0$, which in turn implies $\lambda_1 \leq -\|w\|^2$. In a similar way, we can show that ww^H satisfies the second set of constraints of (21.15). ■

The first argument of the proposition indicates that the convex relaxation (21.37) is not as tight as the non-convex relaxation (21.15) in general (namely, (21.37) always gives a lower-bound of (21.15)). In other words, the corresponding optimal values satisfy $v^\star((21.37)) \leq v^\star((21.15)) \leq v^\star((21.3))$. The second argument of Proposition 21.1 implies that if (21.37) has an optimal rank-one solution $w^\star w^{\star H}$ (together with optimal $\{\mu_m^\star\}$ and $\{\lambda_k^\star\}$), then relaxation problems (21.37), (21.15), and the original optimal beamforming problem (21.3) are equivalent to each other in the sense that they share the same optimal value. Although the relaxation (21.37) appears looser, there is a trade-off: SDP (21.37) can be solved in a single step, unlike (21.15) iteratively solved via the one-dimension optimization problem (21.16).

Further, regarding relaxation (21.37), we has the following observations: (i) if we get a rank-one optimal solution $w^\star w^{\star H}$ for (21.37), then there is no gap between (21.3) and (21.37), and, thus, we do not have to solve (21.15) (via (21.16)); (ii) in case of getting a solution $W^\star$ of rank two or higher for (21.37), the optimal value of (21.37) can serve as a new t_0 for solving (21.16), namely, update $t_0 := \max\{t_0, \sqrt{I \bullet W^\star}\}$; (iii) in order to generate an approximate solution for (21.7) (or equivalently, (21.3)), it is possible to use $W^\star$ as a covariance, according to (21.17) and (21.18). The third observation motivates us to design a new approximation algorithm for the robust secondary downlink beamforming problem (21.7).

21.4.4 Another Randomized Approximation Algorithm via Convex SDP Relaxation

In this subsection, we wish to establish another Gaussian randomization algorithm to solve the beamforming problem (21.7), taking advantage of efficiently solving SDP (21.37).

To proceed, let us assume that $W^\star$ is an optimal solution for (21.37). Upon $W^\star$, we can employ (21.17) to randomly generate a beamforming vector w fulfilling (21.7b), (21.7c), thus, satisfying (21.3b), (21.3c) (which, in turn, means that ww^H is feasible for (21.37b) and (21.37c) with some $\{\mu_m\}$ and $\{\lambda_k\}$). Based on such an observation, we conclude a randomized approximation algorithm for (21.7) via the SDP relaxation (21.37), which consists of solving the SDP (21.37) (obtaining an optimal solution $(W^\star, \{\mu_m^\star\}, \{\lambda_k^\star\})$), and steps 2–4 of Algorithm 1. Algorithm 10 summarizes the mentioned procedure producing an approximate solution for the robust beamforming problem (21.7).

Algorithm 10 Gaussian randomization procedure via SDP (21.37) for robust beamforming problem (21.7)

Require: H_m, G_k, σ_m, τ_m, ε_m, η_k, ε'_k, I;

Ensure: a randomized approximate solution w of (21.7);

1: solve SDP (21.37), obtaining an optimal solution $(W^\star, \{\mu_m^\star\}, \{\lambda_k^\star\})$;

2: if Rank $(W^\star) = 1$, then output $w^\star$ with $w^\star w^{\star H} = W^\star$ and terminate;

3: draw random vectors $w_i \in \mathcal{N}_{\mathbb{C}}(0, W^\star)$, $i = 1, \ldots, I$, and compute $\lambda(w_i)$ by (21.17);

4: pick up w_{i_0}, such that $i_0 = \arg\min\{\|w_i\| : \|w_i\| \leq \lambda(w_i), i = 1, \ldots, I\}$.

Observe that the complexities of the two approximation algorithms are dominated by solving the respective SDP relaxation problems. In Algorithm 1, the cost of outputting a solution by t-search is about 20 times empirically of solving an SDP feasibility problem, which has worst-case complexity of $O((\max\{M+K,N\})^4 N^{0.5} \log(1/\zeta))$ for a given accuracy ζ (cf. [32]); in the algorithm via (21.37), the computational cost is higher since the sizes of the involved SDP cones are quite large; for instance, in the particular case of $N_m = N_k' = 1$ $\forall m,k$, the worst-case complexity is up to of $O(N^{6.5} \log(1/\zeta))$ for a small $(M+K)$ (cf. [29]).

21.5   Simulation Results

We consider a spectrum sharing CR network with a three-antenna secondary transmitter, five single-antenna secondary receivers, and two single-antenna primary receivers (i.e., $N = 3$, $M = 5$, $K = 2$, and $N_m = N_k' = 1$, $\forall m,k$). The elements of the channels (from the secondary transmitter to either the primary or the secondary users) are assumed to be iid complex Gaussian distributed with mean 0 and variance 1. We fix the secondary receivers' noise variance $\sigma_m^2 = 1$ for all m, and set $\tau_m = 10$ dB for all m and $\eta_k = 0$ dB for all k. The same channel perturbation level is assumed for all primary and secondary channels, i.e., $\varepsilon_m = \varepsilon_k' = \varepsilon$, $\forall m,k$. A total of 3000 channel realizations (each with 10000 Gaussian randomizations) are tested.

In the simulation, we wish to examine how the average transmit power is affected by the radius of the channel perturbation set. We compare our two proposed robust beamforming designs, namely t-search (problem (21.8)) and S-lemma (problem (21.37)) designs, with an existing robust design provided by problem (17) of [4]. Moreover, as the robust power minimization problem with QoS constraints could be intrinsically infeasible, the average transmit power in Figure 21.2 (a) is obtained by averaging only those channel realizations for which all the three robust designs are feasible, i.e., at least one feasible beamforming solution can be found for each design after randomization procedure. In the legend, the "beamformer" stands for the result after randomization, while "SDP relaxation value" means the optimal value of the SDP relaxations corresponding to the three robust designs.

As shown in Figure 21.2 (a), we see a result that higher transmit power is required to assure larger radius of the channel error set (i.e., provide more robust beamformer). Figure 21.2 (a) also shows that the average transmit powers by our proposed robust beamformers are lower than that by (17) of [4] in general. This means that the former methods are less conservative than the latter. Let us compare the performance of our two robust proposed designs. In Figure 21.2 (a), for the SDP relaxation values, we note that S-lemma yields a slightly lower value than t-search, which is consistent with our claim in Proposition 21.1, i.e., the relaxation (16) (S-lemma) is looser than (12) (t-search). For the beamformer's power, we see that S-lemma leads to slightly better performance than t-search. As observed, the performance gap of our two algorithms however is not big. This phenomenon may be caused by the precision of the relaxed solution, the approximation procedure employed and the simulation settings.

To get a better understanding of the conservativeness, Figure 21.2 (b) plots the feasibility rate of the three designs with the same setup as Figure 21.2 (a). Here the feasibility rate is denoted as the ratio of the number of channel realizations, for which we can generate a feasible beamformer via randomization, over the total 3000 channel trials. It can be seen from Figure 21.2 (b) that the proposed two robust designs have much higher feasibility rates than that of [4] over the whole perturbation radii tested.

In Figure 21.2 (b), we also observe that t-search method yields slightly higher feasibility rate than S-lemma. In contrast, as Figure 21.2 (a) shows, S-lemma design has superior performance in terms of the SDP relaxation values and the transmit power of beamformers. In other words, there is a trade-off between the two proposed robust designs.

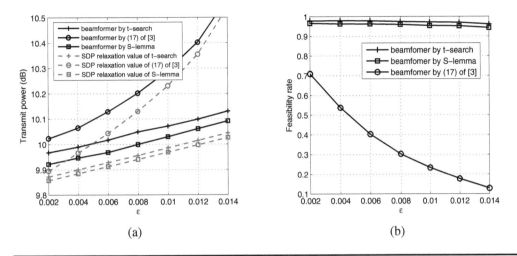

Figure 21.2: (a) Average transmit power vs. the radius of channel perturbation set. (b) Feasibility rate vs. perturbation radius ε.

21.6   Conclusion

We have considered a robust secondary multicast beamforming design problem in a MIMO spectrum sharing CR network. Two efficient algorithms for the robust problem have been proposed: One algorithm includes solving a one-dimensional optimization problem, checking the feasibility of SDPs, and a Gaussian randomized procedure; the other algorithm contains solving an LMI relaxation problem for a QMI problem formulated via S-lemma, and a randomization procedure. Particularly, in the former algorithm, we have shown the closed-form optimal value for a norm-constrained quadratic optimization problem, which is a key to formulate the one-dimensional optimization problem. Further, we have proved that the robust optimal beamforming problem can be solved efficiently up to the global optimality for the special cases of "not many" primary and secondary receivers (cf. Section 21.3.3). In the latter algorithm, we have provided an alternative (new) proof for the complex-valued S-lemma. The performance of the proposed beamforming designs has been demonstrated by simulations.

21.7   Acknowledgment

The author would like to thank Dr. Qiang Li, Professor Wing-Kin Ma, and Professor Shuzhong Zhang for their great helps to this work.

References

[1] Y.-C. Liang, K. Chen, Y. Li, and P. Mahonen, "Cognitive radio networking and communications: An overview," *IEEE Transactions on Vehicular Technology*, vol. 60, no. 7, pp. 3386–3407, September 2011.

[2] R. Zhang, Y.-C. Liang, and S. Cui, "Dynamic resource allocation in cognitive radio networks," *IEEE Signal Processing Magazine*, vol. 27, no. 3, pp. 102–114, May 2010.

[3] Y. J. Zhang and A. M.-C. So, "Optimal spectrum sharing in MIMO cognitive radio networks via semidefinite programming," *IEEE Journal on Selected Areas in Communications*, vol. 29, no. 2, pp. 362–373, February 2011.

[4] K. T. Phan, S. A. Vorobyov, N. D. Sidiropoulos, and C. Tellambura, "Spectrum sharing in wireless networks via QoS-aware secondary multicast beamforming," *IEEE Transactions on Signal Processing*, vol. 57, no. 6, pp. 2323–2335, June 2009.

[5] Y. Huang, Q. Li, W.-K. Ma, and S. Zhang, "Robust multicast beamforming for spectrum sharing-based cognitive radios," *IEEE Transactions on Signal Processing*, vol. 60, no. 1, pp. 527–533, January 2012.

[6] N. Sidiropoulos, T. D. Davidson, and Z.-Q. Luo, "Transmit beamforming for physical-layer multicasting," *IEEE Transactions on Signal Processing*, Vol. 54, No. 6, pp. 2239–2251, June 2006.

[7] A. B. Gershman, N. D. Sidiropoulos, S. Shahbazpanahi, M. Bengtsson, and B. Ottersten, "Convex optimization-based beamforming: From receive to transmit and network designs," *IEEE Signal Processing Magazine*, vol. 27, no. 3, pp. 62–75, May 2010.

[8] E. Karipidis, N.D. Sidiropoulos, and Z.-Q. Luo, "Convex transmit beamforming for downlink multicasting to multiple co-channel groups," *Proceedings of IEEE ICASSP*, pp. 973–976, 2006.

[9] Y. Huang and D. P. Palomar, "Rank-constrained separable semidefinite programming with applications to optimal beamforming," *IEEE Transactions on Signal Processing*, vol. 58, no. 2, pp. 664–678, February 2010.

[10] Y. Huang and D. P. Palomar, "A dual perspective on separable semidefinite programming with applications to optimal downlink beamforming," *IEEE Transactions on Signal Processing*, vol. 58, no. 8, pp. 4254–4271, August 2010.

[11] G. Zheng, K.-K. Wong, and B. Ottersten, "Robust cognitive beamforming with bounded channel uncertainties," *IEEE Transactions on Signal Processing*, vol. 57, no. 12, pp. 4871-4881, December 2009.

[12] G. Zheng, K.-K. Wong, and T.-S. Ng, "Robust linear MIMO in the downlink: A worst-case optimization with ellipsoidal uncertainty regions," *EURASIP Journal on Advances in Signal Processing*, vol. 2008, Article ID 609028, 15 pages, 2008.

[13] A. De Maio, Y. Huang, and M. Piezzo, "A Doppler robust max-min approach to radar code design," *IEEE Transactions on Signal Processing*, vol. 58, no. 9, pp. 4943–4947, September 2010.

[14] A. Ben-Tal, L. E. Ghaoui, and A. Nemirovski, *"Robust Optimization,"* Princeton University Press, Princeton, New Jersey, 2009.

[15] Y. Huang and D. P. Palomar, "Randomized algorithms for optimal solutions of double-sided QCQP with applications in signal processing," *IEEE Transactions on Signal Processing*, vol. 62, no. 5, pp. 1093–1108, March 2014.

[16] S.A. Vorobyov, A.B. Gershman, and Z.-Q. Luo, "Robust adaptive beamforming using worst-case performance optimization: A solution to the signal mismatch problem," *IEEE Transactions on Signal Processing*, vol. 51, no. 2, pp. 313–324, February 2003.

[17] R. G. Lorenz and S. P. Boyd, "Robust minimum variance beamforming," *IEEE Transactions on Signal Processing*, vol. 53, no. 5, pp. 1684–1696, May 2005.

[18] S.A. Vorobyov, A.B. Gershman, Z.-Q. Luo, and N. Ma, "Adaptive beamforming with joint robustness against mismatched signal steering vector and interference nonstationarity," *IEEE Signal Processing Letters*, vol. 11, no. 2, pp. 108–111, 2004.

[19] Y. Huang and S. Zhang, "Approximation algorithms for indefinite complex quadratic maximization problems," *Science in China Series A: Mathematics*, vol. 53, no. 10, pp. 2697–2708, October 2010.

[20] T. G. Kolda, R. M. Lewis, and V. Torczon, "Optimization by direct search: new perspectives on some classical and modern methods," *SIAM Review*, vol. 45, no. 3, pp. 385–482, 2003.

[21] A. R. Conn, K. Scheinberg, and L. N. Vicente, *"Introduction to Derivative-Free Optimization,"* MPS-SIAM Series on Optimization, Philadelphia, 2009.

[22] A. De Maio, S. De Nicola, Y. Huang, D. P. Palomar, S. Zhang, and A. Farina, "Code design for radar STAP via optimization theory," *IEEE Transactions on Signal Processing*, vol. 58, no. 2, pp. 679–694, February 2010.

[23] W. Ai, Y. Huang, and S. Zhang, "New results on Hermitian matrix rank-one decomposition," *Mathematical Programming: Series A*, vol. 128, no. 1-2, pp. 253–283, June 2011.

[24] J.F. Sturm and S. Zhang, "On cones of nonnegative quadratic functions," *Mathematics of Operations Research*, vol. 28, no. 2, pp. 246–267, 2003.

[25] Y. Huang and S. Zhang, "Complex matrix decomposition and quadratic programming," *Mathematics of Operations Research*, vol. 32, no. 3, pp. 758–768, 2007.

[26] Y. Huang, A. De Maio, and S. Zhang, "Semidefinite programming, matrix decomposition, and radar code design," in *Convex Optimization in Signal Processing and Communications*, D. P. Palomar and Y. Eldar, Eds. Cambridge, U.K.: Cambridge University Press, 2010, chapter 6.

[27] Y. Huang, D. P. Palomar, and S. Zhang, "Lorentz-positive maps and quadratic matrix inequalities with applications to robust MISO transmit beamforming," *IEEE Transactions on Signal Processing*, vol. 61, no. 5, pp. 1121–1130, March 2013.

[28] R. T. Rockafellar, *Convex Analysis*, Princeton University Press, Princeton, New Jersey, 1970.

[29] A. Ben-Tal and A. Nemirovski, "Lectures on modern convex optimization," *Class Notes, Georgia Institute of Technology*, Fall 2013. [Online]. Available: http://www2.isye.gatech.edu/~nemirovs/Lect_ModConvOpt.pdf.

[30] W. Ai, Y. Huang, and S. Zhang, "On the low rank solutions for linear matrix inequalities," *Mathematics of Operations Research*, vol. 33, no. 4, pp. 965–975, November 2008.

[31] M. Medra, Y. Huang, W.-K. Ma, T. N. Davidson, "Low-complexity robust MISO downlink precoder design under imperfect CSI," manuscript, July 2015.

[32] Z.-Q. Luo, W.-K. Ma, A. M.-C. So, Y. Ye, and S. Zhang, "Semidefinite relaxation of quadratic optimization problems: From its practical deployments and scope of applicability to key theoretical results," *IEEE Signal Processing Magazine*, vol. 27, no. 3, pp. 20–34, May 2010.

POWER CONTROL V

Chapter 22

Interference-Aware Power Allocation in Spectrum Sharing Cognitive Radio

Gosan Noh, and Daesik Hong

CONTENTS

In a spectrum sharing environment, interference is one of the main factors that disrupts the transmission of a secondary user. While the transmission of the primary user can be protected by the interference constraint at the primary user receiver, there is no mechanism to protect the transmission of the secondary user from the interference. Accordingly, the interference from the primary user impairs the signal-to-interference-plus-noise ratio (SINR) of the secondary user and thus degrades the ergodic capacity of the secondary user.

As a solution to deal with these impairments, we employ an interference cancellation technique and investigate the best strategy for balancing the interference cancellation capability and secondary user capacity. In this regard, we propose an interference-aware power allocation approach which can maximize the ergodic capacity of the secondary user while guaranteeing the interference constraints to the primary user. Analysis and simulation show that the ergodic capacity can be increased by up to 130% with the proposed interference-aware power allocation over what is possible with conventional power allocation. The obtained results can be used when predicting how much capacity gain is obtained by employing interference cancellation in a spectrum sharing environment.

22.1   Introduction

22.1.1   Objective of the chapter

There has recently been a surge in demand for radio spectrum arising from the rapid growth of wireless applications. As a result, the radio spectrum available for wireless communication is fast approaching exhaustion, a situation referred to as the *spectrum scarcity problem* [1]. Meanwhile, a large portion of the licensed spectrum is unused at any given time and location, according to actual spectrum usage measurement [2].

The concurrence of spectrum scarcity and inefficient spectrum usage arises from a fixed spectrum allocation, where each frequency band is exclusively allocated to licensed users. In order to resolve the problems caused by fixed spectrum allocation, the Federal Communications Commission (FCC) Spectrum Policy Task Force has decided to make spectrum access more efficient, so that the unused portion of the licensed spectrum can be used by unlicensed users [2]. It is, therefore, possible for unlicensed users to access the temporally and locally unused licensed spectrum, with the aid of cognitive radio technology.

Cognitive radio enables the unlicensed user to learn from and adapt to the external radio environment [3]. With the ability to utilize the spectrum more flexibly, cognitive radio can be realized in one of two scenarios: spectrum sharing and spectrum sensing. In both scenarios, hierarchical spectrum reuse is allowed between primary and secondary users. The primary user is a licensed user in possession of a license to exclusively use the spectrum. On the other hand, the secondary user is an unlicensed user who has no spectrum license. Coexistence of the primary and secondary users is possible only if the secondary user is allowed to use the spectrum while the transmission of the primary user is sufficiently protected.

In the spectrum sharing scenario, which assumes simultaneous transmission by the primary and secondary users, the secondary user can transmit at a low enough power, such that the interference to the primary user does not exceed a predefined threshold [1]. In this regard, much of the research has focused on allocating power in such a way that it satisfies the interference constraint at the primary user receiver and the corresponding capacity analysis.

Although the primary user can be sufficiently protected by the interference constraint, the secondary user cannot be protected by the interference from the primary user. This is because there is no mechanism to mitigate the interference from the primary user that the secondary user experiences in the spectrum sharing scenario. Hence, the primary user interference degrades the SINR of the secondary user and its corresponding capacity [8].

Research on spectrum sharing among hierarchical systems was motivated by the *interference temperature* concept, introduced by the FCC [2]. The interference temperature indicates the tolerable interference level at the primary user receiver. As long as the interference power received by the primary user is less than the received-power constraint (i.e., a metric for the interference temperature), the primary and secondary user can coexist at the same frequency band, enhancing the overall spectral efficiency.

22.1.2 Prior Works

Gastpar provided capacity analysis with the average received-power constraint in additive white Gaussian noise (AWGN) channels [4]. Ghasemi *et al.* derived the capacity of the secondary user assuming fading channels with both average and peak received-power constraints [5]. Kang *et al.* proposed optimal power allocation to maximize the ergodic and outage capacities under various power constraints and fading channels [6]. Wang *et al.* analyzed the capacity gain achieved by exploiting the selection diversity over the multiple primary channels and multiple secondary users [7]. Although these works derived the secondary user capacity, they did not consider the interference from the primary user, which can severely degrade the SINR of the secondary user.

The effect of primary user interference on secondary user transmission was investigated in [8], [9], [10], [11]. Cho *et al.* derived an upper bound on the capacity under the interference-to-signal ratio (ISR) constraint when the interference from a primary transmitter to a secondary receiver is considered [8]. Suraweera *et al.* obtained the secondary user capacity with imperfect channel information when the interference term from the primary user is included in the analysis [9]. Kim *et al.* provided upper- and lower-bounds for the ergodic capacity considering the effect of outdated channel information [10]. Almalfouh *et al.* investigated power allocation when considering the effect of primary user interference, but their work was limited to a spectrum sensing environment where the source of the interference is the sensing failure [11].

22.1.3 Motivation

All of the prior works focused only on protecting the primary user from secondary user transmission, which can be guaranteed by the interference constraint. However, no means of protecting the secondary user transmission from the interference of the primary user (i.e., interference-ignorant transmission) has been proposed to date. If the interference from the primary user is not properly limited or managed, secondary user transmission will be difficult, especially when the primary user signal is strong, such as in the case of a large-sized cellular network. Therefore, a new transmission strategy for overcoming primary user interference and enhancing secondary user performance is required.

Interference cancellation has been shown to be an effective approach to solving the interference problem in conventional interference-limited wireless networks, such as multiple- input-multiple-output (MIMO) networks [12], [13] and ad-hoc networks [14], [15]. Unlike these conventional schemes that cancel interference from users with equal priority, our spectrum sharing-based cognitive radio system decodes and cancels the interference from the primary user (having higher priority) so as to improve the quality of the transmission of the secondary user (having lower priority).

22.1.4 Contribution and Outline

Motivated by the foregoing, we are proposing interference-aware power allocation in a spectrum sharing environment. Since the interference cancellation performance is sensitive to the transmit power level, the transmit power of the secondary user is allocated so as to optimally balance the interference cancellation capability and the secondary user capacity while satisfying the interference constraint, i.e., the average and peak received-power constraint. We then analyze how much gain in secondary user ergodic capacity can be achieved by the proposed interference-aware power allocation.

The rest of this chapter is organized as follows. Section 22.2 presents the system model. In Section 22.3, we present our interference-aware power allocation scheme as an optimal transmission strategy in a spectrum sharing environment. Section 22.4 provides the capacity analysis when the proposed scheme is employed. Numerical results are given in Section 22.5, and Section 22.6 concludes this chapter.

22.2 System Model and Assumptions

Consider a spectrum sharing model where the primary and secondary systems coexist on the same frequency band, as shown in Figure 22.1. The primary system consists of a primary transmitter (PU-Tx) and a primary receiver (PU-Rx), and the secondary system consists of a secondary transmitter (SU-Tx) and a secondary receiver (SU-Rx). The links from PU-Tx to PU-Rx and from SU-Tx to SU-Rx include the intended signals, which are multiplied by the channel gains $h_p(k)$ and $h_s(k)$, while the links from PU-Tx to SU-Rx and from SU-Tx to PU-Rx include the interference signals, which are multiplied by $h_{ps}(k)$ and $h_{sp}(k)$. The received signals at the k-th time instant $r_p(k)$ and $r_s(k)$ are given by

$$\begin{aligned} \text{At PU-Rx:} \quad r_p(k) &= h_p(k)s_p(k) + h_{sp}(k)s_s(k) + v_p(k) \\ \text{At SU-Rx:} \quad r_s(k) &= h_s(k)s_s(k) + h_{ps}(k)s_p(k) + v_s(k), \end{aligned}$$

(22.1)

where $s_p(k)$ and $s_s(k)$ are the signals transmitted from the primary and secondary transmitters and are assumed to have mean powers P_p and P_s, respectively. The noises $v_p(k)$ and $v_p(k)$ are AWGN, with a common power spectral density (PSD) N_0.

Ignoring time indices for simplicity, we assume that each channel gain is independent and identically distributed (IID) Rayleigh flat faded, i.e., $\mathbf{h} = (h_p, h_{ps}, h_{sp}, h_s) \sim \mathcal{CN}(\mathbf{0}, \mathbf{1})$ (a zero-mean,

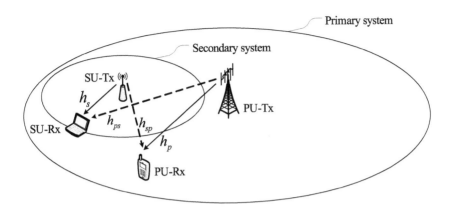

Figure 22.1: System model for spectrum sharing.

unit-variance circularly symmetric complex Gaussian random vector). The channel power gain for each link is denoted by $\mathbf{g} = (g_p, g_{ps}, g_{sp}, g_s) = |\mathbf{h}|^2 = (|h_p|^2, |h_{ps}|^2, |h_{sp}|^2, |h_s|^2)$.

We also assume that the channel power gains g_s, g_{ps}, and g_{sp} are available at the secondary transmitter. The channel power gains h_s and h_{ps} can be estimated by the secondary receiver and then fed back to the secondary transmitter. The channel gain h_{sp} can be obtained by direct estimation [16], [17] or an indirect method with a band manager [18]. It is assumed that the optimal Gaussian codebook is used for both the primary and secondary users. Since the secondary system is deployed within an area where the primary system is operating, it is reasonable to assume that the secondary user can decode the primary user signal with prior information of the primary user, such as the modulation scheme, codebook, and bandwidth [19]. In order to avoid system-specific implementations and provide a more generalized capacity limit, our analysis employs SINR thresholding as a means for evaluating the success of the interference cancellation.

We consider two basic types of interference constraint: the average received-power constraint and the peak received-power constraint. Under the average received-power constraint, the average received power at PU-Rx is limited to a predefined value, i.e., $E[g_{sp}P_s(\mathbf{g})] \leq Q_{\text{th}}$. The average received-power constraint is used to satisfy a delay-insensitive primary user having a long-term QoS requirement [20]. On the other hand, under the peak received-power constraint, the peak received power at PU-Rx must be below a predefined threshold, i.e., $g_{sp}P_s(\mathbf{g}) \leq Q_{\text{th}}$. The peak received-power constraint can be applied when the primary user has an instantaneous QoS requirement [20].

22.3 Interference-Aware Power Allocation for Capacity Maximization

In this section, we present an interference-aware power allocation scheme, where the interference from the primary user is canceled and then the secondary user adjusts its transmit power based upon the success of the interference cancellation in order to maximize the secondary user capacity. We first show how the interference cancellation can be employed in the spectrum sharing environment and then provide the optimal power allocation for both the average and peak received-power constraints. We also provide the optimal transmit power when considering the transmit power constraint in addition to the received-power constraints.

22.3.1 *Employing Interference Cancellation In a Spectrum Sharing Environment*

The main problem in a spectrum sharing scenario is the unintended interference that occurs when the primary and secondary users simultaneously transmit their own data on the same frequency band. From the primary user's perspective, interference is not quite as big a problem, because the interference from the secondary user is limited to an acceptable level by a received-power constraint (e.g., average or peak) [5]. However, without any corresponding similar interference mitigation mechanism, the secondary receiver may experience strong interference from the primary user, especially when the primary user transmit power is high. As seen in Figure 22.1, it is assumed that the primary user is in a large-scale cellular network with high power, while the secondary user is in a small-scale hotspot with low power.

As a way of reducing the interference from the primary user, we employ an interference cancellation technique. Instead of directly decoding the secondary user signal, the secondary receiver first decodes the primary user signal $s_p(k)$ and cancels it from the secondary user received signal $r_s(k)$ in (22.1). The received signal after interference cancellation $r_s^{IC}(k)$ is then given by

$$r_s^{IC}(k) = h_s(k)s_s(k) + h_{ps}(k)\big(s_p(k) - \hat{s}_p(k)\big) + v_s(k), \tag{22.2}$$

where $\hat{s}_p(k)$ is the decoded signal of $s_p(k)$. If $s_p(k)$ is successfully decoded, i.e., $\hat{s}_p(k) = s_p(k)$, the interference term is perfectly cancelled. Otherwise, the secondary user does not cancel the interference and treats it as noise[1].

Whether or not the primary user signal has been successfully decoded is determined by the SINR thresholding [22], [21], [23], [24]. According to the SINR thresholding, successful decoding of the primary user signal depends on whether the SINR[2] of the primary user measured at the SU-Rx exceeds a predefined threshold γ_{th}.

The SINR of the primary user signal at the SU-Rx is given by

$$SINR_p = \frac{g_{ps}P_p}{g_sP_s + N_0}. \tag{22.3}$$

The successful decoding of the primary user is, thus, guaranteed when the $SINR_p \geq \gamma_{th}$ (i.e., the link from PU-Tx to SU-Rx is not in outage). On the other hand, the decoding of the primary user is considered failed when the $SINR_p < \gamma_{th}$ (i.e., the link is in outage).

From the secondary user's perspective, SINR improvement is achieved by cancelling out the interference from the primary user. With interference cancellation, the SINR of the secondary user can be written as

$$SINR_s = \frac{g_sP_s}{g_{ps}P_p \cdot 1(SINR_p < \gamma_{th}) + N_0}, \tag{22.4}$$

where $1(\cdot)$ is the indicator function that indicates whether or not the interference from the primary user exists and depends on the decoding success of the primary user signal. Then, we expect not only the SINR improvement but also the capacity enhancement of the secondary user by employing interference cancellation.

Note also that the degree of capacity enhancement depends on the transmit power of the secondary user P_s. In general, the secondary user can achieve higher capacity with higher transmit power. However, at the same time, the primary user signal decoding at the interference cancellation stage begins to fail as the transmit power of the secondary user increases. Therefore, it is expected that there will be some optimal transmit power that can maximize not only the SINR but also the secondary user capacity. Since P_s is a controllable system parameter, we will provide a method

---

[1]It is assumed that an error detection mechanism, such as parity checking, is in place so that the secondary user can know whether the primary user signal (i.e., interference) is successfully decoded for each sample of the primary user signal [21].

[2]Since the primary user signal is the target signal to decode in this case, the secondary user signal is regarded as noise.

for obtaining the optimal P_s in the next subsections for both the average and peak received-power constraint cases.

22.3.2 Optimal Power Allocation with Average Received-Power Constraint

Using the interference cancellation technique discussed above, the ergodic capacity maximization problem with the average received-power constraint Q_{th} can be formulated as

$$C_{avg} = \max_{P_s(\mathbf{g})} E_{\mathbf{g}}\left[\log\left(1 + \frac{g_s P_s(\mathbf{g})}{\chi_p(\mathbf{g}) \cdot g_{ps}P_p + N_0}\right)\right] \quad \text{s.t.} \quad E\left[g_{sp}P_s(\mathbf{g})\right] \leq Q_{th}, \tag{22.5}$$

where $E[\cdot]$ is an expectation operator, and $\chi_p(\mathbf{g})$ is an indicator function for the success of the interference cancellation, defined as

$$\chi_p(\mathbf{g}) = \begin{cases} 0, & \frac{g_{ps}P_p}{g_s P_s(\mathbf{g}) + N_0} \geq \gamma_{th} \\ 1, & \text{otherwise.} \end{cases} \tag{22.6}$$

The optimization problem (22.5) can be solved using the Lagrangian method [25], as follows:

$$L(P_s, \lambda) = E_{\mathbf{g}}\left[\log\left(1 + \frac{g_s P_s(\mathbf{g})}{\chi_p(\mathbf{g}) \cdot g_{ps}P_p + N_0}\right)\right] - \lambda\left(E_{\mathbf{g}}\left[g_{sp}P_s(\mathbf{g})\right] - Q_{th}\right), \tag{22.7}$$

where λ is a Lagrangian multiplier. By solving the equation $\frac{\partial L(P_s, \lambda)}{\partial P_s(\mathbf{g})} = 0$, we obtain the following water-filling power allocation:

$$P_s^{wf}(\mathbf{g}) = \left[\frac{1}{\lambda g_{sp}} - \frac{\chi_p(\mathbf{g}) \cdot g_{ps}P_p + N_0}{g_s}\right]^+, \tag{22.8}$$

with $[x]^+ = \max(0, x)$. Note that $P_s^{wf}(\mathbf{g})$ depends on the success of the interference cancellation. The interference term $\chi_p(\mathbf{g}) \cdot g_{ps}P_p$ in (22.8) becomes zero when the interference cancellation is successful. The result is then the interference-free water-filling power allocation $P_s^{wf-if}(\mathbf{g}) = \left[\frac{1}{\lambda g_{sp}} - \frac{N_0}{g_s}\right]^+$. If the interference term is not cancelled, we have the interference-rich water-filling power allocation $P_s^{wf-ir}(\mathbf{g}) = \left[\frac{1}{\lambda g_{sp}} - \frac{g_{ps}P_p + N_0}{g_s}\right]^+$.

Due to the nature of the indicator function, there is a critical point "A" with power $P_s^A(\mathbf{g})$ that divides the interference-free ($\mathcal{R}_1$) and interference-rich ($\mathcal{R}_2$) regions, as seen in Figure 22.2. In the interference-free region, the secondary user experiences no primary user interference since the condition $\text{SINR}_p = \frac{g_{ps}P_p}{g_s P_s(\mathbf{g}) + N_0} \geq \gamma_{th}$ holds with smaller transmit power P_s. In the interference-rich region, however, the primary user interference is not cancelled because the transmit power P_s is too high to satisfy the above condition. Hence, $P_s^A(\mathbf{g})$ is the maximum transmit power possible while still ensuring successful interference cancellation, which can be obtained by manipulating the equation $\frac{g_{ps}P_p}{g_s P_s(\mathbf{g}) + N_0} = \gamma_{th}$, as follows:

$$P_s^A(\mathbf{g}) = \frac{g_{ps}P_p/\gamma_{th} - N_0}{g_s}. \tag{22.9}$$

If we further increase P_s beyond the point "B" with power $P_s^B(\mathbf{g})$, the secondary user capacity will be higher than the maximum interference-free capacity even with the primary user interference. We denote this region as a high-power region ($\mathcal{R}_3$). $P_s^B(\mathbf{g})$ can be obtained by solving the following equation

$$\log\left(1 + \frac{g_{ps}P_p/\gamma_{th} - N_0}{N_0}\right) = \log\left(1 + \frac{g_s P_s^B(\mathbf{g})}{g_{ps}P_p + N_0}\right), \tag{22.10}$$

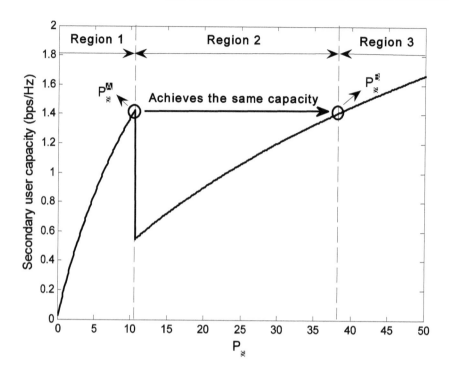

Figure 22.2: Snapshot of the secondary user capacity as a function of its transmit power for the following channel realizations: $g_s = 0.16$, $g_{ps} = 0.27$, **and** $g_{sp} = 1$. **The SINR threshold is assumed to be** $\gamma_{th} = 1$. **There is a critical point "A," where the maximum transmit power with interference cancellation is attained. Beyond point "B," higher capacity can be obtained even in the face of primary user interference due to the higher power transmission.**

which yields

$$P_s^B(\mathbf{g}) = \frac{g_{ps}P_p + N_0}{g_s}\left(\frac{g_{ps}P_p}{\gamma_{th}N_0} - 1\right). \tag{22.11}$$

Using the above characteristics between the capacity and the transmit power of the secondary user, we will now show how the optimal transmit power can be obtained for each region.

- **Interference-free region** ($\mathcal{R}_1$): If the secondary user has a very low power budget (mainly due to a low Q_{th} value), its transmit power should be lower than $P_s^A(\mathbf{g})$ but still within the interference-free region. In this case, the interference-free water-filling solution $P_s^{wf-if}(\mathbf{g})$ according to the current channel realization is optimal.

- **Interference-rich region** ($\mathcal{R}_2$): If the secondary user can use the power between $P_s^A(\mathbf{g})$ and $P_s^B(\mathbf{g})$, the optimal solution is to use $P_s^A(\mathbf{g})$, which can achieve higher capacity than any transmit power within $[P_s^A(\mathbf{g}), P_s^B(\mathbf{g})]$.

- **High-power region** ($\mathcal{R}_3$): In this case, the optimal transmit power is the water-filling solution $P_s^{wf-ir}(\mathbf{g})$. Even though the primary user interference is not cancelled, the capacity degradation due to interference can be overcome by using a higher transmit power. However, even though the secondary user transmits with higher power, the received-power constraint must be maintained.

In summary, the optimal transmit power of the secondary user in the interference cancellation-based spectrum sharing environment under an average received-power constraint is given by

$$
P_s^{avg}(\mathbf{g}) = \begin{cases} \left[\frac{1}{\lambda g_{sp}} - \frac{N_0}{g_s}\right]^+, & [P_s^{wf-if}(\mathbf{g})]^+ \le P_s^A(\mathbf{g}) \\ \frac{g_{ps}P_p/\gamma_{th}-N_0}{g_s}, & \{[P_s^{wf-if}(\mathbf{g})]^+ > P_s^A(\mathbf{g})\} \cap \{P_s^B(\mathbf{g}) \ge \max[P_s^{wf-ir}(\mathbf{g})]^+\} \\ \left[\frac{1}{\lambda g_{sp}} - \frac{g_{ps}P_p+N_0}{g_s}\right]^+, & P_s^B(\mathbf{g}) < [P_s^{wf-ir}(\mathbf{g})]^+, \end{cases}
$$

$$(22.12)$$

where the boundaries for λ are obtained by solving the equations $P_s^{wf-if}(\mathbf{g}) = P_s^A(\mathbf{g})$ and $P_s^B(\mathbf{g}) = P_s^{wf-ir}(\mathbf{g})$, respectively. Note that λ is determined by the average received-power constraint Q_{th}. With higher values for Q_{th}, a lower λ can be determined, such that more transmit power can be allocated.

22.3.3 Optimal Power Allocation with Peak Received-Power Constraint

The ergodic capacity maximization problem with the peak received power constraint can be formulated as follows:

$$
C_{peak} = \max_{P_s(\mathbf{g})} E_{\mathbf{g}}\left[\log\left(1 + \frac{g_s P_s(\mathbf{g})}{\chi_p(\mathbf{g}) \cdot g_{ps}P_p + N_0}\right)\right] \quad \text{s.t.} \quad g_{sp}P_s(\mathbf{g}) \le Q_{th}, \tag{22.13}
$$

where the objective function is the same as in (22.5), but the peak received-power constraint is employed instead of the average received-power constraint.

In conventional spectrum sharing schemes without interference cancellation, the transmission strategy for the peak received-power constraint is to transmit using the maximum power that still satisfies the constraint, i.e., $P_s^{mp}(\mathbf{g}) = Q_{th}/g_{sp}$ [5], [6]. However, as discussed for the average received-power constraint case, the higher the transmit power, the more difficult it becomes to correctly decode the primary user signal and cancel it from the received signal. Hence, as seen in Figure 22.2, the secondary user needs to transmit with power $P_s^A(\mathbf{g})$ instead of $P_s^{mp}(\mathbf{g})$ in $\mathcal{R}_2$. Since the secondary user can increase its transmit power as Q_{th} increases, the boundaries dividing the regions are determined by the value of Q_{th}.

The optimal transmit power of the secondary user, with the peak received-power constraint, is given by

$$
P_s^{peak}(\mathbf{g}) = \begin{cases} \frac{Q_{th}}{g_{sp}}, & Q_{th} \le \frac{g_{ps}P_p/\gamma_{th}-N_0}{g_s/g_{sp}} \\ \frac{g_{ps}P_p/\gamma_{th}-N_0}{g_s}, & \frac{g_{ps}P_p/\gamma_{th}-N_0}{g_s/g_{sp}} < Q_{th} \le \frac{g_{ps}P_p+N_0}{g_s/g_{sp}}\left(\frac{g_{ps}P_p}{\gamma_{th}N_0}-1\right) \\ \frac{Q_{th}}{g_{sp}}, & Q_{th} > \frac{g_{ps}P_p+N_0}{g_s/g_{sp}}\left(\frac{g_{ps}P_p}{\gamma_{th}N_0}-1\right), \end{cases}
$$

$$(22.14)$$

where the boundaries for Q_{th} are obtained by solving the equations $P_s^{mp}(\mathbf{g}) = P_s^A(\mathbf{g})$ and $P_s^B(\mathbf{g}) = P_s^{mp}(\mathbf{g})$, respectively.

22.3.4 Consideration of Transmit Power Constraint

In this subsection, we further investigate the effect of the transmit-power constraint as well as the received-power constraints. The transmit power constraint originates from the hardware restrictions, such as radio frequency (RF) non-linearity, and can significantly affect the ergodic capacity of the secondary user when the channel power gain g_{sp} or the received-power constraint Q_{th} becomes large [9].

Since the transmit power of the secondary user cannot exceed the transmit power constraint P_{th}, the optimal transmit power $P_s^{avg}(\mathbf{g})$ and $P_s^{peak}(\mathbf{g})$ is modified as follows:

$$
P_{s,tpc}^{avg}(\mathbf{g}) =
\begin{cases}
\min\left(\left[\frac{1}{\lambda g_{sp}} - \frac{N_0}{g_s}\right]^+, P_{th}\right), & [P_s^{wf-if}(\mathbf{g})]^+ \le P_s^A(\mathbf{g}) \\[2mm]
\frac{g_{ps}P_p/\gamma_{th} - N_0}{g_s}, & \{[P_s^{wf-if}(\mathbf{g})]^+ > P_s^A(\mathbf{g})\}\cap\{P_s^B(\mathbf{g}) \ge \max[P_s^{wf-ir}(\mathbf{g})]^+\} \\[2mm]
\min\left(\left[\frac{1}{\lambda g_{sp}} - \frac{g_{ps}P_p+N_0}{g_s}\right]^+, P_{th}\right), & P_s^B(\mathbf{g}) < [P_s^{wf-ir}(\mathbf{g})]^+,
\end{cases}
$$

$$
P_{s,tpc}^{peak}(\mathbf{g}) =
\begin{cases}
\min\left(\frac{Q_{th}}{g_{sp}}, P_{th}\right), & Q_{th} \le \frac{g_{ps}P_p/\gamma_{th} - N_0}{g_s/g_{sp}} \\[2mm]
\frac{g_{ps}P_p/\gamma_{th} - N_0}{g_s}, & \frac{g_{ps}P_p/\gamma_{th} - N_0}{g_s/g_{sp}} < Q_{th} \le \frac{g_{ps}P_p+N_0}{g_s/g_{sp}}\left(\frac{g_{ps}P_p}{\gamma_{th}N_0} - 1\right) \\[2mm]
\min\left(\frac{Q_{th}}{g_{sp}}, P_{th}\right), & Q_{th} > \frac{g_{ps}P_p+N_0}{g_s/g_{sp}}\left(\frac{g_{ps}P_p}{\gamma_{th}N_0} - 1\right),
\end{cases}
$$

$$(22.15)$$

where we note that the transmit power at the critical point "A" $P_s^A(\mathbf{g})$ is not affected by the transmit power constraint. If $P_s^A(\mathbf{g})$ is limited by P_{th}, the interference-free region ($\mathcal{R}_1$) will be selected instead of the critical point "A."

22.4 Capacity Analysis

In this section, the ergodic capacity achieved by the proposed interference-aware power allocation is derived both with the average and peak received-power constraints. Since the ergodic capacity depends greatly on which transmission region is selected, we obtain the ergodic capacity as a combination of conditional capacity and selection probability for each region.

22.4.1 *Ergodic Capacity with Average Received-Power Constraint*

The optimal transmit power is determined along with the three different regions (i.e., $\mathcal{R}_1$, $\mathcal{R}_2$, and $\mathcal{R}_3$), as in (22.12) and (22.14). We first obtain the probability that each transmission region is selected with the average received-power constraint in the following lemma.

Lemma 22.1 (region selection probability with average received-power constraint) *The probability that a specific transmission region is selected out of three regions is given by*

$$
\begin{aligned}
P(\mathcal{R}_1^{avg}) &= \exp\left(-\frac{\gamma_{th}N_0}{P_p}\right)\left[1 + \frac{\gamma_{th}}{\lambda P_p}F\left(\frac{\gamma_{th}}{P_p}\left(N_0 + \frac{1}{\lambda}\right)\right)\right], \\
P(\mathcal{R}_2^{avg}) &= 1 - P(\mathcal{R}_1^{avg}) - P(\mathcal{R}_3^{avg}), \\
P(\mathcal{R}_3^{avg}) &= 1 - e^{-\gamma_{th}N_0/P_p} + \int_{\frac{\gamma_{th}N_0}{P_p}}^{\infty} \frac{e^{-w}}{\frac{\lambda P_p^2}{\gamma_{th}N_0}w\left(w + \frac{N_0}{P_p}\right) + 1}\,dw.
\end{aligned}
$$

$$(22.16)$$

where $F(x) = \exp(x)\mathrm{Ei}(-x)$.

Proof 22.1   See Appendix A.   ■

Unfortunately, the integral in $P(\mathcal{R}_3^{avg})$ can only be calculated numerically. However, since Region 3 is an interference-rich environment (i.e., $P_p \gg N_0$), the noise density N_0 can be ignored, and its approximated value can be obtained in a closed-form expression as follows:

$$
\int_{\frac{\gamma_{th}N_0}{P_p}}^{\infty} \frac{e^{-w}}{\frac{\lambda P_p^2}{\gamma_{th}}w\left(w+\frac{N_0}{P_p}\right)+1}dw \approx \frac{\gamma_{th}N_0}{\lambda P_p^2}\int_0^{\infty}\frac{e^{-w}}{w^2+\frac{\gamma_{th}N_0}{\lambda P_p^2}}dw
$$

$$
=\sqrt{\frac{\gamma_{th}N_0}{\lambda P_p^2}}\left[\operatorname{ci}\left(\sqrt{\frac{\gamma_{th}N_0}{\lambda P_p^2}}\right)\sin\left(\sqrt{\frac{\gamma_{th}N_0}{\lambda P_p^2}}\right)-\operatorname{si}\left(\sqrt{\frac{\gamma_{th}N_0}{\lambda P_p^2}}\right)\cos\left(\sqrt{\frac{\gamma_{th}N_0}{\lambda P_p^2}}\right)\right],
$$

$$(22.17)$$

where $\operatorname{ci}(x) = -\int_x^{\infty}\cos(t)/t\,dt$ and $\operatorname{si}(x) = -\int_x^{\infty}\sin(t)/t\,dt$ are cosine and sine integral functions, respectively.

Using the results obtained for the region selection probabilities $P(\mathcal{R}_1^{avg})$, $P(\mathcal{R}_2^{avg})$, and $P(\mathcal{R}_3^{avg})$, we can evaluate the ergodic capacity of the secondary user when the corresponding optimal transmit power is allocated in the following theorem.

Theorem 22.1 (ergodic capacity with average received-power constraint) *According to the law of total probability, the ergodic capacity of the secondary user can be calculated by summing the products of the conditional capacity of each region and the probability that each region is selected, as follows:*

$$
C_{avg} = C(\mathcal{R}_1^{avg})P(\mathcal{R}_1^{avg}) + C(\mathcal{R}_2^{avg})P(\mathcal{R}_2^{avg}) + C(\mathcal{R}_3^{avg})P(\mathcal{R}_3^{avg}), \tag{22.18}
$$

with the conditional capacity for each region:

$$
C(\mathcal{R}_1^{avg}) = \log\left(1+\frac{1}{\lambda N_0}\right)
$$

$$
C(\mathcal{R}_2^{avg}) = \log\left(\frac{P_p}{\gamma_{th}N_0}\right) - \omega \tag{22.19}
$$

$$
C(\mathcal{R}_3^{avg}) = \log\left(1+\frac{1}{\lambda N_0}\right) + F\left(\frac{N_0}{P_p}\right) - F\left(\frac{N_0}{P_p}+\frac{1}{\lambda P_p}\right),
$$

where ω is Euler's constant [26], defined as

$$
\omega = \lim_{s\to\infty}\left(\sum_{m=1}^{s}\frac{1}{m}-\log s\right) = 0.577215\ldots \tag{22.20}
$$

Proof 22.2 See Appendix B. ■

We can expect that $C(\mathcal{R}_1^{avg})$ and $C(\mathcal{R}_3^{avg})$ are decreasing functions of λ, while $C(\mathcal{R}_2^{avg})$ is not affected by λ. As discussed before, the power budget for the secondary user becomes tighter with higher value of λ. Hence, the ergodic capacity of the secondary user is a decreasing function of λ.

This principle can be used in evaluating the optimal λ^* that achieves the ergodic capacity. The optimal λ^* can be obtained by solving the equation: $E_{\mathbf{g}}[g_{sp}P_s(\mathbf{g})] = Q_{th}$ [5]. We do this by deriving the average received-power $E_{\mathbf{g}}[g_{sp}P_s(\mathbf{g})]$ that encompasses the three transmission regions in the following theorem.

Theorem 22.2 (average received-power at a primary user) *The average received-power at a primary user is given by*

$$
E_{\mathbf{g}}[g_{sp}P_s(\mathbf{g})] = E_{\mathbf{g}}[g_{sp}P_s^{wf-if}(\mathbf{g})]P(\mathcal{R}_1^{avg}) + E_{\mathbf{g}}[g_{sp}P_s^{A}(\mathbf{g})]P(\mathcal{R}_2^{avg})
$$

$$
+ E_{\mathbf{g}}[g_{sp}P_s^{wf-ir}(\mathbf{g})]P(\mathcal{R}_3^{avg}), \tag{22.21}
$$

where the average received-power for each region is:

$$E_{\mathbf{g}}\left[g_{sp}P_s^{wf-if}(\mathbf{g})\right] = \frac{1}{\lambda} - N_0 \log\left(1 + \frac{1}{\lambda N_0}\right),$$

$$E_{\mathbf{g}}\left[g_{sp}P_s^A(\mathbf{g})\right] = \left(2\alpha - \frac{1}{\lambda}\right) - \left(\alpha - N_0\left(1 + \frac{1}{\gamma_{th}}\right)\right)F(\beta) - \left(\alpha - \left(\frac{1}{\lambda} + N_0\right)(1-\gamma)\right)F(\gamma)$$

$$- \alpha \int_0^\infty (w-\delta)e^{-w}\left(\frac{1}{1 + \frac{1}{\beta\delta}w(w+\beta)} - \log\left[1 + \frac{1}{\beta\delta}w(w+\beta)\right]\right)dw$$

$$E_{\mathbf{g}}\left[g_{sp}P_s^{wf-ir}(\mathbf{g})\right] = \frac{1}{\lambda} - P_p\left[(1+\gamma)\log\left(1 + \frac{1}{\lambda N_0}\right) + F(\beta) - \left(1 - \frac{1}{\lambda P_p}\right)F\left(\beta + \frac{1}{\lambda P_p}\right)\right],$$

$$\tag{22.22}$$

where $\alpha = \frac{P_p}{\gamma_{th}}$, $\beta = \frac{N_0}{P_p}$, $\gamma = \frac{\gamma_{th}}{\lambda P_p}$, and $\delta = \frac{\gamma_{th}N_0}{P_p}$.

Proof 22.3 The proof is given in Appendix C. ■

Similar to the case for evaluating the region selection probability, the integration in $E_{\mathbf{g}}\left[g_{sp}P_s^A(\mathbf{g})\right]$ can be solved using the interference-rich environment assumption ($P_p \gg N_0$):

$$\int_0^\infty (w - \delta)e^{-w}\left(\frac{1}{1 + \frac{1}{\beta\delta}w(w+\beta)} - \log\left[1 + \frac{1}{\beta\delta}w(w+\beta)\right]\right)dw$$

$$\approx \beta\delta\int_0^\infty \frac{we^{-w}}{w^2 + \beta\delta}dw - 2\int_0^\infty we^{-w}\log w\, dw + \log(\beta\delta)\int_0^\infty we^{-w}dw \tag{22.23}$$

$$= 2\omega - 1 + \sqrt{\beta\delta}\left[\text{ci}\left(\sqrt{\beta\delta}\right)\sin\left(\sqrt{\beta\delta}\right) - \text{si}\left(\sqrt{\beta\delta}\right)\cos\left(\sqrt{\beta\delta}\right)\right].$$

Since $E_{\mathbf{g}}\left[g_{sp}P_s(\mathbf{g})\right]$ is a monotonically decreasing function of λ, λ^* can be determined by numerically solving the equation $E_{\mathbf{g}}\left[g_{sp}P_s(\mathbf{g})\right] = Q_{th}$, as in [5]. Substituting the obtained λ^* into (22.18) yields the ergodic capacity of the secondary user achievable by the proposed interference-aware power allocation scheme under the average-received power constraint.

22.4.2 Ergodic Capacity with Peak Received-Power Constraint

Similar to the average received-power constraint case, we first calculate the region selection probabilities for each region ($\mathcal{R}_1$, $\mathcal{R}_2$, and $\mathcal{R}_3$) under the peak received-power constraint in the following lemma.

Lemma 22.2 (region selection probability with peak received-power constraint) *The probability of selecting one of three transmission regions is given by*

$$P(\mathcal{R}_1^{peak}) = \exp\left(-\frac{\gamma_{th}N_0}{P_p}\right)\left[1 + \frac{\gamma_{th}Q_{th}}{P_p}F\left(\frac{\gamma_{th}Q_{th}}{P_p}\right)\right],$$

$$P(\mathcal{R}_2^{peak}) = 1 - P(\mathcal{R}_1) - P(\mathcal{R}_3), \tag{22.24}$$

$$P(\mathcal{R}_3^{peak}) = 1 - \exp\left(-\frac{\gamma_{th}N_0}{P_p}\right) + \int_{\frac{\gamma_{th}N_0}{P_p}}^\infty \frac{e^{-w}}{\frac{P_p}{Q_{th}}\left(\frac{P_p}{\gamma_{th}N_0}w^2 + \frac{1-\gamma_{th}}{\gamma_{th}}w - \frac{N_0}{P_p}\right) + 1}dw.$$

Proof 22.4   See Appendix D.   ■

Although $P(\mathcal{R}_3^{peak})$ can be calculated numerically, its approximated value can be obtained in a closed-form expression with the assumption of $P_p \gg N_0$, as follows:

$$
\int_{\frac{\gamma_{th} N_0}{P_p}}^{\infty} \frac{e^{-w}}{\frac{P_p}{Q_{th}}\left(\frac{P_p}{\gamma_{th}N_0}w^2 + \frac{1-\gamma_{th}}{\gamma_{th}}w - \frac{N_0}{P_p}\right)+1}dw \approx \frac{\gamma_{th}Q_{th}N_0}{P_p^2}\int_0^\infty \frac{e^{-w}}{w^2 + \frac{\gamma_{th}Q_{th}N_0}{P_p^2}}dw
$$

$$
= \frac{\sqrt{\gamma_{th}Q_{th}N_0}}{P_p}\left[\mathrm{ci}\left(\frac{\sqrt{\gamma_{th}Q_{th}N_0}}{P_p}\right)\sin\left(\frac{\sqrt{\gamma_{th}Q_{th}N_0}}{P_p}\right) - \mathrm{si}\left(\frac{\sqrt{\gamma_{th}Q_{th}N_0}}{P_p}\right)\cos\left(\frac{\sqrt{\gamma_{th}Q_{th}N_0}}{P_p}\right)\right].
$$

$$(22.25)$$

We then evaluate the ergodic capacity of the secondary user under the peak received-power constraint in the following theorem.

Theorem 22.3 (ergodic capacity with peak received-power constraint) *Similar to the counterpart of the average received-power constraint case, the ergodic capacity can be calculated as*

$$
C_{peak} = C(\mathcal{R}_1^{peak})P(\mathcal{R}_1^{peak}) + C(\mathcal{R}_2^{peak})P(\mathcal{R}_2^{peak}) + C(\mathcal{R}_3^{peak})P(\mathcal{R}_3^{peak}),
\qquad (22.26)
$$

with the conditional capacity for each region:

$$
C(\mathcal{R}_1^{peak}) = \frac{\log(Q_{th}/N_0)}{1 - N_0/Q_{th}},
$$

$$
C(\mathcal{R}_2^{peak}) = \log\left(\frac{P_p}{\gamma_{th}N_0}\right) - \omega,
\qquad (22.27)
$$

$$
C(\mathcal{R}_3^{peak}) = \int_0^\infty \frac{e^{-w}\log\left(\frac{P_p w + N_0}{Q_{th}}\right)}{\frac{P_p w + N_0}{Q_{th}} - 1}dw.
$$

Proof 22.5   See Appendix E.   ■

Although the exact value of $C(\mathcal{R}_3^{peak})$ can be obtained by numerical integration, its approximated value can be derived in a closed-form solution, in the following corollary.

Corollary 22.1 (Approximated capacity for region 3) *The approximation of $C(\mathcal{R}_3^{peak})$ can be calculated as*

$$
C(\mathcal{R}_3^{peak}) \approx \log N_0 - F(N_0/P_p).
\qquad (22.28)
$$

Proof 22.6   In (22.26), since Q_{th} has very high value in Region 3, the following approximation can be used:

$$
\log\left(1 + Q_{th}\frac{g_s/g_{sp}}{g_{ps}P_p + N_0}\right) \approx \log\left(Q_{th}\frac{g_s/g_{sp}}{g_{ps}P_p + N_0}\right)
$$

$$
= \log Q_{th} + \log g_s - \log g_{sp} - \log(g_{ps}P_p + N_0).
\qquad (22.29)
$$

The triple integral then becomes the sum of single integrals, as follows:

$$C(\mathcal{R}_3^{peak}) = \log Q_{th} + \int_0^\infty \log x e^{-x} dx - \int_0^\infty \log y e^{-y} dy - \int_0^\infty \log(P_p z + N_0) e^{-z} dz, \quad (22.30)$$

with $x = g_s$ and $y = g_{sp}$. The integrals $\int_0^\infty \log x e^{-x} dx$ and $\int_0^\infty \log y e^{-y} dy$ cancel each other out. The integral in the last term of (22.30) can be solved using (4.337.1) in [26], as follows:

$$\int_0^\infty \log(P_p z + N_0) e^{-z} dz = \log N_0 - e^{N_0/P_p} \text{Ei}(N_0/P_p), \quad (22.31)$$

which yields the desired result (22.29). ■

According to the capacity results in (22.26) and (22.27), we can expect the ergodic capacity to increase as Q_{th} increases. In addition, it is important to note that the selection of transmission region affects the transmit power and the corresponding ergodic capacity.

22.5 Numerical Results

This section provides numerical results showing the ergodic capacity gain of the proposed interference-aware power allocation over conventional interference-ignorant power allocation in a spectrum sharing environment, which was investigated in our previous paper [27]. It is assumed that all the channel gains are independent and IID with unit gains, all of which follow Rayleigh distribution. In the analysis, both the exact ergodic capacity from numerical integration and the approximated ergodic capacity are given. In the simulation, the result is averaged from 10,000 iterations.

22.5.1 Effect of Received-Power Constraint

The effect of the received-power constraint on the ergodic capacity of the secondary user is shown in Figure 22.3 for the average received-power constraint and in Figure 22.4 for the peak received-power constraint. We assume the following parameters: $P_p = 10$ and $\gamma_{th} = 0.5$. It is shown that the ergodic capacity increases as the received-power constraint Q_{th} increases for all cases. This is because the secondary user can transmit at a higher power as Q_{th} increases.

Note that the interference-aware power allocation approach proposed here achieves higher capacity than conventional interference-ignorant power allocation due to the effectiveness of the interference cancellation and the optimal power allocation between the interference-free and interference-rich signals. However, the ergodic capacity is upper-bounded by the interference-free case, where there is no primary user interference. Note also that the results from both the exact analysis and the simulation coincide, and the gap between the exact and approximated capacities is tiny. Finally, we can see the ergodic capacity is degraded by the transmit power constraint, especially with higher Q_{th}.

22.5.2 Effect of Interference Cancellation Capability

The effect of the interference cancellation capability is determined by the SINR threshold γ_{th}, which is shown in Figure 22.5. The figure shows that the ergodic capacity of the proposed interference-aware

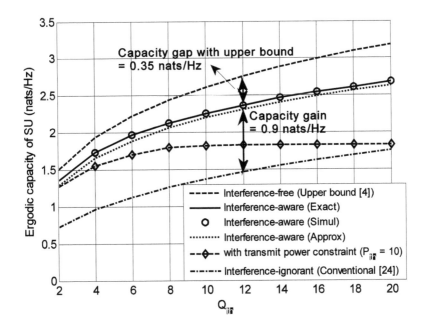

Figure 22.3: Ergodic capacity of the secondary user vs. received-power constraint (Q_{th}) of the proposed interference-aware power allocation, with the average received-power constraint ($P_p = 10$, $\gamma_{th} = 0.1$). Ergodic capacities for both the interference-free and interference-ignorant (conventional) schemes are also depicted for comparison.

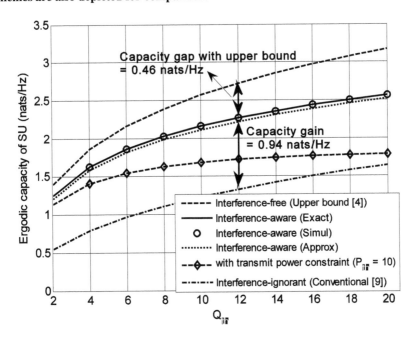

Figure 22.4: Ergodic capacity of the secondary user vs. received-power constraint (Q_{th}), with the peak received-power constraint ($P_p = 10$, $\gamma_{th} = 0.1$).

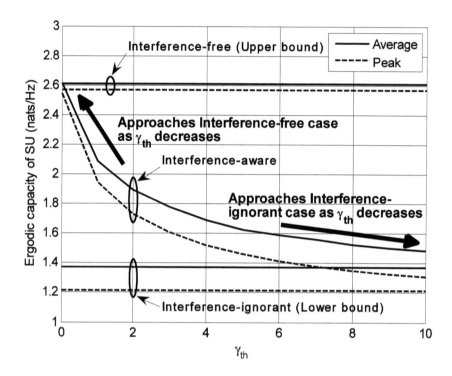

Figure 22.5: Ergodic capacity of the secondary user vs. SINR threshold (γ_{th}) with both the average and peak received-power constraints. The ergodic capacity of the proposed interference-aware power allocation decreases as γ_{th} increases.

scheme decreases with γ_{th}. Recall that a higher γ_{th} is assumed in the case of vulnerable transmission, as occurs in higher modulation systems. On the other hand, a lower γ_{th} is assumed with robust transmission. Hence, we conclude that the interference cancellation capability improves with lower γ_{th}, enhancing the ergodic capacity. Meanwhile, the ergodic capacities of the interference-free and interference-ignorant schemes do not vary with γ_{th} because they do not perform interference cancellation.

22.5.3 Effect of primary user transmit power

The effect of the primary user transmit power on the ergodic capacity of the secondary user is shown in Figure 22.6. We can see that the ergodic capacity of the proposed interference-aware scheme decreases with P_p when P_p is low but increases when P_p is high. In a low P_p region, as P_p increases the primary user interference also increases, degrading the ergodic capacity. However, in a high P_p region, the interference cancellation capability improves with higher values of P_p. If the interference cancellation is successful, the effect of the primary user interference can be ignored. Hence, the ergodic capacity increases along with P_p due to the improved interference cancellation capability.

The conventional interference-ignorant scheme, on the other hand, which cannot cancel the primary user interference, experiences a continuous decrease in ergodic capacity as P_p decreases. This is because there is no way of preventing the SINR degradation that results from the interference.

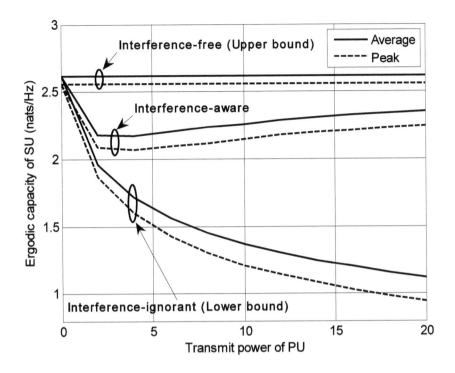

Figure 22.6: Ergodic capacity of the secondary user vs. transmit power of the primary user (P_p), with both the average and peak received-power constraints. The ergodic capacity of the proposed interference-aware scheme decreases when P_p is low but increases when P_p is high, while that of the conventional interference-ignorant scheme decreases continuously with P_p.

22.6 Conclusion

This chapter investigated the interference-aware power allocation in a spectrum sharing scenario. By employing interference cancellation, the interference from the primary user can be eliminated based upon successfully decoding the primary user signal, thus, improving the SINR of the secondary user. Using this fact, we proposed an interference-aware power allocation approach that optimally balances the interference cancellation capability and the ergodic capacity of the secondary user, considering both the average and peak received-power constraint at the primary receiver. We then provided an ergodic capacity analysis to quantify the degree to which capacity can be enhanced by the proposed interference-aware power allocation.

Numerical results show that there are significant capacity gains compared to the conventional interference-ignorant power allocation, which satisfies only the received-power constraints. The obtained results can be used when predicting whether interference cancellation should be used and how much capacity gain is obtained by employing interference cancellation, thus, providing a criterion for developing practical interference cancellation techniques in realistic spectrum sharing environments.

22.7 Appendix A: Proof of Lemma 1

A.1 Derivation of $P(\mathcal{R}_1^{avg})$

$P(\mathcal{R}_1^{avg})$ is the probability that the interference-free water-filling transmit power $P_s^{wf-if}(\mathbf{g})$ is lower than the maximum possible transmit power while the interference cancellation is successful $P_s^A(\mathbf{g})$. Thus, $P(\mathcal{R}_1^{avg})$ can be expressed as

$$
\begin{aligned}
P(\mathcal{R}_1^{avg}) &= \Pr\left[\max\left(0, P_s^{wf-if}(\mathbf{g})\right) \leq P_s^A(\mathbf{g})\right] \\
&= \Pr\left[0 \leq P_s^A(\mathbf{g}), 0 > P_s^{wf-if}(\mathbf{g})\right] + \Pr\left[P_s^{wf-if}(\mathbf{g}) \leq P_s^A(\mathbf{g}), 0 \leq P_s^{wf-if}(\mathbf{g})\right] \\
&= \Pr\left[g_{ps} \geq \frac{\gamma_{th}N_0}{P_p}, \frac{g_s}{g_{sp}} < \lambda N_0\right] + \Pr\left[\lambda N_0 \leq \frac{g_s}{g_{sp}} \leq \frac{\lambda P_p}{\gamma_{th}}g_{ps}\right].
\end{aligned}
\tag{22.32}
$$

Suppose that $z = g_s/g_{sp}$ and $w = g_{ps}$. As discussed before, z is a log-logistic random variable, and w is an exponential random variable. Since z and w are independent of each other, the first term of (22.32) can be calculated as

$$
\begin{aligned}
\Pr\left[w \geq \frac{\gamma_{th}N_0}{P_p}, z < \lambda N_0\right] &= \Pr\left[w \geq \frac{\gamma_{th}N_0}{P_p}\right] \Pr\left[z < \lambda N_0\right] \\
&= \exp\left(-\frac{\gamma_{th}N_0}{P_p}\right)\left(1 - \frac{1}{1 + \lambda N_0}\right).
\end{aligned}
\tag{22.33}
$$

The second term of (22.32) is derived as

$$
\begin{aligned}
\Pr\left[\lambda N_0 \leq z \leq \frac{\lambda P_p}{\gamma_{th}}w\right] &= \int_{\frac{\gamma_{th}N_0}{P_p}}^{\infty} e^{-w} \int_{\lambda N_0}^{\frac{\lambda P_p}{\gamma_{th}}w} \frac{dz}{(z+1)^2}dw \\
&= \int_{\frac{\gamma_{th}N_0}{P_p}}^{\infty} e^{-w}\left(\frac{1}{1 + \lambda N_0} - \frac{1}{1 + \lambda P_p w/\gamma_{th}}\right)dw \\
&= \exp\left(-\frac{\gamma_{th}N_0}{P_p}\right)\left[\frac{1}{1 + \lambda N_0} + \frac{\gamma_{th}}{\lambda P_p}\exp\left(\frac{\gamma_{th}}{P_p}\left(N_0 + \frac{1}{\lambda}\right)\right)\mathrm{Ei}\left(-\frac{\gamma_{th}}{P_p}\left(N_0 + \frac{1}{\lambda}\right)\right)\right].
\end{aligned}
\tag{22.34}
$$

With (22.33) and (22.34), we can obtain the final results for $P(\mathcal{R}_1^{avg})$ as in (22.16).

A.2 Derivation of $P(\mathcal{R}_2^{avg})$

$P(\mathcal{R}_2^{avg})$ is the probability that the secondary user transmits with a power P_s^A. Since the events for the regions $\mathcal{R}_1$, $\mathcal{R}_2$, and $\mathcal{R}_3$ are mutually exclusive and collectively exhaustive [28], i.e., $P(\mathcal{R}_1^{avg}) + P(\mathcal{R}_2^{avg}) + P(\mathcal{R}_3^{avg}) = 1$, $P(\mathcal{R}_2^{avg})$ can be given by

$$
P(\mathcal{R}_2^{avg}) = 1 - P(\mathcal{R}_1^{avg}) - P(\mathcal{R}_3^{avg}).
\tag{22.35}
$$

A.3 Derivation of $P(\mathcal{R}_3^{avg})$

$P(\mathcal{R}_3^{avg})$ is the probability that the interference-rich water-filling power $P_s^{wf-ir}(\mathbf{g})$ is higher than $P_s^B(\mathbf{g})$, as follows:

$$
\begin{aligned}
P(\mathcal{R}_3^{avg}) &= \Pr\left[P_s^B(\mathbf{g}) < \max\left(0, P_s^{wf-ir}(\mathbf{g})\right)\right] \\
&= \Pr\left[P_s^B(\mathbf{g}) < 0, 0 > P_s^{wf-ir}(\mathbf{g})\right] + \Pr\left[P_s^B(\mathbf{g}) < P_s^{wf-ir}(\mathbf{g}), 0 \le P_s^{wf-ir}(\mathbf{g})\right] \\
&= \Pr\left[\frac{1}{\lambda P_p}\frac{g_s}{g_{sp}} - \frac{N_0}{P_p} < g_{ps} < \frac{\gamma_{th}N_0}{P_p}\right] \\
&\quad + \Pr\left[\frac{g_s}{g_{sp}} > \frac{\lambda P_p^2}{\gamma_{th}N_0}\left(g_{ps}^2 + \frac{N_0}{P_p}g_{ps}\right), \frac{g_s}{g_{sp}} \ge \lambda P_p\left(g_{ps} + \frac{N_0}{P_p}\right)\right].
\end{aligned}
\tag{22.36}
$$

Suppose that $z = g_s/g_{sp}$ and $w = g_{ps}$. The first term of (22.36) can be calculated as

$$
\begin{aligned}
\Pr\left[\frac{1}{\lambda P_p}z - \frac{N_0}{P_p} < w < \frac{\gamma_{th}N_0}{P_p}\right] &= \int_0^{\frac{\gamma_{th}N_0}{P_p}} \int_0^{\lambda P_p(w+\frac{N_0}{P_p})} \frac{e^{-w}}{(z+1)^2} dz\, dw \\
&= \int_0^{\frac{\gamma_{th}N_0}{P_p}} e^{-w}\left(1 - \frac{1}{\lambda P_p w + \lambda N_0 + 1}\right) dw \\
&= 1 - e^{-\gamma_{th}N_0/P_p} + \frac{\exp\left(\frac{1+\lambda N_0}{\lambda P_p}\right)}{\lambda P_p}\left[\text{Ei}\left(-\frac{1+\lambda N_0}{\lambda P_p}\right) - \text{Ei}\left(\frac{1+(1+\gamma_{th})\lambda N_0}{\lambda P_p}\right)\right],
\end{aligned}
\tag{22.37}
$$

where the integration is solved with the help of (3.352.1) in [26].

The second term of (22.36) is given by

$$
\begin{aligned}
&\Pr\left[z > \frac{\lambda P_p^2}{\gamma_{th}N_0}\left(w^2 + \frac{N_0}{P_p}w\right), z \ge \lambda P_p\left(w + \frac{N_0}{P_p}\right)\right] \\
&= \underbrace{\int_0^{\frac{\gamma_{th}N_0}{P_p}} \int_{\lambda P_p(w+\frac{N_0}{P_p})}^{\infty} \frac{e^{-w}}{(z+1)^2} dz\, dw}_{A_1} + \underbrace{\int_{\frac{\gamma_{th}N_0}{P_p}}^{\infty} \int_{\frac{\lambda P_p^2}{\gamma_{th}N_0}(w^2+\frac{N_0}{P_p}w)}^{\infty} \frac{e^{-w}}{(z+1)^2} dz\, dw}_{A_2}.
\end{aligned}
\tag{22.38}
$$

We can solve A_1 as follows:

$$
\begin{aligned}
A_1 &= \int_0^{\frac{\gamma_{th}N_0}{P_p}} \frac{e^{-w}}{1 + \lambda N_0 + \lambda P_p w} dw \\
&= \frac{\exp\left(\frac{1+\lambda N_0}{\lambda P_p}\right)}{\lambda P_p}\left[\text{Ei}\left(\frac{1+(1+\gamma_{th})\lambda N_0}{\lambda P_p}\right) - \text{Ei}\left(-\frac{1+\lambda N_0}{\lambda P_p}\right)\right],
\end{aligned}
\tag{22.39}
$$

where the integration is solved using (3.352.1) in [26].

By solving the inner integral, A_2 is given by

$$
A_2 = \int_{\frac{\gamma_{th}N_0}{P_p}}^{\infty} \frac{e^{-w}}{1 + \frac{\lambda P_p^2}{\gamma_{th}N_0}w\left(w + \frac{N_0}{P_p}\right)} dw.
\tag{22.40}
$$

Substituting A_1 and A_2 into (22.37) yields $P(\mathcal{R}_3^{avg})$ in (22.16).

22.8 Appendix B: Proof of Theorem 1

B.1 Derivation of $C(\mathcal{R}_1^{avg})$

$C(\mathcal{R}_1^{avg})$ is the ergodic capacity averaged over all the fading blocks when water-filling power allocation is used on condition that there is no primary user interference due to the successful decoding and cancellation (i.e., interference-free environment). This environment is similar to that of [5]. Thus, we have

$$C(\mathcal{R}_1^{avg}) = \iint_{\frac{g_s}{g_{sp}} \geq \lambda N_0} \log\left(\frac{1}{\lambda N_0}\frac{g_s}{g_{sp}}\right) f_{g_s}(g_s) f_{g_{sp}}(g_{sp}) dg_s dg_{sp}. \tag{22.41}$$

Under the Rayleigh fading assumption, g_s and g_{sp} follow the exponential distributions with unit-means. Then, g_s/g_{sp} becomes the log-logistic distribution [5], with a probability density function (PDF) of $f_z(z) = \frac{1}{(z+1)^2}$. The calculation of (22.41) is similar to that in [5], as follows:

$$C(\mathcal{R}_1^{avg}) = \log\left(1 + \frac{1}{\lambda N_0}\right). \tag{22.42}$$

B.2 Derivation of $C(\mathcal{R}_2^{avg})$

$C(\mathcal{R}_2^{avg})$ is the ergodic capacity when the maximum possible power satisfying the condition of successful interference cancellation $P_s^A = (g_{ps}P_p/\gamma_{th} - N_0)/g_s$. Since there is no primary user interference in this case, the capacity is given by

$$C(\mathcal{R}_2^{avg}) = \int_{g_s}\int_{g_{ps}} \log\left(1 + \frac{g_s\frac{(g_{ps}P_p/\gamma_{th}-N_0)}{g_s}}{N_0}\right) f_{g_s}(g_s) f_{g_{ps}}(g_{ps}) dg_s dg_{ps}$$

$$= \int_{g_{ps}} \log\left(\frac{g_{ps}P_p}{\gamma_{th}N_0}\right) f_{g_{ps}}(g_{ps}) dg_{ps} \tag{22.43}$$

$$= \int_0^\infty \log\left(\frac{P_p}{\gamma_{th}N_0}w\right) e^{-w} dw,$$

with $w = g_{ps}$. From (4.331.1) in [26], the solution of the integral (22.43) is given by

$$C(\mathcal{R}_2^{avg}) = \log\left(\frac{P_p}{\gamma_{th}N_0}\right) - \omega. \tag{22.44}$$

B.3 Derivation of $C(\mathcal{R}_3^{avg})$

$C(\mathcal{R}_3^{avg})$ is the ergodic capacity when the power $P_s^{wf-ir}(\mathbf{g})$ is used in the interference-rich environment. Thus, the capacity is given by

$$C(\mathcal{R}_3^{avg}) = \iiint_{\frac{g_s}{g_{sp}} \geq \lambda(g_{ps}P_p+N_0)} \log\left(\frac{g_s}{g_{sp}}\cdot\frac{1}{\lambda(g_{ps}P_p+N_0)}\right) \times f_{g_s}(g_s) f_{g_{sp}}(g_{sp}) f_{g_{ps}}(g_{ps}) dg_s dg_{sp} dg_{ps}, \tag{22.45}$$

This triple integral was solved in our previous research [27], yielding the closed-form expression (22.27).

22.9 Appendix C: Proof of Theorem 2

C.1 Derivation of $E_g[g_{sp}P_s^{wf-if}(g)]$

For Region 1, the average received-power with transmit power $P_s^{wf-if}(\mathbf{g})$ can be obtained by calculating the following integration:

$$E_{\mathbf{g}}[g_{sp}P_s^{wf-if}(\mathbf{g})] = \iint_{\frac{g_{sp}}{g_s} \le \frac{1}{\lambda N_0}} \left(\frac{1}{\lambda} - N_0 \frac{g_{sp}}{g_s} \right) f_{g_s}(g_s) f_{g_{sp}}(g_{sp}) dg_s dg_{sp}. \tag{22.46}$$

Substituting $y = g_{sp}/g_s$, (22.46) can be calculated as

$$E_{\mathbf{g}}[g_{sp}P_s^{wf-if}(\mathbf{g})] = \int_0^{\frac{1}{\lambda N_0}} \left(\frac{1}{\lambda} - N_0 y \right) \frac{1}{(y+1)^2} dy$$
$$= \frac{1}{\lambda} - N_0 \log\left(1 + \frac{1}{\lambda N_0} \right), \tag{22.47}$$

where the integral can be solved using (2.113.2) in [26].

C.2 Derivation of $E_g[g_{sp}P_s^A(g)]$

The average received-power for Region 2 is defined as

$$E_{\mathbf{g}}[g_{sp}P_s^A(\mathbf{g})] = \iiint_{\frac{\lambda P_p}{\gamma_{th}} g_{ps} < \frac{g_s}{g_{sp}} \le \frac{\lambda P_p^2}{\gamma_{th} N_0} g_{ps}(g_{ps}+\frac{N_0}{P_p})} \frac{g_{sp}}{g_s} \left(\frac{P_p}{\gamma_{th}} g_{ps} - N_0 \right) f_{g_s}(g_s) f_{g_{sp}}(g_{sp}) f_{g_{ps}}(g_{ps}) dg_s dg_{sp} dg_{ps}. \tag{22.48}$$

By substituting $z = g_s/g_{sp}$ and $w = g_{ps}$, (22.48) can be rewritten as

$$E_{\mathbf{g}}[g_{sp}P_s^A(\mathbf{g})] = \frac{P_p}{\gamma_{th}} \int_0^\infty \left(w - \frac{\gamma_{th} N_0}{P_p} \right) e^{-w} \underbrace{\int_{\frac{\lambda P_p}{\gamma_{th}} w}^{\frac{\lambda P_p^2}{\gamma_{th} N_0} w(w+N_0/P_p)} \frac{dz}{z(z+1)^2}}_{B} dw. \tag{22.49}$$

The integration of B can be solved using (2.119.1) in [26], yielding

$$B = \frac{1}{1 + \frac{\lambda P_p^2}{\gamma_{th} N_0} w\left(w + \frac{N_0}{P_p}\right)} - \frac{1}{1 + \frac{\lambda P_p}{\gamma_{th}} w} + \log\left[\left(1 + \frac{P_p}{N_0}w\right)\left(1 + \frac{\lambda P_p}{\gamma_{th}}w\right) \right]$$
$$- \log\left[1 + \frac{\lambda P_p^2}{\gamma_{th} N_0} w\left(w + \frac{N_0}{P_p}\right) \right]. \tag{22.50}$$

Then, after some manipulations, $E_{\mathbf{g}}[g_{sp}P_s^A(\mathbf{g})]$ becomes

$$E_{\mathbf{g}}[g_{sp}P_s^A(\mathbf{g})] = \frac{P_p}{\gamma_{th}} \left[\underbrace{\int_0^\infty \left(w - \frac{\gamma_{th} N_0}{P_p} \right) e^{-w} \log\left[\left(1 + \frac{P_p}{N_0}w\right)\left(1 + \frac{\lambda P_p}{\gamma_{th}}w\right) \right] dw}_{B_1} - \underbrace{\int_0^\infty \frac{\left(w - \frac{\gamma_{th} N_0}{P_p}\right) e^{-w}}{1 + \frac{\lambda P_p}{\gamma_{th}} w} dw}_{B_2} \right.$$
$$\left. + \underbrace{\int_0^\infty \left(w - \frac{\gamma_{th} N_0}{P_p} \right) e^{-w} \left(\frac{1}{1 + \frac{\lambda P_p^2}{\gamma_{th} N_0} w\left(w + \frac{N_0}{P_p}\right)} - \log\left[1 + \frac{\lambda P_p^2}{\gamma_{th} N_0} w\left(w + \frac{N_0}{P_p}\right) \right] \right) dw}_{B_3} \right], \tag{22.51}$$

We first integrate B_1 as follows:

$$B_1 = \underbrace{\int_0^\infty w e^{-w} \log\left(1 + \frac{P_p}{N_0}w\right) dw}_{B_{11}} - \underbrace{\frac{\gamma_{th}N_0}{P_p} \int_0^\infty e^{-w} \log\left(1 + \frac{P_p}{N_0}w\right) dw}_{B_{12}}$$
$$+ \underbrace{\int_0^\infty w e^{-w} \log\left(1 + \frac{\lambda P_p}{\gamma_{th}}w\right) dw}_{B_{13}} - \underbrace{\frac{\gamma_{th}N_0}{P_p} \int_0^\infty e^{-w} \log\left(1 + \frac{\lambda P_p}{\gamma_{th}}w\right) dw}_{B_{14}},$$

(22.52)

where B_{11} and B_{12} can be obtained using integration by parts and (4.337.2) in [26]:

$$B_{11} = 1 - \left(1 - \frac{N_0}{P_p}\right)\exp\left(\frac{N_0}{P_p}\right)\text{Ei}\left(\frac{N_0}{P_p}\right)$$
$$B_{12} = -\exp\left(\frac{N_0}{P_p}\right)\text{Ei}\left(\frac{N_0}{P_p}\right).$$

(22.53)

Since B_{13} and B_{14} can be obtained in similar ways, the solution for B_1 is given by

$$B_1 = 2 - \left(1 - \frac{N_0}{P_p}(1 + \gamma_{th})\right)\exp\left(\frac{N_0}{P_p}\right)\text{Ei}\left(-\frac{N_0}{P_p}\right) - \left(1 - \frac{\gamma_{th}}{P_p}\left(\frac{1}{\lambda} + N_0\right)\right)\exp\left(\frac{\gamma_{th}}{\lambda P_p}\right)\text{Ei}\left(-\frac{\gamma_{th}}{\lambda P_p}\right).$$

(22.54)

We can obtain B_2 with the help of (3.352.4) and (3.353.5) in [26], as follows:

$$B_2 = \int_0^\infty \frac{w e^{-w}}{1 + \frac{\lambda P_p}{\gamma_{th}}w} dw - \frac{\gamma_{th}N_0}{P_p} \int_0^\infty \frac{e^{-w}}{1 + \frac{\lambda P_p}{\gamma_{th}}w} dw$$
$$= \frac{\gamma_{th}}{\lambda P_p}\left[1 + \frac{\gamma_{th}}{P_p}\left(\frac{1}{\lambda} + N_0\right)\exp\left(\frac{\gamma_{th}}{\lambda P_p}\right)\text{Ei}\left(-\frac{\gamma_{th}}{\lambda P_p}\right)\right].$$

(22.55)

Unfortunately, B_3 cannot be solved in a closed-form. Combining B_1, B_2, and B_3, we can obtain the desired result in (22.22).

C.3 Derivation of $E_g[g_{sp}P_s^{wf-ir}(g)]$

The average received-power for Region 3 is given by

$$E_\mathbf{g}[g_{sp}P_s^{wf-ir}(\mathbf{g})] = \iiint_{\frac{g_{sp}}{g_s} \le \frac{1}{\lambda(g_{ps}P_p + N_0)}} \left(\frac{1}{\lambda} - \frac{g_{sp}}{g_s}(g_{ps}P_p + N_0)\right)$$
$$\times f_{g_s}(g_s)f_{g_{sp}}(g_{sp})f_{g_{ps}}(g_{ps})dg_s dg_{sp} dg_{ps}.$$

(22.56)

This triple integral was solved in our previous research [27], resulting in $E_\mathbf{g}[g_{sp}P_s^{wf-ir}(\mathbf{g})]$ in (22.22).

22.10   Appendix D: Proof of Lemma 3

D.1 Derivation of $P(\mathcal{R}_1^{peak})$

$P(\mathcal{R}_1^{peak})$ is the probability that the maximum possible transmit power that satisfies the received-power constraint $P_s^{mp}(\mathbf{g})$ is lower than the optimal transmit power for the interference-free case

P_s^A (i.e., maximum possible transmit power while interference cancellation is successful). Thus, $P(\mathcal{R}_1^{peak})$ can be expressed as

$$P(\mathcal{R}_1^{peak}) = \Pr\left[P_s^{\lim}(\mathbf{g}) \le P_s^{\max}(\mathbf{g})\right] = \Pr\left[\frac{g_s}{g_{sp}} \le \frac{P_p}{\gamma_{th}Q_{th}}g_{ps} - \frac{N_0}{Q_{th}}\right]. \tag{22.57}$$

Suppose that $z = g_s/g_{sp}$ and $w = g_{ps}$. z is a log-logistic random variable, and w is an exponential random variable as discussed before.

We can evaluate (22.57) by integrating $f_{z,w}(z,w)$ over the region $\{z \le P_p/(\gamma_{th}Q_{th})w - N_0/Q_{th}\}$, as follows:

$$
\begin{aligned}
P(\mathcal{R}_1^{peak}) &= \int_{\frac{\gamma_{th}N_0}{P_p}}^{\infty} \int_0^{\frac{P_p}{\gamma_{th}Q_{th}}w - \frac{N_0}{Q_{th}}} \frac{e^{-w}}{(z+1)^2} dz dw \\
&= \int_{\frac{\gamma_{th}N_0}{P_p}}^{\infty} e^{-w} dw - \frac{\gamma_{th}N_0}{P_p} \int_{\frac{\gamma_{th}N_0}{P_p}}^{\infty} \frac{e^{-w}}{w + \frac{\gamma_{th}}{P_p}(Q_{th} - N_0)} dw \\
&= \exp\left(-\frac{\gamma_{th}N_0}{P_p}\right) + \frac{\gamma_{th}Q_{th}}{P_p} \exp\left(\frac{\gamma_{th}}{P_p}(Q_{th} - N_0)\right) \mathrm{Ei}\left(-\frac{\gamma_{th}Q_{th}}{P_p}\right),
\end{aligned}
\tag{22.58}
$$

where the integral on the second term can be solved with the help of (3.352.4) in [26].

D.2 Derivation of $P(\mathcal{R}_2^{peak})$

Similar to the case for the average received-power constraint, we can obtain $P(\mathcal{R}_2^{peak})$ using the property that the event for the regions $\mathcal{R}_1$, $\mathcal{R}_2$, and $\mathcal{R}_3$ are mutually exclusive and collectively exhaustive. Thus, we have

$$P(\mathcal{R}_2^{peak}) = 1 - P(\mathcal{R}_1^{peak}) - P(\mathcal{R}_3^{peak}). \tag{22.59}$$

D.3 Derivation of $P(\mathcal{R}_3^{peak})$

$P(\mathcal{R}_3^{peak})$ is the probability that $P_s^{mp}(\mathbf{g})$ is higher than $P_s^B(\mathbf{g})$, as follows:

$$
\begin{aligned}
P(\mathcal{R}_3^{peak}) &= \Pr\left[P_s^B(\mathbf{g}) < P_s^{mp}(\mathbf{g})\right] \\
&= \Pr\left[\frac{g_s}{g_{sp}} > \frac{P_p}{Q_{th}}\left(\frac{P_p}{\gamma_{th}N_0}g_{ps}^2 + \frac{1-\gamma_{th}}{\gamma_{th}}g_{ps} - \frac{N_0}{P_p}\right)\right].
\end{aligned}
\tag{22.60}
$$

Suppose that $z = g_s/g_{sp}$ and $w = g_{ps}$. Using the joint PDF of z and w, (22.60) can be calculated as

$$
\begin{aligned}
P(\mathcal{R}_3^{peak}) &= \int_0^{\infty} \frac{1}{(z+1)^2} \int_0^{\frac{\gamma_{th}N_0}{P_p}} e^{-w} dw dz \\
&\quad + \int_{\frac{\gamma_{th}N_0}{P_p}}^{\infty} e^{-w} \int_{\frac{P_p}{Q_{th}}\left(\frac{P_p}{\gamma_{th}N_0}w^2 + \frac{1-\gamma_{th}}{\gamma_{th}}w - \frac{N_0}{P_p}\right)}^{\infty} \frac{1}{(z+1)^2} dz dw.
\end{aligned}
\tag{22.61}
$$

Solving the integrals, (22.61) is calculated as

$$
\begin{aligned}
P(\mathcal{R}_3^{peak}) &= 1 - \exp\left(-\frac{\gamma_{th}N_0}{P_p}\right) \\
&\quad + \int_{\frac{\gamma_{th}N_0}{P_p}}^{\infty} \frac{e^{-w}}{\frac{P_p}{Q_{th}}\left(\frac{P_p}{\gamma_{th}N_0}w^2 + \frac{1-\gamma_{th}}{\gamma_{th}}w - \frac{N_0}{P_p}\right) + 1} dw.
\end{aligned}
\tag{22.62}
$$

22.11   Appendix E: Proof of Theorem 2

E.1 Derivation of $C(\mathcal{R}_1^{peak})$

$C(\mathcal{R}_1^{peak})$ is the ergodic capacity averaged over all the fading blocks when the maximum possible power satisfying the received-power constraint $P_s^{mp}(\mathbf{g}) = Q_{th}/g_{sp}$ is used on condition that there is no primary user interference due to the successful decoding and cancellation (i.e., interference-free environment). This environment is similar to that of [5], yielding the result $C(\mathcal{R}_1^{peak})$ in (22.27).

E.2 Derivation of $C(\mathcal{R}_2^{peak})$

$C(\mathcal{R}_2^{peak})$ is actually the same as the counterpart for the average received-power constraint case, i.e., $C(\mathcal{R}_2^{peak}) = C(\mathcal{R}_2^{avg})$. This is because the transmit powers for both cases are the same for Region 2.

E.3 Derivation of $C(\mathcal{R}_3^{peak})$

$C(\mathcal{R}_3^{peak})$ is the ergodic capacity when $P_s^{mp}(\mathbf{g})$ is used when primary user interference exists (i.e., interference-rich environment). Thus, the ergodic capacity is given by

$$C(\mathcal{R}_3^{peak}) = \int_{g_{ps}} \int_{g_{sp}} \int_{g_s} \log_2\left(1 + Q_{th}\frac{g_s/g_{sp}}{g_{ps}P_p + N_0}\right)$$
$$\times f_{g_s}(g_s)f_{g_{sp}}(g_{sp})f_{g_{ps}}(g_{ps})dg_s dg_{sp} dg_{ps}. \tag{22.63}$$

Let us suppose that $z = g_s/g_{sp}$, $w = g_{ps}$. Then, since z follows the log-logistic distribution, $C(\mathcal{R}_e^{peak})$ is given as

$$C(\mathcal{R}_3^{peak}) = \int_0^\infty e^{-w} \int_0^\infty \frac{\log_2\left(1 + \frac{Q_{th}z}{P_p w + N_0}\right)}{(z+1)^2} dz dw. \tag{22.64}$$

The inner integral can be solved using integration by parts:

$$\int_0^\infty \frac{\log_2\left(1 + \frac{Q_{th}z}{P_p w + N_0}\right)}{(z+1)^2} dz = \frac{\log\left(\frac{P_p w + N_0}{Q_{th}}\right)}{\frac{P_p w + N_0}{Q_{th}} - 1}, \tag{22.65}$$

which yields the desired result $C(\mathcal{R}_3^{peak})$ in (22.27).

References

[1] Q. Zhao and B. M. Sadler, "A survey of dynamic spectrum access: Signal processing, networking, and regulation policy," *IEEE Signal Process. Mag.*, vol. 24, no. 3, pp. 79–89, May 2007.

[2] FCC Spectrum Policy Task Force, "Report of the spectrum efficiency working group," ET Docket 02-135, Nov. 2002.

[3] J. Mitola and G. Q. Maguire, "Cognitive radio: Making software radios more personal," *IEEE Personal Commun. Mag.*, vol. 6, no. 4, pp. 13–18, Aug. 1999.

[4] M. Gastpar, "On capacity under received-signal constraints," in *Proc. Allerton Conf. Commun. Control Comput.*, Sep. 2004, pp. 1322–1331.

[5] A. Ghasemi and E. S. Sousa, "Fundamental limits of spectrum-sharing in fading environments," *IEEE Trans. Wireless Commun.*, vol. 6, no. 2, pp. 649–658, Feb. 2007.

[6] X. Kang, H. K. Garg, Y.-C. Liang, and R. Zhang, "Optimal power allocation for OFDM-based cognitive radio with new primary transmission protection criteria," *IEEE Trans. Wireless Commun.*, vol. 9, no. 6, pp. 2066–2075, Jun. 2010.

[7] H. Wang, J. Lee, S. Kim, and D. Hong, "Capacity of secondary users exploiting multispectrum and multiuser diversity in spectrum-sharing environments," *IEEE Trans. Veh. Technol.*, vol. 59, no. 2, pp. 1030–1036, Feb. 2010.

[8] H. Cho and J. G. Andrews, "Upper bound on the capacity of cognitive radio without cooperation," *IEEE Trans. Wireless Commun.*, vol. 8, no. 9, pp. 4380–4385, Sep. 2009.

[9] H. A. Suraweera, P. J. Smith, and M. Shafi, "Capacity limits and performance analysis of cognitive radio with imperfect channel knowledge," *IEEE Trans. Veh. Technol.*, vol. 59, no. 4, pp. 1811–1822, May 2010.

[10] H. Kim, H. Wang, S. Lim, and D. Hong, "On the impact of outdated channel information on the capacity of secondary user in spectrum sharing environments," *IEEE Trans. Wireless Commun.*, vol. 11, no. 1, pp. 284–295, Jan. 2012.

[11] S. M. Almalfouh and G. L. Stuber, "Interference-aware power allocation in cognitive radio networks with imperfect spectrum sensing," in *Int. Conf. Commun. (ICC 2010)*, pp. 1–6, May 2010.

[12] A. Zanella, M. Chiani, and M. Z. Win, "MMSE reception and successive interference cancellation for MIMO systems with high spectral efficiency," *IEEE Trans. Wireless Commun.*, vol. 4, no. 3, pp. 1244–1253, May 2005.

[13] T. Tang and R. W. Heath Jr., "Space-time interference cancellation in MIMO-OFDM systems," *IEEE Trans. Veh. Technol.*, vol. 54, no. 5, pp. 1802–1816, Sep. 2005.

[14] S. P. Weber, J. G. Andrews, X. Yang, and G. De Veciana, "Transmission capacity of wireless ad hoc networks with successive interference cancellation," *IEEE Trans. Inf. Theory*, vol. 53, no. 8, pp. 2799–2814, Aug. 2007.

[15] J. Lee, J. G. Andrews, and D. Hong, "The effect of interference cancellation on spectrum-sharing transmission capacity," *IEEE Trans. Commun.*, vol. 61, no. 1, pp. 76–86, Jan. 2013.

[16] R. Zhang and Y.-C. Liang, "Exploiting hidden power-feedback loops for cognitive radio," in *Proc. IEEE Symp. New Frontiers Dynamic Spectr. Access Netw. (DySPAN 2008)*, pp. 1–5, Oct. 2008.

[17] K. Hamdi, W. Zhang, and B. K. Letaief, "Power control in cognitive radio systems based on spectrum sensing side information," in *Proc. IEEE Int. Conf. Commun. (ICC 2007)*, Jun. 2007, pp. 5161–5165.

[18] J. M. Peha, "Approaches to spectrum sharing," *IEEE Commun. Mag.*, pp. 10–12, Feb. 2005.

[19] K. Nishimori, H. Yomo, and P. Popovski, "Distributed interference cancellation for cognitive radios using periodic signals of the primary system," *IEEE Trans. Wireless Commun.*, vol. 10, no. 9, pp. 2971–2981, Sep. 2011.

[20] X. Kang, Y.-C. Liang, A. Nallanathan, H. K. Garg, and R. Zhang, "Optimal power allocation for fading channels in cognitive radio networks: Ergodic capacity and outage capacity," *IEEE Trans. Wireless Commun.*, vol. 8, no. 2, pp. 940–950, Feb. 2009.

[21] S.-M. Cheng, S.-Y. Lien, F.-S. Chu, and K.-C. Chen, "On exploiting cognitive radio to mitigate interference in macro/femto heterogeneous networks," *IEEE Wireless Commun. Mag.*, pp. 40–47, Jun. 2011.

[22] K. Huang, J. G. Andrews, R. W. Heath, D. Guo, and R. A. Berry, "Spatial interference cancellation for mobile ad hoc networks: Perfect CSI," in *Proc. IEEE Global Commun. Conf. (GLOBECOM 2008)*, pp. 1–5, Nov. 2008.

[23] E. Gelal, K. Pelechrinis, I. Broustis, S. V. Krishnamurthy, and B. Rao, "Topology control for effective interference cancellation in multi-user MIMO networks," in *Proc. INFOCOM 2010*, pp. 1–9, Mar. 2010.

[24] D. Halperin, T. Anderson, and D. Wetherall, "Taking the sting out of carrier sense: Interference cancellation for wireless LANs," in *Proc. ACM Int. Conf. Mobile Comput. Netw. (MobiCom 2008)*, pp. 339–350, Sep. 2008.

[25] S. Boyd and L. Vandenberghe, *Convex Optimization*. Cambridge, UK: Cambridge University Press, 2004.

[26] I. S. Gradshteyn and I. M. Ryzhik, *Tables of Integrals, Series, and Products*, 7th ed. London, UK: Academic Press, 2007.

[27] G. Noh, S. Lim, and D. Hong, "Exact Capacity Analysis of Spectrum Sharing Systems: Average Received-Power Constraint," *IEEE Commun. Lett.*, vol. 17, no. 5, pp. 884–887, May 2013.

[28] A. Papoulis and S. U. Pillai, *Probability, Random Variables and Stochastic Processes*, 4th ed. New York, NY: McGraw-Hill, 2002.

Chapter 23

Energy-Efficient Power Control for Spectrum Sharing in Next-Generation Wireless Networks

Xiangping Zhai, Liang Zheng, and Chee Wei Tan

CONTENTS

Energy efficiency is particularly important in next-generation wireless mobile networks that can support a large number of battery-powered mobile terminals. An upcoming trend in mobile network operation has been the decoupling of wireless carrier infrastructure from mobile services that enables the multiplexing of services offered by different mobile network operators (MNOs), e.g., Google's Project Fi. This wireless access by multiple heterogeneous MNOs presents new challenges to wireless resource sharing and interference management. In this chapter, we study the design of energy-efficient power control in a network with multiple MNOs. We provide a new and novel perspective to the design of single-MNO iterative power control algorithm that also provides the basis to design fast-convergent iterative power control algorithms for the more complex multiple-MNOs case. We also discuss a joint energy-efficient power control and admission control using the perspective of optimization-theoretic feasibility. Lastly, we provide numerical simulation to demonstrate the energy efficiency gain and the performance of the algorithms.

23.1 Introduction

The demand for mobile data services has grown significantly in recent years. This has spurred the growth of the mobile network operators (MNOs) and various new wireless services. Unlike the traditional wireless carrier operators, the MNOs do not own the physical wireless carrier infrastructures, but instead lease these carrier infrastructures from the carrier operators to offer new wireless services independently. An example is the Google's Project Fi. As such, many new devices today are operating in a shared wireless spectrum that is utilized by different users belonging to the traditional wireless carrier operators as well as the MNOs [1]. Figure 23.1 illustrates an example of this wireless access by different mobile users with different technological connectivities to three different MNOs. Due to the decentralized management of wireless resources, wireless resource sharing can be far from perfect thus making resource control and interference management especially important. Power control allows each mobile user to transmit enough power to achieve the required quality of service without causing unnecessary interference to other users. An energy-efficient power allocation must thus be able to support a large number of battery-powered mobile terminals with reduced energy consumption.

Due to the uncoordinated decentralized access to multiple MNOs and the broadcast nature of the wireless medium, multiuser interference can be a major source of performance impairment. A non-adaptable wireless system can suffer from deteriorating quality due to fixed resource allocation that fails to consider this multiuser interference. Also, power control schemes that are designed primarily for a single wireless carrier operator can lead to a higher energy consumption whenever multiuser interference increases. There are a number of related works on energy efficiency in wireless networks. The authors in [2–4] studied interference management algorithms for energy efficiency according to the dynamics of mobile usage. The authors in [5] proposed switching off base stations to reduce energy consumption by exploiting the spatio-temporal traffic fluctuations. The authors in [6–8] addressed the problem of maximizing the energy efficiency and user utility satisfaction. The authors in [9] proposed robust power control to study the fundamental tradeoff between energy consumption and data requirements. The authors in [10–13] analyzed several representative utility fairness problems using the nonlinear Perron-Frobenius theory and designed energy-efficient power control for utility maximization. Also, various power control algorithms have been proposed for the wireless throughput maximization problem [14–16].

In this chapter, we study the problem of energy minimization subject to data rate requirements in a wireless network with MNOs, and focus on the design of energy-efficient power control algorithms. In essence, this problem is similar to the energy minimization problem that assumes a multi-carrier system model (see, e.g., [17]). However, unlike a multi-carrier model, the MNOs typically cannot cooperate in wireless resource sharing and interference management. In addition, this class of energy minimization problems is challenging to solve due to the inherent nonconvexity in the signal-to-interference-and-noise ratio (SINR)-related function used to model data rates. As such,

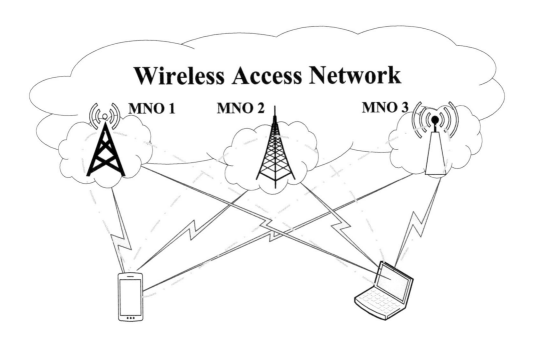

Figure 23.1: Illustration of mobile users transmitting information through three different MNOs to maximize resource utilization.

the algorithms that we have developed in this chapter can also be useful for other problems that assume a multi-carrier system model, e.g., in [17]. Another practical challenge is that the decentralized access to multiple MNOs can cause system infeasibility whenever it is not possible to simultaneously meet the data rate requirements of all the users. As such, we also discuss how to resolve this system infeasibility issue by proposing a joint energy-efficient power control and admission control algorithm that uses the sum-of-infeasibilities heuristic in optimization theory.

The rest of the chapter is organized as follows. We first study the single MNO problem using the standard interference function framework that exploits the problem structure directly to design power control algorithms. Secondly, we provide a formulation of the power control problem for the multi-user multiple MNOs with data rate requirements, and develop an iterative distributed power control algorithm that utilizes the standard interference function to compute a feasible solution. Lastly, we address the system infeasibility issue by considering a vector-cardinality optimization formulation and its relaxation. Based on this formulation, we propose a joint power control and admission control algorithm using the sum-of-infeasibilities heuristic. We conclude the chapter by listing some open issues in this research area.

We adopt the following notations in this chapter. Lowercase boldface and uppercase boldface are used for vectors and matrices, respectively. We use $\mathbf{x}_1 \circ \mathbf{x}_2$ to represent the Schur product of two vectors $\mathbf{x}_1$ and $\mathbf{x}_2$. The spectral radius of a matrix $\mathbf{A}$ is denoted by $\rho(\mathbf{A})$. The super-script $(\cdot)^{\top}$ denotes the transpose. We use $\mathbf{1}$ to represent the vector, with all the entries being one, $\mathbf{1}_l$ to represent the vector, with all entries being zero, except its l-th entry being one. Let $\mathbf{I}$ and $\mathrm{diag}(\mathbf{x})$ denote the identity matrix and the diagonal matrix with the entries of $\mathbf{x}$ on the diagonal, respectively. Let $e^{\mathbf{x}}$ and $\log \mathbf{x}$ denote $(e^{x_1}, \ldots, e^{x_n})^{\top}$ and $(\log x_1, \ldots, \log x_n)^{\top}$, respectively.

23.2 Power Control in Single MNO

In this section, we study the total power minimization problem subject to given data rate requirements of all the users in a single MNO network. We will first review the well-known solution to this problem using the Foschini-Miljanic algorithm [18], and then use the standard interference function framework to propose a new distributed power control algorithm to solve it optimally. This new power control algorithm is later leveraged in the design of a power control algorithm for the general case of multiple MNOs in Section 23.3.

23.2.1 System Model

Consider a single MNO communication system with a finite number of mobile users. There are L users (transmitter-receiver pairs) that want to communicate simultaneously through a common MNO. Denote the set of users by $\mathcal{L} = \{1, 2, \cdots, L\}$. For any $l \in \mathcal{L}$, suppose $s_l \in \mathbb{C}$ to be the symbol that transmitter l wishes to send to receiver l, then the received signal $\hat{s}_l$ at receiver l can be expressed by:

$$\hat{s}_l = \sum_{j=1}^{L} h_{lj} s_j + z_l, \tag{23.1}$$

where $h_{lj} \in \mathbb{C}$ is the channel coefficient between the j-th transmitter and the l-th receiver, and $z_l \in \mathbb{C}$ is the additive white Gaussian noise (AWGN), with distribution $\mathcal{CN}(0, \sigma_l)$. Denoting the power of s_l by p_l, i.e., $p_l = |s_l|^2$, the received power at receiver l is given by:

$$\sum_{j=1}^{L} G_{lj} p_j + \sigma_l, \tag{23.2}$$

where $G_{lj} = |h_{lj}|^2$ stands for the channel gain between the l-th transmitter and the j-th receiver. The SINR of receiver l is given by:

$$\text{SINR}_l(\mathbf{p}) = \frac{G_{ll} p_l}{\sum_{j \neq l} G_{lj} p_j + \sigma_l}, \tag{23.3}$$

where $\mathbf{p} = (p_1, p_2, \ldots, p_L)^\top$. Assuming a fixed bit error rate at the receiver, the achievable data rate r_l of the l-th transmitter can be computed by the Shannon capacity formula [19]:

$$r_l(\mathbf{p}) = \log\left(1 + \frac{\text{SINR}_l(\mathbf{p})}{\Gamma}\right) \text{ nats/symbol}, \tag{23.4}$$

where Γ is the SINR gap to capacity, which is always greater than 1. In this chapter, we absorb $1/\Gamma$ into G_{ll} for all l, and write the achievable rate as:

$$r_l(\mathbf{p}) = \log\left(1 + \text{SINR}_l(\mathbf{p})\right). \tag{23.5}$$

The problem to minimize the total energy consumption subject to given data rate requirements in the single MNO communication system is formulated as follows:

$$\text{minimize} \quad \sum_{l=1}^{L} p_l$$

$$\text{subject to} \quad \log \left(1 + \frac{G_{ll} p_l}{\sum_{j \neq l} G_{lj} p_j + \sigma_l} \right) \geq \bar{r}_l, \quad l = 1, \ldots, L \quad (23.6)$$

$$0 \leq p_l \leq \bar{p}_l, \quad l = 1, \ldots, L$$

$$\text{variables}: \quad \mathbf{p},$$

where $\bar{p}_l$ is the upper-bound of the transmit power for the l-th users, and $\bar{r}_l$ is a given minimum rate threshold that represents the l-th user's rate requirement in the single MNO network. Denote a vector $\bar{\mathbf{r}} = (\bar{r}_1, \ldots, \bar{r}_L)^T$. In other words, we are interested to find the optimal power allocation in (23.6), such that the achieved transmission rates of all the users are at least larger than $\bar{\mathbf{r}}$.

23.2.2 Problem Reformulation and Analysis

In this section, we reformulate (23.6) into an equivalent problem (23.7), with SINR constraints instead of the original rate constraints:

$$\text{minimize} \quad \sum_{l=1}^{L} p_l$$

$$\text{subject to} \quad \text{SINR}_l(\mathbf{p}) \geq \bar{\gamma}_l, \quad l = 1, \ldots, L \quad (23.7)$$

$$0 \leq p_l \leq \bar{p}_l, \quad l = 1, \ldots, L$$

$$\text{variables}: \quad \mathbf{p},$$

where $\bar{\gamma}_l = e^{\bar{r}_l} - 1$ for all l can be interpreted as the minimum SINR requirement.

It is possible to give a more compact representation to (23.7). Let us define the nonnegative vector:

$$\mathbf{v} = \left(\frac{\sigma_1}{G_{11}}, \frac{\sigma_2}{G_{22}}, \cdots, \frac{\sigma_l}{G_{LL}} \right)^\top, \quad (23.8)$$

and the nonnegative matrix $\mathbf{F}$ with entries:

$$F_{lj} = \begin{cases} 0, & k = j, \\ \dfrac{G_{lj}}{G_{ll}}, & l \neq j, \end{cases} \quad . \quad (23.9)$$

Moreover, we shall assume that the matrix $\mathbf{F}$ is irreducible, i.e., each user has at least an interferer. Then, we can rewrite (23.7) as the following linear program [20]:

$$\text{minimize} \quad \mathbf{1}^\top \mathbf{p}$$

$$\text{subject to} \quad (\mathbf{I} - \text{diag}(\bar{\boldsymbol{\gamma}})\mathbf{F})\mathbf{p} \geq \text{diag}(\bar{\boldsymbol{\gamma}})\mathbf{v}, \quad (23.10)$$

$$\mathbf{0} \leq \mathbf{p} \leq \bar{\mathbf{p}},$$

$$\text{variables}: \quad \mathbf{p}.$$

In general, (23.10), or, equivalently, (23.6), may be feasible or it may not be. This means that it may not be possible to have the transmitting rates of all the users be larger than $\bar{\mathbf{r}}$ in (23.6). We shall address the infeasibility issue of (23.6) later in Section 23.4. Let us suppose that (23.6) is feasible. We first review how to compute the optimal solution of (23.7) using the well-known Foschini-Miljanic algorithm [18].

Lemma 23.1
If (23.7) is feasible, then the SINR constraints in (23.7) are tight at optimality:

$$\text{SINR}_l(\mathbf{p}^\star) = \bar{\gamma}_l, \quad l = 1, \ldots, L. \tag{23.11}$$

Proof 23.1   Suppose that the rate constraint of a specific user l is not tight at optimality, i.e.,

$$\text{SINR}_l(\mathbf{p}) > \bar{\gamma}_l. \tag{23.12}$$

It is easy to verify that the left-hand side of inequality (23.12) is a strictly increasing function in p_l, and is a strictly decreasing function in p_j for all $j \neq l$. We can reduce the power p_l by a sufficiently small amount $\varepsilon > 0$ so that (23.12) is still satisfied using new transmit power for user l, i.e., $\hat{p}_l = p_l - \varepsilon$. By doing so, the other users' SINR increase and their SINR requirements can still be satisfied. Thus, we can further reduce the total energy consumption by ε, which is a contradiction to the assumption that the powers are optimal.   ■

Using Lemma 23.1 and the standard interference function framework[1] introduced in [18, 21], we can solve (23.7) using the following fixed-point iterative algorithm (and, thus, solving (23.6)). Let k index the iteration number.

Algorithm 11 Distributed Power Control Algorithm

$$p_l(k+1) = \min\left\{ \frac{e^{\bar{r}_l} - 1}{\text{SINR}_l(\mathbf{p}(k))} p_l(k), \bar{p}_l \right\}. \tag{23.13}$$

Theorem 23.1
When (23.6) is feasible, Algorithm 11 converges geometrically fast to the optimal solution of (23.7), equivalently, (23.6) from any feasible initial point $\mathbf{p}(0)$.

We refer the readers to [18, 21] for the details of the proof of Theorem 23.1 using the standard interference function framework.

23.2.3   *Iterative Power Control Algorithm*

In this section, we propose another alternative to solve (23.6) that also utilizes the standard interference function framework in [21] but leading to a different algorithm that solves (23.6) more directly. In particular, we give another fixed-point characterization at the optimality of (23.6) in the following. This leads to a new distributed algorithm to solve (23.6) instead of Algorithm 11.

Lemma 23.2
If (23.6) is feasible, then the rate constraints in (23.6) are tight at optimality:

$$\log\left(1 + \text{SINR}_l(\mathbf{p}^\star)\right) = \bar{r}_l, \quad l = 1, \ldots, L. \tag{23.14}$$

---

[1] See Definition 22.1 and Lemma 22.3 on the standard interference function.

Proof 23.2   Suppose that the rate constraint of a specific user l is not tight at optimality, i.e.,

$$\log\left(1 + \text{SINR}_l(\mathbf{p})\right) > \bar{r}_l. \tag{23.15}$$

It is easy to verify that the left-hand side of inequality (23.15) is a strictly increasing function in p_l, and is a strictly decreasing function in p_j for all $j \neq l$. We can reduce the power p_l by a sufficiently small amount $\varepsilon > 0$ so that (23.15) is still satisfied using new transmit power for user l, i.e., $\hat{p}_l = p_l - \varepsilon$. By doing so, the other users' rates increase and their data rate requirements can still be satisfied. Thus, we can further reduce the total energy consumption by ε, which is a contradiction to the assumption that the powers are optimal.   ■

Now, using the fixed-point characterization in Lemma 23.2, we propose the following iterative fixed-point algorithm that solves (23.6) more directly. Let k index the iteration number.

Algorithm 12 Rate Update Algorithm

$$p_l(k+1) = \min\left\{\frac{\bar{r}_l}{\log\left(1 + \text{SINR}_l(\mathbf{p}(k))\right)} p_l(k), \bar{p}_l\right\}. \tag{23.16}$$

Theorem 23.2
When (23.6) is feasible, Algorithm 12 converges geometrically fast to the optimal solution of (23.6) from any feasible initial point $\mathbf{p}(0)$.

Proof 23.3   The burden of proof lies in the standard interference function framework of [21], and so we first introduce it before showing how it is applied to prove Theorem 23.2.

Definition 23.1   An interference function $\mathbf{I}(\mathbf{p})$ is standard if, for all $\mathbf{p} \geq 0$, the following properties are satisfied:[2]

- Monotonicity: If $\mathbf{p}_1 \geq \mathbf{p}_2$, then $\mathbf{I}(\mathbf{p}_1) \geq \mathbf{I}(\mathbf{p}_2)$.
- Scalability: For all $\alpha > 1, \alpha\mathbf{I}(\mathbf{p}) > \mathbf{I}(\alpha\mathbf{p})$.

Lemma 23.3
If $\mathbf{p}$ is a feasible power vector, then $\mathbf{I}(\mathbf{p})$ is a monotone increasing sequence of feasible power vector in a fixed-point iteration that converges to the unique fixed point $\mathbf{p}^\star$ that satisfies:

$$\mathbf{p} = \mathbf{I}(\mathbf{p}). \tag{23.17}$$

Next, we state the following lemma.

Lemma 23.4
Let $D \subseteq \mathbb{R}^K$ and $f : D \to \mathbb{R}$ be such that for all $x \in D, \frac{\partial f}{\partial x_l}, l = 1, \dots, L$ exist on D. Then f is monotonically increasing on D if and only if $\frac{\partial f}{\partial x_l} \geq 0, l = 1, \dots, L$.

---

[2]Notice that, even though $\mathbf{p} \geq 0$ was required in [21], the results hold equally for just $\mathbf{p} > 0$. Moreover, notice that positivity (i.e., $\mathbf{I}(\mathbf{p}) > 0$) was required explicitly in [21], but can actually be implied by monotonicity and scalability, since the latter two yield $\alpha\mathbf{I}(\mathbf{p}) > \mathbf{I}(\alpha\mathbf{p}) \geq \mathbf{I}(\mathbf{p})$, for $\alpha > 1$.

Proposition 23.1

Let:

$$I_l(\mathbf{p}) = \frac{\bar{r}_l}{\log(1 + \text{SINR}_l(\mathbf{p}))} p_l. \tag{23.18}$$

The interference function

$$\mathbf{I}(\mathbf{p}) = (I_1(\mathbf{p}), I_2(\mathbf{p}), \cdots, I_L(\mathbf{p})) \tag{23.19}$$

is a standard interference function.

1. **Monotonicity:** From Lemma 23.4, if we can verify that $\frac{\partial I_l(\mathbf{p})}{\partial p_j} \geq 0$ for all j, then the monotonicity of $I_l(\mathbf{p})$ is guaranteed, and, furthermore, the monotonicity of $\mathbf{I}(\mathbf{p})$ is guaranteed.

   - If $j = l$, we have:

   $$\frac{\partial I_l(\mathbf{p})}{\partial p_l} = \frac{\log(1 + \text{SINR}_l(\mathbf{p})) - \text{SINR}_l(\mathbf{p})/(1 + \text{SINR}_l(\mathbf{p}))}{\log^2(1 + \text{SINR}_l(\mathbf{p}))} \bar{r}_l. \tag{23.20}$$

   We use y to substitute $1 + \text{SINR}_l(\mathbf{p})$ and define

   $$g(y) = \log(1 + \text{SINR}_l(\mathbf{p})) - \text{SINR}_l(\mathbf{p})/(1 + \text{SINR}_l(\mathbf{p})) = \log y + \frac{1}{y} - 1, \tag{23.21}$$

   where $y \geq 1$, and it is obvious to verify that $g(1) = 0$ and $g'(y) = 1 - 1/y^2 > 0$ for any positive power allocations, i.e., $y > 1$. Therefore, we have $g(y) \geq 0$, and then we have $\partial I_l(\mathbf{p})/\partial p_l \geq 0$.

   - If $j \neq l$, similarly, we have

   $$\frac{\partial I_l(\mathbf{p})}{\partial p_j} = \frac{G_{lj}G_{ll}p_l^2}{(1 + \text{SINR}_l(\mathbf{p}))\log^2(1 + \text{SINR}_l(\mathbf{p}))\left(\sum_{j \neq l} G_{lj}p_j + \sigma_l\right)^2} \bar{r}_l \geq 0. \tag{23.22}$$

   Therefore, $\mathbf{I}(\mathbf{p})$ is a monotonically increasing function in $\mathbf{p}$.

2. **Scalability:** We scale each user's power by $\alpha > 1$, and then we have:

$$
\begin{aligned}
I_l(\alpha\mathbf{p}) &= \alpha\bar{r}_l p_l / \log\left(1 + \frac{\alpha G_{ll}p_l}{\alpha \sum_{j \neq l} G_{lj}p_j + \sigma_l}\right) \\
&= \alpha\bar{r}_l p_l / \log\left(1 + \frac{G_{ll}p_l}{\sum_{j \neq l} G_{lj}p_j + \sigma_l/\alpha}\right) \\
&< \alpha\bar{r}_l p_l / \log\left(1 + \frac{G_{ll}p_l}{\sum_{j \neq l} G_{lj}p_j + \sigma_l}\right) = \alpha I_l(\mathbf{p}).
\end{aligned}
\tag{23.23}
$$

Since scaling up the power can suppress the noise σ_l/α, this increases the rate of user l. Thus, the denominator of $I_l(\alpha\mathbf{p})$ is larger than that of $\alpha I_l(\mathbf{p})$, while the numerator remains the same, and, therefore, we have $\alpha I_l(\mathbf{p}) > I_l(\alpha\mathbf{p})$ and $\alpha\mathbf{I}(\mathbf{p}) > \mathbf{I}(\alpha\mathbf{p})$.

Leveraging the standard interference function results of [21], the function $\hat{\mathbf{I}}(\mathbf{p}) = \min\{\mathbf{I}(\mathbf{p}), \bar{\mathbf{p}}\}$ is still standard and, therefore, the convergence of Algorithm 12 is guaranteed. ∎

In summary, the standard interference function of the more commonly-known Foschini-Miljanic Algorithm 11 and our Algorithm 12 are linear and nonlinear in the powers, respectively. Also, Algorithm 12 demonstrates the practical feasibility of directly using the achieved data rate for power updates, as opposed to an indirect treatment in the SINR domain.

23.2.4 Numerical simulations

In this section, we consider the single MNO and compare the performance of our proposed Algorithm 12 with the Foschini-Miljanic Algorithm 11. Especially of interest is how the linear and nonlinear standard interference functions affect the convergence behavior to the optimal solution.

Example 23.1 *Consider a single MNO with 6 users whose transmitter locations are randomly drawn on a 2 km × 2 km square. For each transmitter location, the corresponding receiver location is drawn randomly in a disc of radius 400 meters. The upper-bounds of the transmit power and the minimum data rate threshold are the same for all l, i.e., $\bar{p}_l = 8$ W and $\bar{r}_l = 0.5$, respectively. The receiver noise is set as -60 dBm. The channel gain is adopted from the well-known model $G_{lj} = d_{lj}^{-4}$, where d_{lj} is the Euclidean distance between the j-th transmitter and the l-th receiver, and is such that (23.6) is feasible.*

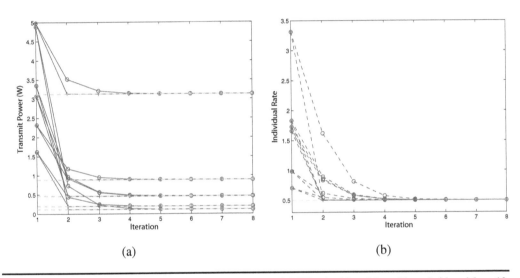

(a)   (b)

Figure 23.2: Comparison of the evolution of Foschini-Miljanic Algorithm 11 and Algorithm 12. The green lines illustrate the optimal solution of (22.6) in (a) and the minimum rate thresholds in (b), respectively. The blue lines illustrate the evolution of individual transmit power in (a) and rate in (b) for Algorithm 11, respectively. The red lines illustrate the evolution of individual transmit power in (a) and rate in (b) for Algorithm 12, respectively.

Figure 23.2 shows the comparison of the power and rate iterate evolution between Foschini-Miljanic Algorithm 11 and Algorithm 12. The numerical example in Figure 23.2 (a) plot the evolution of each transmit power that runs Algorithm 11 and Algorithm 12. Both Algorithm 11 and Algorithm 12 converge fast to the optimal solution of (22.6). Figure 23.2 (b) verifies Lemma 23.2 that each user transmits at its minimum rate to achieve the minimum energy consumption of the system. Figure 23.3 (a) shows that the total power consumptions of Algorithm 11 and Algorithm 12 achieve the optimal value of (22.6). Figure 23.3 (b) plots the topology of the 6 random transmitter-receiver pairs and their corresponding optimal power solution. The width of the connection line identities the strength of the optimal transmit power for each transmitter-receiver pair.

Example 23.2 *Considering a single MNO network with more users, we compare Algorithm 11 and Algorithm 12 by Monte-Carlo simulations by averaging 300 instances. The transmitter-receiver pairs are randomly generated in the region, and the channel gains are generated to make (22.6)*

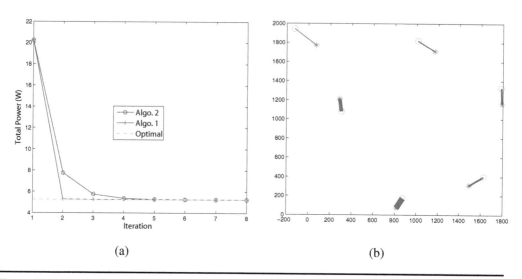

(a)   (b)

Figure 23.3: Illustration of the system energy consumption and the topology for the single MNO. (a) The evolution of the total energy consumption for Algorithm 11 and Algorithm 12. (b) The blue stars denote the location of the transmitters, and the red circles denote the location of the receivers.

feasible. Figure 23.4 shows that Algorithm 11 and Algorithm 12 have comparable convergence rates, even when the number of users is large.

23.3   Power Control in Multiple MNOs

In this section, we consider the total power minimization problem subject to the data rate constraints of all the users in a multiple MNOs network. Unlike the single-MNO case, this problem is more challenging to solve optimally. We study how to design low-complexity algorithms that can yield a feasible solution, particularly leveraging Algorithm 12 developed in the previous section.

23.3.1   System model

Consider a multiple MNOs system, where there are L users (transmitter-receiver pairs) sharing M discrete MNOs. Denote the set of MNOs by $\mathcal{M} = \{1, 2, \ldots, M\}$, respectively. For any $l \in \mathcal{L}$ and $m \in \mathcal{M}$, the AWGN has a distribution $\mathcal{CN}(0, \sigma_l^m)$. Let us denote the transmit power of l-th user through m-th MNO as p_l^m. The received power at receiver l through the m-th MNO is given by

$$\sum_{j=1}^{L} G_{lj}^m p_j^m + \sigma_l^m, \tag{23.24}$$

where G_{lj}^m stands for the channel gain between the l-th transmitter and the j-th receiver through the m-th MNO. We let the SINR of receiver l through the m-th MNO be

$$\mathsf{SINR}_l^m(\mathbf{p}^m) = \frac{G_{ll}^m p_l^m}{\displaystyle\sum_{j \neq l} G_{lj}^m p_j^m + \sigma_l^m}, \tag{23.25}$$

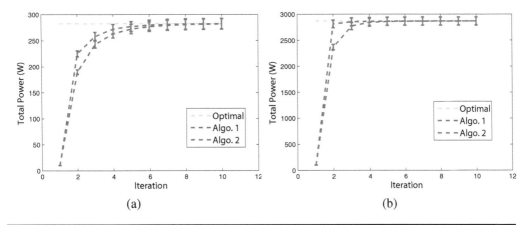

Figure 23.4: Comparison of the convergence behavior between Algorithm 11 and Algorithm 12, with the same environmental parameters as Figure 23.2. (a) There are 100 users sharing one common MNO, and the transmitter locations are random on a 10 km × 10 km square. (b) There are 1000 users sharing one common MNO, and the transmitter locations are random on a 40 km × 40 km square.

where $\mathbf{p}^m = (p_1^m, p_2^m, \cdots, p_L^m)^\top$, $m = 1, \ldots, M$, and the l-th achievable data rate r_l^m (nats/sec) through the m-th MNO is:

$$r_l^m(\mathbf{p}^m) = \log\left(1 + \mathsf{SINR}_l^m(\mathbf{p}^m)\right). \tag{23.26}$$

The problem to minimize the total energy consumption subject to given data rate requirements in the multi-user multiple MNOs communication system is formulated as follows:

$$
\begin{aligned}
&\text{minimize} && \sum_{l=1}^{L}\sum_{m=1}^{M} p_l^m \\
&\text{subject to} && \sum_{m=1}^{M} \log\left(1 + \mathsf{SINR}_l^m(\mathbf{p}^m)\right) \geq \bar{r}_l, \quad l = 1, \ldots, L \\
& && \sum_{j=1}^{L}\sum_{n=1}^{M} a_{lj}^{mn} p_j^n \leq \bar{p}_l^m, \quad l = 1, \ldots, L; m = 1, \ldots, M \\
& && p_l^m \geq 0, \quad l = 1, \ldots, L; m = 1, \ldots, M \\
&\text{variables}: && \mathbf{p}^m, \quad m = 1, \ldots, M,
\end{aligned}
\tag{23.27}
$$

where $\bar{p}_l^m$ and a_{lj}^{mn} are, respectively, the upper-bound and the non-negative weight vectors of the weighted power constraints for the l-th user through the m-th MNO, and $\bar{r}_l > 0$ is the minimum transmission rate requirement of the l-th user. We assume that (23.27) is feasible for the given parameters $\{\bar{p}_l^m, \bar{r}_l\}$, $l = 1, \ldots, L$ and $m = 1, \ldots, M$.

We illustrate two special cases of (23.27) with commonly used power constraints. Suppose we let the power weight be $a_{ll}^{mm} = 1$ and $a_{lj}^{mn} = 0$ for all $j \neq l$ and $m \neq n$. Then, we have the optimization problem with individual power constraints

$$
\begin{aligned}
\text{minimize} \quad & \sum_{l=1}^{L}\sum_{m=1}^{M} p_l^m \\
\text{subject to} \quad & \sum_{m=1}^{M} \log\left(1 + \mathsf{SINR}_l^m(\mathbf{p}^m)\right) \geq \bar{r}_l, \quad l = 1,\ldots,L \\
& p_l^m \leq \bar{p}_l^m, \quad l = 1,\ldots,L; m = 1,\ldots,M \\
& p_l^m \geq 0, \quad l = 1,\ldots,L; m = 1,\ldots,M \\
\text{variables}: \quad & \mathbf{p}^m, \quad m = 1,\ldots,M.
\end{aligned}
\tag{23.28}
$$

On the other hand, suppose the power budget is given by $\bar{p}_l^m = \bar{p}_l, m = 1,\ldots,M$, and let the power weight be $a_{lj}^{mn} = 1$ for all $j = l$ and $a_{lj}^{mn} = 0$ for all $j \neq l$. Then, we have the optimization problem with total power constraints

$$
\begin{aligned}
\text{minimize} \quad & \sum_{l=1}^{L}\sum_{m=1}^{M} p_l^m \\
\text{subject to} \quad & \sum_{m=1}^{M} \log\left(1 + \mathsf{SINR}_l^m(\mathbf{p}^m)\right) \geq \bar{r}_l, \quad l = 1,\ldots,L \\
& \sum_{m=1}^{M} p_l^m \leq \bar{p}_l, \quad l = 1,\ldots,L \\
& p_l^m \geq 0, \quad l = 1,\ldots,L; m = 1,\ldots,M \\
\text{variables}: \quad & \mathbf{p}^m, \quad m = 1,\ldots,M.
\end{aligned}
\tag{23.29}
$$

In general, (23.27) is hard to solve, and so we mainly focus on one of the above special cases, namely, (23.28). The optimal solution in (23.28) is hard to compute due to the nonconvexity in the data rate constraints, since these are nonlinear and nonconvex SINR functions of the powers in (23.28). In addition, a practical concern is that the power solution of each user be determined without any cooperation between the multiple MNOs. As in the previous, our approach exploits a fixed-point characterization of the primal constraints associated with the data rate requirements in (23.28) that leads to the design of low-complexity iterative power control algorithms.

23.3.2 Weighted Rate Update Algorithm

Similar to the single-MNO case, we state a fixed-point optimality condition related to (23.28) in the following result that will be leveraged for power control algorithm design.

Lemma 23.5
If (22.28) is feasible, then the rate constraints in (22.28) are tight at optimality

$$
\sum_{m=1}^{M} (r_l^m)^\star = \sum_{m=1}^{M} \log\left(1 + \mathsf{SINR}_l^m((\mathbf{p}^m)^\star)\right) = \bar{r}_l, \quad l = 1,\ldots,L.
\tag{23.30}
$$

Proof 23.4   The proof is similar to that of Lemma 23.2. Suppose that the rate constraint of a specific user l is not tight at optimality, i.e.,

$$
\sum_{m=1}^{M} r_l^m = \sum_{m=1}^{M} \log\left(1 + \mathsf{SINR}_l^m(\mathbf{p}^m)\right) > \bar{r}_l.
\tag{23.31}
$$

Note that we can reduce the power p_l^m on a particular m-th MNO by a sufficiently small amount $\varepsilon > 0$ to satisfy (23.31), and we can obtain a new transmit power for the user l through this m-th MNO that is given by $\hat{p}_l^m = p_l^m - \varepsilon$. In fact, this reduced power leads to an increase in the achieved data rates of the other users, thus, still satisfying the data rate requirements, as well as the power constraints (i.e., still satisfying the constraint set). Thus, we can further reduce the total energy consumption (optimal value) by ε, which is a contradiction to the assumption that the powers in (23.31) are optimal. ■

By leveraging Lemma 23.5, we propose the following power control algorithm that yields a feasible solution (whenever it exists) to (23.28).

Algorithm 13 Weighted Rate Update Algorithm

---

1. Initialize an arbitrarily feasible $\mathbf{p}^m(0)$ and a small positive ε.

2. Compute the corresponding initial weight of the l-th rate for the m-th MNO:

$$\omega_l^m(0) = \frac{\log(1 + \mathsf{SINR}_l^m(\mathbf{p}^n(0)))}{\displaystyle\sum_{n=1}^{M} \log(1 + \mathsf{SINR}_l^n(\mathbf{p}^n(0)))}. \tag{23.32}$$

3. Repeat until convergence at each m-th MNO:

$$p_l^m(k+1) = \min\left\{ \frac{\bar{r}_l}{(1 - \omega_l^m(t))\bar{r}_l + \log(1 + \mathsf{SINR}_l^m(\mathbf{p}^m(k)))} p_l^m(k), \bar{p}_l^m \right\} \tag{23.33}$$

for all l, i.e., $\|\mathbf{p}^m(k+1) - \mathbf{p}^m(k)\|_2 \leq \varepsilon$, or the iterations exceed a predefined threshold K, where k is the discrete iteration index at each MNO and $p_l^m(t+1) = \lim_{k \to K} p_l^m(k)$.

4. Normalization of rates:

$$\omega_l^m(t+1) = \frac{r_l^m(t+1)}{\displaystyle\sum_{n=1}^{M} r_l^n(t+1)}, \tag{23.34}$$

for all l and m, where t is the discrete iteration index for updates between all MNOs, and

$$r_l^m(t+1) = \log(1 + \mathsf{SINR}_l^m(\mathbf{p}^m(t+1))). \tag{23.35}$$

5. Go to Step 3 until $\|\mathbf{r}^m(t+1) - \mathbf{r}^m(t)\|_2 \leq \varepsilon$.

Remark 22.1 In general, Algorithm 13 yields only a feasible local optimal solution.

Remark 22.2 At Step 3 of Algorithm 13, we have leveraged our previous result (Theorem 23.2) to show the convergence of $p_l^m(k)$ for each MNO. This is because (23.33) is equivalent to

$$
\begin{aligned}
p_l^m(k+1) &= \min\left\{ \frac{\frac{r_l^m(t)}{\|\mathbf{r}^m(t)\|_1}\bar{r}_l}{\log\left(1 + \mathsf{SINR}_l^m(\mathbf{p}^m(k))\right)} p_l^m(k), \bar{p}_l^m \right\} \\
&= \min\left\{ \frac{\omega_l^m(t)\bar{r}_l}{\log\left(1 + \mathsf{SINR}_l^m(\mathbf{p}^m(k))\right)} p_l^m(k), \bar{p}_l^m \right\},
\end{aligned}
\tag{23.36}
$$

given a fixed MNO updating matrix $\boldsymbol{\omega}(t)$. In other words, the inner loop (23.33) corresponds to solving M separate power control problems (each one corresponding to a MNO). Thus, the inner loop (23.33) converges to a local optimal solution whenever (23.28) is feasible. Lastly, at Step 4 of Algorithm 13, i.e., the outer loop, the rate iterates obtained from the inner loop are normalized to satisfy (23.30). In particular, Lemma 23.5 shows that this normalization yields a feasible solution to (23.28) at each iteration.

23.3.3 Numerical Simulations

In this section, we consider the multiple MNOs system and illustrate the convergence performance of Algorithm 13 numerically.

Example 23.3 *Consider a multiple MNOs system with 2 users transmitting through 2 MNOs, whose transmitter locations are random on a 2 km × 2 km square. For each transmitter location, the corresponding receiver location is drawn randomly in a disc of radius 1000 meters. The upper-bounds of the transmit power and the receiver noise are the same for all l and m, i.e., $\bar{p}_l^m = 8$ W and $\sigma_l^m = -60$ dBm, respectively. The minimum data rate thresholds are fixed as $\bar{\mathbf{r}} = (0.8, 0.6)^\top$. The parameters are chosen to make (22.28) feasible.*

Figure 23.5 illustrates the convergence of Algorithm 13, and the evolution of the individual transmit power and the corresponding rates. Figure 23.5 (a) shows that Algorithm 13 converges fast to an equilibrium solution. Figure 23.5 (b) verifies Lemma 23.5 that each user transmits at its minimum rate threshold. Figure 23.6 illustrates that Algorithm 13 achieves a solution close to the solution obtained by exhaustive search.

Example 23.4 *Consider a larger multiple MNOs system with 10 users transmitting through 3 MNOs. All the upper-bounds of transmit power are the same for all l and m, i.e., $\bar{p}_l^m = \infty$ to ensure (22.28) is feasible. The minimum rate thresholds are the same for all l, i.e., $\bar{r}_l = 0.5$. The other problem parameters remain the same as Example 22.3. Figure 23.7 (a) shows the individual rate evolution of Algorithm 13 and that each user achieves the minimum rate threshold. Figure 23.7 (b) shows the evolution of the total power consumption.*

23.4 Joint Power Control and Admission Control for Feasibility

In the previous sections, we assume that the optimization problems are always feasible. In this section, we address the issue of infeasibility related to the energy efficiency optimization problems using the approach in [9, 22]. Addressing the infeasibility issue is as challenging as overcoming the nonconvexity hurdle in the multiple-MNO case. As such, we primarily focus on the infeasibility issue for a single-MNO case.

23.4.1 Energy-Infeasibility Optimization Model

Consider the optimization problem in (23.6) when it is infeasible to satisfy the data rate requirements of all the users simultaneously. As such, some users may have to be removed from accessing the

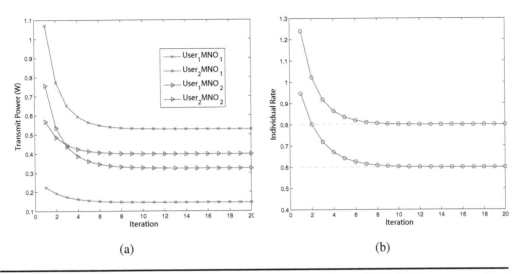

Figure 23.5: Illustration of the convergence of Algorithm 13 for multiple MNOs system with 2 mobile users transmitting through sharing 2 MNOs. The red and blue lines show the evolution of each user, respectively. The green lines show the minimum individual rate thresholds. The evolution of transmit powers and individual rate are shown in (a) and (b), respectively.

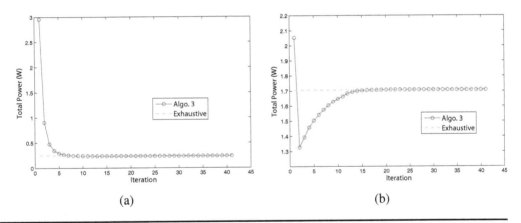

Figure 23.6: Comparison of the solution between Algorithm 13 and exhaustive search. (a) Exhaustive search grid is 0.01. (b) Exhaustive search grid is 0.005.

MNO. However, how to retain the largest set of users whose rate thresholds can all be satisfied in (23.6) whenever it is infeasible is an NP-hard combinatorial problem [23]. The problem becomes intractable when the number of mobile users is large. In the following, we study an approximation methodology to find a feasible set to (23.6) with the maximal cardinality. First, we formulate an optimization problem related to (23.6) by adding auxiliary variables s_l to the right-hand side of the rate constraint for each l-th user:

$$
\begin{aligned}
& \text{minimize} && \|\mathbf{s}\|_0 \\
& \text{subject to} && \frac{e^{\bar{r}_l} - 1}{\text{SINR}_l(\mathbf{p})} \leq 1 + s_l, \quad l = 1,\dots,L \\
& && \mathbf{0} \leq \mathbf{p} \leq \bar{\mathbf{p}}, \\
& \text{variables}: && \mathbf{p}, \mathbf{s},
\end{aligned}
\tag{23.37}
$$

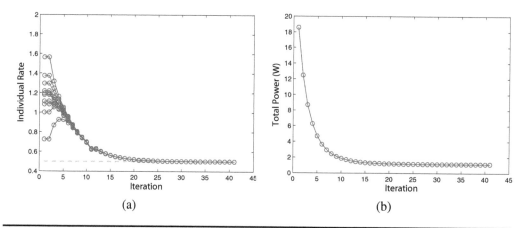

Figure 23.7: Illustration of the evolution of Algorithm 13 for multiple MNOs system, with 10 mobile users transmitting through 3 MNOs. The red lines show the evolution of each user's rate in (a) and the total power consumption in (b), respectively. The green line shows the common minimum rate threshold.

where $\mathbf{s}$ can be interpreted as an indicator of infeasibility that also has a physical meaning of rate margins being added to the rate thresholds, and the objective function $\|\mathbf{s}\|_0$ is the ℓ_0 norm that measures the cardinality of $\mathbf{s}$. For brevity, we call $\mathbf{s}$ the rate margin variable.

Lemma 23.6

If $\mathbf{s}$ is a feasible solution of (22.37), we have:

$$\rho\left(\text{diag}\left(\frac{e^{\bar{\mathbf{r}}}-1}{1+\mathbf{s}}\right)\left(\mathbf{F}+\frac{1}{\bar{p}_l}\mathbf{v}\mathbf{e}_l^\top\right)\right) \le 1, \quad l=1,\ldots,L. \tag{23.38}$$

Proof 23.5   From the constraint set of (23.37), we have

$$\begin{cases} p_l \le \bar{p}_l \Rightarrow \frac{1}{\bar{p}_l}\mathbf{e}_l^\top\mathbf{p} \le 1, \quad l=1,\ldots,L_m+L_s, \\ \frac{e^{\bar{r}_l}-1}{\text{SINR}_l(\mathbf{p})} \le 1+s_l \Rightarrow \text{diag}\left(\frac{e^{\bar{\mathbf{r}}}-1}{1+\mathbf{s}}\right)(\mathbf{F}\mathbf{p}+\mathbf{v}) \le \mathbf{p}, \\ \Rightarrow \text{diag}\left(\frac{e^{\bar{\mathbf{r}}}-1}{1+\mathbf{s}}\right)\left(\mathbf{F}+\frac{1}{\bar{p}_l}\mathbf{v}\mathbf{e}_l^\top\right)\mathbf{p} \le \mathbf{p}, \quad l=1,\ldots,L, \end{cases}$$

where $\mathbf{e}_l$ denotes the l-th unit coordinate vector. Let $\mathbf{H}_l = \text{diag}\left(\frac{e^{\bar{\mathbf{r}}}-1}{1+\mathbf{s}}\right)\left(\mathbf{F}+\frac{1}{\bar{p}_l}\mathbf{v}\mathbf{e}_l^\top\right)$ for all l. Note that $\mathbf{H}_l$ is a nonnegative matrix that is irreducible whenever $\mathbf{F}$ is for all l. Using Theorem 1.6 in [24] (Subinvariance Theorem), we deduce the following: Suppose that $\mathbf{H}_l$ is an irreducible nonnegative matrix, and there is a vector $\mathbf{p} \ge \mathbf{0}$, with $\mathbf{p} \ne \mathbf{0}$ satisfying $\mathbf{H}_l\mathbf{p} \le \mathbf{p}$ (implying that (23.37) is feasible), then $\mathbf{p} > \mathbf{0}$ and $\rho(\mathbf{H}_l) \le 1$.   ■

From Lemma 23.6, (23.6) is feasible if and only if the optimal value of (23.37) is zero. We have $s_l > 0$ if the rate threshold of the l-th secondary user cannot be achieved. Intuitively, a feasible set of users for (23.6) can be obtained by removing all the mobile users satisfying $s_l > 0$ at the optimality of (23.37). However, (23.37) is still a computationally hard problem due to the nonsmooth and nonconvex ℓ_0 norm function.

23.4.2   Sum-of-Infeasibilities Based Convex Relaxation Heuristic

We consider the following optimization problem by replacing the ℓ_0 norm objective function of (23.37), with the sum of $\mathbf{s}$, i.e., using the sum-of-infeasibilities[3] heuristic, given by

$$
\begin{aligned}
\text{minimize} \quad & \mathbf{1}^\top \mathbf{s} \\
\text{subject to} \quad & \frac{e^{\bar{r}_l} - 1}{\text{SINR}_l(\mathbf{p})} \leq 1 + s_l, \quad l = 1, \ldots, L \\
& \mathbf{0} \leq \mathbf{p} \leq \bar{\mathbf{p}}, \\
& \mathbf{s} \geq \mathbf{0}, \\
\text{variables}: \quad & \mathbf{p}, \mathbf{s}.
\end{aligned}
\tag{23.39}
$$

Let us denote the optimal solution $\mathbf{p}$ and $\mathbf{s}$ in (23.39) by $\mathbf{p}^\star$ and $\mathbf{s}^\star$, respectively.

Remark 22.3   Since the nonnegative rate margin variable $\mathbf{s}$ satisfies $1 - s_l \leq \dfrac{1}{1 + s_l}$ for all l, the objective function of (23.39) satisfies

$$
\sum_{l=1}^{L} s_l \geq \sum_{l=1}^{L} \left(1 - \frac{\text{SINR}_l(\mathbf{p})}{e^{\bar{r}_l} - 1} \right).
\tag{23.40}
$$

The inequality in (23.40) is tight if (23.6) is feasible. Otherwise, minimizing the left-hand side of (23.40) has the effect of minimizing the differences between the rate thresholds and the achieved rates of all the mobile users. This viewpoint, thus, motivates the sum-of-infeasibilities heuristic as a viable way to approximate the maximum feasible set of mobile users.

Although (23.39) is still nonconvex, we can transform it to a convex problem by using a logarithmic transformation on the transmit power, i.e., $\tilde{\mathbf{p}} = \log \mathbf{p}$. Then, we obtain the following equivalent convex optimization problem:

$$
\begin{aligned}
\text{minimize} \quad & \mathbf{1}^\top \mathbf{s} \\
\text{subject to} \quad & \log \frac{e^{\bar{r}_l} - 1}{\text{SINR}_l(e^{\tilde{\mathbf{p}}})} \leq \log(1 + s_l), \quad l = 1, \ldots, L \\
& e^{\tilde{\mathbf{p}}} \leq \bar{\mathbf{p}}, \\
& \mathbf{s} \geq \mathbf{0}, \\
\text{variables}: \quad & \tilde{\mathbf{p}}, \mathbf{s}.
\end{aligned}
\tag{23.41}
$$

Note that the optimal solution $\tilde{\mathbf{p}}^\star$ in (23.41) is related to $\mathbf{p}^\star$ in (23.39) by $\tilde{\mathbf{p}}^\star = \log \mathbf{p}^\star$. Next, we present results on the optimality of (23.39) that will be used to design a price-driven algorithm to solve (23.37) in Section 23.4.3.

Theorem 23.3
The optimal solution $\mathbf{p}^\star$, $\mathbf{s}^\star$, and the dual solution $(\mathbf{v}^\star, \boldsymbol{\lambda}^\star)$ of (23.41) satisfy:

$$
\mathbf{p}^\star = \text{diag}\left(\frac{e^{\bar{r}} - 1}{1 + \mathbf{s}^\star} \right) (\mathbf{F}\mathbf{p}^\star + \mathbf{v}),
\tag{23.42}
$$

$$
v_l^\star = p_l^\star \left(\sum_{i \neq l} \frac{G_{il} v_i^\star}{\sum_{j \neq i} G_{ij} p_j^\star + n_i} + \lambda_l^\star \right), \quad l = 1, \ldots, L
\tag{23.43}
$$

---

[3]The sum-of-infeasibilities method is routinely used in the first phase of many convex programming algorithms, e.g., interior-point method, to find a feasible point. It often violates only a small number of inequalities, and this interesting phenomenon is under active research in sparse recovery, e.g., basis pursuit and ℓ_1 norm regularization (cf. Chapter 11.4 in [25]).

$$\lambda_l^\star(p_l^\star - \bar{p}_l) = 0, \quad l = 1, \dots, L \tag{23.44}$$

and

$$s_l^\star = \max\{v_l^\star - 1, 0\}, \quad l = 1, \dots, L \tag{23.45}$$

where $v_l \in \mathbb{R}_+$ is the dual variable associated with the l-th rate constraint and $\lambda_l \in \mathbb{R}_+$ is the dual variable associated with the l-th power constraint. Interestingly, v_l can be interpreted as the admission price of the l-th mobile user (once admitted into the system, the l-th mobile user pays this price to maintain his or her rate requirement in co-existence with the other users in the network). In particular, from (22.45), the mobile users with the largest rate margin pays the highest price at the optimality of (22.41). Furthermore, by introducing an auxiliary variable $x_l^\star = v_l^\star/p_l^\star$ for each l, we can rewrite (22.43) as:

$$\mathbf{x}^\star = \mathbf{F}^\top \text{diag}\left(\frac{e^{\bar{\mathbf{r}}} - 1}{1 + \mathbf{s}^\star}\right)\mathbf{x}^\star + \boldsymbol{\lambda}^\star. \tag{23.46}$$

Proof 23.6 Since (23.41) is a convex optimization problem, we derive its Karush-Kuhn-Tucker (KKT) optimality conditions. We introduce nonnegative dual variables $(\boldsymbol{v}, \boldsymbol{\lambda}, \boldsymbol{\mu})$ and write the Lagrangian function of (23.41):

$$L(\tilde{\mathbf{p}}, \mathbf{s}, \boldsymbol{v}, \boldsymbol{\lambda}, \boldsymbol{\mu}) = \mathbf{1}^\top \mathbf{s} - \sum_{l=1}^{L} v_l \log \text{SINR}_l(e^{\tilde{\mathbf{p}}}) - \boldsymbol{\mu}^\top \mathbf{s}$$

$$+ \sum_{l=1}^{L} v_l \log(e^{\bar{r}_l} - 1) + \sum_{l=1}^{L} \lambda_l(e^{\tilde{p}_l} - \bar{p}_l) - \sum_{l=1}^{L} v_l \log(1 + s_l). \tag{23.47}$$

It is easy to obtain the KKT optimality conditions:

$$\begin{cases} \boldsymbol{v}^\star \geq 0, \boldsymbol{\lambda}^\star \geq 0, \boldsymbol{\mu}^\star \geq 0, \mathbf{s}^\star \geq 0, \\ \log(e^{\bar{r}_l} - 1) - \log \text{SINR}_l(e^{\tilde{\mathbf{p}}^\star}) - \log(1 + s_l^\star) \leq 0, \quad l = 1, \dots, L, \\ e^{\tilde{p}_l^\star} - \bar{p}_l \leq 0, \quad l = 1, \dots, L, \\ v_l^\star(\log(e^{\bar{r}_l} - 1) - \log \text{SINR}_l(e^{\tilde{\mathbf{p}}^\star}) - \log(1 + s_l^\star)) = 0, \quad l = 1, \dots, L, \\ \lambda_l^\star(e^{\tilde{p}_l^\star} - \bar{p}_l) = 0, \quad l = 1, \dots, L, \\ \mu_l^\star s_l^\star = 0, \quad l = 1, \dots, L, \\ \frac{\partial L}{\partial s_l} = 1 - \mu_l^\star - \frac{v_l^\star}{1 + s_l^\star} = 0, \quad l = 1, \dots, L, \\ \frac{\partial L}{\partial \tilde{p}_l} = \lambda_l^\star e^{\tilde{p}_l^\star} - v_l^\star + \left(\sum_{i \neq l} \frac{G_{il} v_i^\star e^{\tilde{p}_l^\star}}{\sum_{j \neq i} G_{ij} e^{\tilde{p}_j^\star} + n_i}\right) = 0, \quad l = 1, \dots, L. \end{cases} \tag{23.48}$$

In particular, from the transformation $p_l^\star = e^{\tilde{p}_l^\star}$ and by defining a new auxiliary variable $x_l^\star = v_l^\star/p_l^\star$ for all l, we obtain (23.42)–(23.45). ■

Remark 1.4 Theorem 23.3 is deduced by applying the KKT optimality conditions (cf. Chapter 5.5 in [25]) to (23.41). From the KKT complementarity slackness condition, the dual variable $\lambda_l^\star$ is equal to zero whenever $p_l^\star < \bar{p}_l$ at the optimality of (23.41). If the optimal value of (23.37) is greater than zero, the dual variables satisfy $\boldsymbol{v}^\star > \mathbf{0}$ and $\boldsymbol{\lambda}^\star \neq \mathbf{0}$. In general, $\mathbf{x}$ can be regarded as an auxiliary variable to assist in the computation of the optimal primal and dual solution of (23.41).

23.4.3 Price-Driven Spectrum Access Algorithm Design

In this section, we propose a price-driven algorithm for joint power and admission control by leveraging the admission price and the fixed-point equations established in Theorem 23.3 to solve the energy-infeasibility optimization problem. We propose a joint power and admission control algorithm that determines the spectrum access of mobile users iteratively through admission control to identify a subset of mobile users that is feasible in (23.6). The key idea is to compute the transmit power (primal solution of (23.41)) and the admission prices (dual solution of (23.41)) iteratively, and then remove secondary users based on the admission prices in a greedy fashion.

Theorem 23.4
Let us define a locally asymptotically stable solution in the Lyapunov sense to be one such that all solutions starting near the stable solution remain near it and tend toward it as $k \to \infty$ [26]. Algorithm 14 converges to a locally asymptotically stable solution that is feasible in (22.6).

We refer the readers to [22] for the details of the proof of Theorem 23.4 using the Lyapunov stability theory.

Remark 22.5 The computation of (23.50) and (23.54) can be made distributed by message passing. We may have more than one user satisfying (23.54). In this case, we remove users by breaking ties uniformly at random. The limit point of $\lim_{k\to\infty} \mathbf{s}(k)$ and its condition that $\lim_{k\to\infty} \mathbf{1}^\top \mathbf{s}(k) = 0$ implies that $\lim_{k\to\infty} \mathbf{p}(k)$ is a feasible solution to (23.6).

Remark 22.6 Theorem 23.1 only characterizes the local convergence behavior of Algorithm 14, and its global convergence is an open problem. Our numerical evaluation in Section 23.4.4, however, demonstrates that Algorithm 14 has good empirical convergence behavior, even when the iterates are far from the fixed-point solution.

From the condition that $\mathsf{SINR}_l(\mathbf{p}^\star) = \dfrac{e^{\bar{r}_l} - 1}{1 + s_l^\star}$, $s_l^\star = 0$ implies that the l-th user can achieve its rate threshold. Otherwise, $s_l^\star > 0$ implies that the l-th user cannot reach its rate threshold, and it can possibly be removed. Now, if we remove all the users that satisfy $s_l^\star > 0$ for all l, then (23.6), with a reduced number of constraints, is guaranteed to be feasible. However, some users may be unnecessarily removed, since we have used the optimality conditions in (23.39) instead of that in (23.37). An educated guess to reduce the sum of infeasibilities is to remove the user corresponding to $\arg\max_{l\in\mathcal{A}(k)} v_l(k+1)$ at the k-th iteration. This is implemented in Step 14. This user removal criterion is motivated by (23.45) in Theorem 23.3, namely, that the user with the largest rate margin variable pays the highest price. This user is removed to reduce the interference to other users in subsequent iterations. Upon convergence, the total energy consumption is minimized on the set of the remaining users, whose rate constraints are all satisfied. In general, other user removal criterion based on the admission price can also be considered.

23.4.4 Numerical Simulations

In this section, we provide experimental results to illustrate that our proposed algorithms outperform other known alternatives in terms of energy consumption and system capacity.

Example 23.5 *We compare our methods with the distributed power control algorithm with temporary removal and feasibility check (DFC) in [23]. Although the model in [23] is the special case for single-cell that the channel gains for one link are the same $G_{1j} = G_{jj}$, we use the same environment for the convenience of comparison. The AWGN at the receiver, i.e., $n = \sigma^2$, is assumed to be 5×10^{-15} W. The channel gain is adopted from the well-known model $G_{jj} = kd_j^{-4}$, where d_j is the distance between the jth transmitter and its receiver, and $k = 0.09$ is the attenuation factor*

Algorithm 14 Sum-of-Infeasibilities-Based Joint Power and Admission Control

---

1. **Initialization:**

   - Initialize the set of users $\mathcal{A}(0) = \{1, \dots, L\}$.

2. **Update by each user $l \in \mathcal{A}(k)$:**

   - Update the transmitter power $p_l(k+1)$ at the $(k+1)$th step for all the mobile users $l \in \mathcal{A}(k)$:

   $$p_l(k+1) = \min\left\{\frac{e^{\bar{r}_l} - 1}{\max\{v_j(k), 1\}\mathsf{SINR}_l(\mathbf{p}(k))}p_l(k), \bar{p}_l\right\}. \qquad (23.49)$$

3. **Update by each user $l \in \mathcal{A}(k)$:**
   If $p_l(k+1) < \bar{p}_l$

   - Update the auxiliary variable $x_l(k+1)$:

   $$x_l(k+1) = \sum_{j \in \mathcal{A}(k)} \frac{F_{jl}(e^{\bar{r}_j} - 1)x_j(k)}{\max\{v_j(k), 1\}}. \qquad (23.50)$$

   - Update the admission price $v_l(k+1)$:

   $$v_l(k+1) = x_l(k+1)p_l(k+1). \qquad (23.51)$$

   else

   - Update the admission price $v_l(k+1)$:

   $$v_l(k+1) = \frac{e^{\bar{r}_l} - 1}{\mathsf{SINR}_l(\mathbf{p}(k+1))}. \qquad (23.52)$$

   - Update the auxiliary variable $\mathbf{x}(k+1)$:

   $$x_l(k+1) = v_l(k+1)/p_l(k+1). \qquad (23.53)$$

   end

4. **Inner loop stopping condition:**

   - If $\|\mathbf{p}(k+1) - \mathbf{p}(k)\|_2 < \varepsilon$, go to Step 14.
   - Otherwise, go to Step 14.

5. **User admission control:**

   - Let $s_l(k+1) = \max\{v_l(k+1) - 1, 0\}$ for all the users $l \in \mathcal{A}(k)$. If $\mathbf{1}^\top \mathbf{s}(k+1) > 0$, then remove a user z satisfying:

   $$z = \arg\max_{l \in \mathcal{A}(k)} v_l(k+1). \qquad (23.54)$$

   - Update the set $\mathcal{A}(k+1) \leftarrow \mathcal{A}(k) - z$ and go to Step 14.

---

that represents power variations due to path loss. The upper-bounds of the transmit power for all users are the same $\bar{p}_l = 1$ W for all l. There are 5 links indexed from 1 to 5 in a single-cell environment, where the distance vector is $d = [300, 530, 740, 860, 910]^\top$ m, in which each element is the distance of the corresponding receiver from its transmitter. The minimum rate threshold vector is $\bar{\mathbf{r}} = [0.3365, 0.2624, 0.3, 0.2231, 0.2231]^\top$, which is equivalent to the SINR threshold vector $\bar{\boldsymbol{\gamma}} = [-4, -5.2, -4.6, -6, -6]^\top dB$.

Figure 23.8 shows the same simulation results of DFC as [23]. As the system is infeasible, DFC sets $p_5 = 0$ to switch off Link 5 and then other links reach their rate thresholds with minimum total transmit power. The solution of power vector is $\mathbf{p}^\star = [0.0061, 0.0483, 0.2063, 0.2904, 0]^\top$, whereas, the performance of DFC depends on the initial point. Although it performs well when the initial point is proper, the evolution may fall into the oscillation more often based on different initial points.

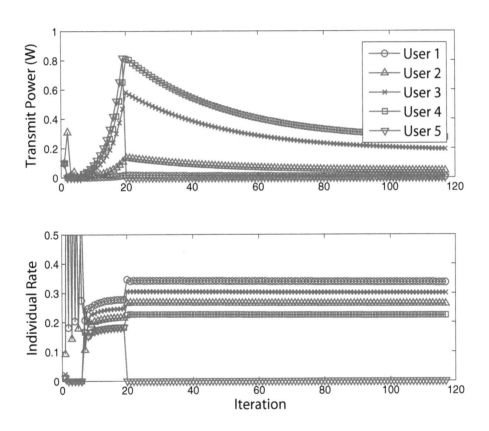

Figure 23.8: Evolution of transmit power and individual rate for DFC with proper initial point. The blue lines are 4 supported users. The red line is the removed user.

Figure 23.9 shows that Algorithm 14 obtains the same feasible set in terms of cardinality. Whereas, Algorithm 14 removes Link 1 instead of Link 5 because the admission price is $\mathbf{v}^\star = [1.3590, 1.2233, 1.2954, 1.1406, 1.1668]$, and the solution of power vector is $\mathbf{p}^\star = [0, 0.0109, 0.0463, 0.0653, 0.0818]^\top$. Hence, our feasible set saves the energy consumption $(0.5511 - 0.2043)/0.5511 = 63\%$. The main reason is that DFC temporally removes the user that

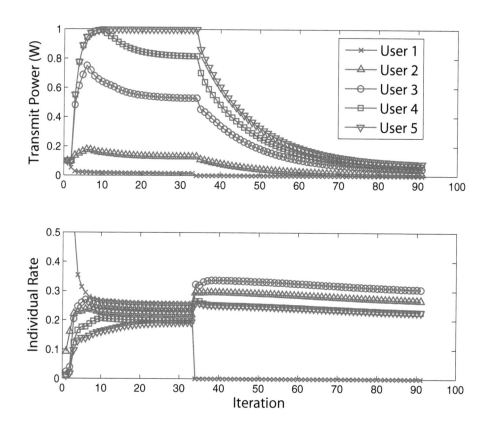

Figure 23.9: Evolution of transmit power and SINR for Algorithm 14. The blue lines are 4 supported users. The red line is the removed worst user.

first achieves the upper-bound of each link's power. Meanwhile, our method predicts the worst user with an educated guess according to the margin variable.

Example 23.6 *It is possible to obtain different maximum feasible sets from different algorithms, as the maximum feasible set may not be unique. Hence, we compare the system capacity obtained and the energy consumption based on different algorithms. This example reports the Monte-Carlo (MC) average results for at least 300 MC runs. For each MC run, transmitter locations are uniformly drawn on a 2Km × 2Km square. For each transmitter location, a receiver location is drawn uniformly in a disc of radius 400 meters, excluding a radius of 10 meters. All upper-bounds of transmit power are fixed $\bar{p}_l = 1$ W. The channel gains are calculated by $G_{lj} = d_{lj}^{-4}$, where d_{lj} is the Euclidean distance between the j-th transmitter and the l-th receiver. The receiver noise is set as −60 dBm.*

In Figure 23.10, Algo. 4 is our proposed Algorithm 14 in Section 23.4.1, Cent is the centralized removal algorithm in [27], and Exce is the algorithm where we use the heuristic that considers the removal metric [28] with the worst secondary user j,

$$j = \arg\max_{a \in \mathcal{A}} \sum_{l \neq a} G_{la} p_a^e + \sum_{l \neq a} G_{al} p_l^e, \tag{23.55}$$

where $\mathcal{A}$ is the set of current secondary users in the system, and p_l^e is the excess transmission power

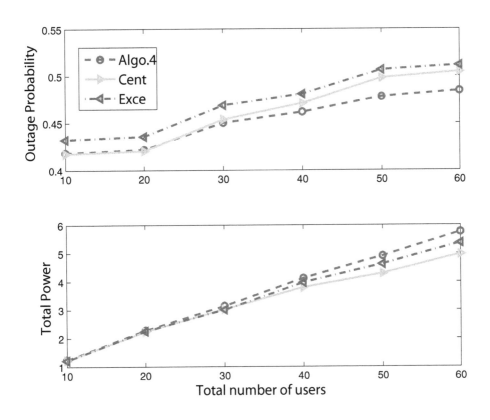

Figure 23.10: Average outage probability and average total energy consumption vs. total number of users. The lower-bound of rate thresholds are the same $\bar{r}_l = 0.2231$ for all l.

needed for User l to attain its SINR threshold, instead of our metric based on sensitivity analysis. Figure 23.10 shows that our Algorithm 14 outperforms the centralized greedily removal algorithm in [27] and the algorithm that uses the removal heuristic in (22.55) for admission control. Although the centralized algorithm and the algorithm that uses the removal heuristic in (22.55) have an overall smaller total energy consumption than those obtained by Algorithm 14, this is due to the fact that they support fewer users, thus, yielding a lower system capacity. When the outage probabilities are the same, Algorithm 14 achieves smaller total energy consumption than the Exce algorithm. This demonstrates the value of optimizing the admission price, as compared to the metric in (22.55).

23.5   Open Issues

We have introduced several relevant optimization problems related to energy consumption minimization in a multiple-MNO network and the solution methodology, with an emphasis on designing energy-efficient power control algorithms. There are, however, many interesting open issues, and we list them down in the following:

- What are other possible methodologies that can compute the globally optimal solution to the general problem in (23.27)? Can we leverage convex relaxation to bound the global optimal value or other convex approximation to obtain feasible solutions with good quality?

- How do we design a computationally efficient algorithm where there are coupling power constraints, e.g., interference temperature constraints and nonlinear power budget constraints, between users in the single-MNO and multiple-MNOs case? For example, the other special case of (23.27) that is given by (23.29) has a single total power constraint coupling across all the MNOs for each mobile user. An efficient power control algorithm to this special case problem will also be useful to other problem setting, e.g., the multi-carrier system model for digital subscriber line networks [17].

- Whenever the energy minimization problems are infeasible, finding the maximum system capacity, i.e., the maximum number of mobile users whose data rate requirements can be simultaneously supported, is still an open problem in both the single-MNO case and the multiple-MNOs case.

- There can be various wireless performance Pareto efficiency tradeoff curves, e.g., the energy-robustness tradeoff studied in [9], and the energy-infeasibility tradeoff studied in [22] and in this chapter. What, then, is the fundamental Pareto efficiency tradeoff curve between energy efficiency and data rate in a multiple-MNO network?

References

[1] J. G. Andrews, S. Buzzi, W. Choi, S. V. Hanly, A. Lozano, A. C. K. Soong, and J. C. Zhang, "What will 5G be?" *IEEE Journal on Selected Areas in Communications*, vol. 32, no. 6, pp. 1065–1082, Jun. 2014.

[2] S. He, Y. Huang, H. Wang, S. Jin, and L. Yang, "Leakage-aware energy-efficient beamforming for heterogeneous multicell multiuser systems," *IEEE Journal on Selected Areas in Communications*, vol. 32, no. 6, pp. 1268–1281, Jun. 2014.

[3] L. Al-Kanj, H. V. Poor, and Z. Dawy, "Optimal cellular offloading via device-to-device communication networks with fairness constraints," *IEEE Transactions on Wireless Communications*, vol. 13, no. 8, pp. 4628–4643, Aug. 2014.

[4] Z. Ding, Z. Yang, P. Fan, and H. V. Poor, "On the performance of non-orthogonal multiple access in 5G systems with randomly deployed users," *IEEE Signal Process. Lett.*, vol. 21, no. 12, pp. 1501–1505, 2014.

[5] S. Han, C. Yang, and A. F. Molisch, "Spectrum and energy efficient cooperative base station doze," *IEEE Journal on Selected Areas in Communications*, vol. 32, no. 2, pp. 285–296, Feb. 2014.

[6] L. Venturino, A. Zappone, C. Risi, and S. Buzzi, "Energy-efficient scheduling and power allocation in downlink OFDMA networks with base station coordination," *IEEE Transactions on Wireless Communications*, vol. 14, no. 1, pp. 1–14, Jan. 2015.

[7] Y. Wu, Y. Chen, J. Tang, D. So, Z. Xu, C.-L. I, P. Ferrand, J.-M. Gorce, C.-H. Tang, P.-R. Li, K.-T. Feng, L.-C. Wang, K. Borner, and L. Thiele, "Green transmission technologies for balancing the energy efficiency and spectrum efficiency trade-off," *IEEE Communications Magazine*, vol. 52, no. 11, pp. 112–120, Nov. 2014.

[8] L. Zheng and C. W. Tan, "Maximizing sum rates in cognitive radio networks: Convex relaxation and global optimization algorithms," *IEEE Journal on Selected Areas in Communications*, vol. 32, no. 3, pp. 667–680, Mar. 2014.

[9] C. W. Tan, D. P. Palomar, and M. Chiang, "Energy-robustness tradeoff in cellular network power control," *IEEE/ACM Transactions on Networking*, vol. 17, no. 3, pp. 912–925, Jun. 2009.

[10] C. W. Tan, M. Chiang, and R. Srikant, "Fast algorithms and performance bounds for sum rate maximization in wireless networks," *IEEE/ACM Transactions on Networking*, vol. 21, no. 3, pp. 706–719, Jun. 2013.

[11] L. Zheng and C. W. Tan, "Cognitive radio network duality and algorithms for utility maximization," *IEEE Journal on Selected Areas in Communications*, vol. 31, no. 3, pp. 500–513, Mar. 2013.

[12] C. W. Tan, S. Friedland, and S. H. Low, "Spectrum management in multiuser cognitive wireless networks: Optimality and algorithm," *IEEE Journal on Selected Areas in Communications*, vol. 29, no. 2, pp. 421–430, Feb. 2011.

[13] C. W. Tan, M. Chiang, and R. Srikant, "Maximizing sum rate and minimizing MSE on multiuser downlink: Optimality, fast algorithms and equivalence via max-min SINR," *IEEE Transactions on Signal Processing*, vol. 59, no. 12, pp. 6127–6143, Dec. 2011.

[14] W. Yu, G. Ginis, and J. Cioffi, "Distributed multiuser power control for digital subscriber lines," *IEEE Journal on Selected Areas in Communications*, vol. 20, no. 5, pp. 1105–1115, Jun. 2002.

[15] W. Yu and R. Lui, "Dual methods for nonconvex spectrum optimization of multicarrier systems," *IEEE Transactions on Communications*, vol. 54, no. 7, pp. 1310–1322, Jul. 2006.

[16] J. Papandriopoulos and J. Evans, "Scale: A low-complexity distributed protocol for spectrum balancing in multiuser DSL networks," *IEEE Transactions on Information Theory*, vol. 55, no. 8, pp. 3711–3724, Aug. 2009.

[17] P. Tsiaflakis, Y. Yi, M. Chiang, and M. Moonen, "Dynamics spectrum management for green DSL," *EURASIP Journal of Wireless Communications*, pp. 1–14, Oct. 2011.

[18] G. J. Foschini and Z. Miljanic, "A simple distributed autonomous power control algorithm and its convergence," *IEEE Transactions on Vehicular Technology*, vol. 42, no. 4, pp. 641–646, Nov. 1993.

[19] T. M. Cover and J. A. Thomas, *Elements of Information Theory*. USA: John Wiley & Sons, 1991.

[20] M. Chiang, P. Hande, T. Lan, and C. W. Tan, "Power control in wireless cellular networks," *Foundations and Trends in Networking*, vol. 2, no. 4, pp. 381–533, 2008.

[21] R. D. Yates, "A framework for uplink power control in cellular radio systems," *IEEE Journal on Selected Areas in Communications*, vol. 13, no. 7, pp. 1341–1347, Sep. 1995.

[22] X. Zhai, L. Zheng, and C. W. Tan, "Energy-infeasibility tradeoff in cognitive radio networks: Price-driven spectrum access algorithms," *IEEE Journal on Selected Areas in Communications*, vol. 32, no. 3, pp. 528–538, Mar. 2014.

[23] M. Rasti, A. R. Sharafat, and J. Zander, "Pareto and energy-efficient distributed power control with feasibility check in wireless networks," *IEEE Transactions on Information Theory*, vol. 57, no. 1, pp. 245–255, Jan. 2011.

[24] E. Seneta, *Non-negative Matrices and Markov Chains*. New York: Springer Verlag, 1981.

[25] S. Boyd and L. Vandenberghe, *Convex Optimization*. UK: Cambridge University Press, 2004.

[26] R. M. Murray, Z. Li, and S. S. Sastry, *A Mathematical Introduction to Robotic Manipulation*. USA: CRC Press, 1994.

[27] H. Mahdavi-Doost, M. Ebrahimi, and A. Khandani, "Characterization of SINR region for interfering links with constrained power," *IEEE Transactions on Information Theory*, vol. 56, no. 6, pp. 2816–2828, Jun. 2010.

[28] I. Mitliagkas, N. D. Sidiropoulos, and A. Swami, "Joint power and admission control for ad-hoc and cognitive underlay networks: Convex approximation and distributed implementation," *IEEE Transactions on Wireless Communications*, vol. 10, no. 12, pp. 4110–4121, Dec. 2011.

SECURITY

Chapter 24

Spectrum Sharing Vulnerability and Threat Assessment

Timothy X. Brown and Douglas C. Sicker

CONTENTS

24.1 Introduction

Security is a concern in any communication system. When compared to conventional wireless communication systems, spectrum sharing adds functionality and complexity that raises additional security concerns. These additional security concerns are the subject of this chapter. The chapter is written for security experts who want to understand how they might contribute to spectrum sharing security and also written for spectrum sharing designers who want to incorporate security into their spectrum sharing designs. It is based on several more detailed surveys of spectrum sharing security [Arkoulis et al., 2008; Baldini et al. 2012; Brown and Sethi, 2008; Clancy and Goergen, 2008].

We contribute a broad discussion of the vulnerabilities that are introduced by spectrum sharing and potential threats that can exploit these vulnerabilities. We also discuss specific security controls that can mitigate these threats. We start with a brief review of spectrum sharing to highlight what is different about this technology and to introduce the components and functionality that complicates securing these systems. For a more detailed description of spectrum sharing, the reader is directed to other chapters.

24.1.1   Steps In Spectrum Sharing

Spectrum sharing radios dynamically choose spectrum bands to communicate while avoiding harmful interference to primary spectrum users. Primary users are the existing incumbent users, such as television broadcasters and navigation radars. Such primary users have an expectation that they can operate without undue interference from spectrum sharing.

Spectrum sharing consists of four basic steps. The first is spectrum awareness. In this step, the transmitter and receiver seek information on what spectrum is available for its use. The second is channel selection. Here, the transmitter and receiver select from among the available spectrum a mutually agreeable channel over which to communicate. The third step is media access. The selected channel may have rules for coordinating users, and within these rules the transmitter seeks an opportunity to communicate. The final step is spectrum handoffs. During a communication session, the available spectrum may change and require the transmitter and receiver to reassess the available spectrum and select a new channel.

Spectrum awareness is the key new feature of spectrum sharing. The transmitter and receiver must gather data either by direct sensing, information from other radios, or information from some centralized source. So-called policies may be defined to proscribe what spectrum can and cannot be used. These policies may be limited by time, geographic location, and type of communication.

Channel selection in spectrum sharing is more complicated. The transmitter and receiver have different views on what spectrum is available, since they differ in their location, capabilities, privileges, and gathered information. Spectrum sharing radios may rely on a common control channel or so-called rendezvous protocols to find each other in the radio space.

Media access in spectrum sharing depends on the spectrum dynamics and licensing model. Cases where the transmitter and receiver are given a long-term exclusive license to the spectrum reduces to conventional wireless access. But, other models may require the communication to avoid primary users or may have multiple non-cooperative radio systems sharing the same band. These require additional sensing and media coordination protocols.

Spectrum handoffs are triggered by mobility, expiration of policies, appearance of primary users, or other dynamics that make the existing channel unavailable. These handoffs are above and beyond the ordinary radio management protocols that handle deterioration in channel conditions or congestion. Spectrum sharing radios may negotiate fallback channels and, if necessary, start over with the spectrum awareness or channel selection process.

24.1.2   Components of a Shared Spectrum Radio

To support these steps in spectrum sharing, the radio architecture has fundamental differences as, shown in Figure 23.1.

An ordinary radio might consist of an operating system controlling a radio. Here, the radio has additional features that enable the spectrum sharing. These include sensing, policies, geolocation and time, cooperation with other devices, and the radio device itself.

Sensing includes the detectors and detection algorithms used to measure if there are existing primary users in the spectrum. The sensor can be a separate receiver, but more likely uses the existing radio receiver for a fraction of the time. To avoid harmful interference to primary users, the sensor may be designed to detect the presence of even weak signals.

A policy is an electronic document that specifies a spectrum band and who, where, when, and how it can be used. We use the term policy to broadly include so-called time-limited leases, contracts

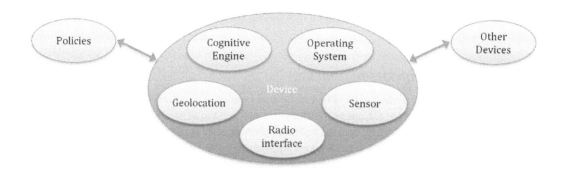

Figure 24.1: Components of spectrum sharing radio

from spectrum brokers, and so on. The source of policies can be from sources, such as a long-term permission downloaded from a database, a short-term lease purchased from a spectrum license owner, or radio characteristics loaded into a device at time of manufacture.

Geolocation and time are used to identify what policies can be applied. Geolocation sources include satellite navigation, such as GPS, other spectrum sharing radios, by hand for fixed radios, cellular system base stations, or indirect means, such as the pattern of channel occupancy. Time can come from similar sources or from Internet sources, if connected. These sources have more or less accuracy. Even imprecise sources can be useful if margins are included to account for the precision.

Cooperation is used in spectrum sharing to select channels, to coordinate spectrum handoffs, and to share sensing, policy, time, and location information. The cooperation can be peer-to-peer or mediated by centralized agents. Cooperation increases the awareness of individual radios that increases the amount of spectrum they can use and reduces the problem of causing harmful interference to primary radios.

The radio device for spectrum sharing is flexible. It has the ability to tune over multiple channels. It senses signals with different modulation. It may need to cooperate with other users using different communication protocols. As such, it is supported by more software and has broader operating range than other radios

In this brief description of the features of spectrum sharing, we focus on the features most relevant to spectrum sharing. For this reason we do not dwell on specific architectures, such as whether this is an ad hoc network, point-to-point link, or base station infrastructure. Nor do we discuss the types of users; whether they are individuals, government organizations, or commercial service providers. Nor is the regulatory framework (e.g., licensed vs. unlicensed) central to this discussion.

By setting aside such details, we can keep our attention on the overall picture of spectrum sharing issues. In the next section, we discuss general security issues with spectrum sharing.

24.2   Security in Spectrum Sharing

The main challenge with security is to protect resources from attacks. We first describe the resources that are being protected, then the elements of an attack and finish with a discussion of the attacker threat model.

24.2.1 What Is Being Secured?

In any communication system, we seek three security properties: confidentiality, integrity, and availability.

Confidentiality is protecting the content of messages from disclosure to unauthorized recipients. This is a general problem not specific to spectrum sharing and is solved using various encryption techniques. Independent of the message confidentiality, traffic confidentiality protects the information on who is talking to whom, at what time, and where. Message confidentiality is reduced by spectrum sharing, as described in later sections.

Integrity is protecting messages from undetected alteration, insertion, or loss. This is again a general problem. While we do not consider direct attacks on communication integrity, communication integrity is a central challenge to the correct operations of spectrum sharing mechanisms.

Availability is ensuring that users are able to communicate. Denial of service is when communication is prevented or degraded. This is the key challenge for spectrum sharing. Spectrum sharing denial of service attacks can affect either the spectrum sharing radio or, by inducing harmful interference, it can affect the primary user.

24.2.2 What Is An Attack?

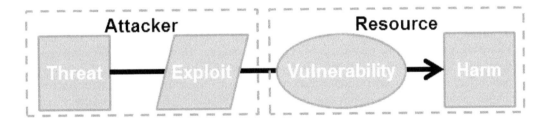

Figure 24.2: Elements of a security attack.

An attack consists of a threat that exploits a vulnerability to cause harm to a resource, as shown in Figure 23.2. The harm as described in the previous section is a loss of confidentiality or availability to communication. The threat is the cause of harm. It may be an intentional actor or an unintentional act, such as a component failure. We focus on intentional threats. The vulnerability is the weakness that allows the attack to happen. Weakness can be in the design of a system, its implementation, or the operation of the system. We focus here on spectrum sharing design weaknesses that affect communication security. Controls are safeguards put in place to minimize the risk of an attack. They can be used to prevent an attack, detect an attack, or mitigate the harm caused by the attack.

24.2.3 Threat Model

An attack consists of all three elements: a threat, a vulnerability, and an exploit. All must be present for the attack to be successful. In this chapter we describe shared spectrum vulnerabilities and the threats that can be applied to those vulnerabilities. A summary description of the exploits will also be given, although without details.

In order to understand the risk associated with each attack, it is important to understand the resources and motivation of the attacker. Some attacks require specialized knowledge and equipment. Other attacks can be accessible to anyone with limited skill. In military uses of spectrum sharing, we might expect motivated, highly capable, and resource-rich attackers. In a residential setting, attackers will be less motivated and capable. It might be useful to think of different types of attackers and their motives. The list below is illustrative of the types of attackers that spectrum sharing user might need to consider.

Determined, skilled, resource-rich attacker with a goal of denial of service: Such an attacker will have access to specialized radio equipment, such as software-defined radios that are able to be programmed to emulate different signals under the control of the attacker. They can also have spectrum analyzers and other measurement tools for assessing user behavior. As well, they can have high-gain antennas and radio front ends capable of high-power transmission. In addition, they have network resources to try to attack databases and other resources used to support spectrum sharing. This hypothetical attacker is motivated to deny legitimate users ability to communicate via shared spectrum.

Note that such an attacker has the resources to send strong jamming signals and directly deny communication to the target radio. Such direct attacks can be simple brute force jamming or more subtle intelligent jamming that leverages the communication protocols as surveyed here [Pelechrinis et al., 2011]. In either case, it is outside the scope of this chapter. Instead, we focus on the attacks that are enabled by the spectrum sharing itself. It may be more difficult to detect the source of such attacks may be more difficult to detect the and potentially enables leveraged attacks (e.g. by attacking access to a database, the attacker may be able to deny many users' communication). The target of this attack can be the spectrum sharing users or a primary user.

Determined, skilled, resource-rich attacker with a goal of violation of privacy: Such an attacker will have access to the same resources as the previous attacker. However, here, the attacker seeks to learn about the radio user themselves. Examples include the user's location, with whom do they communicate, and what type of communication do they use. Protecting such information is considered traffic security. Message security, protecting the confidentiality of message content, is a more general problem and outside the scope of this chapter.

Individual seeking more than their fair share of bandwidth: In this case, the attacker is familiar with the operation of their shared spectrum radio and able to change its configuration and operation settings. They have access to low-cost radio measurement tools (e.g.. USB spectrum analyzers that work in the 2.4 GHz industry, scientific, medical (ISM) band). They may have basic hacking tools that allow them to attack network resources. The goal of the attacker is to access more than their fair share of spectrum to get more data bandwidth. Like denial of service, this is a more general problem. For instance, outside of spectrum sharing, by violating the IEEE 802.11 protocol, it is possible to shorten so-called contention windows to get priority access to the spectrum when there are many users sharing the same unlicensed bands. Here, we focus on the strategies that leverage the spectrum sharing mechanisms themselves.

We can propose other attacks. For instance, an attacker that sells access to spectrum might manipulate the perceived available spectrum so that a spectrum sharing radio pays more for the spectrum than it would otherwise. Or, the same attacker may drive customers away from competitors. Or, an attacker can try to deny access to the spectrum near the attacker's spectrum in order to reduce interference. Or, an attacker can try to sell policy database services without having access to legitimate policies. Or, a military attacker may drive an enemy transmitter to operate in a band where the attacker has the equipment to perform other attacks (jamming, eavesdropping, etc.). This list is suggestive of the types of spectrum sharing attackers and their motives.

For all these attackers, they can either be participants or non-participants to a spectrum sharing protocol. There may be a set of spectrum sharing users that have a protocol to work together. The attacker may be one of these participants. As a result, they are able to directly subvert the protocol to achieve their goal. The protocol may authenticate users and encrypt control messages in order to exclude non-participants. Or the attacker may not be capable of participating in the protocol because of limitations in their radios. In this case, the attacker must use indirect means of attack.

With the background provided in the first two sections, we turn to understanding the vulnerabilities exposed by spectrum sharing.

24.3   Spectrum Sharing Vulnerabilities

This section describes the weaknesses introduced by the components that are used in spectrum sharing, as described in Section 23.2. Not all components are used in every spectrum sharing design. The designers of spectrum sharing systems can consider these weaknesses as they decide the best spectrum sharing approach.

24.3.1   Sensing

Sensing is the process used to gain awareness of what spectrum is being used or not used. In general, a spectrum sharing radio depends on sensing to know what spectrum is available. Any loss or degradation of sensing will reduce the spectrum it can use. Any false sensing can reduce the spectrum it can use, or lead it to cause harmful interference.

Sensing can come from internal detectors or external sources. We address each in turn.

24.3.1.1   Internal Detectors

Internal detectors can be a general-purpose power detector, such as a spectrum analyzer or a detector specific to the type of signal. There are fundamental limits to how weak a signal is detectable according to the noise floor, which depends on the measurement bandwidth, interfering signals present and detector noise figure. Higher quality detectors have lower noise figures and are, thus, able to discern weaker signals from noise. Signal specific filters and signal processing improve detection sensitivity. Higher gain antennas increase sensitive but are more focused in a narrower set of directions.

The primary weakness with the detector approach is the presence of detection errors where the detector either fails to detect the presence of a primary signal or gives false detections when no primary signal is present. The latter can be caused by noise and other signals. Or, atmospheric conditions and sensitive detectors can detect primary signals hundreds of kilometers from their normal coverage area. A particular challenge comes once an available channel is identified and then used by a spectrum sharing user. Subsequent measurements may see this as an occupied band and no longer identify it as unused by the primary user.

24.3.1.2   External Sources

Alternatively, external sources can tell the spectrum sharing radio what spectrum is being used or not used. A beacon signal could broadcast availability in a specific area. A network connected database could enable queries of spectrum usage in specific bands at specific locations at specific times. Radios may be connected to a centralized controller (e.g., like a mobile phone base station) that gives connected radios spectrum information.

The weakness here then becomes the availability and integrity of the information from these sources. Further, these sources themselves depend on other sources. The information from these other sources must be correct and up-to-date.

24.3.2   Policies

Policies are used to express spectrum availability information. A single spectrum radio may receive policies from several types of sources: regulatory, operational, and hardware. Regulatory agencies may indicate the existence of licensed users or what spectrum bands are allowed for spectrum sharing. The policy may also express power limits and other usage restrictions. The spectrum sharing radio may also be part of a service and the service provider may use policies to guide the operation of individual radios. Or, a spectrum sharing user may pay a spectrum broker who provides policies to indicate the purchased spectrum access. The spectrum sharing radio device manufacturer may include policies at time of manufacture to indicate on which bands the radio is physically capable of operating.

The weakness is that the policies must be accurate and the provenance of the policy must be clear. For instance, when purchasing access from a spectrum broker, how does the radio determine that the broker is indeed a spectrum broker who has a right to sell the indicated spectrum access? Another weakness is that the interaction between policies can be complex and potentially conflicting and requires careful reasoning to achieve the intended result.

24.3.3   Geolocation and Time

Since policies have restrictions on when and where they can be used, it is critical that the spectrum sharing user knows its location and the time. There are many possible sources for this information. Satellite navigation systems, such as GPS and GLONASS, provide accurate time and position but rely on receiving very weak signals and on visibility of a sufficient number of satellites. Both become problems among tall buildings in cities. Other time and position information can be used, such as broadcast channels from mobile base stations. Potentially, indirect radio information, such as the pattern and usage of TV channels or call signs from FM radio stations, can be used to identify location as well. Location can also be entered by hand. For instance, a professional installer at time of setup could enter the location of a fixed spectrum sharing radio. Finally, the location may simply be implicit by market: If it is sold in the United States, then its location is the United States.

The weakness is that most notions of dynamic spectrum sharing require location and time to work. Spectrum sharing may not need location and time if it uses a beacon or a central controller to distribute policies since a policy received in this way is implicitly applicable to the reception area of the beacon and timing may be relative to the time that the policy was transmitted. However, this pushes the problem of location and time one step back to the beacon or the controller.

24.3.4   Cooperation

Cooperation is part of every step in spectrum sharing. Spectrum sharing users share sensing, geolocation, time, and policy data to improve the accuracy of spectrum awareness. Channel selection requires both transmitter and receiver to converge on a common channel. Channel access requires different spectrum access radios to cooperatively share the spectrum resources.

The primary weakness with cooperation is that the spectrum sharing radio depends on the other radios providing accurate information and correctly following protocols. The other radios may intentionally mislead or could have poor quality or faulty information.

24.3.5   Radio Devices

The spectrum sharing radio is a more capable device than conventional radios. It depends on correct software to follow the spectrum sharing algorithms and protocols. It depends on accurate hardware to correctly transmit in the correct band at the correct power, modulation, and time. Detectors need to measure correctly. Non-ideal behavior, such as intermods, spurious emissions, and non-ideal filters, need to be characterized, since the spectrum sharing radio is operating among primary radios.

24.4 Spectrum Sharing Threats

For each of the vulnerabilities in the previous section, we can identify corresponding threats that would seek to exploit these vulnerabilities.

24.4.1 Sensing

We identify threats to the internal detectors and external sources of spectrum usage information.

24.4.1.1 Primary Radio Emulation

Here, an attacker sends a signal similar to a primary transmitter in a given band so that the internal detector marks this channel as already used and unavailable. For example, an attacker transmits a TV-like signal in a TV band to convince others that the TV channel is being used.

This is a serious threat to shared spectrum receivers and subject to dedicated surveys, such as [Nguyen-Thanh et al., 2015]. There are three different attack ranges. The shortest range is the jamming range, the range that an attacker needs to be to overwhelm a desired signal. The next range is the reception range, the range that an attacker needs to be to send spoofed messages that are correctly received by the legitimate receiver. The furthest range is the detection range; this is the range at which the sensitive detectors will detect the presence of the attacker's bogus primary signal. Thus, a bogus primary signal sent by an attacker can deny access to spectrum sharing radios over a wide area.

24.4.1.2 Fabricated External Data Sources

Data from other external sources may be shared measurement data, policies, or beacons. They can be fabricated and sent to the spectrum sharing radio. If accepted, the attacker can arbitrarily deny access to any part of the spectrum. Depending how the shared spectrum radio interprets conflicting reports, the attacker may be able to guide spectrum sharing users to transmit harmful interference to primary users. Conflicting data may also lead to delays as the spectrum radio seeks to process the conflict or seeks additional data to resolve the conflict.

The data can be fabricated and inserted at several points. It can be sent directly to the spectrum sharing radio. It can be entered into the database from which the data is forwarded to spectrum sharing radios. Existing messages can also be altered in transit. The database itself may be compromised and fabricate, alter, or remove data.

24.4.1.3 Replayed External Data Sources

This is similar to fabricated external data sources. Here the threat is that the attacker captures earlier valid data and repeats it at a later time or at a different location in order to mislead a spectrum sharing radio. This attack is different from the previous attack in that the data appears valid and was generated by valid sources. For this reason it may be more difficult for the shared spectrum radio to discern that the reports are not valid.

24.4.1.4 Denial of Service on External Data Sources

External data is delivered to the shared spectrum radio over a radio channel or via a wired network connection. An attacker can prevent the delivery of external data to the radio by jamming the radio channel, e.g., jamming a beacon, common control channel, or channel for sharing measurements. The radio jamming is effective if the attacker signal is sufficiently stronger than the radio channel used by the external source. This means the attacker is either close to the receiver or is transmitting at a higher power.

The attacker can alternatively mount an attack on either the connection or the server (or database) that is sending the data. This may be any number of known attacks to deny or degrade access to the external data via channel overload or attacks on the communication protocols. If effective, for instance, by denying a beacon station access to the information it needs to broadcast, the attack can affect many users.

24.4.1.5   Loss of Privacy from External Sources

An attacker who eavesdrops on the communication between a spectrum sharing radio and an external source may be able to see the identity of the spectrum sharing radio, their location, type of communication, and so on. For example when a spectrum sharing radio makes a request to a centralized database for policies, the request may contain the identity of the radio, its geographic location, and the type of spectrum it is seeking. The identity may be required to validate the request comes from a legitimate user of the data. If the database itself is compromised, it can track users.

When radios share information, there is an inherent tension between privacy and trust. A spectrum sharing radio may not accept measurements or data from other radios unless it can identify the information source. As a result, radios may be required to reveal their identity to other nearby radios. The radio identity can be correlated with user identities and, thus, a spectrum sharing radio can know who is in their vicinity.

24.4.2   Policies

The previous section described basic attacks to information gathered from external sources, including policies. In this section, we develop two more subtle threats specific to policies.

24.4.2.1   Policies Issued without Authority

In this attack, an attacker distributes policies without authority. The attacker may have authority to distribute policies for some spectrum but may not have authority for the spectrum in a given policy. Establishing the provenance of a policy may require several steps of checking. Further, authority is not simply for a given band but also with other constraints on power, bandwidth, spectrum etiquette, and so on. A policy may violate any of these dimensions, and so all must be checked for the policy to be deemed legitimate.

24.4.2.2   Policy Overload

An attacker can issue many interrelated policies. The policies may be legitimate, in which case the spectrum sharing radio would be overloaded trying to determine what is the available spectrum. Or, the policies may be illegitimate, in which case the spectrum sharing radio would be overloaded processing the provenance of the policies. Either way, the spectrum sharing radio would be overwhelmed processing the many policies and unable to determine the available spectrum. Along the same lines, an attacker could load many legitimate policies into a database. In this case, queries to the database will return a large number of responses, potentially overwhelming the policy distribution channel and each spectrum sharing radio that makes a query.

24.4.3   Geolocation and Time

Since geolocation and time are essential to the interpretation of policies, any attack that denies, degrades, or corrupts a device's concept of location or time will lead the radio to use or not use policies inappropriately. If successful, this attack may deny access to spectrum or cause the spectrum sharing radio to cause harmful interference. Section 4.1 already described attacks to geolocation and time reports from external sources. This section describes further attacks.

24.4.3.1   Satellite Jamming and Spoofing

Satellite navigation signals are weak and easily jammed [Volpe, 2001]. More sophisticated attacks can cause the location estimate to drift. Ground-based transmitters can emulate satellite signals to introduce arbitrary location estimate errors. If successful, the attack could affect spectrum sharing radios over a large area.

24.4.3.2   False Radio Environment

The attacker can send false signals representing FM radio call signs from different locations. Patterns of spectrum usage can be created that represent different areas. For instance, in a rural area where there are few television stations, the attacker could transmit a pattern of television stations representing a specific urban market. The urban market would then dictate a much more limited spectrum availability.

24.4.4   Cooperation

Cooperation can be used to improve available information, to enable a transmitter and receiver to select a common channel, and to manage access to the spectrum. We describe attacks to each of these.

24.4.4.1   Information-Sharing Attacks

Section 23.4.1 describes general attacks on information from external sources. In cooperative sensing, users share sensing data. This data is not inherently validated and is subject to measurement noise. To solve this, a spectrum sharing radio may take averages or other statistics of shared sensing data. To arbitrarily skew the statistics, the attacker can send many false measurements using different identities (possibly from only one or a few attacker radios). If successful, the attacker's version of the measurements may prevail and, further, may undermine the confidence in the legitimate spectrum sharing radios.

24.4.4.2   Subverting the Channel Selection Process

Two peer spectrum sharing radios find a common channel for each to communicate via a process of searching through each radio's available spectrum. An attacker that participates in this process can keep another radio busy for a long time by never completing the process. An attacker observing the process can jam two participants as they settle on a common channel and force the process to continue. If successful, an attacker with little effort can keep these spectrum sharing radios busy and not actually communicating.

24.4.4.3   Subverting Un-shared Media Access Protocol

A spectrum sharing radio that is accessing the spectrum must contend with three types of other users. The first is other spectrum sharing radios that are participating in a common protocol. The second is spectrum sharing radios that are not cooperating via a common protocol. Finally, there are primary users who have priority access to the channel and the spectrum sharing radio must avoid harmful interference. The complexity of coordinating these different users opens an opportunity for an attacker to degrade others' access to spectrum and increasing their own access to spectrum.

24.4.5 Radio Devices

Changes to the software or hardware in spectrum sharing radio devices open the possibility of the following attacks.

24.4.5.1 Malicious Device Software

The spectrum sharing radio depends on correct implementations of the spectrum sharing protocols and algorithms. An attacker that can install malicious software can cause harmful interference to primary users, send false reports, keep users busy by violating protocols, and so on. The extent of this attack depends on whether one radio or many radios are affected by the malicious software. If malicious software became installed widely, then these problems could become widespread and opens the possibility of different radio devices colluding in their attacks. Collusion might coordinate harmful interference to reduce detection on any one radio or it might coordinate false sensing reports.

24.4.5.2 Altered Radio Hardware

The spectrum radio hardware depends on correct implementations of the device hardware. The radio receiver can be modified to use higher power or higher gain antennas. Transmitted signals might be shifted to other bands. Radio circuitry might be damaged or modified so that it transmits more spurious emissions. As a result of these changes, the radio transmits outside of policy limits. Sensing measurements can be corrupted by disabling related antennas and receivers. An attacker can disable satellite navigation antennas or receivers on the spectrum sharing device denying location information. Unlike malicious software, altering the radio hardware is likely restricted to one or a small number of devices, unless it is a systematic flaw manufactured into many radios.

24.4.5.3 Changing Device Settings

An attacker with access to the device configuration can enter an incorrect location or set the clock incorrectly. They might alter the type of user (e.g., from a government to civilian user). They can reset the device, deleting any stored policy, measurement, or key information. They might change the default locations used to download policies or alter keys used to access or verify policies. Such attacks can lead the device to incorrectly use policies or deny the user access to valid policies. This attack is limited to the individual devices to which an attacker can gain access. Devices with remote configuration options are more susceptible to this attack.

24.4.5.4 Steering Non-ideal Device Characteristics

Spurious signals of radio devices, such as intermodulation products, adjacent channel interference, or other out-of-band signals, if known, can potentially be guided to cause interference with other radios. This attack would be through one of the means described earlier to generate the sensor, policy, or other data necessary to guide the radio. The difference in this attack is that since it is generating the interference indirectly, it may not be recognized as an attack.

24.4.5.5 Device Cloning

An attacker can clone a legitimate device in order to access the services and privileges of the original device. The owner of the original device would be charged for paid spectrum used by the attacker or become liable for the malicious actions carried out by the attacker. The attacker may access services, such as access to a policy database that has already been paid for by the legitimate user.

24.4.6 Summary

The threats are summarized in Table 23.1. The first column is the threat. The second is a list of the technical skills needed to execute the attack. The third column indicates the extent of a successful attack. The fourth column indicates the effect of a successful attack. The technical skills that are needed are categorized into the radio, network, crypto, and participant.

Radio means the attacker understands radio communication systems. The attacker has the ability to analyze and generate radio waveforms, usually in a wide range of bands. This will require specialized hardware, like spectrum analyzers, signal generators, or software defined radios.

Network means the attacker understands network protocols. The attacker has the ability to receive and send packets using existing protocols. The bits of the packets can be read and, if not encrypted, can be interpreted according to the protocols being used. To eavesdrop on packets, this requires access to so-called packet sniffers for wireless communication or to have access to the switches, routers, or cables of a wired network. To insert packets, the attacker needs to have access to the network on which the target resource lies. For some resources, this means the attacker must be a participant. Other resources are accessible over the Internet. The inserted packets may include information of the attacker's choosing.

Crypto means that the attacker has the ability to analyze cryptographically secured information. As an example, they have the resources to crack passwords or otherwise gain access to a database. They also understand cryptographic protocols and can generate appropriate signatures, certificates, and so on.

Participant means that the attacker has legitimate access to a resource. The attacker abuses its access privilege to achieve the attack goal. As examples, the attacker is another spectrum sharing radio or has access to enter data into an external database, or is a spectrum broker.

The attack effect lists the primary effect of the attack. The ultimate effect is implied. For instance, many attacks are used to manipulate or degrade a spectrum sharing radio's awareness of which spectrum is available. This enables an attacker to deny access to spectrum that can either clear out spectrum for the use of the attacker or prevent the spectrum sharing radio to communicate. It also enables the attacker to steer the affected radio to cause interference to other radios.

24.5 Security Controls

Security controls are used to reduce the risk of a successful attack. Controls address vulnerabilities, increase the technical skill and resources needed by the attacker, and mitigate the loss of a successful attack. In this chapter, we focus on controls that affect the threats to spectrum sharing mechanisms. We describe a representative set of controls. For each we identify how it addresses threats and any new vulnerabilities it introduces.

24.5.1 Primary Radio Verification

A sensor that detects a primary radio signal will want to distinguish between a valid primary radio and an attacker attempting primary radio emulation. True primary signals, such as a TV signal or a radar, have characteristics that can distinguish them from an attacker signal. They may be high power or have a specific physical profile, such as the radar sweeping period or have a specific content, such as the TV programming. These characteristics can be part of the detection. An isolated sensor will find it difficult to distinguish between a distant high-power signal and a nearby

Table 24.1   Summary of Spectrum Sharing Threats

Threat	Technical Skill Needed	Attack Extent	Attack Effect
Primary Radio Emulation	Radio	Widespread in an area around the attacker	Denies access to chosen spectrum bands
Fabricated External Data Sources	Network, Crypto, or Participant	Depends on extent of fabricated data recipients	Manipulates spectrum awareness
Replayed External Data Sources	Network	Depends on extent of replayed data recipients	Manipulate spectrum awareness
DoS on External Data Sources	Radio or Network	Depends on recipients denied access to data	Degrades spectrum awareness
Loss of Privacy from External Data Sources	Network or Participant	Depends on participants that can be monitored	Learn who communicates where, when, and how
Policies Issued Without Authority	Crypto or Participant	Depends on recipients of unauthorized policies	Manipulates spectrum awareness
Policy Overload	Crypto or Participant	Depends on recipients of denied policy source	Degrades spectrum awareness
Satellite Jamming and Spoofing	Radio	Widespread in an area around the attacker	Manipulates spectrum awareness
False Radio Environment	Radio	Vicinity of attacker	Manipulates spectrum awareness
Information Sharing Attacks	Participant	Participants cooperating with the attacker	Manipulates spectrum awareness
Subverting the Channel Selection Process	Radio or Participant	Participants cooperating with attacker	Prevents or degrades communication
Subverting Unshared Media Access Protocol	Radio, Network, or Participant	Attackers sharing in a media access domain	Prevent or degrade communication
Malicious Device Software	Network, Crypto, or Participant	Depends on devices that use attacker software	Arbitrary access
Altered Radio Device Hardware	Radio	Depends on devices physically accessed	Degraded performance to device and in its vicinity
Changing Device Settings	Network or Participant	Depends on devices that settings can be accessed	Manipulate or degrade spectrum awareness
Steering Non-ideal Device Characteristics	Radio, Network, or Participant	In area around radios manipulated by attacker	Degrade performance in affected bands
Device Cloning	Radio or Crypto	Depends on radios that can be cloned	Access to services and networks of cloned radio

low-power signal. Verification is improved by two or more physically separated sensors comparing measurements, as described in a later section.

Another approach tries to localize the source of the primary user transmitter and compare this with the location of known primary users. However, this adds additional steps that can be threatened. Further, a spectrum sharing radio that knows its own location and the primary radio transmitter location already knows enough for spectrum awareness and may not gain significantly from the sensing.

24.5.2 Graceful degradation to Missing Information

Denial of service or other degradations in performance can deny the spectrum sharing radio access to sensing, position, or time information. The spectrum sharing algorithms should be designed to use degraded information with varying levels of error by including appropriate margins in their harmful interference estimates. Even crude location information with many kilometers of error is sufficient to identify large swaths of land that work in a marine radar band. Time errors of days, if included in the spectrum awareness algorithm, will enable use of most spectrum policies.

Direct spectrum sensing is more problematic, as it is difficult to set decision thresholds. A threshold conservative enough to prevent harmful interference in primary user bands excludes access to almost all unused bands, as well. While it is enough to identify one communication band, adding additional margins because of measurement errors will make it even more difficult to identify usable bands.

24.5.3 Physical Layer Protection

An attacker can try to obtain private information about a user by eavesdropping on wireless connections. Transmission power control can lower transmit power to a minimum level necessary for reception by the desired receiver. This minimizes the range that an attacker can overhear the conversation.

Robust communication via spread spectrum techniques makes the signal more difficult to intercept and to even detect that it is present. It also has the benefit that the signal is more difficult to jam in a denial of service attack.

Directional antennas, electronic beam steering, and similar techniques focus the desired communication in specific directions. Signals from other directions are less likely to jam communication and eavesdroppers in other directions are less likely to intercept messages. Directional antennas are generally larger but may work for fixed or larger mobile platforms.

24.5.4 Cryptographic Techniques

In order to avoid eavesdropping and message corruption that the spectrum sharing can incorporate encryption and cryptographic message hashes. Encrypted communication can use shared secret keys in a centralized scheme that manages participant keys. Or, it can use a public key infrastructure between peers. The encryption makes it difficult for eavesdroppers to read messages and to follow protocols for targeted jamming. The cryptographic hashes make it more difficult for messages to be changed or corrupted without being detected by the recipient.

A general design question is which communication is protected with cryptographic techniques. Is it used just for private communications, for information sharing, or for each step of a protocol? Cryptography adds additional steps to setting up a connection in a process that is already complex and that adds overhead to the communications process. As an example, it is difficult to incorporate cryptography before two radios find a common communication channel. So, some protocols may inherently be unable to incorporate shared cryptography. Furthermore, cryptography may require trusted third party participation and the infrastructure associated with this trust.

24.5.5   Challenge-Response Information Sharing

One attack is to replay messages from earlier times. It is not enough for the messages to be signed, as the signature can be valid, even though the message was created earlier. An approach to avoid this problem is to use a protocol where a spectrum sharing radio requests specific information (sensing measurements, geolocation, time, or policies), and the external sources reply with the requested information. To avoid a replay, the requester includes a random string in the request, which is included in the reply. Since the reply with the random string is signed, the requester knows the information is from a time after the random string was created. This approach works in a request-response information model. It is more difficult to incorporate in broadcast models, such as beacons.

24.5.6   Reputation-Based Mechanisms

Reputation-based mechanisms try to associate a level of trustworthiness to participants in a spectrum sharing scheme. The trustworthiness can be assigned. For instance, a validated central controller may be more trusted than a client spectrum sharing radio. Alternatively, trustworthiness can be based on historical performance at delivering accurate information to other radios. More trustworthy sources will be given higher weight in spectrum awareness decisions than less trustworthy sources. As an extreme, sources that are not sufficiently trustworthy are simply ignored. But, the power of this approach is that even less trustworthy sources can be included in decisions.

Such approaches require messages to attach a unique identifier of the message source. To avoid attackers from simply copying legitimate IDs into their messages, messages need to be cryptographically signed. Unique identifiers, if present with each transmission, can help to identify legitimate transmissions from rogue transmissions, or they can help to associate problematic transmissions with specific radios.

The reputation scheme itself becomes a target of attack. Attackers play a game of trying to remain trustworthy while still degrading shared information. Or a group of attackers may generate similar false measurements that overwhelm legitimate measurements, lowering the reputation of the legitimate measurements. Similarly, an individual attacker may take advantage of botnets or other means of replicating identities to launch what is referred to as a Sybil attack.

24.5.7   Radio Monitoring and Enforcement

In general, spectrum sharing radios can misbehave by incorrectly following protocols, transmitting in prohibited bands, causing interference to other radios, and forwarding incorrect information. Reputation-based schemes help identify the latter. The other misbehaviors can be identified via monitoring and reporting. An area that is subject to spectrum sharing may have dedicated monitor radios that attempt to identify misbehavior. For instance, if a source is sending jamming information, the monitors can attempt to identify or localize the misbehaving radio. A source that does not correctly follow a protocol can be reported by other participants. Identifying a misbehaving radio requires unique IDs, as described under reputation or a localization method. Localization is difficult to do precisely in most environments unless there are many monitoring devices used for the localization or a mobile direction finding approach is used to zero in. If the device is mobile or changing transmission characteristics, then localization may not be possible.

A misbehaving device that is identified or localized can have a variety of possible enforcement mechanisms applied. The devices reputation could be lowered. The device could be blacklisted. Blacklisted devices are refused services and cooperation. Devices may have a mechanism to allow authorities to shut down misbehaving devices remotely. Such a mechanism may or may not be circumvented by the attacker. The device could be seized by authorities. Finally, other devices may actively punish the misbehaving radio by jamming or otherwise disrupting its communication. These enforcement mechanisms themselves are avenues for attack, so they must be applied after thorough consideration.

24.5.8   Policy Certificate Chains

A challenge is to establish the provenance of a given policy. A chain of trust can be established between a known trusted authority, such as a national regulator, and the final entity issuing a policy. A regulator can issue a policy that allows a second entity to control policies for a specific spectrum at a specific place for a specific time. For spectrum under its control, this second entity can issue a sub-policy to a third entity, and so on. By cryptographically signing these policies, the chain of trust can be created.

24.5.9   Distributed Databases

A database may store policies, measurements, and other data to support spectrum sharing devices. Having data replicated in multiple databases reduces the threat of denial of service attacks on any given database. The replication mechanism needs to be robust so that false information is not redistributed across the databases.

24.5.10   Trusted Radio Module

A trusted radio module is a combination of hardware and software controls that ensure radio operation within the defined spectrum policies and laws as well as maintaining the security and integrity of on-board user and device specific information. It also enables a secure download, installation, and configuration of radio software. Using embedded hardware keys installed at time of manufacture, the trusted radio module works with monitoring devices to detect device tampering and cloning.

24.5.11   Architecture Controls

Security can be controlled by the choice of the spectrum sharing architecture. Below is a selection of choices to illustrate the interaction between security and the architecture. These choices are not necessarily mutually exclusive. Isolated radios might be policy-based. A trusted coordinator might be connected to the Internet, and so on.

24.5.11.1   Isolated Operation

The spectrum sharing radio can operate on its own without access to any external data sources, such as cooperating nodes or policy databases. It uses policies that are loaded at time of manufacture or via occasional updates by a professional installer. It depends on its own sensing, localization, and timing measurements. Such a choice precludes some controls but avoids many threats to the radio. The loss of real-time access to policies reduces its spectrum sharing potential. The radio itself may inadvertently be misbehaving but not be able to detect it. The result is a greater potential for harmful interference. An example of this scenario is a military handheld radio.

24.5.11.2   Cooperative Information Sharing

Cooperative information sharing enables more dynamic and precise spectrum sharing. It enables the radios to better coordinate and check their activities. These advantages are offset by the new

Table 24.2 Threats Addressed by Each Control

Threat	Primary Radio verification	Graceful degradation	Physical layer protection	Cryptographic techniques	Challenge-response	Reputation mechanisms	Radio monitoring/enforcement	Policy certificate chains	Distributed databases	Trusted radio module	Isolated operation	Cooperative information sharing	Trusted coordinator	Internet connected	Policy-based radio
Primary Radio Emulation	X											X	X	X	X
Fabricated External Data Sources		X		X		X		X		X			X	X	X
Replayed External Data Sources		X		X	X	X				X			X		X
DoS on External Data Sources		X	X						X		X				
Loss of Privacy			X	X							X				
Policies Issued Without Authority				X		X		X						X	
Policy Overload									X						
Satellite Jamming and Spoofing		X										X	X		
False Radio Environment	X	X										X	X		
Information Sharing Attacks		X		X		X			X		X		X		X
Subverting Channel Selection			X				X						X		
Subverting Unshared MAC Protocol							X						X		
Malicious Device Software				X		X	X			X					
Altered Radio Device Hardware						X	X			X					
Changing Device Settings				X			X								
Steering Non-ideal Characteristics							X								
Device Cloning										X					

vulnerabilities and complexity of operation that are introduced. An example of this operation is a set of laptop computers communicating with an access point.

24.5.11.3   Trusted Coordinator

In this model, there exists a special spectrum sharing device that collects and validates sensing information, is a reliable source of spectrum awareness information, and helps coordinate channel selection, media access, and spectrum handoff. The trusted coordinator can manage keys for secure communication and help monitor for misbehaving radios. The other spectrum sharing radios need to establish a single initial connection with the trusted coordinator, possibly over well-known preselected channels. Once established, subsequent communication can be managed via the trusted coordinator. A security concern is that the trusted coordinator becomes a target for denial of service and is a single point of failure. An example of this operation is a base station controlling the subscriber units in a mobile telephone system. A spectrum broker is a more limited version that only coordinates requests for spectrum but does not coordinate other aspects of communication.

24.5.11.4   Internet Connected

Spectrum sharing radios connected to the Internet or other communication networks have access to resources, such as policies, spectrum brokers, public keys, blacklisted radios, software updates, time, and other information. This enables more dynamic spectrum access and supports many of the other controls. One question is whether the Internet access is continuous or intermittent. The frequency of connectivity determines the frequency of updates and dynamism of spectrum access. From a security perspective, depending on the Internet exposes the spectrum sharing to attacks from a much broader set of sources that could be located anywhere on the Internet.

24.5.11.5   Policy Based Radios

Sensing is vulnerable to several threats and requires cooperation to accurately assess spectrum awareness. Policy-based radios eliminate or minimize the role of sensing. Instead, they rely on policies, combined with location and time, for spectrum awareness. Since no direct measurements are made, this approach is more conservative and may not be able to exploit every spectrum sharing opportunity.

24.5.12   Summary

Spectrum sharing has many security vulnerabilities that need addressed. This section presents a set of security controls that can address these vulnerabilities and reduce the risk. Table 23.2 summarizes the relationship between each threat and control. An "X" indicates that the control in that column partially or fully addresses the threat in that row. A blank means that the control has minimal or no effect.

24.6   Conclusion

This chapter describes the relationship between spectrum sharing and security. The additional functionality necessary for spectrum sharing introduces vulnerabilities that an attacker can exploit to deny communication for the spectrum sharing radio, reveal personal information about the spectrum sharing user, or cause harmful interference to primary users. The spectrum sharing architecture itself becomes a target to degrade or provide free access to paid spectrum sharing services. The vulnerabilities, threats that exploit these vulnerabilities, and controls that can help reduce the risk of an attack are briefly described to give a broad picture of how these elements interact. The reader is encouraged to review the papers in the references that provide more details.

References

[1] S. Arkoulis, L. Kazatzopoulos, C. Delakouridis, and G.F. Marias, "Cognitive Spectrum and Its Security Issues," in *The Second International Conference on Next Generation Mobile Applications, Services and Technologies* (NGMAST '08)., pp. 565–570, 16–19 Sept. 2008, Cardiff, UK.

[2] G. Baldini, T. Sturman, A.R. Biswas, R. Leschhorn, G. Godor, M. Street, "Security aspects in software defined radio and cognitive radio networks: A survey and a way ahead," *Communications Surveys & Tutorials, IEEE*, vol.14, no.2, pp. 355–379, Second Quarter 2012

[3] T. X Brown and A. Sethi, "Potential Cognitive Radio Denial-of-Service Vulnerabilities and Protection Countermeasures: A Multi-dimensional Analysis and Assessment," *Journal Mobile Networks and Applications*, v. 13, n. 5, October 2008, pp. 516–532, 17 p.

[4] T.C. Clancy, N. Goergen, "Security in Cognitive Radio Networks: Threats and Mitigation," *Cognitive Radio Oriented Wireless Networks and Communications, 2008. CrownCom 2008. 3rd International Conference on*, pp. 1–8, 15-17 May 2008, Singapore

[5] N. Nguyen-Thanh, P. Ciblat, A. Pham, V. Nguyen, "Surveillance Strategies against Primary User Emulation Attack in Cognitive Radio Networks," *Wireless Communications, IEEE Transactions on,* to appear.

[6] K. Pelechrinis, M. Iliofotou, S.V. Krishnamurthy, "Denial of service attacks in wireless networks: The case of jammers," *Communications Surveys & Tutorials, IEEE* , vol.13, no.2, pp. 245–257, Second Quarter 2011

[7] J. A. Volpe, *Vulnerability assessment of the transportation infrastructure relying on the global positioning system*, Final Report for the National Transportation Systems Center, U.S. Dept. of Trans., Aug. 29, 2001 (www.navcen.uscg.gov/gps/geninfo/pressrelease.htm).

Chapter 25

Security Measures for Efficient Spectrum Sharing

Ethan Gaebel and Wenjing Lou

CONTENTS

Security is now widely recognized as being a major issue in all electronic devices and systems. Each domain has its own set of security issues, and each domain's set of issues can vary to the point of being totally distinct from another's. Spectrum sharing includes many security issues that may not be immediately obvious because they are not present in the everyday domain of desktop, server, mobile security, or even typical wireless security. This can be attributed to the still recent status of spectrum sharing in comparison to other areas.

One useful aspect of spectrum sharing's newness is that we have the opportunity to embed security into the initial design process of spectrum sharing systems. This is a stark contrast to most technologies, which are developed to maturity and have security bolted on afterward. Easily recognizable examples of this include the World Wide Web where many sites are still migrating from HTTP to HTTPS; databases where SQL injection is still a viable attack vector; and others.

Spectrum sharing must be different if it is to succeed. The result of attacks on spectrum sharing systems will usually be lack of service or poor service quality, as opposed to loss of data, although compromised data is also a very real result. If spectrum sharing systems lack security, they simply will not work and, thus, will not be adopted.

Many techniques and protocols have been proposed to facilitate spectrum sharing and many more will likely be proposed before widespread adoption occurs. It is important to understand the security implications that each technique brings as baggage before selecting a set of techniques to implement in a real system. This paper seeks to provide guidance as to how certain techniques fit together and how others cannot.

The rest of this chapter is organized as follows: Section 24.1 lays out regulatory limitations set by the Federal Communications Commission (FCC), Section 24.2 discusses potential consequences of not implementing proper security measures in spectrum sharing schemes, Section 24.3 goes over various network architecture choices, Section 24.4 reviews some current IEEE standards relating to spectrum sharing, Section 24.5 covers the many attack vectors and defenses the research community has examined thusfar, Section 24.6 concludes.

25.1   Regulatory Limitations and Terminology

The FCC has had several exploratory reports prepared on spectrum sharing, and as a result, has already made several rulings on how it believes spectrum sharing should be carried out in the United States. The push for spectrum sharing in the United States was kick started by an FCC report showing that spectrum utilization in the various allocated bands in the United States varied from 15% utilization to 85% utilization [26]. Many bands are approaching saturation, while many others remain underused.

After these findings emerged, the FCC prepared a report that laid out rules for spectrum sharing and opened the TV whitespaces that correspond to the UHF (300 MHz–3 GHz) and VHF (30 MHz–300 MHz) bands [16]. These bands are unused in many locations due to the fact that TV stations will buy a block of spectrum and not necessarily use all of it; they hold a lease on vacant property.

The owners of a particular block of spectrum are referred to as the primary users, while users who share the spectrum owned by the primary users are referred to as secondary users. Secondary users participate in opportunistic access on the available bands by using software defined radios (SDRs), which are capable of changing their operating parameters on the fly. Further specialized devices that automatically adjust their operating parameters by obtaining information about the current wireless environment are referred to as cognitive radios (CRs). These CRs are driven by a cognitive engine, a piece of software running several algorithms designed to deliver the best quality of service as defined by the user.

The FCC has stated that secondary users are responsible for not interfering with primary user operation [16, 22, 23], a feat that can be approached in numerous ways, which will be discussed in later sections. The FCC defines not interfering with primary user operation with a series of time requirements. Secondary users may enter a band only after monitoring it for a minimum of 30 seconds and verifying that no primary users are in operation [23]. Secondary users must monitor the band every 60 seconds to detect the reemergence of the primary user [23]. After a secondary user detects a primary user in the band it is operating on, the secondary user must vacate the band within two seconds [23]. The FCC has also ruled that primary users should not require modification to allow spectrum sharing to occur [12, 16]. These regulations add many constraints to the already rich field of problems we face in designing secure, efficient, and fair spectrum sharing systems.

25.2   What's at Stake

Spectrum sharing offers unique challenges through a combination of technological, economic, and usability constraints.

The main problems we will be concerned with as follows:

- Ensuring priority for primary users

- Ensuring primary users are not interfered with

- Determining if primary users are currently operating in a geographic area

- Making requisite infrastructure changes minimal (preferably non-existent)

- Providing a high quality of service to secondary users

- Providing a persistent connection to the Internet

- Fairly scheduling secondary user spectrum usage

These problems leak across the decisions and schemes necessary to facilitate spectrum sharing. To sum them up, spectrum sharing must be transparent for primary users, and secondary users must be able to obtain the level of service they desire, so long as it does not disrupt the primary users' service.

Why are these problems so central? Primary users remain the owners of their respective bands and, therefore, should retain priority in their property. In addition, if their service is infringed upon, they will likely take action to put a stop to spectrum sharing. Furthermore, secondary users must receive a quality of service comparable to that which they enjoy now in their respective bands for them to be willing to use the shared spectrum. There can be no regression in quality for either primary users or secondary users. How can we ensure this happens? By introducing adequate security measures from the beginning to protect all users from attackers who would disrupt service. Spectrum sharing is concerned with correctly scheduling users in time on specific bands. If this scheduling is allowed to be disrupted, then spectrum sharing will simply not work correctly or at all. It is necessary to devise secure schemes that prevent as many attacks as possible while remaining robust in the face of attacks that are able to get by security measures.

25.3   Architectural Considerations

At this current juncture, there are myriad approaches to facilitate spectrum sharing. At the highest level, there is the question of how to obtain information regarding the current spectrum usage in a secondary user's area. This question reduces to geolocated databases vs. spectrum sensing, both of which we discuss in Section 24.3.1. After determining how spectrum information can be obtained, we turn to the question of how users operating in this spectrum sharing system are going to connect to the rest of the world via the Internet. Communicating point-to-point over wireless is useful, but, realistically, Internet access must be provided, as well. At a high level, we can reduce this to two choices, centralized vs. decentralized. Centralized Internet access is a replication of our current cell tower and wireless access point architecture, many users connect to a single point over the wireless medium, which acts as a gateway to the Internet. The decentralized approach is more diverse; there are many specific schemes that can be used, but all of them share the common infrastructure of self-organizing nodes, where one or more secondary users provide Internet access to other users. We briefly discuss the merits and demerits of the approaches to each problem, as well as raising some high-level security issues, which will be addressed at a more granular level in the sections that follow.

25.3.1   Acquiring Reliable Spectrum Information

Spectrum information acquisition is the most critical issue in any spectrum sharing system. Without information about spectrum availability, there will be no spectrum sharing, so it is important to broach this topic, if only to discuss the security implications. Geolocated databases and spectrum sensing seem to be two opposing paradigms, but, in practice, we are likely to see both in use for different scenarios or, as proposed by the FCC, both may be used to provide a higher quality of spectrum information [4,16,29,42]. This approach may complicate implementation of such systems, as well as securing them, but should provide a richer environment for users of these systems.

Spectrum information itself is great; however, if we cannot validate its authenticity, then our spectrum sharing systems will be open to attack. Complicating this issue is the fact that devices may be receiving spectrum information from many sources. We see a requirement to discern which sources are reliable, which are untrustworthy, and which are in-between. Now we discuss the two dominant approaches to acquiring spectrum data.

25.3.1.1   Geolocated Databases

Geolocated databases are to be accessed either over the Internet or through co-located stations. These databases provide spectrum information to authorized secondary users about the current spectrum availability in their area of operation. The databases can obtain spectrum data in a few ways. First, primary users can register to use the spectrum in specific time blocks (this is particularly practical for small devices used at long events) [42]. Second, a model of wave propagation can be run that takes into account primary users and all of their specific parameters (height, power, frequency), landscape (particularly features that block radio frequency (RF)), and other factors [42]. In [42] the Longley–Rice propagation model [3] is used to model primary user signal propagation due its increased complexity that takes into account climatic effects, soil conductivity, permittivity, the curvature of the Earth, as well as surface refractivity. The FCC also opted to use the Longley–Rice model to compute TV contours [3]. It is also possible to combine geolocated database lookup with spectrum sensing to provide a more dynamic approach [42]. The access protocol described in [42] is as follows: The base station providing connectivity to the area broadcasts a single beacon on each currently available channel containing all channel availability in all regions of its coverage area, which is repeated every second. Once every minute, the base station listens on each of the channels broadcasted and listens for secondary users who would like to join the network. After secondary users join the network, they must transmit their location to the base station so the correct spectrum information can be applied to them.

25.3.1.2 Spectrum Sensing

Spectrum sensing is facilitated by having cognitive radios that passively sense various channels' usage and cycle through channels, usually based on some intelligent scheme. If a primary user is detected in a channel, then that channel is unavailable. This technique is more power intensive than a geolocated database lookup, but will provide more accurate insight into real-time spectrum usage, especially for low power devices, and there exists much research into reducing power demands through intelligent sensing strategies [27]. Another advantage of spectrum sharing is its robustness; no preexisting infrastructure is necessary for a single radio to deploy spectrum sensing. This will be significant for many hobbyists trying to participate in opportunistic spectrum access as well as those in under-served areas.

In spectrum sensing, we have one unavoidable security issue that is shared across any system that provides spectrum information: authenticity. Secondary users sense spectrum in search of primary user signals, which indicate an occupied channel, but how do we know that a primary user's signal is actually a primary user's signal? If malicious users can forge a primary user's signal characteristics, then they can obtain unobstructed, unfair spectrum access. This specific issue is called primary user emulation (PUE) and is one of the most studied issues in spectrum sharing. The solution is to devise a secure authentication scheme for primary users. This will be thoroughly discussed in Section 24.5.1.

25.3.1.3 Spectrum Sensing with Information Sharing

On its own, directly sensing spectrum to gain availability information is more decentralized; a secondary user need not rely on infrastructure to obtain spectrum information. It can be obtained anywhere at anytime. However, through cooperation, there can be performance perks. Secondary users can share sensing information in a localized group. This group can be centralized, with all secondary users communicating with a central entity, referred to as a fusion center, that combines (or fuses) information together and reports spectrum availability back to the group. Alternatively, this group can be decentralized, with secondary users broadcasting spectrum information on a control channel to surrounding users. Sharing spectrum information allows each device to save power by not sensing on every channel and can also decrease the time required to determine channel availability [63].

Allowing secondary users to share spectrum sensing information would undoubtedly increase the efficiency of the system through fewer overall channel sweeps [63], if we assume that all the information shared is true. Secondary users may transmit false spectrum information and cause primary users to be interfered with, cause secondary users to be interfered with, or cause total failure of the spectrum sharing system [47]. Here, we find an issue of trust and trust management among users or between users and a centralized entity collecting and distributing spectrum information. For both cases, we must introduce an additional layer of security. Just like with the geolocated database and the primary user, a device must be able to authenticate all communications from a fusion center and vice-versa. This issue will be discussed in great detail in Section 24.5.2.

25.3.2 Securely Connecting to the Internet

It would be naive to think that users in spectrum sharing systems would be satisfied with only communicating with nearby users vis-a-vis wireless communication; the Internet must be linked to spectrum sharing systems if such systems are to be implemented and used.

25.3.2.1 Centralized

The current cell tower and wireless access point infrastructure is a perfect example of how a centralized Internet access system is structured. The only difference is that the access points will need to accommodate more access channels to account for all of the possible bands given the spectrum available for sharing in a particular geographic area. Other than that, this architecture is identical to the cell tower infrastructure used everyday.

Centralizing Internet access is quite appealing from a security perspective. It will be necessary to authenticate access points for the fullest security guarantees, and relying on static, centralized infrastructure eases authentication significantly, in comparison to the decentralized case.

25.3.2.2 Decentralized

A decentralized or ad-hoc network is built on many point-to-point connections. This architecture is another way that secondary users can achieve Internet access in spectrum sharing systems. This approach relies on at least one secondary user having some form of Internet connectivity already, from 3G, 4G, a hard line, etc., and also on that user sharing their Internet with the remainder of the network by acting as a gateway. This scheme remains advantageous if the channel being used to provide Internet access is saturated or unavailable to other secondary users.

Much research has been done on security issues in ad-hoc and mesh networks, and that research can directly be applied here, so we will not discuss it in detail. However, a few fundamental points are worth mentioning. Decentralized Internet access involves significant trust issues, since the intentions of the other secondary users participating in a mesh network are unclear. Attackers can easily perform denial of service (DoS) attacks by simply not passing on messages from other secondary users. The same goes for eavesdropping on unsecured channels. If channels are secured, we encounter the age-old issue of key management; for a large network of secondary users, the number of keys required to facilitate secure communication between each pair of users is 2^N, where N is the number of secondary users participating in the network. Realistically, it is unlikely that every secondary user will be in close enough proximity to communicate with every other user, but this gives us a hard upper-bound.

Despite the implementation issues and security issues involved with such decentralized networks, they can be useful when secondary users are grouped in an area with poor Internet coverage from traditional sources and no centralized infrastructure, but they are not ideal.

25.4 Current Standards

IEEE has already begun work on establishing protocols to support spectrum sharing. These standards are 802.11af and 802.22, respectively. Both refer to operation explicitly in the TV whitespaces (UHF and VHF). An advantage of using these bands over traditional WiFi (802.11a,b,n,ac) is that the attenuation and fading performance when many building materials, such as brick and concrete, are encountered is superior.

25.4.1 IEEE 802.11af

Standard 802.11af falls under the wireless local area network (WLAN) classification, and as such, is designed for relatively short range communication, within 1 km [4]. Its physical layer uses a variation of orthogonal frequency domain multiplexing (OFDM), which is very similar to the one used in 802.11ac. Topological and protocol considerations are what make the 802.11af standard interesting in this context. The 802.11af standard stipulates that various access points will operate in TV whitespace bands whose availability is obtained from geolocated databases that are assumed to be operated by regulatory entities. Secondary users may connect to these access points by listening on each TV whitespace channel in turn for an enabling signal from the access point. Once an enabling signal is obtained, the secondary user responds with an enabling response frame. The access point responds with a successful enablement response, and the secondary user can now communicate over the access point on the channel on which it received the initial enabling signal [4].

25.4.2 IEEE 802.22

Standard 802.22 is still in the draft stage, but it shows quite a bit of promise in being a robust definition of how spectrum sharing schemes should be implemented. It falls under the wireless regional area network (WRAN) classification, and, as such, is designed to support communication over 100 km areas. In the 802.22 standard, the topology revolves around base stations and secondary users. Base stations are stationary points that provide various accessibility services to secondary users in the area. Base stations obtain spectrum information from spectrum information databases, perform spectrum information fusion using data sent by secondary users in the surrounding area, provide Internet access, and perform scheduling for secondary users. Secondary users must be able to detect a primary user in ≤ 2 seconds, as stipulated by the FCC in [23], with a $\geq 90\%$ probability of detection and a false alarm rate of $\leq 10\%$ [2]. Base stations in 802.22 only communicate with secondary users on a single channel at once and use the spectrum information from sensing and databases to decide when to change channels. Secondary users communicate with base stations using orthogonal frequency-division multiple access (OFDMA) [11].

25.5 Attacks and Defenses

The central theme of attack and defense models in spectrum sharing is authentication. It cuts across almost every security issue in spectrum sharing and is definitely present in all of the biggest problems. For reasons we will discuss in the following sections, authenticating users via the wireless medium is difficult, and many techniques have been developed to increase its efficacy and robustness.

25.5.1 Primary User Emulation (PUE)

Primary user emulation emerged as one of the first security issues to be identified when dealing with dynamic spectrum access [14, 34, 44, 51, 65, 67]. As stated earlier, the FCC mandated that radios engaging in spectrum sharing must not interfere with the primary user of a particular band [16]. This opens up a lucrative attack vector for users trying either to gain unfair access to the spectrum or launch DoS attacks. In both attacks, a malicious user masquerades as the primary user to trick other secondary users monitoring the channel. The secondary users will detect the malicious user as the primary user and will not attempt to access the channel. Entities attempting to enforce spectrum fairness will be unable to take action against the malicious user due to the need for secondary users to respect the priority of the primary user, who is believed to be operating. How exactly malicious users trick others into believing they are the primary user depends on the security measures (if any) that are in place. This attack demonstrates the need for primary user authentication mechanisms.

There are two important limitations to consider when thinking about primary user authentication. First, performing authentication in the upper layers of the protocol stack, i.e., MAC, Network, Transport, Application, is problematic for performance and compatibility reasons. Many radios do not share common components at each of these layers, so developing a widely used standard at the upper layers would be practically impossible. For secondary users to manage all the different interfaces would be arduous at best and practically infeasible at worst. Performance is also a problem in this scenario, since as we travel up the stack, we necessarily introduce more overhead; to reach an upper layer only to discard the signal as illegitimate would be wasteful. Hence, the discussion of authenticating primary users is grounded in the physical layer.

Second, the FCC has ruled that primary users should not be modified to accommodate spectrum sharing. This ruling boils down to minimizing cost to the primary users; primary users are not likely to accommodate spectrum sharing if it means bearing additional costs. This means we are in need of some scheme that can be added on the secondary users' side or a scheme that incurs a very small cost to the primary users. In the following subsections we discuss methods to authenticate primary users to guard against the PUE attack.

25.5.1.1 Energy Detection

Much research has focused on identifying the primary user by both signal and channel features [14, 30, 44, 67, 68].

Detecting signal energy was proposed in [65, 68] to identify primary users with distinct energy levels. A secondary user receives a signal assumed to be composed as follows: $y(n) = s(n) + w(n)$, where $s(n)$ is a transmitted signal and $w(n)$ is additive white Gaussian noise. The secondary user now how two hypotheses: $H_1 : y(n) = w(n) \rightarrow$ no primary user is present, and $H_2 : y(n) = s(n) + w(n) \rightarrow$ a primary user is present. The secondary user determines which hypothesis is true by gathering N samples, $y_i(n)$, where $0 \leq i \leq N$ and summing the square of each received signal $M = \sum_{i=1}^{N} y(n)^2$. The secondary user then compares M to some threshold, P, such that if $M > P$, the primary user is present, and if $M < P$, the primary user is absent. P is determined using expected noise power and signal power.

Determining P for dynamic scenarios is not robust, as it is dependent on distance from the primary user and fading properties, which are unpredictable. In practice, determining P becomes a balancing act between the probability of detection, P_d, and the probability of false positive, P_f [68].

Specific research targeting scenarios where TV towers are the primary users has tried using received signal strength, as measured at different locations in a sensor network (or alternatively a network of secondary users with sensing abilities) in an attempt to localize the statically located TV towers and use this location to authenticate primary users [14]. If this technique were dealing with a lower power primary user, it would be feasible for attackers to physically place themselves nearby the primary user and adjust their operating parameters to cause sensors to measure a received signal strength in the same range as the primary user.

For primary users with low energy signals, these methods of energy detection are totally inadequate, since such signals can easily be spoofed or rendered undetectable by other devices. Chen et al. give a description and perform an evaluation of just such an attack in [67]. The attack uses maximum likelihood estimation to determine the best estimates for the primary user's transmission power and variance, and the attacker adjusts their operating parameters accordingly. The authors evaluate their attack in terms of probability of false positive (identifying the primary user as illegitimate) and false negative (identifying the attacker as the primary user) for the secondary users trying to sense the primary user. The probability of a secondary user correctly detecting a primary user is given as $1 - P_{FN}$, where P_{FN} is the probability of false negative. Their results show that even when the variance of the attacker and the variance of the primary user differ, which may result if the two not transmitting from the same location, that the attacker is able to emulate the primary user with a very high probability. Chen et al. are able to establish a stronger defense by taking more signal samples to obtain a better estimate of channel variance, but are unable to devise a defense strategy when attacker and primary user variances are the same. Furthermore, this advanced defense strategy relies upon prior knowledge of the variance in the channel between primary user and receiver, which is not knowledge that can be easily disseminated prior to deployment, since it is dependent on channel characteristics and relative location to the primary user.

Identifying primary users by energy level and received signal strength works well for these high power primary users, since a TV tower's transmitter output power is usually on the order of hundreds of thousands of watts [14]. Such high energy levels cannot be realistically replicated by conventional attackers. For general purpose applications, where the primary user is not a high power entity, these authentication mechanisms will be highly insecure, since attackers can tweak operating parameters to work at specific energy levels [67]. For the current state of spectrum sharing, where TV whitespace bands are the only ones open for spectrum sharing, energy detection is a feasible technique to provide primary user authentication, but it is not a reliable method of authentication for future lower power applications of dynamic spectrum sharing.

25.5.1.2 Feature Detection

Wireless signals have features more diverse and more discriminating than energy. These features have been used for years in radio fingerprinting for identifying devices in cellular networks [50], by the military for identifying radios [35], and other applications involving wireless transceivers [9, 24, 59]. Many techniques from radio fingerprinting find application in primary user authentication. One powerful technique from radio fingerprinting relies upon manufacturing defects in the analog wireless components. These defects affect several measurable features that are part of the modulation process and endow each wireless radio with a unique signature with respect to particular features. In [59] Vladimir et al. identify the following features:

- Frame frequency error, the difference between ideal and observed carrier frequency

- Frame SYNC correlation, the correlation of I/Q values forms ideal and observed SYNC, which is a short frame sent before the data for synchronization

- I/Q offset, the distance between the origin of the ideal I/Q plane and the observed I/Q plane

- Frame magnitude error, the average distance between ideal symbol magnitude and observed signal magnitude

- Phase error, the average difference between the ideal symbol angle and observed symbol angle

These are referred to as modulation features, since they are identified and created in the demodulation and modulation processes, respectively. These modulation features are placed into a vector and normalized to be between 0 and 1 before being passed to a classification algorithm for matching. In [59], the authors examined using k-nearest neighbors (kNN) and a support vector machine (SVM) to perform classification. Their results show that SVM achieves near perfect accuracy, having an overall error rate of 0.34%, although SVM performed over 5 times slower than kNN. kNN saw very high false reject rates for certain devices, peaking at more than 60%, but the average false reject rate was only 3%. Evaluated over worst-case similarity, how likely two devices are to be mixed up, kNN had an average score of 3% again, although it also performed very badly for certain devices. This shows that kNN can perform well enough for certain applications, and its performance can perhaps be increased by the inclusion of additional features. Further research into effective classifiers for radiometric identification remains an open research topic, which is constrained by various performance requirements and feature selection.

In [59] Vladimir et al do not explore the robustness of their radiometric identification scheme with respect to security and how easily it can be defeated. Luckily, Danev et al. do explore techniques to defeat these forms of radio fingerprinting in [19]. Two attack varieties are examined, signal replay and feature replay. In signal replay, the attacker captures a signal and replays it exactly as captured using a waveform generator, while in feature replay, the attacker captures a signal, processes it to identify the values of the features, and then transmits a signal with those feature values replicated. Feature replay attacks were performed using the Universal Software Radio Platform (USRP) in conjunction with gyeongsang national university (GNU) Radio [61]. These attacks were both found to be quite effective. Danev et al were able to impersonate all features except for frame SYNC correlation and frame magnitude error, but it was found that these features were not as discriminating as the rest, and, thus, they obtained a 98% success rate in their impersonation attacks. Signal replay was even more successful, making it is nearly impossible to differentiate between legitimate signals and replayed ones. In the context of primary user emulation, signal replay provides only a mechanism for attackers to perform DoS attacks, since it requires exact signal replay, with no opportunity to modify the signal content. For an attacker to impersonate a primary user and transmit data of their own, an attacker must rely on feature replay attacks.

Another signal feature with discriminating properties is the signal transient. The signal transient is the part of the signal where the amplitude rises from channel noise levels to full transmit power levels. As with the modulation features, the signal transient is a unique identifier due to the subtle manufacturing defects in the analog radio components. In a regular transmission, the transient occurs before each transmitted packet, making it a regularly occurring authentication mechanism, which can only speed up primary user authentication. Danev et al. explored using the signal transient to uniquely identify devices [18] in sensor networks. To use the transient in signal identification, the signal is put through a four-step process: capture, transform, extract, and match. First, a number of samples from the signal are captured, then they are passed through a transformation. In [18], various transformations were proposed: using the raw samples (no transformation), the Hilbert transformation, the raw samples +/- the Fast Fourier Transform (FFT) spectra of the samples, the Hilbert transformation +/- the FFT spectra of the samples, and the relative differences between adjacent FFT spectra of the raw samples. Their tests found the relative differences between adjacent FFT spectra transformation to be superior. After the transformation is performed, the features to identify the unique transient must be extracted. This is achieved by using Fisher Linear Discriminant Analysis (LDA), a technique used to classify human biometrics [8]. The samples are matched by comparing the similarities between transient fingerprints using a distance metric. This scheme exhibited quite good results in the author's experiments, performing with under 4% error rates under all parameters tested. It is important to note that these results include several measurements from nodes in the sensor network and only extend to a maximum distance of 40 meters. For application to primary user authentication, further work should be conducted to examine the maximum distance at which a signal transient is usable for authentication.

Danev et al. examined the resiliency of signal transient based identification to attack in both [18] and [19]. In [18], a hill climbing attack was evaluated in simulation, and it was found that for under five devices in the sensor network, the signal transient was theoretically vulnerable to impersonation, which applies to our PUE case, since we cannot guarantee a minimum number of users participating in spectrum sensing at one time, nor can we even guarantee that a spectrum sharing system will support collaborative spectrum sensing. However, while signal transient impersonation may be possible in theory, Danev et al. find in [19] that it is very difficult in practice. The experiment used a waveform generator to replay a captured transient to impersonate a device. It was found that the replayed transient failed in impersonating the device due to the replayed transient signals' alteration in the wireless channel. Since the transient was captured with interference from the wireless channel, further interference from retransmission corrupted its unique signature beyond tolerable authentication levels. The authors suggest that capturing the signal transient directly from the device being emulated or replaying the transient from the location of the emulated device may overcome the problems in impersonation, but these suggestions remain unverified.

While the transient has been shown to be quite robust to impersonation attacks, it is not as robust against DoS attacks [18]. Danev et al. demonstrated using an USRP and GNU Radio [61] that the transient can be selectively jammed in low power devices, making authentication impossible. For high energy devices like TV towers, the transient will not exhibit the same vulnerability, however, and, thus, may prove to be an effective tool in verifying the identity of these high energy devices, especially if used in conjunction with energy detection or other features.

Signal cyclostationarity is another feature that shows some promise in identifying distinct signals. Cyclostationarity is a property of a stochastic process, $x(t)$, where the mean $M_{x(t)}$ and autocorrelation $R_{x(t)}$ are periodic with period T_0 [36, 45, 46]. Wireless signals can be modeled as stochatic processes, so if there are any periodically repeating characteristics of a signal, we can use these characteristics to help identify the signal. There are several such periodic characteristics in wireless signals used for communications, arising from modulation protocols, such as OFDM [45], specific modulations, such as quadrature amplitude modulation (QAM), or phase-shift keying (PSK) [36, 46].

Cyclostationarity is extracted from signals by applying various transformations to the signal to obtain some data to be fed to a classifer. Often this culminates in determining the cyclic domain profile, which can be used for generating the spectral correlation function to be used in identifying similarity between signals. In [46], Ramkumar examined using neural networks and hidden Markov models (HMM) to classify signals based on cyclostationarity existing in the following modulations: binary phase-shift keying (BPSK), quadrature phase-shift keying, frequency-shift keying (FSK), and minimum-shift keying (MSK). The neural network approach showed a large variance in performance with respect to the number of samples necessary to achieve a high probability of correct classification. Achieving a greater than 95% chance of correct classification required between 30 and 100 samples, depending on the modulation, with the probability never reaching 100%. The HMM approach was able to achieve a greater than 95% chance of correct classification between 60 and 260 samples, making it much less reliable.

Many other researchers have also studied using cyclostationarity for primary user identification, and the FCC has discussed using cyclostationarity-based detection [16].

In [12], Chen et al. point out that cyclostationary feature detectors can be defeated by attackers altering their transmission properties to match those of an observed primary user. Cyclostationarity on its own does not suffice to reliably authenticate primary users, but these features may prove useful in conjunction with others to make attacks sufficiently difficult.

Feature detection is currently a not-entirely-secure method to authenticate primary users. Due to the over-the-air nature of the features being relied upon for security, attackers are able to observe and spoof them, although some feature detection schemes are secure in specific scenarios due to the resources requirements that would be required to spoof them (e.g., TV towers). If feature detection is used to authenticate primary users, then a good solution to provide usability and higher levels of security should harness as many of the described features as possible to make impersonation attacks very difficult for attackers, although there can be no guarantees when relying on these signal and channel characteristics.

25.5.1.3 Signal Watermarking

The lack of guarantees provided by feature detection has caused much research to be conducted into how to embed some sort of identifying information into a primary user signal that cannot be replicated by an attacker. Such procedures try to do so cheaply, both economically (with respect to necessary hardware and software modifications) and computationally. This process is referred to as signal watermarking, a moniker taken from the process of watermarking currency to prevent counterfeiting. Various methods have been proposed to accomplish this, which focus on embedding during either channel coding or modulation. Due to the non-modification requirement the FCC has included, all of these techniques try to store information in unlikely places during the modulation or coding processes to minimize the alterations necessary in both hardware ans software. As a result, these schemes all have a similar process: the primary user embeds authentication information somewhere in the physical layer using an unconventional procedure; the primary user transmits the message, usually in some altered form due to the authentication information; and the secondary user retrieves the authentication information, separating the authentication information from the original message before the message is passed to the upper layers.

In [57], Kumar et al. propose a scheme they call frame frequency modulation (FFM) and demonstrate it using OFDM. FFM introduces extra information into the encoding by adding frequency offsets to the OFDM frames described by the equation below. Here, f_a is the maximum possible value of the embedded frequency offset, M is the total number of possible frequency offsets for a b bit authentication symbol, that is, $M = 2^b$, and m is the current frequency offset being used so that $1 \le m \le M$.

$$f_m = f_a * \left(1 - 2 * \frac{m-1}{M-1}\right)$$

This will show up as noise to unaware listeners, and users attempting to authenticate the sender will be able to approximate the frequency offsets that can be decoded into the embedding information, which is encoded using error correcting codes prior to being converted to frequency offsets to allow for real noise in the wireless channel that will likely corrupt the authentication message.

In [54], Tan et al. propose two different embedding techniques, one based on adding authentication tags to the modulation process, and one based on adding authentication tags to the error-correcting codes (ECCs). Embedding in modulation adds a slight phase shift, θ, to the constellation of each symbol. This phase shift is either positive or negative depending on whether a 1 or a 0 is being encoded. The authors rely on noise tolerance in the modulation scheme to allow the message and authentication symbols to be separated by secondary users and for the authentication symbols to be treated as negligible noise by the primary users. ECC embedding involves intentionally "corrupting" symbols in the bit stream at particular positions to encode the authentication message, effectively diminishing the error-correcting capabilities of the ECC being used.

Goergen et al. offer another approach to signal watermarking in [28]. Their scheme uses a series of synthesized channel impulses $c_l^H(t)$, introduced to the primary signal $p(t)$ to encode the extra authentication information. The transmitted signal is: $x(t) = c_l^H(t)p(t)$, and total channel impulse response becomes: $h(t) = c_l(t) \oplus u(t)$, where $u(t)$ denotes the already existent channel impulse and $\oplus$ denotes the convolution operation. At the receiver, total measured channel impulse becomes $g(t) = h_l^H(t)p(t) + n(t)$, where $n(t)$ denotes additive white Gaussian noise. The receiving secondary user utilizes an equalizer designed to respond to channel impulse response, and, thus, the authentication symbols can be recovered by feeding the equalizers' response to the channel impulse response into a separate signal processing unit, while the primary signal is fed off into the regular demodulation hardware. This scheme has the advantage of not affecting the final demodulated message, since the authentication information is stripped off in the equalization phase.

If information is going to be used for authentication, it must have cryptographic assurances to prevent forgery and other illegitimate uses. Two cryptographic embeddings have been proposed for this application. The first, a one-way hash chain,

$$h_n \rightarrow h_{n-1} \rightarrow ...h_1 \rightarrow h_0,$$

where h_0 is publicly published (likely on a daily basis) so that any h_i hashes can be verified by performing hash function operations until h_0 is reached [54, 58]. The other proposed technique [28, 57] involves embedding a tuple and the digital signature of the tuple in the following form:

$$\{TS, F, T, C, Sign_C(TS, F, T)\},$$

where
TS = Time stamp;
F = Transmitting frequency;
T = Time slot authorized to transmit in; and
C = Certificate.

This information is all necessary to ensure legitimacy and to prevent replay attacks. The time stamp prevents replay of a captured authentication code, the frequency prevents replay on a different frequency, and the time slot the user is authorized to transmit in is included to prevent replay on the same frequency in a different time slot. Due to its resiliency via the extra information, the certificate method should be preferred to maximize security.

Analysis of radio propagation has shown that watermarking a primary user's signal can have detrimental results for a large number of secondary users attempting to authenticate the primary user's signal [21]. High error rates result from this embedding when ECCs are not utilized, while adding ECCs increases authentication time. It remains necessary for further analysis to be done regarding the secondary user authentication rate when ECCs are used, as well as the percentage of secondary users able to authenticate when using a signal watermarking scheme in comparison to other techniques not involving the extra noise such schemes include. In [54], it is suggested that the authentication rate should be much more favorable, but direct analysis remains necessary, as [54] relies only on a simulation for verification of their results. Techniques for cryptographic embedding remain a promising option for authenticating primary user signals, and there exists potential for many research problems to explore ways to optimize the secondary user authentication rate while keeping primary user bit error rate low.

25.5.1.4 Helper Node

A method related to watermarking that does not require any modification to existing systems was proposed in [38]. The authors' scheme involves placing a secondary device in close proximity to the primary user and allowing it to broadcast channel availability when the primary user is dormant, thus, notifying secondary users that the channel is available. The secondary node must also sense the primary user's signal to avoid interfering; the authors rely on the wireless characteristics induced by the helper node's close proximity to the primary user to authenticate the primary user's signal.

Although there have been no attacks exhibited against this scheme explicitly, there is much skepticism on the efficacy of security techniques reliant on the characteristics of the wireless environment, as well as much work to back up this skepticism.

25.5.1.5 Reminder of Non-modification

When discussing signal watermarking schemes and helper nodes, we must remember the FCC's ruling stipulating that primary users should not be modified. The watermarking techniques discussed above rely on the assumption that if the cost to modify existing systems is low enough, then the FCC will acquiesce to the proposed techniques. However, the FCC is not the primary concern. It is unlikely that primary users will feel incentivized enough to pay to alter their existing schemes until dynamic spectrum access shows serious economic potential for primary users in the form of leases or some other legal framework whereby primary users can profit by renting out their spectrum while they are not using it. In addition to a profit framework being in place, if the activities of corporate entities in other areas involving security are any indication, primary users will likely not pay to alter their systems until they begin to lose substantial amounts of spectrum-leasing revenue to attackers performing primary user emulation attacks against their unsecured systems. To sum all of this up, it is unlikely that signal watermarking will be used in practice until spectrum sharing becomes a very mature and in-use technology. Until that time, secondary users will have to rely upon radio fingerprinting and energy detection to identify primary users' signals. In light of this, it is important for further research to be conducted in the area of radio fingerprinting to determine which techniques are the most secure, which are easily defeated, and which features have a synergistic effect on increasing attack difficulty.

25.5.2 Spectrum Data Falsification

If we are using a dynamic spectrum access scheme where cognitive radios sense spectrum, it seems natural to combine the spectrum sensing efforts of all cognitive radios in an area to improve spectrum awareness and decrease the total amount of spectrum sensing. The process of combining pieces of spectrum information is referred to as fusion. As mentioned earlier, there are two approaches to performing fusion, the centralized approach and the decentralized approach. The centralized approach utilizes infrastructure called fusion centers, while the decentralized approach involves a point-to-point network of secondary users.

   In both cases, the fusion process from the perspective of the fusion system is composed of three essential steps:

1. Collect spectrum information from cognitive radios in the geographic area

2. Determine which bands are unoccupied

3. Schedule secondary users into bands

The central security concern here is ensuring that spectrum information accurately reflects the current state of spectrum availability. If a system relies upon spectrum information fusion, then attackers may decide to attack the fusion center or the decentralized equivalent instead of emulating the primary user, since the effect of a successful attack will be the same. Transferring decision responsibility to a fusion center or to the decentralized equivalent, while providing a performance increase, also serves to create another security hole, which must be filled. In addition to attackers sending false data to trick the system for opportunistic access, we can see a whole new problem from examining the above three-step process. A new problem to add to the list from the "What's at Stake" section. How can we ensure that secondary users perform their fair share of sensing work? Users may send false data, not to gain opportunistic access through convincing all other users that a primary user is present, but to reap the rewards of cooperative spectrum sensing systems without performing any resource-consuming spectrum sensing of their own. With these problems in mind, let's examine the two possible schemes, centralized spectrum data fusion via fusion centers and decentralized mesh network communication between secondary users, which has several possible schemes of its own.

25.5.2.1 Centralized Spectrum Information Sharing (Fusion Centers)

Problems similar to those experienced with PUE attacks can be experienced in fusion center schemes as well. Users may falsify data to convince fusion centers that no spectrum is available so that they can gain unfair access or simply to ruin the fusion center's operation. This class of attackers is referred to as Byzantine Attackers. The problem of detecting this attack is exacerbated by the possibility of users legitimately returning false data due to sensing problems or other issues. Several initial schemes for data fusion considered each vote from a secondary user as a binary variable and combined them together using AND, OR, or majority rules fusion. More complex schemes included use of the Neyman-Pearson test, which was shown in [66] to provide an optimal solution (assuming honest secondary users), and Wald's sequential probability ratio test (SPRT) [56]. These techniques give no consideration to security and [13] showed that these schemes all fail when under attack by a relatively small number of Byzantine Attackers. So how can we punish malicious providers of false data while sparing innocent providers of false data? In [13], Chen et al. use a model that takes reputation into account using a weighted sequential probability ratio test (WSPRT). Here, each user is assigned a weight, w_i, initially 1, and this weight is decreased every time a user reports a value that is not consistent with the global decision made by the WSPRT system. The WSPRT system makes this decision by taking the likelihoods of each reported observation, raising each one to the w_i^{th} power, taking the product of all of these expressions, and then comparing the result to a threshold determined by the tolerated false alarm and false negative probabilities. This differs from

SPRT only in raising each likelihood to the w_i^{th} power. The addition of a weight is powerful, since it encodes a reputation assigned to each user. This gives the WSPRT the ability to thwart persistent attackers, ensuring they can only confuse the system for a short amount of time until their reputation is reduced to a negligible level.

25.5.2.2 Decentralized Spectrum Information Sharing

The architecture in decentralized data fusion is fundamentally different from centralized. Secondary users must self-organize to communicate among each other, and relevant information must be propagated throughout the network to come to a global decision. These complications are not without benefit; decentralized data fusion is more robust, since it is not reliant on a single point of failure, and it can be more scalable, under certain conditions, since there is no central communications bottleneck [20]. The unique security problems facing decentralized spectrum fusion schemes are based in the network architecture and, thus, belong to the ad-hoc and mesh networking research communities, who have extensively researched these topics. Specific research applying decentralized communications techniques to data fusion has also been conducted in [7, 20]. This research focuses on efficiency as opposed to security, but still may be useful for building security schemes. The problems facing centralized data fusion also face decentralized data fusion, correctly identifying attackers and handling their input accordingly, and maintaining normal operation under duress from as many attackers as possible.

25.5.3 Jamming

Jamming is an old problem in channel availability, and, as a result, it is well studied. Spectrum sharing, however, changes the dynamics involved in a jamming attack by giving the user being jammed some flexibility: They can simply change channels. But a change in channel by a user can then be followed by a channel change by the attacker. Back and forth this exchange will go, without any clear advantage for either side. Indeed, Clancy et al. [62] have shown that this interplay can be modeled as a game, and a Nash equilibrium can be derived. Furthermore, they show that this exchange reduces to an arms race; given equal resources, neither the jammer nor the user being jammed can gain a clear advantage. This is much better than the traditional jamming case, where the user being jammed has no recourse against a jammer. This can be particularly useful for military communications where messages may be short so channel hopping to deliver an important message is a very real alternative.

25.5.4 Spectrum Information Database Attacks

The use of geolocated databases introduces a whole host of potential database attacks to the spectrum sharing world.

Privacy issues are especially prevalent here, as some primary users may wish to conceal their operational privacy, and databases are large sources of information that can be leveraged to unveil private information. Such an attack would be particularly concerning for government entities such as the Department of Defense (DoD). Bahrak et al. discuss in [6] how database inference attacks can be used to identify primary users' operational details. Attackers perform inference attacks by repeatedly querying the database for spectrum availability information. Attackers gather a large number of query results from various locations to determine the geographic boundaries between channel availability and occupancy. Using this information, attackers can discern the location of static primary users, the path of movement of mobile primary users, and the time of operation [6]. The solution suggested by the authors is to perturb data records, causing the database to not accurately report the channel usage of the primary users. This turns into a tradeoff between spectrum sharing efficiency and primary user privacy. It is likely, however, that most primary users will not be concerned with such issues, so a small number of primary users perturbing their channel availability records should not cause significant detriment to the spectrum sharing schemes overall.

The other side of the privacy issue concerns the location privacy of the secondary users. Both the 802.11af [1] and 802.22 [2] draft stipulate that secondary users must supply their geographic location via GPS or alternative means so that they can be supplied with localized spectrum information. Under current U.S. law, if "customer location information is used for performance of network communications or to authenticate user identity, no notice to the customer is required" [31]. These privacy concerns are more regulatory in nature, since it is hard to argue that geolocated databases should transmit all spectrum information for all locations to users. Obfuscation could be performed, similar to the suggested schemes for primary users in [6], but the legality of extended storage of secondary user data is likely to be the primary concern here.

25.5.5 Cognitive Engine Attacks

In many spectrum sharing schemes, a cognitive radio is the pervasive tool used to facilitate automatic radio parameter adjustments. Cognitive radios are driven by cognitive engines, a collection of algorithms that optimize radio operation with respect to regulatory requirements and quality of service [32]. These cognitive engines can either follow a static set of rules, which may be updated periodically, or they may learn continuously from their environment, after being trained extensively prior to deployment. These cognitive engine varieties are referred to as policy radios and learning radios, respectively [10, 15]. We shall be mostly concerned with learning radios, since they present the most dynamic behavior and, thus, the greatest security risks.

A unique caveat of cognitive engines driving decisions about radio parameters is that they are susceptible to being misled when in close proximity to attacking devices. If an attacker can determine the state of the various software components in the cognitive engine, then they may be able to create specific conditions through their own interaction with the wireless medium to cause another cognitive radio to reach a specific state of the attacker's choosing. Such attacks have been demonstrated to be feasible in the area of adversarial learning and adversarial reverse engineering by performing queries on the learning systems [17, 39, 40]. Here, a query refers to an input to the learning system with a corresponding response to the input. In the wireless context, this could be an attacker shifting their own radio parameters and observing their victim's reactions. An attacker might want to launch such an attack to perform denial of service, gain access to a channel their victim was trying to occupy, or alter some specific part of the victim's radio parameters for a very specific gain.

In cognitive engine attacks, an attacker targets a learning radio with the goal of retraining the cognitive engine to put the engine in a state of the attacker's choosing, which will persist after the attacker ceases the attack. This attack will require either long-term interaction between the attacker and a user or repeated shorter-term interactions. Due to the complex nature of the software in a cognitive engine, as well as the current lack of standardization, the term cognitive engine attack is ambiguous. An example of a cognitive engine attack is the case of an attacker targeting a classifier that decides how to classify signals for identifying primary users. An attacker can construct signals to move the decision boundary of the classifier to where they want it to disrupt the cognitive engine's ability to correctly identify primary users and other types of users. In [43] and [52], Clancy et al. describe how an attacker can craft such a signal. In [52], the authors also describe how to defend against such attacks, advocating for the use of the X-means* algorithm over K-means. In [5], Bahrak et al. formulate optimal attacks and defenses when considering cognitive engine algorithms for optimal channel selection. Specifically, two attack strategies, myopic [5] and softmax [5], are analyzed for security, with attack and defense phrased as optimization problems for attackers and defenders to solve, respectively. These are just specific cognitive engine attacks, there are many techniques that can be used in cognitive engines, and all of them must be analyzed by security researchers if we are to ensure these systems operate with minimal obstruction.

Many researchers see this move to more automation through intelligent algorithms as a trend that will eventually make its way to the upper levels of the protocol stack [15, 32, 37, 41]. As algorithms take over operation and optimization of many different layers and protocols, we should expect to see better and better performance in general, since these algorithms will be much more capable and available, with more data to use in optimizing performance than we have now. But these performance gains do not necessarily come without cost.

Performance boosts from cognition may bring new attack vectors, if designers and implementers do not take care to consider the relevant security issues. Most machine learning algorithms are not hardened against attackers, and researchers implicitly assume a non-adversarial environment [25, 48, 49, 55, 64], which cannot always be expected to be the case in the wireless environment. As intelligence creeps further up the protocol stack care must be taken to harden the security of these systems that we are handing control over to; otherwise, attacks targeting the decision-making systems could cripple the future communications infrastructure. Securing machine learning algorithms remains an open research area that should be of great interest to researchers interested in securing CRs and dynamic spectrum sharing.

25.5.6   *Cross-Layer Attacks*

Cross-layer attacks utilize two attack vectors to provide attackers with a unified attack that is either more powerful than the two executed separately, more cost effective, less likely to be detected, or a combination of the three [33, 60]. This class of attacks encompasses an immense number of attacks, since combining any of the two attacks previously discussed together constitutes a cross-layer attack. Thus, we will only consider a few examples here.

The Lion cross layer attack [33] leverages a primary user emulation attack at the physical layer in conjunction with an attack on the transmission control protocol (TCP) at the transport layer to create a cheaper denial of service attack. TCP is sensitive to high variations in delay and bandwidth, so transmission interruption can cause various problems. If an attacker causes a temporary disruption to service by performing a PUE attack, the TCP layer will be unaware of the interuption and will continue to send data to be queued for transmission at lower layers. These outstanding TCP segments pile up and eventually overflow the queue, requiring the transport layer to re-send data. Furthermore, TCP keeps a retransmission timer for each outstanding segment whose value is set using round-trip time measurements. If the retransmission timer times out, then the segment is considered lost and is retransmitted. With each failed retransmission, the retransmission timer doubles in length. The attacker leverages these properties of TCP and only emulates the primary user at select intervals to cause an accumulation in delay, saving the attacker power. Thanks to standards like IEEE 802.22, these retransmission timers can be easily looked up by attackers [2]. The Lion attack can be mitigated by either solving the primary user emulation problem, which we have discussed in detail, or by augmenting the TCP protocol to be more delay tolerant. The authors in [33] suggest the alternative Freeze-TCP [53] protocol designed for TCP use in mobile environments.

In [60], Wang et al. describe two possible cross-layer attacks using attacks in the physical layer (PHY) and medium access control (MAC) layers, respectively. The first of these attacks aims to reduce channel utilization by performing a PUE attack at the PHY layer and simultaneously performing a denial of service attack on the MAC layer via either exploiting small back-off windows (as in Lion) or pure jamming. These attacks are synergistic in nature. Performed independently, they can achieve the same ends, but when used together, they achieve those ends more efficiently. The second attack described aims to cause a secondary user to interfere with a primary user. To achieve cross-layer attacks, the attacker(s) deceive the secondary user into believing the primary user is not present through attacks on a data fusion system or channel manipulation. Then attackers forward data to the secondary users so they have data to transmit that can interfere with the primary user. The secondary user forwards the data on and interferes with the primary user. This attack can be particularly effective if the secondary users are selected to be nearby the primary user. To help mitigate the effects of these cross-layer attacks, Wang et al. suggest creating trust evaluation systems in both the PHY and MAC layers that evaluate and update other secondary users' trust values. In addition, these two trust evaluation systems would communicate with a trust fusion component that combines the trust evaluations at each layer together so that cross-layer attacks can be identified. The authors found via simulation that this system was effective at repelling the combined effect, finding that the attacker's optimal choice was to perform a single attack, which has less of an effect than cross-layer attacks.

25.6 Conclusion

Spectrum sharing brings together many disciplines, creating a security environment that requires knowledge spanning seemingly disparate security areas. Researchers and implementers must gather information from radio fingerprinting, machine learning, protocol design, network architecture, databases, signal analysis, and general purpose security to ensure their work is of the highest quality. The security issues associated with all of these areas is staggering, not to mention the security issues caused by the convergence of all of these areas. With data flowing between so many users, verification and authentication must be performed to ensure maliciously incorrect data is not entered into secondary user systems as well as infrastructure systems.

Spectrum sharing is still in its infancy; we have the opportunity to ensure it flourishes and is secured from the beginning. This is an opportunity that should not be passed by.

References

[1] Part 11: Wireless LAN medium access control (MAC) and physical layer (PHY) specifications amendment 5: Television whitespaces (TVWS) operation, 2013.

[2] Part 22: Cognitive wireless ran medium access control (MAC) and physical layer (PHY) specifications: Policies and procedures for operation in the TV bands amendment 1: Management and control plane interfaces and procedures and enhancement to the management information base (MIB), 2014.

[3] FCC OET Bulletin No. 69. Longley-Rice methodology for evaluating TV coverage and interference. Technical report, FCC, February 2004.

[4] Edward W. Knightly Peter Ecclesine Adriana B. Flores, Ryan E. Guerra, and Santosh Pandey. IEEE 802.11af: A standard for TV white space spectrum sharing. *IEEE Communications Magazie*, 51(10):92–100, October 2013.

[5] Benham Bahrak and Jung-Min "Jerry" Park. Security of spectrum learning in cognitive radios. *SK Telecom Telecommunications Review*, 22(6):850–864, December 2012.

[6] Abid Ullah Jung-Min "Jerry" Park Jeffery Reed Behnam Bahrak, Sudeep Bhattarai, and David Gurney. Protecting the primary users' operational privacy in spectrum sharing. In *2014 IEEE International Symposium on Dynamic Spectrum Access Networks*, pages 236–247. IEEE, April 2014.

[7] S. ; Chakrabarti-I. ; Pathak S.S. Bera, D. ; Maheshwari. Decentralized cooperative spectrum sensing in cognitive radio without fusion centre. In *2014 Twentieth National Conference on Communications (NCC)*, pages 1–5, February 2014.

[8] C. Bishop. *Pattern Recognition and Machine Learning*. Springer, 2006.

[9] Davide Zanetti Boris Danev, and Srdjan Capkun. On physical-layer identification of wireless devices. *ACM Computing Surveys*, 45(1), November 2012.

[10] E. Stuntebeck C. Clancy, J. Hecker, and T. O'Shea. Applications of machine learning to cognitive radio networks. *IEEE Wireless Communications*, 14(4):47–52, August 2007.

[11] Kran Challapali, Carlos Cordeiro, and Dagnachew Birru. IEEE 802.22: An introduction to the first wireless standard based on cognitive radios. *IEEE Journal of Communications*, 1(1):38–47, April 2006.

[12] Ruiliang Chen and Jung-Min Park. Ensuring trustworthy spectrum sensing in cognitive radio networks. In *1st IEEE Workshop on Networking Technologies for Software Defined Radio Networks*, pages 110–119, September 2006.

[13] Ruiliang Chen, Jung-Min Park, and Kaigui Bian. Robust distributed spectrum sensing in cognitive radio networks. Technical report, TR-ECE-06-07, Department of Electrical and Computer Engineering, Virginia Tech, July 2006.

[14] Ruiliang Chen, Jung-Min Park, and Jeffrey H. Reed. Defense against primary user emulation attacks in cognitive radio networks. *IEEE Journal on Selected Areas in Communications*, vol., 26(1), January 2008.

[15] T. Charles Clancy, and Nathan Goergen. Security in cognitive radio networks: Threats and mitigation. *Cognitive Radio Oriented Wireless Networks and Communications*, 2008.

[16] Federal Communications Commision. Facilitating opportunities for flexible, efficient, and reliable spectrum use employing spectrum agile radio technologies. *ET Docket*, (0, pages 3–108, December 2003.

[17] Nilesh Dalvi, Pedro Domingos, Sumit Sanghai Mausam, and Deepak Verma. Adversarial classification. In *Conference on Knowledge Discovery and Data Mining (KDD)*. ACM, August 2004.

[18] Boris Danev and Srdjan Capkun. Transient-based identification of wireless sensor nodes. In *International Conference on Information Processing in Sensor Networks (IPSN)*, pages 25–36. IEEE, April 2009.

[19] Boris Danev, Heinrich Luecken, Srdjan Capkun, and Karim El Defrawy. Attacks on physical-layer identification. In *WiSec*, March 2010.

[20] High Durrant-Whyte and Mike Stevens. Data fusion in decentralised sensing networks. *Proceeding on Information Fusion*, pages 302–307, 2006.

[21] Tingting Jiang et al. On the limitation of embedding cryptographic signature for primary transmitter authentication. *IEEE Wireless Communications Letters*, 1(4), August 2012.

[22] FCC. Revision of parts 2 and 15 of the commissions rules to permit unlicensed national information infrastructure (u-nii) devices in the 5 GHZ band. Technical report, FCC, 2003.

[23] FCC. Small entity compliance guide part 15 TV bands devices. Technical report, FCC, April 2012.

[24] Chowdhury Shahriar, Chang-Tien Lu, Wenjing Lou, Feng Chen, Qiben Yan, and T. Charles Clancy. On passive wireless device fingerprinting using infinite hidden Markov random field.

http://www.contrib.andrew.cmu.edu/ fchen1/device-fingerprinting-detection.pdf.

[25] T.W. ; Bin Le Feng Ge ; Qinqin Chen ; Ying Wang ; Bostian, C.W. ; Rondeau. Cognitive radio: From spectrum sharing to adaptive learning and reconfiguration. In *IEEE Aerospace Conference*, pages 1–10. IEEE, March 2008.

[26] FCC Spectrum Policy Task Force. Report of the spectrum efficiency working group. Technical report, FCC, November 2002.

[27] Andrey Garnaev and Wade Trappe. Bandwidth scanning involving a bayesian approach to adapting the belief of an adversary's presence. In *2014 IEEE Conference on Communications and Network Security (CNS)*, pages 35–43, IEEE, October 2014.

[28] Clancy T.C. Newman T.R. Goergen, N. Physical layer authentication watermarks through synethetic channel emulation. In *IEEE Symposium on New Frontiers in Dynamic Spectrum*, April 2010.

[29] Google. Google spectrum database.

[30] Nikhil Gulati, Rachel Greenstadt, Kapil R. Dandekar, and John M. Walsh. Gmm based semi-supervised learning for channel-based authentication scheme. In *IEEE 78th Vehicular Technology Conference*, September 2013.

[31] Janine S. Hiller and Jung-Min Park. Spectrum sharing and privacy: A research agenda. In *The Research Conference on Communication and Internet Policy*, September 2014.

[32] Joseph Mitola Iii and Gerald Q. Maguire Jr. Cognitive radio: Making software radios more personal. *IEEE Personal Communication*, August 1999.

[33] Olga Leon Juan Hernandez-Serrano and Miguel Soriano. Modeling the lion attack in cognitive radio networks. *EURASIP Journal on Wireless Communications and Networking*, August 2011.

[34] A. A. Beex T. Charles Clancy Vireshwar Kumar, Jung-Min (Jerry) Park, Jeffrey H. Reed and Benham Bahrak. Security and enforcement in spectrum sharing. *Proceedings of the IEEE*, 192(3), March 2014.

[35] P.R. Duley K.I. Talbot and M.H. Hyatt. Specific emitter identification and verification. *Technology*, 2003.

[36] Kyung K. Bae, Jung-sun Um, Chad M. Spooner, Kyouwoong Kim, Ihsan A. Akbar, and Jeffrey H. Reed. Cyclostationary approaches to signal detection and classification in cognitive radio. In *IEEE International Symposium on New Frontiers in Dynamic Spectrum Access Networks (DySPAN)*, pages 212–215, April 2007.

[37] Bin Le, Thomas W. Rondeau, and Charles W. Bostian. Cognitive radio realities. *Wireless Communications and Mobile Computing*, 7(9):1037–1048, May 2007.

[38] Yao Liu, Peng Ning, and Huaiyu Dai. Authenticating primary users' signals in cognitive radio networks via integrated cryptographic and wireless link signatures. In *2010 IEEE Symposium on Security and Privacy (SP)*. IEEE, May 2010.

[39] Daniel Lowd. Good word attacks on statistical spam filters. In *In Proceedings of the Second Conference on Email and Anti-Spam (CEAS)*, 2005.

[40] Daniel Lowd and Christopher Meek. Adversarial learning. In *Conference on Knowledge Discovery and Data Mining (KDD)*, August 2005.

[41] David Maldonado, Bin Le, Akilah Hugine, Thomas W Rondeau, and Charles W Bostian. Cognitive radio applications to dynamic spectrum allocation: A discussion and an illustrative example. In *New Frontiers in Dynamic Spectrum Access Networks, 2005. DySPAN 2005. 2005 First IEEE International Symposium on*, pages 597–600, IEEE, 2005.

[42] R; Moscibroda T; Bahl P; Murty, R; Chandra. Senseless: A database driven white spaces network. *IEEE Transactions on Mobile Computing*, 11(2):189–203, September 2010.

[43] Timothy R. Newman and T. Charles Clancy. Security threats in cognitive radio signal classifiers.

[44] Nam Tuan Nguyen, Rong Zheng, and Zhu Han. On identifying primary user emulation attacks in cognitive radio systems using nonparametric bayesian classification. *IEEE Transactions on Signal Processing*, 60(3), March 2012.

[45] Keith E. Nolan, Paul D. Sutton, and Linda E. Doyle. Cyclostationary signatures in practical cognitive radio applications. *IEEE Journal on Selected Areas in Communications*, 26(1):13–24, January 2008.

[46] Barathram Ramkumar. Automatic modulation classification using cyclic feature detection. *IEEE Circuits and Systems Magazine*, 9(2):27–45, 2009.

[47] Ankit Singh Rawat, Priyank Anand, Hao Chen, and Pramod K. Varshney. Collaborative spectrum sensing in the prescence of byzantine attacks in cognitive radio networks. *IEEE Transactions on Signal Processing*, 59(2), February 2011.

[48] Christian James Rieser. Biologically Inspired Cognitive Radio Engine Model Utilizing Distributed Genetic Algorithms for Secure and Robust Wireless Communications and Networking. PhD thesis, Virginia Polytechnic Institute and State University, Blacksburg, Virginia, August 2004.

[49] T.W. ; Bostian C.W. ; Gallagher T.M. Rieser, C.J. ; Rondeau. Cognitive radio testbed: further details and testing of a distributed genetic algorithm based cognitive engine for programmable radios. In *IEEE Military Communications Conference, 2004. MILCOM 2004*, vol. 3. IEEE, 2004.

[50] Michael J. Riezenman. Cellular security: Better, but foes still lurk. *IEEE Spectr*, 37(6):39–42, June 2000.

[51] Prasant Mohapatra, Shaxun Chen, and Kai Zeng. Hearing is believing: Detecting mobile primary user emulation attack in white space. *IEEE Transactions on Mobile Computing*, 12(3):401–411, December 2013.

[52] Awais Khawar, T. Charles Clancy, and Timothy R. Newman. Robust signal classification using unsupervised learning. *IEEE Transactions on Mobile Computing*, 10(4):1289–1299, April 2011.

[53] D.S. Phatak T. Goff, J. Moronski and V. Gupta. Freeze-tcp: A true end-to-end tcp enhancement mechanism for mobile environments.pdf. In *Nineteenth Annual Joint Conference of the IEEE Computer and Communications Societies (INFOCOM)*, pages 1537–1545, March 2000.

[54] Xi Tan, Kapil Borle, Wenliang Du, and Biao Chen. Cryptographic link signatures for spectrum usage authentication in cognitive radio. In *WiSec*, 2011.

[55] Alexander M. Wyglinski, Arvin Agah, Joseph B. Evans, Tim R. Newman, Brett A. Barker, and Gary J. Minden. Cognitive engine implementation for wireless multicarrier transceivers. *Wireless Communications and Mobile Computing*, 7(9):1129–1142, November 2007.

[56] Pramod K. Varshney. *Distributed Detection and Data Fusion*. Springer-Verlag, 1996.

[57] Jung-Min "Jerry" Park, Vireshwar Kumar, and Kaigui Bian. Blind transmitter authentication for spectrum security and enforcement. In *SIGSAC Conference on Computer and Communications Security*, pages 787–798. Conference on Computer and Communications Security, 2014.

[58] T Charles Clancey, Vireshwar Kumar, Jung-Min "Jerry" Park, and Kaigui Bian. PHY-layer authentication by introducing controlled inter symbol interference. In *IEEE Conference on Communications and Network Security (CNS)*. IEEE, 2013.

[59] Marco Gruteser, Vladimir Brik, Suman Banerjee, and Sangho Oh. Wireless device identification with radiometric signatures. In *14th ACM International Conference on Mobile Computing and Networking*, pages 116–127, 2008.

[60] Wenkai Wang, Yan Sun, Huseng Li, and Zhu Han. *Cross-Layer Attack and Defense in Cognitive Radio Networks*. IEEE Globecom, 2010.

[61] GNU Radio Website, accessed May 2015.

[62] Yongle Wu, K. J. Ray, Liu Beibei Wang, and T. Charles Clancy. Anti-jamming games in multi-channel cognitive radio networks. *IEEE Journal on Selected Areas in Communications*, 30(1), January 2012.

[63] Tevfik Yucek and Huseyin Arslan. A survey of spectrum sensing algorithms for cognitive radio applications. *IEEE Communications Surveys and Tutorials*, 11(1):116–130, 2009.

[64] Hong Jiang Yuqing Huang, and Jiao Wang. Model of learning inference and decision-making engine in cognitive radio. In *2010 Second International Conference on Networks Security Wireless Communications and Trusted Computing (NSWCTC)*, vol. 2, pages 24–25, April 2010.

[65] S. Anaand Z. Jin and K.P. Subbalakshmi. Detecting primary user emulation attacks in dynamic spectrum access networks. In *IEEE International Conference on Communications (ICC)*, pages 1–5, June 2009.

[66] Sepideh Zarrin and Teng Joon Lim. Belief propagation on factor graphs for cooperative spectrum sensing in cognitive radio. In *3rd IEEE Symposium on New Frontiers in Dynamic Spectrum Access Networks (DySPAN)*, pages 1–9. IEEE, October 2008.

[67] Chao Chen, Zesheng Chen, Todor Cooklev, and Carlos Pomalaza-Raez. Modeling primary user emulation attacks and defenses in cognitive radio networks. In *IEEE 28th International Performance Computing and Communications Conference (IPCCC)*. IEEE, December 2009.

[68] Pan Jianguo and Zhai Xuping. Energy-detection based spectrum sensing for cognitive radio. In *IET Conference on Wireless, Mobile and Sensor Networks*, pages 944–947. IET, December 2007.

Chapter 26

Pragmatic Security Issues and Solutions for Spectrum Sharing in Wireless Networks

Lei Li and Chunxiao Chigan

CONTENTS

26.1 Introduction

The explosions in the number of mobile users and the use of diverse bandwidth-hungry wireless applications have resulted in an exponential growth in the demand for wireless access to the Internet. As a consequence, not only the volume of mobile data traffic is overwhelming the capacity of existing wireless communication networks, but the required high data rates also drains the limited radio resources. The spectrum sharing paradigm has drawn considerable attention for improving the network performance such as the network capacity and data rate. A variety of networks have been designed with spectrum sharing capability, among which the cognitive radio network (CRN) is a promising technique that has been widely researched.

The dynamic spectrum allocation/access (DSA) mechanism enables the cognitive radio network (CRN) to share the spectrum resources that assigned to the licensed users (primary users or PUs) [1]. Attributed to the spectrum sharing strategy, the spectrum utilization efficiency is improved drastically. To realize the DSA mechanism, the CRN users (secondary users or SUs) are necessitated to be aware of their environment, and configure their radio parameters (such as operating frequency, bandwidth, transmission power etc.) to accommodate the current needs of surrounding environment, which are called spectrum sensing and spectrum mobility [2].

Nevertheless, the cognitive features such as spectrum sensing and spectrum mobility induce more vulnerabilities of CRNs to the specific security threats. For example, Primary User Emulation Attack (PUEA) [3][4], Spectrum Sensing Data Falsification (SSDF) attack [5][6] and Lion Attack [7][8] and so on. These issues have been widely studied theoretically. However, since the cognitive featured networks (such as cognitive long term evolution (LTE) network, [9] etc.) are being commercialized in the market [10][11], new security issues relevant to pragmatic aspects of CRNs have been explored recently. The major security threats and vulnerabilities in the future cognitive spectrum sharing wireless networks can be summarized as follows:

(1) Security threats relevant to the deployment of relay

Due to the requirement of interference avoidance with PUs, the SUs transmission power is restrained to be under a threshold [12][13]. Therefore, relay nodes are necessary for forwarding the SUs message to the destination [14–17]. For efficient implementation, relay nodes are proposed to be selected from the SUs with high performance [18]. However, this provides the opportunity for the adversaries to compromise SUs who are selected as relay nodes to launch kaleidoscopic attacks.

The pollution attack described in [19] is one example caused by the relay deployment. The compromised relay nodes may manipulate the messages that they received before forwarding them to the destination. Consequently, the detection error rate at the destination will be increased due to the pollution attack. Authors in [20] introduce a routing toward primary user (RTPU) attack. The adversaries with RTPU attack intend to compromise the routing protocol and force the network to select the route with relays that are close to the PUs. In this sense, the PUs might be interfered by the SUs.

(2) Security threats relevant to the ineligible users

Most of existing works on the dynamic spectrum access technique concentrate on the improvement of the network performance (e.g., network throughput and channel utilization, etc.). There are few works paying attention to the eligibility of SUs' spectrum access requests. If the ineligible SUs gain the access to the network with privilege, many advanced severe attacks, such as PUEA, SSDF, etc., can be launched. This will cause dramatic degradation of network performance. Hence, it is crucial to grant the spectrum resource only to the eligible users, which are referred to as network access control (NAC) mechanism [21].

(3) Security threats in the network coexistence scenarios

The principle of the CRNs is to coexist with the primary network. However, with the increasing demands on the wireless network services, a large number of cognitive featured networks are being developed. In this sense, the coexistence among the cognitive featured networks is inevitable [22]. Different from the coexistence scenarios with PUs, the priority of the coexisting networks may be identical. Therefore, these networks compete with each other to obtain the privilege of spectrum utilization. As a consequence, the fairness is a critical aspect in the design of coexistence mechanisms.

Intuitively, the deployment of a coordinator who is responsible for the management of coexistence of networks is an easy way to achieve the fairness. However, this may result in a lot of message exchanges between the coordinator and each network, which would increase the network overhead. The high expense is another bottleneck of coordinator-based mechanisms [23]. Alternatively, the non-coordinator-based mechanisms are provided in [24][25]. Sophisticated strategies or protocols are proposed for each coexisting network to achieve the fairness. However, the high network latency and implementation complexity due to the sophisticated approaches impedes the deployment of non-coordinator-based mechanisms.

In this chapter, the aforementioned three categories of security threats will be discussed in detail, and the corresponding representative solutions will be elaborated.

26.2 Security Issues Relevant to the Relaying Strategy

Relaying strategy has been widely studied and adopted in the design of CRNs in order to improve the network performance. There are mainly three types of relaying strategies, i.e., amplitude and forward (AF), compress and forward (CF), and decode and forward (DF). With the AF or CF strategy, the relay node amplifies or compresses the signals it received, respectively, and directly forwards them to the destination. On the contrary, the relay node with DF strategy decodes its received signal first and encodes again before forwarding to the destination. No matter which strategy is adopted, the signal transmitted from source node will be processed by the relay node. In the scenarios with relaying strategy, the adversaries can easily launch attacks by compromising the relay nodes due to the openness of the relay node mentioned above.

In this section, we first introduce the necessities of relaying strategy in the design of CRNs. Then, several security threats relevant to the relay strategy will be discussed in the section 2.2. Finally, we will elaborate some approaches countering against the aforementioned attacks.

26.2.1 Necessities of Relay Nodes In the Cognitive Radio Networks

Considering the implementation of the CRNs in reality, the relaying strategy is realized as a necessary and effective way to enhance the network performance. It can be described from the following three aspects:

(1) *Increasing the network serving capacity*

The power consumption of the portable devices draws much attention recently. Reducing the transmission power is one of the efficient schemes. Consequently, the transmission range of the portable devices is very limited. This restrains the serving capacity of the base station (the number of users can be served by a base station) in the single-hop-based centralized networks. Relaying strategy is an effective method for extending the user devices' transmission range by deploying the relay nodes. For example, the IEEE 802.22 standard [26–27] (the first CR-based wireless regional area network standard) specifies the CRN as a centralized network. In order for serving more users, the transmission range of the CRN base station is supposed to be around 30 km to 40 km, with a high transmission power (about 4 W). However, due to the transmission power limit of user device, the messages sent from user devices cannot reach the CRN base station with only one hop. Hence, the relay nodes are necessary for forwarding the users' messages to the CRN base station.

(2) *Improving the network performance*

Cooperative communication was proposed to achieve the spatial and multiuser diversities. Wherein the relay nodes are deployed in the network and act as a virtual antenna array to help the source nodes forward their messages to the destination nodes. In this sense, the network throughput and transmission reliability can be increased. A novel cooperative relay-based CRN paradigm was proposed in the literature recently [28]. Figure 25.1 shows an example achieving high network throughput credited to the spectrum diversity.

Figure 25.1(a) depicts the original network model without cooperative relay deployment and Figure 25.1(b) and 25.1(c) shows two consecutive time slots with deployment of cooperative relay. The source node allowed to occupy channels 1 and 4 wants to send messages to the destination node with the available channels 1 and 6. However, the relay node is allowed to access channels 4 and 6. As shown in Figure 25.1(a), the transmission on the direct link between source and destination nodes cannot satisfy the demanded data rate. However, with the assist of the relay node shown in Figure 25.1(b) and 25.1(c), the different transmitted messages are allocated on the channel 1 and channel 4 (this can be done by orthogonal frequency division multiple access (OFDM) technique) by the source node and sent to the destination

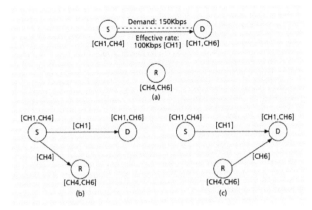

Figure 26.1: Cooperative relay-based CRN [15].

node and relay, respectively, at the first time slot. Thus, the relay node sends the received messages to the destination node on channel 6 at the second time slot. It is obvious that the network data rate is increased by parallel transmission through two independent links.

(3) *Interference control by relays*

The CRNs are allowed to access into the spectrum resources assigned to the primary networks while avoiding interference to them. This restrains the SUs to transmit with low power, such that the summation of power of SUs' signals arrived at the primary receiver is lower than the interference temperature. Therefore, the relay nodes are needed to extend the SUs' transmission coverage to reach the destination.

The network performance can be improved by adoption of relaying strategy, as described above. On the contrary, it also brings in a variety of specific attacks that may severely destroy the network. Several typical attack models relevant to the relaying strategy are introduced as follows.

26.2.2   Attack Models

26.2.2.1   Sybil Attack

In the Sybil attack, an individual malicious entity is able to masquerade multiple identities for the purposes of malicious behaviors, such as selfishly obtaining more spectrum resources. Figure 25.2 illustrates an example of the network in the presence of the Sybil attack. The user in dark red is the entity who is launching the Sybil attack (we call it *Sybil entity* in this chapter). The users in purple denote the *Sybil nodes* forged by the Sybil entity. The Sybil entity and Sybil nodes will pretend to be the independent individuals to communicate with the base station in order to achieve their purpose. The Sybil attack has appeared in many forms in both academic work and in the real world [29–32].

In the relay featured CRNs, the adversaries may attempt to compromise the relay node. Once the compromised relay node is asked to forward messages for the users, it may generate a large number of fake messages and transmit those messages received from users along with the generated fake messages together to the destination.

This Sybil attack may cause a severe problem due to the cooperative operation in the CRNs. For example, the cooperative spectrum sensing mechanism is adopted in order to deal with the hidden terminal problem. In the cooperative spectrum sensing, a fusion center collects the SUs

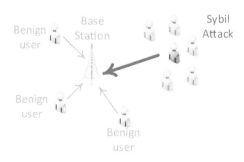

Figure 26.2: Illustration of Sybil attack.

individual spectrum sensing reports and make the final decision of the channel status according to the fusion rules (such as OR rule, AND rule, etc.). Due to the limit of transmission range, the relay nodes are used to forward the cognitive users' sensing report to the fusion center. In this sense, the compromised relay node by the Sybil entity may generate a large number of fake sensing reports and send those sensing reports from benign users along with the fake sensing reports to mislead the fusion center to make wrong decision. This may lead to either the low spectrum utilization efficiency or interference to the PUs.

26.2.2.2 Pollution Attack

The pollution attack is another attack caused by the relaying strategy. The adversary who is launching the pollution attack intentionally manipulates the signals sent from the source node [33]. It is obvious that the pollution attack can be easily launched by the compromised relay node, since the relay node is allowed to process the signals (as shown in AF, DF, and CF relaying strategies).

The manipulation of the messages can be described as follows. The compromised relay node may intentionally "add" malicious signals on top of the signals to be forwarded. As a consequence, the signal received at the destination can be represented as the addition of signal sent from the source node and the interference generated by the adversaries. In this sense, this kind of manipulation can be considered as an additive channel. Additive white Gaussian noise (AWGN) channel is a widely used channel model in the communication theory. Different from the AWGN channel, the "noise" generated by the additive channel caused by the compromised relay nodes may follow any arbitrary distribution, while the Gaussian distribution is assumed in the AWGN channel. The authors in [19] propose a novel forward error correction (FEC)-enabled network coding framework to reduce the effect from this kind of pollution attack.

26.2.2.3 Relay Attack toward Primary Users

The adversaries with the attacks introduced above attempt to reduce the network performance of the CRN. However, since the CRNs are allowed to share the spectrum resources and even cooperate with the primary networks, it is also possible for the adversaries to attack the primary network. Basically, there are two types of attacks toward the PUs caused by the relaying strategy.

26.2.2.4 Routing toward Primary User Attack

First, the adversary attempts to attack the PU indirectly by taking advantage of SUs, for example, the routing toward primary user attack (RPUA) [20]. With RPUA, the adversary always claims that they have optimum route with low costs. In this case, the other honest users will route data packets through that adversary. Then the adversary will select its next hop that is closest to a PU's footprint

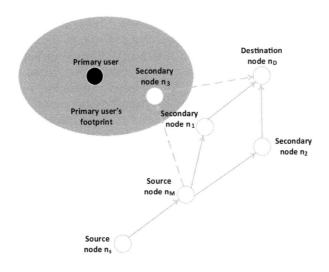

Figure 26.3: Routing toward a primary user attack [20].

(defined as the PU's interference region in which any SU locates will stop transmission to protect the PU). As a result, the transmission of node nearby the footprint may interfere the PU with a high probability.

Figure 25.3 shows an example of RPUA [20]. As shown in this figure n_S, n_D, n_1, n_2, and n_3 are all SUs. Nodes n_S and n_D are source and destination nodes, respectively. The shaded region is the footprint of a PU. Since SU n_3 is inside this region, it will not transmit although it receives anything needed to be forwarded. Considering the scenario that the source node n_S wants to transmit some packets to the destination node n_D, since the malicious node n_M claims that it has the shortest path to destination node n_D, source node n_S will send all its packets to the node n_M. Then, node n_M tries to forward the data to node n_1, which is closer to the PU compared to node n_2, even though node n_M knows clearly that node n_2 can also help to forward the packets.

According to the description above, the adversary with RPUA may interfere with the PU with a high probability. Nevertheless, due to the dynamic environment (such as fast fading), the footprint of PU may change. For example, as shown in the Figure 25.3, the footprint of PU may be enlarged, such that the node n_1 may be covered in the footprint. In this sense, the node n_1 will stop its transmission, and there will be no interference to the PU. However, in this case, the node n_S needs to select another available route and retransmit the messages, which may increase the latency of the CRN. A believe propagation-based RPUA defending mechanism is proposed in [20]. The detailed description of this approach is referred to section 2.3.1.2.

26.2.2.5 *Cooperative Networking Attack*

Since the relay strategy is able to improve the network performance, as described above, a PU may select a SU with high performance as its relay [34]. In this sense, the PU has to lease its own spectrum resources to that SU for a fraction of time as a reward. If all the SUs selected as the relay for PU are well-behaved, both PU and SU can benefit from the cooperation. However, when there are some dishonest SUs, the performance of both primary network and secondary network will be degraded. Specifically, the following possible security issues arising in CRN need to be addressed [34].

To secure the cooperation described above, the primary system needs to meet the following basic security requirements: confidentiality, integrity, and authentication, which can be provided by suitable cryptography approaches (e.g., encryption and decryption, digital signature, authentication, message authentication code, etc.). However, a legitimate SU may be compromised and misbehaves when it is selected to cooperate with the PU. A dishonest SU may not obey the cooperation rule during the cooperative transmission to pursue more self-benefits, e.g., it may transmit its own packets instead of relaying the packets from the PU. Moreover, considering the mobility of SUs, the malicious or dishonest SUs may misbehave at one place then move to other places. Since there is no record of the past behaviors in a geographical area, these users can have the same opportunity to be selected to cooperate with the PU, and then continue to harm the system. In a nutshell, without considering these security threats, the PU may choose an untrustworthy SU for cooperation, which may cause the failure of the cooperation and degrade its quality of service (QoS).

26.3 Detection And Mitigation of Relay Featured Security Threats

In this section, the detection and mitigation mechanisms for the attacks described in the Section 25.2.2 are elaborated. These approaches can be mainly categorized as reputation-based scheme, network coding-based schemes, on game theory-based schemes. The reputation-based scheme is a straightforward way to detect the users with abnormal behaviors. In the reputation-based schemes, the reputation value is calculated and maintained for each user. This value is then used to make decisions on the user's validity. Most of the schemes for the calculation of reputation values are based on the statistics of a user's behavior [38]. Therefore, to achieve more accurate reputation value, the number of samples should be large enough. Hence, a long time is needed to collect the user's behavior. Intuitively, this method would lead to severe network latency. In order to reduce the network latency, the network coding-based security threat defending mechanisms were proposed. Among the network coding-based approaches [35–37], the physical-layer network coding schemes are shown to be able to provide the attack resilience with low network latency. In the physical-layer network coding design, there are a variety of factors need to be considered-such as transmission power, etc. In order to get the optimal solution under those realistic conditions, game theory-based mechanisms were proposed. The details of each type of mechanism will be elaborately introduced in the following subsections.

26.3.1 Reputation-Based Secure Network Design

To ensure the robustness of spectrum sensing, a reputation-based mechanism for identifying misbehavior and mitigating their harmful effect on the sensing performance is studied in [52]. In the reputation system, a reputation value, indicating the reliability of a message sent from a user, is calculated according to that user's historical behaviors. The reputation value can be adopted to decide with which users to cooperate (the users with higher reputation values) and which users to avoid (the users with lower reputation values). Once the reputation values are calculated, the network nodes can be classified and the network management policies can be enforced.

The major issue of a reputation system is the representation of reputation for a network node, i.e., how the reputation is built and updated, and for the latter, how the ratings of others are considered and integrated. In the following, we will introduce several reputation systems that are used to detect and mitigate the misbehavior users in the CRNs.

26.3.1.1 Bayesian Reputation Computation-Based Secure CRN

The Beta Bayesian framework

In the basic Bayesian reputation systems, the reputation of a node is based on the collection of

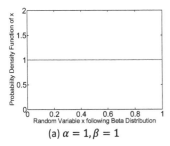
(a) $\alpha = 1, \beta = 1$

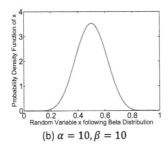
(b) $\alpha = 10, \beta = 10$

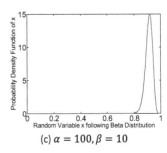
(c) $\alpha = 100, \beta = 10$

Figure 26.4: Probability density function of variable following Beta distribution with different parameters.

ratings about this node maintained by the others. For example, node i models the behavior of node j with the following principles. Node i considers that node j misbehaves with a probability θ. Since the probability θ is unknown, node i considers that θ as a random variable. Therefore, the goal of a Bayesian-based scheme is to estimate the probability density function (PDF) of θ with current and historical observations. Usually, θ is considered to be following the Beta distribution. The Beta PDF can be used to describe probability distributions of binary events. There are two parameters in the representation of Beta distribution, i.e., α and β. The α and β corresponds to how frequently that two events happen. The shape of the Beta function tends to be a Delta function with more samples (observations). Therefore, the variable θ can be estimated eventually as the value indicated by the Delta function corresponds to the PDF of θ. The PDF of variable θ can be expressed as $f(\theta; \alpha, \beta) = \frac{1}{B(\alpha, \beta)} \theta^{\alpha-1} (1-\theta)^{\beta-1}$. The PDF of θ with different parameters are shown in the Figure 25.4.

In the Bayesian algorithm, the parameters α and β are updated according to the historical observations. With enough samples, the ratio of α to β will be converged, which means the shape of the PDF of variable θ is almost stable. In the standard Bayesian process, the α and β are configured as 1 initially, which means the variable θ follows the distribution with beta function $Beta(1, 1)$. In this sense, the θ follows uniform distribution on $[0,1]$ is shown in Figure 25.4(a). This represents there is no knowledge about θ is known. In the Bayesian algorithm-based misbehavior detection system, when a new observation is made, say, with s observed misbehaviors (may be referred to as packet drop) and f observed correct behaviors (referred to as successful packet delivery), the parameters α and β are updated according to $\alpha = \alpha + s$ and $\beta = \beta + f$. Assuming θ is constant, then after a large number (e.g., denoted by n) of observations, $\alpha \sim n\theta$ (in expectation), $\beta \sim n(1-\theta)$ and $Beta(\alpha, \beta)$ becomes close to a Dirac at θ, as shown in Figure 25.4(c). In summary, one node can evaluate another node's reputation value by following steps: (1) assume the reputation value of a node as a random variable following Beta distribution; (2) update the parameters of PDF of the variable (reputation value), i.e., α and β, with the observations; (3) decide the reputation value of the node according to its PDF.

26.3.1.2 Believe Propagation-Based Trust and Reputation System

As described above, node i can evaluate node j, which is adjacent to node i, using the mechanism shown in the Beta reputation system. However, in a multi-hop network, at the path selection stage (relevant to the routing algorithm), the tradition Beta reputation system cannot be adopted directly for the source node to decide whether there is any malicious node in a link between the source and destination nodes. This is because that there are lots of non-adjacent nodes in the middle of source and destination nodes.

In order for the source node to evaluate reputation values of non-adjacent intermediate relay nodes, a believe propagation mechanism is proposed in [20]. The basic idea of this believe propagation bade scheme can be described as following. Along a predetermined path, each node sends the test packets to its next hop (child node). The feedback message, indicating how reliable the child node is, is required from each node to its previous hop (parent node) that has direct connection with it and the source node. The feedback message is calculated and sent at each node from the destination to source along the path sequentially. The feedback message is composed of two terms. One is the local information, and the other is external information. For example, node w is requested to send feedback message to node v. Then the local information can be calculated by the concept of basic Beta reputation algorithm, in terms of how many test packets are successfully received by node w. The external information is the feedback message come from the child node of node w. Eventually, the source node will have the knowledge about the feedback information of all the nodes along the predetermined path. Therefore, the source node can determine which node is misbehaved, according to their feedback information.

26.3.1.3 Reinforcement Learning-Based Trust and Reputation Model

As shown above, the general reputation-based mechanisms consider the behaviors of network nodes are static. In this sense, the reputation value for each network node can be calculated by evaluating their historical behaviors in a statistic manner. However, it is not necessary for a network node to be always honest or malicious for a long period. For example, with the cooperative networking attack in Section 2.2.3.2, the intelligent adversaries may compromise the SUs and intentionally perform better than the other nodes in order to be selected as relay for the primary users with high chance. Once they are chosen, the compromised SUs may attempt to launch the attack on the primary network. In this case, the traditional reputation system may fail due to the adversaries' dynamic behaviors. Hence, it is critical to develop a trust and reputation model that is robust and adaptable to the operating environment in order to increase the detection efficiency of malicious SUs.

In order to deal with the dynamic threat scenario described above, the reinforcement learning-based trust and reputation model was proposed. With the reinforcement learning, the system is able to observe and learn about the static or dynamic operating environment in the absence of guidance. In [39], the Q-learning, a typical reinforcement learning algorithm, was adopted to learn and update the reputation values of network nodes and make final decision on the strategy selection. The Q-learning is an online algorithm in the reinforcement learning, which is able to observe, learn, and act simultaneously in real-time.

In the Q-learning-based trust and reputation systems, the state and action pair (s_t, a_t) is defined. The states represent the environment situations (corresponds to the user i is estimated to be malicious or honest). The action represents the choice taken by the system (corresponds to the user i is claimed as malicious or honest, according to the observation) in order to maximize the benefits (corresponds to the network performance) it gains. By taking one action, the system will gain a corresponding reward immediately, which is called short-term reward. Meanwhile, the environment situation may change, which means the system goes to the next state. The summation of rewards by all possible actions at the new state is defined as long-term reward. There are multiple combinations of state and action with different rewards. The Q-learning-based reputation system learns and collects the states corresponding to the environment situation by taking different actions. Assuming after a long period, the reputation system has perfect knowledge about the environment, it can then choose an optimal action to perform to get the highest reward, including both short-term reward and long-term reward.

The summation of short-term and long-term reward corresponding to a state-action pair is represented by Q-values. The short-term reward gained at time t corresponding to the state s_t is represented as $\gamma_t(s_t)$. The long-term reward $\gamma \cdot \max_{a \in A} Q_t(s_{t+1}, a)$ represents the cumulative rewards received in the future (next state). In the Q-learning algorithm, the Q-value is updated in each itera-

Table 26.1   Q-Value Update Procedural

For each random initial state (s_t, a_t), initialize Q-table: $Q_t(s_t, a_t) \leftarrow 0$
Observe current state s_t
For each time step t:
    Select an action a_t and execute it
    Receive delayed reward $\gamma_{t+1}(s_{t+1})$,
    Observe the new state (s_{t+1}),
    Update the Q-table entry for $Q_t(s_t, a_t)$
    t = t + 1;
End For

tion, as shown in the following equation:

$$Q_{t+1}(s_t, a_t) \leftarrow (1 - \alpha) \cdot Q_t(s_t, a_t) + \alpha \cdot [\gamma_{t+1}(s_{t+1}) + \gamma max_{a \in A} Q_t(s_{t+1}, a)]\, 1. \tag{26.1}$$

The long-term reward factor γ emphasizes on the importance of future rewards. If $\gamma = 1$, the system considers the same weightage for both short-term and long-term rewards. If $\gamma = 0$, the system only considers the short-term reward. The update of the Q-value can be also represented Table 25.1.

Figure 25.5 depicts the comparison of reinforcement learning-based trust and reputation system with the traditional trust and reputation system [39]. According to the simulation results shown in Figure 25.5, the detection performance of reinforcement learning-based trust and reputation system outperforms the traditional model. The increased in detection performance is due to the capacity of reinforcement learning-based system to learn and re-learn from its dynamic operating environment. However, it is noticed that there is only slight performance improvement in reinforcement learning-based mechanism when the percentage of malicious users are lower and higher than a threshold. This is because that when the malicious users is low, the reinforcement learning-based mechanism would have less previous learned knowledge of each node's behavior, hence, its performance is similar to that of the traditional schemes. Similarly, when the number of malicious SUs is high, the learning process of reinforcement learning-based scheme would masquerade the traditional one since higher a number of malicious SUs indicates easier detection.

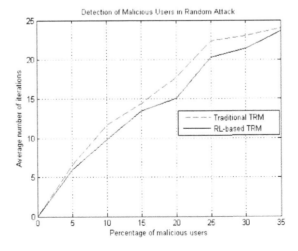

Figure 26.5: Performance comparison of RL-based and traditional TRM [39].

26.3.1.4 Indirect Reciprocity Game-Based Trust and Reputation Model

[40] provides an indirect reciprocity game modeling-based trust and reputation system for secure wireless networks. In this chapter, the author states that the trust/reciprocity mechanism is a powerful tool to improve the security and stimulate cooperation in wireless networks. The author classified the trust/reciprocity mechanism as direct and indirect reciprocity principles.

The main ideas for indirect reciprocity schemes are "I help you, and somebody else helps me." The indirect reciprocity game is promising to stimulate cooperation in cognitive networks, and can be used to improve the Sybil-resistance for the accounting of peer contributions in peer-to-peer networks. The reputation propagation mechanism in the indirect reciprocity system allows attackers to be known and punished by a larger population of nodes in the network. Compared with the direct reciprocity system, the indirect reciprocity system provides stronger security protection, especially for the large-scale networks with node mobility.

The indirect reciprocity game is promising to stimulate cooperation in cognitive networks, and can be used to improve the Sybil-resistance for the accounting of peer contributions in peer-to-peer networks. In the indirect reciprocity game-based secure system, in each transmission period, the intended receiver and other observing nodes evaluate the behavior of each node in this area, reduce the reputations of the attackers, and propagate the new reputations to the whole network through the gossip channels. More specifically, the author in [40] built a public social norm and reputation updating process to assign to the attackers low (bad) reputations, due to which most nodes reject their requests for network service over a long time.

In this reputation mechanism, each node obtains a reputation vector according to its past and current actions. Each time slot consists of the message transmission stage and the performance evaluation stage. In the second stage, the reputation of each source node is updated and broadcast to the other nodes via the gossip channels. The network determines whether to accept its future transmission requests based on its reputation. In general, the reputation of a node decreases, if it attacks the network or declines the request from a transmitter with a good reputation. In order to punish attackers, the reputation of the node decreases, if it is helping a node with a bad reputation. Its reputation improves in other cases. A reputation updating process was designed to compute the reputation vector for each node, based on its last reputation, and the instant reputation resulting from its current action. The forgetting factor is used to weight the last reputation in the calculation, and it is related to the instant reputation. In this system, attackers are not only rejected by their direct victims, but also by most other nodes in the network during a long time. The punishment period is determined by the forgetting factors. The nodes are forgiven and regain the network access, if following the network social norm during the punishment period.

No matter the static (Bayesian reputation system) or dynamic (Q-learning-based reputation system) trust and reputation system, the reputation system detects the abnormal behaviors based on the statistics of the user's historical behaviors. The network latency caused by the reputation system is inevitable. However, the network latency may result in the dramatic degradation of network performance, such as throughput. Therefore, the design of mechanisms considering low latency is necessary. Network coding is a promising technique that has been widely researched recently. The network coding allows operation on the bit level of the messages. Compared with the mechanisms dealing with the messages in the upper layer, the processing latency might be reduced with the network coding schemes.

26.3.1.5 Network Coding-Based Secure Network Design

Network coding was first proposed to be implemented in the upper layer (such as network layer) to improve the network throughput, reduce the network congestion, and enhance the network robustness. Instead of simply relaying the packets of information they receive, the nodes of the network take several packets and combine them together for transmission. This can be used to attain the maximum possible information flow in a network. Except for improvement of network throughput, the network coding can also provide secrecy for the network if the network coding scheme is used in the physical layer, which is called physical-layer network coding (PNC).

In the following, we will introduce the network coding-based security threats defending mechanisms categorized by upper-layer mechanism and physical-layer mechanism, respectively.

(1) Upper-layer network coding-based mechanism

Consider a system that acts as information relay, such as a router, a node in an ad-hoc network, or a node in a peer-to-peer distribution network. Traditionally, when forwarding an information packet destined to some other node, it simply repeats it. With network coding, the node is allowed to combine a number of packets it has received or created into one or several outgoing packets. The following figure shows a typical network architecture, provides the comparison of a traditional network without network coding mechanism and one with network coding.

As shown in Figure 25.6, assuming the source node wants to send message b_1 and b_2 to destination node F and E, respectively, the source node broadcast messages b_1 and b_2 to the relay nodes A and B, respectively. Then the relay nodes A and B forward their received messages b_1 and b_2 to E and F, respectively, which are the destination nodes. Additionally, the messages received from source node are also forwarded to relay node C. Node C will then forward the messages b_1 and b_2 to the next hop node D sequentially in the network without network coding scheme. Then the node D forwards the received b_1 and b_2 to destination node E and F, respectively. On the contrast, in the network coding-based network, the messages b_1 and b_2 are combined together using eXclusive OR (XOR) operation at node C and forward the combined version of b_1 and b_2 to node E and F through node D. Since node E and F have the knowledge of message b1 and b2, node E and F can get the messages b_2 and b_1 by simply using XOR operation on the received combined message and b_1 and b_2 respectively. It is obvious that the network coding scheme reduces the network propagation delay, compared with traditional network.

Many researchers have adopted the data flow of network coding-based network, as shown in Figure 25.6 to construct the security threat defending mechanisms. In [42], a novel network coding-based malicious detection approach called DENNC for the wireless networks was proposed. Instead of only sending the data packet, as shown in Figure 25.6, the source nodes in [42] are designed to send their data packet along with the result of hash function on their data packet. Messages received by the intermediate nodes are exclusive ORed together, as in the conventional network coding scheme. As a result, each receiver can calculate the original messages and their hash results

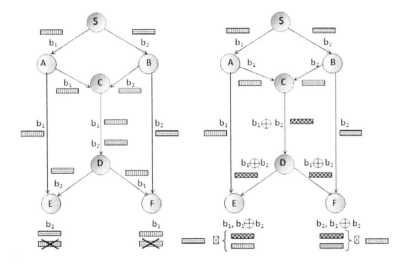

Figure 26.6: Network model with network coding scheme.

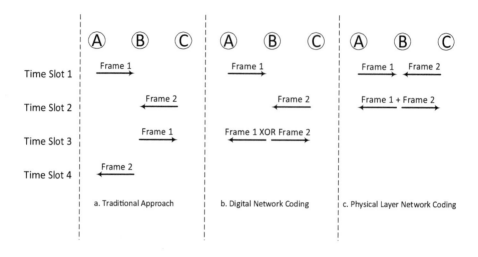

Figure 26.7: Physical-layer network coding.

by using exclusive OR operation. Then, the integrity of the messages can be verified using the original packet with its hash result. This approach does not need additional hardware and secret key encryption; it can detect the malicious node in highly probability.

(2) Physical-layer network-coding-based mechanism

Traditional network coding is conducted on the network layer wherein the packet operations are involved. In order to further improve the network throughput, the physical layer network coding was proposed [43]. It is obvious from Figure 25.7 that the network throughput with physical-layer network coding is improved, since only two time slots are needed for transmission between node A and node C, while 4 and 3 time slots are needed in the traditional and digital network coding approaches, respectively. In the physical-layer network coding-based mechanisms, by taking advantage of bit operation, more secrecy can be achieved [41].

In [44], the author proposed the Sybil attack defending mechanism based on the physical-layer network coding. This mechanism identifies the malicious nodes by evaluating their locations through the network coding mechanism. The example shown in Figure 25.8 demonstrates the principle of location estimation. As shown in the figure, node C and D send messages to node A and node B, respectively. Assume that node C starts sending at $T_C = 0$ and D starts sending at T_D, which is later than node C. According to [44], the time of messages collision due to node C and D happens at node A is t_{diffA}, and the time of messages collision happens at node B denotes t_{diffB}. The difference between these two instant collision time points can then be denoted as: $t_{diffB} - t_{diffA} = \frac{(d_{BD} - d_{AD}) + (d_{AC} - d_{BC})}{s}$, where d_{XY} indicates the distance between node X and node Y. s denotes the radio wave propagation speed.

According to the above description, the difference between the collision time points happen at node A and B becomes larger with the increase of distance of node A and node B. This collision time points difference happens at the Sybil nodes is 0, since those Sybil nodes are implemented in one physical device. Therefore, the Sybil attack can be distinguished by this collision time point difference theoretically. However, the t_{diffA} and t_{diffB} reported by the receivers cannot be adopted directly, since malicious nodes will lie about the values. Therefore, [44] also proposed a mecahnism to defense the Sybil attack. In the proposed mechanism, the messages sent from source nodes can be recovered if the distance between two received nodes is greater than 0.

Figure 25.9 shows an example illustrating how the proposed mechanism in [44] works. Without losing generality, it is assumed that the collisions at node A and B happen at the fourth and seventh bits of sequence C, respectively. If the interference results can be viewed as the sum of the two

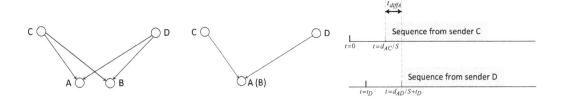

Figure 26.8: Network coding-based location estimation [44].

signals, Figure 25.8 also shows the received sequences at A and B. If the interfered signal is "0" or "2", the corresponding bits in both sequences are "0" or "1". However, if the interfered signal is "I", the receiver cannot tell which sequence contains the bit "I." The receiver can take a wild guess, but it has only 50% chance to guess correctly. Therefore, when the received sequences are long enough, a single receiver cannot recover the two sequences. However, if nodes A and B combine their information, they can accomplish the data recovery task. As illustrated in Figure 25.2, since B already knows that the fourth bit in sequence C is a "1," it can help A to figure out that the first bit of sequence D is "0." This will then help B to determine that the seventh bit from C is "1." This procedure will continue, and A and B will recover the two sequences.

This network coding-based Sybil attack defending mechanism adopts the signal processing method to resist the message recovery capability of Sybil nodes. However, this scheme requires the cooperative processing of two receivers. In this sense, the extra network overhead will be introduced, and the receiver selection should be another optimization issue need to be addressed.

In [45], the author proposes a PUE attack detection mechanism based on the PNC technique. PNC uses the additive nature of the electromagnetic waves to serve as the coding procedure. In this approach, the position of a wireless node is estimated by letting its radio signals interfere with a reference sender. These interfered sequences will be captured by multiple secondary users. Combining the starting points of signal interference results with their positions, the secondary users will determine a group of hyperbolas on which the wireless sender resides. Then they will compare the intersection point of these hyperbolas with the known position of the PU to detect the PUE attack.

In [46], the author proposed the physical-layer network coding in a manner of scheduling scheme for DF-based relay mechanism in the multi-hop networks. This scheme considers the network model, as shown in Figure 25.10. As shown in this figure, node A and B want to send mes-

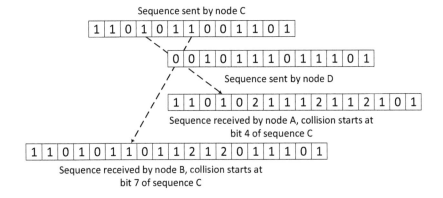

Figure 26.9: Physical-layer network coding-based Sybil attack defending algorithm [44].

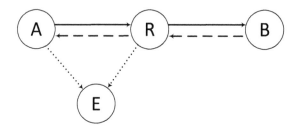

Figure 26.10: Network model considered in [46].

sages to each other through the assist of node R. Node E represents the eavesdropper who is intend to overhear the massages sent from node A and node B in order to decode their messages.

The network coding proposed in this chapter can be represented as the following scheduling scheme:

Considering that the time has been partitioned into equal length slots, and let $x_1(i)$ and $x_2(i)$ denote the ith message from A to B and from B to A, respectively, consequently, we can express the transmitted signals by using $x_1(i)$ and $x_2(i)$. The scheduling policy can then be described as follows.

In the first time slot, nodes A and B send messages $x_1(1)$ and $x_2(1)$ to node R.

In the $(2k) - th$ time slot $(k \geq 1)$, node R sends message $x_1(k) + x_2(k)$ to nodes A and B.

In the $(2k+1) - th$ time slot $(k \geq 1)$, nodes A and B send messages $x_1(k+1) + x_2(k)$ and $x_2(k+1) + x_1(k)$ to node R. In this manner, the relay node R will receive $x_1(k+1) + x_2(k) + x_2(k+1) + x_1(k)$ in time slot $2k + 1$. Note that the node R has $x_1(k) + x_2(k)$, it can decode $x_1(k+1) + x_2(k+1)$.

This scheme makes it difficult for the eavesdropper to decode node A and node B's messages unless he can overhear all of messages sent from node A and B from the beginning.

Considering there are multiple entities that want to enlarge their own benefits from the cooperation, an alternative effective way to model this scenario is game theory, which is described in the following.

26.3.2   Game-Based Secure Network Design

Physical layer (PHY) security was first introduced in Wyner's seminal work [47] over the wire-tap channel, and it was then extended to the wireless and multi-user channels [48]. The main idea behind PHY security is to exploit the wireless channel characteristics, such as noise and fading, so as to improve the reliability of wireless transmission. This reliability is quantified through the notion of secrecy rate, which is defined as the maximum rate of reliable information from the source to the destination, with no information obtained by the eavesdroppers. According to the information theory, the channel capacity is defined as the maximum data rate between a transceiver and is determined by the transmitter's transmission power. Therefore, the power allocation is one of the critical factors that should be optimized for a network to achieve the maximum data rate. For the relay featured cognitive radio networks, however, different from the non-relay featured networks, the total secrecy rate is intensively affected by the power allocation strategies of the relay. Therefore, the power allocation is the critical factor that affects the system performance of CRNs.

Game theory has been used in many resource allocation problems. Auction theory, as a kind of game theory, was pioneered by the paper of William Vickrey, who first gives an analysis from the perspective of an incomplete information game. In the game-based power allocation mechanisms, the power is considered as the merchandise, while the user ends are considered as the bidder to bid

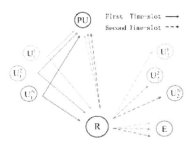

Figure 26.11: Network model for Vickrey game-based mechanism.

for the channel bands to gain the higher data rate. Some game-based power allocation mechanisms are shown as follows:

In [49], the authors consider the scenario with multiple pairs of SUs. There is one source node and one destination node in each pair of SUs. The source node sends messages to the destination node, with the assist of one relay node that helps forward the messages from source node to the destination node. One PU node is considered in the network model in this work. The network structure can be depicted Figure 25.11.

In this work, the channel is assumed as flat fading with single antenna. Therefore, the relationship between messages at the source node and the destination node for a pair of SUs can be described as $y = ha + w$, where $w \sim CN(0, N_0)$, and $h = cr^{-2}$ with $c \sim CN(0, 1)$ and r the propagation distance. It assumes that the users have perfect channel state information (CSI) of all channels, which remains constant during the relaying. The transmit power of U_1^i is P_t^i. The total transmit power of the relay is P_r, which are divided into K equal units for auction, with each unit $Punit = Pr/K$. For SU pair i with k^i power units P_{unit}) obtained, the relay transmit power for it is $k^i \cdot P_{unit}$. The channel gains from U_1^i to PU, from U_1^i to relay node R, from R to PU, from R to U_2^i, and from R to E are denoted by $h_{1,p}^i, h_1^i, h_{2,p}^i, h_2^i, h_{2,e}^i$, respectively. The message transmission from source node to destination node can be described as the following procedure:

In the first slot, U_1^i transmit its corresponding signal a^i to R. The received signal at the relay node is $y_r^i = \sqrt{P_t^i} h_1^i a^i + n_1^i$. In the second slot, the relay transmits it with power $k^i \cdot P_{unit}$ to the destination U_2^i, the received signal at U_2^i is $y^i = \sqrt{k^i P_{unit} P_t^i} h_1^i h_2^i \beta^i a^i + \sqrt{k^i P_{unit}} h_2^i \beta^i n_1^i + n_2^i$, where $\beta^i = (P_t^i |h_1^i|^2 + N_0)^{-1/2}$ is the normalizing factor. Then, the capacity from U_1^i to U_2^i, with k^i power units is given by:

$$C^i(k^i) = \frac{W}{2} log_2 \left(1 + \frac{k^i P_{unit} P_t^i |h_1^i h_2^i \beta^i|^2}{k^i P_{unit} |h_2^i \beta^i|^2 N_0 + N_0} \right) 2, \qquad (26.2)$$

where W is the bandwidth for each user pair. Similarly, the capacity of the wire-tap channel from U_1^i to E is given by $C^i(k^i) = \frac{W}{2} log_2 (1 + \frac{k^i P_{unit} P_t^i |h_1^i h_{2,e}^i \beta^i|^2}{k^i P_{unit} |h_{2,e}^i \beta^i|^2 N_0 + N_0})$. Then, the secrecy rate of i is given by $C_s^i(k^i) = max[C^i(k^i) - C_e^i(k^i), 0]$ and the marginal secrecy rate of the kth unit for SU pair i is given by $\Delta C_k^i = max[C_s^i(k) - C_s^i(k-1), 0]$.

In order to model the network structure as an auction model, the channel bands are considered as the merchandise. The source nodes in each pair of SUs are considered as bidders. The transmission power subject to the interference to the primary user is considered as the bidder's cost, and the secrecy rate is then the payoff of the source nodes (corresponding to each pair of SUs).

The Vickrey auction scheme was adopted to allocate power for secondary users in order to maximize the secrecy rate $C_s^i(k^i)$. The basic idea of the Vickrey auction is that the bidder who bid with the highest price will win the merchandise by paying the second highest bid price. The Vickrey auction based power allocation can be described in Table 25.2.

Table 26.2   Algorith-Based on the Vickrey Auction

* **Common Knowledge:** P_1 and CSI of all channels.
* **Mobile User Pair:**
(1) Wait for an auction and set $P_t^i = P_{inter}/|h$.
(2) Calculate the marginal secrecy rate of each unit given by (3).
(3) Calculate the largest number of units allowed
    $k_{max}^i = P_{inter}/(P_{unit}|h_1^i$.   Set $\Delta C_k^i = 0, k > k$.
(4) Submit the bids of the only weakly dominant-strategy equilibrium
    $b^i = (\Delta C_1^i, \ldots, \Delta$, where the marginal capacity is deemed the value of
    correcponding unit. Go to (1).

Relay node:
(1) Announce an auction with units with each be P.
(2) Allocate kP power to the idder who wins non-zero bids of the highest bids. Go to (1).

Table 26.3   Algorithm Based on the Sequential First-Price Auction with Uniform Distribution Assumption

***Common Knowledge:** P_1 and CSI of all channels.
***Mobil User Pair:**
(1) Wait for an announcement of an allocation and then set
    $P_t^i = P_{inter}/|h$.
(2) Calculate the marginal capacity of each power unit given by (5).
(3) Calculate the largest number of units allowed
    $k_{max}^i = P_{inter}/(P_{unit}|h_1^i$.   Set $\Delta C_k^i = 0, k > k$.
(4) Wait for an announcement of an auction with units, and each be i, and then determine the
    value for this auction x : if he has gotten n units in this allocation.
(5) Submit the bids of all rounds based on the symmetric equilibrium
    $\beta_k^*(x) = \frac{N-K}{N-k+1} x, k = 1, 2, \ldots$.
(6) Get notification, update the number of units obtained in this allocation.
    Go to (4) unless the allocation is over.
(7) Go to (1).

Relay node:
(1) Announce an allocation.
(2) Group all units, with each set contains K units.
(3) Announce an auction with units, with each be P.
(4) Remove the highest bidder in each round immediately after allocating one
    power unit to him. Stop at the -th round or the round with no non-zero bids submitted.
    Notify the result.
(5) Go to (2) unless all units are auctioned off.
(6) Announce the end of this distribution and allocate kP power
    to the bidder who wins units in total. Go to (1).

However, as it is known, the critical drawback of the Vickrey auction is that it may not bring the optimal payoff for the auctioneer due to the unreasonable second-highest bid price when the difference between the highest price and the second highest price is large- and the second-highest price is unreasonable low. In this sense, the transmission power of the source node will be fairly low which may result in insufficient network performance.

The sequential first-price auction avoids the auctioneer's loss. The authors in [49] also adopt the sequential first-price auction to solve the security issue shown above. The sequential first-price auction is an auction where the K identical units are sold to N > K bidders using a series of first-price sealed-bid auctions. Specifically, one of the units is auctioned off at one go, where the unit is sold at the price of the highest bid to the corresponding bidder, and the knock-down price is announced immediately. After K first-price sealed-bid auctions, the sequential first-price auction ends. For simplicity, the work in [49] only considers the situations in which each bidder requires at most one unit—the case of single-unit demand. The sequential first-price auction-based power allocation mechanism can be described in Table 25.3.

As we discussed before, the cooperative relay [50] is an emerging promising technology that can significantly improve the system throughput. In the cooperative relay scheme, the SU will be selected to help the PUs forward their messages to gain transmission opportunities as a reward. However, in reality, the selected relay node in the cooperative relay scheme may not be honest. Those malicious or compromised nodes may launch attacks, such as black or a hole attack, etc. Moreover, the dishonest nodes may not obey the cooperation rule during the cooperative transmission to pursue more self-benefits [34]. For example, it may transmit its own packets instead of relaying the packet from the primary users. Several game theory-based mechanisms were proposed [34][51].

In these mechanisms, the interaction between PUs and SUs are modeled using the game theory. The equilibrium solution is obtained to maximize the system performance of both primary network and secondary network simultaneously.

For example, in [34], the author proposed a cooperative framework in the CRNs which addresses the energy efficiency of the PU and the trustworthiness of the SUs. In this cooperative strategy, the PUs select the most suitable SUs as the cooperative relay and lease the spectrum resources to the SUs rewards. Based on the PUs' strategy and corresponding information, the SU should determine its own transmission power in order to maximize the data rate for the PUs' transmission and avoid the interference to the other PUs. If all the SUs are well-behaved, both PU and SU can benefit from the cooperation. However, when there are some dishonest or malicious SUs, the normal operation of CRN will not be guaranteed.

Based on the above assumptions, the authors used the Stackelberg game to model the interactions between PUs and SUs which is a typical leader-follower game.

In this model, the PUs pay more attention on its energy efficiency. In other word, the PUs try to minimize their power consumption due to the message transmission. Therefore, in [34], the utility of the PUs are expressed as the multiplication of secondary relay's trust value, with the difference between the power consumption before and after the cooperative relay mechanism provided by the secondary relay used. The trust value of the SU is evaluated by the Bayesian framework. The difference between the power consumption before and after the cooperative relay mechanism provided by the secondary relay is used can be represented using the time slot and transmission power assigned to the secondary relays. Therefore, the utility of the PU is indicated as the transmission power saving achieved from the cooperative relay, while considering the cooperative relay's trust value.

SUs are selected as the relay to help the PUs forward their messages. The SUs are allowed to transmit their secondary messages over the same spectrum bands as the rewards after they finish the transmission for the PUs. Therefore, the utility for the secondary in [34] is given by the difference between the channel capacity that the SU can achieve over the spectrum bands leased by the PU and the total cost (in terms of energy used to transmit PU's messages) caused by helping PUs for their transmission.

26.3.3 Other Solutions to Security Issues by Relay Feature

Considering that the reputation and network coding-based solutions are two effective ways to counter against the security threats in the CRNs, a novel forward error correction—driven (FEC-driven) network coding scheme was proposed by integrating the network coding and the reputation scheme to countering against the pollution attack [19]. In this work, the uplink of centralized CRNs is considered.

The FEC-driven network coding scheme considers the effect of the pollution attack as the chan-

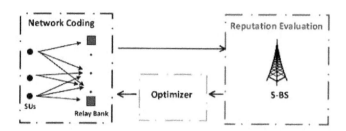

Figure 26.12: Network coding-based pollution attack defending framework.

nel with additive noise (as described in the section 25.2.2.2). Hence, the effect of pollution attack can be overcome by the FEC code. The traditional FEC code is operated in each individual device. However, this leads to high implementation complexity. Therefore, the work in [19] proposes to realize the FEC in the network level, which is called network error correct code. For simplicity of implementation, the convolutional code is adopted in this work.

The overview architecture of FEC-driven network coding mechanism is described in Figure 25.12. Assuming the relay nodes are distributed around the SUs densely, whenever SUs notify the secondary base station for data delivery, the base station will group several relay nodes and SUs, and optimize the parameters of coding schemes for each group of users. In each group, the relay nodes cooperatively encode the SUs' messages, and forward the encoded messages to the secondary base station. Assuming the base station knows what coding scheme is used, it can recover the original messages with errors inserted by the pollution attack.

There are three challenges in this mechanism: (1) how to realize the convolutional encoding in the network level in each group, (2) how to group SUs and relay nodes, and (3) how to optimize the parameters of coding scheme for each group of SUs and relays. These three questions will be solved in the following explanations, respectively.

(1) Convolutional code based network error correction coding

Figure 25.13 illustrates the basic idea of conventional convolutional code. As shown in this figure, the message bits are input into the convolutional encoder sequentially. The output of the convolutional encoder is composed of convolution of some input signals.

In order to realize the convolutional code in the network level, each exclusive OR operation is conducted by a relay node. Assuming the messages sent from every secondary user are reachable

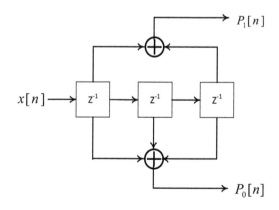

Figure 26.13: Structure of convolutional encoding.

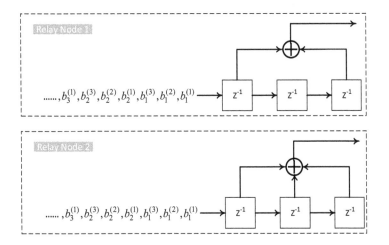

Figure 26.14: Network encoding procedure.

at all the selected relay nodes within a group, each relay node then organizes the received bits from the SUs in the order shown in Figure 25.14, and exclusive OR them sequentially.

Assume there are three secondary users that send messages with the assists of two relay nodes, and all bits sent from SUs are correctly received by relay nodes. Denote the bits from the ith SU arriving at each relay node in the order of time sequence can be represented as $b_n^{(i)}$. n is the time index. At each relay node, the received bits will be organized in the order as the following pattern, $b_1^{(1)}, b_1^{(2)}, b_1^{(3)}, b_2^{(1)}, b_2^{(2)}, b_2^{(3)}, \cdots\cdots$. Considering the convolutional encoding structure shown in Figure 25.13 is adopted, then, the output bits sent from the first relay node can be represented as $b_1^{(1)} \oplus b_1^{(3)}, b_1^{(2)} \oplus b_2^{(1)}, \ldots\ldots$, and the output bits sent from the second relay node will be $b_1^{(1)} \oplus b_1^{(2)} \oplus b_1^{(3)}, b_1^{(2)} \oplus b_1^{(3)} \oplus b_2^{(1)}, \ldots\ldots$.

Since the secondary base station has the knowledge about parameters of the convolutional code that the network is using, it can easily adopt the traditional convolutional decoding algorithms, such as Viterbi algorithm, to decode the bits for each secondary user.

(2) Grouping strategy

The SUs and the relay nodes that are suitable for construction of the FEC-driven network coding should be grouped first. The major goal of the grouping strategy is to ensure the signal sent from the SUs is achievable at every relay node within a group. There are many options to indicate the connectivity of transceivers. Among with the channel quality indicator (CQI) is a metric that are commonly used. The CQI value is calculated based on the received signal strength (RSS) at the receiver.

(3) Optimization of parameters of coding scheme

After grouping relay nodes and SUs, the parameters of convolutional encoding, such as the number of registers, the number of exclusive OR operations, and so on, should be optimized to achieve better error correction performance. The optimization problem can be represented as shown below.

$$\begin{cases} f_1 = \max\left\{f_{d_{free}}\right\} \\ f_2 = \min\left\{f_{eff}\right\} \\ f_3 = \min\left\{f_{delay}\right\} \end{cases} . \tag{26.3}$$

Equation (25.3) shows the objective function of this optimization problem from the following three aspects: (1) maximization of the free distance of convolutional code; (2) minimization of the efficiency rate which is defined as the ratio of number of relay node to the number of SUs served by those relay nodes; and (3) minimization of the delay caused by the convolutional encoding.

The constraints of this optimization problem can be represented in the following equation:

$$\begin{cases} c_1 : \frac{r}{n} < 1 \\ c_2 : \eta_i > \eta_{th} \\ c_3 : P_{tx}^{pu} < P_{th} \\ c_4 : P_{tx}^{su} < P_{max}^{su} \\ c_5 : P_{int}^{su} < P_{int}^{th} \end{cases} . \tag{26.4}$$

In Equation (25.4), c1 ensures that a relay is not configured to simple AF wherein only one SU message is forwarded. Constraint c2 makes sure that every relay node has a certain level of trustworthiness based on its history, where η_{th} and η_i represent required reliability threshold and individual reliability values, respectively. Channel availability due to the PU activities is pointed out in c3, where primary network (PN) transmission power (P_{tx}^{pu}) should be below certain threshold (P_{th}) for SU to decide as idle for SU relay-user pair to be able to communicate via utilizing subjected licensed channel. c3 points out that SU transmit power (P_{tx}^{su}) is limited by peak transmit power (P_{max}^{su}) due to hardware limitation. Another constraint inherited from CR is power limitation on SU, such that the interference level it contributes to the media should be less than a certain threshold (P_{int}^{th}) so that primary network performance is not degraded.

Reputation for data transmission from relay to the base-station (BS) is evaluated in order to quantify the trustworthiness of a relay node. All relay nodes begin with equal reputation values in BS and it updates over the time according to their performance. Intuitively, reputation value of a relay node is increased by one once it helps a successful data transmission; otherwise, the reputation value is decreased by one as mentioned in [52]:

$$\eta_i(t) = \eta_i(t-1) + (-1)^{\gamma_i(t) + \gamma_g(t)}, \tag{26.5}$$

where $\eta_i(t-1)$ shows reputation value up to current time, $\eta_i(t)$ indicates current reputation, whereas $\gamma_i(t)$ and $\gamma_g(t)$ represents individual relay data and global data that is decoded by relay, respectively. However, Equation (25.5) can be further improved by introducing the importance of the history, as mentioned in [16], such that:

$$\eta_i(t) = \sum_{m=1}^{M} \left(\eta_i(t-m) e^{-(t-m)/\vartheta} + (-1)^{\gamma_i(t) + \gamma_g(t)} \right), \tag{26.6}$$

where M represents the oldest time instance of interest, and ϑ indicates time decaying factor. One should configure the ϑ in such a way that malicious behavior could be caught on time while intermittent errors are not over penalized. According to (25.6), the smaller the value of ϑ, the larger the rate of decay is, which means the history becomes less important, and vice versa.

26.4 Secure Coexistence Issues In Cognitive Radio Networks

The capability of coexistence with licensed networks enables the secondary networks to improve the spectrum utilization efficiency. However, with the proliferation of cognitive radio featured wireless systems (e.g., WiMAX, 5G, etc.), the coexistence issue of competing multiple secondary networks is unavoidable. In order to effectively realize the coexistence among multiple networks, the knowledge of information of other networks is significant for optimizing the individual or overall network performance. Whereas, due to the variety of cognitive featured wireless networks, the cooperation among heterogeneous networks is a challenge for the deployment of coexisting networks.

In the coexisting scenarios with multiple secondary networks, the spectrum resources are shared by the multiple secondary networks. Here, each individual network intends to selfishly occupy as much spectrum resources as possible to provide high-quality (high data rate) services to its users. However, due to the limit of available spectrum resources, the high-quality services of one network may cause the severe performance degradation of the other secondary networks. In the severe case, this may lead to the selfish attack in the co-existing secondary networks. Therefore, it is critical for the coexisting mechanisms to achieve high overall network performance with fairness toward individual secondary networks.

To achieve the fairness among the coexisting networks, the fairness concept should be aware in every spectrum resource allocation related mechanisms. The major spectrum allocation related mechanisms in the secondary networks are dynamic spectrum allocation (DSA) process and the media access control (MAC) process. The DSA mechanisms attempt to assign optimal bandwidth and allocate corresponding spectrum bands to the coexisting networks. By doing DSA, the overall network performance is aimed to be maximized subject to the constraints, such as the data rate, provided by the network should be sufficient for its users' minimum QoS requirements, and the interference to the PUs caused by the SUs are avoided. Intuitively, the fairness should be another constraint for the optimization problems in DSA mechanisms.

In general, the network resource is not only in the spectrum domain, but also in time and other domain depending on the multiplexing access mechanism (e.g., time-division multiplexing access (TDMA), frequency division multiple access (FDMA), code division multiple access (CDMA) etc.). The MAC protocol specifies how to assign the network resources to the users such as assigning frequency to the users in FDMA scheme and assigning the time slots to the users in the TDMA scheme, etc. Considering the coexistence with other secondary networks fairly, the chance for the other networks to obtain the network resource should be considered in the design of MAC protocol for a network. For example, in a TDMA-based system, the "quiet period" should be kept in which the users in this network do not transmit anything and give the transmission chance for the other networks.

Furthermore, in the modern CRNs, the SUs are capable of sensing their surrounding environment in order to help to maintain the available channel bands for the cognitive network. However, if the channel status is only considered as "occupied" and "unoccupied" by the primary users, it will not be fair for the other SUs if one SU always occupies the channel, since the other SUs consider the channel is occupied by the PU. To achieve the fairness among the SUs, new channel status model ("occupied by primary users," "occupied by secondary users," and "unoccupied") should be considered in the spectrum sensing process. In the following sub-sections, the fairness aware designs in the coexisting network scenarios are introduced from the aforementioned three aspects (DSA, MAC, spectrum sensing).

26.4.1 Definition of Fairness

(a) The most popular definition of fairness was called Jain's fairness index, which was proposed in [t]. Jain's fairness index is widely used to evaluate the TCP fairness [53]. The definition of Jain's fairness index can be represented as following equation:

$$J(x_1, x_2, \cdots, x_n) = \frac{\left(\sum_{i=1}^{n} x_i\right)^2}{n \cdot \sum_{i=1}^{n} x_i^2} 7, \tag{26.7}$$

where x_i is the throughput for the ith connection, and n is the number of users in the network. The result ranges from $\frac{1}{n}$ (worst case) to 1 (best case), and it is maximum when all users receive the same allocation. This index is $\frac{k}{n}$ when k users equally share the resource, and the other n-k users receive zero allocation. The Jain's fairness index was adopted in [u] as one of the metrics to evaluate the proposed MAC protocol.

(b) In the Jain's fairness index, the fairness is achieved when every connection is allocated with identical amount of resources. However, the users in a network may be served with different services. The required QoS of different services are also different. From this perspective, the Jain's fairness index may not be able to evaluate the fairness of network accurately. In order to tackle this problem, [54] proposed to utilize ratio of allocated data rate to the traffic demand for each user as the indicator to evaluate the fairness.

(c) The aforementioned two fairness indicator evaluates the network from the network resource allocation perspective. However, in the design of MAC layer protocol for cognitive radio network, the latency of resource access is one of the major concerns, since the spectrum sensing and spectrum mobility may need a long time to finish. Considering the cognitive users with the same priority, the duration of network access for a cognitive user can be considered as the representation of fairness [55].

Considering the fairness as one of metric, the design of coexistence mechanism can be categorized into following groups.

26.5 Fairness-Based Coexistence Network Design

26.5.1 Fair Spectrum Sensing

Spectrum sensing is one of the important features of the CRNs. With the spectrum sensing scheme, the cognitive users are capable of being aware the surrounding environment dynamically. In the traditional spectrum sensing scenarios, only two states of channel are considered, i.e., the channel is occupied (H_1) and it is idle (H_0). The simple energy detection algorithm is adopted as the spectrum sensing mechanism to detect the channel status.

According to the nature of CRNs, whenever the SU detects the presence of PUs over a specific spectrum band, it has to vacant the spectrum bands and switch to the other available bands. Whereas, according to the two state channel assumptions, the SU does not know who is occupying the channel bands. Therefore, it has to vacate the channel and switch to the other available channels. However, if it is the SU is occupying the spectrum bands, then the fairness of SUs will be violated, since the SUs are not competing with each other under the same condition.

In practice, multiple CRNs often coexist together. The two-state sensing model is insufficient for such a system. Consider a scenario that a channel might be occupied by an SU from one CRN, and therefore, SUs in other CRNs perceive this channel as occupied by the PU. When there are a large number of available idle channels, whether a channel is occupied by a PU or an SU is not important. SUs can simply switch to other idle channels. However, with the expected rapid proliferation of CRNs, it is often true that several CRNs have to co-exist on the same channel. In this case, it is essential to determine whether a PU or an SU is using the channel. The reason is that if SUs in one CRN are accessing a channel, SUs in other CRNs would detect it as busy and, hence, be starved.

In order to achieve the fairness between the SUs, [56] proposed the three-state model for the spectrum sensing. Intuitively, the detection between the state H0 and the other two states can be conducted using the simple energy detection scheme straightforwardly. However, it is difficult to distinguish between state H1 and H2. Feature sensing seems a promising way to achieve this objective. However, this method strongly depends on specific signals, and it is not trivial to get such a feature. Currently, only TV band signal provides detailed signature information. For other bands being released, it is possible to get the signature information, but more investigation is needed. Therefore, a distance estimation-based technique was proposed in this chapter aiming at distinguishing the state H_1 from H_2. In this mechanism, the received signal at the secondary user is utilized to estimate the corresponding transmitter's location. By assuming the PU transmitter's location is known by the secondary network, it is possible to distinguish the transmitter of a signal received at the SUs.

26.5.2   Fair Dynamic Resource Allocation

Dynamic resource allocation is one of the major functionalities for the secondary networks. Due to the varying environment, including the mobility of both PUs and SUs, and the changing utilization of spectrum resources, the SUs are allowed to opportunistically access into the licensed spectrum bands with proper transmission configurations, such as transmission power, etc. In the coexistence scenarios, the configurations for the SUs should be optimized in order to achieve the better overall network performance.

The existing dynamic resource allocation mechanisms can be categorized into two groups from the network architecture perspective, i.e., non-coordinator-based and coordinator-based schemes. In the non-coordinator-based mechanisms, the coexisting networks operate individually in a distributed manner wherein there is no third party to coordinate the operations of each network, such as spectrum allocation, power allocation, and time slot assignment, etc. It is obvious that the advantage of these mechanisms is that no extra hardware, e.g., the coordinator, is needed, which results in the low-cost implementation. However, the complex algorithm or protocols are necessary for these scenarios, since the status of networks are unknown among the coexisting networks. This may limit the network performance, such as network throughput. On the contrary, in the coordinator-based mechanisms, a coordinator is deployed as the manager of the coexisted networks, who is responsible for allocating the spectrum resources to the coexisting networks and control the transmission power of the subscribers in each network. The coordinator is supposed to have the knowledge of all networks. Therefore, the optimal strategies can be made by the coordinator to optimize the overall network performance of coexisting networks.

26.5.2.1   Non-coordinator-Based Mechanisms

A non-coordinator-based coexisting scheme is proposed in [57]. This approach filled the technical void of achieving efficient co-channel sharing in the uplink. This work proposed an uplink soft frequency reuse mechanism to enable the efficient co-channel spectrum sharing targeting on the global power-efficiency and local fairness. In view of potentially mobile/portable devices in CRNs, power saving is necessary for battery-powered uplink transmitters. In addition, fairness guarantee is also important, because user terminals, either close to or far away from their home BS, consume largely different amounts of power for the same level of signal-to-interference plus-noise ratio (SINR).

In this mechanism, the uplink resource allocation (URA) in each network cell is formulated as an optimization problem. Considering the scenario that there are total of N cells coexisting on a common channel, which consists of K subchannels. In each cell n, for $n \in N \equiv \{1, \cdots, N\}$, there are $M^{(n)}$ active sessions. Let $U^{(a)}$ be a matrix with binary entries represent the channel assignment result (i.e., 1 at the ith row and the jth column indicates the sub-channel i is assigned to the jth user). $P^{(a)}$ is the power allocation matrix that denotes the corresponding allocated transmission power and $P^{(-a)} \equiv \dot{X}_{n \in N, n \# a} P^{(n)}$, where $\dot{x}$ represents the Cartesian product. Hence, the optimization problem for the scenario described above can be formulated as following:

Find: $U^{(a)}, P^{(a)}, P^{(-a)}$;

Minimize: $L^{(a)}$;

Subject to:

(1) $\begin{cases} U_{m_n,k}^{(n)} \widehat{Q}_{m_n,k}^{(n)} \leq P_{m_n,k}^{(n)} \leq U_{m_n,k}^{(n)} \bar{Q}_{m_n,k}^{(n)} \\ for\, n \in \mathcal{N}; m_n \in \mathcal{M}^{(n)}; k \in \mathcal{K}, \end{cases}$

(2) $\begin{cases} \sum_{k=1}^{k} \log\left(1 + P_{m_n,k}^{(n)} H_{m_n}^{(n,n)}\right) \geq \theta_{m_n}^{(n)} \\ for\, n \in \mathcal{N}; m_n \in \mathcal{M}^{(n)}, \end{cases}$

(3) $\sum_{m_a=1}^{M^{(a)}} U_{m_a,k}^{(a)} \leq 1\, for\, k \in \mathcal{K},$

where $L^{(a)}$ is power consumption, which is defined as

$$L^{(a)} \triangleq \sum_{m_a=1}^{M^{(a)}} \omega_{m_a}^{(a)} \sum_{k=1}^{K} P_{m_a,k}^{(a)}. \tag{26.8}$$

Each $\omega_{m_a}^{(a)}$ denotes session m_a's weight or priority, which is defined as

$$\omega_{m_a}^{(a)} \triangleq \frac{\sum_{n=1,n\neq a}^{N} H_{m_a}^{(a,n)}}{H_{m_a}^{(a,a)}}, \tag{26.9}$$

where each $H_{m_a}^{(a,n)}$ denotes the propagation gain from session m_a to BS n.

In the first constraint, if $U_{m_a,k}^{(a)} = 1$, then $P_{m_a,k}^{(a)}$ should be lower-bounded by $\widehat{Q}_{m_a,k}^{(a)}$, the minimum power for meeting session m_a's SINR requirment, denoted by $\gamma_{m_a}^{(a)}$. Moreover, $P_{m_a,k}^{(a)}$ should also be upper-bounded by $\bar{Q}_{m_a,k}^{(a)}$, the maximum power of session m_a on each subchannel k. But if $U_{m_a,k}^{(a)} = 0$ then $P_{m_a,k}^{(a)} = 0$. The BS a cannot make decisions for the other cells to change $\mathcal{U}^{-a}$, so we say $\mathcal{U}^{-a} \equiv \overline{\mathcal{U}}^{-a}$, where $\overline{\mathcal{U}}^{-a}$ is a fixed strategy matrix set. But $P^{(a)}$d P^{-a} may interact with each other due to the change of inter-cell interference. Hence, these bounds should be satisfied in each cell n, and we have constraint (1). In constraint(1), each $\widehat{Q}_{m_n,k}^{(n)}$ is written as

$$\widehat{Q}_{m_n,k}^{(n)} \triangleq \frac{\gamma_{m_n}^{(n)} \left(\sum_{n'=1,n'\neq n} \sum_{m_{n'}=1}^{M^{(n')}} P_{m_{n'},k}^{(n')} H_{m_{n'}}^{(n',n)} + N_0 \right)}{H_{m_n}^{(n,n)}}, \tag{26.10}$$

In which N_0 denotes average noise power.

In the second constraint, $I_k^{(n)}$ denotes the interference requirement on subchannel k, which is written as

$$I_k^{(n)} \triangleq \sum_{n'=1,n'\neq n}^{N} \sum_{m_{n'}=1}^{M^{(n')}} P_{m_{n'},k}^{(n')} H_{m_{n'}}^{(n',n)}. \tag{26.11}$$

$\theta_{m_n}^{(n)}$ denotes QoS requirement in second constraint. Since each session $m_{n'}$ aggregated uplink capacity in cell n should meet its corresponding QoS requirement, we have constraint (2).

As above, the BS a should mot assign more than one session in the same cell to any subchannel k, and we have constraint (3).

Finally, we can decouple the complex URA problem in the cell a into two subproblems:

- SCA by adapting $\mathcal{U}^{(a)}$ given fixed $\mathcal{P}^{(a)}$ and $\mathcal{P}^{-a}$;

- TPC by adapting $\mathcal{P}^{(a)}$ and $\mathcal{P}^{-a}$ given fixed $\mathcal{U}^{(a)}$.

The above problem is formulated as a mixed-integer non-linear program, which is NP-hard in general. Therefore, this problem is decoupled into two subproblems: subchannel allocation (SCA) and transmit power control (TPC). Solving the former one requires global knowledge, whereas solving the latter one does not. The multi-cell TPC is framed as a non-cooperative game, and prove that the Nash equilibrium can be established in the TPC game without inter-cell coordination. For the SCA subproblem requiring global knowledge, therefore, a low-complexity heuristic is present. The authors also frame multi-cell SCA as a non-cooperative game. After that, the TPC and SCA games are integrated and formulate a two-level game-theoretic approach that is heuristic yet distributed.

Due to the distributed nature of CRNs, each cell in the multi-cell system has to conduct local

URA individually. In view of possible conflicts in coexisting cells' local optimal strategies, the game theory is used to study the global performance of multi-cell URA problem. The self-coexistence of uncoordinated CRNs can be modeled as a non-cooperative game, in which each network cell acts as a player. In the URA game, each cell solves Problem 1 independently. Then, minimizing $L^{(n)}$ is equivalent to optimizing cell n's utility. According to the decoupled SCA and TPC subproblems, a two-level game-theoretic approach is adopt to generate globally power-efficient and locally fair coexistence patterns in a distributed manner. Specifically, the URA game can be regarded as two levels of non-cooperative games for SCA and TPC, respectively. In the two-level frame-work, each acting cell plays the SCA game on the first level. Given a strategy taken by any cell in the SCA game, the Nash equilibrium is achieved in the TPC game on the second level, which will be proved in the next section. The optimal utility gain by taking this SCA strategy is shown accordingly. Based on the utility gain, the acting cell is able to know whether this two-level URA strategy is beneficial. As soon as nobody can find an improving strategy, a stabilized coexistence pattern is commonly agreed by all the cells in the URA game.

The approach shown above considers coexistence of centralized networks wherein the frequency allocation for the uplink of each network is optimized. However, this approach fails when relay nodes are necessary in the network. The authors in [58] proposed the spectrum allocation solution for the coexistence of multi-hop networks.

In [58], the author convert the spectrum allocation, problem as the rate allocation since the data rate of the user end is determined by the bandwidth of its allocated spectrum bands. Therefore, an optimization problem is formulated with the objective as maimizing the summation of end users' data rate. While this optimization problem is constrained by maximizing the demand satisfaction factor (DSF) for each end user which is defined as the ratio of assigned data rate to their demand data rate to achieve the fairness. For example, if r_i denotes the assigned data rate for the ith and user and its demand data rate is represented as d_i, then the demand satisfaction factor of the ith end user can be represented as $\alpha_i = r_i/d_i$. Thus, the optimization problem can be represented as, $\max \sum_{i=1}^{N} r_i$ where r_i is the spectrum allocation for the ith end user, such that for any other allocation strategy r_i', and the following inequality holds, i.e., $min\{\alpha_i|1 \leq i \leq N\} \geq min\{\alpha_i'|1 \leq i \leq N\}$.

In order to obtain the optimal rate allocation for this problem, [58] proposed a multi-channel contention graph based approach. The multi-hop multi-channel network is first modeled, as shown in Figure 25.15.

In an multi-channel contention graph (MCCG) $G_C(V_C, E_C)$, every vertex corresponds to a user-channel pair in A. There is an undirected edge connecting two nodes in V_C if their corresponding user-channel pairs interfere with each other, which can be determined based on conditions described in Section 25.3. Note that if two users ik are incident to each other, then there will be undirected edges between every two user-channel pairs that contain i and k, respectively, because they always interfere with each other no matter which channels are considered. Next, a simple example is used to illustrate how to construct an MCCG. In this example, there are five users (transmitter–receiver

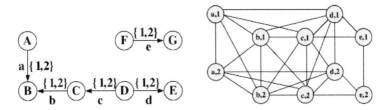

Figure 26.15: Network structure Figure 25.16b illustration of MCCG [58].

pairs), $a;b;c;d;e$, and two channels, channel 1 and channel 2, available to each user, which are shown in Figure 25.15a. In the figure, we have $d(A,B) = d(B,C) = d(C,D) = d(D,E) = d(F,G) = d(D,F) = d(E,G) = R = 0.5I$, where R and I are the transmission and interference range of each user, respectively. We can obtain the corresponding MCCG, which is shown in Figure 25.15b. In the figure, each vertex corresponds to a user-channel pair, for example, vertex $(a,2)$ corresponds to user-channel pair $(a,2)$. Here, we can see that there are edges between nodes $(a,1)$ and $(b,1)(a,1)$ and $(b,2)(a,2)$ and $(b,1)$, and $(a,2)$ and $(b,2)$, because user a is incident to user b. Moreover, there is an edge between node $(a,1)$ and $(a,2)$, because any user can only work on one channel at one time. Considering the data rate for a pair of transceivers can be easily calculated by the function of summation of bandwidth of channel bands assigned to them. Then, the optimal solution to the rate allocation problem described above can be solved assisted with the help of MCCG graph with existing algorithms [59].

26.5.2.2 Coordinator-Based Mechanisms

As described above, the non-coordinator-based mechanisms adopt complex algorithms for each network to achieve the fairness. These complex algorithms may lead to severe network latency and degrade the network performance. Alternatively, the coordinator-based mechanism controls the operations of each network by deploying the third-party "manager." In this sense, the complex algorithms or protocols are moved to the coordinator.

[23] proposes a fair spectrum allocation considering the selfish of CRN using the 0-1 Multiple Knapsack. The coexistence between IEEE 802.22 and IEEE 802.11af is taken as an example to show the network scenario considered in this work and the solution to the problem modeled. In the coexistence scenario mentioned in [23], the IEEE 802.22 network, is considered as a proactive network relative to the IEEE 802.11af network, since the cognitive users in the 802.22 network is able to actively and dynamically sense the channel status and decide to occupy the spectrum that is vacant. However, the 802.11af networks have to passively jump to the channel bands that are shown as free from the spectrum database, which might be constructed by the sensing report from the 802.22 users. There is a specific scenario that the transmission starting time for 802.11af network is later than 802.22 network ($t_1 > t_2$), as illustrated in Figure 25.16. In this scenario, 802.22 users have already received their channel assignment at t_2, so they will start to transmit over the assigned available spectrum bands. Meanwhile, the 802.11af users will keep quiet according to carrier sense multiple access (CSMA) protocol, since the spectrum is occupied. Therefore, the 802.11af users will have no opportunity to access the spectrum under the conditions with selfish DSA in the 802.22 networks.

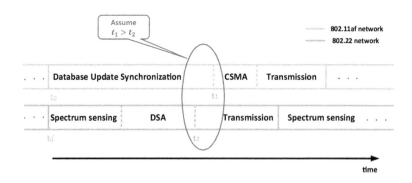

Figure 26.16: 802.11af and 802.22 network working cycle.

In order to avoid the selfish scenario, as described above, a coordinator based network coexistence framework is proposed. In this framework, the spectrum allocation is divided into two steps, i.e., bandwidth allocation and channel allocation. The spectral efficiency of the 802.22 and 802.11af coexistence network is regarded as the performance to be maximized.

In [23], the coexistence networks consists of one 802.22 network and M 802.11af subnetworks, as well as a centralized spectrum coordinator, which collects the bandwidth requirement from $1 + M$ subnetworks and makes the decision on the availability of the spectrum. The authors assume there is a common channel for the communications between each subnetwork and spectrum coordinator. It also assumes that all the bandwidth requests from $1 + M$ subnetworks arrive at the coordinator at the same time. Thus, the coordinator collects the requests and allocates the available spectrum optimally and, hence, the overall spectral efficiency is maximized.

Assume the bandwidth is equally distributed with unit bandwidth b_w, and suppose the total available bandwidth is denoted by B, thus, the total assigned channels from coordinator N can be calculated by $[B/b_w]$ where $[B/b_w]$ is rounded up of B/b_w. Assume each secondary user's transmission power P_T and noise power P_N is the same as other secondary users', as a result, the distribution of channel gain g determines the distribution of signal to noise ratio (SNR). At a decision time, each secondary subnetwork requests a bandwidth b$_i$ to the spectrum coordinator for the current transmission. Among all $1 + M$ requests, the coordinator decides which requests are accepted. The authors in [23] assume the maximum bandwidth that coordinator assigns to subnetwork is B_i, and there is an average channel gain $\bar{g}_{i,j}$ of each network i, which can be calculated by the average of channel gain of each user in i^{th} subnetwork.

The system spectral efficiency is represented as Equation (25.12), which is a nonnegative and nondecreasing function with respect to $\bar{g}_{i,j}$. Denote a variable of $x_{i,j}$, which equals 1 if the bandwidth request from the ith subnetwork is accepted by the spectrum coordinator through assigning a frequency segment in the jth channel band. Then, it is reasonable to formulate the decision problem of the spectrum coordinator as:

$$
\begin{aligned}
max \quad & \eta = \sum_{i=1}^{1+M} \sum_{j=1}^{N} log_2(1 + \bar{g}_{i,j} x_{i,j} \tfrac{P_T}{P_N}) \\
s.t. \quad & b_w \sum_{j=1}^{N} x_{ij} \leq B_i \\
& min(g_{k,j} x_{i,j} P_{T_{k,j}}) \geq \tau,
\end{aligned}
\tag{26.12}
$$

where the first constraint ensures that the assigned bandwidth to i^{th} subnetwork should be less than or equal to the maximum bandwidth assigned by spectrum coordinator. The lower-bound of the power constraint is set by the second inequality, where each power spectral density allocated for each user at channel j should be higher than the sensitivity limit. τ denotes transmission power on channel j (equal to P_T). $g_{k,j}$ stands for the channel gain on the jth channel between the kth user and the base station in 802.22 networks or AP (access point) in the 802.11af network. Define that user k is a cognitive user in the ith subnetwork. Instead of satisfying all secondary users' allocated power constraint, the user with the minimum received power and is selected. The optimization objective is to let the user with the minimum value larger than the threshold τ, thus, all the users in the subnetwork i will meet the power constraint. In conclusion, the spectrum coordinator needs to find the optimum value, denoted by $x_{i,j}^* = [x_{1,j}^*, x_{2,j}^* \ldots, x_{1+M,j}^*]$ which maximizes the overall throughput while the constraints are satisfied.

(a) Bandwidth allocation

According to the aforementioned assumptions, each subnetwork i requests bandwidth $b_i(i = 1, \ldots 1 + M)$ from spectrum coordinator. Let r denotes the ratio factor, which is the ratio between the ith subnetwork requested bandwidth b_i and the total requested bandwidth $\sum_{i=1}^{1+M} b_i$. Since the total available bandwidth is B, then the rule of maximum allocated bandwidth B_io i^{th} subnetwork can be represented as follows:

$$B_i = rB = \frac{b_i}{\sum_{i=1}^{1+M} b_i} B. \tag{26.13}$$

Fairness is taken into consideration in (25.13) so that each subnetwork is assigned to corresponding proportional bandwidth based on its request. Nevertheless, it's likely that the allocated bandwidth to i^{th} subnetwork B_i is not an integer multiple of the unit bandwidth b_w. In such case, allocating bandwidth B_i to subnetwork is inappropriate. Therefore, define the bandwidth B_i', which considers the factor that the total available bandwidth is integer multiples of the unit bandwidth:

$$B_i' = \frac{B_i}{b_w} \cdot b_w = \left[\frac{\frac{b_i}{\sum_{i=1}^{1+M} b_i} B}{b_w} \right] b_w. \tag{26.14}$$

On the other hand, if the bandwidth is allocated in accordance of formula (26.14), the total allocated bandwidth to subnetworks $\sum_{i=1}^{1+M} B_i'$ may larger than the total available bandwidth B due to the property of rounding function. As a consequence, there is redundant bandwidth $\sum_{i=1}^{1+M} B_i' - B$ to be eliminated.

The authors discuss the fairness of bandwidth allocation of each subnetwork in our coexistence system using Jain's fairness index. The fairness index provides a fairness criterion which takes all the subnetworks into account. The formular of Jain's fairness index (FI) is as follows:

$$FI = \frac{(\sum_{i=1}^{n} x_i)^2}{n(\sum_{i=1}^{n} x_i^2)}. \tag{26.15}$$

To make our coexistence system fairest, the aforemented problem can be expressed as:

$$\begin{aligned} max \quad & FI = \frac{(\sum_{i=1}^{1+M} x_i)^2}{(1+M)(\sum_{i=1}^{1+M} x_i^2)} \\ s.t. \quad & \sum_{i=1}^{1+M} z_i = \sum_{i=1}^{1+M} B_i' - B. \end{aligned} \tag{26.16}$$

Define x_i as the ratio between requested bandwidth b_i and allocated bandwidth $B_i' - z_i$ of subnetwork i, where z_i is the redundant bandwidth of ith subnetwork to be eliminated. According to formular (26.17), x_i can be calculated by:

$$x_i = \frac{b_i}{B_i' - z_i} = \frac{b_i}{\frac{B_i}{b_w} \cdot b_w - z_i}. \tag{26.17}$$

So far, authors haven't focused on the complexity analysis of the algorithm. Brute-force computation is used to solve the fairness optimization problem. Other methods that reduce the complexity will be used in the future work.

(b) Channel allocation

With the result of bandwidth allocation from the bandwidth allocation section, centralized coordinator will then make the decision which channels are assigned to each subnetwork. It is apparent to note that the decision problem formulated in (25.12) can be modeled as a 0-1 MKP, where the available channels are regarded as the items and subnetworks as knapsacks. Thus, the optimization problem can be represented as follows: Given N total available channels and $(1+M)$ subnetworks, we need to find out a channel assignment algorithm to maximize the coexistence system spectral efficiency, which corresponds to the profit in the knapsack problem.

Because the function $log_2(1 + \bar{g}_{i,j}x_{i,j}\frac{P_T}{P_N})$n (12) has the same monotony as $\bar{g}_{i,j}\frac{P_T}{P_N}x_{i,j}$, we can derive the maximum value of $\sum_{i=1}^{1+M}\sum_{j=1}^{N}\bar{g}_{i,j}\frac{P_T}{P_N}x_{i,j}$ instead. Then we can obtain $x_{i,j}$ rom the maximum and substitute $x_{i,j}$ to formular (12) to obtain the maximal spectral efficiency. Let $f(\bar{g}_{i,j}) = \bar{g}_{i,j}\frac{P_T}{P_N}x_{i,j}$ Without loss of generality, we assume in each subnetwork, the channels are sorted so that:

$$\frac{f\left(g_{i,1}\right)}{b_w} \geq \frac{f\left(g_{i,2}\right)}{b_w} \geq \ldots \geq \frac{f\left(g_{i,N}\right)}{b_w}. \tag{26.18}$$

And the maximum allocated bandwidth B_i to i^{th} subnetwork is sorted as follows:

$$B_1 \leq B_2 \leq \ldots \leq B_N. \tag{26.19}$$

An initial feasible solution is determined by applying the Algorithm 1 to the first subnetwork, then to the second subnetwork by using only the remaining channels, and so on. This can be obtained by calling N times the following procedure, where η is the desired maximum spectral efficiency. Let's define a vector $y_i (i = 1, 2 \ldots, 1 + M)$ as follows:

$$y_i = \begin{cases} 0, if\,the\,i^{th}\,subnetwork\,is\,currently\,unassigned \\ index\,of\,the\,channel\,it\,is\,assigned\,to,\,otherwise \end{cases}. \tag{26.20}$$

Based on the description above, it is easy to follow the polynomial-time approximation algorithm proposed by Martello and Toth [89]. The main idea of the algorithm is as follows: After calling Greedy $1 + M$ times, the algorithm will improve on the solution through local exchanges. First, it considers all pairs of channels assigned to different subnetworks and, if possible, interchanges them to allow a new channel to be inserted. When all pairs have be considered, the approximation algorithm tries to exclude in turn each channel currently in the solution and to replace it with one or more channels not in the solution, such that the total throughput is increased.

Both coordinator and non-coordinator based DSA mechanisms have been introduced in this section wherein the spectrum resources are allocated to the coexisting networks optimally. However, spectrum is not the only network resource for the networks. For example, in the TDMA-based networks, the time is the resource which is allocated to the users for their accessing. Therefore, the fairness should also be considered in the design of MAC protocols, wherein the multiplexing access mechanisms such as FDMA, TDMA, etc., are realized.

26.5.3 Fair MAC Protocol

The CRNs offer the flexibility to utilize the spectrum resources efficiently. Consequently, the design of MAC layer protocol differs from the traditional networks significantly in terms of spectrum sensing, spectrum mobility, and so on. In particular, coexisting scenario brings in more difficulties in the design of MAC protocol since the unknown of other networks. Among those existing MAC protocols for the CRNs, fairness is one of the most significant metrics considered in the design of MAC protocol. For example, in [60], a fair opportunistic spectrum access scheme is proposed that, based on a fast catch up strategy, manages to reduce the amount of time after which all SUs have equal access rights to the available licensed channels (LCs). In [61], a Homo Egualis-based learning model was proposed to achieve fairness among dissimilar SUs, while in [58] the authors proposed heuristic channel allocation algorithms based on multi-channel contention graphs and linear programming aiming at achieving a good trade-off between throughput and fairness, while ensuring interference-free transmissions. In [62], the authors derive the optimal access probabilities for two independent SUs focusing on achieving a good trade-off between spectrum efficiency and fairness. However, all these proposals assume that an LC that is occupied by an SU cannot be accessed by another SU. In particular, the LC appears as being busy to the SU and, thus, it is avoided. Hence, most coexistence schemes in the literature totally overlook the case where several secondary networks (SNs) coexist and share the same PU resources. To overcome the aforementioned problem, in [55], the authors propose fair MAC protocol (FMAC), a MAC protocol that utilizes a three-state sensing model. Specifically, FMAC uses a spectrum sensing algorithm [56] to distinguish whether a busy channel is occupied by a PU or by an SU and, in the latter case, gives the option to the SU to share the channel with the SUs of other SNs that are currently using it. Nevertheless, in [55] a simple system model consisting only of one LC is considered, while, more importantly, the scheme employs a constant back-off window. As a result, unlike the proposed coexistence scheme, it shows low adaptability to any changes in the number of contending SUs in an LC.

Following passages will illustrate several CRN MAC protocol designs considering the fairness as one of their targets. These CRN MAC protocols can be categorized into two groups. In one group, the SU does not care whether the spectrum resources are occupied by PU or other SU. In this sense, whenever the SU detects the spectrum band is occupied according to the simple energy detection mechanism, then this SU will stop transmission and switch to the other available spectrum bands. These mechanisms resort to the simple energy detection spectrum sensing scheme which leads to the implementation with low complexity.

However, from the fairness perspective, these mechanisms are not fair to the SUs. For example, if the spectrum bands are occupied by a SU. This may lead to the unsuccessful transmission for the other SUs since the other SU have to vacate the spectrum band, which they detect as occupied regardless of occupied by primary user or other SU. Therefore, the SUs who successfully access the spectrum bands will have more "privilege" than the others.

For example, the authors in [55] claims that it is not fair for the SUs using traditional spectrum sensing algorithms in which only two states of channel are assumed, i.e., the channel is occupied (H_1) and the channel is not occupied (H_0). This is because that if a MAC-based on the two-state sensing model is used, then when the SUs of one CRN are accessing the channel, the SUs of other CRNs often starve. Hence, such a MAC protocol is not coexistence friendly and results in poor fairness. For a MAC to be coexistence friendly, the SUs of one CRN must be able to share a channel with the SUs of another CRN. Based on this, [55] introduces a novel concept for spectrum sensing wherein the channel has three states, i.e., the channel is idle (H_0), channel is occupied by PU (H_1) and the channel is occupied by the SU (H_2). In this sense, when the channel is claimed as occupied, the SU will further detect if the channel is occupied by PU or SU. If (H_2) is claimed, this SU will switch to the contention mode with the SU who is occupying the channel. Intuitively, the SUs will have the same chance to access into the spectrum by contending with each other. For instance, if there are N users competing for a channel, the ideal fairness is achieved if each user accesses the channel for 1/N of the total period that the channel is available for SU access.

This MAC protocol can be described as a state machine with three states in it, i.e., H0: channel is idle, H1: channel is occupied by PU and H2: the channel is occupied by the SU. When the SU has traffic to transmit. There are three possible states for a channel at any time, i.e., H0, H1, and H2. An SU that has traffic for transmission takes distinct actions based on the detected state of the channel. Specifically, if the channel state is H1, the SU does not simply switch to another channel. Instead, it keeps silent and continues to monitor the channel. If the channel state is H0, the SU accesses the channel immediately. In contrast, if the channel state is H2, the SU knows that the channel is occupied by another SU, which may be from a different CRN, and can participate in competition for channel access. During SU transmission, the SUs keep sensing the channel. If the sensing result is H0 or H2, SUs continue accessing the channel. However, they have to vacate the channel whenever the PU comes back. The operation of the FMAC protocol is illustrated in Figure 25.17(b), which is compared with a MAC with the two-state model in Figure 25.17(a).

In the FMAC, the channel access scheme by SUs under state H2 is based on the IEEE 802.11 MAC protocol. Specifically, an SU monitors the channel activity when it has a packet to transmit. The SU starts transmitting only after an idle period equal to a distributed inter-frame space (DIFS).

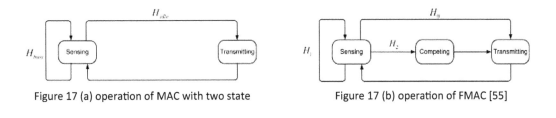

Figure 17 (a) operation of MAC with two state   Figure 17 (b) operation of FMAC [55]

Figure 26.17: Operation of MAC with two-state Figure 25.17 (b) operation of FMAC [55].

In the case that the channel is busy, i.e., another SU is currently occupying the channel, the SU randomly selects a back- off interval from [0, W-1], where W represents the size of a contention window. The backoff time counter is decremented whenever the channel is sensed idle, stopped when a transmission is detected, and reactivated when the channel is sensed as idle again for a DIFS. The SU transmits when the backoff time counter reaches 0. In case that a collision occurs, i.e., two or more SUs transmit simultaneously, the same backoff mechanism is repeated. SUs continue sensing the channel during transmission or a backoff period.

Although FMAC is designed based on IEEE 802.11, the cognition or spectrum sensing capability that is required for operation in a scenario of co-existing CRNs is added, as well as design mechanisms, to achieve fairness. As a legacy protocol, IEEE 802.11 does not have the cognition or spectrum sensing capability. Spectrum sensing is the unique feature of CRNs and it is fully considered in the design of FMAC. A novel renewal process is proposed for the situation whenever the PU comes back. Furthermore, FMAC is based on the three-state sensing model and responds distinctly when the channel is used by the PU or another SU. This is critical to improve the fairness for coexistence of CRNs. Recent studies pointed out that the fairness performance of IEEE 802.11 is not satisfactory because of the binary exponential backoff technology [15]. As studied in [16], users with different contention window sizes have different channel access probabilities, which then results in poor fairness among users. Therefore, to achieve optimal fairness among SUs, the binary exponential backoff technology is not adopted by FMAC. Instead, the same contention window size is used for all SUs, and, hence, the channel access probability is the same for all SUs, which results in optimal fairness.

However, only one licensed channel is considered in the FMAC protocol shown above, while, more importantly, the scheme employs a constant back-off window. As a result, unlike the proposed coexistence scheme, it shows low adaptability to any changes in the number of contending SUs in an licensed channel. To tackle this problem, the [60] presents an energy efficient contention aware channel selection-based MAC protocol to deal with the random spectrum access problem for distributed CRNs. Since the distributed cognitive networks are assumed in this mechanism, there is no centralized entity, such as base station, in the infrastructure-based networks in the cognitive networks. Hence, this MAC protocol regulates a coordinator for each secondary network. This coordinator is assigned to each secondary user in the secondary network in a round-robin manner.

By using this MAC protocol, the secondary network is assumed to be initially located in the highly congested unlicensed channels (ISM bands). Initially, the coordinator contends with other users operating over the unlicensed bands using CSMA with back-off mechanism. Whenever the coordinator obtains the unlicensed bands, it sends spectrum sensing request to other secondary users over the unlicensed bands. The sensing requirements for the other SUs such as (i) which licensed bands will be sensed; (ii) the order in which the SUs will send their sensing result to the coordinator and (iii) how often the sensing procedure will be triggered will be included in the spectrum sensing request. In the second period, the SUs sense the licensed channels assigned by the coordinator in the first period. If the channel is claimed as occupied, then the cyclostationary feature detection algorithms will be launched to decide whenever the channel is occupied by the PU or other SUs. In the last period, all the individual sensing results from SUs will be sent to the coordinator and the cooperative spectrum sensing with OR fusion rule is adopted at the coordinator to obtain the final spectrum sensing result.

This mechanism is able to achieve a higher fairness level in terms of fairness index due to the fact that for short contention periods, i.e., in the case that the PU resumes its activity in the LC shortly after the SN under study has hopped to it, the sensory nerve conduction studies (SNCS) achieves much better fairness among the SUs than in FMAC, as an SU that is involved in a collision defers its transmission for a longer time, and, thus, the transmissions opportunities are more equally distributed among the contending SUs.

On the contrary, [60] claims that even if the two-state channel model is used, the fairness can also be achieved due to the interference channel. Additionally, a contention-free MAC protocol was also proposed, which is called fair opportunistic spectrum access. The following example shows the case that the fairness is achieved due to the interference channel.

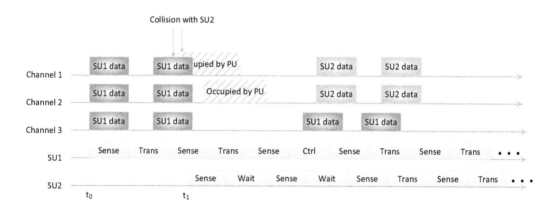

Figure 26.18: Example of contention based MAC protocol [60].

As shown in the Figure 25.18, SU1 accessed the channels first, and it sensed that all the 3 channels are available to access. So, it took all the 3 channels to transmit. When SU2 came, not any channel was left for it. It had to wait. Soon, SU1 found the presence of PUs at channels 1 and 2. It had to vacate them. After the absence of PUs at channels 1 and 2, SU2 found the two idle channels first; so it took it and started to transmit data.

However, in the traditional CRNs, three states can be used to represent the cognitive uses' statuses, which are control state, sense state, and transmission state. These three states can make up two cycles, i.e., control cycle composed of control state and sense state; and the sending cycle, which consists of sense state and transmission state. In the sense state, the SUs sense their surrounding environment to check if a particular channel is occupied or not. If the channel is claimed as occupied (which indicates there is collision with PU), then the SU switches to the control state. In the control state, the SUs look for the other available channel bands and switch to the other available channel bands. After successful switching to the other available channels, the SUs switch to the sense state again and start to sense the required channel status. On the contrary, if the SU claims that the channel it is sensing is idle (which demonstrates that there is no collision with PU), then the SU will switch to the transmission state and start to transmit its user data. After transmitting the user data, the secondary user will switch back to the sense state again. The aforementioned state machine can be represented in Figure 25.19:

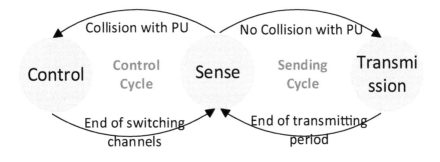

Figure 26.19: State machine for MAC protocol without FCS strategy.

However, since each SU is allowed to maintain a list of available channels for their occupation. When an incoming SU wants to access the spectrum, it may obtain a few channels first. The incoming SU will have more opportunity to transmit data in the a few channels, but not to actively occupy more idle channels which are found in the sensing state. This problem may damage the fairness of performance. Therefore, in order to solve this problem, [60] proposed fast catch-up strategy (FCS). With FCS strategy, whenever the SUs find the existence of PU in a particular channel or there are more channels available to access, this SU will switch to the control state. Otherwise, the SU switches to the transmission state. In this way, the collision of SU with other SU with more available channels will be avoided. The state transition of FCS scheme can be depicted Figure 25.20.

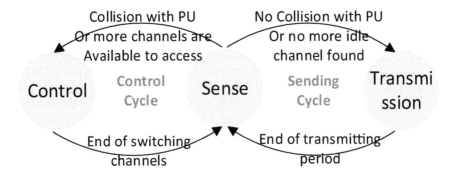

Figure 26.20: State machine for MAC protocol with FCS strategy.

26.6 Security Issues Relevant to Ineligible Access of Cognitive Radio Networks

Although having been actively studied, most existing work on the dynamic spectrum access technology focuses on the improvement of network performance alone (e.g., network throughput and channel utilization, etc.). There is only a few work of DSA paying attention to the eligibility of SUs' dynamic spectrum access requests in the CRNs. If the ineligible SUs gain access to the CRN network with privilege, many advanced severe attacks such as spectrum sensing data falsification, etc., can be launched. This will cause dramatic degradation of network performance. Hence, the network access control (NAC) mechanism is significantly critical for the CRN performance with which only the eligible users are granted with the spectrum resources dynamically. Generally, the network access control procedure can be divided into the following three steps, i.e., identification, authentication, and authorization.

Identification is the process that a new user presents his identity to the network. It is done in the first stage of gaining access to the network. In the identification process, the user is asked to provide the evidence to show their validity. The evidence can be in any form, such as the IP address, the specific characteristic of the user etc., as long as these evidences are known by the authority that is responsible for granting the resource to the users.

Authentication is the process of validating the user who requests for the privilege of network access. The authentication is done to ensure the authenticity of the user who is requesting the spectrum resources. For example, the adversaries may pretend to be a valid users by stealing a valid user's identifications. By authentication, the adversaries are identified and their requests for accessing the network are rejected. Usually, the authentication is conducted by verifying the unique characteristics of the user mentioned above, such as user ID, password, etc. After the validity of a user is verified, the authorization phase will start by the authority to grant the resources to the corresponding users.

According to the description above, the authentication mechanism is the core of the NAC process. Traditionally, the cryptography-based authentication is widely adopted in the existing networks. It can be mainly categorized into two groups from architecture perspective. They are symmetric key based authentication and public-key-based authentication [63].

In the symmetric key scenario, the user shares a single, secret key with an authentication server (normally the key is embedded in a token). The user is authenticated by sending to the authentication server his/her username together with a random challenge message that is encrypted by the secret key. The user is considered as authenticated if the server can match the received encrypted message using its share of the secret key. In the public key authentication mechanisms, the public key-based protocol always combines the digital signature to realize the authentication, which can be described as follows. The sender calculates the hash corresponds to the messages he wants to send first. The digital signature is calculated by encrypting the hash result by the sender's private key. Then the sender's original messages, along with the digital signature, are sent to the receiver. The sender can then be authenticated at the receiver by comparing the decrypted digital signal using the sender's public key with the hash operation of sender's original messages. The users' public key might be obtained from key distribution center (KDC).

However, due to the unique feature of CRNs such as dynamic spectrum mobility, the direct adoption of conventional cryptographic authentication in the CRNs will incur high network overhead. Furthermore, since the SUs are required to be able to dynamically accommodate the RF configuration that is suitable for the surrounding environment, the new users have no knowledge about their correct configurations, including operating frequency, modulation scheme, and coding schemes, etc. Consequently, this necessitates the new paradigm of authentication in CRNs, such as mechanisms assisted with common control channel (CCC).

Considering the unique features of the cognitive radio networks, there are several specific designs for the network access control mechanism which concentrate on the authentication mechanism, network access control with or without common control channel, and mechanisms for confidentiality. The rest of this section will introduce the details of network access control design considering the specific features of cognitive radio network from these three aspects.

26.7   Authentication Mechanisms

Several approaches were proposed to provide the authentication function for CRNs [64–68]. Generally, these works can be categorized into two groups: trust-based authentication and physical-layer-based authentication.

In the trust-based mechanisms, the trust values, which indicate the reliability of the user, are assigned for each user according to their behaviors. These values are updated in real-time and used to compare with a threshold to authenticate the user. However, the trust-based mechanisms are conducted on the upper layer (above the PHY layer). This requires the users to communicate with each other by assuming the configuration of the RF frontend and digital signal processing schemes (such as modulation and coding schemes) of every user is identical. Whereas, due to the dynamic spectrum access in the CRNs, those configurations should be optimized according to the surrounding environment. In this sense, it is difficult for the users to have the same configuration especially at the beginning stage of NAC when a new user attempts to request the spectrum resources.

Intuitively, to overcome the drawback of trust-based mechanisms, the NAC mechanisms for the CRN should be deployed at the PHY layer in which the radio waveforms are dealt with. The PHY-layer authentication mechanisms have been well studied. In those works, the transmitters are authenticated by their specific characteristics from the following aspects: (1) unique authentication messages; (2) unique link (channel) characteristics; and (3) unique location information. By sending the specific authentication messages at the transmitter, the receiver is able to verify if the signal is sent from the desired transmitter. The problem of how to integrate authentication messages with user data affects the network performance significantly. The link signature-based algorithm takes advantage of the unique characteristic (e.g., channel fading) of a link due to the environment, such as multipath, etc., between the transmitter and receiver to verify the authenticity of the transmitter, while the transmitter's location information is adopted in the location-based mechanisms. Obviously, these schemes are very sensitive to the environment, especially when the channel status and user's location change fast. Several examples of these mechanisms will be introduced in the following Subsection.

26.7.1   Trust Based Authentication

The trust-based authentication schemes build up the secondary networks' trust model with which the trust values for the SUs can be calculated and disseminated within the secondary network. One of the critical issues in the trust-based system is the updating of the trust value for the SU. Aiming at the different targets, the trust value updating models are different. In this section, several trust-based authentication mechanisms will be introduced.

In [64], a trust-based network was built, as shown in the Figure 25.21. The primary base station (PUBS) and secondary base station (SUBS) are in charge of the PUs and SUs respectively. The PUBS and SUBS are able to connect with a certificate authority (CA) containing a trust repository. The trust repository records the trust values for all users (including both PUs and SUs). In this proposed mechanism, there are two trust values for each user, i.e., a public trust value and a private trust value. The public trust value is visible to every node in the network, while the private value is only able to be accessed by the CA. This value is actually preserved for security purposes. If there are any hackers or attackers in the network and they intentionally alter the trust value, then the CA can check the private value of trust and obtain information about which node has been attacked. Then the CA broadcasts one message to revoke the hacked node from the network.

The trust values can be used as follows. When one SU tries to access one PU's free spectrum, the PUBS at first checks the SU's trust value from the CA's trust repository. If the value is greater than the predefined threshold, then the PUBS assigns free spectrum to the requested SU. If this is not the case, then the PUBS checks the reference trust value with which the SU already has a connection. The PUBS computes the average of the reference values of the trust value and checks the trust value.

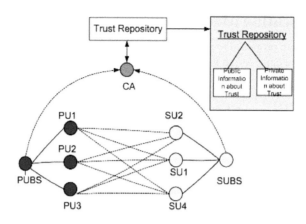

Figure 26.21: System model of trust-based network model.

If it is not an acceptable trust value, then the PUBS declines the request. But if a new node wants to access the PU's free spectrum, at first the joining node should meet the agreement with the base station. The SUBS, PUBS, and the member nodes assign the trust value to the new joining node by seeing its past reports

Trust value management is a critical procedural in the trust-based systems. It is mainly responsible for the trust value calculation and updating. In [64], every node in the network (including PUs, SUs, PUBS and SUBS) is assigned with a trust value. The trust value of a user is calculated by its $l - hop$ neighbor. For example, the trust value calculated by its ith hop neighbor of user j is denoted as $TV_i(j)$. Then, the averaged trust value for the user j is given by $TV(j) = \frac{\sum_{i \in T} TV_i(j)}{|J|}$, where T is the node set containing the nodes that have l hops from the user j. This averaged trust value will be sent and saved in the trust repository.

This service provides the assurance that the requesting entity is the one that it claims to be. An authentication is proposed by establishing trust value of every CR node that is stored by the CA. Whenever an SU wants to access the PU's free spectrum band, the SU shows its good manners in order to gain spectrum access. Then the PU accesses the trust table from the CA, and then the PU makes the decision on whether or not the SU can have access to the free spectrum.

26.7.2 Physical-Layer authentication

Physical-layer (PHY) authentication mechanisms can effectively authenticate the transmitters, since the authentication messages represented by bits are indeed embedded with the user data and sent to the receivers. The PHY-layer authentication mechanisms can be categorized into three groups, i.e., link (channel) signature-based authentication, localization-based authentication, and authentication message-based mechanisms. In the design of authentication messaged-based mechanisms, the problem of concern is how to effectively and efficiently embed the authentication messages with the user data.

26.7.2.1 Authentication Message-Based Mechanisms

In [65], the authors proposed a PHY-layer authentication scheme called *hierarchically modulated duobinary signaling for authentication* (HM-DSA), which is based on duobinary signaling, a waveform shaping technique that has been traditionally used to increase the bandwidth efficiency. The basic idea is to utilize the redundancy induced in the message signal due to the addition of inter-symbol interference (ISI) to embed the authentication signal.

Assuming that the message signal is transmitted in blocks of binary sequences, each with length N and represented by $\{d_n\}$ $n = 1, 2 \cdots, N$. Using non-return-to-zero (NRZ) encoding, a bipolar state sequence, $\{w_n\}$, is generated from $\{d_n\}$. Further, a duobinary sequence, $\{y_n\}$, is generated by adding the delayed and weighted states of $\{w_n\}$. It is achieved by using a digital filter represented by $y_n = w_n + \delta \cdot w_{n-1}$, where $0 < \delta < 1$. Hence, the ISI introduced to each y_n, corresponding to the state w_n, comes only from the preceding state, w_{n-1}. Moreover, the extent of ISI is controlled by δ. For $w_n = \pm 1$, we can obtain a four-level hierarchically modulated output, i.e., y_n has one of four possible values: $+1 + \delta, +1 - \delta, -1 + \delta$, or $-1 - \delta$, as shown in the following figure. We note that the four-level output of y_n is used to express one of the two binary values of the message signal, and, hence, there is an inherent redundancy in this process. We also observe that the encoded signal, y_1, is given by $y_1 = w_1 + \delta \cdot w_0$, where w_1 and w_0 are the bipolar states for d_1 and d_0, respectively. Hence, an extra bipolar state, w_0, and a corresponding bit, d_0, are required to start the encoding of the message signal, $\{d_n\}, n = 1, 2 \cdots, N$. Bit d_0 is called an initialization bit, and the bipolar state w_0 is called an initialization state. The received HM-DSA-based signal can be easily decoded using the regular maximum likelihood detection algorithm.

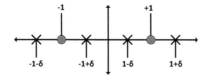

Figure 26.22: Modulation scheme for authentication messages [65].

Since the receiver needs to know the authentication messages that the transmitted user used, the authentication messages should be known by both transmitter and receiver in advance. This results in the key dissemination mechanism, which is widely discussed in the cryptography area. Furthermore, this approach requires the receiver has the knowledge about the signal processing parameters in the transmitter, such as the modulation and coding schemes. However, this is not always true in the spectrum sharing scenarios, since heterogeneous networks might coexist with each other, and they may use different signal processing techniques. In order to tackle this problem, [66] proposed a novel approach in which the receiver does not need to have any knowledge of the transmitter, which is called blind transmitter authentication (BTA).

The basic idea behind the BTA can be explained as follows. The transmitter embeds the authentication signal in the form of embedded frequency offset (EFO) in each frame of the message signal in the baseband. The embedded signal in the baseband is sent to the oscillator, where it gets up-converted and transmitted along with the inherent carrier frequency offset (CFO) due to the inaccurate oscillator. This overall frequency offset does not affect the decoding procedure of the message signal by the intended receivers. These intended receivers estimate and correct any frequency offset present in the received signal, with the help of the preamble symbols and the pilot samples. This scheme can be described as two main steps, i.e., generation and integration of the user message and authentication message, and extraction of the authentication messages.

Message generation: The message data to be transmitted is assumed to be a sequence of quadrature amplitude modulated (QAM) samples, which are statistically independent and identically distributed with zero mean and average power represented by σ_s^2. For each OFDM symbol, the transmitter generates N_f samples by taking the inverse fast fourier transform (IFFT) of N_u QAM samples corresponding to N_u non-zero sub-carriers loaded with data or pilot samples. The last N_c samples out of the N_f samples are repeated at the beginning of the N_f samples as the cyclic prefix (CP) to generate an OFDM symbol of $N_o = N_f + N_c$ samples. The message signal is transmitted in frames, and each frame contains $N_s = N_p + N_d$ FDM symbols, where N_p represents the number of symbols carrying the preamble, and N_d represents the symbols carrying data. The samples of a frame is represented by $\{s(n)\}$ where $n = 0, 1, \cdots N_s \cdot N_o - 1$.

The authentication signal contains three pieces of information: frequency, location, and time (represented by F, L, and T, respectively) at which the message signal is authorized to be transmitted. A timestamp, represented by TS, is also used to prevent replay of the authentication signal. This information, represented by $A_m = \{TS, F, L, T\}$, is digitally signed using a privacy preserving group signature scheme, and the signature of A_m, is represented by $sign(A_m)$. In this work, it assumes that a unique membership certificate has been issued by a designated CA to each member (including Alice) of the group of secondary users, and Dave has access to all the information available at the designated CA. Hence, one authentication sequence is given by $A_s = \{A_m, sign(A_m)\}$. Each A_s is channel-coded with an error-correcting code (e.g., convolution code), and synchronization and guard bits are appended to generate the authentication message with K bits.

Message combination: Frame frequency modulation (FFM) is used to embed the authentication messages into the user messages. In the FFM, the frequency offset of each frame of the message signal is modified (modulated) according to the authentication signal. FFM of order M (M-FFM) is represented by a set of M possible frequency offsets corresponding to $M = 2^b$ possible b-bit authentication symbols. Here, an authentication symbol is defined as a set of b authentication bits and is obtained by using b-bit Gray code. The set of frequency offsets in M-FFM can be represented by $\{f_m\}$ such that $f_m = f_a \cdot (1 - 2 \cdot \frac{m-1}{M-1})$, where $m = 1, 2, \cdots, M$ and f_a is the maximum positive frequency offset that can be used to embed the authentication signal into a frame of the message signal.

In the kth frame of the message signal, the authentication symbol is embedded, represented by a_k, by embedding a frequency offset, f_k. Hence, for $n = 0, 1 \cdots N_s \cdot N_o - 1$, each sample of a frame of the embedded signal in the baseband is given by $x(n) = s(n) \cdot e^{j2\pi \frac{f_k}{F_s} n}$, where F_s is the sampling frequency. The embedded signal is up-converted to the carrier frequency (F_c) and transmitted. Assuming that CFO due to the inaccurate oscillator at Alice is f_t, the total frequency offset of the transmitted signal is $f_k + f_t$.

Authentication message extraction: Since the authentication messages are integrated in the manner of frequency offset, these authentication messages can be extracted by the frequency detection mechanisms, such as phase locked loop (PLL) and carrier offset estimation schemes.

26.7.2.2 Localization-Based Authentication

In the localization-based authentication mechanisms, the location of the valid transmitters are pre-known. Therefore, an invalid user can be easily claimed when the distance between the invalid user's location and the desired valid user's location is greater than a predefined threshold. Hence, the critical problem in the localization-based authentication mechanisms is to extract the location information. For instance, [67] and [68] proposed two localization-based defense strategies against the PUEA. In [67], the received signal strength (RSS)-based localization is used to determine the location of the attacker by deploying an additional sensor network. In this work, the sensors are assumed to be uniformly deployed in an area that covers the transmitter who is being authenticated (called SU_x). Furthermore, the sensors are also assumed to be dense and can receive the signal transmitted by SU_x. It is obvious that the signal strength measured by the sensor, which is closest to the SU_x, is higher than any other sensors. Hence, the SU_x's location is estimated as the location of the sensor with highest RSS value.

Since the "best" sensor's position is considered as the estimated location of SU_x, the inaccuracy of estimation may be the drawback of this mechanism. Although this work also discusses the optimization of deploying the sensors, the optimal sensor deployment solution is highly related to the statistics of the channel model in which only the path loss model is considered. However, the channel characteristics may keep changing with time in reality due to the dynamic environment. Therefore, this approach may not be able to performance well in the dynamic environment.

In [68], a joint position verification method using both time difference of arrival (TDOA) and frequency difference of arrival (FDOA) was proposed to enhance the positioning accuracy. However, extra expensive hardware for the TDOA and FDOA are needed to be deployed in the network. Among the above techniques, time of arrival (TOA) is a receiver-localization technique and needs to be enhanced to support transmitter localization so that it can be applied to the primary signal transmitter (PST) localization problem. Such an enhancement is not trivial, especially when one considers the possibility that a malicious transmitter may craft its transmitted signal. TDOA and angle-of-arrival (AOA) techniques can both be used for transmitter localization and have relatively high localization precision. To apply them to the PST localization problem, special care must be taken to consider the situations where multiple transmitters or an attacker equipped with a directional antenna exists. The common drawback of both techniques in [67] and [68] is the requirement of expensive hardware, preventing them from a large-scale deployment. To tackle the implementation problems as mentioned above, [69] proposed an RSS-based approach, which does not need to deploy a sensor network and avoids the deployment of extra expensive hardware.

In [69], the authors consider the primary transmitter is the TV tower, which is fixed located, and the location of TV tower is assumed to be known by the SUs. The basic idea of this approach is to compare the calculated location of unknown SUs based on the corresponding RSS value. If the calculated location of an SU is deviated from the known primary transmitter's location, this SU will be claimed as adversary.

According the RSS value, two benign SUs (SU_1 and SU_2) are able to determine the trace of an unknown SU's (SU_x) location, assuming the SUs are equipped with GPS, i.e., the SUs are able to know their accurate locations. Denoting the two benign user's location as (x_1, y_1) and (x_1, y_1) respectively, based on the signal propagation model, then the trace of the unknown user's location is a circle whose center is $(\frac{\eta^2 x_1 - x_2}{\eta^2 - 1}, \frac{\eta^2 y_1 - y_2}{\eta^2 - 1})$, and radius is $\frac{\eta d_{12}}{\eta^2 - 1}$, where $\eta = \frac{d_{SU_x, SU_2}}{d_{SU_x, SU_1}} = 10^{\frac{RSS_1 - RSS_2}{\alpha}}$, an RSS_1 and RSS_2 are RSS values measured at SU_1 and SU_2, respectively. α is the path loss constant, and d_{12} is the distance between SU_1 and SU_2. Intuitively, three traces (circles) are needed to determine the unique location of an unknown SU. In another words, four benign SUs cooperate with each other to estimate an unknown user's location. An example is shown in the Figure 25.23.

26.7.2.3 Link (Channel) Signature-Based Authentication

As aforementioned, in the spectrum sharing scenarios, when a selfish SU wants to access into the whitespace, it is possible for him to transmit the signals to masquerade the PUs over the whitespace in order to get the privilege to access into the whitespace for their message transmission. According to the communication theory, the link (channel) characteristics between different pair of transceivers are different. Therefore, if the link (channel) characteristics between the valid transmitter and receiver are pre-known, it can be adopted to identify the invalid users who masquerade the valid users' signal to access into the spectrum resources.

However, in the spectrum sharing scenarios, there are three main critical problems that should be notified when using link-signature-based authentication: (1) the channel (link) characteristics keep changing due to the mobility of the users. Hence, the link characteristic between the transmitter and receiver cannot be adopted directly. (2) The Federal Communications Commission (FCC) states that "no modification to the incumbent system (i.e., PU) should be required to accommodate opportunistic use of the spectrum by SUs" [71]. Therefore, the channel characteristics are hard to be extracted without any cooperative processing between the transceiver. (3) Traditionally, the link-signature-based authentication needs the receiver to have the knowledge of channel pattern between the transmitter and it as reference. However, it is impossible for the SUs to store the link patterns of

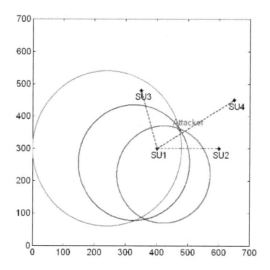

Figure 26.23: Illustration of the proposed location detection strategies by interactions between neighboring users [69].

all primary transmitters, especially when the SU is mobile user. To overcome the aforementioned problems, the link-signature-based channel utilization authentication mechanisms were proposed.

In [70], authors assume the primary transmitter is fixed. Based on this assumption, the authors proposed a two-step link-signature-based authentication mechanism. Step 1 is authentication of spectrum utilization by a helper node that is deployed near by the primary transmitter. In this step, the helper node located near by the primary transmitter authenticates if the specific channel bands are occupied by the PU or malicious SUs by verifying the estimated channel impose response at the helper node. This scheme takes advantages of multipath fading in reality, which can be described in the following figure.

As shown in Figure 25.24, T, R, and B represent the primary transmitter, helper node, and obstacle, respectively. The signal sent from primary transmitter can arrive at the helper node in two ways, i.e., direct link from T to R (link 1) and the link reflected by the obstacle B (link 2). Due to the different distance that the signal travels, the power of received signals along these two links at the helper node are different at different time instances. Figure 25.25 shows several examples of received signal power with different distances.

Since the helper node is very close to the primary transmitter, the received signal power from link 1 should be much more than link 2, intuitively. Therefore, the malicious users can be identified by comparing the ratio of received signal power along link 1 to the received signal power along

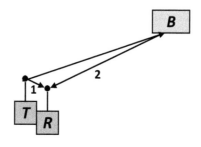

Figure 26.24: Illustration of multipath [70].

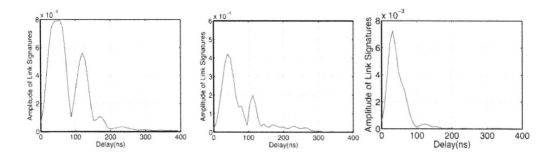

Figure 26.25: Received signal power at the helper node with two links [70].

link 2 with a threshold. Due to the randomness and uncertainty of the surrounding environment, this ratio measured for primary transmitter may not always be larger than the pre-determined threshold. Hence, there may be two types of possible errors: false alarm and false negative. With a false alarm, the PU's signal is incorrectly identified as the attacker's signal, while the attacker's signal is incorrectly identified as the primary transmitter's signal in the false negative. With the required false alarm and false negative, the threshold can be optimized.

The helper node may periodically trigger the SUs to calculate the channel characteristics. At the same time, the helper node sends its channel utilization detection result to the SUs to train them a what the primary transmitter's signal should look like. In this sense, the SUs will have enough channel patterns about the primary transmitter that can be used for the following spectrum utilization authentication, even when the helper node is in the sleeping mode.

Since the link signature-based authentication mechanisms authenticate the primary transmitters based on the unique (uncorrelated) channel characteristics between the primary transmitter and the secondary receiver, and the channel is open to everyone, the traditional link signature-based approaches are vulnerable to several specific attacks, such as mimicry attack [71] and correlation attack [72].

In [71], the mimicry attack was identified for the scenario with link signature-based authentication. Let y_t and y_a denote the received symbols from the transmitter and the attacker, respectively. The attacker's goal in the mimicry attack is to make y_a approximately the same as y_t. Thus, when the receiver attempts to extract the link signature from the attacker's symbols y_a, it will get a link signature similar to the one estimated from y_t. There are two ways for the attacker to transmit the similar signal as primary transmitter, i.e., by forging or forwarding. Note that it is possible to simply add digital signatures or Message Integrity Code (MIC) into each frame. As a result, the frames forged by the attacker can be easily detected through authentication of message content. Thus, the remaining threat is from the frames that are originally generated by the transmitter but forwarded by the attacker. To overcome the mimicry attack, the time-synched link signature approach was proposed.

Moreover, with replay attack detection mechanism such as sequence numbers, if the verifier can receive the original frames sent by the transmitter, it can easily identify frames forwarded by the attacker as duplicates and discard them. Thus, the unresolved threats are from the following two cases: (1) when the attacker can jam and replay the Transmitter's frames (jam-and-replay attack [73]), and (2) when the transmitter and the verifier are out of communication range, but the jammer forwards frames from the Transmitter to the Verifier.

For the case 1, in order to overcome the jam-and-replay attack, [71] proposed to bring "time" into the scheme. It assumes that the transmitter and the verifier have synchronized clocks. The transmitter may include a timestamp in the transmitted frame, which indicates the time when a particular bit or byte called the anchor (e.g., the start of frame delimiter (SFD) field) is transmitted over the air. The transmitter is assumed to use authenticated timestamping techniques (e.g., [74]) to ensure that the timestamp precisely represents the point in time when the anchor is transmitted. Upon receiving a frame, the verifier can use this timestamp and the frame receiving time to estimate the frame traverse time. An overly long time indicates that the frame has been forwarded by an intermediate attacker.

To tackle the threat in case 2, the location of the training sequence is made unpredictable until the end of the frame transmission. Specifically, the authors insert the training sequence at a random location in the payload, and place this location, which can be represented as the offset from the start of the frame header, at the end of the frame. In order for a PHY-layer symbol repeater to mimic the link signature of the transmitter, she has to manipulate the PHY-layer symbols corresponding to the training sequence in a frame. If the location of the training sequence is not revealed until the end of the frame, the attacker will have to wait until the end of the transmission to learn it. This forces a PHY-layer symbol repeater attack to degenerate into a frame repeater attack.

Paper [72] also introduces a potential thread in the scenarios with link-signature-based authentication. The success of these schemes relies crucially on the uniqueness of link signatures resulting from the assumed fast spatial decorrelation of wireless channels; in particular, it is widely accepted that half a wavelength separation is sufficient for security assurance.

However, two critical questions remain unclear. First, does the common "half-wavelength decorrelation" assumption hold in all circumstances? Since the spatial channel correlation is significantly influenced by the angular spread (AS) of the incoming signal, when two receivers are located with rich scatters, their corresponding AS is usually large and the half-wavelength decorrelation conclusion holds. But when a line-of-sight (LOS) component exists or the waveguide propagation effect dominates, the AS is small and will induce high spatial channel correlation. In fact, high spatial channel correlations have already been observed in real-world experiments. Second, when the half-wavelength decorrelation assumption is violated, is the current link signature technique still able to provide security protection to wireless applications? Therefore, according to the aforementioned statements, the link-signature-based authentication is still in the middle of design stage. There are lots of potential threats need to be overcome.

26.7.2.4 Re-Authentication Problems

Nowadays, with the rapid enhancement of integrated circuit design, more and more radios are able to be equipped on one user device for different network services, such as WLAN, Bluetooth, and cellular network, and so on. This provides the chance for the user end to realize the seamless network connection service by the cooperation of different networks. For example, for a user device equipped with WLAN and Cellular network ratios, if this user end moves into the "dead zone" of the cellular network while this user is being served, he can immediately switch on the WLAN radio and continue the service through the WLAN network, considering the access point of WLAN network is available in that area. It is obvious that the QoS for the users can be improved in this sense.

However, whenever a user tries to access into a new network, the mutual authentication between this user and the new network is needed. Furthermore, different networks specify their own authentication mechanism, which may quite different from each other. This requires the frequent authentication when the user dives into the area with very poor channel characteristics (e.g., deep fading, etc.), whereas, the frequent authentication induces severe network overhead and may reduce the network throughput. Therefore, a fast radio independent re-authentication mechanism is needed in this scenario to improve the network performance.

The authors in [85] proposed a radio independent re-authentication mechanism based on the mobile user's location trail information. In this work, the authors assume that the mobile user's location information is available for both the user himself and also the network. The mobile user's

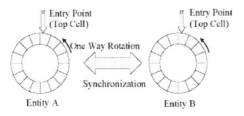

Figure 26.26: Carousel structure for shared key generation.

location information is then used to generate the shared keys for the user and network. The shared keys are generated by a carousel structure, which is shown in the Figure 25.26.

As shown in Figure 25.26, there are two carousels called entity A and entity B that are maintained in the mobile user and network, respectively. Each carousel is composed of several cells that contains the location information of the mobile user. Whenever the mobile user changes its location (i.e., its location information is updated), the new location information will be recorded in both of the carousels at the entry points. Then the carousel rotates by a random number of cells. If the entry point has the old location information for the user, this old location information will be overwritten by the new location information. If these two entities are identical, they could be adopted to generate the same shared keys for the mobile user and the network.

The factor that results in the difference between two entities is the random rotation. Therefore, the synchronization is needed in which entity B rotates in order to ensure the rotation number of entity B is same as entity A (assuming entity A is maintained by the mobile user and entity B is maintained by the network. Hence, the rotation of entity A caused by the new location information will trigger the rotation of entity B). The synchronization is achieved by the following steps: (1) the first entity generates an authentication key from the carousel and sends a challenge to the second, and (2) the second entity generates a key by rotating the carousel, attempts to decode the challenge, and continues to rotate until decoding is successful.

The work in [85] provides an effective radio independent re-authentication mechanism based on the mobile user's location trail information. However, this work assumes there is a location server deployed in the network, which is responsible for collecting the location information from the mobile users and disseminating each user's location information to the neighbor networks. It is obvious that with a large number of users, the mechanism proposed in [85] might lead to a very crowded network environment between the mobile user and the network. Therefore, the effective and efficient radio independent re-authentication mechanism is still needed.

26.7.3   Jamming Attack on the Common Control Channel and Solutions

Once a new user wants access into a network, he/she needs to provide the identity information for authentication, and authorization etc., which is called NAC, as mentioned in the previous section. Due to the dynamic spectrum allocation mechanisms in the modern networks (such as IEEE802.22, IEEE 802.11af, etc.), the user does not know which channel bands are used in the network. Therefore, a dedicated channel, CCC, is adopted for users to send their network access requests at this stage. However, due to the openness of the predetermined CCC, the jamming attack, wherein the adversaries intentionally send the jamming signals to interfere with the CRN subscribers, can be easily launched on the CCC. With the jamming attacks, the SINR, drops dramatically, which may jeopardize the whole network [79–80]. Therefore, the protection of the CCC is crucial for the overall performance of CRNs.

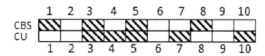

Figure 26.27: Channel availability at the BS and SU.

26.7.4   CCC Avoidance Mechanisms

The jamming attack on the CCC of CRNs has attracted researchers' great attention recently. Several anti-jamming strategies for the CCC of CRNs have been proposed in the literature, which can be grouped into two categories. In one category, the utilization of CCC is attempted to be avoided [81, 82]. In this sense, there is no control messages need to be exchanged. The specific frequency hopping criteria are designed for the source and destination nodes to hop to the same channel for their user data communication. However, the large time-to-rendezvous (i.e. the time needed for source and destination nodes to find a common channel) is the bottleneck of this approach. The detail description of this kind of approach can be further grouped into two classes according to network architecture, i.e., for centralized network and decentralized network.

26.7.4.1   For Centralized Networks

In a centralized architecture, a manager (usually the base station—BS) maintains a timer that counts to T_S seconds. It initially starts its search from the channel with the lowest frequency and starts its timer, T_S. It shifts to the next channel when the timer expires. In each time slot, the channel is scanned for the presence of a PU. If the channel is not free, then BS will immediately shift to the next channel and reset the timer. If the channel is free, a beacon is sent indicating its presence in that channel. It will wait for a response for the rest of the time slot until the T_S timer expires and then tune to the next channel, starting its timer again. If in the meantime a response is received from a SU, a different CR is assigned the task of carrying on the negotiations with the SU and the BS continues its search for other potential users. After all the channels are searched, it will restart from the lowest frequency again. If all the N channels were free, core banking solution (CBS) would take $N \times T_S$ seconds to complete a cycle of searching all the channels.

Every SU maintains a wait timer, T_W, which is set to $N \times T_S$. It initially starts from the channel with lowest frequency and scans for the availability. If the channel is not free, it shifts to the next channel and resets its timer. If the channel is free, it waits for a beacon from the CBS until the timer T_W expires. Since BS will search all the channels at-least once in T_W seconds, the SU can be sure of receiving a beacon if the channel it was listening to is free with BS. The total process is illustrated using Figure 25.27. Each block in the figure represents a channel. So, there are a total of 10 channels with each having a BS and SU. A shaded block means that PU is active in that channel. BS starts its search from the channel by setting its timer, T_S. Since the first channel is not available, it will reset its timer and shift to the second channel. As the BS scans and sees that channel 2 available, it sends beacons in this channel and waits until the timer expires for a response. Similarly, an SU starts from the first channel and waits for T_W seconds and will not receive any beacon because BS does not send beacon in that channel. After the T_W timer expires, the SU shifts to the next channel, where it will receive a beacon from the CBS and respond to the beacon and request a connection. It should be observed that an SU will receive a beacon in a maximum time of $N^2 \times T_S$ seconds if at least one channel is free with both BS and SU.

26.7.4.2 For Decentralized Networks

Compared with the centralized network scenarios, the rendezvous process in the decentralized is more difficult, since at the beginning stage, every user does not have any knowledge of the others, and there is no controller (such as BS) to help them jump into the same frequency band. Therefore, the critical problem for the frequency hopping-based rendezvous process can be summarized as follows: (1) to guarantee the periodic overlap between any pair of frequency hopping sequences so that a pair of nodes that wish to establish a link can rendezvous; and (2) to guarantee that any two frequency hopping sequences will rendezvous in more than one channel within a sequence period.

For example, in [83], the author proposed the use of non-orthogonal sequences to attain rendezvous, while still not requiring any synchronization between radios. For the purposes of rendezvous, the use of non-orthogonal sequences is proposed so as to maximize the probability that two radios looking for each other will eventually be searching on the same channel.

The use of pre-defined sequences by each radio is proposed to determine the order in which potential channels are to be visited. These sequences are constructed in such a way to minimize the maximum or the expected time-to-rendezvous even when radios are not synchronized to each other. For instance, consider radio 1 starting to look for a peer at time t_1, and radio 2 doing the same at time t_2. In our method, each radio follows a pre-defined sequence in visiting the potentially-available channels in search of each other. The properties of the time to rendezvous depend on the sequence selected.

A concrete example is provided by describing one method for building these sequences below. Consider again a set of N potentially-available channels, numbered 1 through N. A visiting sequence $a = (a_1, a_2, a_3, \ldots)$ describes the order in which a radio visits channels in search of other radios with which to rendezvous. It is notable to design the sequences that are periodic and that, for fairness reasons, contain in each period the same number of instances of each channel.

One method for building such a sequence is to select a permutation of the N channels (there are $N!$ such permutations) and building the sequence. The selected permutation appears (N+1) times in the sequence: N times the permutation appears contiguously, and once the permutation appears interspersed with the other N permutations. The average time-to-rendezvous of this approach is given by $E[TTR] = \frac{N^4 + 2N^2 + 6N - 3}{3N(N+1)}$.

The most existing rendezvous mechanisms consider only one radio equipped on each user. However, as the cost of wireless transceivers is dropping, this feature can be exploited to significantly improve the rendezvous performance at low cost. In particular, when an SU is equipped with multiple radios, the time-to-rendezvous can potentially be reduced by a large amount, while the additional cost (i.e., cost of the extra radios) is low.

In [83], the author proves that the parallel sequence strategy for channel hopping can achieve smaller mean time-to-rendezvous. The basic idea of the parallel sequence strategy is to assign multiple radios with two roles: general radio and dedicated radio. There is only one dedicated radio, and the remaining radios are general radios. Users hop on available channels in the general radios while staying on a specific channel in the dedicated radio. The rendezvous is expected to be achieved between the general radios of one user and the dedicated radio of the other. Suppose that a user is equipped with m radios. Our algorithm, role-based parallel sequence (RPS), is described as follows:

(i) All radios are divided into two groups, $(m - 1)$ general radios and one dedicated radio.

(ii) A starting index i is randomly selected from $[1, P - 1]$. A step-length r is randomly selected from $[1, P - 1]$. P is the smallest prime number, which is not smaller than Q.

(iii) The $(m - 1)$ general radios in parallel hop on P channels with step-length r in the round-robin fashion.

(iv) The dedicated radio stays on one channel for $\frac{P}{m-1}$ time slots and switches to next channel for the same duration. The stay channel is taken from $[1, Q]$ in the round-robin fashion.

(v) If the channel is not available to the user, a random available channel will be selected to replace it.

26.7.4.3 Frequency Hopping-Based CCC Mechanisms

The dynamic frequency hopping and location mobility [84, 85] strategies were proposed as the other category. In these mechanisms, the jamming detection phase, is followed by the operation phase wherein the channel bands for CCC as well as the locations of the SUs are dynamically moved. This leads to the unprotected period of jamming detection phase of the CCC. The legal user signal might be interfered in this period. Channel or location switching time is another drawback of this type of approach.

Figure 25.28 depicts the block diagram of the traditional frequency hopping strategy. A main limitation with this design structure is the strong requirement on PN acquisition, as exact frequency synchronization has to be kept between transmitter and receiver. Synchronization dominates the complexity and the performance of the system. Slow hopping systems, therefore, have been popular due to their relaxed synchronization requirement. On the other hand, due to their resistance to hostile jamming and interception, fast hopping systems are highly desired in classified information transmission. This raises a big challenge in transmitter and receiver design. In addition to strict synchronization requirement, traditional frequency hopping systems are also being challenged to transport more information with little or no increase in allocated bandwidth. Meeting these challenges requires advanced signaling techniques.

To tackle the aforementioned problem, [85] proposed a typical example of frequency hopping based anti-jamming CCC design. The basic idea is that part of the message will be acting as the PN sequence for carrier frequency selection. Taking the original modulation technique (such as frequency-shift keying (FSK) or phase-shift keying (PSK)) into consideration, transmission of information through frequency control in, fact, adds another dimension to existing constellations, and the resulting coding gain increases the spectral efficiency significantly. At the same time, the receiver is designed to be able to detect the transmission frequency automatically, hence, relaxes the burden on PN acquisition.

26.7.5 Mechanisms for Confidentiality

The confidentiality is one of the most critical metrics in the design of a network system. It is more important in the NAC procedure. In the NAC processes, the network nodes may need to send the network access request along with their passwords and their configuration parameters. If these messages are eavesdropped by the malicious users, they may be able to access into the network with privilege and launch severe attacks, such as spectrum sensing data falsification (SSDF), and other relay featured attacks, as shown in the first section of this chapter. Therefore, the design of NAC mechanisms should be aware of the confidentiality of exchanged messages.

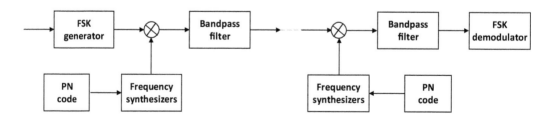

Figure 26.28: The block diagram of the conventional frequency hopping scheme.

The traditional approach is the key-based cryptography [63]. However, the implementation complexity is the bottleneck of this kind of approach. Another potential approach protects the confidentiality of the communication between honest nodes by producing an interference signal toward the eavesdropper [76–78]. Similarly, this kind of approach also induces many extra signal processes to generate the interference to the eavesdropper, while ensuring the honest communication is not affected by this artificial interference. The third kind of confidentiality protection mechanism is the most promising one, based on the dynamic spectrum allocation/access (DSA). Since the nature of DSA is similar to frequency hopping, DSA also provides security capability. Further, DSA is an indispensable module used in the CRNs. Therefore, the adoption of DSA for confidentiality does not need any extra devices and processing. Unfortunately, the DSA-based confidentiality protection mechanism has not been studied well so far. In the following sub-section, we will elaborate the DSA-based mechanisms.

26.7.6 *DSA-Based mechanism*

The spectrum sharing is credit to the dynamic spectrum allocation/access mechanism supported by the cognitive functionalities, such as spectrum sensing, spectrum mobility, etc. In the spectrum allocation mechanism, the available spectrum resources are optimized to be allocated to the shared networks and their users. Due to the changing spectrum resource utilization in the dynamic spectrum allocation scheme, the DSA is able to provide confidentiality, which is similar with frequency hopping schemes. Since the DSA is indispensable for a spectrum sharing system, it is a good chance to design the DSA scheme with considering the confidentiality.

For example, in [75], the authors proposed a DSA-based confidentiality for the CRNs. In this work, a confidentiality oriented DSA was designed. Compared with the traditional DSA schemes, in the confidentiality oriented DSA, each active user should randomly hop over channels in one transmission session to prevent entire messages from being intercepted. This is different from the traditional frequency hopping schemes, wherein during one transmission session, there is no frequency hopping that occurs. The number of channels that a user should access during one transmission depends on the number of eavesdroppers nearby the user. Without knowing the channel hopping pattern of benign CRs, the best action for an adversary is to consecutively eavesdrop over one channel. If there are n attackers eavesdropping on a benign user, the user must at least hop over n+1channels to keep entire data from being possibly exposed. Moreover, hopping over more channels can increase the diversity of the spectrum access pattern. Therefore, a user should access all available channels during one transmission session. Another conclusion is that the frequency hopping DSA-based cryptography can only work when the number of adversaries around any benign user is less than the channel number. This is referred to as the first law of frequency hopping DSA-based cryptography. Without loss of generality, in the following, the number of attacker around a benign user is considered to be only one. For other cases, as long as the first law is satisfied, only the confidentiality degree is degenerated, while the cryptography method itself is not affected.

With each user accessing spectrum multiple times, the DSA process is modeled as a fractional coloring process. Such a fractional coloring process turns out to be a constraint satisfaction problem (CSP) with both confidentiality goal and spectrum utility goal. However, the traditional search-based CSP solutions are incompetent for the DSA CSP due to myriad possible coloring schemes for fractional coloring. To keep the cost for confidentiality low, a Latin square-based (LS) confidentiality oriented DSA algorithm is proposed, which utilizes the art of graph expansion and special properties of the LS to give an efficient CSP solution. The graph model for this confidentiality oriented DSA is described as follows.

Consider N active CR users $N = \{1,\ldots,N\}$ sharing access of orthogonal channels $K = \{1,\ldots,K\}$, with a schedule containing T time slots. Each user u_i corresponds to a pair of transmitting and receiving CRs. Then, the user u_i can be considered as the vertex in the graph, and the edge between the vertices corresponding to u_i and u_j denotes the interference between user u_i and u_j. Since the confidentiality oriented DSA requires each CR user to access multiple channels during one transmission session, the DSA process is no longer modeled as traditional graph coloring, where each vertex is assigned with only one color. Rather, it is modeled as a fractional graph coloring, where a set of colors is assigned to each vertex. For a user u_i allocated with l channel-time grids, v_i is colored by a color vector, $vi = (c_{i1}, t_{i1}, c_{i2}, t_{i2}, \cdots, c_{il}, t_{il})$, where $c_{im}, t_{im}, m = 1, 2, ..., l$ is a 2-dimension color denoting a channel-time grid, and $t_{i1} \leq t_{i2} t_{i1} \leq \cdots \leq t_{il}$. The fractional coloring process has three constraints: (1) a user cannot access more than one channel during a time slot, hence, for v_i, it is required $t_{i1} < t_{i2} t_{i1} < \cdots < t_{il}$; (2) adjacent vertices cannot be colored with the same channel-time grid, i.e., if $e_{ij} \in E, v_i \neq v_j$. Here, $v_i \neq v_j$ is defined as: for $\forall c_{im}, t_{im}$ and c_{in}, t_{in} if $t_{im} = t_{in}$, then $c_{im} \neq c_{in}$; (3) each user should use all the available channels, i.e., $K \subseteq C_i$, where $C_i = \{c_{i1}, c_{i2}, \cdots, c_{il}\}$ is the sequence of channels accessed by u_i. Hence, in essence, the graph coloring problem is a CSP that can be formulized as:

Goal: (1) confidentiality and (2) original goal of DSA process.

Variables: V-Vertices of G

Domain of variable v_i:$D_i = \{v_i\} v_i$ is value for variable v_i

Constraints: (1) $t_{i1} < t_{i2} t_{i1} < \cdots < t_{il}, \forall v_i \in V$

(2) $v_i \neq v_j, \forall e_{ij} \in E$

(3) $K \subseteq C_i, \forall v_i \in V$

The optimization problem above can be solved by a number of methods even including the exhaust search. However, for keeping low cost of implementation, [75] proposed a LS-based algorithm to solve this optimization problem, which takes advantages of graph extension algorithm. The details of this solution are beyond the scope of this chapter. The reader who is interested in it can refer to reference [75].

The simulations in [75] show the of proposed confidentiality oriented DSA algorithm. To achieve the same goal, the authors in [85] show the theoretical analysis of confidentiality provided by the regular DSA. This theoretical analysis is based on the following observation. Since some of the channel bands within a wide range of spectrum resources are selected for the SUs, it is very costly for the eavesdroppers to overhear all the possible spectrum resources to obtain all the messages exchanged within the secondary network. Hence, the communication between the SU and eavesdropper is modeled as the binary erasure channel in [85]. According to the communication theory, the forward error correction code is the only method that can be used to recover the sender's original messages from the received incomplete message signals (under the binary erasure channel). Therefore, the confidentiality of the secondary network, defined as the probability that the eavesdropper cannot correctly recover the sender's original message, can be quantified as the function in terms of: (1) the total number of available sub-channels for SUs N; (2) the number of sub-channels assigned to the ith SU m_i; (3) total number of sub-channels occupied by the adversary M; and (4) probability that the eavesdropping ratio (refers to as the ratio of the number of common sub-channels that are occupied by this SU and the adversary simultaneously to the total number of this SU's sub-channels), with respect to the ith SU, is $a\%$, which can be represented as: $P_{er}^i = \dfrac{A_{m_i \times a\%}^{m_i} \cdot A_{M-m_i \times a\%}^{N-m_i}}{A_M^N}$, where $A_y^x = \dfrac{x!}{y! \cdot (x-y)!}$.

Additionally, from the network performance perspective, [85] also modeled the DSA as a M/G/k queuing system. In this queuing system, the channel bands occupied by users are considered as "queue servers". The queue server's serving capability is determined by its spectrum bandwidth. By assuming the available channel bands are sufficient for users' requests with respect to their QoS requirements, the number of queue servers in the queuing system equals to the number of users being served simultaneously. The interval between two consecutive requests is a random variable and is assumed to be following Poisson distribution, and the queue servers' serving times are assumed to be identical and independently distributed with unknown distribution. Based on these assumptions, the overall cost function of a queuing system can be described as $P^i_{er} = \frac{A^{m_i}_{m_j \times a\%} \cdot A^{N_i}_{M-m_j \times a\%}}{A^N_M}$, where $A^y_x = \frac{x|}{y! \cdot (x-y)}$.

$$G_i(\mu_i) = E(w) \cdot h(\mu_i, p_i) = \left(\frac{1}{\mu_i} + \frac{\rho_i^2 + \lambda_i^2 \sigma_v^2}{2\lambda_i(1-\rho_i)} \right) \cdot h(\mu_i, p_i),$$

and $c_i(\mu_i)$ is the service cost rate, which measures the average cost per unit time associated with the operating facility at rate μ_i.

Accordingly, the security-oriented DSA in [85] is modeled as a multi-objective optimization problem, which is given by

$$\min f_1 = \sum_{i \in I} \left[c_i(\mu_i) + \lambda_i \cdot \left(\frac{1}{\mu_i} + \frac{\rho_i^2 + \lambda_i^2 \sigma_v^2}{2\lambda_i(1-\rho_i)} \right) \cdot h(\mu_i, p_i) \right] \quad \min f_2 = \sum_{i \in I} \left[1 - P^i_{de} \right]$$

s.t.

$m_i \in z, where\, Z\, is\, intege\, domain$

$$\sum_{i \in I} m_i \leq N \qquad \text{(C1)}$$

$$(B_{Total}/N) \cdot m_i/K \geq \lambda_i \qquad \text{(C2)}$$

$$b^{QoS}_i \leq (B_{Total}/N) \cdot m_i. \qquad \text{(C3)}$$

The N in inequality (C1) indicates the total number of available sub-channels. So, this inequality makes sure the total spectrum resources assigned to the SUs is less than the total available spectrum resource. The constraint (C2) illustrates that the serving rate should not be less than the customer's arrival rate in the queuing system in order to make the system stable. Constraint (C3) shows the bandwidth of spectrum resource assigned to each user should not be less than the one minimal value for each user's application, according to its QoS requirement.

26.8 Simulator for Cognitive Radio Networks and Its Security

To facilitate the deployment of CRNs, it is desirable to have effective simulators to verify the efficiency of various design schemes in light of pragmatic challenges with low cost and short development cycle. However, most widely used existing network simulators, such as NS-2, NS-3, OPNET, and QUALNET, have no cognitive radio feature imbedded. Therefore, new cognitive radio protocols and algorithms cannot be adequately verified with these simulators. Hence, there is a demand to extend existing simulators to support CR features.

NS-2 is widely used in academia, since it is a free open source network simulator. In addition, NS-2 supports the simulation of MAC layer protocols and the routing algorithms in the network layer. Users can build their own network by configuring the network architecture and specifying the MAC protocol and routing algorithms. Furthermore, there are many exemplary implementations of MAC protocols and routing algorithms in NS-2 for current popular networks, such as WLAN, WPAN, and so on. Attributed to these advantages of NS-2, a CR cognitive network (CRCN) simulator using NS-2 was developed (http://faculty.uml.edu/Tricia_Chigan/Research/CRCN_Simulator.htm). In this NS-2-based CRCN simulator, the basic cognitive features for PHY-layer, MAC layer and routing protocols were provided. Several application programming interfaces (APIs) were also provided for the users to embed their own design into this simulator for verification.

NS-3 shares many advantages with NS-2, for example: (1) NS-3 is an open source software, thus any contributions to NS-3 are freely accessible; (2) NS-3 provides many radio models, such as 802.11, 802.16, 802.15.3, 802.15.4. Users can use these radio models for cognitive radio network simulations; and (3) NS-3 has incorporated different topology and traffic generators, which enable users to create different simulation scenarios, etc. In addition, compared with NS-2, NS-3 has the following advantages:

(1) A simulation script can be written as a C++ program, which is not possible in NS2.

(2) With modern hardware capabilities, the compile time of NS-3 is not an issue, as in NS-2.

(3) NS-3 supports the testbed-based experiments with novel protocol stacks, to emit/consume network packets over real device drivers or virtual local area networks (VLANs).

(4) NS3 performs better than NS2 in terms of memory management.

(5) The aggregation system prevents unnecessary parameters from being stored, and packets don't contain unused reserved header space.

Consequently, the NS-3 is more suitable for the simulation of the current research on CRNs, since the NS-3-based CRN simulator can be easily downloaded into the hardware to construct realistic scenarios. In addition, to support the research of security issues relevant to pragmatic aspects of CRNs, as discussed in this chapter, it is necessary to further extend the CRCN simulator with security features. Therefore, the NS-2, based simulator was extended to NS-3, based CRCN simulator with security modules.

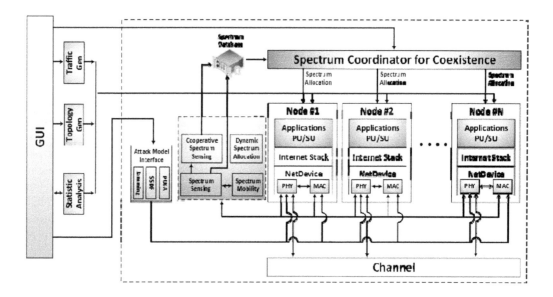

Figure 26.29: Architecture of CRCN simulator with NS3.

This NS-3, based CRCN simulator will be able to support performance evaluations for the proposed dynamic spectrum resource allocation, power control algorithms, coexistence mechanisms, and the adaptive CR networking protocols, such as the CR MAC protocols. The effects of attack models can also be evaluated using this CRCN simulator. This simulator uses NS-3 to generate realistic traffic and topology patterns. For each node in this simulator, a reconfigurable multi-radio multi-channel PHY layer is available by customizing the spectrum parameters, such as transmission power, etc. The design architecture of this CRCN simulator is shown in Figure 25.30. This simulator is accessible at http://faculty.uml.edu/Tricia_Chigan/Research/CRCN_NS3.html.

26.9   Potential Security Issues

Since the CR technique has been proved as a promising technology that can effectively improve the spectrum utilization efficiency, the CR features have been developed in a variety of areas. For example, IEEE802.22 is the first standard aiming at development of CRN in the world. Although several practical issues still exist in the development and deployment of the IEEE802.22 network [86], there are already a number of wireless systems developed for some specific areas based on the concepts in the IEEE802.22 standard, e.g., the monitoring system for protection of forest, smart grid, etc. [84]. These systems operate over the TV whitespace following the FCC regulation. Therefore, the TV set should be protected to avoid interference. However, since the TV set is a passive receiver, compared with the primary transmitter, it is even easier for the adversaries to emulate the behavior of passive receiver. Therefore, the passive receiver emulation attack and its kind (e.g., medical sensor device, earphone, etc.) [90] are even more difficult to be identified, since there is no signal emission.

Due to the advantages of CR, it has been considered to be integrated into various existing networks in order to improve their network performance. For example, cognitive-long term evolution (LTE) [87] or LTE in the unlicensed band (LTE-U) [91] was proposed by integrating the cognitive radio concept into the LTE network. Furthermore, the 5G cellular network has already embedded the cognitive radio feature into its standard [88]. However, small cell communication is the trend of those networks for lower interference and higher network throughput. On the one hand, this makes these networks more vulnerable to the security threats. This is because that in the small cell networks, there are only a few number of network nodes that can cooperate with each other in a sub-network. Therefore, it is easy for a malicious node to attack most of network nodes in a sub-network, which may destroy this whole sub-network.

In addition, with various emerging commercial wireless technologies with expanded unlicensed spectrum sharing capabilities, such as of IEEE 802.11 (e.g., ac, af, ax), IEEE 802.15, IEEE 802.16, and IEEE 802.22, new security vulnerabilities and threats need to be further addressed in this heterogeneous wireless spectrum sharing paradigm [92][93].

References

[1] S. Haykin, "Cognitive radio: Brain-empowered wireless communications," *IEEE Journal on Selected Areas in Communications*, vol. 23, no. 2, pp. 201–220, February 2005.

[2] W. Wenkai, L. Husheng, S. Yan and H. Zhu, "Securing collaborative spectrum sensing against untrustworthy secondary users in cognitive radio networks", *Eurasip Journal on Advances in Signal Processing, special issue on Advanced Signal Processing for Cognitive Radio Networks*, vol. 2010, Article ID 695750, 15 pages, 2010.

[3] Jin, Z.; Anand, S.; Subbalakshmi, K.P., "Impact of primary user emulation attacks on dynamic spectrum access networks", *IEEE Transactions on Communications*, Volume: 60, Issue: 9, 2012.

[4] Feijing Bao; Huifang Chen; Lei Xie, "Analysis of primary user emulation attack with motional secondary users in cognitive radio networks", *Personal Indoor and Mobile Radio Communications (PIMRC), 2012 IEEE 23rd International Symposium on*, Pages: 956–961, 2012.

[5] Attar, A.; Tang, H.; Vasilakos, A.V.; Yu, F.R.; Leung, "A survey of security challenges in cognitive radio networks: Solutions and future research directions", *Proceedings of the IEEE*, Volume: 100, Issue: 12, Page(s): 3172–3186, 2012.

[6] Rawat, A.S.; Anand, P.; Hao Chen; Varshney, P.K. "Countering byzantine attacks in cognitive radio networks", *IEEE International Conference on Acoustics Speech and Signal Processing (ICASSP)*, Page(s): 3098–3101, 2010.

[7] Olga León, Juan Hernandez-Serrano and Miguel Soriano, "A new cross-layer attack to TCP in cognitive radio networks", *Proceedings of the 2nd International Workshop on Cross Layer Design (IWCLD '09)*, Palma, Spain, June, 2009, pp.1–5.

[8] Nanthini SB, Hemalatha M, Manivannan D, Devasena L. "Attacks in cognitive radio networks (CRN) - A survey". *Indian Journal of Science and Technology*. 2014 Apr; 7(4):530-6.

[9] Bartoli, G.; Fantacci, R.; Marabissi, D.; Pucci, M., "Resource allocation schemes for cognitive LTE-A femto-cells using zero forcing beamforming and users selection", *Global Communications Conference (GLOBECOM), 2014 IEEE*, Pages: 3447–3452, 2014.

[10] Hyoil Kim; Shin, K.G., "Optimal admission and eviction control of secondary users at cognitive radio hotspots", *Sensor, Mesh and Ad Hoc Communications and Networks, 2009. SECON '09. 6th Annual IEEE Communications Society Conference on*, Papers (12), 2009.

[11] Saha, N.; Mondal, R.K.; Yeong Min Jang, "Opportunistic channel reuse for a self-organized visible light communication personal area network", *Ubiquitous and Future Networks (ICUFN), 2013 Fifth International Conference on*, Pages: 131–134, 2013.

[12] Xin Kang; Ying-Chang Liang; Nallanathan, A.; Garg, H.K.; Rui Zhang, "Optimal power allocation for fading channels in cognitive radio networks: Ergodic capacity and outage capacity", *Wireless Communications, IEEE Transactions on*, Volume: 8, Issue: 2, 2009.

[13] Taheri, Z.; Taft, A.A.T.; Hoseini, S.M.M., "Cooperative spectrum sensing, power and throughput tradeoffs in cognitive radio systems", *Telecommunications (IST), 2012 Sixth International Symposium on*, Pages: 261–265, 2012.

[14] Zhipeng Cai; Shouling Ji; Jing He; Lin Wei; Bourgeois, A.G., "Distributed and asynchronous data collection in cognitive radio networks with fairness consideration", *Parallel and Distributed Systems, IEEE Transactions on*, Year: 2014, Volume: 25, Issue: 8, Pages: 2020–2029.

[15] Qian Zhang; Juncheng Jia; Jin Zhang, "Cooperative relay to improve diversity in cognitive radio networks", *Communications Magazine, IEEE*, Volume: 47, Issue: 2, 2009.

[16] Lusina, P.; Schober, R.; Lampe, L., "Diversity-multiplexing trade-off of the hybrid nonorthogonal amplify-decode and forward protocol", *Information Theory, 2008. ISIT 2008. IEEE International Symposium on*, Pages: 2375–2379, 2008.

[17] Pengyu Zhang; Jian Yuan; Jianshu Chen; Jian Wang; Jin Yang, "Analyzing amplify-and-forward and decode-and-forward cooperative strategies in wyner's channel model", *Wireless Communications and Networking Conference, 2009. WCNC 2009. IEEE*, Pages: 1–5, 2009.

[18] J. Jia, J. Zhang, and Q. Zhang, "Cooperative relay for cognitive radio networks," to appear, *IEEE INFOCOM*, 2009.

[19] Demirdogen, I., Li, L., Chigan, C., "FEC-driven network coding based pollution attack defense in cognitive radio networks," *2015 IEEE Wireless Communications and Networking Conference (WCNC) - Workshop on Smart Spectrum*, March 2015.

[20] Z. Yuan, Z. Han, Y. Sun, H. Li, and J. B. Song, "Routing-toward-primary-user attack and belief propagation-based defense in cognitive radio networks," *IEEE Transactions on Mobile Computing*, vol. 12, no. 9, pp. 1750–1760, Sep. 2013.

[21] Jarecki, S.; Saxena, N., "On the insecurity of proactive RSA in the URSA mobile ad hoc network access control protocol", *IEEE Transactions on Information Forensics and Security*, Volume:5, Issue: 4, Page(s): 739–749, 2010.

[22] Huaizhou Shi; Prasad, R.V.; Niemegeers, I.G.M.M.; Ming Xu; Rahim, A., "Self-coexistence and spectrum sharing in device-to-device WRANs", *Communications (ICC), 2014 IEEE International Conference on*, Pages: 1651–1656, 2014.

[23] Yuan, S., Li, L., Chigan, C. (2015), "A selfishness-aware coexistence scheme for 802.22 and 802.11af networks", *2015 IEEE Wireless Communications and Networking Conference (WCNC) - Workshop on Smart Spectrum*, March 2015.

[24] Boram Choi; Hyuk Lim; Hyunduk Kang; Byung-Jang Jeong, "Dynamic priority scheduling for heterogeneous cognitive radio networks", *Sensor, Mesh and Ad Hoc Communications and Networks (SECON), 2012 9th Annual IEEE Communications Society Conference on*, Pages: 62–64, 2012.

[25] Amjad, M.F.; Chatterjee, M.; Zou, C.C., "Inducing cooperation for optimal coexistence in cognitive radio networks: A game theoretic approach", *Military Communications Conference (MILCOM)*, 2014 IEEE, Pages: 955–961, 2014.

[26] 802.22 Working Group, "IEEE 802.22 D1: draft standard for wireless regional area networks", http://grouper.ieee.org/groups/802/22/.

[27] C. Stevenson, G. Chouinard, Z. D. Lei, W. D. Hu, S. Shellhammer, and W. Caldwell, "IEEE 802.22: the first cognitive radio wireless regional area network standard," *IEEE Communications Magazine*, vol. 47, no. 1, pp. 130–138, 2009.

[28] Q. Zhang, J. Jia, and J. Zhang, "Cooperative relay to improve diversity in cognitive radio networks," *IEEE Commun. Mag.*, vol. 47, no. 2, pp. 111–117, 2009.

[29] Hyoungsuk Jeon; McLaughlin, S.W.; Jeongseok Ha, "Secure communications with untrusted secondary users in cognitive radio networks", *Global Communications Conference (GLOBECOM), 2012 IEEE*, Pages: 1072–1078, 2012.

[30] P. Judge. ZDNet: .net vote rigging illustrates importance of web services. http://news.zdnet.co.uk/software/ 0,39020381,2102244,00.htm, 2002.

[31] J. Bulgatz. "More extraordinary popular delusions and the madness of crowds." *Three Rivers Press*, 1992.

[32] A. Viglucci, J. Tanfani, and L. Getter. "Herald special report: Dubious tactics tilted mayoral votes". *Miami Herald*, February 8, 1998.

[33] Oggier, F., Fathi, H.: An authentication code against pollution attacks in network coding. CoRR abs/0909.3146 (2009).

[34] Ning Zhang; Ning Lu; Rongxing Lu; Mark, J.W.; Xuemin Shen, "Energy-efficient and trust-aware cooperation in cognitive radio networks", *Communications (ICC), 2012 IEEE International Conference on*, Pages: 1763–1767, 2012.

[35] Lijuan Geng; Ying-Chang Liang; Chin, F., "Network coding for wireless ad hoc cognitive radio networks", *Personal, Indoor and Mobile Radio Communications, 2007. PIMRC 2007. IEEE 18th International Symposium on*, Pages: 1–5, 2007.

[36] C. Karlof and D. Wagner, "Secure routing in wireless sensor networks: Attacks and countermeasures," *Ad Hoc Networks*, vol. 1, no. 23, pp. 293–315, Sep. 2003.

[37] Kejie Lu; Shengli Fu; Yi Qian; Tao Zhang, "On the security performance of physical layer network coding", *Communications, 2009. ICC '09. IEEE International Conference on*, Pages: 1–5, 2009.

[38] Minsoo Lee; Xiaohui Ye; Johnson, S.; Marconett, D.; Chaitanya, V.S.K.; Vemuri, R.; Ben Yoo, S.J., "Cognitive security management with reputation based cooperation schemes in heterogeneous networks", *Computational Intelligence in Cyber Security, 2009. CICS '09. IEEE Symposium on*, Pages: 19–23, 2009.

[39] Mee Hong Ling; Yau, K.-L.A., "Reinforcement learning-based trust and reputation model for spectrum leasing in cognitive radio networks", *IT Convergence and Security (ICITCS), 2013 International Conference on*, Pages: 1–6, 2013.

[40] Liang Xiao; Lin, W.S.; Yan Chen; Liu, K.J.R., "Indirect reciprocity game modelling for secure wireless networks", *Communications (ICC), 2012 IEEE International Conference on*, Pages: 928–933, 2012.

[41] Zhiwei Li; Di Pu; Weichao Wang; Wyglinski, A., "Node localization in wireless networks through physical layer network coding", *Global Telecommunications Conference (GLOBE-COM 2010), 2010 IEEE*, Pages: 1–5, 2010.

[42] Hong Song; Xiao Xiao; Weiping Wang; Luming Yang, "DENNC: A wireless malicious detection approach based on network coding", *Trust, Security and Privacy in Computing and Communications (TrustCom), 2011 IEEE 10th International Conference on*, Pages: 160–165, 2011.

[43] Alnabelsi, S.H.; Kamal, A.E.; Jawadwala, T.H., "Uplink channel assignment in cognitive radio WMNs using physical layer network coding", *Communications (ICC), 2011 IEEE International Conference on*, Pages: 1–5, 2011.

[44] Weichao Wang; Di Pu; Wyglinski, A., "Detecting sybil nodes in wireless networks with physical layer network coding", *Dependable Systems and Networks (DSN), 2010 IEEE/IFIP International Conference on*, Pages: 21–30, 2010.

[45] X. Xie, W. Wang, Detecting primary user emulation attacks in cognitive radio networks via physical layer network coding, *Procedia Comput. Sci.* 21 (2013) 430–435. Special Issue: The 4th International Conference on Emerging Ubiquitous Systems and Pervasive Networks (EUSPN-2013) and the 3rd International Conference on Current and Future Trends of Information and Communication Technologies in Healthcare (ICTH) Edited By Elhadi Shakshuk.

[46] Kejie Lu; Shengli Fu; Yi Qian; Tao Zhang, "On the security performance of physical-layer network coding", *Communications, 2009. ICC '09. IEEE International Conference on*, Pages: 1–5, 2009.

[47] A. Wyner, "The wire-tap channel," *Bell System Technical Journal*, vol. 54, no. 8, pp. 1355–1387, 1975.

[48] Y. Liang, H. Poor, and S. Shamai, "Secure communication over fading channels," *IEEE Transactions on Information Theory*, vol. 54, no. 6, pp. 2470–2492, June 2008.

[49] Tianyu Wang; Lingyang Song; Zhu Han; Xiang Cheng; Bingli Jiao, "Power allocation using vickrey auction and sequential first-price auction games for physical layer security in cognitive relay networks", *Communications (ICC), 2012 IEEE International Conference on*, Pages: 1683–1687, 2012.

[50] Q. Zhang, J. Jia, and J. Zhang, "Cooperative relay to improve diversity in cognitive radio networks," *IEEE Comm. Magazine*, vol. 47, no. 2, pp. 111–117, Feb. 2009.

[51] Sheng Zhong; Haifan Yao, "Towards cheat-proof cooperative relay for cognitive radio networks", *Parallel and Distributed Systems, IEEE Transactions on*, Pages: 2442–2451, 2014.

[52] K. Zeng, P. Paweczak, and D. Cabric, "Reputation-based cooperative spectrum sensing with trusted nodes assistance," *Communications Letters*, IEEE, vol. 14, no. 3, pp. 226–228, 2010.

[53] R. Jain, W. Hawe, D. Chiu, "A quantitative measure of fairness and discrimination for resource allocation in shared computer systems," *DEC-TR-301*, September 26, 1984.

[54] J. Tang, S. Misra and G. Xue, "Joint spectrum allocation and scheduling for fair spectrum sharing in cognitive radio wireless networks," *Computer Networks*, vol. 52, no. 11, pp. 2148–2158. 2008.

[55] Yanxiao Zhao; Min Song; ChunSheng Xin, "FMAC: A fair MAC protocol for coexisting cognitive radio networks", *INFOCOM, 2013 Proceedings IEEE*, Pages: 1474–1482, 2013.

[56] Y. Zhao et al., "Spectrum sensing based on three-state model to accomplish all-level fairness for co-existing multiple cognitive radio networks," in *Proc. IEEE INFOCOM*, Mar. 2012.

[57] Gao, B.; Jung-Min Park; Yang, Y., "Uplink soft frequency reuse for self-coexistence of cognitive radio networks operating in white-space spectrum", *INFOCOM, 2012 Proceedings IEEE*, Pages: 1566–1574, 2012.

[58] J. Tang, S. Misra, and G. Xue, "Joint spectrum allocation and scheduling for fair spectrum sharing in cognitive radio wireless networks," *Comput. Netw. J.*, vol. 52, no. 11, pp. 2148–2158, Aug. 2008.

[59] M.S. Bazaraa, J.J. Jarvis, H.D. Sherali, Linear programming and network flows, third ed., John Wiley & Sons, 2005.

[60] Z. Ma et al., "A fair opportunistic spectrum access (FOSA) scheme in distributed cognitive," in *Proc. IEEE ICC*, May 2008.

[61] S. Shankar, "Efficiency and coexistence strategies for cognitive radio," *in Cognitive Radio, Software Defined Radio, and Adaptive Wireless Systems*, H. Arslan, Ed. Boston/Dordrecht/London: Springer, 2007, ch. 7, pp. 189–234.

[62] B. Wang et al., "Primary-prioritized markov approach for dynamic spectrum access," *in Proc. IEEE DySPAN*, Apr. 2007.

[63] Wang, Fengjiao; Zhang, Yuqing, "A new provably secure authentication and key agreement mechanism for SIP using certificateless public-key cryptography", Pages: 809–814, 2007.

[64] Parvin, S.; Song Han; Biming Tian; Hussain, F.K., "Trust-based authentication for secure communication in cognitive radio networks", *Embedded and Ubiquitous Computing (EUC), 2010 IEEE/IFIP 8th International Conference on*, Pages: 589–596, 2010.

[65] Kumar, V.; Jung-Min Park; Clancy, T.C.; Kaigui Bian, "PHY-layer authentication using hierarchical modulation and duobinary signaling", *Computing, Networking and Communications (ICNC), 2014 International Conference on*, Pages: 782–786, 2014.

[66] Vireshwar Kumar, Jung-Min Park, Kaigui Bian, "Blind transmitter authentication for spectrum security and enforcement", *Proceedings of the 2014 ACM SIGSAC Conference on Computer and Communications Security*, 2014.

[67] R. Chen, J. M. Park, and J. Reed, "Defense against primary user emulation attacks in cognitive radio networks", *IEEE Journal on Selected Areas in Communications*, vol. 26, pp. 25–37, 2008.

[68] L. Huang, L. Xie, H. Yu, W. Wang, and Y. Yao, "Anti-PUE attack based on joint position verification in cognitive radio networks", in *Proc. 2010 International Conference on Communications and Mobile Computing (CMC)*, Shenzhen, China, April 2010.

[69] Zhou Yuan; Niyato, D.; Husheng Li; Zhu Han, "Defense against primary user emulation attacks using belief propagation of location information in cognitive radio networks", *Wireless Communications and Networking Conference (WCNC), 2011 IEEE*, Pages: 599–604, 2011.

[70] Yao Liu; Peng Ning; Huaiyu Dai, "Authenticating primary users' signals in cognitive radio networks via integrated cryptographic and wireless link signatures", *Security and Privacy (SP), 2010 IEEE Symposium on*, Pages: 286–301, 2010.

[71] Federal Communications Commission. Facilitating opportunities for flexible, efficient, and reliable spectrum use employing spectrum agile radio technologies. ET Docket, (03-108), Dec. 2003.

[72] S. Ganeriwal, S. Capkun, C. Han, and M. B. Srivastava. Secure time synchronization service for sensor networks. In *Proceedings of 2005 ACM Workshop on Wireless Security (WiSe 2005)*, pages 97–106, September 2005.

[73] K. Sun, P. Ning, C. Wang, A. Liu, and Y. Zhou. TinySeRSync: Secure and resilient time synchronization in wireless sensor networks. In *Pro ceedings of 13th ACM Conference on Computer and Communications Security (CCS '06)*, pages 264–277, October/November 2006.

[74] Chao Zou; Chunxiao Chigan, "Dynamic spectrum allocation based confidentiality for cognitive radio networks", *Global Telecommunications Conference (GLOBECOM 2010), 2010 IEEE*, Pages: 1–6, 2010.

[75] Y. Liang, H. V. Poor, and S. Shamai, "Information theoretic security", *MA: Now Publishers*, vol. 5, no. 4-5, pp. 355–580, 2008.

[76] F. Oggier and B. Hassibi, "The secrecy capacity of the MIMO wiretap channel," *IEEE Trans. Inf. Theory*, vol. 57, no. 8, pp. 4961–4972, Oct. 2007.

[77] S. Goel and R. Negi, "Guaranteeing secrecy using artificial noise," *IEEE Trans. Wireless Commun.*, vol. 7, no. 6, pp. 2180–2189, July 2008.

[78] A. Mukherjee amd A. Swindlehurst, "Robust beamforming for security in MIMO wiretap channels with imperfect CSI," *IEEE Trans. Signal Process.*, vol. 59, no. 1, pp. 351–361, Jan. 2011.

[79] G. Safdar and M. O'Neill, "Common control channel security framework for cognitive radio networks," in *Vehicular Technology Conference 2009*, 2009.

[80] J. W. C. X. Ying Dai, "Efficient virtual backbone construction without a common control channel in cognitive radio networks," *IEEE Transaction on Parallel and Distributed Systems*, vol. 25, no. 12, 2014.

[81] J. Zhang and Z. Zhang, "Initial link establishment in cognitive radio networks without common control channel," in *IEEE Wireless Communications and Networking Conference (WCNC)*, 2011, 2011.

[82] Lu Yu; Hai Liu; Yiu-Wing Leung; Xiaowen Chu; Zhiyong Lin, "Multiple radios for effective rendezvous in cognitive radio networks", *Communications (ICC), 2013 IEEE International Conference on*, Pages: 2857–2862, 2013.

[83] X. He, H. Dai and P. Ning, "Dynamic adaptive anti-jamming via controlled mobility," *IEEE Transactions on Wireless Communications*, vol. 13, no. 8, pp. 4374–4388, 2014.

[84] Honggang Wang; Yi Qian; Sharif, H., "Multimedia communications over cognitive radio networks for smart grid applications", *Wireless Communications, IEEE*, Volume: 20, Issue: 4, Volume: 20, Issue: 4, 2013.

[85] Lei Li and Chunxiao Chigan, "Security-oriented DSA for network access control in cognitive radio networks", submitted to GlobalCom 2015.

[86] Chang-Woo Pyo; Xin Zhang; Chunyi Song; Ming-Tuo Zhou; Harada, H, "A new standard activity in IEEE 802.22 wireless regional area networks: Enhancement for broadband services and monitoring applications in TV whitespace", *Wireless Personal Multimedia Communications (WPMC), 2012 15th International Symposium on*, Pages: 108–112, 2012.

[87] Beluri, M.; Bala, E.; Yuying Dai; Di Girolamo, R.; Freda, M.; Gauvreau, J.; Laughlin, S.; Purkayastha, D.; Touag, A., "Mechanisms for LTE coexistence in TV white space", *Dynamic Spectrum Access Networks (DYSPAN), 2012 IEEE International Symposium on*, Pages: 317–326, 2012.

[88] Mavromoustakis, C.X.; Mastorakis, G.; Bourdena, A.; Pallis, E.; Kormentzas, G.; Dimitriou, C.D., "Joint energy and delay-aware scheme for 5G mobile cognitive radio networks", *Global Communications Conference (GLOBECOM)*, 2014 IEEE, Pages: 2624–2630, 2013.

[89] Martello and P. Toth, "Knapsack problems: Algorithms and computer implementations", John Wiley and Sons, New York, 1990.

[90] Ahmad, A.; Ahmad, S.; Rehmani, M.H.; Hassan, N.U., "A survey on radio resource allocation in cognitive radio sensor networks", *Communications Surveys & Tutorials, IEEE*, Volume: 17, Issue: 2, Pages: 888–917, 2015.

[91] Liu, F.; Bala, E.; Erkip, E.; Beluri, M.; Yang, R., "Small cell traffic balancing over licensed and unlicensed bands", *Vehicular Technology, IEEE Transactions on, Volume: PP*, Issue: 99, Pages: 1–1, 2015.

[92] https://www.google.com/url?sa=t&rct=j&q=&esrc=s&source=web&cd=1&cad=rja&uact=8& ved=0CCAQFjAA&url=https%3A%2F%2Fwww.qualcomm.com%2Fmedia%2Fdocuments% 2Ffiles%2Flte-unlicensed-coexistence-whitepaper.pdf&ei=uBSJVe_uO4v7sAWHu4DoDw& usg=AFQjCNHHP-iRjiB9M82LfxcwppRzym9tnw&bvm=bv.96339352,d.b2w

[93] http://www.nsf.gov/pubs/2015/nsf15550/nsf15550.htm

MILITARY
APPLICATIONS

Chapter 27

Spectrum in Defense: From Commodity to Maneuver Space

Jesse Bourque Jr.

CONTENTS

27.1 Introduction

The Electromagnetic Spectrum (EMS) is a fundamental feature and a strategic resource for National defense, civil infrastructure, and international commerce. International presence in the world within commercial, policy, and security contexts will be challenged by state and non-state adversaries attempting to restrict freedom of access to and action within the EMS in the interests of inducing instability, disruption, asymmetric cost-imposing effects, and regional anti-access/area-denial stressors. In response, for example, the United States Defense policy and Joint military doctrine is now evolving to acknowledge the converging responsibilities of Electronic Warfare (EW) and Spectrum Management (SM) in order to form a future-proof solution [1–4], known as Electromagnetic Spectrum Operations (EMSO). This evolution will reduce process inefficiency and latency, consolidating expertise and capability in pursuit of vital *EMS Control*. Although it may appear as if things are going well within current paradigms and operating models, or at least that all would be going well *if only we spent more*, they are not. National security challenges in the Spectrum are not incremental; they are structural, resulting from a decades-old, increasingly false sense of "elbow room" now exacerbated by increasing Spectrum complexity, density, demand, and contest. To procure enduring capabilities, organizations must transition from a "break glass" model of quick-reaction capability (QRC) *expenditure* to a stable architecturally-based *investment* strategy. In haste to provide so many perishable materiel solutions dependent upon spectrum, an "owner" has yet to be empowered to represent the enduring DOTMLPF-P problem set (Doctrine, Organization, Training, materiel, Leadership and education, Personnel, Facilities, and Policy), an actual *opportunity* space comprised of *all radiated electromagnetic energy*. There is no such thing as radiated *digital* energy or radiated "data" per se, only digital and informational modulation transported within narrow regions of radiated analog EM energy, which must be *deliberately* provided and protected. That energy is divided into individual electromagnetic operating environments (EMOE), which are shared and measurable operational maneuver spaces aligned with multi-national operating areas wherein commanders require reliable control to plan and achieve campaign objectives.

In this chapter, all instances of Electromagnetic Spectrum (e.g., "Spectrum") are capitalized as a matter of convention. In contrast to the generic use of "*spectrum*" to denote a *range* of operations or conflict; *the physical maneuver space comprised of all radiant EM energy* comprises a diverse and expansive *place*, like Australia. This treatment will serve not only the useful purpose of distinguishing the two concepts, but over time will also help leaders, decision makers, and stakeholders to internalize the reality of Spectrum as a *domain of operations* and help us to unlearn the counterproductive notion of the Spectrum as a simple commodity of fragments. The Spectrum is a physical, continuous, operational *maneuver space*, the only physical space which unites all others. Regarding the Spectrum as a simple utility of fragments keeps national planners and strategists mentally locked into an isolated "*threat*" paradigm and sustains the resulting disarray in our menagerie of disparately-created EMS capabilities. In contrast, recognition of the Spectrum as a domain equal to the *other* natural domains of Land, Maritime, Air, and Space will drive broad adoption of an "*opportunity*" paradigm, enabling force-wide coherence across all EMS-dependent capabilities and by extension, *all* capabilities.

We are now revisiting a historical pattern wherein technologies, capabilities and responsibilities within a uniform physical context have massed to demonstrate its existence as a new domain. The preceding physical domains "discovered" throughout history have *always* existed and followed this same pattern. The issue for each was simply that at some point in time, our technology and abilities became mature enough to plainly illustrate the practicality of their existence. Logic and reality now compel us to acknowledge the totality of radiant EM energy comprising the EMS in similar terms. In the case of the EMS, a '*meta-domain*' due to its unique unifying quality, it should be regarded simply as *the physical domain comprised of all radiant electromagnetic energy*.

27.2 Situation

The issue of Spectrum sharing across the "Defense-commerce" boundary has two identities. Although well intended by those with commercial interests, the innocent notion of sharing translates to a de facto condition of *forfeiture* for Defense and first responder organizations expected to train, sustain, and project operational capability within the current framework of static Spectrum technologies in reduced regions of Spectrum. This is mostly driven by the governmental motivation to decrease disruption or impediments to full realization of commercial Spectrum access for fear of upsetting the substantial revenue stream it generates. To address increasing demand, commercial stakeholders will continue their successful efforts to increase their allocated Spectrum, dislocating current government occupants from the claimed regions. In sharp contrast to this *revenue* motive,

Defense and first responder participants in the Spectrum share a motivation of *expediency* and *freedom of action*, logically compelling a prioritization mandate for use of any Spectrum available, beyond their currently approved but dwindling allocations. Regarding military operations, the EMS no longer tolerates isolated behaviors or intentions; our twin mandates of *global reach* and *combat agility* require freedom in the EMS. Spectrum demand, density, complexity, and contest are increasing globally while international agreements must remain intact as conflict is resolved. These EMS conditions underscore a systemic state of *"fractured dependence,"* the consequences of which will induce intolerable risk via low decision speed, inefficient EMS maneuver, unavoidable EM fratricide, and inability to assure EMS Control for universally EMS-reliant forces.

Under *Title 10* authorities in the US, the purchase of EMS capabilities remains within the purview of the individual Armed Service components, in contrast to the fact that warfighting is inescapably Joint (i.e., multi-Service, cross-domain). In no other domain is this truer than in the Spectrum since a Service may physically sequester a ship, tank, or aircraft within articulated and well-defined geospace, while all Spectrum participants within the broad effects of EM energy share the effects of all participation, everywhere and immediately. As each of the Services sets out to solve the challenge commonly characterized as *the* Defense Spectrum problem, intending to present their solution as *the* solution, they are instinctively solving *only* for the portions and attributes of the problem apparent from their institutional perspective. To characterize Services' typical Spectrum participation: the Army deploys as a large, dense, 'stationary' enterprise with tens of thousands of EM transmit/receive apertures; the Navy during operations is medium-sized, spread, mobile, and federated with hundreds of apertures; the Marines in combat are relatively small, light, highly-mobile, and task-organized with thousands of apertures; the Air Force in the fight is essentially localized into small formations, moving at hundreds of knots and engaged *remote* from friendly force concentrations with tens of apertures; while Special Operations Forces act in covert very small teams, discreet and isolated with very few apertures. Although simplistic, these *iconic* modes of employment have dictated disparate responses to *"the"* Spectrum problem and have driven EMS capability development approaches optimized for employment beyond the Spectral influence of other Services' capabilities. This misalignment between combatant command (CCMD) / Joint task force (JTF) warfighting capability *requirements* for aggregation and Service resourcing actions that assume culmination via localized influence brings us back to reality. We must confidently achieve *"compatibility"* across EMS capabilities *before* we can expect "interoperability" across operations. To address this systemic challenge and its complex and cascading consequences, it is necessary to raise the corporate level of interest to the EMS itself above those of the individual capabilities operating *within* it. Spectrum policy, doctrine, organization, and acquisition must be harmonized; frequency allocation, management and enforcement will converge; and a unified and empowered EMS Governance feature must be established to champion a new future for Defense, underwritten by *"Decisive Aggregation."*

27.3   Electromagnetic Spectrum Control

EMS Control can be thought of as *a condition of reliable freedom of action and adversary capability denial for the times, locations, frequencies, and durations necessary to achieve mission objectives at acceptable risk to friendly personnel, facilities, and equipment.* Adversary action degrades this requisite condition, *further* challenged by the chaotic combination of friendly interference, civil-commercial use, technology proliferation, and environmental unknowns. While some may dismiss the requirement for EMS Control as fleeting or unnecessary, we should consider the following simple justification framework which demonstrates the need for EMS Control and the specific activities uniquely able to *provide* it. Due to the omnipresence and immediately shared impact of radiant EM energy, inter/intra-domain freedom of action is at risk where that EM energy is not controlled.

Freedom of action within each domain results from contextual integration of awareness, *attack*, *defense*, and *management*. In the Spectrum, that group is comprised of EW and SM operations, together known as EMS Operations (EMSO), since expecting adherence to process (SM) without an enforcement mechanism (EW) is as senseless as conducting attack (EW) beyond the context of a larger maneuver plan (SM). To this end, EW's internal feature of *Electronic Warfare Support* (ES) represents a historically untapped *trove* of capability and authority (commonly treated as an Intelligence responsibility), central to this discussion. In the continental US, the FCC and NTIA provide enforcement by protecting *friendly* EMS use and penalizing interference. In combat – far away from FCC or NTIA protection – EW performs ES to shape *Electronic Protection* (EP) for ensuring *friendly* capability, and performs ES to inform *Electronic Attack* (EA) for penalizing *adversary* capability, resulting in a state of *operationally enforced* EMS Control. But the required coordination between EW and SM experts cannot move at the speed of phone calls or emails; process latency induced by "as required" coordination between EMS allocation, management, sensing, attack, and interference remediation activities costs time, money, mission success, and *lives*. The focus must shift to EMS Control "outcomes" *beyond* individual EW inputs for the entire discussion to be culminated; governance will address this process latency by institutionally combining its EMS-*controlling* components of EW and SM to comprise EMSO.

It is important to understand the defining difference between 'operations *in* the Spectrum' (e.g., PNT, ISR, C2, Radar, UAS, Communications, Cyberspace Operations and Space Operations) and 'operations *for* the Spectrum' (EW and SM, or EMSO). The former is the very valuable "*user group*" of beneficiaries or customers of EMS Control, while the latter is the critically essential "*control group*" for the purposes of performing that function and providing a single point of responsibility to *guarantee dedicated effort toward that result*. We no longer live in a world wherein EMS Control can simply be assumed, it must be deliberately made to happen and this can only be reliably done by leveraging focused EMSO acquisition, investment, organization and effort. Taken together, the *user group* and *control group* form an ecosystem of sorts, in which a sustainable '*engine*' for EMS Control is established. First, participants in the *user group* - comprising all friendly force EMS traffic - articulate their Spectrum operating requirements. The EMS requirements for these operations are then planned, deconflicted, allocated, and *enforced* by the integrated efforts of the EMSO *control group*, creating localized and measurable conditions of EMS Control. Finally, this 'improved condition' is recycled back out into the *user group*, setting up a continuous dynamic refinement of Spectrum requirements to be presented again to the EMSO *control group*, feeding a reiterative, self-sustaining process. In contrast to this ecosystem view of EMS Control, many still believe that RF jamming can occur without due consideration for the consequences to capabilities, intelligence, operations, and decision making. Although reasonable when considered in isolation, that thinking defeats the accelerating, clarifying intent of EMSO and actually *opposes* shared objectives of EMS Control.

In currently published doctrine, EMSO consists of EW and SM, with the product/result of that focused activity benefitting Communications, C2, Cyberspace Operations, Space Operations, etc., just as the benefits of law enforcement extend to those of us who aren't responsible for policing, merely 'living within the law.' In the larger context of managed sharing, the current definition of EMSO also makes the EMS Control objective *achievable* by its avoidance of excessive scope. The alternative view of "*everything that happens in the EMS is EMSO*" leaves us to expect EMS Control as a fickle *coalition of the willing* instead of a tightly controlled and manageable group, which must remain an easily identifiable target for resourcing and sustainment. Although the *results* of EMSO extend to non-EMSO stakeholders, expecting everyone to be equally responsible for *creating* that state (beyond compliance with EMSO C2, i.e. '*living within the law*') is unmanageable and growing more so, as EMS complexity, demand, and contest globally escalate.

Although the use of *EMS Control* in this discussion may be seen to compete with the labels "*EMS superiority / EMS dominance*" to convey the same meaning, logically it does not. "Dominance" in the global commons of radiant energy and information – implying *operations with impunity* – causes unrealizable expectations of our capability and capacity. As "Spectrum or information *dominance*" create hyperbolic expectations, *localized and measurable* situational control remains a warfighting requirement. U.S. Joint analysis and Service doctrine documents prominently mention EMS Control as an objective, and for good cause. The term "*superiority*" maintains usefulness as an abstract *comparative* state conveying success in deterrence, whereas "control" is an *objective* condition denoting a state of *practical reality*. Interceptors wield *superiority*, convincing potential adversaries to leave their aircraft parked on the ramp (i.e., deterrence), but when challenged they demonstrate *control* over the airspace. No such luxury of deterrence inducement exists in Spectrum contests due to the omnipresence of radiant EM energy, only *demonstration* in every case. Therefore, efficacy must be demonstrated – *Control* – as the measurable ability *to access, attack, sense, communicate, and outmaneuver the adversary* across the Spectrum.

27.4 Electromagnetic Spectrum Operations

EMSO is defined as *"those activities consisting of Electronic Warfare and Spectrum Management operations used to exploit, attack, protect and manage the electromagnetic operational environment to achieve the commander's objectives"* [1]. The sense of urgency in pursuing EMSO is tied to the need to *fight/adapt/decide faster* by aggregating capability and *mitigating environmental chaos* in a way currently unavailable to forces as organized. EMSO includes SM expertise, since SM provides systemic protection against the effects of radiant EM energy via organization and protocols. The evolution toward convergence of EMS *warfare* and *management* mission responsibilities and technologies must be pursued aggressively. In contrast to cyberspace experts' necessary focus on application of executable *code* (CO) and protection of *data* in use (IA), the actual convergence of EW and SM is based upon a uniquely shared, *completely overlapping* interest in the manipulation, protection, and weaponization of radiant analog EM energy for control of their objective maneuver space (EMOE) and adversary capabilities within it. Technologies for sensing, attack, protection, management, and visualization with respect to radiant EM energy must take this true convergence into account to maximize force efficiency and effectiveness. If deliberately pursued, this new strategy of coherence will yield force-wide benefits to the larger *user group* of participants within the EMS Domain and doctrinal adoption of EMSO will catalyze that paradigm shift. EMSO is not simply an operational model, it also responds to future needs with an *inclusive framework* for evolved engagement. Deliberate pursuit of EMSO will drive and normalize changes in EMS technology, organization, requirements, coordination, acquisition, and process enforcement. Although capabilities are still acquired under competing Service level authorities, EMSO is intrinsically multi-domain ("Joint") in *every instance* of its application, due to cross-domain impact regularly beyond that of its localized intent. Use of the term "JEMSO" (to include the "J") creates the counterproductive expectation that any organization may acquire capabilities and conduct EMSO without respect to other operations in any other domain. EMSO capability becomes EMSO operations at the CCMD / JTF levels. From the open ocean surface combatant interfering with space capability, to the convoy interfering with close air support platforms, to the airborne EA platform interfering with cross-domain C2, the profound and immediate effects of EMS activity are inevitable unless considered prior; there are *no 'un-Joint' instances* of EMSO, notwithstanding their individual intent.

27.5 The "Shift"

Spectrum energy is ubiquitous, as the only truly shared physical space common to all operations. As a matter of fact, digital code/data extends only to the physical limits of cyberspace infrastructure before it depends entirely on the EMS for extended, global transport. As a physical space so fundamentally shared, *coherent development effort* is an increasingly logical and critical requirement across EMS policy, strategy, capabilities, acquisition, fielding and employment. A legacy model of isolated acquisition of EMS capabilities, although appropriate in previous decades, no longer constitutes an effective or sustainable approach to meeting increasingly interdependent Spectrum maneuver requirements. Elevated cross-organizational attention will help to ensure effectiveness of the defense Spectrum enterprise across all DOTMLPF-P areas of responsibility. Failure to *durably* empower such an orchestrator to assist and inform multi-domain users will ensure that EMS policy, strategy and capability gains remain unlikely, episodic, and unsustainable. Cost-effective, future-proof solutions will depend upon integrated S&T, R&D, and acquisition effort across the enterprise.

The evolution toward EMSO acknowledges the reality that Spectrum warfighters and planners are engaged in the *same conversation*, no longer able to plan separately, since it is no longer reasonable to conduct offensive operations in the EMS with disregard for the broader consequences. As a systemic remedy to this challenge, *maneuver* in the EMS is a simple concept denoting the ability to manipulate **a)** *Parametric* (frequency, modulation, phase, polarity, etc); **b)** *Spatial* (directionality, shape, volume, etc); and **c)** *Competitive* (ERP, gain, LPD, etc) - three "dimensions" - over *Time*, delivering campaign-level effects through simultaneous action in order to achieve EMS Control, to include neutralization of all relevant adversarial capabilities within *disruptive influence* of EM energy. Ultimately, the most important thing about EMSO is the acknowledgement of the *shift* in thinking its implementation will require: from "*EMS as a utility*" to "*EMS as a maneuver space*"; from focus on *means* (electronics) to focus on *ends* (EMS Control); from the entropy of *self-interest* (measure of performance) to the efficiency of *shared interest* (measure of effectiveness); from isolated focus on *threat* to the creation of *opportunity* by aggregation; from *tech-defined* capabilities to *requirements-driven* technologies; and from *episodic* EA to *holistic* Spectrum Maneuver. This shift should come as no surprise, since a litany of recent, high-level studies and analyses [3] (and references therein) have demonstrated the merits of adopting such a framework, underscored by multi-national level security and defense strategy mandates to 'defend the global commons,' *itself* built upon and completely underwritten by the EMS.

A future approach based upon a *proactive* posture, for which EMS Control will prove foundational, enables growth beyond historical focus on kinetic "threats" to shift instead toward non-kinetic "opportunities." A proactive engagement strategy of *Decisive Aggregation* based upon the universality of radiant EM energy would go beyond merely *deterring a tech-peer adequately*, it would convey the strategic objective of *fighting any state or non-state actor decisively*, anywhere and without caveat. Combat at the speed of coherently aggregated EM energy - versus lumbering at the speed of assembled Mass - would provide the potential to out-turn, out-reach, and *out-know* any adversary, no matter their size, density, organization, sophistication, or pace of operations.

27.6 Compliance and Chaos Models of Spectrum

Civil-commercial use of Spectrum relies on and benefits from predictability ("compliance model") whereas effective and adaptive military use relies upon the mitigation – or *creation* – of uncertainty ("chaos model"). Although some technologies may be used in both contexts, it is critical to remember that their objectives are sharply distinguished, especially as they pertain to transitional and escalatory phases of conflict. A future of EMS Control based upon civil-commercial patterns of use holds defense at risk, since those patterns depend upon voluntarily compliant behaviors of all participants and a profit model for systemic optimization. This is very different from the chaotic, disorienting, violently transitional, and *expedient* use of Spectrum during conflict. A challenge is pursuing military solutions that do not completely fall prey to *non-military* assumptions, since EMSO comprises a larger group of opportunities than the diminutive notion of Spectrum as merely a layer in the OSI Model or a mere collection of tactical materiel. Civil-commercial access to the Spectrum is based on a 'path to revenue' focus inspiring confidence, grouping, planning, and sharing, resulting in efficient predictability. Warfare challenges those assumptions via introduction of chaos, requiring a mission-expedient 'freedom of action' focus characterized by tolerance of disorientation, spreading, escalation, and denial, resulting in un-predictability. In the context of chaos-driven combat operations, compliance-based expectations and technologies are dangerously inappropriate. Unfortunately, mission-driven military Spectrum requirements have not fared well against revenue-yielding commercial Spectrum interests, causing a systemic impediment to EMS-dependent military efficacy. Transcending this challenge requires evolution to a paradigm of dynamic maneuver, living comfortably yet unpredictably in EMS "white space" (allocated but momentarily unoccupied) in order to regain the initiative and fundamentally complicate adversary targeting. Cognitive EM technologies enabled by software-defined radios (SDRs) and emerging Dynamic Spectrum *Access* (DSA) schema provide a means by which temporarily "empty" Spectrum can be dynamically used and vacated by synchronized devices. Non-military stakeholders may continue their routine use of fixed (FCC/NTIA) Spectrum allocations, but military command, control, and communications (C3) and EMSO capabilities must leverage operational Spectrum contest as *license* to evolve to this new paradigm of Spectrum maneuver. Due to obvious EMS use trends, this approach will be increasingly required across our EMS-reliant portfolio to the maximum extent possible. It will relieve domestic Spectrum congestion, ensure viability of current capabilities, reduce stress on foreign Spectrum environments, and most importantly, remain *unpredictable* to future adversaries possessing increasingly advanced EMS capabilities.

It is useful to pause and reconsider the term "military technology." We have broadly forgotten the intent of this term in the Spectrum context, as evidenced by the broad substitution of combat-specialized government-off-the-shelf (GOTS) technologies by increasingly popular commercial (COTS) counterparts. Logically, this cannot be considered appropriate in *every* case. The term "military technology" although not exclusively applicable to GOTS, identifies technologies as *appropriate in performance and survivability* for combat conditions. In current cost/schedule-driven acquisition programs however, they may increasingly forego the "appropriate for combat" standard. Commercial technologies provide attractive options, but they *are* so because they are intended for use in compliant environments, requiring *voluntary adherence* to civilized behavioral standards in order to capture the efficiencies/value inherent to them. However, when these technologies and their tolerance *maximized* for remedial congestion is subjected to the chaos of *deliberate attack* or sabotage in combat, their efficiencies become overshadowed by their relative fragilities. FCC-compliant, COTS-based Spectrum devices – as a matter of practical fact – assure every user (or *exploiter*) of precisely the frequency ranges or "Spectral terrain" to be used. In doing so, the essential warfighting tenets of *Surprise, Offensive, Maneuver*, and *Security* may all be held at risk for prioritized economy, potentially conferring systemic, *enterprise-wide advantage* to potential adversaries, be they nation-state technology peers or non-state insurgents.

While commercial Spectrum providers will continue development of dynamic *allocation* technologies (e.g., cellular telecommunication), military and first responder operations at home and abroad must develop Dynamic Spectrum *Access* capabilities in order to **a)** facilitate *live* training at home; **b)** better facilitate competing commercial EMS demand; **c)** minimize impact to Spectrum sovereignty of nation-states in conflict; and **d)** maintain operational initiative/surprise with respect to rapidly advancing and proliferating adversary counter-EMS capabilities. Predictable or known applications of EMS technology can insidiously create conditions of "net-centric *vulnerability*" (versus *net-centric warfare*). Military information assurance (IA) activities and measures must acknowledge the greatly increased EP requirements upon which information transport depends in contested electromagnetic operating environments, since IA alone does not accomplish EP. Existence of insufficiently adaptive or survivable links across the Spectrum can invite cascading failure; links must be survivable under *deliberate operational stress*, and not simply tolerant of routine civil congestion. Subjecting technologies to chaotic military EMS conditions with compliant commercial expectations in mind can have the effect of multiplying exposure to adversary cost-imposing strategies, with true potential for uncontained risk to a larger capability portfolio.

27.7 From Electronic Warfare to Spectrum Warfare

As a core component of EMSO, EW has been defined well [2], but is labeled narrowly and has been executed incoherently as a result of the narrow, *tactical* expectations created. The findings of the 2015 Defense Science Board study on "*21ˢᵗ Century Military Operations in Complex Electromagnetic Environments*" essentially support this assertion [3]. Although the EW mission area and each of its subdivisions carry the subordinate "electronic" label and it is often couched as an "information related capability", EW's enduring doctrinal and definitional identity clearly conveys a focus on control of *energy and capability, versus* 'electronics' or 'information.' An individual target may be *electronic,* among the doctrinal targets of "*personnel, facilities and equipment*" [4], but culminating the entirety of the discussion requires elevating the concept of warfare from focusing on the transient materiel *capabilities* that support it to the enduring operational *responsibilities* that define it in order to achieve freedom of action across the *maneuver space* in which it occurs. This follows an historic line of precedence: Sea Control, Airspace Control, Space Control, and now Spectrum Control. In order to gain and maintain this EMS Control, EMSO drives a shift from *electronics*-based to *energy*-based thinking. To paraphrase the brilliant 19ᵗʰ Century inventor Michael Faraday, our attention should not be captivated by the magnets, but instead by *the important space between* them. As with Land Warfare, Maritime Warfare, Air Warfare – and now *Spectrum* Warfare – the objective is to control participation across mission-relevant portions of the EMS Domain, in the same way that interceptors exert control of the air, and warships exert control at sea.

In technology acquisition, one should transcend the proscriptive, defensive mindset of EW; Spectrum Warfare within the EMSO framework serves a *sword* as well as a shield. EA in fact goes beyond traditional jamming to include the entire continuum of radiant energy for offensive purposes, specifically high-power microwave, laser, and electromagnetic pulse applications. As Spectrum sharing strategies evolve beyond current RF-centric applications, these capabilities will yield greater influence on those strategies. The true scope of ES is also *much* larger than we have been conditioned to appreciate, since all collection of radiant EM energy arguably *begins* as ES, or *collection for action.* It is only post-collection processing under special authorities that turns this energy into intelligence, or *collection for analysis.* Finally, EP includes protection from *any effects* of EM energy (e.g., multi-Spectral low observability "stealth", laser and DE hardening, EMP hardening, and anti-jam measures, etc), *not* merely from the effects of adversary jamming. Warfighting doctrine supports these assertions, as it has for many years.

Elevating the focus from capabilities to Spectrum *won't* marginalize the capabilities acting within it as some might fear. It will instead cause a systemic "pull" on the *right* mix of the *right* capabilities, instead of allowing forces to erroneously imagine their isolated employment. The contemporary choice of "stealth or EW" provides a litmus test of this thesis, since the de facto condition presents an opportunity to balance "EP and EA" investments, respectively, as a unified portfolio consideration of EMS Control. The key to coherent acquisition of EMS capability is not simply access to the finest technology, but instead in truly *understanding the dynamics* of the Spectrum contest. Spectrum *attack, protection, sensing,* and *management* will require balance to achieve mission success within an affordable portfolio.

27.8   Training for Effective EMS Operations

Emerging technologies will create great opportunities for a wider Spectrum Warfare portfolio of capabilities. However, benefits of these new technologies are not held by any single nation. Global availability of high quality foreign-manufactured, high-grade chipsets and proliferation of competitive military technologies will produce normalizing pressures across national military capabilities, creating a *de facto* state of technology parity. That condition *now* exists regarding the COTS technologies broadly incorporated – since anyone may obtain them – and is rapidly approaching among proliferating foreign GOTS technologies as well. Advanced information technology (IT) systems now available will enable an EMSO *center of gravity* shift from EW's tactical roots to beyond the operational level of engagement. A modular open systems approach (MOSA) for IT-enhanced EMS warfare will increase effectiveness, scalability, adaptability, and efficiency, reducing resource requirements to achieve a better end product at a net savings in manpower. As IT-enabled EMSO capabilities, payloads, and processes are carried aboard flying, rolling, walking and floating host platforms, deployed EMS Operations Cells (EMSOCs) will become C2-capable weapon systems *versus* mere staff coordination groups.

The approaching COTS/GOTS technology parity is telling, since it underscores the significance of the *remaining* component of capabilities: the expert Operator. Decreasing opportunities now exist for realistic EMS training in *real space* due to government restrictions, Spectral disruption in/near civil environments, and increasing potential for EMS capability compromise. As these challenges chase military operations out of real space prior to large scale development and implementation of DSA (*Access*) capabilities, an increasing proportion of EMS capability training, rehearsal, experimentation, testing, validation, and assessment will require access to live, virtual, constructive and distributed (LVC-D) environments. EMS warfare experts – aviators, ground troops, and seamen – are forged by chaotic *immersion*, not simply by technical expertise. Allowing reduced opportunities to experience and overcome chaos in the EMS will systematically erode the operator's ability to tolerate and adapt to chaotic conditions, an essential component of any EMS sharing *or* warfighting strategy. Civil norms of EMS use are incompatible with military readiness; training must be facilitated according to operational plans. Ensuring EMSO viability requires accelerated investment in LVC-D simulation, training, and rehearsal capabilities. In the end, the product of this approach will be an *EMSO Professional*: integrator, critical thinker, planner, EM battlespace controller, strategist, acquisition advisor, innovator, educator, lesson keeper ... and *Warfighter*.

27.9   The Way Ahead

The contest for control of the Spectrum is experiencing a watershed period of transition as sophisticated power handling and sensor technologies, antenna applications, DE and laser innovations, multi-Spectral low observability technologies, complex waveforms, and tailored IT capabilities are concurrently maturing to support it. The indicators all around defense point to a future requirement for scalable, dynamic, multi-functional, distributed, adaptive, surgical, collaborative, and platform/waveform-agnostic EMS capabilities in a MOSA framework. This approach will use hardware and software commonality to control cost, but must do so in a way that *avoids predictability and fragility* in its implementation. Uniting EMS systems in this chaos-tolerant manner will require a system of spread-Spectrum (frequency-agile/diverse) software defined translation modules for collaborative employment of all proximate/opportune EMS systems. The requisite architecture will not be simply interconnected, but instead form the foundation of a swarming, end-to-end *Spectrum maneuver force* for delivering decisive surgical effects under deliberate operational Spectrum stress. Based upon this common foundation of universal architecture and data exchange protocols within a shared operational framework, technology developments can pursue the following near term objectives to prepare the future force:

- to ***orchestrate*** operations: agile EM battle management (EMBM) tools and processes, supported by purpose-built EMBM architecture and data structures, employed within Joint policy-based Spectrum operations (PBSO) frameworks;

- to ***visualize*** operations: high-fidelity, many-on-many modeling and simulation capabilities, expressed within an EMS user-defined operating picture (UDOP) framework with scalable simultaneous volumetric electromagnetic propagative visualization;

- to ***adapt*** operations effectively within chaotic EMS environments: DSA (*Access*) technologies supported by collaborative platform-agnostic sensing with real-time exchange features, enabled by advanced aperture technologies; *and*

- to ***engage*** adversaries decisively with EM energy: high-energy weapons enabled by deliberate *size, weight, and power* (SWaP) reduction efforts for high-power systems; while incorporating platform and personnel *protection* against adversary weapons of similar type.

Friendly forces will be exposed to an increasing range of power projection and regional stability missions, necessitating a broad shift to EMS multi-functionality. Common architectures, protocols, and technical standards will form critical enablers to the growth and flexibility necessary for meeting diverse global security challenges *affordably* and ensuring our Spectrum capabilities are adaptable and *future proof*. Transitioning broadly from a box/pod/QRC-centric model to an *architectural* approach will drive collaborative capabilities that potentially outperform and even *anticipate* a broader range of unanticipated EMS threats. The force will benefit from accelerated inclusion of high-efficiency (e.g., gallium nitride based) technologies for increased power at reduced form factor, to allow deployment of swarm-capable EMS attack and sensing capabilities aboard unmanned vehicles. Dynamic Spectrum *Access* in concert with prudently incorporated (commercial) dynamic *allocation* technologies will form the basis of sufficiently flexible *and* affordable future military operations. Advanced physics-based modeling and simulation (M&S) will be essential to high-confidence, high-fidelity LVC-D environments, EMS capability development, and effective test, training, and experimentation. Cost-per-shot, portability, precision, and multiple/dynamic lethal targeting requirements will drive inclusion of high-energy EW weapon systems. Multi-Spectral sensing systems must become adaptive, "aware" of each other, offer instant precision signal geolocation, and be scalable in order to suit dynamic operational requirements. They will also require provisions for *ad hoc* cooperation with external systems and architectures, real-time status reporting, and graceful degradation. EMS capabilities will share near-real-time access to the same meta-repository of (or gateway to) relevant data and supporting intelligence. Spectrum warfighting actions and capabilities will share dynamic EMBM nodes and hierarchy, be capable of push- *and* pull-based asset allocation and tasking, and maintain a filterable EMS UDOP down to the tactical decision level. These UDOPs will include awareness of EMS-dependent systems and activities as well as the ability to rapidly support those operators with sensing, interference mitigation, and synergistic attack via collaborative optimization.

Mission success across all domains now depends on EMS Control. As with previous evolutions of Land, Maritime, and Air Warfare, logic drives prioritizing control of the *EMS* Domain *above* the capabilities operating within it in order to field an effective force. Policy, doctrine, TTPs, and exercises supporting the EMS Control objective are now in development and assumption of adoptive technology risk should be encouraged to expedite that revolution. As Spectrum energy rests at the foundation of what force application efforts hope to accomplish, future success in conflict will be defined by the speed and reach of EM energy but dictated by the ability to reduce process latency and mitigate chaos in combat while denying adversary maneuver and capability. While failure to acquire and plan employment of EMS capability at that tempo and scale during operations could progressively eliminate the potential for *mission success at acceptable levels of risk*, continuing a systemic pursuit of EMS Control will inevitably impose unsustainable costs on any adversary who chooses to oppose EMSO-equipped friendly forces. This truism is driving the convergence of EW and SM experts and capabilities - beyond that of a mere *enabling* force - to become the cadre of decisive future EMS Operations and *guarantors* of EMS Control.

References

[1] U.S. Department of Defense (Various Authors), DoD EMS Strategy. Washington DC: DoD Press, 2013, pg. 2.

[2] U.S. Department of Defense (Various Authors), "Executive Summary," in Joint Publication 3-13.1, Electronic Warfare. Washington DC: DoD Press, 2012, pg. viii.

[3] U.S. Defense Science Board (Various Authors), Report: 21st Century Military Operations in a Complex Electromagnetic Environment. Washington DC: DoD Press, 2015, pg. 6.

[4] U.S. Department of Defense (Various Authors), "Executive Summary," in Joint Publication 3-13.1, Electronic Warfare. Washington DC: DoD Press, 2012, pg. viii.

Milton Keynes UK
Ingram Content Group UK Ltd.
UKHW052031071024
449327UK00027B/2513